Theory of Stochastic Integrals

In applications of stochastic calculus, there are phenomena that cannot be analyzed through the classical Itô theory. It is necessary, therefore, to have a theory based on stochastic integration with respect to these situations.

Theory of Stochastic Integrals aims to provide the answer to this problem by introducing readers to the study of some interpretations of stochastic integrals with respect to stochastic processes that are not necessarily semimartingales, such as Volterra Gaussian processes, or processes with bounded p-variation among which we can mention fractional Brownian motion and Riemann-Liouville fractional process.

Features

- Self-contained treatment of the topic
- Suitable as a teaching or research tool for those interested in stochastic analysis and its applications
- Includes original results.

Jorge A. León studied the PhD in equivalence of solutions to stochastic evolution equations at the Department of Mathematics of Cinvestav-IPN, Mexico. He has carried out joint research work with internationally recognized researchers. He has mainly contributed to the development of stochastic calculus and integration, and their applications to stochastic differential equations with different interpretations of stochastic integral. He has co-organized several conferences on stochastic analysis among which we can mention the Latin American Congress of Probability and Mathematical Statistics (CLAPEM), the joint meeting between USA and Mexico, Bernoulli-IMSWorld Congress, Symposium on Probability and Mathematics Statistics, which was the most important meeting in probability at Mexico, etc. He has taught at Cinvestav-IPN for 35 years.

Theory of Stochastic Integrals

Jorge A. León

CRC Press is an imprint of the
Taylor & Francis Group, an **informa** business
A CHAPMAN & HALL BOOK

Designed cover image: Shutterstock 1162365019

First edition published 2025
by CRC Press
2385 NW Executive Center Drive, Suite 320, Boca Raton FL 33431

and by CRC Press
4 Park Square, Milton Park, Abingdon, Oxon, OX14 4RN

CRC Press is an imprint of Taylor & Francis Group, LLC

ISBN: 978-1-032-77810-5 (hbk)
ISBN: 978-1-032-77812-9 (pbk)
ISBN: 978-1-003-48491-2 (ebk)

DOI: 10.1201/9781003484912

Typeset in CMR10 Font
by KnowledgeWorks Global Ltd.

Publisher's note: This book has been prepared from camera-ready copy provided by the authors.

To my wife Edith
and my son Jorge Aarón.

Contents

Foreword

The study of stochastic processes and their complex behaviours has become a fundamental aspect of contemporary mathematical and statistical analysis. At the heart of this exploration is Harold Edwin Hurst, who gained prominence with his seminal 1951 paper titled **"The Behaviour of the Long-Term Storage in Reservoirs"**. In this influential work, he introduced the concept of the Hurst parameter, which describes the long-term dependence and self-similarity in time series data. This paper laid the foundation for his subsequent contributions to the study of stochastic processes and their applications in various fields.

The Hurst parameter (H) is instrumental in characterizing the long-term memory of a time series, revealing whether the series exhibits persistent, anti-persistent, or random behaviour. When H exceeds 0.5, it suggests a persistent, trending behaviour, whereas values below 0.5 indicate anti-persistent or mean-reverting behaviour. A value of $H=0.5$ corresponds to the standard Brownian motion, known for its lack of long-term dependence. An extension of the classical Brownian motion framework is the Fractional Brownian motion (fBm), which incorporates the Hurst parameter.

One of the seminal works on fractional Brownian motion (fBm) was done by **Benoît B. Mandelbrot** and **J. W. van Ness** were also pivotal in extending the concept of Brownian motion.

Unlike standard Brownian motion, which features independent increments, fBm exhibits long-range dependence and self-similarity, maintaining its statistical properties under rescaling of time and space. The parameter H influences the degree of roughness or smoothness of the fBm path, thereby affecting the correlation of the process's increments. This characteristic makes fractional Brownian motion a valuable model in various fields, including finance, hydrology, and telecommunications, where data frequently display long-term dependencies.

To analyze processes like fBm, stochastic integrals are crucial. These integrals generalize the concept of integration to address the inherent randomness in processes. Traditional methods such as Itô calculus fall short when dealing with fBm's long-range dependence. Instead, fractional calculus and the theory of fractional stochastic integrals, such as the Riemann-Liouville, Forward and Stratonovich integration, among others, provide the necessary tools. These integrals facilitate the rigorous definition and computation of integrals with respect to fBm, making them indispensable for modeling and analyzing phenomena with complex statistical dependencies.

An interesting book covers the study of fractional Brownian motion (fBm) and associated integrals is **Stochastic Calculus for Fractional Brownian Motion and Applications** (2008) by Francesca Biagini, Yaozhong Hu, Bernt Øksendal, and Tusheng Zhang. This book provides a comprehensive overview of fBm theory as well as the related integration techniques and analysis, addressing both theoretical aspects and practical applications of fBm in various fields.

The second book is **Stochastic Integration with Respect to Fractional Brownian Motion** by Vladas Pipiras and Murad S. Taqqu, published in 2017, provides a comprehensive and authoritative text on the subject. This book digs into the theory of stochastic integration in the context of fractional Brownian motion (fBm), presenting advanced techniques and methods for dealing with fBm's unique properties. It explores fractional calculus, fractional stochastic integrals, and their applications in various fields. The authors provide rigorous mathematical frameworks and practical insights, making it an essential resource for researchers and practitioners working with fBm and related stochastic processes.

The book we present here provides a broad exploration of stochastic integration techniques applicable to a range of stochastic processes beyond fBm. This book also explores alternative interpretations and methods, equipping readers with the skills to navigate and understand stochastic integrals in a broader context. The focus is on expanding readers' ability to handle complex stochastic processes through advanced methods such as fractional Brownian motion and other non-semimartingale frameworks.

Furthermore, this book highlights the practical applications of stochastic differential equations (SDEs) in these extended settings, demonstrating how they can model and analyze real-world phenomena with intricate statistical dependencies. By bridging classical theories with modern applications, this text aims to enhance readers' analytical capabilities and provide a deeper understanding of stochastic processes in diverse and challenging contexts.

The author of the book, Jorge A. León, is a researcher at Cinvestav (Centro de Investigación y de Estudios Avanzados del Instituto Politécnico Nacional) in Mexico. He is widely recognized for his major contributions to the study of stochastic processes and their applications. His research explores into advanced topics in stochastic calculus, including stochastic differential equations and their interpretations in non-standard contexts. León's work extends traditional stochastic analysis methods to complex scenarios where classical approaches, like Itô calculus, may not be suitable. His contributions advance our understanding of stochastic integrals and their practical applications in finance, physics, engineering, and beyond.

At Cinvestav, León engages in cutting-edge research and collaborations, bolstering the institution's reputation as a leading center for scientific and technological research. His extensive publications and international collaborations, including guidance and advice to researchers from Colombia and Chile, underscore his commitment to advancing the global scientific community. León's efforts not only bridge theoretical advancements with practical applications but also enhance problem-solving capabilities on an international scale.

I met Jorge A. León during one of his early visits to Chile in 2000, in Valparaíso. Following this visit, we began a collaboration that has lasted for more than 20 years.

M. Soledad Torres D.
Valparaíso, August 2024

Preface

In applications of stochastic calculus, there are phenomena that cannot be analyzed through the classical theory of Itô. Among them, we can mention problems governed by stochastic processes with memory. That is, their future outcomes not only depend on the present state but also the past history. Therefore, it is necessary to have a theory based on stochastic integration with respect to these types of processes, which are not semimartingales in general. Motivated by this fact, the main aim of this book is to introduce the reader to the study of some interpretations of stochastic integral with respect to stochastic processes that are not necessarily semimartingale such as Volterra Gaussian processes, or processes with bounded p-variation among which we can mention fractional Brownian motion and Riemann-Liouville fractional process.

In 2010, I have been invited to contribute with a course on stochastic integration with respect to the fractional Brownian motion and some applications to SDE's at the Spring School "Stochastic Control in Finance", which took place in Roscoff (France). Then, the idea to write this book arose from an invitation to give a series of lectures on integration with respect to functions at CIMPA School 2014 held at Universidad de Valparaíso, Chile. Later, these lectures were improved in an invited course on stochastic integration given at Jornadas de Probabilidad & Procesos Estocásticos at the Universidad Nacional de Colombia in 2015. I would like to thank Liliana Blanco, Rainer Buckdahn and Soledad Torres for their invitations. In all these series of lectures, two important facts were to establish properties of integrals and to prove that, in general, the difference between two interpretations of the stochastic integral is a term that depends on the derivative operator in the Malliavin calculus sense.

The Malliavin calculus is an extension of the calculus of variations for functions in the case of stochastic processes. Started by Paul Malliavin in the 70s, Malliavin calculus has been the basis of numerous scientific works and, year after year, the range of its possible applications has been expanded. At the beginning of this calculus, its usefulness was to characterize the conditions under which a random variable has a smooth density. Although for many years it was considered that this was its only application, this situation changed in the mid-1990s when Ocone stated an explicit expression of the Clark's formula in terms of the derivative operator. Subsequently, many researchers have utilized the techniques of Malliavin calculus in several contexts and applications since these not only allow us to consider problems involving processes that are not adapted to the underlying filtration, but also allow us to deal with problems that either are driven by non-semimartingales, or should only be studied by means of the classical Itô's calculus, such as in the examination of the existence and uniqueness of solutions to stochastic evolution equations with adapted random coefficients to the underlying filtration, or in the calculation of closed-form expressions for the at-the-money behaviour of the implied volatility in stochastic volatility model, where the volatility could be driven by fractional Brownian motion (see Sections 1.5 and 7.6 below). In this way, the Malliavin calculus becomes an important tool in the theory of stochastic processes and its application via stochastic differential equations. In consequence, we introduce some important features of Malliavin calculus for Gaussian processes with values in Hilbert space needed to understand some results in this book.

We suppose that the reader is familiar with a basic background on measure, martingale and probability theories. However, The book includes an appendix that makes it self-contained.

In general, for the definitions of integral considered in this monograph, we mainly deal with dominated convergence-type theorems, maximal inequalities, relations between two integrals, formulas of Itô type, Fubini theorems, the existence and uniqueness for different interpretations of the solution to stochastic differential equations, local properties and substitution formulas, among another facts.

The monograph is organized as follows: In Chapter 1, we present the framework and some basic elements of stochastic calculus that we utilize. In particular, here we introduce the canonical Wiener space, the derivative operator for Gaussian processes with values in Hilbert space, fractional Brownian motion, fractional integrals and derivatives, multiple Wiener-Itô integrals. The Riemann-Stieltjes integral is studied in Chapter 2. This integral is analyzed using different approaches among which are an algebraic method, fractional calculus and p-variation theory. The classical integral in the Itô sense is given in Chapter 3, where we apply localization arguments similar to those of Malliavin calculus. The divergence operator for Gaussian processes and some of its extensions are examined in Chapter 4. The purpose of Chapters 5 and 6 is to deal with the forward and Stratonivich integrals, mainly, using the Russo and Vallois's definitions. In Chapter 7, as an application, we study several existence and uniqueness results for the solutions to stochastic differential equations with different interpretations of integrals. Finally, the auxiliary results needed in this book are presented in Chapter 8.

I am very grateful to the people who encouraged me to finish writing this book. I especially want to express my gratitude to Elisa Alòs and Soledad Torres. Moreover, I would like to thank Héctor Araya for his helpful suggestions to improve the presentation of this book.

August 31, 2024 Jorge A. León

Symbol Description

Symbol Description

A^* adjoint of the operator A.
β beta function.
$\mathcal{B}(X)$ Borel σ-algebra of X.
Bm Brownian motion.
C generic constant.
$C(\alpha)$ constant depending on α.
$C^n(\mathbb{R}^m)$ space of all the real-valued functions on $\mathbb{R}^m$ with continuous partial derivatives up to order $n \in \mathbb{N}$.
$\mathcal{C}_1^\mu([a,b])$ space of all the $\mathbb{R}$-valued μ-Hölder continuous functions on $[a,b]$.
$\mathcal{C}_k^\mu([a,b];V)$ $(k-1)$-increments that are μ-Hölder continuous.
$\mathcal{C}_3^{1+}([a,b];V)$ set $\cup_{\mu>1}\mathcal{C}_3^\mu([a,b];V)$.
$C_0([0,T])$ $\mathbb{R}$-valued continuous functions f defined on $[0,T]$ such that $f(0)=0$.
$C_b^\infty(\mathbb{R}^n)$ $\mathbb{R}$-valued bounded functions on $\mathbb{R}^n$ with bounded derivatives.
$C_p^\infty(\mathbb{R}^n)$ $\mathbb{R}$-valued functions f on $\mathbb{R}^n$ such that f and all its partial derivatives have polynomial growth.
$\mathrm{Dom}\,\delta_x^-$ domain of forward integral.
$\mathrm{Dom}\,\delta_x^{\mathrm{S}}$ domain of Stratonovich integral.
$\mathcal{D}(\delta_S^{B^H})$ domain of weak Stratonovich integral.
$Dom(\int_a^b \cdot d^Y x)$ domain of Riemann-Stieltjes integral.
D derivative operator in the Malliavin calculus sense.
$D_{\mathcal{T}}$ $(\mathcal{T}\otimes I_{\mathcal{K}})^*(\mathcal{T}\otimes I_{\mathcal{K}})D$.
$\mathbb{D}^{n,p}(\mathcal{K})$ $\mathcal{K}$-valued random variables in $L^p(\Omega;\mathcal{K})$ with n derivatives in the Malliavin calculus sense.
$\mathbb{D}^{n,p}$ space $\mathbb{D}^{n,p}(\mathbb{R})$.
$\mathbb{D}^{1,\infty}(\mathcal{K})$ domain of the operator D.
$\mathbb{D}_{\mathcal{T}}^{1,2}(\mathcal{K})$ domain of $D_{\mathcal{T}}$.
$\mathbb{D}_{loc}^{n,p}(\mathcal{K})$ local domain of D.
D_{a+}^α left-sided Riemann-Liouville fractional derivative of order α.
D_{b-}^α right-sided Riemann-Liouville fractional derivative of order α.
δ_p divergence operator.
δ operator on $C_k(V)$ into $C_{k+1}(V)$, or extension of δ_2.
$\hat{\delta}$ extension of δ_2.
$\{e_1,\ldots,e_d\}$ canonical basis of $\mathbb{R}^d$.
$E(X)$ expectation of X.
$E[X|\mathcal{G}]$ conditional expectation of X with respect to the σ-algebra $\mathcal{G}$.
$\mathcal{E}([a,b])$ family of simple, or step functions on $[a,b]$.
$\mathcal{E}_n$ family of the step function in $L^2\left(([0,T]\times\{i,\ldots,d\})^n,(\lambda\times\delta)^n\right)$.
$\mathcal{E}(|\mathcal{H}|)$ stochastic processes of the form $\sum_{j=0}^{n-1} F_i 1_{[t_i,t_{i+1}]}$ with $F_i = f_i(B(h_{1,i}),\ldots,B(h_{n_i,i})) \in \mathcal{S}$, $h_{j,i}\in|\mathcal{H}|$ and $0=t_0<t_1<\ldots<t_n=T$.
$e(\alpha)$ $\mathrm{Symm}(e_{j_1}^{\otimes\alpha_1}\otimes\ldots\otimes e_{j_{\hat{\alpha}}}^{\otimes\alpha_{\hat{\alpha}}})$, for $\alpha=((\alpha_1,j_1),\ldots,(\alpha_{\hat{\alpha}},j_{\hat{\alpha}}))$.
$f^{(\ell)}$ ℓ-th component of the fuction $f: X\to\mathbb{R}^n$.
$\mathcal{F}_A$ σ-algebra generated by $\{W(1_B): B\subset A\}$.
F_X distribution function of the random vector X.
fBm fractional Brownian motion.
$\bar{\mathcal{F}}_t$ σ-algebra generated by $\mathcal{F}_t$ and all the set of $\mathcal{F}$ with probability zero.
$\mathcal{F}_t^Y$ σ-algebra generated by $\{Y_s: s\in[0,t]\}$.
g^{b-} function $g(b-)-g$.
Γ gamma function.
$h\otimes_r h_1$ contraction of $h\in\mathcal{H}^{\otimes n}$ and $h_1\in\mathcal{H}^{\otimes m}$.
$\mathcal{H}\otimes\mathcal{K}$ tensor product of the Hilbert spaces $\mathcal{H}$ and $\mathcal{K}$.
$A\otimes B$ tensor product of the bounded linear operator A and B.

$H_n(\cdot,\lambda)$ Hermite polynomial of order n with parameter λ.
$\mathcal{H},\mathcal{K}$ real separable Hilbert spaces.
$|\mathcal{H}|$ measurable functions $\varphi:[0,T]\to\mathbb{R}$ such that $\|\varphi\|_{|\mathcal{H}|}<\infty$.
$\mathcal{H}_K$ completion of $\mathcal{E}([0,T])$ with respect to the seminorm $\|\cdot\|_K$.
K kernel of a Gaussian volterra process.
K^* isometry from the involved reproducing kernel Hilbert space to $L^2([0,T])$.
I^α_{a+} left-sided Riemann-Liouville fractional integral of order α.
I^α_{b-} right-sided Riemann-Liouville fractional integral of order α.
$\langle\cdot,\cdot\rangle_{1,2,\mathcal{K}}$ inner product in $\mathbb{D}^{1,2}(\mathcal{K})$.
$\langle\cdot,\cdot\rangle_{\mathcal{K}}$ inner product of $\mathcal{K}$.
$\langle\cdot,\cdot\rangle_{\mathcal{H}\otimes\mathcal{K}}$ inner product of $\mathcal{H}\otimes\mathcal{K}$.
$I^\alpha_{a+}(L^p)$ image of $L^p([a,b])$ by I^α_{a+}.
$I^\alpha_{b-}(L^p)$ image of $L^p([a,b])$ by I^α_{b-}.
$I_n(h)$ multiple Wiener-Itô integral of h of order n.
$L^p(\Omega)$ family of random variables with finite p-moment.
Λ operator from $\mathcal{ZC}^{1+}_3([a_1,a_2];V)$ to $\mathcal{C}^{1+}_2([a_1,a_2];V)$.
$\mathcal{L}(X,Y)$ linear and continuous operators from X to Y.
$\mathcal{L}^2_a(\Omega\times[a,b])$ family of square-integrable and $\mathcal{F}_t$-adapted processes.
$\mathcal{L}^2_a(\Omega;L^2([a,b]))$ space of all measurable and $\mathcal{F}_t$-adapted processes with square-integrable paths.
$L^p(\Omega;\mathcal{K})$ $\mathcal{K}$-valued ramdom variables X such that $E(|X|^p_{\mathcal{K}})<\infty$.
$\mathbb{L}^{1,2,f}$ stochastic processes that are differentiable in the future sense.
$\mathbb{L}^{1,2,f}_{q-}$ processes X in $\mathbb{L}^{1,2,f}$ such that $D^-X\in L^q(\Omega\times[0,T];\mathbb{R}^{d\times d})$.
$\mathbb{L}^{1,p}_H$ processes $u\in\mathbb{D}^{1,p}(|\mathcal{H}|)$ such that $E\left(\|u\|_{L^{1/H}([0,T])}+\|Du\|_{L^{1/H}([0,T]^2)}\right)<\infty$.
$\mathbb{L}^F$ stochastic processes that are twice differentiable in the two dimensional future.
$\mathbb{L}^F_{q-}$ family of processes in $\mathbb{L}^F\cap\mathbb{L}^{1,2,f}_{q-}$
$\mathbb{N}$ family of all positive integers.
$\|A\|$ norm of the operator A.
$(\Omega,\mathcal{F}.P)$ probability space.
$(C_0([0,T]),\mathcal{F}^*,P^{*,W})$ canonical Wiener space.
μ^* outer measure.
P_ε function $x\mapsto e^{-\frac{x^2}{2\varepsilon}}/\sqrt{2\pi\varepsilon}$.
$L^p(X)-\lim$ limit in $L^p(X)$.
$P-\lim$ limit in probability.
$\mathcal{P}(X)$ family of all the subsets of X.
$\mathcal{P}([a,b])$ predictable σ-algebra on $\Omega\times[a,b]$.
R_H covariance function of fBm with Hurst parameter $H\in(0,1)$.
RLfp Riemann-Liouville fractional process.
$\mathrm{RS}(f,g,\pi,\tilde\pi)$ Riemann-Stieltjes sum.
φ_X characteristic function of the random vector X.
$\mathcal{W}_p([a,b])$ space of functions with finite p-variation on $[a,b]$.
$||X||_{L^p(\Omega)}$ seminorm $(E(|X|^p))^{1/p}$.
$\|\cdot\|_{1,2,\mathcal{T}}$ seminorm in $\mathbb{D}^{1,2}_{\mathcal{T}}(\mathcal{K})$.
$\|\cdot\|_{[p]}$ seminorm

$$\|\cdot\|_{\infty,[a,b]}+(\mathrm{var}_p(\cdot;[a,b]))^p.$$

$\|f\|_{(p),a}$ seminorm

$$|f(a)|+(\mathrm{var}_p(f;[a,b]))^{1/p}.$$

2^B family of all subsets of B.
$\Sigma_{2,n}$ elements of the form

$$((\alpha_1,j_1),\ldots,(\alpha_{\hat\alpha},j_{\hat\alpha}))$$

such that $\alpha_i,j_i,\hat\alpha\in\mathbb{N}$, $\sum_{i=1}^{\hat\alpha}\alpha_i=n$ and $j_1<\ldots<j_{\hat\alpha}$.
$\mathbb{R}^{\mathcal{H}}$ family of functions from $\mathcal{H}$ into $\mathbb{R}$.
$\mathcal{S}(\mathcal{K})$ $\mathcal{K}$-valued smooth random variables.
$\mathcal{S}$ space $\mathcal{S}(\mathbb{R})$.
S_n permutation group of $\{1,\ldots,n\}$.
$\|\cdot\|_{\infty,[a,b]}$ supremum norm on $[a,b]$.
$\|\cdot\|_{\mu,[a,b]}$ seminorm in the space of μ-Hölder continuous functions on $[a,b]$.
$\|\cdot\|_{n,p,\mathcal{K}}$ seminorm in $\mathbb{D}^{n,p}(\mathcal{K})$.
Symm Symmetrization operator.
$\tilde h$ $\mathrm{Symm}\otimes I_{\mathcal{K}}(h)$, for $h\in\mathcal{H}^{\otimes n}\otimes\mathcal{K}$.
T fixed constant bigger than zero.
$\mathrm{var}_p(f;[a,b])$ p-variation of f on $[a,b]$.
w.p.1 with probability 1.
Ξ_h bounded linear operator related to $h\in\mathcal{H}^{\otimes n}$.
$\int_a^b\cdot d^Yg$ Young integral on $[a,b]$ with respect to g.
ζ Riemann zeta function.

Chapter 1

Basic Tool and Concepts

In this chapter, we describe the basic framework that we need in this book. For this, we assume that the reader is familiar with the elementary facts of measure theory and classical probability theory.

Throughout this book, $T > 0$ and $(\Omega, \mathcal{F})$ represents a measurable space. That is, Ω is a non-empty set and $\mathcal{F}$ is a σ-algebra on Ω. It means $\mathcal{F}$ is a family of subsets of Ω such that:

i) For every $A \in \mathcal{F}$, we have A^c (i.e., the complement of A) also belongs to $\mathcal{F}$.

ii) If $\{A_n \subset \Omega : n \in \mathbb{N}\} \subset \mathcal{F}$, then $\cup_{n\in\mathbb{N}} A_n \in \mathcal{F}$.

Remark 1 *If we change Statement ii) by*

ii$'$) For $\{A_i \in \mathcal{F} : i \in \{1, \ldots, n\}\}$, we have $\cup_{i=1}^n A_i \in \mathcal{F}$,

then $\mathcal{F}$ is called an algebra.

Note that $\Omega \in \mathcal{F}$ in both cases. Also, we consider a probability measure P on the measurable space $(\Omega, \mathcal{F})$. That is, a function $P : \mathcal{F} \to [0, 1]$ such that $P(\Omega) = 1$ and, for $\{A_n \subset \Omega : n \in \mathbb{N}\} \subset \mathcal{F}$ such that $A_n \cap A_m = \emptyset$, $n \neq m$, we have

$$P(\cup_{n\in\mathbb{N}} A_n) = \sum_{n=1}^{\infty} P(A_n).$$

In this case, we say that $(\Omega, \mathcal{F}, P)$ is a probability space. Henceforth, we assume as given this probability space that it is complete. In other words, if $A \subset B$, with $B \in \mathcal{F}$ and $P(B) = 0$, then A also belongs to $\mathcal{F}$.

The elements of $\mathcal{F}$ are called *events*, and, in general, they stand for the available information that we have of a certain random phenomenon (i.e., it has different possible realizations). The sample space Ω is the set of all possible outcomes, and P indicates how probable that occurs an event with respect to another one.

In general, a random phenomenon evolves in time. So, we need the following:

Definition 1 *A filtration is a family $(\mathcal{F}_t)_{t\in[0,T]} = \{\mathcal{F}_t \subset \mathcal{F} : t \in [0, T]\}$ of increasing sub-σ-algebras on Ω. That is, $\mathcal{F}_s \subset \mathcal{F}_t$, for $s < t$.*

The elements of $\mathcal{F}_t$ represent the events that may or may not occur up to time t. Also, it is natural to have an increasing family of sub-σ-algebras because the amount of information of a random experiment increases in time.

The following two properties for the filtration $(\mathcal{F}_t)_{t\in[0,T]}$ are known as the *usual conditions.*

i) The filtration $(\mathcal{F}_t)_{t\in[0,T]}$ is right-continuous (i.e., $\mathcal{F}_t = \mathcal{F}_{t+} := \bigcap_{s>t} \mathcal{F}_s$, for every $t \in [0, T)$),

ii) The σ-algebra $\mathcal{F}_0$ contains all the elements of $\mathcal{F}$ with probability zero.

Note that for every $t \in [0, T]$, the probability space $(\Omega, \mathcal{F}_t, P)$ is complete since $(\Omega, \mathcal{F}, P)$ is so.

DOI: 10.1201/9781003484912-1

1.1 Stochastic processes

Now, we introduce a mathematical model for phenomena that evolve in time. Remember that we are considering the complete probability space $(\Omega, \mathcal{F}, P)$.

Definition 2 *A stochastic process X is a map $X : \Omega \times [0,T] \to \mathbb{R}$ such that for every $t \in [0,T]$, X_t is an $\mathcal{F}$-measurable random variable (i.e., $X_t^{-1}(A) \in \mathcal{F}$, for any open set A of $\mathbb{R}$).*

Note that a random process X is a family $X = \{X_t : t \in [0,T]\}$ of random variables. But, we can also look at it as the set $\{X_\cdot(\omega) : \omega \in \Omega\}$ of maps from $[0,T]$ into $\mathbb{R}$. These functions are called the *paths* of X. This motivates the following three definitions.

Definition 3 *Let $X, Y : \Omega \times [0,T] \to \mathbb{R}$ be two stochastic processes. We say that X and Y are indistinguishable if $P(\{\omega \in \Omega : X_t(\omega) \neq Y_t(\omega)$ for some $t \in [0,T]\}) = 0$.*

We observe that the last definition implies that X and Y are indistinguishable if there exists $A \in \mathcal{F}$, $P(A) = 0$, such that the paths $X_\cdot(\omega)$ and $Y_\cdot(\omega)$ are the same, for any $\omega \in A^c$.

Other type of equality between two stochastic processes is the following:

Definition 4 *Let X and Y be two stochastic processes. X is called a modification of Y if $P([X_t = Y_t]) = 1$, for every $t \in [0,T]$. Here, $[X_t = Y_t]$ represents the set $\{\omega \in \Omega : X_t(\omega) = Y_t(\omega)\}$.*

Note that if X and Y are indistinguishable, then X and Y are a modification of each other, but the converse is not true in general, as the following example shows.

Example 1 *We consider the probability space $(\Omega, \mathcal{F}, P)$, with $\Omega = [0,1]$, $\mathcal{F}$ the Lebesgue σ-algebra on Ω and P the Lebesgue measure. Then, the processes*

$$X_t(\omega) = \begin{cases} 1, & \text{if } t = \omega, \\ 0 & \text{otherwise.} \end{cases} \quad \text{and} \quad Y \equiv 0, \quad t \in [0,1],$$

are modifications of each other, but they are not indistinguishable. Indeed, for $t \in [0,1]$ fixed, we have that $X_t(\omega) = 0 = Y_t(\omega)$ on the set $\{\omega \in \Omega : \omega \neq t\}$, which has probability 1. But, the paths $X_\cdot(\omega)$ and $Y_\cdot(\omega)$ are different, for any $\omega \in \Omega$.

Definition 5 *Let X be a stochastic process. We say that X is a continuous (resp. right-continuous with left-limits, or càdlàg for short) process if the paths of X are continuous (resp. càdlàg) w.p.1 (i.e., with probability 1). That is, there is $A \in \mathcal{F}$, $P(A) = 0$, such that $X_\cdot(\omega)$ is a continuous (resp. càdlàg) function, for every $\omega \in A^c$.*

Proposition 1 *Let X and Y be two right-continuous (rep. left-continuous) processes. Then, they are indistinguishable if X is a modification of Y.*

Proof. For $r \in \mathbb{Q} \cap [0,T]$, we set $A_r = \{\omega \in \Omega : X_r(\omega) \neq Y_r(\omega)\}$ and $A = \cup_{r \in \mathbb{Q} \cap [0,T]} A_r$. By hypothesis, we have that $P(A) \leq \sum_{r \in \mathbb{Q} \cap [0,T]} P(A_r) = 0$. Finally, choose $\omega \in A^c$. Then, for $r \in \mathbb{Q} \cap [0,T]$, $X_r(\omega) = Y_r(\omega)$, which gives that $X_t(\omega) = Y_t(\omega)$ for all $t \in [0,T]$, due to the right (resp. left)-continuity of the processes X and Y. Thus, the proof is finished. □

Let X be an $\mathbb{R}^n$-valued random vector. That is, its components $X^{(i)}$, $i \in \{1, \ldots, n\}$, are random variables. Henceforth, the expected value of X, denoted by $E(X)$, is given by

$$E(X) := \int_\Omega X(\omega) P(d\omega) = \left(\int_\Omega X^{(1)} dP, \ldots, \int_\Omega X^{(n)} dP \right),$$

whenever $X^{(i)} \in L^1(\Omega; P)$, for $i \in \{1, \ldots, n\}$.

The following guarantees that a stochastic process that satisfies a suitable condition on the moments of its increments has a continuous modification.

Theorem 1 (Kolmogorov-Čentsov continuity theorem) *Let X be a stochastic process such that*

$$E\left(|X_t - X_s|^{\gamma}\right) \le C\,|t-s|^{1+\beta}, \quad s,t \in [0,T], \tag{1.1}$$

for some positive constants C, γ and β. Then, there exists a modification $\tilde{X}$ of X such that

$$E\left(\left(\sup_{s\neq t} \frac{|\tilde{X}_t - \tilde{X}_s|}{|t-s|^{\alpha}}\right)^{\gamma}\right) < \infty,$$

for every $\alpha \in [0, \beta/\gamma)$.

Remark 2 *Note that the stochastic process $\tilde{X}$ is α-Hölder continuous (i.e., it has α-Hölder continuous paths), for any $\alpha \in (0, \beta/\gamma)$.*

Proof. Consider the sequence $(\pi^n)_{n\in\mathbb{N}}$ of dyadic partitions of $[0,T]$. It means,

$$\pi^n = \{0 = t_0^n < t_1^n < \cdots < t_{2^n}^n = T\}, \text{ with } t_i^n = \frac{iT}{2^n}.$$

In the remaining of this proof, we use the conventions

$$\Pi = \cup_{n=0}^{\infty} \pi^n, \qquad \tilde{\Pi}^n = \{(s,t) \in \pi^n \times \pi^n : |t-s| = 2^{-n}T \text{ and } s < t\}$$

and $M_n = \sup_{(s,t)\in\tilde{\Pi}^n} |X_t - X_s|$. By inequality (1.1), we have

$$E\left(M_n^{\gamma}\right) \le \sum_{(s,t)\in\tilde{\Pi}^n} E\left(|X_t - X_s|^{\gamma}\right) \le CT^{1+\beta}2^{-n(1+\beta)}2^n = C_{T,\beta}2^{-n\beta}. \tag{1.2}$$

For $s \in \Pi$, we introduce the sequence $\{s_n \in \pi^n : n \in \mathbb{N}\cup\{0\}\}$, where $s_n = \max\{t \in \pi^n : t \le s\}$. Note that this sequence is non-decreasing, $s_n = s$, for n large enough, and either $s_m = t_m$, or $(s_m, t_m) \in \tilde{\Pi}^m$. Therefore, for $s,t \in \Pi$, we get

$$X_t - X_s = X_{t_m} - X_{s_m} + \sum_{n=m}^{\infty} \left(X_{t_{n+1}} - X_{t_n}\right) + \sum_{n=m}^{\infty} \left(X_{s_n} - X_{s_{n+1}}\right)$$

and, consequently,

$$|X_t - X_s| \le M_m + 2\sum_{n=m+1}^{\infty} M_n \le 2\sum_{n=m}^{\infty} M_n, \quad \text{for } |t-s| \le 2^{-m}T. \tag{1.3}$$

Now, let $\tilde{M}_\alpha = \sup\left\{\frac{|X_t - X_s|}{|t-s|^{\alpha}} : s,t \in \Pi \text{ and } s \neq t\right\}$. Then, inequality (1.3) leads us to write

$$\begin{aligned}
\tilde{M}_\alpha &\le \sup_{m\in\mathbb{N}\cup\{0\}} \left\{ \sup_{2^{-(m+1)}T<|t-s|\le 2^{-m}T} \frac{|X_t - X_s|}{|t-s|^{\alpha}}, t,s \in \Pi \right\} \\
&\le \sup_{m\in\mathbb{N}\cup\{0\}} \left\{ 2^{\alpha(m+1)}T^{-\alpha} \sup_{|t-s|\le 2^{-m}T} |X_t - X_s|, s,t \in \Pi \text{ and } s \neq t \right\} \\
&\le \sup_{m\in\mathbb{N}\cup\{0\}\}} \left\{ 22^{\alpha(m+1)}T^{-\alpha} \sum_{n=m}^{\infty} M_n, m \in \mathbb{N}\cup\{0\} \right\} \le 2^{\alpha+1}T^{-\alpha} \sum_{n=0}^{\infty} 2^{n\alpha} M_n.
\end{aligned}$$

Thus, from this inequality, together with (1.2), we obtain, for $\gamma \geq 1$,

$$||\tilde{M}_\alpha||_{L^\gamma(\Omega)} \leq 2^{\alpha+1}T^{-\alpha}\sum_{n=0}^{\infty} 2^{n\alpha}||M_n||_{L^\gamma(\Omega)} \leq 2^{\alpha+1}T^{-\alpha}C_{T,\beta}^{1/\gamma}\sum_{n=0}^{\infty} 2^{n\alpha}2^{-n\beta/\gamma} < \infty.$$

For $\gamma < 1$, the penultimate inequality gives

$$E\left(\tilde{M}_\alpha^\gamma\right) \leq 2^{\gamma(\alpha+1)}T^{-\alpha\gamma}\sum_{n=0}^{\infty} 2^{n\alpha\gamma}E\left(M_n^\gamma\right) < \infty.$$

In particular, the paths of X are uniformly continuous with probability 1 on Π, which allows us to see that, there is $A \in \mathcal{F}$, $P(A) = 0$, such that for $\omega \in \Omega$ and $t \in [0,T]$,

$$\tilde{X}_t(\omega) = \lim_{s \to t, s \in \Pi} 1_{A^c}(\omega)X_s(\omega)$$

is well-defined.

Finally, inequality (1.1) and Fatou's lemma imply the process $\tilde{X}$ is the modification of X that we were looking for. Hence, the proof is complete. □

In the sequel, we deal with the following type of measurabilities.

Definition 6 *Let $X = \{X_t, t \in [0,T]\}$ be a stochastic process and $(\mathcal{F}_t)_{t\in[0,T]}$ a filtration on $(\Omega, \mathcal{F}, P)$.*

i) X is a measurable process if $X^{-1}(A) \in \mathcal{F} \otimes \mathcal{B}([0,T])$, for any open set A of $\mathbb{R}$. Here, $\mathcal{B}([0,T])$ is the Borel σ-algebra of $[0,T]$. In this case, we use the convention

$$X : (\Omega \times [0,T], \mathcal{F} \otimes \mathcal{B}([0,T])) \to \mathbb{R}.$$

ii) We say that X is $\mathcal{F}_t$-adapted (or adapted to the filtration $(\mathcal{F}_t)_{t\in[0,T]}$) if $X_t^{-1}(A) \in \mathcal{F}_t$, for any open set A of $\mathbb{R}$ and $t \in [0,T]$.

iii) The stochastic process X is called progressively measurable (with respect to the filtration $(\mathcal{F}_t)_{t\in[0,T]}$) if for each $t \in [0,T]$ we have

$$X\big|_{\Omega\times[0,t]} : (\Omega \times [0,t], \mathcal{F}_t \otimes \mathcal{B}([0,t])) \to \mathbb{R}.$$

Remark 3 *i) In Definitions 6.i) and 6.ii), we can write Borel set A of $\mathbb{R}$ instead of open set A of $\mathbb{R}$ ($A \in \mathcal{B}(\mathbb{R})$ for short), which can be proven by the reader.*

ii) Definition 6.i) only depends on the σ-algebra $\mathcal{F}$, while Definitions 6.ii) and 6.iii) depend on the filtration $(\mathcal{F}_t)_{t\in[0,T]}$. Also, Definition 6.ii) establishes that X_t only depends on the information that we have up to time $t \in [0,T]$.

iii) Note that a process can be adapted, but not measurable. Indeed, let X be given by $X_t(\omega) = 1_A(t)1_\Omega(\omega)$, with $A \notin \mathcal{B}(\mathbb{R})$. Then, X is an adapted process with respect to any filtration, but is not measurable.

iv) It is easy to see that a progressively measurable process is also measurable and adapted.

The following result gives an example of progressively measurable processes.

Proposition 2 *Let X be an $\mathcal{F}_t$-adapted process such that all its paths are left (resp. right)-continuous. Then, X is also a progressively measurable process with respect to $(\mathcal{F}_t)_{t\in[0,T]}$.*

Proof. For $t \in [0,T]$, we set

$$X_s^{(n,t)} = X_0 + \sum_{i=0}^{n-1} X_{\frac{it}{n}} 1_{(\frac{it}{n}, \frac{(i+1)t}{n}]}(s), \quad n \in \mathbb{N} \text{ and } s \in [0,t].$$

(resp.

$$X_s^{(n,t)} = X_0 + \sum_{i=0}^{n-1} X_{\frac{(i+1)t}{n}} 1_{(\frac{it}{n}, \frac{(i+1)t}{n}]}(s), \quad n \in \mathbb{N} \text{ and } s \in [0,t].)$$

Thus, the results are an immediate consequence of the fact that

$$X_s(\omega) = \lim_{n\to\infty} X_s^{(n,t)}(\omega), \quad (\omega, s) \in \Omega \times [0,t].$$

□

Concerning Remark 3.*iv*), we have the following converse argument. This result was first established by Chung and Doob [29], under the assumption that the involved process is separable. Some proofs of the following result can be found in Dellacherie and Meyer [37], Kaden and Potthoff [92], and Ondreját and Seidler [170]. For the sake of completeness and the readers convenience, we include the proof given in [170] because of its simplicity. Some details of it are provided in the appendix (Section 8.2) as auxiliary lemmas in order to improve the presentation.

Theorem 2 *Let X be a measurable and $\mathcal{F}_t$-adapted process. Then, it has a progressively measurable modification.*

Proof. The proof is divided into two steps.

Step 1: Here we assume that the stochastic process X has a countable range. It means, there are $N \in \mathbb{N} \cup \{\infty\}$, $\{x_i \in \mathbb{R} : 1 \le i < N\}$, $x_i \ne x_j$ for $i \ne j$ and an $\mathcal{F} \otimes \mathcal{B}([0,T])$-measurable partition $\{B_j : 1 \le j < N\}$ of $\Omega \times [0,T]$ such that $X \equiv x_j$ on B_j. Note that 1_{B_j} is an $\mathcal{F}_t$-adapted stochastic process due to $1_{B_j}(t) = 1_{\{x_j\}}(X_t)$. Hence Corollary 31 implies that there is a family $\{Y^{(i)} : 1 \le i < N\}$ of progressively measurable processes such that $Y^{(i)}$ is a modification of 1_{B_i}. Set

$$C_j = \{Y^{(j)} = 1\} \text{ and } \Gamma_j = C_j \setminus \cup_{i=1}^{j-1} C_i, \quad 1 \le j < N.$$

Now, it is easy to see that the process

$$Y_t(\omega) = \begin{cases} x_j, & \text{if } (\omega,t) \in \Gamma_j, 1 \le j < N, \\ 0, & \text{otherwise} \end{cases}$$

is a progressively measurable modification of X.

Step 2: By Lemma 84, we have a sequence $\{Y^{(n)} : n \in \mathbb{N}\}$ of measurable processes with countable rang such that $\sup_{(\omega,t)\in\Omega\times[0,T]} |X_t(\omega) - Y_t^{(n)}(\omega)| < \frac{1}{n}$, $n \in \mathbb{N}$. Thus, from Step 1, we can obtain a sequence $\{\tilde{Y}^{(n)} : n \in \mathbb{N}\}$ of progressively measurable processes such that $\tilde{Y}^{(n)}$ is a modification of $Y^{(n)}$ and

$$|X_t - \tilde{Y}_t^{(n)}| < \frac{1}{n}, \quad \text{w.p.1, for each } n \in \mathbb{N}.$$

Therefore, the progressively measurable modification of X can be defined as

$$Y_t = \begin{cases} \lim_{n\to\infty} \tilde{Y}_t^{(n)}, & \text{if the limit exists,} \\ 0, & \text{otherwise,} \end{cases}$$

since $1_{\{(\omega,t)\in\Omega\times[0,T]:\lim_{n\to\infty}\tilde{Y}_t^{(n)}(\omega)\text{ exists}\}}$ is a progressively measurable process (see the proof of Lemma 83). Thus, the proof is finished. □

In general, the filtration considered in Definitions 6.*ii*) and 6.*iii*) is the filtration $(\mathcal{F}_t^X)_{t\in[0,T]}$ generated by the process X, which is defined by

$$\mathcal{F}_t^X = \sigma\{X_s : 0 \le s \le t\},$$

where $\sigma\{X_s : s \le t\}$ is the smallest *sigma*-algebra that contains the family $\mathcal{G} = \{X_s^{-1}(A) : 0 \le s \le t \text{ and } A \in \mathcal{B}(\mathbb{R})\}$. It is easy to see that

$$\mathcal{F}_t^X = \cap_{\tilde{\mathcal{F}}\in\mathcal{G}(X)}\tilde{\mathcal{F}},$$

with $\mathcal{G}(X)$ the family of all the σ-algebras that contain the set $\mathcal{G}$, due to the arbitrary intersection of σ-algebras being also a σ-algebra.

We may think that the filtration $\mathcal{F}^X$ should be right-continuous if the process X is continuous. However, we have the following elementary example.

Example 2 *Consider a car that moves to the right along a straight line with speed 1. At time $t = 1$, we toss a coin. Then, we continue our way if the outcome is head, and we come back otherwise.*

In this case, $\Omega = \{head, tail\}$ and the position of the car at time $t \in [0,2]$, denoted by X_t, has the paths

$$X_t(head) = t \quad \text{and} \quad X_t(tail) = \begin{cases} t, & \text{if } t \le 1, \\ 2-t & \text{if } t > 1. \end{cases}$$

Therefore, X is a continuous process, but

$$\mathcal{F}_t^X = \begin{cases} \{\Omega,\emptyset\}, & \text{if } t \le 1, \\ 2^\Omega, & \text{if } t > 1, \end{cases} \quad \text{and} \quad \mathcal{F}_{t+}^X = \begin{cases} \{\Omega,\emptyset\}, & \text{if } t < 1, \\ 2^\Omega, & \text{if } t \ge 1. \end{cases}$$

Thus $\mathcal{F}^X$ does not satisfiy the usual conditions.

In order to state the following theorem, we need to introduce the following:

Definition 7 *Let $\mathcal{I}$ be an index set. We say that the random variables $\{X_i : i \in \mathcal{I}\}$ are independent if for every finite subset of indices $\{i_1,\dots,i_n\} \subset \mathcal{I}$ and every Borel measurable subsets $B_1,\dots,B_n \in \mathcal{B}(\mathbb{R})$, we have*

$$P([X_{i_1} \in B_1,\dots,X_{i_n} \in B_n]) = \prod_{j=1}^n P([X_{i_j} \in B_j]).$$

In particular, we obtain that if X and Y are two independent random variables in $L^1(\Omega)$, then XY also belongs to $L^1(\Omega)$ and $E(XY) = (EX)(EY)$, which is easily proven since X and Y can be approximated in $L^1(\Omega)$ by simples process of the form $\sum_{k=1}^m a_k 1_{[Z\in B_i]}$, where $a_k \in \mathbb{R}$, $B_k \in \mathcal{B}(\mathbb{R})$, and Z is X or Y, respectively. The proof is left to the reader as an exercise.

In the following result, $\bar{\mathcal{F}}_t^X$ (resp. $\bar{\mathcal{F}}_{t+}^X$) stands for the σ-algebra generated by $\mathcal{F}_t^X$ (resp. $\mathcal{F}_{t+}^X$) and all the sets of $\mathcal{F}$ with probability zero.

Theorem 3 *Let X be a d-dimensional stochastic process with right continuous paths such that the random variables $X_0, X_{t_1} - X_0, \dots, X_{t_n} - X_{t_{n-1}}$ are independent, for any $n \in \mathbb{N}$ and $0 < t_1 < \dots < t_n \le T$ (i.e., X has independent increments). Then, $\bar{\mathcal{F}}_{t+}^X = \bar{\mathcal{F}}_t^X$, for all $t \in [0,T]$.*

Proof. To simplify the notation, we assume that $d = 1$. We first observe that we only need to show that

$$P(A|\mathcal{F}_t^X) = P(A|\mathcal{F}_{t+}^X) \text{ w.p.1}, \quad \text{for each } t \in [0,T] \text{ and } A \in \mathcal{F}_T^X. \tag{1.4}$$

Indeed, let $t \in [0,T]$ and $A \in \mathcal{F}_{t+}^X$. Then, (1.4) implies that $1_A = P(A|\mathcal{F}_t^X)$ w.p.1. In other words, 1_A is equal to an $\mathcal{F}_t^X$-measurable random variable w.p.1. Thus, we have that A also belongs to $\bar{\mathcal{F}}_t^X$, which proves that our claim is true.

Now we observe that it is enough to see that (1.4) is satisfied for

$$A \in \sigma\{X_{u_n}, \ldots, X_{u_1}, X_t, X_{s_1}, \ldots, X_{s_m}\},$$

with $u_n > \ldots > u_1 > t > s_1 > \ldots > s_m$, due to the monotone class lemma in Chapter 8 (Theorem 96) and the fact that the union of these σ-algebras being a π-system. Note that if $t = 0$, we omit the points $s_1, \ldots s_m$. Hence, applying the monotone class lemma (Theorem 96 in Chapter 8) again, we only need to show that (1.4) holds for any A of the form

$$\begin{aligned} A = \{\delta_{u_{n-1},u_n} X \in B_n, \quad \ldots, \quad & \delta_{u_1,u_2} X \in B_2, \delta_{t,u_1} X \in B_1, \delta_{s_1,t} X \in \tilde{B}_1, \\ \ldots, \quad & \delta_{s_m,s_{m-1}} X \in \tilde{B}_m, X_{s_m} \in \tilde{B}_{m+1}\}, \end{aligned}$$

where $\delta_{a,b} X = X_b - X_a$ and $B_i, \tilde{B}_j \in \mathcal{B}(\mathbb{R})$. For this event A, we have that the independence of the increments of X and $u_1 > t$ yield

$$\begin{aligned} P(A|\mathcal{F}_{t+}^X) &= E(E(1_A|\mathcal{F}_{u_1}^X)|\mathcal{F}_{t+}^X) = 1_{\{\delta_{s_1,t} X \in \tilde{B}_1, \ldots, \delta_{s_m,s_{m-1}} X \in \tilde{B}_m, X_{s_m} \in \tilde{B}_{m+1}\}} \\ &\quad \times E(1_{\{\delta_{t,u_1} X \in B_1\}} P(\delta_{u_{n-1},u_n} X \in B_n, \ldots, \delta_{u_1,u_2} X \in B_2 | \mathcal{F}_{u_1}^X)|\mathcal{F}_{t+}^X) \\ &= 1_{\{\delta_{s_1,t} X \in \tilde{B}_1, \ldots, \delta_{s_m,s_{m-1}} X \in \tilde{B}_m, X_{s_m} \in \tilde{B}_{m+1}\}} \\ &\quad \times P(\delta_{t,u_1} X \in B_1 | \mathcal{F}_{t+}^X) \Pi_{k=2}^n P(\delta_{u_{k-1},u_k} X \in B_k). \end{aligned}$$

Similarly, changing $\mathcal{F}_{t+}^X$ by $\mathcal{F}_t^X$ in previous calculations, we can write

$$\begin{aligned} P(A|\mathcal{F}_t^X) &= 1_{\{\delta_{s_1,t} X \in \tilde{B}_1, \ldots, \delta_{s_m,s_{m-1}} X \in \tilde{B}_m, X_{s_m} \in \tilde{B}_{m+1}\}} \\ &\quad \times P(\delta_{t,u_1} X \in B_1 | \mathcal{F}_t^X) \Pi_{k=2}^n P(\delta_{u_{k-1},u_k} X \in B_k). \end{aligned}$$

Therefore, in order to finish the proof, we must see that

$$E(f(X_u, X_t)|\mathcal{F}_{t+}^X) = E(f(X_u, X_t)|\mathcal{F}_t^X) \quad \text{w.p.1}$$

holds, for each bounded Borel-measurable function $f : \mathbb{R}^2 \to \mathbb{R}$ and $u > t$. Consequently, the problem is reduced to show that

$$E(g(X_u)|\mathcal{F}_{t+}^X) = E(g(X_u)|\mathcal{F}_t^X) \quad \text{w.p.1}, \tag{1.5}$$

for each $u > t$ and a bounded continuous function $g : \mathbb{R} \to \mathbb{R}$ because by the monotone class lemma in Chapter 8 (Theorem 97), we can consider f of the form $f(x_1, x_2) = f_1(x_1) f_2(x_2)$.

Finally, we observe that, for a bounded continuous function, $g : \mathbb{R} \to \mathbb{R}$ and $u > t$, the independence of the increments leads us to conclude

$$E(g(X_u)|\mathcal{F}_t^X) = E(g(X_u - X_t + X_t)|\mathcal{F}_t^X) = g_u(t, X_t). \tag{1.6}$$

Here, $g_u(t, x) = Eg(X_u - X_t + x)$, $x \in \mathbb{R}$. Proceeding similarly, for $v \in (t, u)$, we also have

$$E(g(X_u)|\mathcal{F}_{t+}^X) = E(E(g(X_u)|\mathcal{F}_v^X)|\mathcal{F}_{t+}^X) = E(g_u(v, X_v)|\mathcal{F}_{t+}^X),$$

which, together with the dominated convergence theorem, the fact that X has right-continuous paths and (1.6), gives that (1.5) is true. So, the proof is complete. □

Definition 8 *Let $n \in \mathbb{N}$. A function $X : \Omega \times [0,T] \to \mathbb{R}^n$ is a stochastic process with values in $\mathbb{R}^n$ if each component $X^{(i)}$, $i \in \{1,\dots,n\}$, is a real-valued stochastic process.*

Note that X is an $\mathbb{R}^n$-valued stochastic process if and only if $X_t^{-1}(A) \in \mathcal{F}$, for every $t \in [0,T]$ and $A \in \mathcal{B}(\mathbb{R}^n)$ (i.e., A is a Borel set of $\mathbb{R}^n$). Actually, it is enough that the last property holds for any open set A of $\mathbb{R}^n$.

1.1.1 Gaussian processes

In this section, we use some properties of the characteristic function of a random vector. The reader interested in this subject can consult any book of probability (for instance, Kallenberg [94] or Tucker [209]). In particular, we utilize the following characterization theorem. In order to state it, we denote the distribution of a random vector X by F_X.

Theorem 4 *Let X and Y be two $\mathbb{R}^n$-valued random vectors with distribution functions F_X and F_Y, and with characteristic functions φ_X and φ_Y, respectively. Then, $F_X = F_Y$ if and only if $\varphi_X = \varphi_Y$.*

A random variable X is a Gaussian random variable if its characteristic function φ_X satisfies

$$\varphi_X(u) := E\left(e^{iuX}\right) = \exp\left(imu - \frac{\sigma^2 u^2}{2}\right), \quad u \in \mathbb{R}, \tag{1.7}$$

for some $m, \sigma \in \mathbb{R}$. Consequently, differentiating (1.7) with respect to u, we obtain

$$E(X) = m \quad \text{and} \quad Var(X) := E\left((X-m)^2\right) = \sigma^2. \tag{1.8}$$

In this case, we say that X has the normal distribution $\mathcal{N}(m,\sigma^2)$ ($X \sim \mathcal{N}(m,\sigma^2)$ for short).

This definition can be extended to random vectors:

Definition 9 *Let Y be a random vector with values in $\mathbb{R}^n$. We say that Y is a Gaussian random vector if $\langle Y, y\rangle$ is a Gaussian random variable, for any $y \in \mathbb{R}^n$. Here, $\langle\cdot,\cdot\rangle$ is the usual inner product in $\mathbb{R}^n$.*

Notice that Y is an $\mathbb{R}^n$-Gaussian random vector if and only if any linear combination of its components is a Gaussian random variable. So, equalities (1.7) and (1.8) yield that the characteristic function φ_Y of the $\mathbb{R}^n$-Gaussian random vector Y is given by

$$\begin{aligned}\varphi_Y(y) &:= E\left(e^{i\langle Y,y\rangle}\right) = \exp\left(iE(\langle Y,y\rangle) - \frac{Var(\langle Y,y\rangle)}{2}\right)\\ &= \exp\left(imy^t - \frac{y\Sigma_Y y^t}{2}\right), \quad y \in \mathbb{R}^n. \end{aligned}\tag{1.9}$$

Here, $m = (m_1,\dots,m_n)$ and $\Sigma_Y = (\Sigma_{k,j})_{k,j=1}^n$ with

$$m_j = E(Y^{(j)}),\ \Sigma_{k,j} = E\left(\left(Y^{(k)} - m_k\right)\left(Y^{(j)} - m_j\right)\right), \quad k,j \in \{1,\dots,n\},$$

and y^t is the transpose of the vector $y = (y_1,\dots,y_n)$. Σ_Y is a symmetric and positive definite $n \times n$-matrix (i.e., $y\Sigma_Y y^t \geq 0$, for any $y \in \mathbb{R}^n$) due to

$$y\Sigma_Y y^t = E\left(\left(\sum_{j=1}^n y_j(Y^{(j)} - m_j)\right)^2\right), \quad y \in \mathbb{R}^n.$$

m and Σ_Y are called the *mean vector* and the *covariance matrix* of Y, respectively. Thus, from (1.9), the distribution of a Gaussian random vector is characterized by its mean vector and covariance matrix using the characterization property of φ_Y (see Theorem 4).

Other important consequence of (1.7) is the following result.

Proposition 3 *Let Y be an $\mathbb{R}^n$-valued Gaussian random vector such that*

$$E\left(\left(Y^{(k)} - m_k\right)\left(Y^{(j)} - m_j\right)\right) = 0, \quad k \neq j.$$

Then, the random variables $Y^{(1)}, \ldots, Y^{(n)}$ are independent.

Proof. Note that, in this case, (1.9) yields

$$\varphi_Y(y) = \prod_{j=1}^{n} \exp\left(i m_j y_j - \frac{\Sigma_{j,j} y_j^2}{2}\right), \quad y \in \mathbb{R}^n,$$

which implies that the result is true. □

Now, we recall that a matrix $A = (A_{k,j})_{k,j=1}^n$ with real components is strictly positive definite if it is positive definite and $\sum_{k,j=1}^n y_j y_k A_{k,j} = 0$ implies that $(y_1, \ldots, y_n) = 0$. It is well-known that the rank of a strictly positive definite $n \times n$-matrix is n.

Proposition 4 *Let Y be an $\mathbb{R}^n$-valued Gaussian random vector with mean m and strictly positive definite covariance matrix Σ. Then, the distribution of Y is absolutely continuous with respect to the Lebesgue measure on $\mathbb{R}^n$ and has the density function*

$$f(x) = \frac{1}{(2\pi)^{n/2}(\det\Sigma)^{1/2}} \exp\left(-\frac{(x-m)\Sigma^{-1}(x-m)^t}{2}\right), \quad x \in \mathbb{R}^n,$$

where $\det\Sigma$ *stands for the determinant of* Σ.

Proof. In order to prove that the result holds, we only need to see that

$$\varphi_Y(y) = \int_{\mathbb{R}^n} e^{i\langle y,x\rangle} f(x)dx, \quad y \in \mathbb{R}^n,$$

where φ_Y is given by (1.9) because we can use Theorem 4 again.

The change of variables formula and the fact that, in this case, Σ has a symmetric square root $\Sigma^{1/2}$ lead us to get

$$\begin{aligned}
&\int_{\mathbb{R}^n} e^{i\langle y,x\rangle} \exp\left(-\frac{(x-m)\Sigma^{-1}(x-m)^t}{2}\right) dx \\
&= e^{i\langle y,m\rangle} \int_{\mathbb{R}^n} e^{i\langle y,x\rangle} \exp\left(-\frac{x\Sigma^{-1}x^t}{2}\right) dx \\
&= (\det\Sigma)^{1/2} e^{i\langle y,m\rangle} \int_{\mathbb{R}^n} e^{i\langle y\Sigma^{1/2},x\rangle} \exp\left(-\frac{xx^t}{2}\right) dx \\
&= (2\pi)^{n/2}(\det\Sigma)^{1/2} e^{i\langle y,m\rangle} \exp\left(-\frac{y\Sigma y^t}{2}\right).
\end{aligned}$$

Here, in the last equality, we apply (1.7), (1.9) and the fact that

$$\frac{1}{(2\pi)^{n/2}} \int_{\mathbb{R}^n} e^{i\langle y\Sigma^{1/2},x\rangle} \exp\left(-\frac{xx^t}{2}\right) dx$$

is the characteristic function of a Gaussian random vector with zero mean and covariance matrix equal to the identity matrix I_n. □

In the following chapters, we use the following definition.

Definition 10 *Let $X : \Omega \times \mathcal{I} \to \mathbb{R}$ be a stochastic process indexed by the elements of a set $\mathcal{I}$ (i.e., X_i is a random variable for any $i \in \mathcal{I}$). We say that X is a Gaussian stochastic process if $(X_{i_1}, \ldots, X_{i_n})$ is an $\mathbb{R}^n$-valued Gaussian vector, for every $n \in \mathbb{N}$ and $i_1, \ldots, i_n \in \mathcal{I}$.*

The Kolmogorov–Čentsov continuity theorem (i.e., Theorem 1) for Gaussian processes is stated in the following result.

Proposition 5 *Let $X = \{X_t, t \in [0,T]\}$ be a centered Gaussian process (i.e., with zero mean) such that*

$$E\left((X_t - X_s)^2\right) \leq C|t-s|^\beta, \quad s,t \in [0,T],$$

for some positive constants C and β. Then, for every $\alpha \in (0, \beta/2)$, there exists a modification $\tilde{X}$ of X with α-Hölder continuous paths and with Hölder constant in $L^p(\Omega)$, for any $p > 1$. That is,

$$\left|\tilde{X}_t - \tilde{X}_s\right| \leq M_\alpha |t-s|^\alpha, \quad s,t \in [0,T],$$

where $M_\alpha \in L^p(\Omega)$, for every $p > 1$,

Proof. Let $\sigma^2 = E\left((X_t - X_s)^2\right)$ and $n \in \mathbb{N}$. Then, the fact that X is a Gaussian process and integration by parts formula imply

$$\begin{aligned} E\left((X_t - X_s)^{2n}\right) &= \frac{1}{\sqrt{2\pi\sigma^2}} \int_{\mathbb{R}} x^{2n} \exp(-\frac{x^2}{2\sigma^2}) dx = -\frac{\sigma^2}{\sqrt{2\pi\sigma^2}} \int_{\mathbb{R}} x^{2n-1} \frac{d\exp(-\frac{x^2}{2\sigma^2})}{dx} dx \\ &= \frac{(2n-1)\sigma^2}{\sqrt{2\pi\sigma^2}} \int_{\mathbb{R}} x^{2n-2} \exp(-\frac{x^2}{2\sigma^2}) dx. \end{aligned} \tag{1.10}$$

Hence, using induction on n, we can see that (1.1) holds when we write $2n$ and $n\beta - 1$ instead of γ and β, respectively. □

Also, we will need the following auxiliary result later on.

Lemma 1 *Let Y be a Gaussian random variable with zero mean and variance σ^2. Then, for $\alpha \in (0,1)$, there exists a constant $C = C(\alpha) > 0$ such that*

$$E(\frac{1}{|Y|^\alpha}) \leq \frac{C}{\sigma^\alpha}. \tag{1.11}$$

Proof. By means of the change of variables formula $u = \frac{x}{\sigma}$, we obtain

$$\begin{aligned} E\left(\frac{1}{|Y_s|^\alpha}\right) &= \int_{\mathbb{R}} \frac{1}{|x|^\alpha} \frac{e^{-x^2/2\sigma^2}}{\sqrt{2\pi\sigma^2}} dx = \frac{1}{\sigma^\alpha}[\int_{-1}^{1} \frac{1}{|u|^\alpha} \frac{e^{-u^2/2}}{\sqrt{2\pi}} du + \int_{|u|>1} \frac{1}{|u|^\alpha} \frac{e^{-u^2/2}}{\sqrt{2\pi}} du] \\ &\leq \frac{C}{\sigma^\alpha}[\int_{-1}^{1} \frac{1}{|u|^\alpha} du + \int_{|u|>1} \frac{e^{-u^2/2}}{\sqrt{2\pi}} du] \leq \frac{C}{\sigma^\alpha}. \end{aligned}$$

Thus (1.11) holds. □

1.2 Brownian motion

Here, we study some facts of one of the most important processes of stochastic analysis. Namely, the Brownian motion (Bm), or Wiener process. The Bm was first observed by the botanist Robert Brown [23] when observing the motion of pollen grains in water. Since then, this process has been used in different areas of scientific knowledge (e.g., physics, biology, engineering, quantum mechanics, etc.), because it has allowed us to model and analyze systems with random perturbations (see Gorostiza [65]). The mathematical formulation of Bm was initiated by Bachelier [17] and, in the field of physics, by Einstein [47]. A rigorous mathematical treatment was begun by Wiener [212].

Remember that we are assuming that the involved probability space $(\Omega, \mathcal{F}, P)$ is complete.

Definition 11 *Let $(\mathcal{F}_t)_{t\in[0,T]}$ be a filtration on the probability space $(\Omega, \mathcal{F}, P)$. A real-valued continuous and $\mathcal{F}_t$-adapted process $B = \{B_t : t \in [0,T]\}$ is called a (standard) Brownian motion, or Wiener process, with respect to $(\mathcal{F}_t)_{t\in[0,T]}$, if it satisfies the following three conditions:*

i) $B_0 = 0$ *w.p.1.*

ii) For $0 \le s < t \le T$, the random variable $B_t - B_s$ is independent of the σ-algebra $\mathcal{F}_s$.

iii) For $s,t \in [0,T]$, $s < t$, $B_t - B_s$ is normally distributed with mean 0 and variance $t-s$ (i.e., $B_t - B_s \sim \mathcal{N}(0, t-s)$).

Remark 4 *In Subsection 1.2.1, we consider a complete probability space $(\Omega, \mathcal{F}, P)$, where it is defined a Brownian motion. In general, the filtration considered in Definition 11 is the one generated by B augmented with the null-sets of Ω (i.e., $\mathcal{F}_0$ contains all the null-sets). Remember that this filtration satisfies the usual conditions because of Theorem 3. In this case, we only say that B is a Brownian motion, or a Wiener process, defined on $(\Omega, \mathcal{F}, P)$ and Statement ii) is changed by*

ii′) The process B has independent increments.

Note that Condition *ii)* (or *ii′*)) yields that B and $t \mapsto B_t^2 - t$ are $\mathcal{F}_t$-martingales (i.e., $E[B_t|\mathcal{F}_s] = B_s$ and $E[B_t^2 - t|\mathcal{F}_s] = B_s^2 - s$, for $s < t$). For instance,

$$\begin{aligned} E[B_t^2 - B_s^2|\mathcal{F}_s] &= E[(B_t + B_s)(B_t - B_s)|\mathcal{F}_s] = E[B_t(B_t - B_s)|\mathcal{F}_s] + B_s E[B_t - B_s|\mathcal{F}_s] \\ &= E[(B_t - B_s)^2|\mathcal{F}_s] + 2B_s E[B_t - B_s|\mathcal{F}_s] \\ &= E((B_t - B_s)^2) + E(B_t - B_s) = t - s, \end{aligned} \tag{1.12}$$

where, in the last equality, we consider the mentioned condition, together with *iii)*.

Henceforth, $(B, (\mathcal{F}_t)_{t\in[0,T]})$ denotes a Brownian motion B with respect to the filtration $(\mathcal{F}_t)_{t\in[0,T]}$. The reason that we consider the filtration $(\mathcal{F}_t)_{t\in[0,T]}$ is that the integrands of the Itô's integral with respect to B can be $\mathcal{F}_t$-adapted processes satisfying suitable integrability conditions, as we will see later on.

Now, we prove that a Brownian motion is also a Gaussian process.

Proposition 6 *Let $(B, (\mathcal{F}_t)_{t\in[0,T]})$ be a Brownian motion defined on a complete probability space $(\Omega, \mathcal{F}, P)$. Then, it is a Gaussian process. Moreover, for $0 \le t_1 < \ldots < t_n \le T$ and $A \in \mathcal{B}(\mathbb{R}^n)$, we have*

$$P([(B_{t_1}, \ldots, B_{t_n}) \in A]) = \int_A \prod_{i=1}^{n} \left(\frac{1}{\sqrt{2\pi(t_i - t_{i-1})}} \exp\left(-\frac{(u_i - u_{i-1})^2}{2(t_i - t_{i-1})} \right) \right) du_1 \cdots u_n, \tag{1.13}$$

where $t_0 = u_0 = 0$.

Proof. Consider the linear operator $U : \mathbb{R}^n \to \mathbb{R}^n$ defined as $U(r_1, \ldots, r_n) = (r_1, r_1 + r_2, \ldots, r_1 + \cdots + r_n)$, whose inverse is $U^{-1}(\theta_1, \ldots, \theta_n) = (\theta_1, \theta_2 - \theta_1, \ldots, \theta_n - \theta_{n-1})$. Hence $U(I) = J$, with

$$I = (B_{t_1}, B_{t_2} - B_{t_1}, \ldots, B_{t_n} - B_{t_{n-1}}) \quad \text{and} \quad J = (B_{t_1}, \ldots, B_{t_n}).$$

From (1.7), $B_{t_j} - B_{t_{j-1}} \sim \mathcal{N}(0, t_j - t_{j-1})$ and the independence of the increments of the Brownian motion B, we get

$$\varphi_I(y) = \prod_{j=1}^{n} E\left(e^{i(B_{t_j} - B_{t_{j-1}})y_j}\right) = \exp\left(-yD(t_1, \ldots, t_n)y^t\right), \quad y \in \mathbb{R}^n,$$

where $D(t_1, \ldots, t_n) = \text{diag}(t_1, t_2 - t_1 \ldots, t_n - t_{n-1})$. That is, I is a Gaussian random vector with zero mean and covariance matrix $D(t_1, \ldots, t_n)$. Therefore,

$$\varphi_J(y) = E\left(e^{i\langle U(I), y\rangle}\right) = E\left(e^{i\langle I, U^*(y)\rangle}\right) = \exp\left(-\frac{y(UD(t_1, \ldots, t_n)U^*)y^t}{2}\right), \quad y \in \mathbb{R}^n.$$

Thus, by (1.9), J is a Gaussian vector with zero mean and covariance matrix $UD(t_1, \ldots, t_n)U^*$ due to this matrix being symmetric and positive definite. Consequently, Proposition 4 leads us to write, for $A \in \mathcal{B}(\mathbb{R}^n)$,

$$\begin{aligned} P\left([(B_{t_1}, \ldots, B_{t_n}) \in A]\right) &= \frac{1}{\sqrt{(2\pi)^n \det(UD(t_1, \ldots, t_n)U^*)}} \\ &\quad \times \int_A \exp\left(-\frac{1}{2}\left\langle (UD(t_1, \ldots, t_n)U^*)^{-1}\theta, \theta\right\rangle\right) d\theta_1 \cdots d\theta_n, \end{aligned}$$

Finally, $\det(U) = 1$ and

$$\begin{aligned} \left\langle (UD(t_1, \ldots, t_n)U^*)^{-1}\theta, \theta\right\rangle &= \left\langle (U^*)^{-1}D(t_1, \ldots, t_n)U^{-1}\theta, \theta\right\rangle \\ &= \left\langle (D(t_1, \ldots, t_n))^{-1}U^{-1}\theta, U^{-1}\theta\right\rangle = \frac{\theta_1^2}{t_1} + \sum_{i=2}^{n} \frac{(\theta_i - \theta_{i-1})^2}{t_i - t_{i-1}} \end{aligned}$$

imply that (1.13) is also satisfied. Thus, the proof is complete. □

Note that the last proof does not use the continuity of the Brownian motion B. So, a process $\tilde{B}$ that meets the conditions of Definition 11 except continuity is a Gaussian process. Thus, if $\mathcal{F}_0$ contains all the null-sets of Ω, the continuous version of $\tilde{B}$ given by Proposition 5 is a Brownian motion. Hence, by Section 1.1.1, Bm is a continuous version of a zero mean Gaussian process $\tilde{B}$ with covariance function

$$R_{1/2}(s, t) := E(\tilde{B}_t \tilde{B}_s) = t \wedge s, \quad s, t \in [0, T].$$

Actually, this continuous version B has Hölder continuous paths for any exponent less than 1/2. That is, for $\gamma \in (0, 1/2)$ and $p > 1$, there is a random variable M in $L^p(\Omega)$ such that

$$|B_t - B_s| \leq M|t - s|^\gamma, \quad \text{for all } t, s \in [0, T], \text{ w.p.1.}$$

We remark that we write $R_{1/2}$ because, later on, we will consider a Brownian motion as a fractional Brownian motion with Hurst parameter $H = 1/2$. Also, observe that, if B is a Brownian motion and $s < t$, we then have

$$E(B_t B_s) = E\left((B_t - B_s) B_s\right) + E\left(B_s^2\right) = s,$$

where the last equality follows from the independence of $B_t - B_s$ and B_s.

Given a Brownian motion $(B, (\mathcal{F}_t)_{t \in [0,T]})$, the following processes are also Brownian motion ones:

a) Translation invariance: Let $t_0 \in (0,T)$ be fixed. Then, $t \mapsto B_{t+t_0} - B_{t_0}$, $t \in [0, T-t_0]$, is a Brownian motion with respect to the filtration $(\mathcal{F}_{t+t_0})_{t\in[0,T-t_0]}$.

b) Scaling invariance: Let $a > 0$. Then, $t \mapsto a^{-\frac{1}{2}} B_{at}$, for $t \in [0, T/a]$, is a Brownian motion with respect to the filtration $(\mathcal{F}_{at})_{t\in[0,T/a]}$.

c) Here we suppose that the $B = \{B_t : t \geq 0\}$ is a Brownian motion. Then,

$$B_t^{(3)} = \begin{cases} tB_{1/t}, & \text{if } t > 0, \\ 0 & \text{if } t = 0, \end{cases}$$

is also a Brownian motion.

Statements *a*) and *b*) are easy to prove and they are left as an exercise to the reader. For Statement *c*), we have that $B^{(3)}$ is zero-mean Gaussian process with covariance $R(s,t) = t \wedge s$ due to Brownian motion being so. Hence, we only need to show that this process is continuous al zero, w.p.1. Towards this end, we have already pointed out that $B^{(3)}$ has a continuous modification $\tilde{B}^{(3)}$ on the interval $[0,1]$ because of Proposition 5, which allows to set $\Omega_s = [B_s^{(3)} = \tilde{B}_s^{(3)}]$ with $P(\Omega_s) = 1$, for every $s \in [0,1]$. Thus, the continuity of the processes $B^{(3)}$ and $\tilde{B}^{(3)}$ gives that $B_t^{(3)} = \tilde{B}_t^{(3)}$ for all $t \in (0,1]$, on $\cap_{r\in[0,1]\cap\mathbb{Q}}\Omega_r$ w.p.1. Consequently,

$$B_0^{(3)} = \tilde{B}_0^{(3)} = \lim_{t\downarrow 0} \tilde{B}_t^{(3)} = \lim_{t\downarrow 0} B_t^{(3)} \quad \text{on} \quad \cap_{r\in[0,1]\cap\mathbb{Q}} \Omega_r \text{ w.p.1},$$

as we wanted to see.

Note that a consequence of Statement *c*) is that $\lim_{t\to\infty} \frac{B_t}{t} = 0$, w.p.1. Actually, the oscillations of Brownian motion paths have been analyzed near zero and as $t \to \infty$, the so-called law of iterated logarithm. We do not study here this property and another one of Brownian motion paths because they are not the aim of this book. For studying them, the reader can consult Karatzas and Shreve [96].

On the other hand, we will introduce the Itô stochastic integral with respect to Brownian motion. One of the problems to define this integral is that Brownian motion does not have paths of bounded variation on bounded intervals, as the following result shows. The proof that we present here was developed by Dvoretzky et al. [46].

Theorem 5 (Paley, Wiener and Zygmund [172]) *Let $(B, (\mathcal{F}_t)_{t\in[0,T]})$ be a Brownian motion. Then, its paths are nowhere differentiable with probability 1.*

Remark 5 *By Royden [187] (Theorems 5.1.3 and 5.2.5), B cannot have paths of bounded variation. That is, for $0 \leq a < b \leq T$, $var_1(B_\cdot(\omega); [a,b]) = \infty$, for almost all $\omega \in \Omega$ (for its definition, see Corollary 1 below).*

Proof. It is enough to see that the set $D = [B_\cdot$ is nowhere differentiable in $[0,T)]$ is such that D^c is included in a set of probability zero because $(\Omega, \mathcal{F}, P)$ is complete. To do so, we first observe that if $f : [0,T] \to \mathbb{R}$ is differentiable at $t_0 \in [0,T)$, then there exist $\delta, L > 0$ such that $|f(t) - f(t_0)| \leq L|t - t_0|$, for all $t \in (t_0 - \delta, t_0 + \delta)$. Taking this into account, for $n \in \mathbb{N}$ large enough, there is $j \in \{1, \ldots, n\}$ such that

$$\frac{j-1}{n}T < t_0 \leq \frac{j}{n}T \quad \text{and} \quad \frac{j}{n}T, \ldots, \frac{j+3}{n}T \in (t_0 - \delta, t_0 + \delta).$$

Thus, for $i \in \{j+1, j+2, j+3\}$, we are able to get the estimate

$$\begin{aligned}\left|f(\frac{i}{n}T) - f(\frac{i-1}{n}T)\right| &\leq \left|f(\frac{i}{n}T) - f(t_0)\right| + \left|f(t_0) - f(\frac{i-1}{n}T)\right| \\ &\leq L\left(\left|\frac{i}{n}T - t_0\right| + \left|\frac{i-1}{n}T - t_0\right|\right) \\ &\leq L\left(\frac{4T}{n} + \frac{3T}{n}\right) = \frac{7TL}{n}.\end{aligned}$$

Hence, in order to prove that the result holds, we only need to see that the set

$$C_{m,L} := \cap_{n=m}^{\infty} \cup_{j=1}^{(n-3)\vee 1} \cap_{i=j+1}^{j+3} \left[\left|B_{\frac{iT}{n}\wedge T} - B_{\frac{(i-1)T}{n}\wedge T}\right| \leq \frac{7TL}{n}\right]$$

has probability zero, for every $m, L \in \mathbb{N}$, since, by last inequality, $D^c \subset \cup_{L\in\mathbb{N}} \cup_{m\in\mathbb{N}} C_{m,L}$. But, the independence of the increments of B implies that, for $n \in \mathbb{N}$ large enough, we have

$$\begin{aligned}P(C_{m,L}) &\leq P\left(\cup_{j=1}^{n-3} \cap_{i=j+1}^{j+3} \left[\left|B_{\frac{iT}{n}} - B_{\frac{(i-1)T}{n}}\right| \leq \frac{7TL}{n}\right]\right) \\ &\leq \sum_{j=1}^{n-3} P\left(\cap_{i=j+1}^{j+3} \left[\left|B_{\frac{iT}{n}} - B_{\frac{(i-1)T}{n}}\right| \leq \frac{7TL}{n}\right]\right) = \sum_{j=1}^{n-3}\left(P\left(\left[\left|B_{\frac{T}{n}}\right| \leq \frac{7TL}{n}\right]\right)^3\right) \\ &\leq n\left(P\left(\left[\left|B_{\frac{T}{n}}\right| \leq \frac{7TL}{n}\right]\right)^3\right) = n\left(\frac{\sqrt{n}}{\sqrt{2\pi T}}\int_{-7TL/n}^{7TL/n} \exp\left(-\frac{nx^2}{2T}\right)dx\right)^3 \\ &= n\left(\frac{1}{\sqrt{2\pi T}}\int_{-7TL/\sqrt{n}}^{7TL/\sqrt{n}} \exp\left(-\frac{x^2}{2T}\right)dx\right)^3 \leq cn\left(\frac{1}{\sqrt{n}}\right)^3 \to 0 \text{ as } n \to \infty.\end{aligned}$$

In others words, the proof is complete. □

We will use the following result when we deal with the integration by parts formula for Itô's integral with respect to Bm. The so-called Itô's formula. To do so, for $0 \leq a < b \leq T$, we denote by $|\pi|$ the norm of the partition $\pi = \{a = t_0 < t_1 < \ldots < t_n = b\}$ of the interval $[a, b]$. It mean, $|\pi| = \sup_{1\leq i\leq n}(t_i - t_{i-1})$.

Proposition 7 *Let $(B, (\mathcal{F}_t)_{t\in[0,T]})$ be a Brownian motion and $\{\pi^n : n \in \mathbb{N}\}$ a sequence of partitions of $[a, b]$ such that $\lim_{n\to\infty} |\pi^n| = 0$. Then,*

$$L^2(\Omega) - \lim_{n\to\infty} \sum_{i=1}^{n} \left(B_{t_i} - B_{t_{i-1}}\right)^2 = b - a. \tag{1.14}$$

Moreover, if $\sum_{i=1}^{\infty} |\pi^n| < \infty$, then the convergence in (1.14) is also true w.p.1.

Remark 6 *The right-hand side of (1.14) is called the quadratic variation of the Brownian motion B on the interval $[a, b]$. In other words, the quadratic variation of B on $[a, b]$ is $b-a$, the length of the interval. Now, let $[B]_t$ be given by the right-hand side of (1.14) when we write t and 0 instead of b and a, respectively. Then, we have that $t \mapsto [B]_t$ is increasing and $t \mapsto B_t^2 - [B]_t$, $t \in [0, T]$, is a martingale (see (1.12)). Here, the involved filtration is the one for which B is a Brownian motion.*

Proof. We have

$$\begin{aligned}
&\left(\sum_{i=1}^{n}\left|B_{t_i}-B_{t_{i-1}}\right|^2-(a-b)\right)^2\\
&= \left(\sum_{i=1}^{n}\left(\left|B_{t_i}-B_{t_{i-1}}\right|^2-(t_i-t_{i-1})\right)\right)^2\\
&= \sum_{i=1}^{n}\left(\left|B_{t_i}-B_{t_{i-1}}\right|^2-(t_i-t_{i-1})\right)^2\\
&\quad +2\sum_{i<j}\left(\left|B_{t_i}-B_{t_{i-1}}\right|^2-(t_i-t_{i-1})\right)\left(\left|B_{t_j}-B_{t_{j-1}}\right|^2-(t_j-t_{j-1})\right).
\end{aligned}$$

Therefore, the independence of the increments and (1.10) imply

$$\begin{aligned}
E\left\{\left(\sum_{i=1}^{n}\left|B_{t_i}-B_{t_{i-1}}\right|^2-(b-a)\right)^2\right\} &= \sum_{i=1}^{n}E\left\{\left(\left|B_{t_i}-B_{t_{i-1}}\right|^2-(t_i-t_{i-1})\right)^2\right\}\\
&\leq 2\sum_{i=1}^{n}\left\{E\left(\left|B_{t_i}-B_{t_{i-1}}\right|^4\right)+(t_i-t_{i-1})^2\right\}\\
&\leq C\sum_{i=1}^{n}(t_i-t_{i-1})^2\leq C(b-a)|\pi^n| \qquad (1.15)\\
&\to 0 \text{ as } n\to\infty.
\end{aligned}$$

On the other hand, the Chebyshev's inequality and (1.15) allow us to establish that, for $\varepsilon>0$,

$$\begin{aligned}
&P\left(\cap_{k=1}^{\infty}\cup_{n=k}^{\infty}\left[\left|\sum_{i=1}^{n}\left|B_{t_i}-B_{t_{i-1}}\right|^2-(b-a)\right|>\varepsilon\right]\right)\\
&\leq \sum_{n=k}^{\infty}P\left(\left[\left|\sum_{i=1}^{n}\left|B_{t_i}-B_{t_{i-1}}\right|^2-(b-a)\right|>\varepsilon\right]\right)\\
&\leq \frac{1}{\varepsilon^2}\sum_{n=k}^{\infty}E\left\{\left(\sum_{i=1}^{n}\left|B_{t_i}-B_{t_{i-1}}\right|^2-(b-a)\right)^2\right\}\leq\frac{C}{\varepsilon^2}\sum_{n=k}^{\infty}|\pi^n|, \text{ for any } k\in\mathbb{N}.
\end{aligned}$$

Thus, if $\sum_{i=1}^{\infty}|\pi^n|<\infty$, we obtain

$$P\left(\cup_{k=1}^{\infty}\cap_{n=k}^{\infty}\left[\left|\sum_{i=1}^{n}\left|B_{t_i}-B_{t_{i-1}}\right|^2-(b-a)\right|\leq\varepsilon\right]\right)=1,$$

for all $\varepsilon>0$, and the proof is complete. □

The following two results are an immediate consequence of Proposition 7.

Corollary 1 *Let $(B,(\mathcal{F}_t)_{t\in[0,T]})$ be a Bm. Then, we have*

$$\begin{aligned}
\mathrm{var}_p(B;[a,b]) &:= \sup\{\sum_{i=1}^{n}\left|B_{t_i}-B_{t_{i-1}}\right|^p : \pi \textit{ partition of } [a,b]\}\\
&= \begin{cases}\infty, \textit{w.p.1}, & \text{if } p\in(0,2),\\ <\infty, \textit{w.p.1}, & \text{if } p>2.\end{cases}
\end{aligned}$$

Remark 7 *By Remark 5, we have already known that the result holds for $p = 1$. This result means that B has not (resp. has) finite p-variation, for $p \in (0,2)$ (resp. $p > 2$), w.p.1. On the other hand, by Remark 6, we could think that the fact that the quadratic variation of B is finite must imply* $\mathrm{var}_2(B;[a,b]) < \infty$ *wp.1, for each $0 \le a < b \le T$. But this claim is false, as it is proven in Friz and Victoir [58] (Theorem 13.69).*

Proof. Suppose that $p \in (0,2)$. Then, $p = 2-\delta$ for some $\delta > 0$. Thus,

$$\begin{aligned}\sum_{i=1}^{n} \left|B_{t_i} - B_{t_{i-1}}\right|^2 &\le \left(\max_{i\in\{1,\dots,n\}} \left|B_{t_i} - B_{t_{i-1}}\right|^{\delta}\right) \sum_{i=1}^{n} \left|B_{t_i} - B_{t_{i-1}}\right|^{2-\delta} \\ &\le \left(\max_{i\in\{1,\dots,n\}} \left|B_{t_i} - B_{t_{i-1}}\right|^{\delta}\right) \mathrm{var}_{2-\delta}(B;[a,b]).\end{aligned}$$

Consequently, the fact that B is a continuous process and Proposition 7 yield that the result holds for $p \in (0,2)$.

Finally, assume that $p > 2$. In this case we can choose $\alpha \in (1/p, 1/2)$. Therefore, Proposition 5 implies that B has α-Hölder continuous paths, which leads us to establish, for some random variable M,

$$\sum_{i=1}^{n} \left|B_{t_i}(\omega) - B_{t_{i-1}}(\omega)\right|^p \le M(\omega) \sum_{i=1}^{n} (t_i - t_{i-1})^{p\alpha} \le M(\omega)|\pi|^{p\alpha-1}(b-a) \le M(\omega)(b-a)^{p\alpha},$$

which gives that $\mathrm{var}_p(B_\cdot(\omega);[a,b]) \le M(\omega)(b-a)^{p\alpha} < \infty$, w.p.1. Now, the proof is complete. □

Corollary 2 *Let $(B,(\mathcal{F}_t)_{t\in[0,T]})$ be a Bm and $\alpha \ge 1/2$. Then,*

$$P\left([B \text{ is } \alpha - \textit{Hölder continuous on } [0,T]]\right) = 0.$$

Proof. Let $\{a = t_0 < t_1 < \ldots < t_n = b\}$ be a partition of $[a,b]$, with $0 \le a < b \le T$, and $\omega \in \Omega$ such that $B_\cdot(\omega)$ is α-Hölder continuous on $[a,b]$. Then,

$$\sum_{i=1}^{n} \left|B_{t_i}(\omega) - B_{t_{i-1}}(\omega)\right|^2 \le M(\omega) \sum_{i=1}^{n} (t_i - t_{i-1})^{2\alpha} \le M(\omega)|\pi|^{2\alpha-1}(b-a) \le M(\omega)(b-a)^{2\alpha},$$

which implies that $\mathrm{var}_2(B_\cdot(\omega);[a,b]) < \infty$ Thus, from Remark 7, we obtain that the proof is finished. □

1.2.1 Canonical Wiener space

Here, we introduce a probability space $(\Omega, \mathcal{F}, P)$ in which is defined a Brownian motion $W = \{W_t : t \in [0,T]\}$. The so-called canonical Wiener space and canonical Wiener process, respectively. For the convenience of the reader and to show the main ideas to establish these definitions, we study, in Section 8.9, some technical results and introduce some notation related to outer measures because P is an outer measure induced by a set function.

Let $C_0([0,T])$ be the space of all the real-valued continuous functions ω defined on $[0,T]$ such that $\omega(0) = 0$. It is well-known that $(C_0([0,T]), \|\cdot\|_\infty)$ is a Banach space, where $\|\omega\|_\infty = \sup_{t\in[0,T]} |\omega(t)|$. Moreover, this space is separable because the Stone-Weierstrass theorem implies that the family of all the polynomials with rational coefficients is a dense subset of $C_0([0,T])$.

A cylindrical subset I of $C_0([0,T])$ has the form

$$I = \{\omega \in C_0([0,T]) : (\omega(t_1), \dots, \omega(t_n)) \in B\}, \tag{1.16}$$

with $B \in \mathcal{B}(\mathbb{R}^n)$ and $0 < t_1 < \ldots < t_n \leq T$.

Lemma 2 *The family $\mathcal{R}$ of all the cylinder sets of $C_0([0,T])$ is an algebra.*

Proof. Let $I \in \mathcal{R}$ be a cylinder of the form (1.16), Then, I^c is also a cylinder due to

$$I^c = \{\omega \in C_0([0,T]) : (\omega(t_1), \dots, \omega(t_n)) \in B^c\}.$$

On the other hand, note that I in (1.16) can be represented in different ways. For instance,

$$I = \{\omega \in C_0([0,T]) : (\omega(t_1), \dots, \omega(t_n), \dots, \omega(t_m)) \in B \times \mathbb{R}^{m-n}\}.$$

Here, $m > n$ and $t_n < t_{n+1} < \ldots < t_m \leq T$. Therefore, if $\tilde{I}$ is another cylinder, without loss of generality, we can assume that $\tilde{I}$ has the form

$$\tilde{I} = \{\omega \in C_0([0,T]) : (\omega(t_1), \dots, \omega(t_n)) \in \tilde{B}\}, \tag{1.17}$$

for some $\tilde{B} \in \mathcal{B}(\mathbb{R}^n)$. Thus, we have that

$$I \cup \tilde{I} = \{\omega \in C_0([0,T]) : (\omega(t_1), \dots, \omega(t_n)) \in (B \cup \tilde{B})\},$$

which is also an element of $\mathcal{R}$. □

We consider the σ-algebra $\mathcal{C}$ generated by $\mathcal{R}$ and the Borel σ-algebra $\mathcal{B}(C_0)$ of $C_0([0,T])$.

Proposition 8 *The σ-algebras $\mathcal{C}$ and $\mathcal{B}(C_0)$ are the same.*

Proof. Let I be given by (1.16), but with B an open subset of $\mathbb{R}^n$. Then, in this case, I is an open set of $C_0([0,T])$, which, together with the monotone class theorem (Chapter 8, Theorem 96), allows to conclude that $\mathcal{C} \subset \mathcal{B}(C_0)$.

Now, in order to finish the proof, we only need to show that a closed ball $B_{\omega_0}(\varepsilon) = \{\omega \in C_0([0,T]) : \|\omega - \omega_0\|_\infty \leq \varepsilon\}$ is in $\mathcal{C}$ because $C_0([0,T])$ is a separable space and, consequently, it has the Lindelöf property. That is, any open set of $C_0([0,T])$ can be expressed as the countable union of open balls.

Finally, the result follows from the equality

$$B_{\omega_0}(\varepsilon) = \cap_{r \in \mathbb{Q}} \{\omega \in C_0([0,T]) : \omega(r) \in [-\varepsilon + \omega_0(r), \varepsilon + \omega_0(r)]\}.$$

□

Now we deal with the Wiener measure P^W. This is introduced as follows. Let I be given in (1.16). Then, we define $P^W(I)$ as

$$P^W(I) := \int_B \prod_{i=1}^n \left(\frac{1}{\sqrt{2\pi(t_i - t_{i-1})}} \exp\left(-\frac{(u_i - u_{i-1})^2}{2(t_i - t_{i-1})} \right) \right) du_1 \cdots u_n, \tag{1.18}$$

with $t_0, u_0 = 0$ (compare it with (1.13)). For instance, we have that, for $0 \leq s < t \leq T$ and $B \in \mathcal{B}(\mathbb{R})$,

$$P^W(\{\omega \in C_0([0,T]) : (\omega(t) - \omega(s)) \in B\}) = \frac{1}{\sqrt{2\pi(t-s)}} \int_B \exp\left(-\frac{u^2}{2(t-s)} \right) du. \tag{1.19}$$

Indeed, Let $\tilde{B} = \{(x,y) \in \mathbb{R}^2 : (y-x) \in B\}$. Then, Definition(1.18) gives

$$\begin{aligned}
&P^W(\{\omega \in C_0([0,T]) : (\omega(t) - \omega(s)) \in B\}) \\
&= P^W(\{\omega \in C_0([0,T]) : (\omega(s), \omega(t)) \in \tilde{B}\}) \\
&= \frac{1}{\sqrt{2\pi s}} \frac{1}{\sqrt{2\pi(t-s)}} \int_{\tilde{B}} \exp\left(-\frac{u_1^2}{2s}\right) \exp\left(-\frac{(u_2-u_1)^2}{2(t-s)}\right) du_2 du_1 \\
&= \frac{1}{\sqrt{2\pi s}} \frac{1}{\sqrt{2\pi(t-s)}} \int_{\mathbb{R}} \int_{\tilde{B}+u_1} \exp\left(-\frac{u_1^2}{2s}\right) \exp\left(-\frac{(u_2-u_1)^2}{2(t-s)}\right) du_2 du_1.
\end{aligned}$$

Thus, the change of variables formula $v = u_2 - u_1$ applied to the integral with respect to u_2 implies that our claim is satisfied. It means, (1.19) holds.

We observe that $P^W(I)$ is independent of the representation of I as a cylinder due to Lemma 101 (Section 8.9). Also, we have already pointed out that if I and $\tilde{I}$ are two disjoint cylinders, then, without loss of generality, we can suppose that there exist $B, \tilde{B} \in \mathcal{B}(\mathbb{R}^n)$ such that $B \cap \tilde{B} = \emptyset$ and the cylinders have the representation (1.16) and (1.17), respectively. Thus the right-hand side of (1.18) yields that $P^W(I \cup \tilde{I}) = P^W(I) + P^W(\tilde{I})$. It means $P^W : \mathcal{R} \to [0,1]$ is finitely additive. Actually, we will see that P^W can be extended to a probability measure on $(C_0([0,T]), \mathcal{C})$ (see Theorem 6 below). Before providing the proof of this result, we need to state the following lemmas.

In order to establish the auxiliary results, we recall that we are using the convention $\Pi = \cup_{n=0}^{\infty} \pi^n$, where $(\pi^n)_{n \in \mathbb{N} \cup \{0\}}$ is the sequence of dyadic partitions of $[0,T]$. It means,

$$\pi^n = \{0 = t_0^n < t_1^n < \cdots < t_{2^n}^n = T\}, \text{ with } t_i^n = \frac{iT}{2^n}.$$

Thus, for $s \in \Pi$, there are $k, n \in \mathbb{N}$ such that $s = \frac{kT}{2^n}$ with k odd and n the smallest possible for this representation. Also, for $s, t \in \Pi$ such that $s < t$ and either $s \neq 0$ or $t \neq T$, there exists $r \in [s,t] \cap \Pi$ with the smallest n among all the elements of Π belonging to the interval $[s,t]$. Note that $r = \frac{k_0 T}{2^{n_0}}$ could be equal to either s, or t, and that $n_0 = 0$ if $s \in \{0, T\}$. Now assume that we can find a positive odd number $k_1 \in \mathbb{N}$ such that $s \leq \frac{k_0 T}{2^{n_0}} < \frac{k_1 T}{2^{n_0}} \leq t$. Then, there is a positive integer k such that $k_0 < 2k < k_1$, which is impossible because, in this case, $\frac{kT}{2^{n_0 - 1}}$ is also in the interval $[s,t]$. Therefore, we have shown that there is a unique element in $[s,t] \cap \Pi$ with the smallest $n \in \mathbb{N}$.

Lemma 3 *Let $s, t \in \Pi$ be such that $s < t$ and either $s \neq 0$ or $t \neq T$, and $r \in [s,t] \cap \Pi$ with the smallest n among all the elements of Π belonging to the interval $[s,t]$. Then, there exists a strictly increasing finite sequence $\{k_i : i = 1, \ldots, m\}$ (resp.$\{\ell_j : j = 1, \ldots, p\}$) of positive integers bigger than n_0 such that $r - s = T\sum_{i=1}^m 2^{-k_i}$ (resp. $t - r = T\sum_{j=1}^p 2^{-\ell_j}$) if $r \neq s$ (resp.$r \neq t$).*

Proof. Suppose that $s \neq r$, and $s = \frac{k_1 T}{2^{n_1}}$ and $r = \frac{k_0 T}{2^{n_0}}$ with $n_0 < n_1$, which is possible because of the definition of r. We will use induction on n_1 to show that the result is satisfied. So, we first assume that $n_1 = n_0 + 1$. Hence,

$$r - s = \frac{2k_0 - k_1}{2^{n_0+1}} T.$$

Note that if $2k_0 - k_1 = 1$ the result is true in this case. If $2k_0 - k_1$ is an odd positive integer bigger than 1, then $s < s + \frac{T}{2^{n_0+1}} < r$. But,

$$s + \frac{T}{2^{n_0+1}} = \frac{k_1+1}{2^{n_0+1}} T = \frac{(k_1+1)/2}{2^{n_0}} T,$$

which contradicts the definition of r. That is, the result is true for $n_1 = n_0 + 1$.

By induction, assume that $n_1 = m+1$ and the result holds for any $n_1 \in \{n_0+1, \ldots, m\}$. Now we have

$$r - s = \frac{2^{m+1-n_0}k_0 - k_1}{2^{m+1}}T.$$

The number $2^{m+1-n_0}k_0 - k_1$ is odd. As before, if this number is 1, the result is satisfied. Otherwise, we obtain $s < s + \frac{T}{2^{m+1}} < r$, where $s + \frac{T}{2^{m+1}} = \frac{(k_1+1)/2}{2^m}$. Therefore using the induction hypothesis to the numbers $s + \frac{T}{2^{m+1}}$ and r, we get that the result holds for the case $s \neq r$.

Finally, proceeding as above, we can also see that the result is true in the case that $t \neq r$. □

Lemma 4 *Let $c, \alpha > 0$ and $\omega \in C_0([0,T])$ such that*

$$\left|\omega\left(\frac{kT}{2^n}\right) - \omega\left(\frac{(k-1)T}{2^n}\right)\right| \leq c\left(\frac{T}{2^n}\right)^\alpha, \tag{1.20}$$

for $n, k \in \mathbb{N} \cup \{0\}$, with $1 \leq k \leq 2^n$. Then,

$$|\omega(t) - \omega(s)| \leq \frac{2c}{1-2^{-\alpha}}|t-s|^\alpha, \quad s, t \in [0,T]. \tag{1.21}$$

Remark 8 *Actually we will prove that if $\omega : [0,T] \to \mathbb{R}$ is a function satisfying (1.20), then (1.21) holds for any $s, t \in \Pi$. Moreover, this lemma is used in Section 8.9.1.*

Proof. Note that it is enough to prove that (1.21) is true for $s, t \in \Pi$ such that either $s \neq 0$, $t \neq T$, or $s, t \notin \{0, T\}$. So, choose $s_1, s_2 \in \Pi$ such that $0 < s_1 < s_2 < T$. Let $r = \frac{k_0 T}{2^{n_0}}$ with the smallest n_0 among all the elements of Π belonging to the interval $[s_1, s_2]$. Then, Lemma 3 yields that there exist two strictly increasing finite sequences $\{k_i : i = 1, \ldots, m\}$ and $\{\ell_j : j = 1, \ldots, p\}$ of positive integers bigger than n_0 such that

$$r - s_1 = \sum_{i=1}^{m} \frac{T}{2^{k_i}} \quad \text{and} \quad s_2 - r = \sum_{j=1}^{p} \frac{T}{2^{\ell_j}}.$$

We observe that we only consider one of these sums when r is equal to either s_1 or s_2. Hence, we can consider the partitions

$$\{s_1 < r - \sum_{i=1}^{m-1} \frac{T}{2^{k_i}} < \ldots < r - \frac{T}{2^{k_1}} < r\} \tag{1.22}$$

and

$$\{r < r + \frac{T}{2^{\ell_1}} < \ldots < r + \sum_{j=1}^{p-1} \frac{T}{2^{\ell_j}} < s_2\} \tag{1.23}$$

of the intervals $[s_1, r]$ and $[r, s_2]$, respectively.

Finally, we observe that, for instance,

$$\begin{aligned} r + \sum_{j=1}^{q} \frac{T}{2^{\ell_j}} &= \frac{T}{2^{\ell_q}}\left(2^{\ell_q - n_0}k_0 + \sum_{j=1}^{q} 2^{\ell_q - \ell_j}\right) = \frac{T}{2^{\ell_q}}\left(2^{\ell_q - n_0}k_0 + \sum_{j=1}^{q-1} 2^{\ell_q - \ell_j} + 1\right) \\ &= r + \sum_{j=1}^{q-1} \frac{T}{2^{\ell_j}} + \frac{T}{2^{\ell_q}}. \end{aligned}$$

Hence, (1.20), (1.22) and (1.23), lead us to establish

$$\begin{aligned}|w(s_2)-w(s_1)| &\leq 2c\sum_{k=k_1\wedge\ell_1}^{k_m\vee\ell_p}\left(\frac{T}{2^k}\right)^{\alpha} = 2c\left(\frac{T}{2^{k_1\wedge\ell_1}}\right)^{\alpha}\sum_{k=0}^{k_m\vee\ell_p-(k_1\wedge\ell_1)}\left(\frac{T}{2^k}\right)^{\alpha}\\ &\leq 2c|s_2-s_1|^{\alpha}\sum_{k=0}^{\infty}\left(\frac{T}{2^k}\right)^{\alpha},\end{aligned}$$

which gives that (1.21) is satisfied when we write s_1 and s_2 instead of s and t, respectively. Consequently, the result is satisfied. □

Now, we are ready to give the proof that P^W can be extended to a probability measure on $(C_0([0,T]),\mathcal{C})$.

Theorem 6 *The mapping $P^W:\mathcal{R}\to[0,1]$ has a unique extension to a probability measure on the σ-algebra $\mathcal{C}$.*

Proof. We have that $P^W(I\cap C_0([0,T]))=\mu_X(I)$, where μ_X is defined in (8.45) and

$$I=\{x\in\{0\}\times\mathbb{R}^{(0,T]}:(x(t_1),\ldots,x(t_n))\in B\},$$

for some $B\in\mathcal{B}(\mathbb{R}^n)$ and $0<t_1<\ldots<t_n\leq T$. Also, by Lemma 103 and Proposition 140 in Section 8.9, we have that P^W is a well-defined probability measure on the algebra $\mathcal{R}$. Hence, the Carathéodory extension Theorem (Proposition 139) yields that the result is satisfied. □

Remark 9 *By Proposition 139, we have that $(C_0([0,T]),\mathcal{F}^*,P^{*,W})$ is a complete probability space, where $P^{*,W}$ is the outer measure induced by the measure $P^W:\mathcal{R}\to[0,1]$ (see Proposition 138 for definition). Remember that, in this case, $P^{*,W}|_{\mathcal{R}}=P^W$.*

Now, we introduce the so-called Canonical Wiener process defined on the complete probability space $(C_0([0,T]),\mathcal{F}^*,P^{*,W})$. So, let $W:C_0([0,T])\times[0,T]\to\mathbb{R}$ given by

$$W_t(\omega)=\omega(t),\quad \omega\in C_0([0,T])\text{ and }t\in[0,T]. \tag{1.24}$$

Note that $W_0\equiv 0$ because $\omega(0)=0$, for any $\omega\in C_0([0,T])$, and that W is a process due to $W_t^{-1}(B)=\{\omega\in C_0([0,T]):\omega(t)\in B\}$ is a cylinder of $C_0([0,T])$, for every $t\in[0,T]$ and $B\in\mathcal{B}(\mathbb{R})$. Furthermore, Lemma 103, Proposition 140, (1.19) and (8.45) imply that the random variable W_t-W_s has Normal distribution with zero mean and covariance $t-s$, for $0\leq s<t\leq T$. Finally, we have that the process W is a Brownian motion since it has clearly continuous paths and, proceeding as in the proof of equality (1.19), the random variables $W_{t_n}-W_{t_{n-1}},\ldots,W_{t_2}-W_{t_1},W_{t_1}$ are independent. This last fact is left as an exercise for the reader to see if she/he can handle the measure $P^{*,W}$ efficiently.

Finally, we have the following:

Definition 12 *The probability space $(C_0([0,T]),\mathcal{F}^*,P^{*,W})$ introduced in Remark 9 is called the Canonical Wiener space and the process $W=\{W_t:t\in[0,T]\}$ in (1.24) is the so-called Canonical Wiener process.*

1.2.2 Wiener integral

The aim of this section is to provide an introduction of the stochastic integral with respect to Brownian motion that we need to understand some results in Section 1.4. Namely, we deal with the Wiener integral (or multiples integrals of order 1), whose integrands are square-integrable functions (with respect to the Lebesgue measure). A more complete exposition on stochastic integration for stochastic processes is given in Chapter 3.

Here, we assume that we have a Brownian motion $\{B_t : t \in [0,T]\}$ defined on a complete probability space $(\Omega, \mathcal{F}, P)$.

Let $a, b \in [0,T]$, $a < b$, and $f : [a,b] \to \mathbb{R}$. A natural way to define an integral of the function f with respect to the stochastic process B on $[a,b]$ is through Riemann-Stieltjes sums as it is done in (2.2) below. In this case, B not only depends on time $t \in [0,T]$, but also on $\omega \in \Omega$. So, the convergence of the involved Riemman-Stieltjes sums is pathwise (i.e., ω by ω). That is, for $\omega \in \Omega$ fixed, we wish the stochastic integral to be introduced as

$$\int_a^b f(s)dB_s(\omega) = \lim_{|\pi|\to 0} \sum_{i=1}^{n} f(s_i)\left(B_{t_i}(\omega) - B_{t_{i-1}}(\omega)\right), \tag{1.25}$$

where $\pi = \{a = t_0 < t_1 < \ldots < t_n = b\}$ is a partition of $[a,b]$, $|\pi| = \max\{t_i - t_{i-1} : i = 1, \ldots, n\}$ and $s_i \in [t_{i-1}, t_i]$. The limit in (1.25) could not exist even in the case that f is continuous since B has no paths of bounded variation on $[a,b]$ (see Remark 5), as it is established in Proposition 33 below. But, Corollary 1 and Theorem 29 imply the existence of the limit in (1.25), with probability 1, if f has finite q-variation, for some $q < 2$. It means, $\mathrm{var}_q(f; [a,b]) < \infty$ (see (2.19) below). In the case that $q = 1$ (i.e., f has bounded variation on $[a,b]$), the existence of the involved limit follows from Theorems 16 and 18.

Hence, a natural question is: can we define a stochastic integral with respect to B whose domain includes all the continuous functions? For instance, does the function $f(x) = \sqrt{x}\sin(1/x)$, $x \in [0, 2/\pi]$, belong to the domain of such a stochastic integral? Note that this function f does not have finite p-variation, for any $p \le 2$. Indeed, consider the partition

$$\pi_n = \{0 = t_0 < t_1 = \frac{1}{(n-1)\pi + \pi/2} < \ldots < t_n = \frac{1}{(n-n)\pi + \pi/2} = \frac{2}{\pi}\}.$$

Then,

$$\begin{aligned}
\sum_{i=1}^{n} |f(t_i) - f(t_{i-1})|^2 &= f(t_1)^2 + \sum_{i=2}^{n} \left|(-1)^{n-i}\left\{(t_i)^{1/2} + (t_{i-1})^{1/2}\right\}\right|^2 \\
&> \sum_{i=2}^{n}\{t_i + t_{i-1}\} > 2\sum_{i=2}^{n-1} t_i = 2\sum_{i=1}^{n-2} \frac{1}{i\pi + (\pi/2)}.
\end{aligned}$$

Thus, our claim follows from Proposition 35 below and from the fact that $\sum_{i=1}^{n-2} 1/(i\pi + (\pi/2)) \to \infty$, as $n \to \infty$.

On the other hand, by Proposition 6, we have that the Bm B is a Gaussian process. Thus, from Definitions 9 and 10, the sum $\sum_{i=1}^{n} f(s_i)\left(B_{t_i} - B_{t_{i-1}}\right)$, on the right-hand side of (1.25), is a Gaussian random variable with zero mean and variance

$$\begin{aligned}
& E\left(\sum_{i=1}^{n} f(s_i)\left(B_{t_i} - B_{t_{i-1}}\right)\right)^2 \\
&= \sum_{i=1}^{n} f(s_i)^2 E\left(\left(B_{t_i} - B_{t_{i-1}}\right)^2\right) + 2\sum_{1\le i<j\le n} f(s_i)f(s_j)E\left\{\left(B_{t_i} - B_{t_{i-1}}\right)\left(B_{t_j} - B_{t_{j-1}}\right)\right\} \\
&= \sum_{i=1}^{n} f(s_i)^2 (t_i - t_{i-1}) = \int_a^b \left(\sum_{i=1}^{n} f(s_i) 1_{]t_{i-1}, t_i]}(s)\right)^2 ds,
\end{aligned} \tag{1.26}$$

where the penultimate equality follows from Definitions 11.*ii*) and 11.*iii*). This relation suggests that, on the right-hand side of (1.25), we should change the pathwise convergence by convergence in $L^2(\Omega)$. That is, the stochastic integral of a function $f \in L^2([a,b])$ with respect to B must be the $L^2(\Omega)$-limit of stochastic integrals of simple functions, which are introduced in the following:

Definition 13 *We say that $f : [a,b] \to \mathbb{R}$ is a simple function or a step function ($f \in \mathcal{E}([a,b])$ for short) if it has the form*

$$f(s) = \alpha_1 1_{[a,t_1]}(s) + \sum_{i=2}^{n} \alpha_i 1_{]t_{i-1},t_i]}(s), \quad s \in [a,b],$$

where $\alpha_i \in \mathbb{R}$, $i = 1, \dots, n$, and $\pi = \{a = t_0 < t_1 < \dots < t_n = b\}$ is a partition of $[a,b]$. In this case, we define the stochastic integral of f with respect to the Brownian motion B as

$$\int_a^b f(s)dB_s := \sum_{i=1}^{n} \alpha_i \left(B_{t_i} - B_{t_{i-1}}\right). \tag{1.27}$$

Note that this definition of stochastic integral is inspired by the right-hand side of (1.25) because we want the stochastic integral of $f \in L^2([a,b])$ to be the limit in $L^2(\Omega)$ of stochastic integrals of simple functions. Also note that if $f \in \mathcal{E}([0,T])$, then $f 1_{[a,b]} \in \mathcal{E}([0,T])$ and $f|_{[a,b]}$ (the restriction of f to $[a,b]$) belongs to $\mathcal{E}([a,b])$. In this case,

$$\int_0^T f(s) 1_{[a,b]}(s)dB_s = \int_a^b f|_{[a,b]}(s)dB_s.$$

In general, by convention, the right-hand side of this equality is denoted as $\int_a^b f(s)dB_s$.

The integral given in Definition 13 is independent of the representation of f as a simple function, which yields that this integral is well-defined. This is stated in the following:

Proposition 9 *Let $f : [a,b] \to \mathbb{R}$ be a simple function. Then, the stochastic integral $\int_a^b f(s)dB_s$ given by (1.27) is independent of the representation of f as a simple function.*

Proof. Consider the simple function

$$f(s) = \alpha_1 1_{[a,t_1]}(s) + \sum_{i=2}^{n} \alpha_i 1_{]t_{i-1},t_i]}(s), \quad s \in [a,b],$$

where $\alpha_i \in \mathbb{R}$, $i = 1, \dots, n$, and $\pi = \{a = t_0 < t_1 < \dots < t_n = b\}$ is a partition of $[a,b]$.

We first show the way that we can proceed to see that the stochastic integral $\int_a^b f(s)dB_s$ does not change if we consider f as a simple function with respect to a refinement $\tilde{\pi} = \{a = \tilde{t}_0 < \tilde{t}_1 < \dots < \tilde{t}_m = b\}$ of π. It means, $\pi \subset \tilde{\pi}$ and

$$f(s) = \tilde{\alpha}_1 1_{[a,\tilde{t}_1]}(s) + \sum_{j=2}^{m} \tilde{\alpha}_j 1_{]\tilde{t}_{j-1},\tilde{t}_j]}(s), \quad s \in [a,b],$$

with $\tilde{\alpha_j} = \alpha_i$ if $[\tilde{t}_{j-1}, \tilde{t}_j] \subset [t_{i-1}, t_i]$. So, by induction on the number of points of the set $\tilde{\pi} \setminus \pi$, it is easy to see

$$\sum_{i=1}^{n} \alpha_i \left(B_{t_i} - B_{t_{i-1}}\right) = \sum_{j=1}^{m} \tilde{\alpha}_j \left(B_{t_j} - B_{t_{j-1}}\right).$$

Now assume

$$f(s) = \alpha_1 1_{[a,t_1]}(s) + \sum_{i=2}^{n} \alpha_i 1_{]t_{i-1},t_i]}(s) = \beta_1 1_{[a,r_1]}(s) + \sum_{\ell=2}^{k} \beta_\ell 1_{]r_{\ell-1},r_\ell]}(s), \quad s \in [a,b], \tag{1.28}$$

where $\bar{\pi} = \{a = r_0 < r_1 < \ldots < r_k = b\}$ is another partition of $[a,b]$. Set $\rho = \pi \cup \bar{\pi} = \{a = s_0 < s_1 < \ldots < s_m = b\}$. Thus, (1.28) implies

$$f(s) = \tilde{\alpha}_1 1_{[a,s_1]}(s) + \sum_{j=2}^{m} \tilde{\alpha}_j 1_{]s_{j-1},s_j]}(s) = \tilde{\beta}_1 1_{[a,s_1]}(s) + \sum_{j=2}^{m} \tilde{\beta}_j 1_{]s_{j-1},s_j]}(s), \quad s \in [a,b]. \tag{1.29}$$

Here, $\tilde{\alpha}_j = \alpha_i$ if $[s_{j-1}, s_j] \subset [t_{i-1}, t_i]$ and $\tilde{\beta}_j = \beta_\ell$ if $[s_{j-1}, s_j] \subset [r_{\ell-1}, r_\ell]$. Hence, using that ρ is a partition of $[a,b]$, we obtain that $\tilde{\alpha}_j = \tilde{\beta}_j$, for $j = 1, \ldots, m$. Therefore, the first part of this proof allows us to conclude

$$\sum_{i=1}^{n} \alpha_i \left(B_{t_i} - B_{t_{i-1}}\right) = \sum_{j=1}^{m} \tilde{\alpha}_j \left(B_{s_j} - B_{s_{j-1}}\right) = \sum_{\ell=1}^{k} \beta_\ell \left(B_{r_\ell} - B_{r_{\ell-1}}\right).$$

In other words, we have that $\int_a^b f(s) dB_s$ is independent of the representation of f as a simple function. □

An immediate consequence of the last proof is the following:

Corollary 3 *Let $f_1, f_2 : [a,b] \to \mathbb{R}$ be two simple functions and $c_1, c_2 \in \mathbb{R}$. Then, $c_1 f_1 + c_2 f_2$ is also a simple function and*

$$\int_a^b (c_1 f_1(s) + c_2 f_2(s))\, dB_s = c_1 \int_a^b f_1(s) dB_s + c_2 \int_a^b f_2(s) dB_s.$$

Moreover,

$$E\left\{\left(\int_a^b f_1(s) dB_s\right)\left(\int_a^b f_2(s) dB_s\right)\right\} = \int_a^b f_1(s) f_2(s) ds.$$

Proof. Using the argument to see that (1.28) yields (1.29), we can find a partition $\pi = \{a = t_0 < t_1 < \ldots < t_n = b\}$ of $[a,b]$ such that

$$f_1(\cdot) = \alpha_1 1_{[0,t_1]}(\cdot) + \sum_{i=2}^{n} \alpha_i 1_{]t_{i-1},t_i]}(\cdot) \quad \text{and} \quad f_2(\cdot) = \tilde{\alpha}_1 1_{[0,t_1]}(\cdot) + \sum_{i=2}^{n} \tilde{\alpha}_i 1_{]t_{i-1},t_i]}(\cdot),$$

which give

$$c_1 f_1(\cdot) + c_2 f_2(\cdot) = (c_1 \alpha_1 + c_2 \tilde{\alpha}_1)\, 1_{[0,t_1]}(\cdot) + \sum_{i=2}^{n} (c_1 \alpha_i + c_2 \tilde{\alpha}_i)\, 1_{]t_{i-1},t_i]}(\cdot).$$

and

$$\begin{aligned} & E\left\{\left(\int_a^b f_1(s) dB_s\right)\left(\int_a^b f_2(s) dB_s\right)\right\} \\ = \; & E\left(\sum_{i=1}^{n} \alpha_i \tilde{\alpha}_i \left(B_{t_i} - B_{t_{i-1}}\right)^2\right) \\ & + E\left(\sum_{i \neq j} \alpha_i \tilde{\alpha}_j \left(B_{t_i} - B_{t_{i-1}}\right)\left(B_{t_j} - B_{t_{j-1}}\right)\right) = \sum_{i=1}^{n} \alpha_i \tilde{\alpha}_i \left(t_i - t_{i-1}\right). \end{aligned}$$

Now, it is easy to finish the proof. □

The following step to construct the stochastic integral $\int_a^b f(s)dB_s$ for f belonging to $L^2([a,b])$ is to show that there exists a sequence of simple functions that goes to f in $L^2([a,b])$.

Proposition 10 *Let f be a function in $L^2([a,b])$. Then, there exists a sequence $\{f_n \in \mathcal{E}([a,b]) : n \in \mathbb{N}\}$ such that*

$$\|f - f_n\|_{L^2([a,b])} \to 0, \quad \text{as } n \to \infty. \tag{1.30}$$

Proof. We will use the monotone class theorem (see Theorem 97 in the appendix) to show that the result is true. So, consider the π-system $\mathcal{A} = \{]s,t] : s,t \in]a,b], \ s < t\} \cup \{\emptyset\}$ and the linear family of functions

$$\mathcal{G} = \left\{f \in L^2([a,b]) : \text{there is } \{f_n : n \in \mathbb{N}\} \subset \mathcal{E}([a,b]) \text{ such that (1.30) holds}\right\}.$$

Note that the functions $1 = 1_{[a,b]}$ and 1_A, with $A \in \mathcal{A}$, belong to the space $\mathcal{E}([a,b])$. That is, Conditions *i)* and *ii)* of Theorem 97 are satisfied.

Now, let $\{\xi_n \in \mathcal{G} : n \in \mathbb{N}\}$ be a sequence of non-negative function such that $\xi_n \uparrow \xi$, where ξ is a bounded function on $[a,b]$. Then, the dominated convergence theorem implies that $\xi_n \to \xi$ in $L^2([a,b])$ as $n \to \infty$ and, from the definition of $\mathcal{G}$, we can choose a sequence $\{f_n \in \mathcal{E}([a,b]) : n \in \mathbb{N}\}$ such that

$$\|\xi_n - f_n\|_{L^2([a,b])} \le \frac{1}{n}, \quad \text{for } n \in \mathbb{N},$$

which give that $\|\xi - f_n\|_{L^2([a,b])} \to 0$, as $n \to \infty$. Thus, we have proven that $\mathcal{G}$ contains all the bounded and $\sigma\{\mathcal{A}\}$-measurable-functions due to Theorem 97.

Finally, for $f \in L^2([a,b])$, we have that $(-n) \vee (f \wedge n)$ goes to f in $L^2([a,b])$, as $n \to \infty$, by using the dominated convergence theorem again. Consequently, the previous analysis establishes that the result holds. □

The last step to construct the Wiener integral is the following result.

Lemma 5 *Let f be a function in $L^2([a,b])$ and $\{f_n \in \mathcal{E}([a,b]) : n \in \mathbb{N}\}$ a sequence such that*

$$\|f - f_n\|_{L^2([a,b])} \to 0, \quad \text{as } n \to \infty.$$

Then, the sequence $\left\{\int_a^b f_n(s)dB_s : n \in \mathbb{N}\right\}$ converges in $L^2(\Omega)$. Moreover, the limit is independent of the approximating sequence $\{f_n : n \in \mathbb{N}\}$.

Proof. By (1.26) and Corollary 3, we have

$$\begin{aligned} E\left(\left(\int_a^b f_n(s)dB_s - \int_a^b f_m(s)dB_s\right)^2\right) &= E\left(\left(\int_a^b [f_n(s) - f_m(s)]\, dB_s\right)^2\right) \\ &= \int_a^b [f_n(s) - f_m(s)]^2\, ds. \end{aligned}$$

Hence, using that f_n goes to f in $L^2([a,b])$, we get that the sequence $\left\{\int_a^b f_n(s)dB_s : n \in \mathbb{N}\right\}$ is Cauchy in $L^2(\Omega)$ and, therefore, it converges to a square-integrable random variable.

Finally, choose another sequence $\{g_n \in \mathcal{E}([a,b]) : n \in \mathbb{N}\}$ that converges to f in $L^2([a,b])$. Set, for $n \in \mathbb{N}$,

$$h_n = \begin{cases} f_n, & \text{if } n \text{ is odd,} \\ g_n, & \text{otherwise.} \end{cases}$$

Note that this sequence of simple functions also goes to f in $L^2([a,b])$. Thus, proceeding as before, we can conclude that the sequence of stochastic integrals $\left\{\int_a^b h_n(s)dB_s : n \in \mathbb{N}\right\}$ converges in $L^2(\Omega)$. This leads to

$$L^2(\Omega) - \lim_{n\to\infty} \int_a^b f_n(s)dB_s = L^2(\Omega) - \lim_{n\to\infty} \int_a^b g_n(s)dB_s$$

since any subsequence of a converging sequence goes to the same limit. Consequently, the proof is complete. □

Resuming, if f belongs to $L^2([a,b])$, then, by Proposition 10, there exists a sequence $\{f_n \in \mathcal{E}([a,b]) : n \in \mathbb{N}\}$ converging to f in $L^2([a,b])$. Hence, from Lemma 5, we have a square-integrable random variable, denoted by $\int_a^b f(s)dB_s$, such that

$$\int_a^b f(s)dB_s = L^2(\Omega) - \lim_{n\to\infty} \int_a^b f_n(s)dB_s. \tag{1.31}$$

Moreover, the definition of the random variable $\int_a^b f(s)dB_s$ is independent of the choice of the sequence $\{f_n : n \in \mathbb{N}\}$. Hence, we can now define the Wiener integral.

Definition 14 *Let $f \in L^2([a,b])$. The stochastic integral of f with respect to B on the interval $[a,b]$ (or Wiener integral of f), denoted by $\int_a^b f(s)dB_s$, is the square-integrable random variable defined by (1.31).*

The following corollary is an immediate consequence of Definition 14 and the results established in this section for simple functions.

Corollary 4 *Let $f, g \in L^2([a,b])$ and $c_1, c_2 \in \mathbb{R}$. Then,*

i) $\int_a^b \cdot\, dB$ is a linear operator from $L^2([a,b])$ into $L^2(\Omega)$. That is,

$$\int_a^b (c_1 f(s) + c_2 g(s))\, dB_s = c_1 \int_a^b f(s)dB_s + c_2 \int_a^b g(s)dB_s.$$

ii) For $r,t \in [a,b]$, $r < t$, we have

$$\begin{aligned}\int_a^b f(s)dB_s &= \int_a^b f(s)1_{[a,t]}(s)dB_s + \int_a^b f(s)1_{[t,b]}(s)dB_s \\ &= \int_a^t f(s)dB_s + \int_t^b f(s)dB_s\end{aligned}$$

and $\int_r^t f(s)dB_s = \int_a^b f(s)1_{[r,t]}(s)dB_s$.

iii) $\int_a^b f(s)dB_s$ and $\int_a^b g(s)dB_s$ are two Gaussian random variables with zero mean and covariance

$$E\left\{\left(\int_a^b f(s)dB_s\right)\left(\int_a^b g(s)dB_s\right)\right\} = \int_a^b f(s)g(s)ds.$$

In particular, it is satisfied the isometry relation

$$E\left(\left(\int_a^b f(s)dB_s\right)^2\right) = \int_a^b (f(s))^2 ds.$$

Consequently,

$$\int_a^b f(s)dB_s = L^2(\Omega) - \lim_{n\to\infty} \int_a^b f_n(s)dB_s$$

if $\|f - f_n\|_{L^2([a,b])} \to 0$, *as* $n \to \infty$.

Other consequence of the definition of the Wiener integral is the following result.

Corollary 5 *Let* $f : [a,b] \to \mathbb{R}$ *be a continuous function with finite q-variation on* $[a,b]$, *for some* $q < 2$. *Then,*

$$\int_a^b f(s)dB_s = \int_a^b f(s)d^Y B_s, \quad w.p.1,$$

where the integral on the right-hand side is defined pathwise by (1.25). In other words, in this case the Wiener integral $\int_a^b f(s)dB_s$ *of* f *agrees with the Rieman-Stiltjes integral of it with respect to* B, ω *by* ω.

Proof. Let $\pi_n = \{a = t_0^n < t_1^n < \ldots < t_{N(n)}^n = b\}$ be a partition of $[a,b]$ such that $|\pi_n| \to 0$, as $n \to \infty$, and $s_i^n \in [t_{i-1}^n, t_i^n]$. Set

$$f_n(s) = f(s_1)1_{[0,t_1^n]}(s) + \sum_{i=2}^{N(n)} f(s_i^n)1_{]t_{i-1}^n, t_i^n]}(s), \quad s \in [a,b].$$

Then, the dominated convergence theorem and the continuity of f allow us to get that $\|f - f_n\|_{L^2([a,b])} \to 0$, as $n \to \infty$. Consequently, Corollary 4.*iii*) implies

$$\int_a^b f_n(s)dB_s = \sum_{i=1}^{N(n)} f(s_i^n)\left(B_{t_i^n} - B_{t_{i-1}^n}\right)$$

converges to the Wiener integral $\int_a^b f(s)dB_s$ in $L^2(\Omega)$, as $n \to \infty$. Thus, there is a subsequence $\{n_k : k \in \mathbb{N}\} \subset \mathbb{N}$ such that

$$\int_a^b f(s)dB_s = \lim_{k\to\infty} \sum_{i=1}^{N(n_k)} f(s_i^{n_k})\left(B_{t_i^{n_k}} - B_{t_{i-1}^{n_k}}\right), \quad \text{w.p.1.}$$

It means, ω by ω, for almost all ω. Therefore the result is now a consequence of Corollary 1 and Theorem 29 below because they state that the hand-right side of (1.25) converges to $\int_a^b f(s)d^Y B_s$ w.p.1, as $n \to \infty$. □

In order to finish this section, we extend the Wiener integral to $L^2(\mathbb{R})$. Towards this end we introduce the following:

Definition 15 *A real-valued and continuous process* $B = \{B_t : t \in \mathbb{R}\}$ *is called a two-sided Brownian motion, or two-sided Wiener process if it satisfies the following three conditions:*

i) $B_0 = 0$ *w.p.1.*

ii) For $-\infty < t_1 < t_2 < \ldots < t_n < \infty$, *the random variables* $(B_{t_n} - B_{t_{n-1}}), \ldots, (B_{t_2} - B_{t_1})$ *are independent (i.e.,* B *has independent increments).*

iii) For $s,t \in \mathbb{R}$, *the random variable* $B_t - B_s$ *is normally distributed with mean* 0 *and variance* $|t-s|$ *(i.e.,* $B_t - B_s \sim \mathcal{N}(0, |t-s|)$*).*

Note that, as in Proposition 6, the two-sided Wiener process B is a Gaussian one. So, from (1.9), this process is characterized by its mean and covariance functions, which are given by $E(B_t) = 0$, for $t \in \mathbb{R}$, and

$$E\left(B_t B_s\right) = \begin{cases} |t| \wedge |s|, & \text{if } st \geq 0, \\ 0, & \text{otherwise,} \end{cases}$$

respectively. Also, note that if the complete probability space $(\Omega, \mathcal{F}, P)$ is such that there are two independent Wiener processes $W^{(1)} = \{W_t^{(1)} : t \geq 0\}$ and $W^{(2)} = \{W_t^{(2)} : t \geq 0\}$, then

$$W_t = \begin{cases} W_t^{(1)}, & \text{if } t \geq 0, \\ W_{-t}^{(2)}, & \text{if } t < 0 \end{cases}$$

is a two-sided Wiener process due to $W_0^{(1)} = W_0^{(2)} = 0$.

In analogy with Definition 13, we say that $f : \mathbb{R} \to \mathbb{R}$ is a simple function, or a step function on $\mathbb{R}$ if there are $-\infty < t_0 < t_1 < \ldots < t_n < \infty$, $n \in \mathbb{N}$, and $\alpha_i \in \mathbb{R}$, $i = 1, \ldots, n$, such that

$$f(s) = \sum_{i=1}^{n} \alpha_i 1_{]t_{i-1}, t_i]}(s), \quad s \in \mathbb{R}.$$

The family of all the simple functions on $\mathbb{R}$ is denoted by $\mathcal{E}(\mathbb{R})$.

Also, in analogy with (1.27), we define the Wiener integral of this simple function f with respect to the two-sided Brownian motion B on $\mathbb{R}$ as

$$\int_a^b f(s) dB_s := \sum_{i=1}^{n} \alpha_i \left(B_{t_i} - B_{t_{i-1}}\right).$$

We can apply the proof of Proposition 9 again to show that this definition is independent of the representation of f as a step function.

Lemma 6 *Let $f : \mathbb{R} \to \mathbb{R}$ be in $L^2(\mathbb{R})$. Then, there is a sequence $\{f_n : n \in \mathbb{N}\}$ of simple function on $\mathbb{R}$ such that*

$$\|f - f_n\|_{L^2(\mathbb{R})} \to 0, \quad \text{as } n \to \infty.$$

Proof. Let $n \in \mathbb{N}$. Proceeding as in Proposition 10 (i.e., using the monotone class theorem in the appendix), we can see that the space $\{g(\cdot)1_{[-n,n]} : g \in \mathcal{E}(\mathbb{R})\}$ is dense in $L^2([-n, n])$. Thus, we can choose $f_n \in \mathcal{E}(\mathbb{R})$ satisfying

$$\int_{\mathbb{R}} \left|f(s)1_{[-n,n]}(s) - f_n(s)1_{[-n,n]}(s)\right|^2 ds = \int_{-n}^{n} |f(s) - f_n(s)|^2 ds < \frac{1}{n}.$$

Therefore, the fact that

$$\begin{aligned} &\int_{\mathbb{R}} \left|f(s) - f_n(s)1_{[-n,n]}(s)\right|^2 ds \\ &\leq 2\int_{-n}^{n} |f(s) - f_n(s)|^2 ds + 2\int_{[-n,n]^c} (f(s))^2 ds < \frac{2}{n} + 2\int_{[-n,n]^c} (f(s))^2 ds, \end{aligned}$$

together with the dominated convergence theorem, implies that the result is satisfied. □

Now, we can extend the Wiener integral to $L^2(\mathbb{R})$ similarly as before. That is, for a sequence $\{f_n \in \mathcal{E}(\mathbb{R}) : n \in \mathbb{N}\}$ converging to f in $L^2(\mathbb{R})$, we define the Wiener integral of f on $\mathbb{R}$ (denoted by $\int_{\mathbb{R}} f(s) dB_s$) as

$$\int_{\mathbb{R}} f(s) dB_s = L^2(\Omega) - \lim_{n \to \infty} \int_{\mathbb{R}} f_n(s) dB_s$$

because, as before, we have

$$E\left(\int_{\mathbb{R}} (f_n(s))^2 \, dB_s\right)^2 = \int_{\mathbb{R}} (f_n(s))^2 \, ds.$$

Moreover, the definition of the Gaussian random variable $\int_{\mathbb{R}} f(s)dB_s$ is independent of the choice of the sequence $\{f_n : n \in \mathbb{N}\}$. Sometimes, we use the notation

$$\int_a^b f(s)dB_s := \int_{\mathbb{R}} f(s)1_{[a,b]}(s)ds,$$

for $a, b \in \mathbb{R}$ with $a < b$.

1.3 Gaussian Volterra processes I

Fractional Brownian motion (fBm) is a Gaussian process that is not a semimartingale (see Protter [180] for the definition). So, we cannot use the classical Itô's calculus to introduce a stochastic analysis based on a stochastic integral that allows us to represent phenomena that appear in nature via stochastic differential equations driven by fBm. Therefore, it is an interesting challenge to develop a stochastic analysis for Gaussian processes that are not semimartingales. One way to overcome this problem is to observe that Molchan and Golosov [152] have represented fBm as a Wiener integral, which has become the fundamental tool for the theoretical and practical development of fBm. As we will see in this section, the representation of fBm as a Wiener integral (with respect to a Wiener process) permits to consider a transfer principle. That is, by mean of a suitable isometry from the Hilbert space defined as the completion of step functions with respect to the inner product given by the underlying covariance (see Example 1.*ii*)) into a close linear subspace of $L^2([0,T])$, we can transfer the stochastic analysis for fBm to the one for the underlying Brownian motion and vice versa. It is worth mentioning that fBm is considered in Section 1.4 and that the main goal of this section is to construct the basis of Malliavin calculus for Gaussian processes that have an integral representation with respect to the Wiener process.

Definition 16 *A Gaussian process $B = \{B_t : t \in [0,T]\}$ is called a Volterra process if there exist a Brownian motion $W = \{W_t : t \in [0,T]\}$ and a kernel $K = \{K(t,s) : 0 \leq s < t \leq T\}$ such that $K(t,\cdot) \in L^2([0,t])$ and*

$$B_t = \int_0^t K(t,s)dW_s, \quad t \in [0,T]. \tag{1.32}$$

Note that B defined in (1.32) is a centered Gaussian process with covariance function

$$R_B(t,s) := E(B_t B_s) = \int_0^{t \wedge s} K(t,r)K(s,r)dr, \quad s,t \in [0,T]. \tag{1.33}$$

Conversely a Gaussian process $B = \{B_t : t \in [0,T]\}$ with zero mean and covariance given by (1.33) have the integral representation (1.32) for some Wiener process W, as it is shown by Cramer [34] (see also Park [174]).

In the remaining of this section, we assume that B is a Volterra Gaussian process defined by (1.32) with covariance function given by (1.33), where the kernel K is such that

$$\sup_{t \in [0,T]} \int_0^t K(t,s)^2 ds = \sup_{t \in [0,T]} R_B(t,t) < \infty. \tag{1.34}$$

The reproducing kernel Hilbert space $\mathcal{H}$ is introduced as the completion of the linear span $\mathcal{E}([0,T])$ of the family $\{1_{[0,t]} : t \in [0,T]\}$ of indicator functions on $[0,T]$ with respect to the inner product given by R_B (see Section 8.11 below). It means,

$$R(t,s) = \langle 1_{[0,t]}, 1_{[0,s]} \rangle_{\mathcal{H}} = \langle 1_{[0,t]}K(t,\cdot), 1_{[0,s]}K(s,\cdot) \rangle_{L^2([0,T])}, \quad s,t \in [0,T].$$

Now, for $\varphi \in \mathcal{E}([0,T])$ of the form

$$\varphi = \sum_{i=1}^{n} a_i 1_{(s_i,s_{i+1}]}, \tag{1.35}$$

we define the linear mappings $\varphi \mapsto B(\varphi)$ and $\varphi \mapsto K^*\varphi$ as

$$B(\varphi) = \sum_{i=1}^{n} a_i \left(B_{s_{i+1}} - B_{s_i}\right) \tag{1.36}$$

and

$$(K^*\varphi)(s) = \varphi(s)K(T,s) + \int_s^T (\varphi(t) - \varphi(s))\, K(dt,s), \quad s \in [0,T]. \tag{1.37}$$

As in Proposition 9, we can show that the definitions of these linear operators are independent of the representation of φ as a step function in $\mathcal{E}([0,T])$. In consequence, without loss of generality, we can assume that φ is as in (1.35) with $0 = s_1 < s_2 < \ldots < s_{n+1} = T$. Moreover, we have that $B(\varphi)$ is a square-integrable random variable because of its definition and that $K^*(\varphi)$ is a square-integrable function on $[0,T]$ due to

$$\int_\cdot^T (\varphi(t) - \varphi(s))\, K(dt,\cdot) = \sum_{i=1}^{n-1} 1_{(s_i,s_{i+1}]}(\cdot) \sum_{j=i+1}^{n} (a_j - a_i)\left(K(s_{j+1},\cdot) - K(s_j,\cdot)\right),$$

for φ introduced in (1.35). Furthermore, for $t,s \in [0,T]$, we have

$$\begin{aligned}
\left(K^*1_{[0,t]}\right)(s) &= 1_{[0,t]}(s)\left(K(T,s) + \int_s^T \left[1_{[0,t]}(u) - 1_{[0,t]}(s)\right] K(du,s)\right) \\
&= 1_{[0,t]}(s)\left(K(T,s) - \int_t^T K(du,s)\right) = 1_{[0,t]}(s)K(t,s).
\end{aligned} \tag{1.38}$$

The following result allows to extend the operator $B : \mathcal{E}([0,T]) \to L^2(\Omega)$ to the Hilbert space $\mathcal{H}$.

Proposition 11 *Let B and φ be as in* (1.32) *and* (1.35)*, respectively. Then,*

$$B(\varphi) = \int_0^T (K^*\varphi)(s)dW_s. \tag{1.39}$$

Here, W is the Wiener process in (1.32) *and the right-hand side is the Wiener integral of $K^*\varphi$ with respect to W (see Section 1.2.2).*

Remark 10 *Let $t \in [0,T]$. Then, by* (1.32)*, we get*

$$B(1_{[0,t]}) = \int_0^T \left(K^*1_{[0,t]}\right)(s)dW_s = \int_0^T K(t,s)1_{[0,t]}(s)dW_s = B_t.$$

Proof. The definition of the linear operator K^* implies

$$\begin{aligned}\int_0^T (K^*\varphi)(s)dW_s &= \int_0^T K(T,s)\varphi(s)dW_s \\ &\quad + \sum_{i=1}^{n-1}\int_{s_i}^{s_{i+1}} \sum_{j=i+1}^{n} (a_j - a_i)\left(K(s_{j+1},s) - K(s_j,s)\right) dW_s. \quad (1.40)\end{aligned}$$

To obtain a suitable expression for the last integral, we perform the following calculations: for $s \in [0,T]$, we have

$$\begin{aligned}&\sum_{i=1}^{n-1} 1_{(s_i,s_{i+1}]}(s) \sum_{j=i+1}^{n} (a_j - a_i)\left(K(s_{j+1},s) - K(s_j,s)\right) \\ &= \sum_{i=1}^{n-1} 1_{(s_i,s_{i+1}]}(s) \sum_{j=i+1}^{n} \left(K(s_{j+1},s) - K(s_j,s)\right)\left(\sum_{\ell=i}^{j-1}(a_{\ell+1} - a_\ell)\right) \\ &= \sum_{i=1}^{n-1} 1_{(s_i,s_{i+1}]}(s) \sum_{\ell=i}^{n-1}(a_{\ell+1} - a_\ell) \sum_{j=\ell+1}^{n} \left(K(s_{j+1},s) - K(s_j,s)\right) \\ &= \sum_{\ell=1}^{n-1}(a_{\ell+1} - a_\ell)\left(K(T,s) - K(s_{\ell+1},s)\right) \sum_{i=1}^{\ell} 1_{(s_i,s_{i+1}]}(s).\end{aligned}$$

Hence, (1.40) yields

$$\begin{aligned}\int_0^T (K^*\varphi)(s)dW_s &= \int_0^T \Bigg\{K(T,s)\sum_{i=1}^{n} a_i 1_{(s_i,s_{i+1}]}(s) \\ &\quad + \sum_{i=1}^{n-1}(a_{i+1} - a_i)\left(K(T,s) - K(s_{i+1},s)\right) 1_{[0,s_{i+1}]}(s)\Bigg\} dW_s \\ &= \int_0^T \left\{a_n K(T,s) - \sum_{i=1}^{n-1}(a_{i+1} - a_i)K(s_{i+1},s)1_{[0,s_{i+1}]}(s)\right\} dW_s.\end{aligned}$$

Now, using the definition of B (i.e., equality (1.32)), we have established

$$\begin{aligned}\int_0^T (K^*\varphi)(s)dW_s &= a_n B_T - \sum_{i=1}^{n-1}(a_{i+1} - a_i)B_{s_{i+1}} \\ &= a_n B_T - \sum_{i=2}^{n} a_i B_{s_i} + \sum_{i=1}^{n-1} a_i B_{s_{i+1}} = B(\varphi).\end{aligned}$$

Thus, the proof is finished. □

In the proof of the last proposition, we have shown that, for $s \in [0,T]$ and φ given by (1.35),

$$\begin{aligned}(K^*\varphi)(s) &= a_n K(T,s) - \sum_{i=1}^{n-1}(a_{i+1} - a_i)K(s_{i+1},s)1_{[0,s_{i+1}]}(s) \\ &= a_n K(T,s) - \sum_{i=2}^{n} a_i K(s_i,s)1_{[0,s_i]}(s) + \sum_{i=1}^{n-1} a_i K(s_{i+1},s)1_{[0,s_{i+1}]}(s) \\ &= \sum_{i=1}^{n-1} a_i\left(K(s_{i+1},s)1_{[0,s_{i+1}]}(s) - K(s_i,s)1_{[0,s_i]}(s)\right).\end{aligned}$$

Consequently, Corollary 4.*iii*) and Proposition 11 lead to write

$$\begin{aligned}
E\left(B(\varphi)^2\right) &= \langle K^*\varphi, K^*\varphi\rangle_{L^2([0,T])} \\
&= \left\langle \sum_{i=1}^{n-1} a_i \left(K(s_{i+1},\cdot)1_{[0,s_{i+1}]}(\cdot) - K(s_i,\cdot)1_{[0,s_i]}(\cdot)\right), \right. \\
&\quad \left. \sum_{i=1}^{n-1} a_i \left(K(s_{i+1},\cdot)1_{[0,s_{i+1}]}(\cdot) - K(s_i,\cdot)1_{[0,s_i]}(\cdot)\right)\right\rangle_{L^2([0,T])} \\
&= \langle \varphi,\varphi\rangle_{\mathcal{H}} = \|\varphi\|^2_{\mathcal{H}}.
\end{aligned} \tag{1.41}$$

Now, choose ψ in the reproducing kernel Hilbert space $\mathcal{H}$. By definition (see Section 8.11 below), there exists a sequence $\{\psi^{(n)} \in \mathcal{E}([0,T]) : n \in \mathbb{N}\}$ such that

$$\|\psi - \psi^{(n)}\|_{\mathcal{H}} \quad \text{and} \quad \|\psi\|_{\mathcal{H}} = \lim_{n\to\infty} \|\psi^{(n)}\|_{\mathcal{H}}.$$

Hence, we can deduce that $\{B(\psi^{(n)}) : n \in \mathbb{N}\}$ is a Cauchy sequence in $L^2(\Omega)$ since (1.41) implies that, for $n, m \in \mathbb{N}$,

$$E\left(\left(B(\psi^{(m)}) - B(\psi^{(n)})\right)^2\right) = E\left(B(\psi^{(m)} - \psi^{(n)})^2\right) = \|\psi^{(m)} - \psi^{(n)}\|^2_{\mathcal{H}}.$$

Hence, we define $B(\psi)$ as

$$B(\psi) = L^2(\Omega) - \lim_{n\to\infty} B(\psi^{(n)}). \tag{1.42}$$

As in Lemma 5 and Corollary 4, we obtain that the random variable $B(\psi)$ is independent of the choice of the sequence that goes to ψ in $\mathcal{H}$ and that $B : \mathcal{H} \to L^2(\Omega)$ is a linear operator. Actually, it is an isometry from $\mathcal{H}$ into a closed linear subspace of $L^2(\Omega)$, whose elements are Gaussian random variables due to the $L^2(\Omega)$-limit of Gaussian random variables being Gaussian again (see Section 1.1.1). Proceeding similarly, we are able to use (1.41) again to introduce $K^*\psi$ as

$$K^*\psi = L^2([0,T]) - \lim_{n\to\infty} K^*\psi^{(n)}.$$

In this way, the linear operator K^* provides an isometry between $\mathcal{H}$ and a closed linear subspace of $L^2([0,T])$. That is,

$$\|\psi\|_{\mathcal{H}} = \|K^*\psi\|_{L^2([0,T])} \quad \text{and} \quad \mathcal{H} = (K^*)^{-1}(L^2([0,T])). \tag{1.43}$$

Also, we have that (1.39) holds for any $\varphi \in \mathcal{H}$.

1.4 Fractional Brownian motion

Let $T > 0$ and consider the function

$$R_H(t,s) := \frac{1}{2}\left(s^{2H} + t^{2H} - |t-s|^{2H}\right), \quad s,t \in [0,T]. \tag{1.44}$$

Note that if $H = 1/2$, then $R_{\frac{1}{2}}(t,s) = t \wedge s$, which is the covariance function of Brownian motion (see Section 1.2). For $H \in (0,1)$, we have the following result.

Proposition 12 *Let $W = \{W_t : t \in \mathbb{R}\}$ be a Brownian motion on the whole real line. Then, the process*

$$B_t = C_H\left\{\int_{-\infty}^{0}[(t-s)^{H-1/2}-(-s)^{H-1/2}]dW_s+\int_0^t (t-s)^{H-1/2}dW_s\right\},\quad t\in[0,T],\ (1.45)$$

with $C_H = \left(\int_0^\infty \left[(1+s)^{H-1/2}-s^{H-1/2}\right]^2 ds + \frac{1}{2H}\right)^{-1/2}$*, is a zero mean Gaussian process with covariance function* R_H.

Remark 11 *Remember that the Wiener stochastic integrals with respect to W on the right-hand side of the last equality are introduced in Section 1.2.2. In the proof of the result, we see that these stochastic integrals are well-defined and, therefore, Corollary 4 yields that $\{B_t : t \in [0,T]\}$ is a zero mean Gaussian process.*

Proof. Note that, for $s \in (-\infty, 0)$ and $t \in [0,T]$, the mean value theorem implies

$$\left|(t-s)^{H-1/2}-(-s)^{H-1/2}\right|^2 \le \left(H-\frac{1}{2}\right)^2 T^2(-s)^{2H-3},$$

where the right-hand side belongs to $L^1((-\infty,-M))$, for any $M>0$. Thus, now, it is easy to show that B_t is well-defined for all $t \in [0,T]$. Also note

$$B_t = C_H\int_{\mathbb{R}}\left[((t-s)^{H-1/2}\vee 0)-((-s)^{H-1/2}\vee 0)\right]dW_s,\quad t\in[0,T].$$

Hence, the change of variable $u=-s/t$ and Section 1.2.2 give

$$\begin{aligned} E(B_t^2) &= C_H^2\int_{\mathbb{R}}\left[((t-s)^{H-1/2}\vee 0)-((-s)^{H-1/2}\vee 0)\right]^2 ds \\ &= C_H^2 t\int_{\mathbb{R}}\left[((t+ut)^{H-1/2}\vee 0)-((ut)^{H-1/2}\vee 0)\right]^2 ds \\ &= C_H^2 t^{2H}\left(\int_0^\infty\left[(1+u)^{H-\frac{1}{2}}-(u)^{H-\frac{1}{2}}\right]^2 du+\int_{-1}^0(1+u)^{2H-1}du\right)=t^{2H}. \end{aligned}$$

Also, for $s,t\in[0,T]$, $s<t$, the change of variables $v=u-s$ and last equality lead us to

$$\begin{aligned} E(|B_t-B_s|^2) &= C_H^2\int_{\mathbb{R}}\left[((t-u)\vee 0)^{H-\frac{1}{2}}-((s-u)\vee 0)^{H-\frac{1}{2}}\right]^2 du \\ &= C_H^2\int_{\mathbb{R}}\left[((t-s-v)\vee 0)^{H-\frac{1}{2}}-((-v)\vee 0)^{H-\frac{1}{2}}\right]^2 du=(t-s)^{2H}. \end{aligned}$$

Finally, with last two equalities in mind, we observe

$$\begin{aligned} E(B_tB_s) &= E([B_t-B_s]B_s)+E(B_s^2) \\ &= -E\left([B_t-B_s]^2\right)+E(B_t^2)+E(B_s^2)-E(B_tB_s) \qquad (1.46) \\ &= t^{2H}+s^{2H}-|t-s|^{2H}-E(B_tB_s). \end{aligned}$$

□

The process B introduced in Proposition (1.45) is a fractional Brownian motion. More generally, we have the following:

Definition 17 *A zero mean Gaussian process* $B = \{B_t : t \in [0,T]\}$ *is called a fractional Brownian motion (fBm) with Hurst parameter* $H \in (0,1)$ *if it has the covariance function* R_H *given by* (1.44).

Fractional Brownian motion was first introduced by Kolmogorov [103] and studied by Mandelbrot and Van Ness in [138], where the stochastic integral representation (1.45) in terms of a standard Brownian motion was established. This is known as the Mandelbrot-van Ness integral representation of the fBm.

In the proof of Proposition 12, we show that the process B in (1.45) is such that $E\left((B_t - B_s)^2\right) = |t-s|^{2H}$, $s,t \in [0,T]$, which means that this B has stationary increments. This property is also obtained from the covariance function R_H. Indeed, (1.44) implies that, for $s,t \in [0,T]$,

$$E\left((B_t - B_s)^2\right) = E\left(B_t^2 - 2B_tB_s + B_s^2\right) = |t-s|^{2H}. \tag{1.47}$$

Hence, in general, fBm also has stationary increments according to Definition 17. Moreover, (1.9) implies that fBm is a self-similar process with index H. That is, for any constant $c > 0$, the processes $\left\{c^{-H}B_{ct} : t \ge 0\right\}$ and $\{B_t : t \ge 0\}$ have the same finite-dimensional distributions. Remember that the distribution of a Gaussian vector is described by the mean and the covariance matrix (see Section 1.1.1). Reciprocally, assume that $\tilde{B} = \{\tilde{B}_t \in L^2(\Omega) : t \ge 0\}$ is a self-similar process of index $H \in (0,1)$ with stationary increments. Then, for $n \in \mathbb{N}$,

$$n^H E\left(\left|\tilde{B}_0\right|\right) = E\left(\left|\tilde{B}_0\right|\right)$$

and

$$n^H E\left(\tilde{B}_1\right) = E\left(\tilde{B}_n\right) = E\left(\sum_{i=1}^{n}\left[\tilde{B}_i - \tilde{B}_{i-1}\right]\right) = nE\left(\tilde{B}_1\right).$$

Consequently, $\tilde{B}_0 = 0$ and $\tilde{B}$ is a zero mean process since $E\left(\tilde{B}_t\right) = t^H E\left(\tilde{B}_1\right) = 0$. Moreover, proceeding as in (1.46),

$$\begin{aligned} 2E\left(\tilde{B}_t\tilde{B}_s\right) &= -E\left(\left[\tilde{B}_t - \tilde{B}_s\right]^2\right) + E\left(\tilde{B}_t^2\right) + E\left(\tilde{B}_s^2\right) \\ &= E\left(\left[\tilde{B}_{t-s}\right]^2\right) + E\left(\tilde{B}_t^2\right) + E\left(\tilde{B}_s^2\right) \\ &= E\left(-(t-s)^{2H}\tilde{B}_1^2 + t^{2H}\tilde{B}_1^2 + s^{2H}\tilde{B}_1^2\right). \end{aligned} \tag{1.48}$$

Therefore, fBm is the only Gaussian process which has stationary increments and is self-similar (with index H). Last property is an immediate consequence of the fact that the covariance function (1.44) is homogeneous of order $2H$. See [20, 150, 157, 158, 194, 195] for general background. Fractional Brownian motion is employed in many areas of application (see e.g. [43, 93]).

In Definition 17, we have chosen H in the interval $(0,1)$. So, now we explain why the Hurst parameter H has to be in this interval. To do so, consider a self-similar process $\bar{B} = \{\bar{B}_t : t \in [0,T]\}$ of index $H \in \mathbb{R}$, which also has stationary increments.

First suppose that $\bar{B}_1 \in L^1(\Omega)$, T is a positive integer bigger than 1 and $H > 0$. We then have that $\bar{B}_0 = 0$ with probability 1 since $E\left(|\bar{B}_0|\right) = 0^H E\left(|\bar{B}_1|\right)$. Moreover, in this case, we also get

$$T^H E\left(|\bar{B}_1|\right) = E\left(|\bar{B}_T|\right) = E\left(\left|\sum_{i=1}^{T}\left[\bar{B}_i - \bar{B}_{i-1}\right]\right|\right) \le \sum_{i=1}^{T} E\left(|\bar{B}_i - \bar{B}_{i-1}|\right) = TE\left(|\bar{B}_1|\right),$$

which establishes that $H \leq 1$ if $E\left(|\bar{B}_1|\right) > 0$. Note that if $\bar{B}_1 = 0$ with probability 1, then we also have that $\bar{B}_t = 0$ for all $t \in [0,T]$.

Now assume that $H = 0$ and that $\bar{B}$ is continuous in probability. That is,

$$\lim_{h \to 0} P\left(\left[|\bar{B}_{t+h} - \bar{B}_t| > \varepsilon\right]\right) = 0, \quad \text{for all } t \in [0,T] \text{ and } \varepsilon > 0.$$

In this case, we obtain, for $t \in (0,T]$ and $\varepsilon > 0$,

$$P\left(\left[|\bar{B}_t - \bar{B}_0| > \varepsilon\right]\right) = P\left(\left[t^0 \left|\bar{B}_1 - \bar{B}_0\right| > \varepsilon\right]\right)$$

But, the left-hand side goes to zero as $t \downarrow 0$, while the right-hand side is independent of t. This means that $\bar{B}_t = \bar{B}_0$, with probability 1.

It is time to analyze the case that $H < 0$. Now, we also assume that $\bar{B}$ is continuous in probability and that $\bar{B}_1$ is non-degenerated (i.e., it is not deterministic). Now, we get

$$P\left(\left[\bar{B}_t \leq x\right]\right) = P\left(\left[\bar{B}_1 \leq t^{-H} x\right]\right).$$

Hence, using the continuity in probability of $\bar{B}$ and letting t go to zero, we can establish that, for almost all $x \in \mathbb{R}$,

$$P\left(\left[\bar{B}_0 \leq x\right]\right) = \begin{cases} P\left(\left[\bar{B}_1 \leq 0\right]\right), & \text{if } x \geq 0, \\ P\left(\left[\bar{B}_1 < 0\right]\right), & \text{if } x < 0. \end{cases}$$

Note that the last equality is also true for all $x \in \mathbb{R}$ due to the distribution functions being right-continuous. But, the left-hand side of the last equality cannot be the distribution function of $\bar{B}_0$ because $\bar{B}_1$ is non-degenerated. Actually, if we allow $\hat{B}_1$ to be deterministic, then $\hat{B} \equiv 0$ is the unique self-similar process with index $H < 0$.

Finally, we deal with $H = 1$. Using that $\bar{B}$ is a self-similar process of index 1 with stationary increments and proceeding as in (1.48), we have

$$E\left(\bar{B}_t \bar{B}_s\right) = \frac{1}{2} E\left(\bar{B}_1^2\right)\left[t^2 + s^2 - (t-s)^2\right) = E\left(\bar{B}_1^2\right) ts, \quad s,t \in [0,T].$$

In consequence, we can show that $\bar{B}_t = t\bar{B}_1$ with probability 1, for each $t \in [0,T]$, because

$$E\left(\left(\bar{B}_t - t\bar{B}_1\right)^2\right) = E\left(\bar{B}_t^2 - 2t\bar{B}_t\bar{B}_1 + t^2\bar{B}_1^2\right) = E\left(\bar{B}_1^2\right)\left(t^2 - 2t^2 + t^2\right) = 0.$$

Now, we assume that $\bar{B} = \{\bar{B}_t \in L^2(\Omega) : t \geq 0\}$ is a self-similar process of index $H \in (0,1)$, which also has stationary increments. When $H = 1/2$, the increments of $\bar{B}$ in disjoint intervals are not correlated. Moreover, in the case that $\bar{B}$ is also a Gaussian process, these increments are independent. It means, in this case, we know that $\bar{B}_0 = 0$, with probability 1, and it is not difficult to see that

$$E\left\{\left(\bar{B}_{t_2} - \bar{B}_{t_1}\right)\left(\bar{B}_{s_2} - \bar{B}_{s_1}\right)\right\} = 0, \quad s_1 < s_2 < t_1 < t_2,$$

due to (1.48). However, for $H \neq \frac{1}{2}$, the covariance of its increments in disjoint intervals is non-zero and decays asymptotically as a negative power of the distance between the intervals. If this covariance decreases slowly, we say that B exhibits long-range dependence, or it has long-memory. Otherwise, if the correlation between increments has a fast decay as a function of the distance between intervals, we say that B satisfies short-memory, or short-range dependence:

Proposition 13 *Let $\bar{B} = \{\bar{B}_t \in L^2(\Omega) : t \geq 0\}$ be a self-similar process of index $H \in (0,1)$, with stationary increments. Then the covariance between two increments $\bar{B}_{t+h} - \bar{B}_t$ and $\bar{B}_{s+h} - \bar{B}_s$, where $s + h \leq t$, and $t - s = nh$ is*

$$\begin{aligned}\rho_H(n) &= \frac{h^{2H}}{2} E\left(\bar{B}_1^2\right)\left((n+1)^{2H} + (n-1)^{2H} - 2n^{2H}\right) \\ &\approx h^{2H} H(2H-1) n^{2H-2} E\left(\bar{B}_1^2\right) \to 0\end{aligned}$$

as n tends to infinity. Therefore:

i) If $H > \frac{1}{2}$, $\rho_H(n) > 0$ and $\sum_{n=1}^{\infty} \rho_H(n) = \infty$.

ii) If $H < \frac{1}{2}$, $\rho_H(n) < 0$ and $\sum_{n=1}^{\infty} |\rho_H(n)| < \infty$.

Proof. Let s and t be as in the statement of the proposition. Then, (1.48) and Taylor's theorem imply that there exists $\theta_1 \in (n, n+1)$ and $\theta_2 \in (n-1, n)$ such that

$$\begin{aligned}E\left\{\left(\bar{B}_{t+h} - \bar{B}_t\right)\left(\bar{B}_{s+h} - \bar{B}_s\right)\right\} &= \frac{1}{2} E\left(\bar{B}_1^2\right)\left\{(t+h-s)^{2H} + (t-s-h)^{2H} - 2(t-s)^{2H}\right\} \\ &= \frac{h^{2H}}{2} E\left(\bar{B}_1^2\right)\left\{(n+1)^{2H} + (n-1)^{2H} - 2n^{2H}\right\} \\ &= \frac{h^{2H}}{2} H(2H-1) E\left(\bar{B}_1^2\right)\left\{\theta_1^{2H-2} + \theta_2^{2H-2}\right\}.\end{aligned}$$

Now, it is easy to see that the result holds. □

In case *i*) the increments $\bar{B}_{t+h} - \bar{B}_t$ and $\bar{B}_{s+h} - \bar{B}_s$ are positively correlated and the process presents long memory. In *ii*) last increments are negatively correlated and, in this case, we say that there is short-range dependence. In consequence, fBm exhibits long (resp. short) range-dependence for $H > 1/2$ (resp. $H < 1/2$).

Long memory was observed by Hurst [84]. As examples, long dependence appears in hydrology [139], stock price changes [69], rainfall [137, 140], etc. Comte and Renault [31] observe that the long-maturity is able to be considered by long-memory volatilities. Thus, fBm is an important tool in volatility modeling. Short-memory appears in the modeling of the volatility process in finance (see Alòs et al. [8] and Gatheral et al. [62]). In this last area, we also need to consider processes observing long and short-range dependence, which is another example of the importance of fractional Brownian motion. However, fBm is not a semimartingale for $H \neq 1/2$. So, in general, we cannot study phenomena that exhibit range dependence through classical Itô's calculus (see, for instance, [22, 38, 96, 180], among others). The lack of semimartingale property of the fBm for $H \neq 1/2$ can be proven through the ergodic theorem as follows. This proof has been given by Rogers [184].

Let $B = \{B_t : t \geq 0\}$ be a fractional Brownian motion. Then, the sequence $\{B_k - B_{k-1} : k \in \mathbb{N}\}$ is stationary. That is, given a Borel subset C of $\mathbb{R}^m$ and $k_1, \ldots, k_m, n \in \mathbb{N}$, $k_1 < \ldots < k_m$, we have that

$$\begin{aligned}&P\left(\left[\left(B_{k_1} - B_{k_1-1}, \ldots, B_{k_m} - B_{k_m-1}\right) \in C\right]\right) \\ &\quad = P\left(\left[\left(B_{k_1+n} - B_{k_1+n-1}, \ldots, B_{k_m+n} - B_{k_m+n-1}\right) \in C\right]\right),\end{aligned}$$

which is an immediate consequence of (1.9), and the fact that fBm is a Gaussian process and that (1.44) implies

$$\begin{aligned}&E\left((B_{k_\ell+n} - B_{k_\ell+n-1})((B_{k_j+n} - B_{k_j+n-1})\right) \\ &= E\left((B_{k_\ell} - B_{k_\ell-1})((B_{k_j} - B_{k_j-1})\right), \quad \ell, j \in \{1, \ldots, m\}.\end{aligned}$$

Since $E(B_k - B_{k-1}) = 0$, $E\left((B_k - B_{k-1})^2\right) = 1$ and the family $\{B_k - B_{k-1} : k \in \mathbb{N}\}$ is Gaussian and stationary, we have that Proposition 13 yields that this family is also ergodic. Thus, the ergodic theorem allows us to conclude that, for $p > 0$,

$$\lim_{n\to\infty} \frac{1}{n} \sum_{n=1}^{\infty} |B_n - B_{n-1}|^p = E\left(|B_1|^p\right), \tag{1.49}$$

where the limit is with probability 1 and in $L^1(\Omega)$ (for details on the ergodic theorem, the reader can consult Lindgren [128], Section 6, and Shiryayev [197], Chapter 5).

Now choose $K, T, p > 0$. In consequence, using the H-self-similarity of B and (1.49), we obtain

$$\begin{aligned} P\left(\left[\sum_{j=1}^{n} \left|B_{\frac{(j+1)T}{n}} - B_{\frac{jT}{n}}\right|^p > K\right]\right) &= P\left(\left[\frac{T^{pH}}{n^{pH}} \sum_{j=1}^{n} |B_{j+1} - B_j|^p > K\right]\right) \\ &= P\left(\left[\frac{1}{n} \sum_{j=1}^{n} |B_{j+1} - B_j|^p > \frac{n^{pH-1}}{T^{pH}} K\right]\right) \\ &\to \begin{cases} 0, & \text{if } pH - 1 > 0, \\ 1, & \text{if } pH - 1 < 0, \end{cases} \qquad \text{as } n \to \infty. \end{aligned} \tag{1.50}$$

For $H > 1/2$, we have that $2H > 1$. Therefore, if B is a semimartingale, (1.50) states that the quadratic variation of B is null, which shows that B is a process with bounded variation on $[0, T]$, with probability 1 (see Protter [180] for details). In this case, we get that

$$\limsup_{n\to\infty} P\left(\left[\sum_{j=1}^{n} \left|B_{\frac{(j+1)T}{n}} - B_{\frac{jT}{n}}\right| > K\right]\right) < 1,$$

for $K > 0$ large enough. But, this is impossible due to $H < 1$ and (1.50). It means B is not a semimartingale for any filtration $\{\mathcal{F}_t : t \in [0.T]\}$ such that B is $\mathcal{F}_t$-adapted, when $H > 1/2$.

B is not a semimartingale on $[0, T]$ when $H < 1/2$ either. Indeed, otherwise, for $K > 0$ large enough, we obtain that the quadratic variation of B exists. Thus, we can see

$$\limsup_{n\to\infty} P\left(\left[\sum_{j=1}^{n} \left|B_{\frac{(j+1)T}{n}} - B_{\frac{jT}{n}}\right|^2 > K\right]\right) < 1,$$

which contradicts (1.50) again because $2H < 1$ in this case. Thus, we have stated the following result.

Proposition 14 *Let $B = \{B_t : t \in [0, T]\}$ be a fractional Brownian motion with Hurst parameter $H \neq 1/2$. Then, B is not a semimartingale with respect to any filtration $\{\mathcal{G}_t : t \in [0, T]\}$ for which B is $\mathcal{G}_t$-adapted.*

Remark 12 *Remember that B is Brownian motion for $H = 1/2$, which is a square-integrable martingale (see Section 1.2).*

On the other hand, as Brownian motion (see Theorem 5), the paths of fBm are also nowhere differentiable with probability 1. To see that this fact holds, we must do some changes in the proof of Theorem 5 because fBm has no independent increments for $H \neq 1/2$.

Proposition 15 *Let $B = \{B_t : t \in [0,T]\}$ be a fractional Brownian motion. Then, its paths are nowhere differentiable with probability 1.*

Remark 13 *In Mandelbrot and van Ness [138] (Proposition 4.2), it is shown that $t \mapsto B_t(\omega)$ is not differentiable for almost all $t \in [0,T]$, with probability 1.*

Proof. We first fix $p, q \in \mathbb{N}$ such that

$$\lim_{h \downarrow 0} \frac{\left(h^{1-H}\right)^q}{h} = 0. \tag{1.51}$$

and

$$\limsup_{h \downarrow 0} \sup_{|t-s| \geq ph} \frac{E\left((B_{t+h} - B_t)(B_{s+h} - B_s)\right)}{h^{2H}} < \frac{1}{2q}. \tag{1.52}$$

Note that (1.51) holds if we choose $q > (1-H)^{-1}$ and the existence of such a p follows proceeding as in the proof of Proposition 13.

Now, we choose a sequence $\{a_n \in \mathbb{N} : n \in \mathbb{N}\}$ such that $a_n \uparrow \infty$ and we proceed as in the proof of Theorem 5: Note that if $f : [0,T] \to \mathbb{R}$ is differentiable at t_0, then we can find $\delta, K > 0$ such that $|f(t) - f(t_0)| \leq K|t - t_0|$ for $|t - t_0| \leq \delta$. Therefore, for n large enough, there is $i \in \{1, \ldots, a_n - pq + p - 1\}$ such that $\frac{i-1}{a_n}T \leq t_0 \leq \frac{i}{a_n}T$ and

$$\begin{aligned}
&\left| f\left(\frac{i+pk+1}{a_n}T\right) - f\left(\frac{i+pk}{a_n}T\right) \right| \\
\leq\ & \left| f\left(\frac{i+pk+1}{a_n}T\right) - f(t_0) \right| + \left| f(t_0) - f\left(\frac{i+pk}{a_n}T\right) \right| \\
\leq\ & K\left| \frac{i+pk+1}{a_n}T - t_0 \right| + K\left| t_0 - \frac{i+pk}{a_n}T \right| \\
\leq\ & KT\frac{pk+2}{a_n} + KT\frac{pk+1}{a_n} \leq KT\frac{2(pk+2)}{a_n}, \quad k = 0, \ldots, q-1.
\end{aligned}$$

Hence,

$$\mathcal{C}_K = \left[\limsup_{h \to 0} \frac{B_{t+h} - B_t}{h} < K, \text{ for some } t \in [0,T) \right] \subset \bigcap_{m \geq 1} \bigcup_{n \geq m} \bigcup_{i=0}^{a_n - pq + p - 1} B_i^n,$$

where

$$B_i^n = \bigcap_{k=0}^{q-1} \left[\left| B_{(i+kp+1)T/a_n} - B_{(i+kp)T/a_n} \right| \leq KT\frac{2(pk+2)}{a_n} \right].$$

Now, our goal is to estimate $P(B_i^n)$. Towards this end, observe that

$$P(B_i^n) = \frac{1}{(2\pi)^{q/2}\sqrt{detA}} \int \cdots \int_{\{x_k \leq 2a_n^{H-1}KT^{1+H}(pq+2)\}} \exp\left(-\frac{1}{2}xAx^t\right) dx_1 \cdots x_q$$

with $A = (a_{kk'})_{k,k'=0}^{q-1}$. Here,

$$a_{kk'} = (a_n T)^{-2H} E\left((B_{(i+kp+1)T/a_n} - B_{(i+kp)T/a_n})(B_{(i+k'p+1)T/a_n} - B_{(i+k'p)T/a_n}) \right).$$

By (1.52), we have $a_{kk} = 1$ and $a_{kk'} \leq 1/2q$, for $k \neq k'$ and n large enough. Hence, there is a constant $c > 0$ such that $detA > c$. In consequence, there exists a constant $C > 0$ so that

$$P(B_i^n) \leq C\left(2a_n^{H-1}T^{1+H}K(pq+2)\right)^q.$$

Therefore, for any $m \in \mathbb{N}$, we have that $P(\mathcal{C}_K) \leq C \sum_{n=m}^{\infty} \frac{a_n}{(a_n^{1-H})^q}$.

Finally, we can choose the sequence $\{a_n : n \in \mathbb{N}\}$ such that

$$\sum_{n=1}^{\infty} \frac{a_n}{(a_n^{1-H})^q} < \infty$$

due to (1.51). Now, we are able to finish the proof as that of Theorem 5. □

Last proof is given in Kawada and Kôno [97]. Actually, they have proven a more general result. Namely:

Proposition 16 *Let $\{X_t : t \in [0,1]\}$ be a continuous and centered Gaussian process and φ an even, non-negative and non-decreasing function such that*

$$E\left((X_{t+h} - X_t)^2\right) \geq \varphi^2(h), \quad t+h, t \in [0,1].$$

Assume that there exist $p, q \in \mathbb{N}$ such that

$$\lim_{h \downarrow 0} \left(\frac{h}{\varphi(h)}\right)^q \frac{1}{h} = 0$$

and

$$\limsup_{h \downarrow 0} \sup_{|t-s| \geq ph} \frac{E\left[(X_{t+h} - X_t)(X_{s+h} - X_s)\right]}{\sqrt{E\left[(X_{t+h} - X_t)^2\right] E\left[(X_{s+h} - X_s)^2\right]}} \leq \frac{1}{2q}.$$

Then,

$$P\left(\left[\limsup_{h \downarrow 0} \frac{X_{t+h} - X_t}{h} = \infty, \quad \text{for all } t \in [0,1]\right]\right) = 1.$$

Note that, in Proposition 15, we have shown that $\varphi(h) = |h|^H$ satisfies the assumptions of the last result. In [97], it is pointed out that some varying functions satisfy the hypotheses of Proposition 16.

On the other hand, fractional Brownian motion also has Hölder continuous paths. In particular Brownian motion has them because it is an fBm with parameter $H = 1/2$. Indeed, by Kolmogorov-Čentsov continuity criterion (i.e., Theorem 1), Proposition 5 and (1.44), we deduce that fBm has a version with continuous trajectories. Henceforth, without loss of generality, we always deal with this continuous version. Moreover, by Proposition 5, we can deduce the following modulus of continuity for the trajectories of fBm: For all $\varepsilon \in (0, H)$ and $T > 0$, there exists a nonnegative random variable $G_{\varepsilon,T}$ such that $E\left(|G_{\varepsilon,T}|^p\right) < \infty$ for all $p \geq 1$, and

$$|B_t - B_s| \leq G_{\varepsilon,T}|t-s|^{H-\varepsilon}, \quad \text{for all } s, t \in [0,T].$$

In other words, the Hurst parameter H controls the regularity of the paths. That is, they are Hölder continuous of order $H - \varepsilon$, for any $\varepsilon \in (0, H)$.

Our following goal is to consider fractional Brownian motion as a Volterra type process. Towards this end, we first need to recall some general elements of fractional calculus. The reader can see Samko et al. [193] and Podlubny [179] for a detailed exposition.

1.4.1 Elements of fractional calculus

The fractional calculus began with some correspondence of Leibniz with Bernoulli, L'Hôpital and Wallis, where they discussed the possibility of extending the notion of derivative to non-integer order, in particular of order 1/2.

The generalization of the Young's integral in Section 2.3 is based on the fractional calculus in the Riemann-Liouville sense. The Riemann-Liouville integrals were first studied by Abel [1, 2] in order to derive the shape of an unknown curve u where a particle sliding under the influence of gravity takes the time $f(x)$ to reach the origin 0 after being released from rest at a height x. Abel obtained that the Volterra equation

$$f(x) = \int_0^x u(y)(x-y)^{-1/2} dy, \quad x > 0,$$

represents this problem.

Independent of the contribution of Abel, Riemann [183] was inspired by the work of Liouville [129] on applications in geometry to consider the integral

$$\frac{1}{\Gamma(\alpha)} \int_0^{x-a} \varphi(x-y) y^{-\alpha} dy, \quad x > a,$$

plus an extra function, in his definition of fractional integral. The reader interested in the history of fractional calculus can see the papers by Ross [185, 186] and references therein.

We restrict this section to real-valued functions for notational sake. Consider $0 \leq a < b \leq T$ and an $L^1([a,b])$-function f. In order to motivate the definition of fractional integrals, we observe that the iterated integral

$$\left(I_{a+}^n f\right)(t) = \int_a^t \int_a^{t_{n-1}} \cdots \int_a^{t_1} f(t_0) dt_0 \cdots dt_{n-1}, \quad t \in [a,b],$$

can be written as

$$\begin{aligned} \left(I_{a+}^n f\right)(t) &= \frac{1}{(n-1)!} \int_a^t (t-y)^{n-1} f(y) dy \\ &= \frac{1}{\Gamma(n)} \int_a^t (t-y)^{n-1} f(y) dy, \quad t \in [a,b], \end{aligned} \tag{1.53}$$

due to Fubini's theorem, induction on n and the relation $\Gamma(n) = (n-1)!$, where Γ is the gamma function introduced in (8.10). Note that, for $n \in \mathbb{N}$, $I_{a+}^n f$ is an $\mathbb{R}$-valued function on $[a,b]$. Now, it is natural to consider an arbitrary positive number α instead of $n \in \mathbb{N}$ in equality (1.53). So, in a natural way, we have given the first fractional integral: for $\alpha > 0$, the left-sided Riemann-Liouville fractional integral of f of order α is defined as

$$\left(I_{a+}^\alpha f\right)(t) = \frac{1}{\Gamma(\alpha)} \int_a^t (t-r)^{\alpha-1} f(r)\, dr, \quad t \in [a,b]. \tag{1.54}$$

This integral is the one given by Riemann [183] without the extra function and is called Riemann-Liouville fractional integral in honor of Liouville and Riemann. Furthermore, the fractional integral $I_{a+}^\alpha f$ is well-defined almost surely. In fact, if f belongs to $L^p([a,b])$, for some $p \geq 1$, then $I_{a+}^\alpha f$ is also in $L^p([a,b])$ because of the Fubini's theorem and Hölder inequality. Indeed, we have

$$\begin{aligned} \int_a^b \left|\left(I_{a+}^\alpha f\right)(t)\right|^p dt &\leq \frac{1}{\Gamma(\alpha)^p} \int_a^b \left(\int_a^t (t-r)^{\alpha-1} dr\right)^{p-1} \left(\int_a^t (t-r)^{\alpha-1} |f(r)|^p\, dr\right) dt \\ &\leq C \int_a^b \int_a^t (t-r)^{\alpha-1} |f(r)|^p\, dr dt = C \int_a^b |f(r)|^p \int_r^b (t-r)^{\alpha-1} dt dr \\ &= C \int_a^b |f(r)|^p dt < \infty. \end{aligned} \tag{1.55}$$

Notice that this inequality implies that I^α_{a+} is a linear operator on $L^p([a,b])$. Therefore, for $f\in L^p([a,b])$ and $g\in L^q([a,b])$, with $p^{-1}+q^{-1}=1$. we can use Fubini's theorem again to get

$$\int_a^b \left(I^\alpha_{a+}f\right)(t)g(t)dt=\int_a^b f(r)\left(\frac{1}{\Gamma(\alpha)}\int_r^b g(t)(t-r)^{\alpha-1}dt\right)dr.$$

It means, the adjoint of I^α_{a+} is the right-sided Riemann-Liouville fractional integral of order α, which is introduced as

$$\left(I^\alpha_{b-}f\right)(t)=\frac{1}{\Gamma(\alpha)}\int_t^b (r-t)^{\alpha-1}f(r)\,dr,\quad t\in[a,b]. \tag{1.56}$$

Here, we observe that we only need to assume that $f\in L^1([a,b])$ and $\alpha>0$ in order to have $I^\alpha_{b-}f$ is well-defined due to Fubini's theorem. Actually, as before, $I^\alpha_{b-}:L^p([a,b])\to L^p([a,b])$ is a continuous linear operator, which implies that $I^\alpha_{b-}f$ is also well-defined almost surely on $[a,b]$.

Sections 1.4.2 and 2.3 deal only with the case that $\alpha\in(0,1)$. So, we contemplate this case in the remaining of this subsection.

The fractional integrals I^α_{a+} and I^α_{b-} are not only bounded operators on $L^p([a,b])$, but also satisfy the following property.

Theorem 7 *Let $\alpha\in(0,1)$ and $p\in(1,1/\alpha)$. Then, $I^\alpha_{a+},I^\alpha_{b-}:L^p([a,b])\to L^q([a,b])$ are bounded linear operators for q in the interval $[1,\frac{p}{1-\alpha p})$.*

Remark 14 *Note that $p<p/(1-\alpha p)$. On the other hand, the result is also true for $q=p/(1-\alpha p)$. This case has been established by Hardy and Littlewood. Their proof is quite technical and long, so it is omitted in this book. The reader interested in the proof of this case can consult Hardy and Littlewood [72].*

Proof. We only need to consider the case $p<q$ due to (1.55), and the facts that $L^{p_2}([a,b])\subset L^{p_1}([a,b])$ and $\|\cdot\|_{L^{p_1}([a,b])}\le C\|\cdot\|_{L^{p_2}([a,b])}$, for $1\le p_1<p_2$.

Let φ be in $L^p([a,b])$, $q>p$, $\varepsilon=(q^{-1}-\tilde{q}^{-1})/2$, where $\tilde{q}=p/(1-\alpha p)$, and p' such that $1=p^{-1}+p'^{-1}$. Then, Hölder inequality and (1.54) give

$$\begin{aligned}
&\Gamma(\alpha)\left|\left(I^\alpha_{a+}\varphi\right)(x)\right|\\
&\le \int_a^x \left(|\varphi(y)|^{\frac{p}{q}}(x-y)^{\varepsilon-\frac{1}{q}}\right)|\varphi(y)|^{1-\frac{p}{q}}(x-y)^{\varepsilon-\frac{1}{p'}}dy\\
&\le \left(\int_a^x|\varphi(y)|^p(x-y)^{\varepsilon q-1}dy\right)^{\frac{1}{q}}\left(\int_a^x|\varphi(y)|^{\frac{q-p}{q-1}}(x-y)^{\frac{(p'\varepsilon-1)(p-1)q}{p(q-1)}}dy\right)^{\frac{q-1}{q}}\\
&\le \left(\int_a^x|\varphi(y)|^p(x-y)^{\varepsilon q-1}dy\right)^{\frac{1}{q}}\left(\int_a^x|\varphi(y)|^pdy\right)^{\frac{q-p}{pq}}\left(\int_a^x(x-y)^{p'\varepsilon-1}dy\right)^{\frac{1}{p'}}\\
&\le C\|\varphi\|^{1-\frac{p}{q}}_{L^p([a,b])}\left(\int_a^x|\varphi(y)|^p(x-y)^{\varepsilon q-1}dy\right)^{\frac{1}{q}}.
\end{aligned}$$

Consequently, Fubini's theorem implies

$$\|I^\alpha_{a+}\varphi\|_{L^q([a,b])}\le C\|\varphi\|^{1-\frac{p}{q}}_{L^p([a,b])}\left(\int_a^b\int_a^x|\varphi(y)|^p(x-y)^{\varepsilon q-1}dydx\right)^{\frac{1}{q}}\le C\|\varphi\|_{L^p([a,b])}.$$

Similarly, we can establish that there exists a constant $C>0$ such that $\|I^\alpha_{b-}\varphi\|_{L^q([a,b])}\le C\|\varphi\|_{L^p([a,b])}$. Therefore, the proof is complete. □

A consequence of Theorem 7 is the following result.

Corollary 6 *Let $p, q > 1$ be such that $p^{-1} + q^{-1} \leq 1 + \alpha$, $\alpha \in (0,1)$, $f \in L^p([a,b])$ and $g \in L^q([a,b])$. Then,*

$$\int_a^b \left(I_{a+}^{\alpha} f\right)(t) g(t) dt = \int_a^b f(t) \left(I_{b-}^{\alpha} g\right)(t) dt.$$

Remark 15 *Note that $gI_{a+}^{\alpha} f$ and $fI_{b-}^{\alpha} g$ belong to $L^1([a,b])$ since Theorem 7 and , for instance,*

$$\frac{1}{q} + \frac{1-\alpha p}{p} \leq 1 + \alpha - \frac{1}{p} + \frac{1-\alpha p}{p} = 1.$$

Proof. The result is an immediate consequence of Fubini's theorem. □

The following result will be needed later and it is interesting by itself.

Lemma 7 *Let $\alpha \in (0,1)$, $p > 1$ and $f \in I_{T-}^{\alpha}(L^p([0,T]))$. Then,*

$$\left(\int_0^T |f(x+t) - f(x)|^p \, dx\right)^{1/p} = o(|t|^{\alpha}), \quad \text{as } t \to 0.$$

Remark 16 *By hypothesis, there exists $\varphi \in L^p([0,T])$ such that $f = I_{T-}^{\alpha}(\varphi)$. So, in this result we are assuming that $\varphi \equiv 0$ on $[0,T]^c$. In this way, f is also well-defined on $\mathbb{R}$.*

Proof. Let φ as in the remark of the lemma. Then, for $x \in [0,T]$ and $t \in \mathbb{R}_+$, the change of variables formula $u = r - x$ implies

$$\begin{aligned}
& f(x+t) - f(x) \\
&= \frac{1}{\Gamma(\alpha)} \left(\int_t^{T-x} \varphi(u+x)(u-t)^{\alpha-1} du - \int_0^{T-x} \varphi(u+x) u^{\alpha-1} du\right) \\
&= \frac{t^{\alpha-1}}{\Gamma(\alpha)} \left(\int_t^{T-x} \varphi(u+x)(\frac{u}{t}-1)^{\alpha-1} du - \int_0^{T-x} \varphi(u+x)(\frac{u}{t})^{\alpha-1} du\right) \\
&= t^{\alpha-1} \left(\int_0^{T-x} \varphi(u+x) k(\frac{u}{t}) du\right) = t^{\alpha-1} \left(\int_0^{\infty} \varphi(u+x) k(\frac{u}{t}) du\right),
\end{aligned}$$

where last equality follows from the fact that $\varphi \equiv 0$ on (T, ∞) and

$$k(y) = \frac{1}{\Gamma(\alpha)} \left((y-1)^{\alpha-1} 1_{\{y>1\}} - y^{\alpha-1}\right), \quad y \in \mathbb{R}_+.$$

Hence, the change of variables formula $v = u/t$ and (8.5) lead us to obtain

$$\|f(\cdot + t) - f(\cdot)\|_{L^p([0,T])} \leq t^{\alpha} \int_0^{\infty} \|\varphi(vt + \cdot) - \varphi(\cdot)\|_{L^p([0,T])} |k(v)| dv. \tag{1.57}$$

Indeed, note that, on $(2, \infty)$, we have that the mean value theorem implies $|k(y)| \leq (1 - \alpha)(y-1)^{\alpha-2}$ and, therefore $k \in L^1((0,\infty))$. Moreover, for $x > 1$,

$$\left|\int_0^x k(y) dy\right| = \frac{1}{\alpha\Gamma(\alpha)} \left[x^{\alpha} - (x-1)^{\alpha}\right] \leq \frac{1}{\Gamma(\alpha)} (x-1)^{\alpha-1} \to 0, \quad \text{as } x \to \infty.$$

Thus, (1.57) holds due to the dominated convergence theorem and, consequently, it is now easy to finish the proof. □

In order to calculate the inverse of a Riemann-Liouville fractional integral, suppose that, for $f \in L^1([a,b])$, the Abel equation

$$\frac{1}{\Gamma(\alpha)} \int_a^x \varphi(y)(x-y)^{\alpha-1} dy = f(x), \quad x > a, \tag{1.58}$$

has a solution φ in $L^1([a,b])$. Hence, using the Fubini's theorem again, $\alpha \in (0,1)$ and Lemma 87, we have

$$\begin{aligned} &\Gamma(\alpha) \int_a^x \frac{f(y)}{(x-y)^\alpha} dy \\ &= \int_a^x \frac{1}{(x-y)^\alpha} \int_a^y \frac{\varphi(s)ds}{(y-s)^{1-\alpha}} dy = \int_a^x \varphi(s) \int_s^x (x-y)^{-\alpha}(y-s)^{\alpha-1} dyds \\ &= \frac{\Gamma(1-\alpha)\Gamma(\alpha)}{\Gamma(1)} \int_a^x \varphi(s)ds, \quad x \in [a,b], \end{aligned} \tag{1.59}$$

which allows us to show that the function $x \mapsto \int_a^x \frac{f(y)dy}{(x-y)^\alpha}$ is absolutely continuous on the interval $[a,b]$ and

$$\varphi(x) = \frac{1}{\Gamma(1-\alpha)} \frac{d}{dx} \int_a^x \frac{f(y)dy}{(x-y)^\alpha}, \quad \text{for almost all } x \in [a,b].$$

In particular, we have stated that equation (1.58) has at most a solution $\varphi \in L^1([a,b])$. Thus, we have motivated the following definition.

Definition 18 *Let $f \in L^1([a,b])$ and $\alpha \in [0,1)$.*

i) Assume that the function $x \mapsto \int_a^x \frac{f(y)dy}{(x-y)^\alpha}$ is absolutely continuous on $[a,b]$. The left-sided Riemann-Liouville fractional derivative of order α of f is defined as

$$\left(D_{a+}^\alpha f\right)(x) = \frac{1}{\Gamma(1-\alpha)} \frac{d}{dx} \int_a^x \frac{f(y)}{(x-y)^\alpha} dy, \quad x \in [a,b].$$

ii) Suppose that the map $x \mapsto \int_x^b \frac{f(y)dy}{(y-x)^\alpha}$ is absolutely continuous on $[a,b]$.In this case, the right-sided Riemann-Liouville fractional derivative of order α of f is introduced as

$$\left(D_{b-}^\alpha f\right)(x) = \frac{-1}{\Gamma(1-\alpha)} \frac{d}{dx} \int_x^b \frac{f(y)}{(y-x)^\alpha} dy, \quad x \in [a,b].$$

Remark 17 *i) Note that if $\alpha > 1$, then the integral in (1.59) may not be well-defined. Also note that if $f \equiv C$ (i.e., f is a constant function), then*

$$\left(D_{a+}^\alpha C\right)(x) = \frac{C}{\Gamma(1-\alpha)} (x-a)^{-\alpha} \neq 0, \quad x \in (a,b],$$

unlike $\frac{dC}{dx} \equiv 0$.

ii) For $\alpha = 0$, we have that D_{a+}^0 and D_{b-}^0 agree with the identity operator on $L^1([a,b])$.

iii) For $\alpha = 1$, we define $D_{a+}^1 f = D_{b-}^1 f = \frac{df}{dx}$.

For any $p \geq 1$, we denote by $I^{\alpha}_{a+}(L^p)$ the image of $L^p([a,b])$ by I^{α}_{a+}, and similarly for $I^{\alpha}_{b-}(L^p)$. In other words, $f \in I^{\alpha}_{a+}(L^p)$ if there exists $\varphi \in L^p([a,b])$ such that $f = I^{\alpha}_{a+}\varphi$. Remember that there is only one φ satisfying this equality due to the uniqueness of the solution of Abel equation (1.58).

Immediate consequences of Proposition 135 in the appendix are the following two results.

Corollary 7 *Let f be a function in $L^1([a,b])$ and $\alpha \in (0,1)$. Then, f belongs to $I^{\alpha}_{a+}(L^1)$ (resp. $I^{\alpha}_{b-}(L^1)$) if and only if $I^{1-\alpha}_{a+}f$ is absolutely continuous on $[a,b]$ and $\left(I^{1-\alpha}_{a+}f\right)(a) = 0$ (resp. $I^{1-\alpha}_{b-}f$ is absolutely continuous on $[a,b]$ and $\left(I^{1-\alpha}_{b-}f\right)(b) = 0$). Moreover, in this case, $I^{\alpha}_{a+}(D^{\alpha}_{a+}f) = f$ (resp. $I^{\alpha}_{b-}(D^{\alpha}_{b-}f) = f$).*

Remark 18 *Note that, for $\varphi \in L^1([a,b])$, we also have the proof of Proposition 135 implies $D^{\alpha}_{a+}\left(I^{\alpha}_{a+}\varphi\right) = \varphi$ and $D^{\alpha}_{b-}\left(I^{\alpha}_{b-}\varphi\right) = \varphi$. Thus, D^{α}_{a+} (rep. D^{α}_{b-}) is the inverse of the map $I^{\alpha}_{a+} : L^1([a,b]) \to I^{\alpha}_{a+}(L^1)$ (resp. $I^{\alpha}_{b-} : L^1([a,b]) \to I^{\alpha}_{b-}(L^1)$.*

Corollary 8 *Let f be an absolutely continuous function on $[a,b]$ and $\alpha \in (0,1)$. Then, f belongs to the space $I^{\alpha}_{b-}(L^1)$ (resp. $I^{\alpha}_{a+}(L^1)$) and*

$$\left(D^{\alpha}_{b-}f\right)(x) = \frac{1}{\Gamma(1-\alpha)}\left(\frac{f(b)}{(b-x)^{\alpha}} - \int_x^b \frac{f'(y)}{(y-x)^{\alpha}}dy\right), \quad x \in [a,b],$$

(resp.

$$\left(D^{\alpha}_{a+}f\right)(x) = \frac{1}{\Gamma(1-\alpha)}\left(\frac{f(a)}{(x-a)^{\alpha}} + \int_a^x \frac{f'(y)}{(x-y)^{\alpha}}dy\right), \quad x \in [a,b]).$$

Proof. The analysis of $D^{\alpha}_{a+}f$ is quite similar to that of $D^{\alpha}_{b-}f$. So, we only carry out the study of $D^{\alpha}_{b-}f$.

The fact that f is an absolutely continuous function gives $f(y) = f(b) - \int_y^b f'(s)ds$, $y \in [a,b]$. Therefore, for $x \in [a,b]$, Fubini's theorem and (8.11) lead us to get

$$\begin{aligned}
\left(I^{1-\alpha}_{b-}f\right)(x) &= \frac{1}{\Gamma(1-\alpha)}\left(\int_x^b \frac{f(b)}{(y-x)^{\alpha}}dy - \int_x^b \frac{1}{(y-x)^{\alpha}}\int_y^b f'(s)dsdy\right) \\
&= \frac{f(b)(b-x)^{1-\alpha}}{\Gamma(2-\alpha)} - \frac{1}{\Gamma(1-\alpha)}\int_x^b f'(s)\int_x^s (y-x)^{-\alpha}dyds \\
&= \frac{1}{\Gamma(2-\alpha)}\left(f(b)(b-x)^{1-\alpha} - \int_x^b \frac{f'(s)}{(s-x)^{\alpha-1}}ds\right) \\
&= \frac{1}{\Gamma(1-\alpha)}\left(f(b)\int_x^b (b-s)^{-\alpha}ds - \int_x^b f'(s)\int_x^s (s-t)^{-\alpha}dtds\right) \\
&= \frac{1}{\Gamma(1-\alpha)}\left(f(b)\int_x^b (b-s)^{-\alpha}ds - \int_x^b \left[\int_t^b \frac{f'(s)}{(s-t)^{\alpha}}ds\right]dt\right).
\end{aligned}$$

Hence, we have that $I^{1-\alpha}_{b-}f$ is an absolutely continuous function on the interval $[a,b]$ and $(I^{1-\alpha}_{b-}f)(b) = 0$. Finally, the result follows from Definition 18 and Corollary 7. □

For some functions $f : [a,b] \to \mathbb{R}$, the Riemann-Liouville fractional derivatives are given by equalities (1.63) and (1.64) below (see also Theorems 98 and 99 in the appendix). Expressions (1.63) and (1.64) (see also (1.60) and (1.61)) are associated with the Marchaud fractional derivatives. In Marchaud's thesis [142] and in Marchaud [143], these fractional derivatives were introduced in 1927, which were also defined in Weyl [211] in 1917. The reader interested in the contributions of Machaud and Wayl on fractional calculus can consult the paper by Ferrari [50].

Proposition 17 *Let $f \in C^1([a,b])$ and $\alpha \in (0,1)$. Then, f belongs to $I^{\alpha}_{a+}(L^p) \cap I^{\alpha}_{b-}(L^p)$, for any $p \in [1, 1/\alpha)$. Moreover, we have*

$$\left(D^{\alpha}_{a+}f\right)(t) = \frac{1}{\Gamma(1-\alpha)}\left(\frac{f(t)}{(t-a)^{\alpha}} + \alpha\int_a^t \frac{f(t)-f(r)}{(t-r)^{1+\alpha}}dr\right), \quad t \in [a,b], \tag{1.60}$$

and

$$\left(D^{\alpha}_{b-}f\right)(t) = \frac{1}{\Gamma(1-\alpha)}\left(\frac{f(t)}{(b-t)^{\alpha}} + \alpha\int_t^b \frac{f(t)-f(r)}{(r-t)^{1+\alpha}}dr\right), \quad t \in [a,b]. \tag{1.61}$$

Remark 19 *Note that the integrals on the right-hand sides of (1.60) and (1.61) are well-defined for all $t \in [a,b]$ since f is a Lipschitz function on $[a,b]$. Moreover, in Samko et al. [193], it is proven that $I^{\alpha}_{a+}(L^p) = I^{\alpha}_{b-}(L^p)$ if $1 < p < 1/\alpha$.*

Proof. From Corollary 8, we know that $f \in I^{\alpha}_{a+}(L^1) \cap I^{\alpha}_{b-}(L^1)$. Furthermore, the proofs of (1.60) and (1.61) are very similar to each other. So, we only deal with equality (1.61).

In the remaining of the proof, we use the convention $f \equiv 0$ on $[a,b]^c$. Then, by Corollary 8, we get

$$\begin{aligned}\left(D^{\alpha}_{b-}f\right)(t) &= \frac{1}{\Gamma(1-\alpha)}\left(\frac{f(b)}{(b-t)^{\alpha}} - \int_t^b \frac{f'(y)}{(y-t)^{\alpha}}dy\right) \\ &= \frac{1}{\Gamma(1-\alpha)}\lim_{\varepsilon\downarrow 0}\left(\frac{f(b)}{(b-t)^{\alpha}} - \int_{t+\varepsilon}^b \frac{f'(y)}{(y-t)^{\alpha}}dy\right), \quad t \in [a,b],\end{aligned}$$

where last equality follows from the dominated convergence theorem and the fact that f' is a bounded function. Therefore, the integration by parts formula implies

$$\begin{aligned}\left(D^{\alpha}_{b-}f\right)(t) &= \frac{1}{\Gamma(1-\alpha)}\lim_{\varepsilon\downarrow 0}\left(\frac{f(b)}{(b-t)^{\alpha}} + \frac{f(t)-f(b)}{(b-t)^{\alpha}} - \frac{f(t)-f(t+\varepsilon)}{\varepsilon^{\alpha}}\right. \\ &\qquad \left. +\alpha\int_{t+\varepsilon}^b \frac{f(t)-f(y)}{(y-t)^{1+\alpha}}dy\right), \quad t \in [a,b).\end{aligned}$$

Hence, (1.61) holds because f is a Lipschitz function due to the mean value theorem. The proof is now complete. □

Concerning (1.61), using that $f \in C^1([a,b])$, we have that, for $\eta > 0$ small enough,

$$\begin{aligned}&\left|\int_{t+\varepsilon}^b \frac{f(t)-f(y)}{(y-t)^{1+\alpha}}dy\right| \\ &\leq 1_{\{b<t+\varepsilon\}}\int_b^{t+\varepsilon} \frac{|f(t)|}{(y-t)^{1+\alpha}}dy + 1_{\{t+\varepsilon<b\}}C\int_{t+\varepsilon}^b \frac{1}{(y-t)^{\alpha}}dy \\ &\leq C1_{\{b<t+\varepsilon\}}(b-t)^{-\alpha-\eta}\int_b^{t+\varepsilon} \frac{1}{(y-t)^{1-\eta}}dy + C1_{\{b<t+\varepsilon\}} \\ &\leq C\left((b-t)^{-\alpha-\eta}+1\right), \quad t \in [a,b].\end{aligned} \tag{1.62}$$

In consequence, the dominated convergence theorem and (1.61) yield that, for $p \in (1, 1/\alpha)$,

$$D^{\alpha}_{b-}f = L^p([a,b]) - \lim_{\varepsilon \downarrow 0} \frac{1}{\Gamma(1-\alpha)} \left(\frac{f(\cdot)}{(b-\cdot)^{\alpha}} + \alpha \int_{\cdot+\varepsilon}^{b} \frac{f(\cdot)-f(r)}{(r-\cdot)^{1+\alpha}} dr \right). \tag{1.63}$$

because we can find $\eta > 0$ such that $p(\alpha+\eta) < 1$.

Similarly, by means of (1.60), we can also establish

$$D^{\alpha}_{a+}f = L^p([a,b]) - \lim_{\varepsilon \downarrow 0} \frac{1}{\Gamma(1-\alpha)} \left(\frac{f(\cdot)}{(\cdot-a)^{\alpha}} + \alpha \int_{a}^{\cdot-\varepsilon} \frac{f(\cdot)-f(r)}{(\cdot-r)^{1+\alpha}} dr \right). \tag{1.64}$$

Remember that, in (1.63) and (1.64), we utilize the convention $f \equiv 0$ on $[a,b]^c$.

In general, for $p \geq 1$, either [193] (Remark 13.2) or Theorems 98 (resp. Theorem 99) in the appendix states that $f \in I^{\alpha}_{a+}(L^p)$ (resp. $f \in I^{\alpha}_{b-}(L^p)$) if and only if $f \in L^p([a,b])$ and the limit on the right-hand side of (1.64) (resp. (1.63)) exists. In this case, equality (1.64) (resp. (1.63)) holds. Therefore, it is not difficult to see that the fact that $f \in L^p([a,b])$, $\frac{f(\cdot)}{(\cdot-a)^{\alpha}}$ and $\int_a^{\cdot} \frac{f(\cdot)-f_r}{(\cdot-r)^{1+\alpha}}dr$ (resp. $\frac{f(\cdot)}{(b-\cdot)^{\alpha}}$ and $\int_{\cdot}^{b} \frac{f(\cdot)-f_r}{(r-\cdot)^{1+\alpha}}dr$) belong to $L^p([a,b])$ implies that $f \in I^{\alpha}_{a+}(L^p)$ (resp. $f \in I^{\alpha}_{b-}(L^p)$) and (1.60) (resp. (1.61)) is true. Indeed, for instance, we have, for $\varepsilon > 0$ and η small enough,

$$\begin{aligned}
&\left| \int_{t+\varepsilon}^{b} \frac{f(t)-f(y)}{(y-t)^{1+\alpha}} dy \right| \\
&\leq 1_{\{b<t+\varepsilon\}} \int_{b}^{t+\varepsilon} \frac{|f(t)|}{(y-t)^{1+\alpha}} dy + 1_{\{t+\varepsilon<b\}} \left| \int_{t+\varepsilon}^{b} \frac{f(t)-f(y)}{(y-t)^{1+\alpha}} dy \right| \\
&\leq 1_{\{b<t+\varepsilon\}} \frac{|f(t)|}{(b-t)^{\alpha+\eta}} \int_{b}^{t+\varepsilon} \frac{1}{(y-t)^{1-\eta}} dy + 1_{\{b<t+\varepsilon\}} \left| \int_{t}^{b} \frac{f(t)-f(y)}{(y-t)^{1+\alpha}} dy \right| \\
&\leq C_{\eta} \left(\frac{|f(t)|}{(b-t)^{\alpha+\eta}} + \left| \int_{t}^{b} \frac{f(t)-f(y)}{(y-t)^{1+\alpha}} dy \right| \right), \quad t \in [a,b].
\end{aligned}$$

Thus, for $\tilde{p} \in (1,p)$ and choosing η small enough, we can apply the Hölder's inequality to see that $t \mapsto \int_{t+\varepsilon}^{b} \frac{f(t)-f(y)}{(y-t)^{1+\alpha}} dy$ belongs to $L^{\tilde{p}}([a,b])$. Hence, the dominated convergence theorem implies

$$\begin{aligned}
D^{\alpha}_{b-}f &= L^{\tilde{p}}([a,b]) - \lim_{\varepsilon \downarrow 0} \frac{1}{\Gamma(1-\alpha)} \left(\frac{f(\cdot)}{(b-\cdot)^{\alpha}} + \alpha \int_{\cdot+\varepsilon}^{b} \frac{f(\cdot)-f(r)}{(r-\cdot)^{1+\alpha}} dr \right) \\
&= \frac{1}{\Gamma(1-\alpha)} \left(\frac{f(\cdot)}{(b-\cdot)^{\alpha}} + \alpha \int_{\cdot}^{b} \frac{f(\cdot)-f(r)}{(r-\cdot)^{1+\alpha}} dr \right).
\end{aligned} \tag{1.65}$$

Consequently, Theorem 99 states that $f = I^{\alpha}_{b-}(D^{\alpha}_{b-}f)$. But, $D^{\alpha}_{b-}f$ belongs to $L^p([a,b])$ by hypothesis. So our assertion is verified. In particular, notice that $C_1^{\alpha+\varepsilon}([a,b]) \subset I^{\alpha}_{b-}(L^p) \cap I^{\alpha}_{a+}(L^p)$, with $\varepsilon > 0$ and $p \in (1, 1/\alpha)$ since, for $f \in C_1^{\alpha+\varepsilon}([a,b])$, f is a bounded function and $\frac{|f(t)-f(y)|}{(y-t)^{1+\alpha}} \leq C(y-t)^{\varepsilon-1}$. This is a particular case that is considered in Remark 111 .

On the other hand, the adjoints of the fractional derivatives can be figured out as follows:

Proposition 18 *Let $p, q > 1$ be such that $p^{-1}+q^{-1} \leq 1+\alpha$, $\alpha \in (0,1]$, $f \in I^{\alpha}_{a+}(L^p([a,b]))$ and $g \in I^{\alpha}_{b-}(L^q([a,b]))$. Then,*

$$\int_a^b f(t)\left(D^{\alpha}_{b-}g\right)(t)dt = \int_a^b \left(D^{\alpha}_{a+}f\right)(t)g(t)dt.$$

Remark 20 *As in Remark 15, we have that both integrals are well-defined because of Theorem 7. Moreover, for* $\alpha = 1$*, we have that* $f(\cdot) = \int_a^{\cdot} \varphi(s)ds$ *and* $g = \int_{\cdot}^b \psi(s)ds$*, with* $\varphi \in L^p([a,b])$ *and* $\psi \in L^q([a,b])$*. Hence, f and g belong to* $\in L^r([a,b])$*, for all* $r \geq 1$*.*

Proof. By hypothesis, there exist $\varphi \in L^p([a,b])$ and $\psi \in L^q([a,b])$ such that $f = I_{a+}^{\alpha}\varphi$ and $g = I_{b-}^{\alpha}\psi$. Thus, Corollaries 6 and 7, and Remark 18 lead us to write

$$\int_a^b f(t)\left(D_{b-}^{\alpha}g\right)(t)dt = \int_a^b \left(I_{a+}^{\alpha}\varphi\right)(t)\psi(t)dt = \int_a^b \varphi(t)\left(I_{b-}^{\alpha}\psi\right)(t)dt = \int_a^b \left(D_{a+}^{\alpha}f\right)(t)g(t)dt.$$

□

The fractional integrals and derivatives enjoy the following semigroup properties, respectively.

Proposition 19 *Let* $\alpha, \beta > 0$, $\varphi \in L^1([a,b])$, $f \in I_{a+}^{\alpha+\beta}(L^1)$ *and* $g \in I_{b-}^{\alpha+\beta}(L^1)$*. Then,*

i) $I_{a+}^{\alpha+\beta}\varphi = I_{a+}^{\alpha}\left(I_{a+}^{\beta}\varphi\right) = I_{a+}^{\beta}\left(I_{a+}^{\alpha}\varphi\right)$ *and* $I_{b-}^{\alpha+\beta}\varphi = I_{b-}^{\alpha}\left(I_{b-}^{\beta}\varphi\right) = I_{b-}^{\beta}\left(I_{b-}^{\alpha}\varphi\right).$

ii) if $\alpha + \beta \leq 1$*, we get* $D_{a+}^{\alpha+\beta}f = D_{a+}^{\alpha}\left(D^{\beta}f\right) = D_{a+}^{\beta}\left(D^{\alpha}f\right)$ *and*

$$D_{b-}^{\alpha+\beta}g = D_{b-}^{\alpha}\left(D_{b-}^{\beta}g\right) = D_{b-}^{\beta}\left(D_{b-}^{\alpha}g\right).$$

Proof. Fubini's theorem, (1.54) and Lemma 87 allow to establish

$$\begin{aligned}
\left(I_{a+}^{\beta}\left(I_{a+}^{\alpha}\varphi\right)\right)(t) &= \frac{1}{\Gamma(\alpha)\Gamma(\beta)}\int_a^t (t-x)^{\beta-1}\int_a^x (x-y)^{\alpha-1}\varphi(y)dydx \\
&= \frac{1}{\Gamma(\alpha)\Gamma(\beta)}\int_a^t \varphi(y)\int_y^t (t-x)^{\beta-1}(x-y)^{\alpha-1}dxdy \\
&= \frac{1}{\Gamma(\alpha+\beta)}\int_a^t \varphi(y)(t-y)^{\beta+\alpha-1}dy = I_{a+}^{\beta+\alpha}\varphi.
\end{aligned}$$

Thus, proceeding similarly, we can see that Statement *i*) holds.

Now, we deal with Assertion *ii*). By hypothesis, there exist $\varphi_1\, \varphi_2 \in L^1([a,b])$ such that $f = I_{a+}^{\beta+\alpha}\varphi_1$ and $g = I_{a+}^{\beta+\alpha}\varphi_2$. Therefore, Statement *i*) and Remark 18 yield

$$\begin{aligned}
D_{a+}^{\alpha}\left(D_{a+}^{\beta}f\right) &= D_{a+}^{\alpha}\left(D_{a+}^{\beta}\left(I_{a+}^{\beta}\left(I_{a+}^{\alpha}\varphi_1\right)\right)\right) = D_{a+}^{\alpha}I_{a+}^{\alpha}\varphi_1 = \varphi_1, \\
D_{a+}^{\beta}\left(D_{a+}^{\alpha}f\right) &= D_{a+}^{\beta}\left(D_{a+}^{\alpha}\left(I_{a+}^{\alpha}\left(I_{a+}^{\beta}\varphi_1\right)\right)\right) = D_{a+}^{\beta}I_{a+}^{\beta}\varphi_1 = \varphi_1
\end{aligned}$$

and $D_{a+}^{\alpha+\beta}f = D_{a+}^{\alpha+\beta}\left(I_{a+}^{\beta+\alpha}\varphi_1\right) = \varphi_1$, which implies that the first part of *ii*) holds. Finally, similar calculations lead us to finish the proof. □

We observe that the hypotheses $f \in I_{a+}^{\alpha+\beta}(L^1)$ and $g \in I_{b-}^{\alpha+\beta}(L^1)$ cannot be omitted in Statement *ii*) of previous proposition. For example, let $\alpha = \beta = 1/2$ and $f(x) = x^{-1/2}$, $x \in [0,1]$. Then, from Definition 18 and Lemma 87, we get

$$\left(D_{0+}^{1/2}f\right)(x) = \frac{1}{\Gamma(1/2)}\frac{d}{dx}\int_0^x (x-y)^{-1/2}y^{-1/2}dy = \frac{\Gamma(1/2)^2}{\Gamma(1/2)\Gamma(1)}\frac{d}{dx}1 = 0, \quad x \in [0,1],$$

which gives $D_{0+}^{1/2}\left(D_{0+}^{1/2}f\right) \equiv 0$. But $D_{0+}^{1}f = f' \neq 0$. In order to explain why D_{a+}^{α} and D_{a+}^{β} do not commute in general, we consider the following calculations. Definition 18 and integration by parts formula lead us to write, for $x \in [a,b]$,

$$\begin{aligned}
\left(D^{\alpha}_{a+}\left(D^{\beta}_{a+}f\right)\right)(x) &= \frac{1}{\Gamma(1-\alpha)}\frac{d}{dx}\int_a^x \frac{\left(D^{\beta}_{a+}f\right)(y)}{(x-y)^{\alpha}}dy \\
&= \frac{1}{\Gamma(1-\alpha)\Gamma(1-\beta)}\frac{d}{dx}\int_a^x (x-y)^{-\alpha}\frac{d}{dy}\int_a^y \frac{f(t)}{(y-t)^{\beta}}dtdy \\
&= \frac{(1-\alpha)^{-1}}{\Gamma(1-\alpha)\Gamma(1-\beta)}\frac{d^2}{dx^2}\int_a^x (x-y)^{1-\alpha}\frac{d}{dy}\int_a^y \frac{f(t)}{(y-t)^{\beta}}dtdy \\
&= \frac{(1-\alpha)^{-1}}{\Gamma(1-\alpha)\Gamma(1-\beta)}\frac{d^2}{dx^2}\left[\int_a^x \frac{1-\alpha}{(x-y)^{\alpha}}\int_a^y \frac{f(t)}{(y-t)^{\beta}}dtdy\right. \\
&\qquad \left. -(x-a)^{1-\alpha}\left(\int_a^y \frac{f(t)}{(y-t)^{\beta}}dt\right)\Big|_{y=a}\right].
\end{aligned}$$

Hence, for $x \in [a,b]$, Lemma 87, (8.11) and Fubini's theorem imply

$$\begin{aligned}
\left(D^{\alpha}_{a+}\left(D^{\beta}_{a+}f\right)\right)(x) &= \frac{1}{\Gamma(2-\alpha-\beta)}\frac{d^2}{dx^2}\int_a^x f(t)(x-t)^{1-\alpha-\beta}dt \\
&\quad -\frac{(1-\alpha)^{-1}}{\Gamma(1-\alpha)\Gamma(1-\beta)}\frac{d^2(x-a)^{1-\alpha}}{dx^2}\left(\int_a^y \frac{f(t)}{(y-t)^{\beta}}dt\right)\Big|_{y=a} \\
&= \frac{1}{\Gamma(1-\alpha-\beta)}\frac{d}{dx}\int_a^x f(t)(x-t)^{-\alpha-\beta}dt \\
&\quad +\frac{\alpha}{\Gamma(1-\alpha)\Gamma(1-\beta)}(x-a)^{-1-\alpha}\left(\int_a^y \frac{f(t)}{(y-t)^{\beta}}dt\right)\Big|_{y=a} \\
&= D^{\alpha+\beta}_{a+}f + \frac{\alpha}{\Gamma(1-\alpha)\Gamma(1-\beta)}(x-a)^{-1-\alpha}\left(\int_a^y \frac{f(t)}{(y-t)^{\beta}}dt\right)\Big|_{y=a}.
\end{aligned}$$

In the same way, we also have that, for $x \in [a,b]$,

$$\left(D^{\beta}_{a+}\left(D^{\alpha}_{a+}f\right)\right)(x) = D^{\alpha+\beta}_{a+}f + \frac{\beta}{\Gamma(1-\beta)\Gamma(1-\alpha)}(x-a)^{-1-\beta}\left(\int_a^y \frac{f(t)}{(y-t)^{\alpha}}dt\right)\Big|_{y=a}.$$

Thus, the last two equalities, together with Corollary 7, provide us an answer as to why the fractional derivatives D^{α}_{a+} and D^{β}_{a+} do not necessarily commute.

1.4.2 Fractional Brownian motion as a Volterra type process

Now, our aim is to show that fBm is also a Gaussian Volterra process (see Section 1.3). To do so, we introduce the square-integrable kernel $\{K_H(t,s) : 0 \le s < t \le T\}$ given by

$$K_H(t,s) = c_H s^{\frac{1}{2}-H}\int_s^t (u-s)^{H-\frac{3}{2}}u^{H-\frac{1}{2}}du, \quad \text{for } H > 1/2, \tag{1.66}$$

where $c_H = \left[\frac{H(2H-1)}{\beta(2-2H,H-\frac{1}{2})}\right]^{1/2}$, and

$$\begin{aligned}
K_H(t,s) = \tilde{c}_H &\left[\left(\frac{t}{s}\right)^{H-\frac{1}{2}}(t-s)^{H-\frac{1}{2}}\right. \\
&\left. -(H-\frac{1}{2})s^{\frac{1}{2}-H}\int_s^t u^{H-\frac{3}{2}}(u-s)^{H-\frac{1}{2}}du\right], \quad \text{for } H < 1/2,
\end{aligned} \tag{1.67}$$

with $\tilde{c}_H = \sqrt{\frac{2H}{(1-2H)\beta(1-2H,H+1/2)}}$. In both cases, β is the function Beta, which is analyzed in Lemma 87. We can see Lemma 54 to see other form of the last kernel K in (1.67).

Proposition 20 *Let K_H be the kernel given by (1.66) and (1.67), with $H \in (0,1)$. Then,*

$$\int_0^{t\wedge s} K_H(t,u)K_H(s,u)du = R_H(t,s), \quad t,s \in [0,T]. \tag{1.68}$$

Here, R_H is the covarince function (1.44) *of fBm.*

Proof. We proceed as in Nualart [157]. So, we divide the proof into three steps.
Step 1: Here we consider the case $H > 1/2$.

Observe that, for $s \leq t$,

$$\begin{aligned}
\int_0^t \int_0^s |r-u|^{2H-2} dudr &= \int_0^s \int_0^s |r-u|^{2H-2} dudr + \int_s^t \int_0^s (r-u)^{2H-2} dudr \\
&= 2\int_0^s \int_0^r (r-u)^{2H-2} dudr + \int_s^t \int_0^s (r-u)^{2H-2} dudr \\
&= \frac{2}{2H-1}\int_0^s r^{2H-1}dr + \frac{1}{2H-1}\int_s^t \left(r^{2H-1} - (r-s)^{2H-1}\right) dr \\
&= \frac{s^{2H}}{H(2H-1)} + \frac{t^{2H} - s^{2H} - (t-s)^{2H}}{2H(2H-1)} = \frac{1}{H(2H-1)} R_H(t,s).
\end{aligned}$$

It means, the covariance of fBm can be written as

$$R_H(t,s) = H(2H-1)\int_0^t \int_0^s |r-u|^{2H-2} dudr. \tag{1.69}$$

Indeed, the case $t < s$ is proven similarly.

In order to continue with the proof, we use the change of variables $z = \frac{r-v}{u-v}$ to get

$$\begin{aligned}
&\int_0^u v^{1-2H}(r-v)^{H-\frac{3}{2}}(u-v)^{H-\frac{3}{2}}dv \\
&= \frac{1}{r-u}\int_0^u v^{1-2H}\left[\frac{r-v}{u-v}\right]^{H-3/2} (u-v)^{2H-1}\frac{r-u}{(u-v)^2}dv \\
&= \frac{1}{r-u}\int_{\frac{r}{u}}^{\infty} \left(\frac{zu-r}{z-1}\right)^{1-2H} z^{H-3/2}\left(u - \frac{zu-r}{z-1}\right)^{2H-1} dz \\
&= (r-u)^{2H-2}\int_{\frac{r}{u}}^{\infty} (zu-r)^{1-2H} z^{H-\frac{3}{2}}dz.
\end{aligned}$$

Hence, the change of variables formula $x = \frac{r}{uz}$ allows us to write

$$\begin{aligned}
&\int_0^u v^{1-2H}(r-v)^{H-\frac{3}{2}}(u-v)^{H-\frac{3}{2}}dv \\
&= (r-u)^{2H-2}\int_0^1 \left(\frac{r}{x} - r\right)^{1-2H}\left(\frac{r}{ux}\right)^{H-3/2}\frac{u}{r}\left(\frac{r}{ux}\right)^{-2} dx \\
&= (ru)^{\frac{1}{2}-H}(r-u)^{2H-2}\int_0^1 (1-x)^{1-2H} x^{H-\frac{3}{2}}dx \\
&= \beta(2-2H, H-\frac{1}{2})(ru)^{\frac{1}{2}-H}(r-u)^{2H-2}.
\end{aligned}$$

Thus, (1.66), (1.69) and Fubini's theorem lead us to

$$\begin{aligned}
&\int_0^{t\wedge s} K_H(t,u)K_H(s,u)du \\
&= c_H^2 \int_0^{t\wedge s}\left(\int_u^t (y-u)^{H-\frac{3}{2}}y^{H-\frac{1}{2}}dy\right)\left(\int_u^s (z-u)^{H-\frac{3}{2}}z^{H-\frac{1}{2}}dz\right)u^{1-2H}du \\
&= c_H^2 \int_0^t\int_0^s (yz)^{H-1/2}\int_0^{y\wedge z} u^{1-2H}(y-u)^{H-3/2}(z-u)^{H-3/2}dudzdy \\
&= c_H^2\beta(2-2H,H-\frac{1}{2})\int_0^t\int_0^s |y-z|^{2H-2}dzdy = R_H(t,s).
\end{aligned}$$

Consequently, the result holds for $H>1/2$.
Step 2: Now, we deal with the case $H<1/2$ and $s=t$.

In this case, by (1.67), we have

$$\begin{aligned}
&\int_0^t K_H(t,u)^2du \\
&= \tilde{c}_H^2\left[\int_0^t (\frac{t}{u})^{2H-1}(t-u)^{2H-1}du\right. \\
&\quad -(2H-1)\int_0^t t^{H-\frac{1}{2}}u^{1-2H}(t-u)^{H-\frac{1}{2}}\left(\int_u^t v^{H-\frac{3}{2}}(v-u)^{H-\frac{1}{2}}dv\right)du \\
&\quad \left.+(H-\frac{1}{2})^2\int_0^t u^{1-2H}\left(\int_u^t v^{H-\frac{3}{2}}(v-u)^{H-\frac{1}{2}}dv\right)^2du\right].
\end{aligned}$$

Applying the change of variables $u=tx$ in the first integral and using Fubini's theorem in the last two integrals, we get

$$\begin{aligned}
&\int_0^t K_H(t,u)^2du \\
&= \tilde{c}_H^2\left[t^{2H}\beta(2-2H,2H)\right. \\
&\quad -(2H-1)t^{H-\frac{1}{2}}\int_0^t v^{H-\frac{3}{2}}\left(\int_0^v u^{1-2H}(t-u)^{H-\frac{1}{2}}(v-u)^{H-\frac{1}{2}}du\right)dv \\
&\quad \left.+2(H-\frac{1}{2})^2\int_0^t\int_0^v\int_0^w u^{1-2H}(v-u)^{H-\frac{1}{2}}(w-u)^{H-\frac{1}{2}}w^{H-\frac{3}{2}}v^{H-\frac{3}{2}}dudwdv\right].
\end{aligned}$$

On the right-hand side of this equality, we use the changes of variables $u=vx$ and $u=wx$ in the second and third summands, respectively, to obtain

$$\begin{aligned}
&\int_0^t K_H(t,u)^2du \\
&= \tilde{c}_H^2\left[t^{2H}\beta(2-2H,2H)\right. \\
&\quad -(2H-1)t^{H-\frac{1}{2}}\int_0^t\left(\int_0^1 x^{1-2H}(t-vx)^{H-1/2}(1-x)^{H-1/2}dx\right)dv \\
&\quad \left.+2(H-\frac{1}{2})^2\int_0^t\int_0^v\int_0^1 x^{1-2H}(v-wx)^{H-1/2}(1-x)^{H-1/2}v^{H-3/2}dxdwdv\right].
\end{aligned}$$

Now, we make use of the change of variables formulas $v = ty$ and $w = vy$ in the second and third summands again. In this way, we can write

$$\int_0^t K_H(t,u)^2 du = \tilde{c}_H^2 t^{2H} \left[\beta(2-2H, 2H) - (2H-1)(\frac{1}{4H} + \frac{1}{2}) \right. \\ \left. \times \int_0^1 \int_0^1 x^{1-2H}(1-xy)^{H-\frac{1}{2}}(1-x)^{H-\frac{1}{2}} dxdy \right] = t^{2H},$$

where we have utilized (8.11) and Lemma 87.

Step 3: Finally, we suppose that $s < t$ and $H < 1/2$.

From (1.68), the fundamental theorem of calculus and Step 2, we only need to show that

$$H(t^{2H-1} - (t-s)^{2H-1}) = \int_0^s \frac{\partial K_H}{\partial t}(t,u) K_H(s,u) du \tag{1.70}$$

holds in order to finish the proof.

Note that (1.67) leads to

$$\int_0^s \frac{\partial K_H}{\partial t}(t,u) K_H(s,u) du \\ = \tilde{c}_H^2 (H - \frac{1}{2}) \left[\int_0^s \left(\frac{t}{u} \right)^{H-\frac{1}{2}} (t-u)^{H-\frac{3}{2}} \left(\frac{s}{u} \right)^{H-\frac{1}{2}} (s-u)^{H-\frac{1}{2}} du \right. \\ \left. -(H-\frac{1}{2}) \int_0^s \left(\frac{t}{u} \right)^{H-\frac{1}{2}} (t-u)^{H-\frac{3}{2}} u^{\frac{1}{2}-H} \int_u^s v^{H-\frac{3}{2}} (v-u)^{H-\frac{1}{2}} dvdu \right].$$

Using the change of variables $u = sx$ in the first integral, Fubini's theorem and the change of variables $u = vx$ in the second integral, we can establish

$$\int_0^s \frac{\partial K_H}{\partial t}(t,u) K_H(s,u) du \\ = \tilde{c}_H^2 (H - \frac{1}{2})(ts)^{H-\frac{1}{2}} s \int_0^1 (sx)^{1-2H} (t-sx)^{H-\frac{3}{2}} (s-sx)^{H-\frac{1}{2}} dx \\ -\tilde{c}_H^2 (H-\frac{1}{2})^2 t^{H-\frac{1}{2}} \int_0^s v^{H-\frac{3}{2}} \int_0^v u^{1-2H} (t-u)^{H-\frac{3}{2}} (v-u)^{H-\frac{1}{2}} dudv \\ = \tilde{c}_H^2 (H-\frac{1}{2}) \left[(ts)^{H-\frac{1}{2}} \gamma(\frac{t}{s}) - (H-\frac{1}{2}) t^{H-\frac{1}{2}} \int_0^s v^{H-\frac{3}{2}} \gamma(\frac{t}{v}) dv \right],$$

where $\gamma(y) = \int_0^1 x^{1-2H}(y-x)^{H-\frac{3}{2}}(1-x)^{H-\frac{1}{2}} dx$ for $y > 1$. Therefore, by (1.70), the proof reduces to verify that the equality

$$\tilde{c}_H^2 \left[(H-\frac{1}{2}) s^{H-\frac{1}{2}} \gamma(\frac{t}{s}) - (H-\frac{1}{2})^2 \int_0^s v^{H-\frac{3}{2}} \gamma(\frac{t}{v}) dv \right] = H(t^{H-\frac{1}{2}} - t^{\frac{1}{2}-H}(t-s)^{2H-1}) \tag{1.71}$$

is true. Towards this end, we consider the derivative of the left-hand side of equality (1.71) with respect to t, which is equal to

$$\tilde{c}_H^2 (H-\frac{3}{2}) \left[(H-\frac{1}{2}) s^{H-\frac{3}{2}} \delta(\frac{t}{s}) - (H-\frac{1}{2})^2 \int_0^s v^{H-\frac{5}{2}} \delta(\frac{t}{v}) dv \right]. \tag{1.72}$$

Here, for $y > 1$,

$$\delta(y) = \int_0^1 x^{1-2H}(y-x)^{H-\frac{5}{2}}(1-x)^{H-\frac{1}{2}} dx = \beta(2-2H, H+\frac{1}{2}) y^{-H-\frac{1}{2}} (y-1)^{2H-2},$$

where last equality follows from the change of variables $z = \frac{y(1-x)}{y-x}$. Hence, (1.72) is equal to

$$\begin{aligned}
&\tilde{c}_H^2 \beta(2-2H, H+\frac{1}{2})(H-\frac{3}{2})(H-\frac{1}{2}) \left[st^{-H-\frac{1}{2}}(t-s)^{2H-2} \right. \\
&\quad \left. -(H-\frac{1}{2})t^{-H-\frac{1}{2}} \int_0^s (t-v)^{2H-2} dv \right] \\
&= \tilde{c}_H^2 \beta(2-2H, H+\frac{1}{2})(H-\frac{3}{2})(H-\frac{1}{2}) \left[st^{-H-\frac{1}{2}}(t-s)^{2H-2} \right. \\
&\quad \left. +\frac{1}{2}\left(t^{H-\frac{1}{2}}(t-s)^{2H-1} - t^{H-\frac{3}{2}} \right) \right] \\
&= H(1-2H)\left(t^{-H-\frac{1}{2}} s(t-s)^{2H-2} + \frac{1}{2}(t-s)^{2H-1} t^{-H-\frac{1}{2}} - \frac{1}{2} t^{H-\frac{3}{2}} \right),
\end{aligned}$$

where last equality follows from (8.11) and Lemma 87. In consequence, we have that (1.72) is equal to

$$\begin{aligned}
&H(1-2H)\left(t^{-H-\frac{1}{2}} s(t-s)^{2H-2} + (t-s)^{2H-1} t^{-H-\frac{1}{2}} \right. \\
&\quad \left. -\frac{1}{2}(t-s)^{2H-1} t^{-H-\frac{1}{2}} - \frac{1}{2} t^{H-\frac{3}{2}} \right) \\
&= H(1-2H)\left((t-s)^{2H-2} t^{\frac{1}{2}-H} \left[t^{-1}s + t^{-1}(t-s) \right] \right. \\
&\quad \left. -\frac{1}{2}(t-s)^{2H-1} t^{-H-\frac{1}{2}} - \frac{1}{2} t^{H-\frac{3}{2}} \right) \\
&= H(1-2H)\left((t-s)^{2H-2} t^{\frac{1}{2}-H} - \frac{1}{2}(t-s)^{2H-1} t^{-H-\frac{1}{2}} - \frac{1}{2} t^{H-\frac{3}{2}} \right) \\
&= H\frac{d}{dt}\left(t^{H-\frac{1}{2}} - t^{\frac{1}{2}-H}(t-s)^{2H-1} \right).
\end{aligned}$$

In other words, using that (1.72) is the derivative with respect to t of the left-hand side of (1.71), we establish that there exists a constant $C_s \in \mathbb{R}$ for each $s > 0$ such that, for all $t > s$,

$$\tilde{c}_H^2 \left[(H-\frac{1}{2}) s^{H-\frac{1}{2}} \gamma(\frac{t}{s}) - (H-\frac{1}{2})^2 \int_0^s v^{H-\frac{3}{2}} \gamma(\frac{t}{v}) dv \right] - H\left(t^{H-\frac{1}{2}} - t^{\frac{1}{2}-H}(t-s)^{2H-1} \right) = C_s.$$

Finally, we have that $C_s = 0$ due to $\gamma(\frac{t}{s}), \gamma(\frac{t}{v}) \to 0$ and

$$t^{H-\frac{1}{2}} - t^{\frac{1}{2}-H}(t-s)^{2H-1} = t^{H-\frac{1}{2}} - \left(t^{\frac{1}{2}} - st^{-\frac{1}{2}} \right)^{2H-1} \to 0,$$

as $t \to \infty$, together with the fact that $\gamma(\frac{t}{v}) \le \beta(2-2H, H+\frac{1}{2}) v^{\frac{3}{2}-H}$, for t large enough, and the dominated convergence theorem. Therefore, equality (1.71) is satisfied and the proof is finished. □

1.4.2.1 Case $H < 1/2$

We consider the linear isometry $K^* : \mathcal{H} \to L^2([0,T])$ introduced in (1.37), which, by (1.38), satisfies $K^*(1_{[0,t]})(\cdot) = K(t,\cdot)1_{[0,t]}(\cdot)$. Hence, (1.67) and Corollary 8 imply

$$\begin{aligned}
K^*(1_{[0,t]})(s) &= \tilde{c}_H \Gamma(H+\frac{1}{2}) s^{\frac{1}{2}-H} 1_{[0,t]}(s) \left(D_{t-}^{\frac{1}{2}-H} u^{H-\frac{1}{2}} \right)(s) \\
&= \tilde{c}_H \Gamma(H+\frac{1}{2}) s^{\frac{1}{2}-H} \left(D_{T-}^{\frac{1}{2}-H} u^{H-\frac{1}{2}} 1_{[0,t]}(u) \right)(s), \qquad (1.73)
\end{aligned}$$

where last equality follows from Definition 18.*ii*) and the fact that

$$\int_s^T \frac{u^{H-\frac{1}{2}}1_{[0,t]}(u)}{(u-s)^{\frac{1}{2}-H}}du = 1_{[0,t]}(s)\int_0^{t-s}\frac{(v+s)^{H-\frac{1}{2}}}{v^{\frac{1}{2}-H}}dv.$$

Consequently, Corollary 7 yields

$$1_{[0,t]}(s) = \left(\tilde{c}_H\Gamma(H+\frac{1}{2})\right)^{-1} s^{\frac{1}{2}-H} I_{T-}^{\frac{1}{2}-H}\left(u^{H-\frac{1}{2}}K^*(1_{[0,t]})(u)\right)(s).$$

Thus, using the linearity of the involved operators leads us to conclude that, for $\varphi \in \mathcal{E}([0,T])$ and $s \in [0,T]$,

$$\varphi(s) = \left(\tilde{c}_H\Gamma(H+\frac{1}{2})\right)^{-1} s^{\frac{1}{2}-H} I_{T-}^{\frac{1}{2}-H}\left(u^{H-\frac{1}{2}}K^*(\varphi)(u)\right)(s). \tag{1.74}$$

We recall that the reproducing kernel Hilbert space $\mathcal{H}$ of the fractional Brownian motion is defined as the completion of $\mathcal{E}([0,T])$ with respect to the inner product given by R_H. That is, $\mathcal{H}$ is the completion of $\mathcal{E}([0,T])$ with respect to the inner product

$$\langle\psi,\phi\rangle_{\mathcal{H}} = \langle K^*\psi, K^*\phi\rangle_{L^2([0,T])} \tag{1.75}$$

due to Proposition 20 and (1.38). In order to identify $\mathcal{H}$, we show the following result.

Lemma 8 *Let $\phi \in L^2([0,T])$. Then, the function $s \mapsto s^{\frac{1}{2}-H} I_{T-}^{\frac{1}{2}-H}\left(u^{H-\frac{1}{2}}\phi(u)\right)(s)$ also belongs to $L^2([0,T])$ and there is a constant $C_{H,T}$ independent of ϕ such that*

$$\int_0^T \left[s^{\frac{1}{2}-H} I_{T-}^{\frac{1}{2}-H}\left(u^{H-\frac{1}{2}}\phi(u)\right)(s)\right]^2 ds \le C_{H,T}\int_0^T \phi(s)^2 ds.$$

Remark 21 *The function $s \mapsto s^{H-\frac{1}{2}}\phi(s)$ is in $L^1([0,T])$ because of Hölder inequality. So, the fractional integral in the lemma is well-defined.*

Proof. By Hölder inequality, Fubini's theorem and (1.56) give

$$\begin{aligned}
&\int_0^T \left[s^{\frac{1}{2}-H} I_{T-}^{\frac{1}{2}-H}\left(u^{H-\frac{1}{2}}\phi(u)\right)(s)\right]^2 ds \\
&= \int_0^T s^{1-2H}\left(\frac{1}{\Gamma(\frac{1}{2}-H)}\int_s^T \frac{u^{H-\frac{1}{2}}\phi(u)}{(u-s)^{\frac{1}{2}+H}}du\right)^2 ds \\
&\le \frac{1}{\Gamma(\frac{1}{2}-H)^2}\int_0^T s^{1-2H}\left(\int_s^T \frac{du}{(u-s)^{\frac{1}{2}+H}}\right)\left(\int_s^T \frac{u^{2H-1}\phi(u)^2}{(u-s)^{\frac{1}{2}+H}}du\right) ds \\
&\le C_{H,T}\int_0^T s^{1-2H}\int_s^T \frac{u^{2H-1}\phi(u)^2}{(u-s)^{\frac{1}{2}+H}}duds \le C_{H,T}\int_0^T\int_s^T \frac{\phi(u)^2}{(u-s)^{\frac{1}{2}+H}}duds \\
&= C_{H,T}\int_0^T \phi(u)^2\int_0^u \frac{ds}{(u-s)^{\frac{1}{2}+H}}du \le C_{H,T}\int_0^T \phi(u)^2 du.
\end{aligned}$$

The proof is complete. □

Now choose $\psi \in \mathcal{H}$. Then, there is a sequence $\{\psi_n \in \mathcal{E}([0,T]) : n \in \mathbb{N}\}$ that converges to ψ in $\mathcal{H}$. From (1.75), this means that $K^*(\psi_n)$ goes to $K^*(\psi)$ in $L^2([0,T])$, as $n \to \infty$. Moreover, Lemma 8 and (1.74) allow us to establish that ψ_n also goes to ψ in $L^2([0,T])$ and, for $s \in [0,T]$.

$$\psi(s) = \left(\tilde{c}_H \Gamma(H + \frac{1}{2})\right)^{-1} s^{\frac{1}{2}-H} I_{T-}^{\frac{1}{2}-H} \left(u^{H-\frac{1}{2}} K^*(\psi)(u)\right)(s). \tag{1.76}$$

In order to identify the space $\mathcal{H}$ in the case that $H < 1/2$, we have to show that the image of the operator K^* is the whole space $L^2([0,T])$. Towards this end, we only need to prove that the function $s \mapsto s^{\frac{1}{2}-H} 1_{[0,t]}(s)$ is in the image of K^* since the family of functions of the form $s \mapsto s^{\frac{1}{2}-H} \varphi(s)$, with $\varphi \in \mathcal{E}([0,T])$, is dense in $L^2([0,T])$, which is left as an exercise to the reader. The idea of the proof is as follows. For $t \in [0,T]$, we get

$$s^{\frac{1}{2}-H} 1_{[0,t]}(s) = s^{\frac{1}{2}-H} D_{T-}^{\frac{1}{2}-H} \left(I_{T-}^{\frac{1}{2}-H} 1_{[0,t]}\right)(s), \quad s \in [0,T],$$

with

$$\left(I_{T-}^{\frac{1}{2}-H} 1_{[0,t]}\right)(s) = \frac{1}{\Gamma(\frac{3}{2} - H)} (t-s)^{\frac{1}{2}-H} 1_{[0,t]}(s), \quad s \in [0,T],$$

by (1.56). In consequence, for $s \in [0,T]$, we get

$$s^{\frac{1}{2}-H} 1_{[0,t]}(s) = \tilde{c}_H \Gamma(H + \frac{1}{2}) s^{\frac{1}{2}-H} D_{T-}^{\frac{1}{2}-H} \left(u^{H-\frac{1}{2}} \left[\tilde{d}_H u^{\frac{1}{2}-H} (t-u)^{\frac{1}{2}-H} 1_{[0,t]}(u)\right]\right)(s),$$

where $\tilde{d}_H = \left(\tilde{c}_H \Gamma(H + \frac{1}{2}) \Gamma(\frac{3}{2} - H)\right)^{-1}$. Hence, from (1.73), a natural way to see that our claim holds is to approximate the function $\varphi(s) = \tilde{d}_H s^{\frac{1}{2}-H} (t-s)^{\frac{1}{2}-H} 1_{[0,t]}(s)$ by step functions and to prove that the image of these step functions under K^* converges to $s \mapsto s^{\frac{1}{2}-H} 1_{[0,t]}(s)$ in $L^2([0,T])$. This idea was implemented by Pipiras and Taqqu [177] (Lemma 8.1) to state the following result.

Lemma 9 *Let $H < 1/2$. Then, the image of the operator K^* is the space $L^2([0,T])$.*

Proof. Fix $t \in [0,T]$. Consider the sequence of step functions

$$\varphi_n(s) = \tilde{d}_H \sum_{i=2}^{n-2} \left(\frac{it}{n}\right)^{\frac{1}{2}-H} \left(t - \frac{it}{n}\right)^{\frac{1}{2}-H} 1_{[it/n,(i+1)t/n)}(s), \tag{1.77}$$

for $s \in [0,T]$ and $n \in \mathbb{N}$. In order to calculate $K^*\varphi_n$ by means of (1.73) and Definition 18.ii), we observe that the integration by parts formula leads us to write, for $s, t_1 \in [0,T]$,

$$\begin{aligned}
&-\int_s^T u^{H-\frac{1}{2}} (u-s)^{H-\frac{1}{2}} 1_{[0,t_1]}(u) du \\
&= -1_{[0,t_1]}(s) \int_s^{t_1} u^{H-\frac{1}{2}} (u-s)^{H-\frac{1}{2}} du \\
&= 1_{[0,t_1]}(s) \left(\frac{H-\frac{1}{2}}{H+\frac{1}{2}} \int_s^{t_1} u^{H-\frac{3}{2}} (u-s)^{H+\frac{1}{2}} du - \frac{t_1^{H-\frac{1}{2}} (t_1 - s)^{H+\frac{1}{2}}}{H+\frac{1}{2}} \right) \\
&= \frac{H-\frac{1}{2}}{H+\frac{1}{2}} \int_s^{t_1} u^{H-\frac{3}{2}} (u-s)_+^{H+\frac{1}{2}} du - \frac{t_1^{H-\frac{1}{2}} (t_1 - s)_+^{H+\frac{1}{2}}}{H+\frac{1}{2}},
\end{aligned}$$

where $v_+ = v \vee 0$. Thus, Fubini's theorem allows us to obtain

$$\begin{aligned}
&-\int_s^T u^{H-\frac{1}{2}}(u-s)^{H-\frac{1}{2}} 1_{[0,t_1]}(u)du \\
&= \left(H-\frac{1}{2}\right)\int_s^{t_1} u^{H-\frac{3}{2}} \int_s^u (u-v)_+^{H-\frac{1}{2}} dvdu - t_1^{H-\frac{1}{2}} \int_s^{t_1} (t_1-u)_+^{H-\frac{1}{2}} du \\
&= \left(H-\frac{1}{2}\right)\int_s^{t_1}\int_v^{t_1} u^{H-\frac{3}{2}} (u-v)_+^{H-\frac{1}{2}} dudv - t_1^{H-\frac{1}{2}} \int_s^{t_1} (t_1-u)_+^{H-\frac{1}{2}} du.
\end{aligned}$$

Hence, (1.73) and Definition 18.ii) imply

$$K^*\left(1_{[0,t_1]}\right)(s) = \tilde{c}_H s^{\frac{1}{2}-H}\left(\frac{1}{2}-H\right)\int_s^{t_1} u^{H-\frac{3}{2}}(u-s)_+^{H-\frac{1}{2}} du + \tilde{c}_H \left(\frac{t_1}{s}\right)^{H-\frac{1}{2}} (t_1-s)_+^{H-\frac{1}{2}}.$$

Therefore (1.77) yields that, for $s \in [0,T]$,

$$\begin{aligned}
&\Gamma\left(\frac{3}{2}-H\right)\Gamma\left(\frac{1}{2}+H\right)K^*(\varphi_n)(s) \\
&= s^{\frac{1}{2}-H}\left(\frac{1}{2}-H\right)\int_s^T \sum_{i=1}^{n-2}\left(\frac{it}{n}\right)^{\frac{1}{2}-H}\left(t-\frac{it}{n}\right)^{\frac{1}{2}-H} u^{H-\frac{3}{2}}(u-s)_+^{H-\frac{1}{2}} \\
&\qquad \times 1_{[it/n,(i+1)t/n]}(u)du \\
&\quad + s^{\frac{1}{2}-H}\sum_{i=2}^{n-2}\left(\frac{it}{n}\right)^{\frac{1}{2}-H}\left(t-\frac{it}{n}\right)^{\frac{1}{2}-H}\left[\left(\frac{(i+1)t}{n}\right)^{H-\frac{1}{2}}\left(\frac{(i+1)t}{n}-s\right)_+^{H-\frac{1}{2}}\right. \\
&\qquad \left. -\left(\frac{it}{n}\right)^{H-\frac{1}{2}}\left(\frac{it}{n}-s\right)_+^{H-\frac{1}{2}}\right] = s^{\frac{1}{2}-H}\left[I_{1,n}(s)+I_{2,n}(s)\right].
\end{aligned} \tag{1.78}$$

In consequence,

$$\begin{aligned}
I_{2,n}(s) &= -\sum_{i=2}^{n-2}\left(\frac{it}{n}\right)^{H-\frac{1}{2}}\left(\frac{it}{n}-s\right)_+^{H-\frac{1}{2}} \\
&\qquad \times\left\{\left(\frac{it}{n}\right)^{\frac{1}{2}-H}-\left(\frac{(i-1)t}{n}\right)^{\frac{1}{2}-H}\right\}\left(t-\frac{it}{n}\right)^{\frac{1}{2}-H} \\
&\quad -\sum_{i=2}^{n-2}\left(\frac{it}{n}\right)^{H-\frac{1}{2}}\left(\frac{it}{n}-s\right)_+^{H-\frac{1}{2}}\left(\frac{(i-1)t}{n}\right)^{\frac{1}{2}-H} \\
&\qquad \times\left\{\left(t-\frac{it}{n}\right)^{\frac{1}{2}-H}-\left(t-\frac{(i-1)t}{n}\right)^{\frac{1}{2}-H}\right\} \\
&\quad +\left(\frac{(n-2)t}{n}\right)^{\frac{1}{2}-H}\left(t-\frac{(n-2)t}{n}\right)^{\frac{1}{2}-H}\left(\frac{(n-1)t}{n}\right)^{H-\frac{1}{2}}\left(\frac{(n-1)t}{n}-s\right)_+^{H-\frac{1}{2}} \\
&\quad -\left(\frac{t}{n}\right)^{\frac{1}{2}-H}\left(t-\frac{t}{n}\right)^{\frac{1}{2}-H}\left(\frac{2t}{n}\right)^{H-\frac{1}{2}}\left(\frac{2t}{n}-s\right)_+^{H-\frac{1}{2}} \\
&= I_{2,1,n}(s)+\ldots+I_{2,4,n}(s),
\end{aligned} \tag{1.79}$$

where we apply the equality

$$\sum_{i=2}^{n-2} a_i (b_{i+1} - b_i) = -\sum_{i=2}^{n-2} b_i (a_i - a_{i-1}) + a_{n-2}b_{n-1} - a_1 b_2.$$

Remember that, in order to finish the proof, we must show that $s \mapsto s^{\frac{1}{2}-H} [I_{1,n}(s) + I_{2,n}(s)]$ converges to $s \mapsto \Gamma\left(\frac{3}{2} - H\right) \Gamma\left(\frac{1}{2} + H\right) s^{\frac{1}{2}-H} 1_{[0,t]}(s)$ in $L^2([0,T])$, as $n \to \infty$, due to (1.78). Towards this end, we first observe that Lemma 87 implies

$$\int_0^T s^{1-2H} I_{2,4,n}^2(s) ds \leq C \int_0^{2t/n} s^{1-2H} \left(\frac{2t}{n} - s\right)^{2H-1} ds \to 0, \quad \text{as } n \to \infty, \tag{1.80}$$

and

$$\begin{aligned}
&\int_0^T s^{1-2H} I_{2,3,n}^2(s) ds \\
&\leq C \int_0^{(n-1)t/n} s^{1-2H} \left(t - \frac{(n-2)t}{n}\right)^{1-2H} \left(\frac{(n-1)t}{n} - s\right)^{2H-1} ds \\
&\leq \frac{C}{n^{1-2H}} \int_0^{(n-1)t/n} s^{1-2H} \left(\frac{(n-1)t}{n} - s\right)^{2H-1} ds \to 0, \quad \text{as } n \to \infty.
\end{aligned} \tag{1.81}$$

Now, we deal with the convergence of $s \mapsto s^{\frac{1}{2}-H} I_{1,n}(s)$, where $I_{1,n}$ is introduced in (1.78). It is easy to see that there exists a constant only depending on H such that, for $u \in [0,T]$,

$$\begin{aligned}
&\sum_{i=1}^{n-2} \left(\frac{it}{n}\right)^{\frac{1}{2}-H} \left(t - \frac{it}{n}\right)^{\frac{1}{2}-H} u^{H-\frac{3}{2}} (u-s)_+^{H-\frac{1}{2}} 1_{[it/n,(i+1)t/n]}(u) \\
&\leq C \sum_{i=1}^{n-2} \left(\frac{it}{n}\right)^{\frac{1}{2}-H} \left(t - \frac{(i+1)t}{n}\right)^{\frac{1}{2}-H} u^{H-\frac{3}{2}} (u-s)_+^{H-\frac{1}{2}} 1_{[it/n,(i+1)t/n]}(u) \\
&\leq C u^{\frac{1}{2}-H} (t-u)^{\frac{1}{2}-H} u^{H-\frac{3}{2}} (u-s)^{H-\frac{1}{2}} 1_{[s,t]}(u).
\end{aligned}$$

Moreover, we have

$$\begin{aligned}
&\int_s^t u^{-1}(u-s)^{H-\frac{1}{2}} du \\
&= s^{H-\frac{1}{2}} \int_1^{t/s} u^{-1}(u-1)^{H-\frac{1}{2}} du = s^{H-\frac{1}{2}} \int_0^{(t/s)-1} (v+1)^{-1} v^{H-\frac{1}{2}} dv \\
&\leq s^{H-\frac{1}{2}} \left(\int_0^1 v^{H-\frac{1}{2}} dv + \int_1^\infty v^{H-\frac{3}{2}} dv\right) = s^{H-\frac{1}{2}} \left(\frac{1}{\frac{1}{2}+H} + \frac{1}{\frac{1}{2}-H}\right).
\end{aligned}$$

Therefore, the dominated convergence theorem leads us to show that $s \mapsto s^{\frac{1}{2}-H} I_{1,n}(s)$ goes to

$$s \mapsto s^{\frac{1}{2}-H} \left(\frac{1}{2} - H\right) \int_0^T u^{-1}(t-u)^{\frac{1}{2}-H} (u-s)^{H-\frac{1}{2}} 1_{[s,t]}(u) du \tag{1.82}$$

in $L^2([0,T])$, as $n \to \infty$.

The term $s \mapsto s^{\frac{1}{2}-H} I_{2,2,n}(s)$ is analyzed as follows. The mean value theorem yields that there is a constant C such that

$$\begin{aligned}
|I_{2,2,n}(s)| &\le \left(\frac{1}{2}-H\right)\sum_{i=2}^{n-2}\left(\frac{it}{n}\right)^{H-\frac{1}{2}}\left(\frac{it}{n}-s\right)_+^{H-\frac{1}{2}}\left(t-\frac{it}{n}\right)^{-\frac{1}{2}-H} \\
&\quad \times \left(\frac{(i-1)t}{n}\right)^{\frac{1}{2}-H}\frac{t}{n} \le C\sum_{i=2}^{n-2}\left(\frac{it}{n}-s\right)_+^{H-\frac{1}{2}}\left(t-\frac{(i-1)t}{n}\right)^{-\frac{1}{2}-H}\frac{t}{n} \\
&\le C\int_s^t (u-s)^{H-\frac{1}{2}}(t-u)^{-\frac{1}{2}-H}du.
\end{aligned}$$

Hence, proceeding similarly with the term $s \mapsto s^{\frac{1}{2}-H} I_{2,1,n}(s)$ and using the dominated convergence theorem again, together with (1.78)-(1.82), (8.11) and Lemma 87, we get

$$\begin{aligned}
&L^2([0,T]) - \lim_{n\to\infty}\Gamma\left(\frac{3}{2}-H\right)\Gamma\left(\frac{1}{2}+H\right)K^*(\varphi_n)(s) \\
&= s^{\frac{1}{2}-H}\left(\frac{1}{2}-H\right)1_{[0,t]}(s)\int_0^T u^{-1}(t-u)^{\frac{1}{2}-H}(u-s)^{H-\frac{1}{2}}1_{[s,t]}(u)du \\
&\quad + s^{\frac{1}{2}-H}\left(H-\frac{1}{2}\right)1_{[0,t]}(s)\int_0^T u^{-1}(t-u)^{\frac{1}{2}-H}(u-s)^{H-\frac{1}{2}}1_{[s,t]}(u)du \\
&\quad + s^{\frac{1}{2}-H}\left(\frac{1}{2}-H\right)1_{[0,t]}(s)\int_0^T (u-s)^{H-\frac{1}{2}}(t-u)^{-\frac{1}{2}-H}1_{[s,t]}(u)du \\
&= s^{\frac{1}{2}-H}\left(\frac{1}{2}-H\right)1_{[0,t]}(s)\int_0^T (u-s)^{H-\frac{1}{2}}(t-u)^{-\frac{1}{2}-H}1_{[s,t]}(u)du \\
&= s^{\frac{1}{2}-H}\left(\frac{1}{2}-H\right)\Gamma\left(H+\frac{1}{2}\right)\Gamma\left(\frac{1}{2}-H\right)1_{[0,t]}(s).
\end{aligned}$$

Consequently, the proof is complete. □

Now, with the help of Lemma 9, we can identify the space $\mathcal{H}$.

Theorem 8 *Let $H < 1/2$. Then, $\mathcal{H}$ is a close subset of $L^2([0,T])$ and*

$$\begin{aligned}
\mathcal{H} = \{f \in L^2([0,T]) : \text{there is } \phi_f \in L^2([0,T]) \text{ such that} \\
f(s) = s^{\frac{1}{2}-H} I_{T-}^{\frac{1}{2}-H}\left(u^{H-\frac{1}{2}}\phi_f(u)\right)(s)\}.
\end{aligned} \tag{1.83}$$

The inner product in $\mathcal{H}$ is given by

$$\langle f, g\rangle_{\mathcal{H}} = \tilde{c}_H^2\Gamma(H+\frac{1}{2})^2\langle \phi_f, \phi_g\rangle_{L^2([0,T])}. \tag{1.84}$$

Proof. From (1.76), we have already known that $\mathcal{H}$ is a subset of the right-hand side of (1.83).

Now, let ψ be a function in $L^2([0,T])$. Observe that, in order to see that (1.83) holds, we only need to see that ψ is in the image of K^*. That is, there exists a sequence $\{\phi_n : n \in \mathbb{N}\}$ in $\mathcal{H}$ such that $K^*\phi_n$ goes to ψ in $L^2([0,T])$. In fact, in this case, there is $\phi \in \mathcal{H}$ such that $\phi_n \to \phi$ and $K^*\phi = \psi$ since $\mathcal{H}$ is a Hilbert space and K^* is an isometry from $\mathcal{H}$ into $L^2([0,T])$. Consequently, the function $s \mapsto s^{\frac{1}{2}-H} I_{T-}^{\frac{1}{2}-H}\left(u^{H-\frac{1}{2}}\psi\right)(s)$ belongs to $\mathcal{H}$ by (1.76).

Finally, let $t \in [0,T]$. Then, $1_{[0,t]}$ is in the image of K^* due to Lemma 9 and, therefore, $\mathcal{E}([0,T])$ also belongs to the image of K^* because of the linearity of this operator. Thus,

(1.83) is satisfied since $\mathcal{E}([0,T])$ is dense in $L^2([0,T])$. Moreover, (1.84) is an immediate consequence of (1.75). In this way, the proof is complete. □

On the other hand, let $\varphi \in \mathcal{E}([0,T])$ be as in (1.35) and $B(\varphi)$ as is in (1.36). In particular, we have that $B_t = B(1_{[0,t]})$ and that $B(\varphi)$ is a square-integrable random variable. Indeed, Proposition 20 and (1.38) imply

$$\begin{aligned} E\left(B(\varphi)^2\right) &= E\left[\sum_{k,\ell=1}^{n} a_k a_\ell \left(B_{s_{k+1}} - B_{s_k}\right)\left(B_{s_{\ell+1}} - B_{s_\ell}\right)\right] \\ &= \sum_{k,\ell=1}^{n} a_k a_\ell E\left[B_{s_{k+1}} B_{s_{\ell+1}} - B_{s_{k+1}} B_{s_\ell} - B_{s_k} B_{s_{\ell+1}} + B_{s_k} B_{s_\ell}\right] \\ &= \sum_{k,\ell=1}^{n} a_k a_\ell \left[\left\langle K^*(1_{[0,s_{k+1}]}), K^*(1_{[0,s_{\ell+1}]})\right\rangle_{L^2([0,T])}\right. \\ &\quad - \left\langle K^*(1_{[0,s_{k+1}]}), K^*(1_{[0,s_\ell]})\right\rangle_{L^2([0,T])} - \left\langle K^*(1_{[0,s_k]}), K^*(1_{[0,s_{\ell+1}]})\right\rangle_{L^2([0,T])} \\ &\quad \left. + \left\langle K^*(1_{[0,s_k]}), K^*(1_{[0,s_\ell]})\right\rangle_{L^2([0,T])}\right] = \|K^*(\varphi)\|^2_{L^2([0,T])}. \end{aligned}$$

Thus, as in Section 1.3, we can extend the linear operator $B : \mathcal{E}([0,T]) \to L^2(\Omega)$ to the Hilbert space $\mathcal{H}$ again. In this way, we have defined B as a Gaussian stochastic process indexed by the elements of $\mathcal{H}$ (see Definition 10). Moreover, $B(\mathcal{H})$ is a closed linear subspace of $L^2(\Omega)$, whose elements $B(\varphi)$ are Gaussian random variables with zero mean and variance $\|K^*(\varphi)\|^2_{L^2([0,T])}$. In consequence, $\left\{B\left((K^*)^{-1}(1_{[0,t]})\right) : t \in [0,T]\right\}$ is a Gaussian process with zero mean and covariance

$$E\left(B\left((K^*)^{-1}(1_{[0,t]})\right) B\left((K^*)^{-1}(1_{[0,s]})\right)\right) = t \wedge s, \quad s,t \in [0,T].$$

Remember that this process is well-defined because Lemma 9 establishes that $\mathcal{E}([0,T])$ is included in the image of K^*. So, the process

$$W_t := B\left((K^*)^{-1}(1_{[0,t]})\right), \quad t \in [0,T],$$

is a Brownian motion (see Section 1.2 for details) satisfying

$$B\left((K^*)^{-1}(1_{[0,t]})\right) = \int_0^T 1_{[0,t]}(s)dW_s, \quad t \in [0,T].$$

Here, the right-hand side is the Wiener integral of the square-integrable function $1_{[0,t]}$ studied in Subsection 1.2.2. Note that the linearity of B and K^* yields

$$B\left((K^*)^{-1}(\psi)\right) = \int_0^T \psi(s)dW_s, \quad \psi \in \mathcal{E}([0,T]). \tag{1.85}$$

Therefore, the fact that $L^2([0,T])$ coincides with the image of K^* (see the proof of Theorem 8) and that $B\left((K^*)^{-1}(\psi)\right)$ is a Gaussian random variable with variance

$$\left\|K^*\left((K^*)^{-1}\psi\right)\right\|^2_{L^2([0,T])} = \|\psi\|^2_{L^2([0,T])},$$

together with the definition of Wiener integral, leads to show that (1.85) is also satisfied for all $\varphi \in L^2([0,T])$. Now, we are ready to state that the fractional Brownian motion is a Volterra type process.

Theorem 9 *Let $\phi \in \mathcal{H}$. Then,*

$$B(\phi) = \int_0^T (K^*\phi)(s)dW_s,$$

and therefore The fBm B is a Volterra Gaussian process with kernel K and Wiener process W.

Proof. We use that the image of K^* is $L^2([0,T])$ again to see that, given $\psi \in L^2([0,T])$, there is a unique $\phi \in \mathcal{H}$ such that $K^*\phi = \psi$. Thus the result is an immediate consequence of equality (1.85). □

An important consequence of Theorem 9 is the following result.

Corollary 9 *Let $H < 1/2$. Then, the filtrations $(\mathcal{F}_t^B)_{t\in[0,T]}$ and $(\mathcal{F}_t^W)_{t\in[0,T]}$ generated by B and W, respectively, are the same.*

Proof. Fix $t \in [0,T]$. From Theorem 9, we have $B_t = \int_0^t K(t,s)dW_s$, which implies $\mathcal{F}_t^B \subset \mathcal{F}_t^W$. Similarly, Theorem 9 also implies $W_t = B\left((K^*)^{-1})1_{[0,t]}\right)$. Therefore, the proof of Lemma 9 yields that $\mathcal{F}_t^W \subset \mathcal{F}_t^B$. □

1.4.2.2 Case $H > 1/2$

Now, the kernel K is given by (1.66) and $K^* : \mathcal{E}([0,T]) \to L^2([0,T])$ introduced in (1.37) satisfies

$$(K^*\varphi)(s) = \int_s^T \varphi(u)\frac{\partial K}{\partial u}(u,s)du, \quad \text{for } \varphi \in \mathcal{E}([0,T]) \text{ and } s \in [0,T], \tag{1.86}$$

with

$$\frac{\partial K}{\partial u}(u,s) = c_H\left(\frac{u}{s}\right)^{H-\frac{1}{2}}(u-s)^{H-\frac{3}{2}}, \quad 0 \le s \le u \le T.$$

In particular, we have that $\left(K^*1_{[0,t]}\right)(s) = K(t,s)1_{[0,t]}(s)$, $s,t \in [0,T]$, which leads us to deduce

$$\langle\varphi,\psi\rangle_{\mathcal{H}} = \langle K^*\varphi, K^*\psi\rangle_{L^2([0,T])}, \quad \varphi,\phi \in \mathcal{H},$$

since, by Proposition 20, we have

$$\begin{aligned}\left\langle 1_{[0,t]}, 1_{[0,s]}\right\rangle_{\mathcal{H}} &= R_H(t,s) = \int_0^{t\wedge s} K(t,u)K(s;u)du \\ &= \left\langle K(t,\cdot)1_{[0,t]}(\cdot), K(s,\cdot)1_{[0,s]}(\cdot)\right\rangle_{L^2([0,T])} \\ &= \left\langle K^*1_{[0,t]}, K^*1_{[0,s]}\right\rangle_{L^2([0,T])}, \quad s,t \in [0,T].\end{aligned}$$

From (1.56) and (1.86), the isometry K^* is now a fractional integral. That is, for $\varphi \in \mathcal{E}([0,T])$,

$$(K^*\varphi)(s) = c_H\Gamma(H-\frac{1}{2})s^{\frac{1}{2}-H}I_{T-}^{H-\frac{1}{2}}\left(u^{H-\frac{1}{2}}\varphi(u)\right)(s), \quad s \in [0,T]. \tag{1.87}$$

Hence, Hölder inequality and Fubini's theorem imply that, for $\varphi \in \mathcal{E}([0,T])$ and $\hat{c}_H = c_H\Gamma(H-\frac{1}{2})$,

$$\begin{aligned}\|\varphi\|_{\mathcal{H}}^2 &= \|K^*\varphi\|_{L^2([0,T])}^2 = \hat{c}_H^2\int_0^T s^{1-2H}\left(\int_s^T \frac{u^{H-\frac{1}{2}}\varphi(u)}{(u-s)^{\frac{3}{2}-H}}du\right)^2 ds \\ &\le C\int_0^T s^{1-2H}\int_s^T \frac{u^{2H-1}\varphi(u)^2}{(u-s)^{\frac{3}{2}-H}}duds \\ &\le C\int_0^T \varphi(u)^2\int_0^u \frac{s^{1-2H}}{(u-s)^{\frac{3}{2}-H}}dsdu \le C\int_0^T \varphi(u)^2du.\end{aligned} \tag{1.88}$$

Therefore, we have established that $L^2([0,T]) \subset \mathcal{H}$ (compare it with(1.83)).

As in Section 1.4.2.1, we have a linear isometry $B : \mathcal{H} \to L^2(\Omega)$ such that $B(1_{[0,t]}) = B_t$, $t \in [0,T]$. Moreover, $B = \{B(\varphi) : \varphi \in \mathcal{H}\}$ is a zero mean Gaussian process (indexed by $\mathcal{H}$) with covariance

$$E\left(B(\varphi)B(\psi)\right) = \langle \varphi, \psi \rangle_{\mathcal{H}} = \langle K^*\varphi, K^*\psi \rangle_{L^2([0,T])}, \quad \varphi, \psi \in \mathcal{H}.$$

In order to see that fractional Brownian motion is also a Gaussian Volterra process for $H > 1/2$, we need to state the following result. This is an immediate consequence of Lemma 8.1 in Pipiras and Taqqu [177].

Lemma 10 *Let $H > 1/2$. Then, the image of the operator K^* is $L^2([0,T])$.*

Proof. As in Lemma 9, we only need to show that the square-integrable function $s \mapsto s^{\frac{1}{2}-H} 1_{[0,t]}(s)$ belongs to the image of the operator K^*, for any $t \in [0,T]$. So fix $t \in [0,T]$. Note that Definition 18.ii) yields that, for $s \in [0,T]$,

$$\begin{aligned}
& s^{\frac{1}{2}-H} 1_{[0,t]}(s) \\
&= s^{\frac{1}{2}-H} I_{T-}^{H-\frac{1}{2}} \left(D_{T-}^{H-\frac{1}{2}} 1_{[0,t]} \right) = \frac{s^{\frac{1}{2}-H}}{\Gamma(\frac{3}{2}-H)} I_{T-}^{H-\frac{1}{2}} \left((t-u)^{\frac{1}{2}-H} 1_{[0,t]}(u) \right)(s) \\
&= c_H \Gamma(H - \frac{1}{2}) s^{\frac{1}{2}-H} I_{T-}^{H-\frac{1}{2}} \left(u^{H-\frac{1}{2}} \left[u^{\frac{1}{2}-H} d_H (t-u)^{\frac{1}{2}-H} 1_{[0,t]}(u) \right] \right)(s), \qquad (1.89)
\end{aligned}$$

with $d_H = \left(c_H \Gamma(H-\frac{1}{2})\Gamma(\frac{3}{2}-H)\right)^{-1}$ and $u \mapsto (u(t-u))^{\frac{1}{2}-H}$ in $L^2([0,t])$.

Now, consider a partition $\pi_n = \{0 = t_0^n < t_1^n < \ldots < t_{k_n}^n = t\}$ of the interval $[0,t]$ such that $|\pi_n|$ goes to zero as $n \to \infty$ and introduce the step function

$$\varphi_n(s) = d_H \sum_{i=1}^{k_n - 1} \left(t_{i+1}^n\right)^{\frac{1}{2}-H} (t - t_i^n)^{\frac{1}{2}-H} 1_{[t_i^n, t_{i+1}^n)}(s), \quad s \in [0,T].$$

Finally, (1.87)-(1.89) and the dominated convergence theorem allow us to deduce

$$\int_0^T \left| s^{\frac{1}{2}-H} 1_{[0,t]}(s) - (K^*\varphi_n)(s) \right|^2 ds \le C \int_0^t \left| d_H (s(t-s))^{\frac{1}{2}-H} - \varphi_n(s) \right|^2 ds \to 0, \quad \text{as } n \to \infty.$$

Therefore, the proof is complete. □

As in Section 1.4.2.1, for all $t \in [0,T]$, Lemma 10 states that $1_{[0,t]}$ also belongs to the image of the isometry K^* and, consequently, the process

$$W_t = B\left((K^*)^{-1} (1_{[0,t]}) \right), \quad t \in [0,T],$$

is a Brownian motion that satisfies

$$B(\varphi) = \int_0^T (K^*\varphi)(s) dW_s, \quad \varphi \in \mathcal{H}. \qquad (1.90)$$

In particular, B is a Volterra Gaussian process. That is,

$$B_t = B(1_{[0,t]}) = \int_0^T \left(K^* 1_{[0,t]}\right)(s) dW_s = \int_0^t K(t,s) dW_s, \quad t \in [0,T].$$

Finally, observe that, as in Corollary 9, B and W generate the same filtration due to the proof of Lemma 10.

1.5 Malliavin calculus

Malliavin calculus is an extension of the calculus of variations for deterministic functions in the case of stochastic processes. This calculus was started by Malliavin [136] in 70s. This important calculus has been the basis of numerous scientific works and the range of its possible applications has increased over the time. At the beginning, the utility of Malliavin calculus was to characterize conditions that guarantee the existence of a smooth density for a given random variable. Although this was considered as its only application for many years, this situation changed in the middle of the 80s when Ocone figured out an explicit expression of Clark's formula in terms of the Malliavin derivative operator (see for example Ocone and Karatzas [165], Nualart and Pardoux [159], or Proposition 72 below). Subsequently, many researchers have used these techniques in many other contexts and applications.

The Malliavin calculus has three operators as its fundamental tool. Namely, the derivative operator (in the sense of Malliavin calculus), its adjoint (the divergence operator) and the number operator. In this book we only show some applications based on the derivative operator and its adjoint called the divergence operator. It is worth mentioning that an important characterization of the domain of the derived operator was given by Sugita [202] and that the reader can consult Nualart's book [157] for a deeper analysis of these operators.

In the case that the underlying process is Brownian motion (resp. a centered Poisson random measure), the divergence operator agrees with the Skorohod integral introduced by Skorohod [198] (resp. the extension of the stochastic Itô's integral defined by Kabano [91]). So, in this case, the divergence operator is an integral that coincides with Itô's integral when the integrand is adapted and square-integrable and its domain contains integrands not necessarily adapted to the filtration generated by Brownian motion (resp. the Poisson random measure). The fact that Skorohod integral and the divergence operator are equal has been proven by Gaveau and Trauber [63]. In this way, the techniques of Malliavin calculus become an important tool to open the possibility of multiple applications where it is natural to deal with not adapted processes, as in the case of a financial market with an insider (see León et al. [116]), or the study of the implied volatility for jump-diffusion models with stochastic volatility using the future volatility (see Alòs et al. [8]). But, the Malliavin calculus is also useful to facilitate the analysis of problems that we could study through the classical Itô calculus. For instance, Fournié et al. [55] have obtained an important probabilistic method for numerical computations of Greeks (i.e., price sensitivities) in finance markets via the integration by parts formula used in the Malliavin calculus (see (4.2)). This study is continued in [54], where the authors, among other applications, figure out optimal weight functionals in the sense of minimal variance and compute conditional expectations as applications of Malliavin calculus via (4.2). Moreover, other two examples of problems that could be analyzed using Itô calculus are the analytical study of the second derivative of the implied volatility curve as a function of the strike price developed by Alòs and León [5] and the existence of a unique solution for either stochastic evolution equations with random and adapted generators on a Hilbert space driven by cylindrical Wiener process (see León and Nualart [117]), or one-dimensional stochastic parabolic partial differential equations with random and adapted coefficients perturbed by two-parameter white noise (see Alòs et al. [6]). Concerning the application of Malliavin calculus on finance markets, the reader can see Nualart [158].

1.5.1 Basic elements

Let $\mathcal{H}$ be a real and separable Hilbert space, with inner product $\langle \cdot, \cdot \rangle_{\mathcal{H}}$ and norm $\| \cdot \|_{\mathcal{H}}$.
Here, as it is done in Nualart [158], we consider a Gaussian subspace of $L^2(\Omega, \mathcal{F}, P)$, where $(\Omega, \mathcal{F}, P)$ is a complete probability space. The elements of this subspace are used to define the derivative operator and its adjoint (i.e., the divergence operator). This Gaussian space is introduced as follows.

Definition 19 *We say that the family $W = \{W(h) : h \in \mathcal{H}\}$, defined on a complete probability space $(\Omega, \mathcal{F}, P)$, is an isonormal Gaussian process if it is a Gaussian stochastic process indexed by the elements of $\mathcal{H}$ (see Definition 10) such that*

$$E(W(h)) = 0 \quad \text{and} \quad E(W(h)W(g)) = \langle h, g \rangle_{\mathcal{H}}, \quad \text{for } h, g \in \mathcal{H}. \tag{1.91}$$

Examples 1 *i) Let $B = \{B_t : t \geq 0\}$ be a d-dimensional Brownian motion as it is given in Definition 36. In this case, for $\mathcal{H} = L^2(\mathbb{R}_+; \mathbb{R}^d)$, we define*

$$W(h) = \sum_{i=1}^{d} \int_0^\infty h^i(s) dB_s^i, \quad h \in \mathcal{H}.$$

Then, it is easy to show that $W = \{W(h) : h \in \mathcal{H}\}$ is an isonormal Gaussian process on $\mathcal{H}$.

ii) Let $X = \{X_t : t \in [0, T]\}$ be a centered Gaussian process defined on $(\Omega, \mathcal{F}, P)$ with covariance function

$$R(t, s) = E(X_t X_s), \quad s, t \in [0, T].$$

In order to introduce an isonormal Gaussian process associated with X, we consider the real separable Hilbert space $\mathcal{H}$ given as the completion of Span($\{1_{[0,t]} : t \in [0, T]\}$) with respect to the inner product

$$\langle g, h \rangle_R = \sum_{\substack{1 \leq i \leq n, \\ 1 \leq j \leq m}} f_i h_j R(s_i, t_j),$$

where $f = \sum_{i=1}^n f_i 1_{[0,s_i]}$ and $h = \sum_{j=1}^m h_i 1_{[0,t_j]}$, with $f_i, h_j \in \mathbb{R}$ and $s_i, t_j \in [0, T]$. For the definition of $\mathcal{H}$, the reader can see Section 8.11 in the appendix. Continuing with the construction of the isonormal Gaussian process related to X, we set

$$X(h) = \sum_{j=1}^{m} h_i X_{t_j} \quad \text{for } h = \sum_{j=1}^{m} h_i 1_{[0,t_j]} \text{ in Span}(\{1_{[0,t]} : t \in [0, T]\}),$$

and

$$X(h) = L^2(\Omega) - \lim_{n \to \infty} X(h_n), \quad h \in \mathcal{H},$$

where $h_n \in$ Span($\{1_{[0,t]} : t \in [0, T]\}$) and $\|h - h_n\|_{\mathcal{H}} \to 0$, as $n \to \infty$. Note that the definition of $\mathcal{H}$ implies that such a sequence exists and that we are able to proceed as in Lemma 5 to obtain that $X(h)$ is independent of the choice of the sequence $\{h_n \in \text{Span}(\{1_{[0,t]} : t \in [0, T]\}) : n \in \mathbb{N}\})\}$. Consequently, $\{X(h) : h \in \mathcal{H}\}$ is an isonormal Gaussian process on $\mathcal{H}$.

iii) Let $T > 0$, $H \in (0, 1)$ and $B^H = \{B_t^H : t \in [0, T]\}$ a fractional Brownian motion with Hurst parameter H (see Section 1.4). So, by Statement ii), the fractional Brownian motion B^H can be considered as an isonormal Gaussian family on the Hilbert space $\mathcal{H}$, where $\mathcal{H}$ is the completion of all the simple function with respect to the inner product $\langle 1_{[0,t]}, 1_{[0,s]} \rangle_{\mathcal{H}} = R_H(t, s)$, $t, s \in [0, T]$. Here $R_H(t, s)$ is given in (1.44).

Note that, in Definition 19, we can only ask that the random variables $W(h)$, for $h \in \mathcal{H}$, be Gaussian because this fact and (1.91) imply that the process W is also Gaussian. In fact, let $\alpha \in \mathbb{R}$ and $h, g \in \mathcal{H}$, then

$$\begin{aligned} & E\left([\alpha W(h) + W(g) - W(\alpha h + g)]^2\right) \\ &= E\left(\alpha^2 W(h)^2 + W(g)^2 + W(\alpha h + g)^2\right. \\ &\quad \left. + 2\alpha W(h)W(g) - 2\alpha W(h)W(\alpha h + g) - 2W(g)W(\alpha h + g)\right) \\ &= \langle \alpha h, \alpha h\rangle_{\mathcal{H}} + \langle g, g\rangle_{\mathcal{H}} + \langle \alpha h + g, \alpha h + g\rangle_{\mathcal{H}} \\ &\quad + 2\langle \alpha h, g\rangle_{\mathcal{H}} - 2\langle \alpha h, \alpha h + g\rangle_{\mathcal{H}} - 2\langle g, \alpha h + g\rangle_{\mathcal{H}} = 0. \end{aligned} \tag{1.92}$$

That is, we have that $\alpha W(h) + W(g) = W(\alpha h + g)$ with probability 1. Thus, Definitions 9 and 10, together with the fact that $W(\alpha h + g)$ is a Gaussian random variable, give that the process W is a Gaussian process.

In the remaining of this section, $\mathcal{K}$ is another real separable Hilbert space, $W = \{W(h) : h \in \mathcal{H}\}$, defined on $(\Omega, \mathcal{F}, P)$, stands for an isonormal Gaussian process indexed by the real separable Hilbert space $\mathcal{H}$ and the σ-algebra $\mathcal{F}$ is generated by W (i.e., $\mathcal{F} = \sigma(\{W(h) : h \in \mathcal{H}\})$.

1.5.2 Homogeneous chaos

We first study the space $L^2(\Omega; \mathcal{K})$. Remember that the involved σ-algebra $\mathcal{F}$ is generated by the isonormal Gaussian process W.

Let $p \geq 1$. A $\mathcal{K}$-valued function $G : \Omega \to \mathcal{K}$ belongs to $L^p(\Omega; \mathcal{K})$ if G is measurable in the sense that

$$[G \in K] \in \mathcal{F}$$

for every Borel subset K of $\mathcal{K}$ and $\int_\Omega \|G\|_{\mathcal{K}}^p dP < \infty$. In this case, using that $\mathcal{K}$ is a real separable Hilbert space (see Section 8.3), we can find a sequence $\{\varphi_n : n \in \mathbb{N}\}$ of $\mathcal{K}$-valued simple random variables of the form

$$\varphi_n = \sum_{i=1}^{N_n} k_i 1_{A_i}, \quad N_n \in \mathbb{N},\ k_i \in \mathcal{K} \text{ and } A_i \in \mathcal{F},$$

such that φ_n goes to G, P-almost surely, and

$$\int_\Omega \|G - \varphi_n\|_{\mathcal{K}}^p \, dP \to 0, \quad \text{as } n \to \infty. \tag{1.93}$$

Moreover, $L^p(\Omega; \mathcal{K})$ is a Banach space with the norm $\|G\|_{L^p(\Omega;\mathcal{K})}^p = E\left(\|G\|_{\mathcal{K}}^p\right)$ and $L^2(\Omega; \mathcal{K})$ is a real Hilbert space with inner product $\langle G_1, G_2\rangle_{L^2(\Omega;\mathcal{K})} = E\left(\langle G_1, G_2\rangle_{\mathcal{K}}\right)$. Thus, from (1.93), we have that the collection of all the $\mathcal{K}$-valued simple processes is dense in $L^p(\Omega; \mathcal{K})$. For a detailed exposition of the space $L^p(\Omega; \mathcal{K})$, the reader can consult the book by Hytönen et al. [85] (Section 1).

Now, consider two complete orthonormal systems $\{e_i : i \in \mathbb{N}\}$ and $\{g_i : i \in \mathbb{N}\}$ of $\mathcal{H}$ and $\mathcal{K}$, respectively. Let Σ be the family of all the sequences $\alpha = \{\alpha_n \in \mathbb{N} \cup \{0\} : n \in \mathbb{N}\}$ such that $|\alpha| := \sum_{n=1}^\infty \alpha_n < \infty$, which gives that $\alpha_n = 0$ except for a finite number of them. Also, for $\alpha \in \Sigma$ and $j \in \mathbb{N}$, we use the convention

$$H_{\alpha,j} = \frac{g_j}{\sqrt{\alpha!}} \prod_{n=1}^\infty H_{\alpha_n}(W(e_n), 1),$$

where $\alpha! = \prod_{n=1}^{\infty} \alpha_n!$ and $H_n(\cdot, 1)$ is the Hermite polynomial of order n with parameter 1, which is analyzed in Section 8.13.

In order to state one of the main results of this section, we establish the following auxiliary lemma.

Lemma 11 *Let $\{\mathcal{F}_i \subset \mathcal{F} : i \in \mathbb{N}\}$ be a filtration such that $\mathcal{F} = \sigma(\cup_{i=1}^{\infty} \mathcal{F}_i)$ and $X \in L^1(\Omega; \mathcal{F}, P)$. Then, $X = L^1(\Omega) - \lim_{n\to\infty} E\left[X \middle| \mathcal{F}_n\right]$.*

Proof. By the monotone class theorem (i.e., Theorem 97), together with the fact that $\cup_{i=1}^{\infty} \mathcal{F}_i$ is a π-system, there exists a sequence $\{Y_m : m \in \mathbb{N}\}$ of the form

$$Y_m = \sum_{j=1}^{K_m} y_j^{(m)} 1_{G_j^{(m)}}, \quad \text{with } K_m \in \mathbb{N},\ y_j^{(m)} \in \mathbb{R} \text{ and } G_j^{(m)} \in \cup_{i=1}^{\infty} \mathcal{F}_i$$

such that $X = L^1(\Omega) - \lim_{m\to\infty} Y_m$. Hence, given $\varepsilon > 0$, there is $m_0 \in \mathbb{N}$ so that

$$\|X - Y_{m_0}\|_{L^1(\Omega)} \le \frac{\varepsilon}{2}.$$

Also, we can find $N \in \mathbb{N}$ such that $G_j^{(m_0)} \in \mathcal{F}_N$, for all $j = 1, \ldots, K_{m_0}$. Therefore, Jensen's inequality implies that, for $n \ge N$, we have

$$\begin{aligned}
\left\|X - E\left[X \middle| \mathcal{F}_n\right]\right\|_{L^1(\Omega)} &\le \|X - Y_{m_0}\|_{L^1(\Omega)} + \left\|Y_{m_0} - E\left[X \middle| \mathcal{F}_n\right]\right\|_{L^1(\Omega)} \\
&= \|X - Y_{m_0}\|_{L^1(\Omega)} + \left\|E\left[Y_{m_0} - X \middle| \mathcal{F}_n\right]\right\|_{L^1(\Omega)} \\
&\le 2\|X - Y_{m_0}\|_{L^1(\Omega)} \le \varepsilon.
\end{aligned}$$

Since ε is arbitrary, the proof is finished. □

Now, we are ready to study the following result.

Theorem 10 *Let $\mathcal{F} = \sigma(W)$. Then, the family $\{H_{\alpha,j} : j \in \mathbb{N} \text{ and } \alpha \in \Sigma\}$ is a complete orthonormal system of $L^2(\Omega; \mathcal{K})$.*

Proof. Let $X \in L^2(\Omega; \mathcal{K})$ be orthogonal to the family $\{H_{\alpha,j} : j \in \mathbb{N} \text{ and } \alpha \in \Sigma\}$. In order to show the result, for $n \in \mathbb{N}$, we introduce the σ-algebra $\mathcal{F}_n = \sigma(\{W(e_1), \ldots, W(e_n)\})$. Then, for $j \in \mathbb{N}$, the Doob-Dynkin lemma (see, for example, [181]) yields that there exists a measurable function $\Phi_{n,j} : \mathbb{R}^n \to \mathbb{R}$ such that

$$E\left[\langle X, g_j\rangle_{\mathcal{K}} \middle| \mathcal{F}_n\right] = \Phi_{n,j}\left(W(e_1), \ldots, W(e_n)\right).$$

Hence, for $\alpha = (m_1, \ldots, m_n, 0, \ldots)$, with $m_1, \ldots, m_n \in \mathbb{N} \cup \{0\}$, we obtain

$$\begin{aligned}
0 &= E\left(\langle X, H_{\alpha,j}\rangle_{\mathcal{K}}\right) = E\left(E\left[\left\langle X, \frac{g_j}{\sqrt{\alpha!}}\right\rangle_{\mathcal{K}} \middle| \mathcal{F}_n\right] \prod_{i=1}^{n} H_{m_i}(W(e_i), 1)\right) \\
&= \frac{1}{\sqrt{\alpha!}} \int_{\mathbb{R}^n} \Phi_{n,j}\left(x_1, \ldots, x_n\right) \prod_{i=1}^{n} H_{m_i}(x_i, 1)\mu_n(dx),
\end{aligned}$$

where μ_n is the standard Gaussian measure on $\mathbb{R}^n$. Thus, Proposition 142, Theorem 102 and (8.57) lead us to conclude that $\Phi_{n,j} \equiv 0$, and therefore Lemma 11 gives

$$\langle X, g_j\rangle_{\mathcal{K}} = \lim_{n\to\infty} \Phi_{n,j}\left(W(e_1), \ldots, W(e_n)\right) = 0.$$

Finally, the result holds because $\{g_i : i \in \mathbb{N}\}$ is a basis of $\mathcal{K}$. □

For $n \in \mathbb{N} \cup \{0\}$, let $\mathcal{H}_n$ be the closed linear subspace of $L^2(\Omega; \mathcal{K})$ generated by $\{H_{\alpha,j} : j \in \mathbb{N}, \alpha \in \Sigma \text{ and } |\alpha| = n\}$. Note that $\mathcal{H}_0 = \mathcal{K}$ and $X \in \mathcal{H}_1$ if there exists a sequence $\{x_{n,j} \in \mathbb{R} : n, j \in \mathbb{N}\}$ such that

$$\sum_{n,j} x_{n,j}^2 < \infty \quad \text{and} \quad X = \sum_{n,j=1}^{\infty} x_{n,j} W(e_n) g_j \text{ in } L^2(\Omega; \mathcal{K}).$$

Definition 20 *The space $\mathcal{H}_n$ is called $\mathcal{K}$-valued either Wiener chaos of order n, or the homogeneous chaos of order n.*

The following result is an immediate consequence of Theorem 10.

Theorem 11 *Let $\mathcal{F} = \sigma(W)$. Then, $L^2(\Omega; \mathcal{K})$ is the orthogonal direct sum of the $\mathcal{K}$-valued Wiener chaos $\mathcal{H}_n$. Namely,*

$$L^2(\Omega; \mathcal{K}) = \bigoplus_{n=0}^{\infty} \mathcal{H}_n.$$

In other words, for $X \in L^2(\Omega; \mathcal{K})$, there is a sequence $\{X_n \in \mathcal{H}_n : n \in \mathbb{N}\}$ such that $E(X_n X_m) = 0$, for $n \neq m$, and

$$X = \sum_{n=1}^{\infty} X_n \quad \text{in } L^2(\Omega; \mathcal{K}).$$

By equality (8.64) below, together with induction on n, the monomial x^n belongs to $\text{Span}(\{H_0(x, \rho), \ldots, H_n(x, \rho)\})$. Consequently, for $n \in \mathbb{N}$,

$$\mathcal{P}_n = \mathcal{H}_0 \oplus \mathcal{H}_1 \oplus \ldots \oplus \mathcal{H}_n,$$

where $\mathcal{P}_n$ is the closure in $L^2(\Omega; \mathcal{K})$ of finite linear combinations of $\mathcal{K}$-valued random variables of the form $p(W(h_1), \ldots, W(h_m))k$. Here, $m \geq 1$, $k \in \mathcal{K}$, $h_1, \ldots, h_m \in \mathcal{H}$ and p is a $\mathbb{R}$-valued polynomial of order less or equal to n. The proof of this claim is left for the reader as an exercise. Hence, the Hilbert spaces $\mathcal{H}_n$ are independent of the families $\{e_i : i \in \mathbb{N}\}$ and $\{g_i : i \in \mathbb{N}\}$. To see this holds, choose other complete orthonormal systems $\{\tilde{e}_i : i \in \mathbb{N}\}$ and $\{\tilde{g}_i : i \in \mathbb{N}\}$ of $\mathcal{H}$ and $\mathcal{K}$, respectively, $m < n$ and

$$\tilde{H}_{\alpha,j} = \frac{\tilde{g}_j}{\sqrt{\alpha!}} \prod_{n=1}^{\infty} H_{\alpha_n}(W(\tilde{e}_n), 1), \quad \text{with} \quad |\alpha| = m.$$

Then, $\tilde{H}_{\alpha,j}$ is orthogonal to $\mathcal{H}_n$ since it belongs to $\mathcal{P}_m$, which follows from the facts that there exists a sequence $\{c_{i,n} \in \mathbb{R} : i, n \in \mathbb{N}\}$ such that

$$\tilde{e}_n = \sum_{i=1}^{\infty} c_{i,n} e_i, \quad n \in \mathbb{N},$$

$\tilde{H}_{\alpha,j,k} = \frac{\tilde{g}_j}{\sqrt{\alpha!}} \prod_{n=1}^{\infty} H_{\alpha_n}(W(\sum_{i=1}^{k} c_{i,n} e_i), 1)$ goes to $\tilde{H}_{\alpha,j}$ in $L^2(\Omega; \mathcal{K})$, as $k \to \infty$ and $\tilde{H}_{\alpha,j,k} \in \mathcal{P}_m$, for all $k \in \mathbb{N}$. Consequently, $\mathcal{H}_n$ is orthogonal to the closure $\tilde{\mathcal{H}}_m$ of the space generated by $\{\tilde{H}_{\alpha,j} : j \in \mathbb{N} \text{ and } |\alpha| = m\}$, with $m < n$. Similarly, changing the role of the involved complete orthonormal systems (i.e., taking first $\{\tilde{e}_n, \tilde{g}_n : n \in \mathbb{N}\}$ and then $\{e_n, g_n : n \in \mathbb{N}\}$), we are able to get that $\mathcal{H}_n$ is still orthogonal to $\tilde{\mathcal{H}}_m$, also for $n < m$. Thus, we have proven that our affirmation is satisfied. That is, $\tilde{\mathcal{H}}_n = \mathcal{H}_n$, $n \geq 0$.

1.5.3 Multiple Wiener-Itô integrals

Now, we study the multiple Wiener-Itô integrals, which are useful to describe the elements of the $\mathcal{K}$-valued homogeneous chaos $\mathcal{H}_n$ of order n, for any $n \geq 0$ (see Definition 20). The concept of multiple integrals with respect to Brownian motion was first introduced by Wiener [214] with the name of polynomial chaos. However, polynomial chaos of different order are not orthogonal. Later, Itô [87] defined the so-called multiple Wiener-Itô integrals with respect to a isonormal Gaussian process with the underlying separable Hilbert space $\mathcal{H}$ equals to the space $L^2(X, \mathcal{X}, m)$, where $(X, \mathcal{X})$ is a measurable space and m is a measure without atoms. It turns out that these multiple integrals of different orders are now orthogonal and, by Example 1.*i*), Brownian motion is included as an integrator. Moreover, Itô associated these multiple integrals with the Hermite polynomials, which are analyzed in Section 8.13. So, we will use this relation to give the multiple Wiener-Itô integrals with respect to isonormal Gaussian processes.

Remember that $\{e_i : i \in \mathbb{N}\}$ and $\{g_i : i \in \mathbb{N}\}$ are complete orthonormal systems of $\mathcal{H}$ and $\mathcal{K}$, respectively. Let S_n be the permutation group of $\{1, \ldots, n\}$, $n \in \mathbb{N}$. For $\sigma \in S_n$ and

$$h = \sum_{j=1}^{k} h_{1,j} \otimes \ldots \otimes h_{n,j} \in \mathcal{H}^{\otimes n}, \tag{1.94}$$

where $k \in \mathbb{N}$, set $\tilde{\sigma}(h) = \sum_{j=1}^{k} h_{\sigma(1),j} \otimes \ldots \otimes h_{\sigma(n),j}$. Then, it is easy to see $\tilde{\sigma}$ is a well-defined bounded linear operator (i.e., $\tilde{\sigma}(h)$ is independent of the representation (1.94), which follows from Lemma 109) with norm 1. Thus, $\tilde{\sigma}$ can be extended to a bounded linear operator on $\mathcal{H}^{\otimes n}$, also with norm 1. Now, use the convention

$$\text{Symm} = \frac{1}{n!} \sum_{\sigma \in S_n} \tilde{\sigma}, \tag{1.95}$$

which is an orthogonal projection. That is, $\text{Symm}^2 = \text{Symm}$ and it agrees with its adjoint. Note that the triangle inequality implies that $\text{Symm} : \mathcal{H}^{\otimes n} \to \mathcal{H}^{\otimes n}$ is a bounded linear operator with norm 1. Now, we utilize the notations

$$\Sigma_{2,n} = \left\{ \alpha = ((\alpha_1, j_1), \ldots, (\alpha_{\hat{\alpha}}, j_{\hat{\alpha}})) : \hat{\alpha}, j_i, \alpha_i \in \mathbb{N} \text{ such that } \sum_{i=1}^{\hat{\alpha}} \alpha_i = n \text{ and } j_1 < \ldots < j_{\hat{\alpha}} \right\}$$

and, for $\alpha = ((\alpha_1, j_1), \ldots, (\alpha_{\hat{\alpha}}, j_{\hat{\alpha}})) \in \Sigma_{2,n}$,

$$e(\alpha) = \text{Symm}(e_{j_1}^{\otimes \alpha_1} \otimes \ldots \otimes e_{j_{\hat{\alpha}}}^{\otimes \alpha_{\hat{\alpha}}}). \tag{1.96}$$

By Proposition 142, the set $\{e_{j_1} \otimes \ldots \otimes e_{j_n} : j_i \in \mathbb{N}\}$ is a basis of $\mathcal{H}^{\otimes n}$. Hence, $\{e(\alpha) : \alpha \in \Sigma_{2,n}\}$ is a complete orthogonal system of the real separable Hilbert space $\mathcal{H}^{\odot n} := \text{Symm}(\mathcal{H}^{\otimes n})$ with the inner product $\langle \cdot, \cdot \rangle_{\mathcal{H}^{\otimes n}}$. Henceforth, for $h \in \mathcal{H}^{\otimes n} \otimes \mathcal{K}$, we denote $\text{Symm} \otimes I_{\mathcal{K}}(h)$ by $\tilde{h}$. The linear operator $\text{Symm} \otimes I_{\mathcal{K}}$ is introduces in Proposition 143.

Definition 21 *For $h \in \mathcal{H}^{\otimes n} \otimes \mathcal{K}$ such that*

$$\tilde{h} = \sum_{\substack{\alpha \in \Sigma_{2,n} \\ \ell \in \mathbb{N}}} c_{\alpha,\ell}\, e(\alpha) \otimes g_\ell, \quad c_{\alpha,\ell} \in \mathbb{R},$$

we define the multiple Wiener-Itô integral $I_n(h)$ of h of order n with respect to the isonormal Gaussian process W as

$$I_n(h) = I_n(\tilde{h}) = \sum_{\substack{\alpha \in \Sigma_{2,n} \\ \ell \in \mathbb{N}}} c_{\alpha,\ell} \left(\prod_{i=1}^{\hat{\alpha}} H_{\alpha_i}(W(e_{j_i}), 1) \right) g_\ell. \tag{1.97}$$

Remark 22 *i) For* $n = 1$ *and* $h = \sum_{\alpha,\ell=1}^{\infty} c_{\alpha,\ell} e_\alpha \otimes g_\ell$, *we have*

$$I_1(h) = \sum_{\alpha,\ell=1}^{\infty} c_{\alpha,\ell} W(e_\alpha)\, g_\ell.$$

ii) By Theorems 10 and 102, and (1.97), we get

$$\|I_n(h)\|_{L^2(\Omega;\mathcal{K})}^2 = \sum_{\substack{\alpha\in\Sigma_{2,n}\\ \ell\in\mathbb{N}}} c_{\alpha,\ell}^2 \prod_{i=1}^{\hat{\alpha}} \alpha_i! = n! \left\|\tilde{h}\right\|_{\mathcal{H}^{\otimes n}\otimes\mathcal{K}}^2.$$

Actually, for $h_1, h_2 \in \mathcal{H}^{\otimes n} \otimes \mathcal{K}$,

$$\langle I_n(h_1), I_n(h_2)\rangle_{L^2(\Omega;\mathcal{K})} = n! \left\langle \tilde{h}_1, \tilde{h}_2 \right\rangle_{\mathcal{H}^{\otimes n}\otimes\mathcal{K}}.$$

Thus, the series on the right-hand side of (1.97) converges in $L^2(\Omega;\mathcal{K})$. *Remember that is Section 1.5.2, we show*

$$\langle I_n(h_1), I_n(h_2)\rangle_{L^2(\Omega;\mathcal{K})} = E\{\langle I_n(h_1), I_n(h_2)\rangle_{\mathcal{K}}\}.$$

iii) As a consequence of Statement ii), we have I_n *is a linear operator from* $\mathcal{H}^{\otimes n} \otimes \mathcal{K}$ *into* $L^2(\Omega;\mathcal{K})$.

iv) We will see that the definition of $I_n(h)$ *given by (1.97) is independent of the complete orthonormal systems* $\{e_i : i \in \mathbb{N}\}$ *and* $\{g_i : i \in \mathbb{N}\}$ *of* $\mathcal{H}$ *and* $\mathcal{K}$, *respectively.*

v) Proceeding as in Statement ii), we also have that, for $n, m \in (\mathbb{N} \cup \{0\})$,

$$E(\langle I_n(h_1), I_m(h_2)\rangle_{\mathcal{K}}) = 0$$

if $n \neq m$, $h_1 \in \mathcal{H}^{\otimes n} \otimes \mathcal{K}$ *and* $h_2 \in \mathcal{H}^{\otimes m} \otimes \mathcal{K}$. *In particular , this implies*

$$E(I_n(h_1)) = 0, \quad \text{for } n \geq 1.$$

The following result is an immediate consequence of Theorems 10 and 11, (1.97) and Remark 22. Remember that the involved σ-algebra is generated by the Gaussian process W.

Theorem 12 *Let* $X \in L^2(\Omega;\mathcal{K})$. *Then, there exists a sequence* $\{h_n \in \mathcal{H}^{\otimes n} \otimes \mathcal{K} : n \in \mathbb{N}\}$ *such that*

$$X = E(X) + \sum_{n=1}^{\infty} I_n(h_n), \quad \text{in } L^2(\Omega;\mathcal{K}). \tag{1.98}$$

Moreover, this representation is unique if we only consider sequences with $h_n \in \mathcal{H}^{\odot n} \otimes \mathcal{K}$, *for all* $n \geq 1$.

In order to see that Remark 22.*iv*) holds, we introduce the following definition and the following two auxiliary results.

Definition 22 *Let* $n, m \in \mathbb{N}$, $r \in \{0, \ldots, n \wedge m\}$, $h \in \mathcal{H}^{\otimes n}$ *and* $h_1 \in \mathcal{H}^{\otimes m}$. *We introduce the contraction* $h \otimes_r h_1$ *of* h *and* h_1 *of order* r *as the element in* $\mathcal{H}^{\otimes(n+m-2r)}$ *given by*

$$h \otimes_r h_1 = \sum_{j_1,\ldots,j_r=1}^{\infty} \langle h, e_{j_1} \otimes \cdots \otimes e_{j_r}\rangle_{\mathcal{H}^{\otimes r}} \otimes \langle h_1, e_{j_1} \otimes \cdots \otimes e_{j_r}\rangle_{\mathcal{H}^{\otimes r}}.$$

Here,

$$\langle h, e_{j_1} \otimes \cdots \otimes e_{j_r}\rangle_{\mathcal{H}^{\otimes r}} = \Xi_h(e_{j_1} \otimes \cdots \otimes e_{j_r}),$$

where $\Xi_h : \mathcal{H}^{\otimes r} \to \mathcal{H}^{\otimes(n-r)}$ *is the bounded linear operator related to* h *via Proposition 144.*

Note that if $r = 0$, then $h \otimes_0 h_1 = h \otimes h_1$; and that if $r = 1$, we have

$$h \otimes_1 h_1 = \sum_{i=1}^{\infty} \langle h, e_i \rangle_{\mathcal{H}} \langle h_1, e_i \rangle_{\mathcal{H}} = \sum_{i=1}^{\infty} \Xi_{h_1}(\langle h, e_i \rangle_{\mathcal{H}} e_i) = \Xi_{h_1}(h). \tag{1.99}$$

Hence, $h \otimes_r h_1 = \langle h, h_1 \rangle_{\mathcal{H}}$ if $n, m = 1$ and, for $h = g_1 \otimes \cdots \otimes g_n$ and $h_1 = \hat{g}_1 \otimes \cdots \otimes \hat{g}_m$,

$$h \otimes_r h_1 = \langle g_1 \otimes \cdots \otimes g_r, \hat{g}_1 \otimes \cdots \otimes \hat{g}_r \rangle_{\mathcal{H}} g_{r+1} \otimes \cdots \otimes g_n \otimes \hat{g}_{r+1} \otimes \cdots \otimes \hat{g}_m.$$

Moreover, the series in Definition 22 converges in $\mathcal{H}^{\otimes n+m-2r}$ since (8.62) implies

$$\begin{aligned}
&\sum_{j_1,\ldots,j_r=1}^{\infty} \left| \langle h, e_{j_1} \otimes \cdots \otimes e_{j_r} \rangle_{\mathcal{H}^{\otimes r}} \otimes \langle h_1, e_{j_1} \otimes \cdots \otimes e_{j_r} \rangle_{\mathcal{H}^{\otimes r}} \right|_{\mathcal{H}^{\otimes(n+m-2r)}} \\
&\leq \left(\sum_{j_1,\ldots,j_r=1}^{\infty} \left| \langle h, e_{j_1} \otimes \cdots \otimes e_{j_r} \rangle_{\mathcal{H}^{\otimes r}} \right|^2_{\mathcal{H}^{\otimes(n-r)}} \right)^{1/2} \\
&\times \left(\sum_{j_1,\ldots,j_r=1}^{\infty} \left| \langle h_1, e_{j_1} \otimes \cdots \otimes e_{j_r} \rangle_{\mathcal{H}^{\otimes r}} \right|^2_{\mathcal{H}^{\otimes(m-r)}} \right)^{1/2} = |h|_{\mathcal{H}^{\otimes n}} |h_1|_{\mathcal{H}^{\otimes m}}.
\end{aligned}$$

Lemma 12 *Let $h \in \mathcal{H}^{\odot n}$ and $h_1 \in \mathcal{H}$. Then,*

$$I_{n+1}(h \otimes h_1) = I_n(h) I_1(h_1) - n I_{n-1}(h \otimes_1 h_1).$$

Remark 23 *i) In this lemma, we deal with the case that $\mathcal{K} = \mathbb{R}$. Also $I_0(c) = c$ for any $c \in \mathbb{R}$ by convention. Also, as a consequence of the proof, this result is also true if we write $\tilde{I}_{n+1}$, $\tilde{I}_n$ and $\tilde{I}_1$ instead of I_{n+1}, I_n and I_1, respectively. Here, $\tilde{I}_n$ is the multiple integral of order n when, in Definition 21, we change $\{e_i : i \in \mathbb{N}\}$ and $\{g_i : i \in \mathbb{N}\}$ by the complete orthonormal systems $\{f_i : i \in \mathbb{N}\}$and $\{k_i : i \in \mathbb{N}\}$ of $\mathcal{H}$ and $\mathcal{K}$, respectively.*

ii) Using mathematical induction, it is not difficult to see that, for $h \in \mathcal{H}^{\odot n}$ and $h_1 \in \mathcal{H}^{\odot m}$,

$$I_n(h) I_m(h_1) = \sum_{r=0}^{m \wedge n} r! \binom{m}{r} \binom{n}{r} I_{n+m-2r}(h \otimes_r h_1)$$

holds. The proof is left to the reader as an exercise.

Proof. Let $h = \sum_{\alpha \in \Sigma_{2,n}} c_\alpha e(\alpha)$ and $h_1 = \sum_{\ell=1}^{\infty} d_\ell e_\ell$. Then, (1.97) gives that, in $L^1(\Omega)$, we have

$$\begin{aligned}
I_n(h) I_1(h_1) &= \left(\sum_{\alpha \in \Sigma_{2,n}} c_\alpha \prod_{i=1}^{\hat{\alpha}} H_{\alpha_i}\left(W(e_{j_i}), 1\right) \right) \sum_{\ell=1}^{\infty} d_\ell H_1\left(W(e_\ell, 1)\right) \\
&= \sum_{\substack{\alpha \in \Sigma_{2,n} \\ \ell \in \mathbb{N}}} c_\alpha d_\ell H_1\left(W(e_\ell, 1)\right) \prod_{i=1}^{\hat{\alpha}} H_{\alpha_i}\left(W(e_{j_i}), 1\right).
\end{aligned}$$

Therefore, we can apply (1.97) and (8.66) to write

$$\begin{aligned} I_n(h)I_1(h_1) &= I_{n+1}(h\otimes h_1) + \sum_{\substack{\alpha\in\Sigma_{2,n}\\ \ell\in\mathbb{N}}} c_\alpha d_\ell \sum_{i=1}^{\tilde{\alpha}} \alpha_i 1_{\{j_i\}}(\ell) \\ &\quad \times \left(\prod_{\substack{k=1\\ k\neq i}}^{\hat{\alpha}} H_{\alpha_k}\left(W(e_{j_k},1),1\right) \right) H_{\alpha_i-1}\left(W(e_{j_i}),1)\right) \\ &= I_{n+1}(h\otimes h_1) + n \sum_{\substack{\alpha\in\Sigma_{2,n}\\ \ell\in\mathbb{N}}} c_\alpha d_\ell I_{n-1}(e_\ell \otimes_1 e(\alpha)). \end{aligned}$$

Finally, (1.99) and the proof of Proposition 144 lead us to get

$$\begin{aligned} I_n(h)I_1(h_1) &= I_{n+1}(h\otimes h_1) + n\sum_{\alpha\in\Sigma_{2,n}} c_\alpha I_{n-1}(h_1\otimes_1 e(\alpha)) \\ &= I_{n+1}(h\otimes h_1) + nI_{n-1}(h_1\otimes_1 h). \end{aligned}$$

Now, the proof is complete. □

Lemma 13 *Let h, h_1 and r be as in Definition 22. Then, the contraction $h\otimes_r h_1$ is independent of the complete orthonormal system $\{e_i : i\in\mathbb{N}\}$.*

Proof. Let $\Xi_h : \mathcal{H}^{\otimes r}\to\mathcal{H}^{\otimes(n-r)}$ and $\Xi_{h_1} : \mathcal{H}^{\otimes r}\to\mathcal{H}^{\otimes(m-r)}$ be the bounded linear operators associated with h and h_1 defined in Proposition 144, respectively. Fix $\tilde{h}\in\mathcal{H}^{\otimes(n-r)}$ and $\tilde{h}_1\in\mathcal{H}^{\otimes(m-r)}$. Then, by Proposition 144, we obtain

$$\begin{aligned} &\left\langle h\otimes_r h_1, \tilde{h}\otimes\tilde{h}_1\right\rangle_{\mathcal{H}^{\otimes(n+m-2r)}} \\ &= \sum_{j_1,\dots,j_r=1}^{\infty} \left\langle \Xi_h\left(e_{j_1}\otimes\cdots\otimes e_{j_r}\right), \tilde{h}\right\rangle_{\mathcal{H}^{\otimes(n-r)}} \left\langle \Xi_{h_1}\left(e_{j_1}\otimes\cdots\otimes e_{j_r}\right), \tilde{h}_1\right\rangle_{\mathcal{H}^{\otimes r}} \\ &= \sum_{j_1,\dots,j_r=1}^{\infty} \left\langle e_{j_1}\otimes\cdots\otimes e_{j_r}, \Xi_h^*\left(\tilde{h}\right)\right\rangle_{\mathcal{H}^{\otimes r}} \left\langle e_{j_1}\otimes\cdots\otimes e_{j_r}, \Xi_{h_1}^*\left(\tilde{h}_1\right)\right\rangle_{\mathcal{H}^{\otimes r}} \\ &= \left\langle \Xi_h^*\left(\tilde{h}\right), \Xi_{h_1}^*\left(\tilde{h}_1\right)\right\rangle_{\mathcal{H}^{\otimes r}}, \end{aligned}$$

where Ξ_h^* and $\Xi_{h_1}^*$ are the adjoint operators of Ξ_h and Ξ_{h_1}, respectively. Finally, observe that the last equality is independent of the complete orthonormal system $\{e_i : i\in\mathbb{N}\}$, and therefore, the proof is finished. □

Notice that if (X,μ) is a measure space, $\mathcal{H} = L^2(X;\mu)$, $f\in L^2(X^n,\mu^{\otimes n})$ and $g\in L^2(X^m,\mu^{\otimes n})$, then, for $(s_1,\dots,s_{n+m-2r})\in X^{n+m-2r}$, we have

$$\begin{aligned} &(f\otimes_r g)(s_1,\dots,s_{n+n-2r}) \\ &= \int_{X^r} f(x_1,\dots,x_r,s_1,\dots,s_{n-r})g(x_1,\dots,x_r,s_{n-r+1},\dots,s_{n+m-2r})dx_1\cdots x_r. \end{aligned}$$

The proof of this fact is left to the reader as an exercise.

On the other hand, a particular case that satisfies Remark 22.*iv*) is the following result:

Lemma 14 *Let $\alpha\in\Sigma_{2,n}$. Then, $I_n\left(e(\alpha)\otimes g_\ell\right)$ is independent of the complete orthonormal systems $\{e_i : i\in\mathbb{N}\}$ and $\{g_i : i\in\mathbb{N}\}$.*

Proof. Let $\{f_i : i \in \mathbb{N}\}$ and $\{k_i : i \in \mathbb{N}\}$ be another two complete orthonormal systems of $\mathcal{H}$ and $\mathcal{K}$, respectively.

We first assume that $n = 1$. Then, in this case, $e(\alpha) = e_j$, for some $j \in \mathbb{N}$. Therefore, $e(\alpha) = \sum_{i=1}^{\infty} c_i f_i$ and $g_\ell = \sum_{i=1}^{\infty} d_i k_i$, with $c_i, d_i \in \mathbb{R}$, and consequently, $e(\alpha) \otimes g_\ell = \sum_{i,m=1}^{\infty} c_i d_m f_i \otimes k_m$. Thus, we only need to see that

$$I_1\left(e(\alpha) \otimes g_\ell\right) = W\left(e(\alpha)\right) g_\ell = \sum_{i,m=1}^{\infty} c_i d_m W\left(f_i\right) k_m, \quad \text{in } L^2(\Omega;\mathcal{K}).$$

holds to show that the result is satisfied for $n = 1$ and any Hilbert space $\mathcal{K}$ because the last term is the definition of $I_1\left(e(\alpha) \otimes g_\ell\right)$ using the complete orthonormal systems $\{f_i : i \in \mathbb{N}\}$ and $\{k_i : i \in \mathbb{N}\}$ of $\mathcal{H}$ and $\mathcal{K}$, respectively. Towards this end, we observe that the definition of a isonormal Gaussian process and (1.92) yield

$$\begin{aligned}
&\left| W\left(e(\alpha)\right) g_\ell - \sum_{i,m=1}^{\infty} c_i d_m W\left(f_i\right) k_m \right|^2_{L^2(\Omega;\mathcal{K})} \\
&= \left| W\left(e(\alpha)\right) - \sum_{i=1}^{\infty} c_i W\left(f_i\right) \right|^2_{L^2(\Omega)} = \lim_{n\to\infty} E\left(\left| W\left(e(\alpha)\right) - \sum_{i=1}^{n} c_i W\left(f_i\right) \right|^2 \right) \\
&= \lim_{n\to\infty} E\left(\left| W\left(e(\alpha) - \sum_{i=1}^{n} c_i f_i \right) \right|^2 \right) = \lim_{n\to\infty} \left| e(\alpha) - \sum_{i=1}^{n} c_i f_i \right|^2_{\mathcal{H}} = 0.
\end{aligned}$$

Now, we assume that the result is true for any k in $\{1, \ldots, n\}$ for some $n \in \mathbb{N}$. Consider $\alpha \in \Sigma_{2,n+1}$. Then, defining $\tilde{I}_k$ as the multiple Wiener-Itô integral of order k when we use the complete orthonormal systems $\{f_i : i \in \mathbb{N}\}$ and $\{k_i : i \in \mathbb{N}\}$ of $\mathcal{H}$ and $\mathcal{K}$, respectively, we have that Lemma 13 leads us to establish

$$\begin{aligned}
I_{n+1}(e(\alpha) \otimes g_\ell) &= g_\ell H_{\alpha_1}\left(W(e_{j_1}), 1\right) \prod_{i=2}^{\hat{\alpha}} H_{\alpha_i}\left(W(e_{j_i}), 1\right) \\
&= g_\ell \left[H_1\left(W(e_{j_1}), 1\right) H_{\alpha_1 - 1}\left(W(e_{j_1}), 1\right) \right. \\
&\quad \left. -(\alpha_i - 1) H_{\alpha_1 - 2}\left(W(e_{j_1}), 1\right) \right] \prod_{i=2}^{\hat{\alpha}} H_{\alpha_i}\left(W(e_{j_i}), 1\right) \\
&= I_1(e_{j_1} \otimes g_\ell) I_n\left(e\left((j_1, \alpha_1 - 1), (j_2, \alpha_2), \ldots, (j_{\hat{\alpha}}, \alpha_{\hat{\alpha}})\right)\right) \\
&\quad - I_{n-1}\left(\left(e_{j_1} \otimes_1 e\left((j_1, \alpha_1 - 1), (j_2, \alpha_2), \ldots, (j_{\hat{\alpha}}, \alpha_{\hat{\alpha}})\right)\right) \otimes g_\ell\right) \\
&= \tilde{I}_1(e_{j_1} \otimes g_\ell) \tilde{I}_n\left(e\left((j_1, \alpha_1 - 1), (j_2, \alpha_2), \ldots, (j_{\hat{\alpha}}, \alpha_{\hat{\alpha}})\right)\right) \\
&\quad - \tilde{I}_{n-1}\left(\left(e_{j_1} \otimes_1 e\left((j_1, \alpha_1 - 1), (j_2, \alpha_2), \ldots, (j_{\hat{\alpha}}, \alpha_{\hat{\alpha}})\right)\right) \otimes g_\ell\right).
\end{aligned}$$

Consequently, Remark 23.*i*) implies that $I_{n+1}(e(\alpha) \otimes g_\ell) = \tilde{I}_{n+1}(e(\alpha) \otimes g_\ell)$. Therefore, from induction on n, the result is true. □

Now, we are ready to state the following result.

Theorem 13 *Let $n \in \mathbb{N}$. Then, the definition of the multiple Wiener-Itô integral I_n given in (1.97) is independent of the complete orthonormal systems $\{e_i : i \in \mathbb{N}\}$ and $\{g_i : i \in \mathbb{N}\}$ of $\mathcal{H}$ and $\mathcal{K}$, respectively.*

Proof. For $h \in \mathcal{H}^{\odot n} \otimes \mathcal{K}$ such that

$$h = \sum_{\substack{\alpha \in \Sigma_{2,n} \\ \ell \in \mathbb{N}}} c_{\alpha,\ell}\, e(\alpha) \otimes g_\ell, \quad c_{\alpha,\ell} \in \mathbb{R},$$

Definition 21 allows us to write

$$I_n(h) = \sum_{\substack{\alpha\in\Sigma_{2,n}\\ \ell\in\mathbb{N}}} c_{\alpha,\ell}\left(\prod_{i=1}^{\hat{\alpha}} H_{\alpha_i}\left(W(e_{j_i}),1\right)\right) g_\ell = \sum_{\substack{\alpha\in\Sigma_{2,n}\\ \ell\in\mathbb{N}}} c_{\alpha,\ell} I_n\left(e(\alpha)\otimes g_\ell\right).$$

Hence, for $\tilde{I}_n$ given in Remark 23.*i*), we can apply Lemma 14, and Remarks 22.*iii*) and 22.*iv*) to deduce

$$I_n(h) = \sum_{\substack{\alpha\in\Sigma_{2,n}\\ \ell\in\mathbb{N}}} c_{\alpha,\ell}\tilde{I}_n\left(e(\alpha)\otimes g_\ell\right) = \tilde{I}_n(h) \quad \text{in } L^2(\Omega;\mathcal{K}).$$

It means, the theorem is satisfied. □

1.5.4 Derivative operator

We denote by $\mathcal{S}$ the family of $\mathbb{R}$-valued smooth random variables of the form

$$f\left(W(h_1),\ldots,W(h_n)\right), \tag{1.100}$$

with $h_i \in \mathcal{H}$, $i = 1,\ldots,n$, and $f \in C_b^\infty(\mathbb{R}^n)$ (i.e., f and all its partial derivatives are bounded). The derivative of F is the $\mathcal{H}$-valued random variable

$$DF = \sum_{i=1}^{n} \frac{\partial f}{\partial x_i}\left(W(h_1),\ldots,W(h_n)\right) h_i. \tag{1.101}$$

These definitions can be extended to $\mathcal{K}$-valued random variables as follows.

$\mathcal{S}(\mathcal{K})$ is the family of all $\mathcal{K}$-valued smooth random variables of the form

$$F = \sum_{i=1}^{n} F_i k_i, \tag{1.102}$$

with $F_i \in \mathcal{S}$, $k_i \in \mathcal{K}$ and $n \in \mathbb{N}$. In this case, for $p \geq 1$, the derivative operator $D : \mathcal{S}(\mathcal{K}) \subset L^p(\Omega;\mathcal{K}) \to L^p(\Omega;\mathcal{H}\otimes\mathcal{K})$ is defined by

$$DF = \sum_{i=1}^{n} DF_i \otimes k, \tag{1.103}$$

where F is given in (1.102) and the space $\mathcal{H}\otimes\mathcal{K}$ is introduced in Section 8.12 below. Note that this definition uses the chain rule since we will see that $D(W(h)) = h$, for $h \in \mathcal{H}$, in Remark 25.

The operator D is well-defined (i.e., DF is independent of the representation of F as a $\mathcal{K}$-valued smooth functional) due to W being a Gaussian family of random variables. The details of this fact is left to the reader. Moreover, for $p \geq 1$, the monotone class lemma (i.e., Theorem 97) implies that D is densely defined because $\mathcal{S}(\mathcal{K})$ is a dense set of $L^p(\Omega;\mathcal{K})$. Indeed, the family

$$\mathcal{G} = \left\{F \in L^p(\Omega) : \|F - F_n\|_{L^p(\Omega)} \to 0, \text{ for some sequence } \{F_n \in \mathcal{S} : n \in \mathbb{N}\}\right\}$$

is a linear space of random variables such that $1_A \in \mathcal{G}$, for any set A of the form $A = \cap_{i=1}^m [W(h_i) \in (a_i, b_i)]$, where $h_i \in \mathcal{H}$ and $a_i, b_i \in \mathbb{R}$, $a_i < b_i$, because Spivak [200] (Exercise

2-26) yields that, for $n \in \mathbb{N}$ and $a < b$, we can find a function $\psi : \mathbb{R} \to [0,1]$ in $C_b^\infty(\mathbb{R})$ such that

$$\psi(x) = \begin{cases} 1, & \text{if } x \in [a,b], \\ 0, & \text{if } x \notin \left[a - \frac{1}{n}, b + \frac{1}{n}\right]. \end{cases}$$

Now, it is easy to see that our claim is satisfied.

The following integration by parts formula is stated in order to show that the operator D given in (1.103) is closable, which allows us to work with the closed extension of it, also denote by D. It means, now a $\mathcal{K}$-valued random variable $F \in L^p(\Omega; \mathcal{K})$ is in the domain of this closed operator D if and only if there exists a sequence $\{F_n : n \in \mathbb{N}\}$ of $\mathcal{K}$-valued smooth random variables such that $F_n \to F$ in $L^p(\Omega; \mathcal{K})$ and $\{DF_n : n \in \mathbb{N}\}$ is a Cauchy sequence in $L^p(\Omega; \mathcal{H} \otimes \mathcal{K})$. Furthermore,

$$DF = \lim_{n\to\infty} DF_n \quad \text{in } L^p(\Omega; \mathcal{H} \otimes \mathcal{K}), \text{ as } n \to \infty.$$

Lemma 15 *Let $F \in \mathcal{S}$ and $h \in \mathcal{H}$. Then,*

$$E\left(\langle DF, h\rangle_{\mathcal{H}}\right) = E\left(FW(h)\right).$$

Remark 24 *Let F be given by (1.102) and $h \otimes g \in \mathcal{H} \otimes \mathcal{K}$. Then, this result implies*

$$E\left(\langle DF, h \otimes g\rangle_{\mathcal{H}\otimes\mathcal{K}}\right) = E\left(\langle F, W(h)g\rangle_{\mathcal{K}}\right) = E\left(\langle F, I_1(h \otimes g)\rangle_{\mathcal{K}}\right),$$

where I_1 is the multiple Wiener-Itô integral of order 1 introduced in Definition 21 (see also, Remark 22.i)).

Proof. From the Gram-Schmidt orthonormalization, we can assume without loss of generality that $h = h_1$ and $F = f(W(h_1), \dots, W(h_n))$, where $\{h_1, \dots, h_n\}$ is an orthonormal set of $\mathcal{H}$. Then, Proposition 4 and the integration by parts formula allow us to write

$$\begin{aligned} E\left(\langle DF, h\rangle_{\mathcal{H}}\right) &= (\sqrt{2\pi})^{-n/2} \int_{\mathbb{R}^n} \partial_{x_1} f(x) \exp\left(-\frac{1}{2}\sum_{i=1}^n x_i^2\right) dx \\ &= (\sqrt{2\pi})^{-n/2} \int_{\mathbb{R}^n} x_1 f(x) \exp\left(-\frac{1}{2}\sum_{i=1}^n x_i^2\right) dx = E\left(FW(h)\right). \end{aligned}$$

Thus, the result is satisfied. □

Now we are ready to prove that D given in (1.103) is a closable operator from $L^p(\Omega; \mathcal{K})$ into $L^p(\Omega; \mathcal{H} \otimes \mathcal{K})$, for all $p \geq 1$. That is, if $\{F_n \in \mathcal{S}(\mathcal{K}) : n \in \mathbb{N}\}$ is a sequence that goes to zero in $L^p(\Omega; \mathcal{K})$ and DF_n converges to an $\mathcal{H}\otimes\mathcal{K}$-valued random variable Y in $L^p(\Omega; \mathcal{H}\otimes\mathcal{K})$, then $Y = 0$ with probability 1.

Theorem 14 *The operator D introduced in (1.103) is closable from $L^p(\Omega; \mathcal{K})$ into $L^p(\Omega; \mathcal{H} \otimes \mathcal{K})$, for all $p \geq 1$.*

Proof. Fix $p \geq 1$. Let $\{F_n \in \mathcal{S}(\mathcal{K}) : n \in \mathbb{N}\}$ be a sequence such that it goes to zero in $L^p(\Omega; \mathcal{K})$ and DF_n converges to an $\mathcal{H} \otimes \mathcal{K}$-valued random variable Y in $L^p(\Omega; \mathcal{H} \otimes \mathcal{K})$, as $n \to \infty$. Also, let $G \in \mathcal{S}$, $h \otimes g \in \mathcal{H} \otimes \mathcal{K}$ and $n \in \mathbb{N}$. Then, (1.102), (1.103) and Remark 24 yield that $GF_n \in \mathcal{S}(\mathcal{K})$ and

$$\begin{aligned} & E\left(\langle Y, h \otimes g\rangle_{\mathcal{H}\otimes\mathcal{K}} G\right) \\ &= \lim_{n\to\infty} E\left(\langle DF_n, h \otimes g\rangle_{\mathcal{H}\otimes\mathcal{K}} G\right) \\ &= \lim_{n\to\infty} E\left(\langle D(GF_n), h \otimes g\rangle_{\mathcal{H}\otimes\mathcal{K}}\right) - \lim_{n\to\infty} E\left(\langle (DG) \otimes F_n, h \otimes g\rangle_{\mathcal{H}\otimes\mathcal{K}}\right) \\ &= \lim_{n\to\infty} E\left(G\langle F_n, W(h)g\rangle_{\mathcal{K}}\right) = 0 \end{aligned}$$

due to $\langle DG, h\rangle_{\mathcal{H}}$ and GF_n being bounded. Hence, the fact that the span of the set $\{Gh\otimes g : G \in \mathcal{S}$ and $h \otimes g \in \mathcal{H} \otimes \mathcal{K}\}$ is dense in $L^q(\Omega; \mathcal{H} \otimes \mathcal{K})$, with $p^{-1} + q^{-1} = 1$, implies that $Y = 0$, and therefore the proof is complete. □

The domain of the closed extension of D is denoted by $\mathbb{D}^{1,p}(\mathcal{K})$ (see Theorem 14). Henceforth, we identify the operator D with its closed extension. That is, it is also denoted by D. We remark again that it means that, for $p \geq 1$, a random variable $F \in L^p(\Omega; \mathcal{K})$ belongs to the space $\mathbb{D}^{1,p}(\mathcal{K})$ if and only if there exists a sequence $\{F_n : n \in \mathbb{N}\}$ of $\mathcal{K}$-valued smooth functionals such that $\|F - F_n\|_{L^p(\Omega;\mathcal{K})} \to 0$, as $n \to \infty$, and $\{DF_n : n \in \mathbb{N}\}$ is a Cauchy sequence in $L^p(\Omega; \mathcal{H} \otimes \mathcal{K})$, where DF_n is given in (1.103). In this case, $DF = \lim_{n\to\infty} DF_n$, in $L^p(\Omega; \mathcal{H} \otimes \mathcal{K})$. Consequently, we have that $\mathbb{D}^{1,p}(\mathcal{K})$ is the closure of the $\mathcal{K}$-valued smooth functionals $\mathcal{S}(\mathcal{K})$ with respect to the norm

$$||F||_{1,p,\mathcal{K}} = \left\{E\left(\|F\|_{\mathcal{K}}^p\right) + E\left(\|DF\|_{\mathcal{H}\otimes\mathcal{K}}^p\right)\right\}^{1/p}. \tag{1.104}$$

We use the convention $\mathbb{D}^{1,p} = \mathbb{D}^{1,p}(\mathbb{R})$ and $\|\cdot\|_{1,p} = \|\cdot\|_{1,p,\mathbb{R}}$, and we observe that $\mathbb{D}^{1,2}(\mathcal{K})$ is a real Hilbert space with inner product

$$\langle F, G\rangle_{1,2,\mathcal{K}} = E\left(\langle(F, G\rangle_{\mathcal{K}}\right) + E\left(\langle DF, DG\rangle_{\mathcal{H}\otimes\mathcal{K}}\right).$$

In general, for $n \in \mathbb{N}$, we can introduce the n-th iteration of D, represented by D^n, whose domain $\mathbb{D}^{n,p}(\mathcal{K})$ is the Sobolev space obtained as the closure of $\mathcal{S}(\mathcal{K})$ with respect to

$$||F||_{n,p,\mathcal{K}}^p = E(\|F\|_{\mathcal{K}}^p) + \sum_{j=1}^{n} E(||D^j F||_{\mathcal{H}^{\otimes j}\otimes\mathcal{K}}^p).$$

Now we can see that $W(h)$ belongs to $\mathbb{D}^{1,p}$, for all $h \in \mathcal{H}$ and, therefore, $W(h)g$ is in $D^{1,p}(\mathcal{K})$, for $g \in \mathcal{K}$. Indeed, set

$$\psi(x) = \begin{cases} 1, & \text{if } |x| \leq 1, \\ 0, & \text{if } |x| \geq 2 \end{cases}$$

and $\varphi_n(x) = \psi(x/n)x$, $x \in \mathbb{R}$. Then $\varphi_n(W(h)$ is an $\mathbb{R}$-valued smooth random variable belonging to $\mathbb{D}^{1,p}$, for all $p \geq 1$. Moreover, it is easy to get that $\varphi_n(W(h)$ goes to $W(h)$ and $D\varphi_n(W(h))$ goes to h, both in probability as $n \to \infty$. Thus, our claim holds since $|\varphi_n(W(h))| \leq |W(h)|$ and $|D\varphi_n(W(h))|_{\mathcal{H}} \leq (2\|\psi'\|_\infty + 1)\,|h|_{\mathcal{H}}$ because we can choose ψ bounded by 1 .

The derivative operator D also satisfies the following chain rule.

Proposition 21 *Let $\varphi \in C^1(\mathbb{R}^n)$ be a continuously differentiable function, $p \geq 1$, $g \in \mathcal{K}$ and $F = (F_1, \dots, F_n)$ a random vector whose components belong to $\mathbb{D}^{1,p}$. Then, $\varphi(F)g \in \mathbb{D}^{1,p}(\mathcal{K})$ if $\varphi(F) \in L^p(\Omega)$ and $\sum_{i=1}^n \partial_{x_i}\varphi(F)DF_i \in L^p(\Omega; \mathcal{H})$. Moreover, in this case,*

$$D\left(\varphi(F)g\right) = \sum_{i=1}^{n} \partial_{x_i}\varphi(F)(DF_i \otimes g).$$

Remark 25 *Besides (1.104), we also have other way to show if a random variable belongs to the space $\mathbb{D}^{1,p}(\mathcal{K})$, as a consequence of this proposition. For example, if*

$$F = f\left(W(h_1), \dots, W(h_n)\right) g, \quad \text{with } h_i \in \mathcal{H},\ g \in \mathcal{K} \text{ and } f \in C_p^\infty(\mathbb{R}^n)$$

(i.e., f and all its partial derivatives have polynomial growth), then F belongs to $\mathbb{D}^{1,p}(\mathcal{K})$ due to $W(h_i) \in L^p(\Omega)$, for all $p \geq 1$.

Proof. Note that we can assume that $\mathcal{K} = \mathbb{R}$ without loss of generality. We divide the proof into two steps.
Step 1: Here we assume that $\varphi \in C_b^1(\mathbb{R}^n)$. Let η be a function in $C^\infty(\mathbb{R}^n : \mathbb{R}_+)$ with compact support such that $\int_{\mathbb{R}^n} \eta(y)dy = 1$. Set $\eta_\varepsilon = \varepsilon^{-n}\eta(\cdot/\varepsilon)$. Then, it is well-known that

$$\eta_\varepsilon * \varphi(x) = \int_{\mathbb{R}^n} \eta_\varepsilon(x-y)\varphi(y)dy, \quad x \in \mathbb{R}^n,$$

is a function in $C_b^\infty(\mathbb{R}^n)$ such that $\eta_\varepsilon * \varphi(x)$ and $\frac{\partial \eta_\varepsilon * \varphi}{\partial x_i}(x)$ go to $\varphi(x)$ and $\frac{\partial \varphi}{\partial x_i}(x)$, respectively, as $\varepsilon \downarrow 0$, for $x \in \mathbb{R}^n$ and $i \in \{1, \ldots, n\}$. Moreover, we have

$$\|\eta_\varepsilon * \varphi\|_\infty \leq \|\varphi\|_\infty \quad \text{and} \left\| \frac{\partial \eta_\varepsilon * \varphi}{\partial x_i} \right\|_\infty \leq \left\| \frac{\partial \varphi}{\partial x_i} \right\|_\infty.$$

For $i \in \{1, \ldots, n\}$, consider a sequence $\{F_{i,k} \in \mathcal{S} : k \in \mathbb{N}\}$ of smooth functionals such that $\|F_{i,k} - F_i\|_{1,p} \to 0$, as $k \to \infty$. Then, the random variable $\eta_\varepsilon * \varphi(F_{1,k}, \ldots, F_{n,k})$ belongs to $\mathcal{S}$ and

$$D\eta_\varepsilon * \varphi(F_{1,k}, \ldots, F_{n,k}) = \sum_{i=1}^n \frac{\partial \eta_\varepsilon * \varphi}{\partial x_i}(F_{1,k}, \ldots, F_{n,k})DF_{i,k}, \quad k \in \mathbb{N}.$$

Now, it is easy to finish the proof in this case.
Step 2: Let $\psi : \mathbb{R} \to [0,1]$ be a function in $C^\infty(\mathbb{R}^n)$ such that

$$\psi(x) = \begin{cases} 1, & \text{if } |x| \leq 1, \\ 0, & \text{if } |x| \geq 2. \end{cases} \tag{1.105}$$

For $m \in \mathbb{N}$, $x \in \mathbb{R}$ and $y \in \mathbb{R}^n$, set $\psi_m(x) = \psi(x/m)$ and $\varphi_m(y) = \psi_m(\varphi(y))\varphi(y)\prod_{i=1}^n \psi_m(y_i)$. Then, from Step 1, we have that $\varphi_m(F)$ is in $\mathbb{D}^{1,p}$ with

$$\begin{aligned} D(\varphi_m(F)) &= \left(\psi'(\varphi(F)/m)\frac{\varphi(F)}{m} + \psi_m(\varphi(F))\right)\left(\prod_{i=1}^n \psi_m(F_i)\right)\sum_{i=1}^n \partial_{x_i}\varphi(F)DF_i \\ &\quad + \psi_m(\varphi(F))\frac{\varphi(F)}{m}\sum_{j=1}^n \psi'(F_j/m)(DF_j)\prod_{\substack{i=1 \\ i\neq j}}^n \psi_m(F_i). \end{aligned}$$

Hence $\varphi_m(F) \to \varphi(F)$ and $D(\varphi_m(F)) \to \sum_{i=1}^n \partial_{x_i}\varphi(F)DF_i$ both in probability, as $m \to \infty$.

Finally, noting that $|\varphi_m(F)| \leq |\varphi(F)|$ and

$$\|D(\varphi_m(F))\|_{\mathcal{H}} \leq C\left(\left\|\sum_{i=1}^n \partial_{x_i}\varphi(F)DF_i\right\|_{\mathcal{H}} + \sum_{i=1}^n \|DF_i\|_{\mathcal{H}}\right),$$

the result follows from the dominated convergence theorem. □

Now we establish an immediate consequence of Proposition 21.

Corollary 10 *Let $\varphi \in C^1(\mathbb{R}^n)$ be a bounded continuously differentiable function with bounded partial derivatives of order 1, $p \geq 1$, $g \in \mathcal{K}$ and $F = (F_1, \ldots, F_n)$ a random vector whose components belong to $\mathbb{D}^{1,p}$. Then, $\varphi(F) \in \mathbb{D}^{1,p}$ and*

$$D(\varphi(F)g) = \sum_{i=1}^n \partial_{x_i}\varphi(F)DF_i \otimes g. \tag{1.106}$$

In order to deal with the converse of Proposition 21, we need to analyze the local property of the derivative operator D. Towards this end, we study the following auxiliary result.

Lemma 16 *Let $p \geq 1$, $k \in \mathcal{K}$ and $F \in \mathbb{D}^{1,p}(\mathcal{K})$. Then, $\langle F, k\rangle_{\mathcal{K}}$ belongs to $\mathbb{D}^{1,p}$ and*

$$D\left(\langle F, k\rangle_{\mathcal{K}}\right) = D^*F(k).$$

Remark 26 *Here, D^*F is the adjoint of DF when this last $\mathcal{H}\otimes\mathcal{K}$-valued random variable is considered as the bounded linear operator from $\mathcal{H}$ to $\mathcal{K}$ introduced in Proposition 144.*

Proof. From Proposition 144, the result holds for $F \in \mathcal{S}(\mathcal{K})$. Thus, the lemma follows from the proof of Proposition 144 and (1.104), which establishes that there exists a sequence $\{F_n \in \mathcal{S}(\mathcal{K}) : n \in \mathbb{N}\}$ such that $||F - F_n||^p_{1,p,\mathcal{K}} \to 0$, as $n \to \infty$. Also note that (8.62) gives that the norm of the linear operator $D^*(F - F_n)$ is bounded by $\|D(F - F_n)\|_{\mathcal{H}\otimes\mathcal{K}}$. □

The following result is the local property for the derivative operator D.

Proposition 22 *Let $p \geq 1$ and $F \in \mathbb{D}^{1,p}(\mathcal{K})$. Then $1_{\{F=0\}}DF = 0$ with probability 1.*

Proof. We will divide the proof into two steps.
Step 1: Here, we assume that $\mathcal{K} = \mathbb{R}$.

For $\varepsilon > 0$, we use the convention $\psi_\varepsilon = \psi(\cdot/\varepsilon)$, where ψ is introduced in (1.105), and $\phi_\varepsilon(x) = \int_{-\infty}^x \psi_\varepsilon(y)dy$. Then, Corollary 10 yields that $\phi_\varepsilon(F)$ is in $\mathbb{D}^{1,p}$. Thus, we can choose a sequence $\{F_n \in \mathcal{S} : n \in \mathbb{N}\}$ such that $||\phi_\epsilon(F) - F_n||^p_{1,p} \to 0$, as $n \to \infty$. Also, consider the smooth $\mathcal{H}$-valued process

$$u = \sum_{i=1}^m G_i h_i, \quad \text{with } m \in \mathbb{N},\ G_i \in \mathcal{S} \text{ and } h_i \in \mathcal{H}. \tag{1.107}$$

Therefore, Lemma 15 leads us to deduce

$$\begin{aligned}
&E\left(\sum_{i=1}^m G_i \langle D(\psi(F_n)F_n), h_i\rangle_{\mathcal{H}}\right) \\
&= E\left\{\sum_{i=1}^m \left(\langle D(G_i\psi(F_n)F_n), h_i\rangle_{\mathcal{H}} - \psi(F_n)F_n\langle DG_i, h_i\rangle_{\mathcal{H}}\right)\right\} \\
&= E\left\{\sum_{i=1}^m \left(G_i\psi(F_n)F_nW(h_i) - \psi(F_n)F_n\langle DG_i, h_i\rangle_{\mathcal{H}}\right)\right\}.
\end{aligned}$$

Consequently, for $\varepsilon < 1/4$, we have that the fact that $\|\phi_\varepsilon\|_\infty \leq 4\varepsilon$, and letting $n \to \infty$, yield

$$\begin{aligned}
|E(\psi_\varepsilon(F)\langle DF, u\rangle_{\mathcal{H}})| &= \left|E\left(\sum_{i=1}^m G_i\langle D(\phi_\varepsilon(F)), h_i\rangle_{\mathcal{H}}\right)\right| \\
&= \left|E\left\{\sum_{i=1}^m \left(G_i\phi_\varepsilon(F)W(h_i) - \phi_\varepsilon(F)\langle DG_i, h_i\rangle_{\mathcal{H}}\right)\right\}\right| \\
&\leq 4\varepsilon E\left\{\sum_{i=1}^m \left(|G_iW(h_i)| + |\langle DG_i, h_i\rangle_{\mathcal{H}}|\right)\right\}.
\end{aligned}$$

Now, letting $\varepsilon \downarrow 0$, we get

$$E\left(1_{\{F=0\}}\langle DF, u\rangle_{\mathcal{H}}\right) = 0,$$

which is satisfied for any smooth process u of the form (1.107). So the result is true when $\mathcal{K} = \mathbb{R}$.

Step 2: Let $k \in \mathcal{K}$. Then, Lemma 16, together with Step 1 applied to $\langle F, k\rangle_{\mathcal{K}}$, implies

$$1_{\{F=0\}} D^* F(k) = 0, \quad \text{w.p.1.}$$

Hence, the fact that $\mathcal{K}$ is a separable space implies that the proof is complete. □

The local property for D (i.e., Proposition 22) allows us to extend its domain as follows.

Definition 23 *Let $p \geq 1$. A $\mathcal{K}$-valued random variable F belongs to $\mathbb{D}^{1,p}_{loc}(\mathcal{K})$ if there exists a sequence $\{(\Omega_n, F^{(n)}) \in \mathcal{F} \times \mathbb{D}^{1,p}(\mathcal{K}) : n \in \mathbb{N}\}$ such that*

i) $\Omega_n \uparrow \Omega$ w.p.1, as $n \to \infty$ (i.e., $\Omega_n \subset \Omega_{n+1}$ and $P(\Omega \setminus \cup_m \Omega_m) = 0$).

ii) $F = F^{(n)}$ w.p.1 on Ω_n.

In this case, we say that $\{(\Omega_n, F^{(n)}) : n \in \mathbb{N}\}$ localizes the random variable F in $\mathbb{D}^{1,p}(\mathcal{K})$ and define the $\mathcal{H} \otimes \mathcal{K}$-valued random variable DF as

$$DF = DF^{(n)} \quad \text{on } \Omega_n.$$

Proposition 22 gives that DF is well-defined. That is, the definition of the $\mathcal{H} \otimes \mathcal{K}$-valued random variable DF is independent of the sequence $\{(\Omega_n, F^{(n)}) : n \in \mathbb{N}\}$.

In order to give a random variable in $\mathbb{D}^{1,p}_{loc} \setminus \mathbb{D}^{1,p}$, we choose $h \in \mathcal{H}$, $q : \mathbb{R} \to \mathbb{R}$ a polynomial function of degree bigger or equal than 3 and $F = \exp(q(W(h))) - 1$, which is not in $L^p(\Omega)$ for any $p \geq 1$. But, for ψ introduced in (1.105),

$$F^{(n)} = \exp\left(\psi\left(q\left(W(h)\right)/n\right) q\left(W(h)\right)\right) - 1$$

is a smooth functional (i.e., it is in $\mathcal{S}$) such that $F = F^{(n)}$ on $[|q(W(h))| \leq n]$.

The necessity of Proposition 21 is also satisfied.

Proposition 23 *Let φ, g, p and F be as in Proposition 21. Also assume that $\varphi(F)g$ is in $\mathbb{D}^{1,p}(\mathcal{K})$. Then, the chain rule (1.106) is still satisfied.*

Proof. Let φ_m be as in Step 2 of the proof of Proposition 21. Then the local property of the derivative operator D and Corollary 10 yield

$$D\left(\varphi(F)g\right) = D\left(\varphi_m(F)g\right) = \sum_{i=1}^{n} \partial_{x_i}\left(\varphi(F) DF_i \otimes g\right)$$

on $\Omega_m := [|\varphi(F)| \leq m, |F_1| \leq m, \ldots, |F_n| \leq m]$. Thus, the result is true because $\Omega_m \uparrow \Omega$ with probability 1. □

Now, our aim is to characterize the space $\mathbb{D}^{1,2}(\mathcal{K})$ through the chaos decomposition given by (1.98). We recall that the notation utilized in the following result was introduced in Section 1.5.3.

Lemma 17 *Let $\{e_i : i \in \mathbb{N}\}$ be a complete orthonormal system of $\mathcal{H}$, $g \in \mathcal{K}$ and $\alpha \in \Sigma_{2,n}$. Then, $(I_n(e(\alpha) \otimes g) \in \mathbb{D}^{1,2}(\mathcal{K})$ and*

$$D\left(I_n\left(e(\alpha) \otimes g\right)\right) = nI_{n-1}\left(\sum_{i=1}^{\infty} \langle e(\alpha), e_i\rangle_{\mathcal{H}} \otimes (e_i \otimes g)\right).$$

Here, $\langle e(\alpha), e_i\rangle_{\mathcal{H}}$ stands for $\Xi_{e(\alpha)}(e_i)$, where $\Xi_{e(\alpha)}$ is the linear operator associated with $e(\alpha)$ defined in Proposition 144.

Proof. By Definition 21, we have that $I_n(e(\alpha)\otimes g) = \left(\prod_{i=1}^{\hat{\alpha}} H_{\alpha_i}(W(e_{j_i}),1)\right) g$. Thus, from Proposition 21, (8.65) and using the convention

$$\alpha(\ell)_k = \begin{cases} (\alpha_k, j_k), & \text{if } k \neq \ell \\ (\alpha_k - 1, j_k), & \text{otherwise,} \end{cases}$$

for $k = 1, \ldots, \hat{\alpha}$, we obtain

$$\begin{aligned} D(I_n(e(\alpha)\otimes g)) &= \sum_{\ell=1}^{\hat{\alpha}} \alpha_\ell H_{\alpha_\ell - 1}(W(e_{j_\ell}),1) \left(\prod_{\substack{i=1 \\ i\neq \ell}}^{\hat{\alpha}} H_{\alpha_i}(W(e_{j_i}),1) \right) (e_{j_\ell}\otimes g) \\ &= \sum_{\ell=1}^{\hat{\alpha}} \alpha_\ell I_{n-1}(e(\alpha(\ell)))(e_{j_\ell}\otimes g). \end{aligned}$$

Hence, the fact that $\{e_i : i \in \mathbb{N}\}$ is a complete orthonormal system of $\mathcal{H}$ implies

$$D(I_n(e(\alpha)\otimes g)) = n \sum_{\ell=i}^{\hat{\alpha}} I_{n-1}\left(\langle e(\alpha), e_\ell\rangle_{\mathcal{H}} \otimes (e_\ell \otimes g)\right).$$

Therefore, the result is satisfied. □

A consequence of Lemma 17 is the following two results.

Corollary 11 *Let $n \in \mathbb{N}$ and*

$$h = \sum_{\substack{\alpha\in\Sigma_{2,n} \\ \ell\in\mathbb{N}}} c_{\alpha,\ell}\, e(\alpha)\otimes g_\ell \tag{1.108}$$

in $\mathcal{H}^{\odot n}\otimes\mathcal{K}$. Then, $I_n(h)$ belongs to $\mathbb{D}^{1,2}(\mathcal{K})$,

$$D(I_n(h)) = nI_{n-1}\left(\sum_{i,\ell=1}^{\infty} \langle h^*(g_\ell), e_i\rangle_{\mathcal{H}} \otimes (e_i\otimes g_\ell) \right)$$

and

$$E\left(|D(I_n(h))|^2_{\mathcal{H}\otimes\mathcal{K}} \right) = nn!\,|h|^2_{\mathcal{H}^{\otimes n}\otimes\mathcal{K}}. \tag{1.109}$$

Proof. Let h_n be the first n summands of the series on the right-hand side of (1.108). Then, apply equality (1.97) and Lemma 17 to h_n. Finally, the result follows from Theorem 14, Proposition 144 and Remark 22.*ii*). □

Corollary 12 *Let $X \in L^2(\Omega;\mathcal{K})$ have the chaos decomposition*

$$X = E(X) + \sum_{n=1}^{\infty} I_n(h_n),$$

where $h_n \in \mathcal{H}^{\odot n}\otimes\mathcal{K}$, for all $n \geq 1$ and $\sum_{n=1}^{\infty} nn!\,|h_n|^2_{\mathcal{H}^{\otimes n}\otimes\mathcal{K}} < \infty$. Then, X belongs to $\mathbb{D}^{1,2}(\mathcal{K})$, and

$$DX = \sum_{n=1}^{\infty} nI_{n-1}\left(\sum_{i,\ell=1}^{\infty} \langle h_n^*(g_\ell), e_i\rangle_{\mathcal{H}} \otimes (e_i\otimes g_\ell) \right).$$

Proof. The result follows from Corollary 11 and Theorem 14, which establishes that D is a closed operator from $L^2(\Omega;\mathcal{K})$ into $L^2(\Omega;\mathcal{H}\otimes\mathcal{K})$. □

1.5.5 Derivative operator for a class of Gaussian processes

The purpose of this section is to prepare the basic tool that we need to analyze an extension of the divergence operator with respect to fractional Brownian motion with Hurst parameter $H < 1/2$. Namely, with the help of an isometry, we study a derivative-type operator. For details on the extension of the divergence operator, the reader has to study Section 4.3.1 below. For the convenience of the reader, we mention that part of the main tool to understand the results presented here is Section 8.12 below. As a consequence of this chapter, in Section 6.3.3, we will be able to introduce an extension of the Stratonovich integral with respect to fractional Brownian motion also with Hurst parameter $H < 1/2$. Using, these two extensions of integral, we can consider stochastic differential equations driven by an fBm with parameter $H < 1/2$, when the involved stochastic integral is either the extended divergence operator, or the extended Stratonovich integral.

Throughout this segment, we suppose that $\mathcal{H}$ and $\mathcal{H}_0$ are two real-separable Hilbert spaces with inner products $\langle \cdot, \cdot \rangle_{\mathcal{H}}$ and $\langle \cdot, \cdot \rangle_{\mathcal{H}_0}$, respectively. Furthermore, we assume that $\{W(h) : h \in \mathcal{H}\}$ is an isonormal Gaussian process on $\mathcal{H}$ defined on a complete probability space $(\Omega, \mathcal{F}, P)$, $\mathcal{H}$ is densely and continuously embedded in $\mathcal{H}_0$ and that $\mathcal{T} : \mathcal{H} \subset \mathcal{H}_0 \to \mathcal{H}_0$ is a linear operator (whose domain $\mathcal{D}(\mathcal{T})$ is $\mathcal{H}$ and its adjoint is denoted by $\mathcal{T}^*$) satisfying the following conditions:

(H1) $|\mathcal{T}h|_{\mathcal{H}_0} = |h|_{\mathcal{H}}$, for all $h \in \mathcal{H}$.

(H2) $\mathcal{T}_{\mathcal{H}} := \{h \in H : \mathcal{T}h \in \mathcal{D}(\mathcal{T}^*)\}$ is a dense subset of $\mathcal{H}$.

Note that (H1) is satisfied if and only if $\langle \mathcal{T}h_1, \mathcal{T}h_2 \rangle_{\mathcal{H}_0} = \langle h_1, h_2 \rangle_{\mathcal{H}}$, for $h_1, h_2 \in \mathcal{H}$.

Under these two hypotheses, the main purpose here is to introduce a derivative type operator $D_{\mathcal{T}}$ such that $D_{\mathcal{T}}F = \mathcal{T}^*\mathcal{T}DF$, for suitable random variables $F \in L^2(\Omega; \mathcal{K})$. Towards this end, remember that $C_p^\infty(\mathbb{R}^n)$ stands for the set of C^∞-functions $f : \mathbb{R}^n \to \mathbb{R}$ such that f and all its partial derivatives have polynomial growth. For a real separable Hilbert space $\mathcal{K}$, $I_{\mathcal{K}}$ denotes the identity operator on $\mathcal{K}$, $(\mathcal{T} \otimes I_{\mathcal{K}})^*$ is the adjoint of the operator $\mathcal{T} \otimes I_{\mathcal{K}} : \mathcal{H} \otimes \mathcal{K} \subset \mathcal{H}_0 \otimes \mathcal{K} \to \mathcal{H}_0 \otimes \mathcal{K}$ and $\mathcal{S}(\mathcal{K})$ (resp. $\mathcal{S}_{\mathcal{T}}(\mathcal{K})$) represents the class of smooth random variables of the form

$$F = f(W(g_1), \dots, W(g_n))k, \tag{1.110}$$

where $k \in \mathcal{K}$, $\{g_1, \dots, g_n\}$ is in $\mathcal{H}$ (resp. is in $\mathcal{T}_{\mathcal{H}}$) and $f \in C_p^\infty(\mathbb{R}^n)$. Also remember (see Remark 25) that the derivative of the smooth random variable F given by (1.110) is the $\mathcal{H} \otimes \mathcal{K}$-valued random variable DF defined by (1.101) and (1.103). That is,

$$DF = \sum_{i=1}^{n} \frac{\partial f}{\partial x_i}(W(g_1), \dots, W(g_n))g_i \otimes k.$$

Now, we consider the operator

$$D_{\mathcal{T}} : \mathcal{S}_{\mathcal{T}}(\mathcal{K}) \subset L^2(\Omega; \mathcal{K}) \to L^2(\Omega; \mathcal{H}_0 \otimes \mathcal{K})$$

introduced as

$$D_{\mathcal{T}}(F) = (\mathcal{T} \otimes I_{\mathcal{K}})^*(\mathcal{T} \otimes I_{\mathcal{K}})DF, \quad F \in \mathcal{S}_{\mathcal{T}}(\mathcal{K}). \tag{1.111}$$

In particular, observe that $D_{\mathcal{T}}F = \mathcal{T}^*\mathcal{T}DF$ in the case that $\mathcal{K} = \mathbb{R}$.

As in Theorem 14, our aim is to show that $D_{\mathcal{T}}$ has a closed extension from $L^2(\Omega; \mathcal{K})$ into $L^2(\Omega; \mathcal{H}_0 \otimes \mathcal{K})$. This closed extension is also denoted by $D_{\mathcal{T}}$ and its domain is represented by $\mathbb{D}^{1,2}_{\mathcal{T}}(\mathcal{K})$, which is a dense set of $L^2(\Omega;\mathcal{K})$ due to (H2). It means that $\mathbb{D}^{1,2}_{\mathcal{T}}(\mathcal{K})$ is the completion of the $\mathcal{K}$-valued smooth random variables $\mathcal{S}_{\mathcal{T}}(\mathcal{K})$ with respect to the norm

$$||F||^2_{1,2,\mathcal{T}} = E\left(|F|^2_{\mathcal{K}} + |(\mathcal{T}\otimes I_{\mathcal{K}})^*(\mathcal{T}\otimes I_{\mathcal{K}})DF|^2_{\mathcal{H}_0\otimes\mathcal{K}}\right).$$

On the other hand, note that (H1) gives that $\mathcal{T}$ is a closed operator on $\mathcal{H}_0$. Therefore $\mathcal{D}(T^*)$ is also a dense subset of $\mathcal{H}_0$ (see [182], Theorem VIII.1).

In the remaining of this section, we suppose that Conditions (H1) and (H2) hold.

Lemma 18 *Assume that Hypotheses (H1) and (H2) are satisfied. Then, the operator $D_{\mathcal{T}}$ defined in (1.111) is closable from $L^2(\Omega;\mathcal{K})$ into $L^2(\Omega;\mathcal{H}_0\otimes\mathcal{K})$.*

Proof. Let $\{F_n : n\in\mathbb{N}\}\subset\mathcal{S}_{\mathcal{T}}(\mathcal{K})$ be a sequence that goes to zero in $L^2(\Omega;\mathcal{K})$ such that $D_{\mathcal{T}}(F_n)$ converges to Y in $L^2(\Omega;\mathcal{H}_0\otimes\mathcal{K})$. Then (H1) and Lemma 15 imply that for every $F\in\mathcal{S}_{\mathcal{T}}$, $h\in\mathcal{H}$ and $k\in\mathcal{K}$,

$$\begin{aligned}
E\langle Y, Fh\otimes k\rangle_{\mathcal{H}_0\otimes\mathcal{K}} &= \lim_{n\to\infty} E\langle(\mathcal{T}\otimes I_{\mathcal{K}})^*(\mathcal{T}\otimes I_{\mathcal{K}})DF_n, Fh\otimes k\rangle_{\mathcal{H}_0\otimes\mathcal{K}} \\
&= \lim_{n\to\infty}[E\langle(\mathcal{T}\otimes I_{\mathcal{K}})^*(\mathcal{T}\otimes I_{\mathcal{K}})D(F_nF), h\otimes k\rangle_{\mathcal{H}_0\otimes\mathcal{K}} \\
&\quad -E\langle(\mathcal{T}\otimes I_{\mathcal{K}})^*(\mathcal{T}\otimes I_{\mathcal{K}})DF\otimes F_n, h\otimes k\rangle_{\mathcal{H}_0\otimes\mathcal{K}}] \\
&= \lim_{n\to\infty}[E\langle(\mathcal{T}\otimes I_{\mathcal{K}})D(F_nF), (\mathcal{T}\otimes I_{\mathcal{K}})h\otimes k\rangle_{\mathcal{H}_0\otimes\mathcal{K}} \\
&\quad -E\langle(\mathcal{T}^*\mathcal{T}DF)\otimes F_n, h\otimes k\rangle_{\mathcal{H}_0\otimes\mathcal{K}}] \\
&= \lim_{n\to\infty}[E\langle D(F_nF), h\otimes k\rangle_{\mathcal{H}\otimes\mathcal{K}} - E\langle(\mathcal{T}^*\mathcal{T}DF)\otimes F_n, h\otimes k\rangle_{\mathcal{H}_0\otimes\mathcal{K}}] \\
&= \lim_{n\to\infty} E\left\{\langle F_n, k\rangle_{\mathcal{K}}\left(FW(h) - \langle\mathcal{T}^*\mathcal{T}DF, h\rangle_{\mathcal{H}_0}\right)\right\} = 0,
\end{aligned}$$

because $F_n\to 0$ in $L^2(\Omega;\mathcal{K})$. Hence $Y=0$ since $\mathcal{S}_{\mathcal{T}}$ is dense in $L^2(\Omega)$ due to Hypothesis (H2). Thus, the result is satisfied. □

Remember that the closure of the operator $D_{\mathcal{T}}$ given by (1.111) (with domain $\mathbb{D}^{1,2}_{\mathcal{T}}(\mathcal{K})$) will be also represented by $D_{\mathcal{T}}$.

The relation between operators D and $D_{\mathcal{T}}$ is established in the following two results.

Lemma 19 *Suppose that Hypotheses (H1) and (H2) hold, and let $F\in\mathbb{D}^{1,2}_{\mathcal{T}}(\mathcal{K})$. Then $F\in\mathbb{D}^{1,2}(\mathcal{K})$.*

Proof. We first choose $F\in\mathcal{S}_{\mathcal{T}}(\mathcal{K})$ having the form $F=Gk$, with $G\in\mathcal{S}_{\mathcal{T}}$ and $k\in\mathcal{K}$. Therefore, (H1) gives

$$\begin{aligned}
|DF|^2_{\mathcal{H}\otimes\mathcal{K}} &= \langle DG\otimes k, DG\otimes k\rangle_{\mathcal{H}\otimes\mathcal{K}} = \langle DG\otimes k, \mathcal{T}^*\mathcal{T}DG\otimes k\rangle_{\mathcal{H}_0\otimes\mathcal{K}} \\
&\le C|DF|_{\mathcal{H}\otimes\mathcal{K}}|D_{\mathcal{T}}F|_{\mathcal{H}_0\otimes\mathcal{K}},
\end{aligned}$$

due to $\mathcal{H}$ being continuously embedded in $\mathcal{H}_0$. Thus, we have shown

$$|DF|_{\mathcal{H}\otimes\mathcal{K}} \le C|D_{\mathcal{T}}F|_{\mathcal{H}_0\otimes\mathcal{K}}. \qquad (1.112)$$

Now, let F belong to the space $\mathbb{D}^{1,2}_{\mathcal{T}}(\mathcal{K})$. Then, there exists a sequence $\{F_n\in\mathcal{S}_{\mathcal{T}}(\mathcal{K}) : n\in\mathbb{N}\}$ such that $\|F-F_n\|_{1,2,\mathcal{T}}$ converges to zero, as $n\to\infty$. Hence, $F_n\to F$ in $L^2(\Omega;\mathcal{K})$ and (1.112) implies that $\{DF_n : n\in\mathbb{N}\}$ is a Cauchy sequence in $L^2(\Omega;\mathcal{H}\otimes\mathcal{K})$. Consequently, the proof is finished because of Theorem 14. □

Lemma 20 *Let $F \in \mathbb{D}_{\mathcal{T}}^{1,2}(\mathcal{K})$. Then, under (H1) and (H2), we have that $(\mathcal{T} \otimes I_{\mathcal{K}})DF$ belongs to* Dom $(\mathcal{T} \otimes I_{\mathcal{K}})^*$ *and*

$$D_{\mathcal{T}}F = (\mathcal{T} \otimes I_{\mathcal{K}})^*(\mathcal{T} \otimes I_{\mathcal{K}})DF, \quad w.p.1.$$

Proof. Let $\{F_n : n \in \mathbb{N}\} \subset \mathcal{S}_{\mathcal{T}}(\mathcal{K})$ be a sequence that converges to F in $\mathbb{D}_{\mathcal{T}}^{1,2}(\mathcal{K})$ as $n \to \infty$. Then, using (H1), we obtain

$$E\left(|(\mathcal{T} \otimes I_{\mathcal{K}})(DF_n - DF)|^2_{\mathcal{H}_0 \otimes \mathcal{K}}\right) = E\left(|DF_n - DF|^2_{\mathcal{H} \otimes \mathcal{K}}\right) \to 0 \quad \text{as} \quad n \to \infty, \tag{1.113}$$

Indeed, the convergence to zero of the right-hand side of (1.113) follows from the proof of Lemma 19 (see (1.112)) and, therefore, the equality is a consequence of the facts that $\mathcal{T} \otimes I_{\mathcal{K}}$ is a closed operator on $\mathcal{H}_0 \otimes \mathcal{K}$ and that $\mathcal{H}$ is continuously embedded in $\mathcal{H}_0$. Moreover, we have

$$E\left(|(\mathcal{T} \otimes I_{\mathcal{K}})^*(\mathcal{T} \otimes I_{\mathcal{K}})DF_n - D_{\mathcal{T}}F|^2_{\mathcal{H}_0 \otimes \mathcal{K}}\right) \to 0 \quad \text{as} \quad n \to \infty. \tag{1.114}$$

Finally, the result is satisfied since (1.113), (1.114) and $(\mathcal{T} \otimes I_{\mathcal{K}})^*$ is a closed operator (see [182], Theorem VIII.1). □

We will consider the local property of the extension of the divergence operator in Section 4.3.1. To do so, we state the following result, whose proof uses standard arguments (compare it with the proof of Proposition 66). So we only sketch it.

Lemma 21 *Let $\phi \in C^\infty(\mathbb{R})$ be a function with compact support and $u \in \mathbb{D}_{\mathcal{T}}^{1,2}(\mathcal{K})$. Then $\phi(|u|^2_{\mathcal{K}})$ belongs to $\mathbb{D}_{\mathcal{T}}^{1,2}$ and*

$$D_{\mathcal{T}}(\phi(|u|^2_{\mathcal{K}})) = 2\phi'(|u|^2_{\mathcal{K}})\Big((\mathcal{T} \otimes I_{\mathcal{K}})^*(\mathcal{T} \otimes I_{\mathcal{K}})Du\Big)^*(u).$$

Proof. It is easy to see that the result holds for $u \in \mathcal{S}_{\mathcal{T}}(\mathcal{K})$, using only the definition of $D_{\mathcal{T}}$ for smooth $\mathcal{K}$–valued random variables.

Now let $\{F_n : n \in \mathbb{N}\} \subset \mathcal{S}_{\mathcal{T}}(\mathcal{K})$ be a sequence such that $||F_n - u||_{1,2,\mathcal{T}} \to 0$, as $n \to \infty$. Hence

$$\begin{aligned}
&\left|\phi'(|u|^2_{\mathcal{K}})\Big((\mathcal{T} \otimes I_{\mathcal{K}})^*(\mathcal{T} \otimes I_{\mathcal{K}})Du\Big)^*(u) - \phi'(|F_n|^2_{\mathcal{K}})\Big((\mathcal{T} \otimes I_{\mathcal{K}})^*(\mathcal{T} \otimes I_{\mathcal{K}})DF_n\Big)^*(F_n)\right|_{\mathcal{H}_0} \\
&\leq \; |\phi'(|u|^2_{\mathcal{K}})||u|_{\mathcal{K}}\,|(\mathcal{T} \otimes I_{\mathcal{K}})^*(\mathcal{T} \otimes I_{\mathcal{K}})(Du - DF_n)|_{\mathcal{H}_0 \otimes \mathcal{K}} \\
&\quad +|\phi'(|u|^2_{\mathcal{K}})|\,|(\mathcal{T} \otimes I_{\mathcal{K}})^*(\mathcal{T} \otimes I_{\mathcal{K}})DF_n|_{\mathcal{H}_0 \otimes \mathcal{K}}|u - F_n|_{\mathcal{K}} \\
&\quad +|\phi'(|u_{\mathcal{K}}|^2) - \phi'(|F_n|^2_{\mathcal{K}})|\,|(\mathcal{T} \otimes I_{\mathcal{K}})^*(\mathcal{T} \otimes I_{\mathcal{K}})DF_n|_{\mathcal{H}_0 \otimes \mathcal{K}}|F_n|_{\mathcal{K}} \to 0 \quad \text{in probability.}
\end{aligned}$$

Finally, the facts that $D_{\mathcal{T}}$ is a closed operator and there is a constant $C > 0$ such that

$$|\phi'(|F_n|^2_{\mathcal{K}})|\left|\Big((\mathcal{T} \otimes I_{\mathcal{K}})^*(\mathcal{T} \otimes I_{\mathcal{K}})DF_n\Big)^*(F_n)\right|_{\mathcal{H}_0} \leq C\,|(\mathcal{T} \otimes I_{\mathcal{K}})^*(\mathcal{T} \otimes I_{\mathcal{K}})DF_n|_{\mathcal{H}_0 \otimes \mathcal{K}},$$

together with the dominated convergence theorem, yield that the result is true. □

1.5.6 Fractional Brownian motion case

Let $B = \{B_t : t \in [0, T]\}$ be a fractional Brownian motion with Hurst parameter $H \in (0, 1)$ as in Definition 17. Then, Theorem 9 and (1.90) imply that there exists a Brownian motion $W = \{W_t : t \in [0, T]\}$ such that

$$B(\varphi) = \int_0^T (K^*\varphi)(s)dW_s, \quad \varphi \in \mathcal{H}, \tag{1.115}$$

where the right-hand side is the Wiener integral with respect to W and K^* is introduced in (1.73) and (1.87). Hence, when we deal with fractional Brownian motion, we can consider the derivative operators D and D^W with respect to B and W, whose domains, for $p \geq 1$, are denoted by $\mathbb{D}^{1,p}$ and $\mathbb{D}^{1,p}_W$, respectively. The relation between these two operators is given in the following result.

Proposition 24 *Let $H \in (0,1)$ and $p \geq 1$. Then, $\mathbb{D}^{1,p} = \mathbb{D}^{1,p}_W$ and*

$$D^W F = K^*(DF), \quad F \in \mathbb{D}^{1,p}.$$

Proof. Let F be an $\mathbb{R}$-valued smooth functional of B (see (1.100)). That is, there exists $f : \mathbb{R}^n \to \mathbb{R}$ in $C_b^\infty(\mathbb{R}^n)$ and $\varphi_1, \dots, \varphi_n \in \mathcal{H}$ such that

$$F = f(B(\varphi_1), \dots, B(\varphi_n)). \tag{1.116}$$

Thus, (1.115) implies

$$F = f\left(W(K^*(\varphi_1)), \dots, W(K^*(\varphi_n))\right). \tag{1.117}$$

Therefore, we have that F is also an $\mathbb{R}$-valued smooth functional of the Brownian motion W. Similarly, let G be an $\mathbb{R}$-valued smooth functional of W of the form

$$G = g\left(W(h_1), \dots, W(h_m)\right) = g\left(B((K^*)^{-1}h_1), \dots, B((K^*)^{-1}h_m)\right),$$

where last equality also follows from (1.115). In this way, we have proven that the family of $\mathbb{R}$-valued smooth functional of B agrees with that of W. Moreover, (1.116) and (1.117) yield that $D^W F = K^*(DF)$, which, together with the fact that

$$\|DF\|^p_{\mathcal{H}} = \|K^*(DF)\|^p_{L^2([0,T])} = \|D^W F\|^p_{L^2([0,T])},$$

gives that the result holds (see (1.43) and (1.104)). □

1.5.6.1 Case $H < 1/2$

Here, the aim is to see that fractional Brownian motion is a Gaussian process satisfying Hypotheses (H1) and (H2) in Section 1.5.5. So, throughout this section, $B = \{B_t : t \in [0,T]\}$ is a fractional Brownian motion (fBm) with Hurst parameter $H \in (0, 1/2)$.

From Theorem 8, we have already known that fBm B is a Gaussian process on the Hilbert space

$$\mathcal{H} = \left\{ f : [0,T] \to \mathbb{R} : \exists \phi_f \in L^2([0,T]) \text{ s.t. } f(u) = u^\alpha (I^\alpha_{T-}(s^{-\alpha}\phi_f(s)))(u) \right\}$$

equipped with the inner product

$$\langle f, g \rangle_{\mathcal{H}} = \tilde{c}_H^2 \Gamma(H + \frac{1}{2})^2 \langle \phi_f, \phi_g \rangle_{L^2([0,T])}. \tag{1.118}$$

Here $\alpha = \frac{1}{2} - H$, $\tilde{c}_H = \sqrt{\frac{2H}{(1-2H)\beta(1-2H,H+1/2)}}$ and I^α_{T-} is the right-sided fractional Riemann–Liouville integral of order α given by (1.56) (see Theorem 8 and (1.67)).

Concerning Section 1.5.5, the following result yields that $L^2([0,T])$ plays the role of the Hilbert space $\mathcal{H}_0$.

Proposition 25 *The space $\mathcal{H}$ is densely and continuously embedded in $L^2([0,T])$.*

Proof. Let $f \in \mathcal{H}$. Then, the Hölder inequality and Fubini's theorem imply that there exist $\phi_f \in L^2([0,T])$ and a constant $C_{\alpha,T} > 0$ depending on α and T such that

$$\begin{aligned}\int_0^T (f(u))^2 du &\leq C_\alpha \int_0^T \left(\int_u^T (r-u)^{\alpha-1}\phi_f(r)dr\right)^2 du \\ &\leq C_{\alpha,T} \int_0^T \int_u^T (r-u)^{\alpha-1}\phi_f(r)^2 drdu = C_{\alpha,T}\int_0^T \phi_f(u)^2 du,\end{aligned}$$

which implies that $\mathcal{H}$ is continuously embedded in $L^2([0,T])$ because of the definition of the inner product in $\mathcal{H}$.

Finally, from (1.74), we have that the step functions are included in $\mathcal{H}$. Thus, the proof is finished. □

Now, for $\mathcal{H}_0 = L^2([0,T])$, we introduce the linear operator $\mathcal{T} : \mathcal{H} \subset \mathcal{H}_0 \to \mathcal{H}_0$ defined by

$$(\mathcal{T}f)(u) = \tilde{c}_H \Gamma(H + \frac{1}{2}) u^\alpha D^\alpha_{T-}(s^{-\alpha}f(s))(u) \tag{1.119}$$

$$= \tilde{c}_H \Gamma(H + \frac{1}{2}) \phi_f(u), \tag{1.120}$$

where D^α_{T-} is the right-sided fractional derivative of order α (see Section 1.4.1), which is the inverse of I^α_{T-}. Note that the operator $\mathcal{T}$ is nothing else than K^* since equality (1.76). In this way, we have proven that the operator $\mathcal{T}$ satisfies Hypothesis (H1) in Section 1.5.5.

In order to deal with (H2) in Section 1.5.5, we recall you that D^α_{0+} and I^α_{0+} denote the left-sided fractional derivative and integral of order α, respectively. These operators are introduced in Section 1.4.1.

In the following result, we figure out the operator $\mathcal{T}^*$.

Proposition 26 *Let $g : [0,T] \to \mathbb{R}$ be a function such that $u \mapsto u^\alpha g(u)$ belongs to $I^\alpha_{0+}(L^q([0,T]))$ for some $q > \alpha^{-1} \vee H^{-1}$. Then, $g \in$ Dom $\mathcal{T}^*$ and*

$$(\mathcal{T}^*g)(u) = \tilde{c}_H \Gamma(H + \frac{1}{2}) u^{-\alpha} D^\alpha_{0+}(s^\alpha g(s))(u), \quad u \in [0,T]. \tag{1.121}$$

Proof. We first observe that the fact that $q > H^{-1}$ implies that the right-hand side of (1.121) is in $L^2([0,T])$ due to Hölder inequality.

Secondly, $q > \alpha^{-1}$, Corollary 6 and (1.119) imply

$$\int_0^T (\mathcal{T}f)(u)g(u)du = \tilde{c}_H \Gamma(H + \frac{1}{2}) \int_0^a u^{-\alpha} f(u) D^\alpha_{0+}(s^\alpha g(s))(u)du.$$

Therefore, $g \in \text{Dom}\mathcal{T}^*$ and (1.121) holds. □

The main result of this section is the following:

Theorem 15 *The operator $\mathcal{T}$ given by (1.119) satisfies Conditions (H1) and (H2).*

Proof. We have already seen that (H1) holds because of the definitions of the operator $\mathcal{T}$ and the inner product in $\mathcal{H}$ (see (1.120) and (1.118)).

In order to see that (H2) is satisfied, we introduce space

$$\mathcal{H}_* = \{f \in \mathcal{H} : \exists f^* \in L^\infty([0,T]) \text{ s.t. } \phi_f(u) = u^{-\alpha} I^\alpha_{0+}(s^\alpha f^*(s))(u)\}.$$

We claim that $\mathcal{H}_*$ is a dense set of $\mathcal{H}$. Indeed, we only need to show that the family.

$$L^2_* = \{f : [0,T] \to \mathbb{R} : \exists f^* \in L^\infty([0,T]) \text{ s.t. } f(u) = u^{-\alpha} I^\alpha_{0+}(s^\alpha f^*(s))(u)\}$$

is a dense subset of $L^2([0,T])$. For this, let $g \in L^2([0,T])$ such that for any $f^* \in L^\infty([0,T])$,

$$0 = \int_0^T g(u)u^{-\alpha} I_{0+}^\alpha (s^\alpha f^*(s))(u)du.$$

Hence, by Corollary 6, we obtain that $0 = \int_0^T I_{T-}^\alpha (s^{-\alpha} g(s))(u)u^\alpha f^*(u)du$. Note that, as a consequence of (1.56) and Fubini's theorem, we get that the function $u \mapsto I_{T-}^\alpha (s^{-\alpha} g(s))(u)u^\alpha$ belongs to $L^2([0,T])$. Therefore, Proposition 135 leads us to conclude that $g = 0$ because $L^\infty([0,T])$ is dense in $L^2([0,T])$, and therefore L^2_* is dense in $L^2([0,T])$.

Finally, Proposition 26 gives $\mathcal{H}_* \subset \mathcal{T}_\mathcal{H}$ and the result is proven. That is, (H2) also holds. □

1.5.6.2 Case $H > 1/2$

In this section, we suppose that the Hurst parameter H is bigger than one half.

Remember that the reproducing kernel Hilbert space $\mathcal{H}$ is the completion of the family of step function $\mathcal{E}$ on $[0,T]$ with respect to the inner product

$$\langle 1_{[0,t]}, 1_{[0,s]} \rangle_\mathcal{H} = R_H(t,s) = H(2H-1)\int_0^t \int_0^s |r-u|^{2H-2} drdu, \quad s,t \in [0,T],$$

where last equality is established in (1.69) and R_H is give in (1.44). In consequence, we have

$$\langle \varphi, \psi \rangle_\mathcal{H} = H(2H-1)\int_0^T \int_0^T |r-u|^{2H-2} \varphi(r)\psi(u)drdu, \quad \varphi, \psi \in \mathcal{E}([0,T]). \tag{1.122}$$

In this case (i.e., $H > 1/2$), we see that $L^2([0,T]) \subset \mathcal{H}$ in Subsection 1.4.2.2. So, we cannot guarantee that all elements of $\mathcal{H}$ are function on the interval $[0,T]$ (see Pipiras and Taqqu [176]). Thus, last identity is not possible to be satisfied for all $\varphi, \psi \in \mathcal{H}$. However, we can find a suitable space of measurable functions on $[0,T]$ contained in $\mathcal{H}$. Namely, the space $|\mathcal{H}|$ of all measurable functions $\varphi : [0,T] \to \mathbb{R}$ such that

$$||\varphi||^2_{|\mathcal{H}|} = H(2H-1)\int_0^T \int_0^T |\varphi(r)|\,|\varphi(s)|\,|r-s|^{2H-2} drds < \infty. \tag{1.123}$$

Here, as it is usual, $\varphi = \psi$ in $|\mathcal{H}|$ if and only if $\varphi = \psi$ almost surely.

Proposition 27 *Let $H > 1/2$. Then, the space $(|\mathcal{H}|, ||\cdot||_{|\mathcal{H}|})$ is a Banach one, the class $\mathcal{E}([0,T])$ of all the step functions defined on $[0,T]$ is dense in it and $|\mathcal{H}| \subset \mathcal{H}$.*

Proof. We first show that $||\cdot||_{|\mathcal{H}|}$ is a norm on $|\mathcal{H}|$. In order to deal with the triangle inequality, we observe that the Fubini's theorem, Proposition 20, (1.66) and (1.44) imply that, for $\varphi \in |\mathcal{H}|$,

$$\begin{aligned}
&H(2H-1)\int_0^T \int_0^T |\varphi(r)|\,|\varphi(s)|\,|r-s|^{2H-2} drds \\
&= \int_0^T \int_0^T \frac{\partial^2 R_H}{\partial r \partial s}(r,s)\,|\varphi(r)|\,|\varphi(s)|\,drds \\
&= \int_0^T \int_0^T \left(\int_0^{r\wedge s} \frac{\partial K(s,u)}{\partial s} \frac{\partial K(r,u)}{\partial r} du \right) |\varphi(r)|\,|\varphi(s)|\,drds \\
&= \int_0^T \left(\int_u^T |\varphi(r)| \frac{\partial K(r,u)}{\partial r} dr \right) \left(\int_u^T |\varphi(s)| \frac{\partial K(s,u)}{\partial s} ds \right) du \\
&= \int_0^T \left(\int_u^T |\varphi(r)| \frac{\partial K(r,u)}{\partial r} ds \right)^2 du.
\end{aligned} \tag{1.124}$$

Hence, Definition (1.123) allows to see easily that $\|\cdot\|_{|\mathcal{H}|}$ satisfies the triangle inequality. That is, it is not difficult to get that $\|\cdot\|_{|\mathcal{H}|}$ is a norm on $|\mathcal{H}|$ due to (1.123).

Now, in order to see that $(|\mathcal{H}|, \|\cdot\|_{|\mathcal{H}|})$ is a complete space, we consider a Cauchy sequence $\{\varphi_n : n \in \mathbb{N}\}$ in $(|\mathcal{H}|, \|\cdot\|_{|\mathcal{H}|})$ and the inequality

$$\begin{aligned}\|\varphi\|^2_{L^1([0,T])} &= \int_0^T \int_0^T |\varphi(s)|\,|\varphi(r)|\,dsdr \\ &\leq (T)^{2-2H} \int_0^T \int_0^T |\varphi(s)|\,|\varphi(r)|\,|r-s|^{2H-2}\,dsdr,\end{aligned}$$

which implies that $\{\varphi_n : n \in \mathbb{N}\}$ is also a Cauchy sequence in $L^1([0,T])$. Consequently, there exists a subsequence $\{n_k : k \in \mathbb{N}\}$ of $\mathbb{N}$ and $\varphi \in L^1([0,T])$ such that φ_{n_k} goes to φ almost surely, as $k \to \infty$. The function φ also belongs to the space $|\mathcal{H}|$ since the Fatou's lemma allows to show

$$\|\varphi\|_{|\mathcal{H}|} = \left\| \liminf_{k\to\infty} \varphi_{n_k} \right\|_{|\mathcal{H}|} \leq \liminf_{k\to\infty} \|\varphi_{n_k}\|_{|\mathcal{H}|} < \infty.$$

Proceeding similarly (i.e., using the Fatou's lemma again), we have

$$\|\varphi - \varphi_n\|_{|\mathcal{H}|} = \left\| \liminf_{k\to\infty} (\varphi_{n_k} - \varphi_n) \right\|_{|\mathcal{H}|} \leq \liminf_{k\to\infty} \|\varphi_{n_k} - \varphi_n\|_{|\mathcal{H}|}\,.$$

Hence, the fact that $\{\varphi_n : n \in \mathbb{N}\}$ is a Cauchy sequence in $(|\mathcal{H}|, \|\cdot\|_{|\mathcal{H}|})$ yields that φ_n converges to φ in this space, as $n \to \infty$. Thus, we have stated that $(|\mathcal{H}|, \|\cdot\|_{|\mathcal{H}|})$ is a Banach space.

On the other hand, we prove that $\mathcal{E}([0,T])$ is a dense set of $|\mathcal{H}|$. Towards this end, we first consider a bounded function φ in $|\mathcal{H}|$. Then, from Royden [187] (Proposition 3.22), we are able to find a sequence $\{\varphi_n : n \in \mathbb{N}\} \subset \mathcal{E}([0,T])$ such that $\varphi_n \to \varphi$ almost surely, as $n \to \infty$, and $\sup_{s\in[0,T]} |\varphi_n(s)| \leq \sup_{s\in[0,T]} |\varphi(s)|$, for all $n \in \mathbb{N}$. Therefore, (1.123) and the dominated convergence theorem leads to show that $\varphi_n \to \varphi$ in $|\mathcal{H}|$, as $n \to \infty$. For $\varphi \in |\mathcal{H}|$ not necessarily bounded, we introduce $\varphi_n = (\varphi \wedge n) \vee (-n)$, which belongs to $|\mathcal{H}|$. For this bounded measurable function, we know that there exists $\psi_n \in \mathcal{E}([0,T])$ such that

$$\|\varphi - \psi_n\|_{|\mathcal{H}|} \leq \|\varphi - \varphi_n\|_{|\mathcal{H}|} + \|\varphi_n - \psi_n\|_{|\mathcal{H}|} \leq \|\varphi - \varphi_n\|_{|\mathcal{H}|} + \frac{1}{n}.$$

Therefore, we can conclude that $\mathcal{E}([0,T])$ is a dense set of $|\mathcal{H}|$.

Finally, the fact that $|\mathcal{H}| \subset \mathcal{H}$ is an immediate consequence of equalities (1.122) and (1.123), which establish that $\|\cdot\|_{\mathcal{H}} \leq \|\cdot\|_{|\mathcal{H}|}$. □

Pipiras and Taqqu [176, 177] have shown that the inclusion is strict and that $|\mathcal{H}|$ is not complete with respect to the inner product $\langle \cdot, \cdot \rangle_{\mathcal{H}}$. However, the space $|\mathcal{H}|$ contains the space $L^2([0,T])$ since the Hölder inequality and (1.123) allow us to get

$$\begin{aligned}\|\varphi\|^2_{|\mathcal{H}|} &\leq H(2H-1)\left(\int_0^T \int_0^T \varphi(r)^2 |r-s|^{2H-2} drds\right)^{1/2} \\ &\quad \times \left(\int_0^T \int_0^T \varphi(s)^2 |r-s|^{2H-2} drds\right)^{1/2} \leq C_{H,T} \int_0^T \varphi(r)^2 dr. \quad (1.125)\end{aligned}$$

Actually, we have that $L^{1/H}([0,T]) \subset |\mathcal{H}|$. Certainly, we can use that the following integrand is a symmetric function and Hölder inequality to obtain

$$\begin{aligned}\|\varphi\|^2_{|\mathcal{H}|} &= 2H(2H-1)\left(\int_0^T |\varphi(s)|^{1/H} ds\right)^H \\ &\quad \times \left(\int_0^T \left[\int_0^s |\varphi(r)|(s-r)^{2H-2} dr\right]^{1/(1-H)} ds\right)^{1-H}.\end{aligned}$$

Hence, Remark 14 (see also Theorem 7) implies that there exists a constant $C > 0$ such that

$$\|\varphi\|^2_{|\mathcal{H}|} \leq C\left(\int_0^T |\varphi(s)|^{1/H} ds\right)^{2H}, \tag{1.126}$$

which gives that our claim is true.

Now we introduce the space $|\mathcal{H}| \otimes |\mathcal{H}|$ of all the measurable functions $\varphi : [0,T]^2 \to \mathbb{R}$ such that

$$\|\varphi\|^2_{|\mathcal{H}|\otimes|\mathcal{H}|} = [H(2H-1)]^2 \int_{[0,T]^4} |\varphi_{r,\theta}||\varphi_{u,\eta}||r-u|^{2H-2}|\theta-\eta|^{2H-2} drdud\theta d\eta < \infty.$$

Note that the following proposition implies that $|\mathcal{H}| \otimes |\mathcal{H}| \subset \mathcal{H} \otimes \mathcal{H}$ since last norm has Property (8.56). In fact, we have that $\|\cdot\|_{\mathcal{H}\otimes\mathcal{H}} \leq \|\cdot\|_{|\mathcal{H}|\otimes|\mathcal{H}|}$.

As in Proposition 27, $(|\mathcal{H}| \otimes |\mathcal{H}|, \|\cdot\|_{|\mathcal{H}|\otimes|\mathcal{H}|})$ is also a Banach space, which is stated in the following result.

Proposition 28 *Let $H > 1/2$. Then, $(|\mathcal{H}| \otimes |\mathcal{H}|, \|\cdot\|_{|\mathcal{H}|\otimes|\mathcal{H}|})$ is a Banach space and the class $\mathcal{E}([0,T]) \otimes \mathcal{E}([0,T])$ is dense in it.*

Remark 27 *By Section 8.12, we have that $\varphi \otimes \psi = \varphi\psi$, for $\varphi, \psi \in \mathcal{E}([0,T])$. Moreover, from the definition of $\|\cdot\|_{|\mathcal{H}|\otimes|\mathcal{H}|}$, we have that $|\mathcal{H}| \otimes |\mathcal{H}| \subset \mathcal{H} \otimes \mathcal{H}$. In this way, we have given a dense subset of functions on $[0,T]^2$ that is dense in $\mathcal{H} \otimes \mathcal{H}$.*

Proof. Proceeding as in the beginning of the proof of Proposition 27, we get

$$\begin{aligned}&\int_{[0,T]^4} |\varphi_{r,\theta}||\varphi_{u,\eta}||r-u|^{2H-2}|\theta-\eta|^{2H-2} drdud\theta d\eta \\ &= \int_{[0,T]^4} |\varphi_{r,\theta}||\varphi_{u,\eta}| \left(\int_0^{r\wedge u} \frac{\partial K(r,s)}{\partial r}\frac{\partial K(u,s)}{\partial u} ds\right) \\ &\quad \times \left(\int_0^{\eta\wedge\theta} \frac{\partial K(\eta,v)}{\partial \eta}\frac{\partial K(\theta,v)}{\partial \theta} dv\right) drdud\theta d\eta.\end{aligned}$$

Thus, applying Fubini's theorem twice, we obtain

$$\begin{aligned}&\int_{[0,T]^4} |\varphi_{r,\theta}||\varphi_{u,\eta}||r-u|^{2H-2}|\theta-\eta|^{2H-2} drdud\theta d\eta \\ &= \int_{[0,T]^3}\int_0^{\eta\wedge\theta} \left(\int_s^T |\varphi_{r,\theta}| \frac{\partial K(r,s)}{\partial r}\frac{\partial K(\theta,v)}{\partial \theta} dr\right) \\ &\quad \times \left(\int_s^T |\varphi_{u,\eta}| \frac{\partial K(u,s)}{\partial u}\frac{\partial K(\eta,v)}{\partial \eta} du\right) dvdsd\theta d\eta \\ &= \int_{[0,T]^2} \left(\int_v^T\int_s^T |\varphi_{r,\theta}| \frac{\partial K(r,s)}{\partial r}\frac{\partial K(\theta,v)}{\partial \theta} drd\theta\right)^2 dsdv.\end{aligned}$$

Hence, we can now proceed as in the proof of Proposition 27 in order to see that the result is satisfied. □

As in (1.126), we have that there is a constant $C > 0$ such that

$$\|\cdot\|_{|\mathcal{H}|\otimes|\mathcal{H}|} \leq C \|\cdot\|_{L^{1/H}([0,T]^2)}, \tag{1.127}$$

which leads to deduce that $L^{1/H}([0,T]^2)$ is a subset of $|\mathcal{H}| \otimes |\mathcal{H}|$. Indeed, the definition of the norm in $|\mathcal{H}| \otimes |\mathcal{H}|$ yields

$$\begin{aligned} ||\varphi||^2_{|\mathcal{H}|\otimes|\mathcal{H}|} &= [H(2H-1)]^2 \int_{[0,T]^2} |\theta - \eta|^{2H-2} \left(\int_0^T |\varphi_{u,\eta}| \int_0^u |\varphi_{r,\theta}|(u-r)^{2H-2} dr du \right. \\ &\quad \left. + \int_0^T |\varphi_{u,\eta}| \int_u^T |\varphi_{r,\theta}|(r-u)^{2H-2} dr du \right) d\theta d\eta. \end{aligned} \tag{1.128}$$

Therefore, utilizing Remark 14 and Hölder inequality again, we are able to write

$$\begin{aligned} &\int_{[0,T]^2} |\theta - \eta|^{2H-2} \left(\int_0^T |\varphi_{u,\eta}| \int_u^T |\varphi_{r,\theta}|(r-u)^{2H-2} dr du \right) d\theta d\eta \\ &\leq \int_{[0,T]^2} |\theta - \eta|^{2H-2} \left(\int_0^T |\varphi_{u,\eta}|^{1/H} du \right)^H \\ &\qquad \times \left(\int_0^T \left[\int_u^T |\varphi_{r,\theta}|(r-u)^{2H-2} dr \right]^{1/(1-H)} du \right)^{1-H} d\theta d\eta \\ &\leq C \int_{[0,T]^2} |\theta - \eta|^{2H-2} \left(\int_0^T |\varphi_{u,\eta}|^{1/H} du \right)^H \left(\int_0^T |\varphi_{u,\theta}|^{1/H} du \right)^H d\theta d\eta \\ &\leq C \left(\int_{[0,T]^2} |\varphi_{u,\eta}|^{1/H} du d\eta \right)^{2H}, \end{aligned}$$

where last inequality follows from (1.126). Moreover, proceeding similarly, we can state

$$\begin{aligned} &\int_{[0,T]^2} |\theta - \eta|^{2H-2} \left(\int_0^T |\varphi_{u,\eta}| \int_0^u |\varphi_{r,\theta}|(u-r)^{2H-2} dr du \right) d\theta d\eta \\ &\qquad \leq C \left(\int_{[0,T]^2} |\varphi_{u,\eta}|^{1/H} du d\eta \right)^{2H}, \end{aligned}$$

for some $C > 0$. Therefore, the last two inequalities, together with (1.128), give that our claim is satisfied (i.e., (1.127) holds).

On the other hand, for $p > 1$, an important subset of the domain of the divergence operator δ_p with respect to fBm is the space $\mathbb{D}^{1,p}(|\mathcal{H}|)$. The operator δ_p has been interpreted as a stochastic integral (see Section 4.3), The space $\mathbb{D}^{1,p}(|\mathcal{H}|)$ is defined as all the elements $u \in \mathbb{D}^{1,p}(\mathcal{H})$ such that $u \in |\mathcal{H}|$ and $Du \in |\mathcal{H}| \otimes |\mathcal{H}|$ w.p.1, and

$$||u||^p_{\mathbb{D}^{1,p}(|\mathcal{H}|)} := E(||u||^p_{|\mathcal{H}|}) + E(||Du||^p_{|\mathcal{H}|\otimes|\mathcal{H}|}) < \infty.$$

Examples of stochastic processes u in $\mathbb{D}^{1,2}(|\mathcal{H}|)$ is the family of all the smooth step processes of the form $u = \sum_{j=0}^{n-1} F_i 1_{[t_i,t_{i+1}]}$, with $F_i = f_i(B(h_{1,i}), \ldots, B(h_{n_i,i})) \in \mathcal{S}$, $h_{j,i} \in$

$|\mathcal{H}|$ and $0 = t_0 < t_1 < \ldots < t_n = T$. Remember that, by definition, this family $\mathcal{E}(|\mathcal{H}|)$ of step processes is dense in $\mathbb{D}^{1,2}(\mathcal{H})$. Moreover, it is also dense in $\mathbb{D}^{1,2}(|\mathcal{H}|)$. Indeed, let $u \in \mathbb{D}^{1,2}(|\mathcal{H}|)$ be a process such that $\langle u, \xi \rangle_{\mathbb{D}^{1,2}(|\mathcal{H}|)} = 0$, for all $\xi \in \mathcal{E}(|\mathcal{H}|)$, and $A = \{(\omega, s) \in \Omega \times [0,T] : u_s(\omega) \neq 0\}$. Then,

$$E\left(\int_0^T |u_s| ds \int_0^T 1_A(t) dt\right) \leq C_{T,H} E\left(\int_0^T \int_0^T |u_s| 1_A(t) |t-s|^{2H-2} dt ds\right) = 0,$$

where last equality follows from Hölder inequality, (1.124), (1.125) and the fact that we can find a sequence $\{\xi^{(n)} \in \mathcal{E}(|\mathcal{H}|) : n \in \mathbb{N}\}$ that goes to 1_A in $L^2(\Omega \times [0,T])$.

For $p > 1/H$, the divergence operator δ_p with respect to fractional Brownian motion satisfies a maximal $L^p(\Omega)$–estimate. This result has been proved by Alòs and Nualart [13] and it is necessary to restrict the set of integrands, as is done in Theorem 52 (see Section 4.3 below). This set of integrands is introduced as follows. A process $u \in \mathbb{D}^{1,p}(|\mathcal{H}|)$, $p > 1$, belongs to $\mathbb{L}_H^{1,p}$ if

$$||u||^p_{\mathbb{L}_H^{1,p}} = E(||u||^p_{L^{\frac{1}{H}}([0,T])}) + E(||Du||^p_{L^{\frac{1}{H}}([0,T]^2)}) < \infty. \tag{1.129}$$

Note that (1.122)-(1.127) imply

$$||u||^p_{\mathbb{D}^{1,p}(\mathcal{H})} \leq C_{H,1} ||u||^p_{\mathbb{D}^{1,p}(|\mathcal{H}|)} \leq C_{H,2} ||u||^p_{\mathbb{L}_H^{1,p}}, \tag{1.130}$$

for some constants $C_{H,1}, C_{H,2} > 0$ depending on the Hurst parameter H of fractional Brownian motion B. In particular, the definition of above spaces yields

$$\mathbb{L}_H^{1,p} \subset \mathbb{D}^{1,p}(|\mathcal{H}|) \subset \mathbb{D}^{1,p}(\mathcal{H}).$$

Resuming, we have introduced two families of processes in $\mathbb{D}^{1,p}(\mathcal{H})$ because, in general, the elements of this space are not necessarily functions with probability 1.

1.6 Gaussian Volterra processes II

In this section, we consider a Gaussian Volterra process $B = \{B_t : t \in [0,T]\}$ as in Definition 16. That is, this process has the form

$$B_t = \int_0^t K(t,s) dW_s, \quad t \in [0,T].$$

Here, $W = \{W_t : t \in [0,T]\}$ is a Brownian motion and K satisfies condition (1.34). In addition, we suppose that the kernel K satisfies the following hypothesis:

(K1) The function $K(\cdot, s)$ has bounded variation on the interval $(r, T]$, for any $T \geq r > s \geq 0$.

As in Section 1.3, for a kernel K satisfying Condition (K1), we can introduce the space $\mathcal{H}_K$ as the completion of the step functions $\mathcal{E}([0,T])$ (see Definition 13) with respect to the seminorm

$$\|\varphi\|_K^2 = \int_0^T \varphi(s)^2 K(T,s)^2 ds + \int_0^T \left(\int_s^T |\varphi(t) - \varphi(s)| \, |K|(dt,s)\right)^2 ds, \tag{1.131}$$

where $|K|(dt,s)$ stands for the total variation of the bounded variation function $K(\cdot,s)$. It is well-known that, for a function $g:[a,b]\to\mathbb{R}$ of bounded variation, there exist two non-decreasing functions $g_1,g_2:[a,b]\to\mathbb{R}$ such that $g=g_1-g_2$ (see Lemma 106). So, the total variation of g is the measure μ defined on the Borel sets of $[a,b]$ such that $\mu((c,d])=(g_1(d+)-g_1(c+))+(g_2(d+)-g_2(c+))$ (see Remark 30 and Section 8.10 for details).

Note that the definition of $\|\cdot\|_K$ and Section 1.3 imply that, for $\varphi\in\mathcal{E}([0,T])$,

$$\|\varphi\|_{L^2([0,T];K(T,s)^2ds)}\le\|\varphi\|_K \quad \text{and} \quad \|\varphi\|_{\mathcal{H}}\le\sqrt{2}\,\|\varphi\|_K\,. \tag{1.132}$$

Consequently, the elements of $\mathcal{H}_K$ are function in $L^2([0,T];K(T,s)^2ds)$ and the space $\mathcal{H}_K$ is included in the reproducing kernel Hilbert space $\mathcal{H}$. In other words, by (1.131), f belongs to $\mathcal{H}_K$ if and only if f is a function in $L^2([0,T];K(T,s)^2ds)$ and there is a sequence $\{\varphi_n\in\mathcal{E}([0,T]):n\in\mathbb{N}\}$ such that $\|f-\varphi_n\|_K\to 0$, as $n\to\infty$.

Observe that, besides the process B in this section, we also have the Brownian motion W appearing in (1.32). Hence, we can also consider the derivative operator D^W and the space $\mathbb{D}^{p,k}_W(\mathcal{K})$ related to W, with $\mathcal{K}$ a Hilbert space. Remember that these are given in Section 1.5.4.

The following result is an important tool to deal with the divergence operator with respect to the Gaussian Volterra process B given by (1.32) (see Section 4.4).

Lemma 22 *Let $B=\{B_t:t\in[0,T]\}$ be the Gaussian Volterra process introduced in* (1.32) *with kernel K satisfying Conditions* (1.34) *and (K1). Then,*

$$K^*\left(\mathbb{D}^{1,2}(\mathcal{H})\right)\subset\mathbb{D}^{1,2}_W\left(L^2([0,T])\right),$$

where K^* is defined in (1.37).

Proof. Let $F\in\mathcal{S}$ be a smooth functional of the form

$$F=f(B(\varphi_1),\ldots,B(\varphi_n)).$$

Then, from (1.39), we have that $F=f(W(K^*\varphi_1),\ldots,W(K^*\varphi_n))$, with $W(K^*\varphi_i)=\int_0^T(K^*\varphi_i)(s)dW_s$. Hence, (1.43) implies that F also belongs to $\mathcal{S}_W$ and that $K^*DF=D^WF$. In other words, we have proven that $\mathcal{S}\subset\mathcal{S}_W$.

Now, let $u\in\mathbb{D}^{1,2}(\mathcal{H})$. Then, by (1.104), there is a sequence of $\mathcal{H}$-valued smooth functionals $\{\varphi_n:n\in\mathbb{N}\}$ of the form

$$F=\sum_{i=1}^n F_ih_i,$$

where $F_i\in\mathcal{S}$ and $h_i\in\mathcal{H}$, such that

$$E\left(\|u-\varphi_n\|^2_{\mathcal{H}}+\|D(u-\varphi_n)\|^2_{\mathcal{H}\otimes\mathcal{H}}\right)\to 0, \quad \text{as } n\to\infty. \tag{1.133}$$

In particular, (1.43) yields

$$E\left(\|(K^*u)-K^*\varphi_n\|^2_{L^2([0,T])}\right)=E\left(\|u-\varphi_n\|^2_{\mathcal{H}}\right)\to 0, \quad \text{as } n\to\infty. \tag{1.134}$$

Therefore, using (1.43) again, together with Proposition 143 and (1.133), we are able to obtain

$$\begin{aligned}&\int_0^T\int_0^T\left|D^W_r\left((K^*\varphi_n)(s)-((K^*\varphi_m)(s)\right)\right|^2dsdr\\ &=\quad \left\|(K^*)^{\otimes 2}D(\varphi_n-\varphi_m)\right\|^2_{L^2([0,T]^2)}=\|D(\varphi_n-\varphi_m)\|^2_{\mathcal{H}\otimes\mathcal{H}}\to 0, \quad \text{as } n,m\to\infty.\end{aligned}$$

Hence, (1.134) allows us to conclude that K^*u belongs to $\mathbb{D}_W^{1,2}\left(L^2([0,T])\right)$, Consequently, the proof is complete. □

The importance of Lemma 22 is that we can take advantage of Hypothesis (K1) to control the $L^2(\Omega)$-moment of the divergence operator with respect to B on a suitable space of stochastic processes, which is done in Section 4.4 below. Using the proof of Lemma 22 and (K1), this space is defined as follows: Note that in the last proof we have

$$\int_0^T \int_0^T \left|D_r^W\left((K^*\varphi_n)(s) - ((K^*\varphi_m)(s)\right)\right|^2 dsdr = \int_0^T \left\|D_r^W(\varphi_n - \varphi_m)\right\|_{\mathcal{H}}^2 dr.$$

This equality motivates the following definition.

Definition 24 *Let B be a Gaussian Volterra process given by* (1.32) *such that (K1) holds. The space $\mathbb{D}^{1,2}(\mathcal{H}_K)$ is the completion of $\mathcal{H}_K$-valued simple processes with respect to the seminorm*

$$E\left(\|\varphi\|_K^2 + \int_0^T \left\|D_r^W\varphi\right\|_K^2 dr\right).$$

Notice that a process $\{u_t : t \in [0,T]\}$ belongs to the space $\mathbb{D}^{1,2}(\mathcal{H}_K)$ if there exists a sequence $\{\varphi_n : n \in \mathbb{N}\}$ of $\mathcal{H}_K$-valued simple random variable of the form

$$\varphi_n = \sum_{i=1}^n F_{j,n} 1_{(t_{i-1,n},t_{i,n}]},$$

where $F_{j,n} \in \mathcal{S}$ and $0 = t_{0,n} < t_{1,n} < \ldots < t_{n,n} = T$, such that

$$E\left(\|u - \varphi_n\|_K^2 + \int_0^T \left\|D_r^W(u - \varphi_n)\right\|_K^2 dr\right) \to 0, \quad \text{as } n \to \infty.$$

The reason to introduce the space $\mathbb{D}^{1,2}(\mathcal{H}_K)$ is to state the following result, which is a consequence of Lemma 22.

Corollary 13 *Let B the process given by* (1.32) *such that (K1) holds and $u \in \mathbb{D}^{1,2}(\mathcal{H}_K)$. Then, K^*u belongs to the space $\mathbb{D}_W^{1,2}\left(L^2([0,T])\right)$.*

Proof. From (1.37) and (1.131), we have that $\mathbb{D}^{1,2}(\mathcal{H}_K) \subset \mathbb{D}^{1,2}(\mathcal{H})$. Therefore, the result is an immediate consequence of Lemma 22. □

Our next aim is to give conditions that guarantee that the process $f(B)$ belongs to the space $L^2(\Omega;\mathcal{H}_K)$, where $f : \mathbb{R} \to \mathbb{R}$ is smooth enough. Towards this end, we first consider a measurable function f satisfying

$$|f(x)| \le C\exp(\sigma x^2), \quad x \in \mathbb{R}, \tag{1.135}$$

with $0 < \sigma < \left(\sup_{0\le t\le T} R_B(t,t)\right)^{-1}/4$ and C a constant bigger than zero. Note that we can find such a σ due to assumption (1.34).

Lemma 23 *Assume that* (1.135) *holds and that B is a continuous process. Then, $\sup_{t\in[0,T]}|f(B_t)|$ belongs to $L^p(\Omega)$, for any $p \in \left[2, \left(\sup_{0\le t\le T} R_B(t,t)\right)^{-1}/2\sigma\right)$.*

Proof. Note that (1.135) leads us to $\sup_{t\in[0,T]}|f(B_t)|^p \le C^p\exp(p\sigma\sup_{0\le t\le T} B_t^2)$. Hence, it is enough to see that the right-hand side of this inequality is a random variable in $L^1(\Omega)$.

To do so, and to simplify the notation, we utilize the convention $X = \sup_{0\le t\le T}|B_t|$. So, to finish the proof, we denote the distribution of X by F_X and apply Fubini's theorem to get

$$\begin{aligned} E\left(e^{p\sigma X^2}\right) &= \int_{(0,\infty)} e^{p\sigma y^2} F_X(dy) = \int_{(0,\infty)} \left(p\sigma \int_0^{y^2} e^{p\sigma u} du + 1\right) F_X(dy) \\ &= 1 + p\sigma \int_0^\infty e^{p\sigma u} \int_{(\sqrt{u},\infty)} F_X(dy) du = 1 + p\sigma \int_0^\infty e^{p\sigma u} P\left([X > \sqrt{u}]\right) du. \end{aligned}$$

Hence, Nourdin [155] (Theorem 4.2) implies that $m = E\left(\sup_{0\le t\le t} B_t\right) < \infty$ and

$$\begin{aligned} E\left(e^{p\sigma X^2}\right) &\le 1 + p\sigma \int_0^m e^{p\sigma u} du \\ &\quad + p\sigma \int_m^\infty e^{p\sigma u} \exp\left(-(\sqrt{u} - m)^2/2 \sup_{0\le t\le T} R_B(t.t)\right) du < \infty, \end{aligned}$$

where last inequality follows from the fact that $2p\sigma < \left(\sup_{0\le t\le T} R_B(t,t)\right)^{-1}$. Thus, the result is true. □

The importance of Lemma 23 is that it allows us to obtain conditions that guarantee that the process $f(B)$ belongs to $L^2(\Omega; \mathcal{H}_K)$. Indeed, consider a function $f \in C^1(\mathbb{R})$ such that f and f' satisfy (1.135). Also assume that Hypothesis (K1) holds. Then, Hölder inequality, (1.131) and (8.5) lead us to get that, for $p > 2$,

$$\begin{aligned} E\left(\|f(B)\|_K^2\right) &= E\left(\int_0^T f(B_s)^2 K(T,s)^2 ds\right) \\ &\quad + E\left(\int_0^T \left[\int_s^T |f(B_t) - f(B_s)|\, |K|(dt,s)\right]^2 ds\right) \\ &\le E\left(\left[\sup_{t\in[0,T]} |f(B_t)|^2\right] \int_0^T K(T,s)^2 ds\right) \\ &\quad + E\left(\int_0^T \left[\int_s^T \left(\sup_{u\in[0,T]} |f'(B_u)|\right) |B_t - B_s|||K|(dt,s)\right]^2 ds\right) \\ &\le E\left(\left[\sup_{t\in[0,T]} |f(B_t)|^2\right]\right) R_B(T,T) \\ &\quad + \int_0^T \left[\int_s^T \left\| \left(\sup_{u\in[0,T]} |f'(B_u)|\right) |B_t - B_s| \right\|_{L^2(\Omega)} ||K|(dt,s)\right]^2 ds \\ &\le E\left(\left[\sup_{t\in[0,T]} |f(B_t)|^2\right]\right) R_B(T,T) + \left\| \sup_{t\in[0,T]} |f'(B_t)| \right\|_{L^p(\Omega)}^2 \\ &\quad \times \int_0^T \left[\int_s^T \|B_t - B_s\|_{L^{2p/(p-2)}(\Omega)} |K|(dt,s)\right]^2 ds. \end{aligned} \tag{1.136}$$

On the other hand, let $p \in \left[\left(2, \left(\sup_{0\le t\le T} R_B(t,t)\right)^{-1}/2\sigma\right) \cap \mathbb{Q}\right]$, where σ is as in Lemma 23. So, we can assume that $p = r/q$ with $r, q \in \mathbb{N}$. Then, from (1.10) and using

Hölder inequality again, there exists a constant C depending on r such that

$$\begin{aligned}\|B_t - B_s\|_{L^{2p/(p-2)}(\Omega)} &= \|B_t - B_s\|_{L^{2r/(r-2q)}(\Omega)} = \left[E\left(|B_t - B_s|^{2r/(r-2q)}\right)\right]^{(r-2q)/2r} \\ &\leq \left[E\left(|B_t - B_s|^{2r}\right)\right]^{1/2r} \leq C\|B_t - B_s\|_{L^2(\Omega)}.\end{aligned}$$

Consequently, (1.136) implies

$$\begin{aligned}E\left(\|f(B)\|_K^2\right) &\leq E\left(\left[\sup_{t\in[0,T]}|f(B_t)|^2\right]\right)R_B(T,T) + C^2\left\|\sup_{t\in[0,T]}|f'(B_t)|\right\|_{L^p(\Omega)}^2 \\ &\times \int_0^T\left[\int_s^T \|B_t - B_s\|_{L^2(\Omega)}|K|(dt,s)\right]^2 ds. \qquad (1.137)\end{aligned}$$

Note that the right-hand side of the last inequality is finite if, in addition, the following assumption is satisfied:

(K2) Hypothesis (K1) holds and $\int_0^T\left[\int_s^T \|B_t - B_s\|_{L^2(\Omega)}|K|(dt,s)\right]^2 ds < \infty$.

Now, we are ready to state the following result, which establishes that the Gaussian process B is in the space $L^2(\Omega;\mathcal{H}_K)$. Towards this end, we utilize the convention

$$t_i^n = \frac{Ti}{2^n} \quad \text{and} \quad \bar{t}_i^n = t_i^n + \frac{T}{2^{n+1}}, \quad n \in \mathbb{N}.$$

Proposition 29 *Suppose that Hypothesis (K1) holds, B is a continuous process such that*

$$\sum_{i=0}^{2^n-1}\int_{t_i^n}^{\bar{t}_i^n}\left(\int_{\bar{t}_i^n}^{t_{i+1}^n}\left\|B_{t_{i+1}^n} - B_{t_i^n}\right\|_{L^2(\Omega)}|K|(dt,s)\right)^2 ds \to 0, \quad \text{as } n\to\infty,$$

$f \in C^1(\mathbb{R})$ is such that f and f' satisfy (1.135), and that there exists a measurable, continuous at zero and non-decreasing function $g : [0,T] \to \mathbb{R}_+ \cup \{0\}$ such that $g(0) = 0$, $\|B_t - B_s\|_{L^2(\Omega)} \leq g(t-s)$, for $0 \leq s \leq t \leq T$ and $\int_0^T\left[\int_s^T g(t-s)|K|(dt,s)\right]^2 ds < \infty$. Then, the process $f(B)$ belongs to $L^2(\Omega;\mathcal{H}_K)$.

Remark 28 *Observe that, under the assumptions of this proposition, we have that Hypothesis (K2) is fulfilled. Also observe that, in order to show that the result is true, it is not enough that* (1.137) *be finite because $\mathcal{H}_K$ is the completion of the step functions $\mathcal{E}([0,T])$ with respect to the seminorm $\|\cdot\|_K$ (see* (1.131)*). In other words, we must find a sequence of simple processes that goes to $f(B)$ in $L^2(\Omega;\mathcal{H}_K)$.*

Proof. For $n \in \mathbb{N}$, we consider the simple process $\varphi^{(n)} = \sum_{i=0}^{2^n-1}\left[B_{t_i^n}1_{(t_i^n,\bar{t}_i^n]} + B_{t_{i+1}^n}1_{(\bar{t}_i^n,t_{i+1}^n]}\right]$. Hence, we also have that $f(\varphi^{(n)}) = \sum_{i=0}^{2^n-1}\left[f(B_{t_i^n})1_{(t_i^n,\bar{t}_i^n]} + f(B_{t_{i+1}^n})1_{(\bar{t}_i^n,t_{i+1}^n]}\right]$ is a simple process. In order to finish the proof, we will see that $f(\varphi^{(n)})$ goes to $f(B)$ in $L^2(\Omega;\mathcal{H}_K)$, as $n\to\infty$. Towards this end, we first consider the first addend corresponding to the right-hand side of (1.131).

Let $s \in [0,T]$, then, proceeding as in (1.136) and (1.137), and applying the assumptions on the function g, we have that, for $p \in \left(2, \left(\sup_{0\leq t\leq T} R_B(t,t)\right)^{-1}/2\sigma\right) \cap \mathbb{Q}$, there exists a constant $C > 0$ such that

$$\begin{aligned}
\left\| f(\varphi^{(n)}(s)) - f(B_s) \right\|^2_{L^2(\Omega)} &\le \left\| \left(\sup_{t\in[0,T]} |f'(B_t)| \right) \left(\varphi^{(n)}(s) - B_s \right) \right\|^2_{L^2(\Omega)} \\
&\le \left\| \sup_{t\in[0,T]} |f'(B_t)| \right\|^2_{L^p(\Omega)} \left\| \varphi^{(n)}(s) - B_s \right\|^2_{L^{2p/(p-2)}(\Omega)} \\
&\le C \left\| \sup_{t\in[0,T]} |f'(B_t)| \right\|^2_{L^p(\Omega)} \left\| \varphi^{(n)}(s) - B_s \right\|^2_{L^2(\Omega)} \\
&\le C \left\| \sup_{t\in[0,T]} |f'(B_t)| \right\|^2_{L^p(\Omega)} g(T/2^{n+1})^2 \to 0 \quad \text{as } n \to \infty,
\end{aligned}$$

where last convergence follows from Lemma 23. Consequently,

$$E\left(\int_0^T \left| f(\varphi^{(n)}(s)) - f(B_s) \right|^2 K(T,s)^2 ds \right) \to 0 \quad \text{as } n \to \infty. \tag{1.138}$$

Now, it is time to analyze the second addend corresponding to the right-hand side of equality (1.131). To do so, we consider the inequality

$$\begin{aligned}
&\int_0^T \left[\int_s^T \left(f(\varphi^{(n)}(t)) - f(\varphi^{(n)}(s)) - (f(B_t) - f(B_s)) \right) |K|(dt,s) \right]^2 ds \\
&\le 2 \int_0^T \left[\int_s^T \sum_{i=0}^{2^n-1} \left(f(\varphi^{(n)}(t)) - f(\varphi^{(n)}(s)) - (f(B_t) - f(B_s)) \right) \right. \\
&\qquad \left. \times 1_{(t_i^n, t_{i+1}^n]^2}(s,t) |K|(dt,s) \right]^2 ds \\
&\quad + 2 \int_0^T \left[\int_s^T \sum_{i=0}^{2^n-2} \left(f(\varphi^{(n)}(t)) - f(\varphi^{(n)}(s)) - (f(B_t) - f(B_s)) \right) \right. \\
&\qquad \left. \times 1_{(t_i^n, t_{i+1}^n]}(s) 1_{(t_{i+1}^n, T]}(t) |K|(dt,s) \right]^2 ds.
\end{aligned}$$

Therefore, using that $\{0 = t_0^n < t_1^n < \ldots < t_{2^n} = T\}$ is a partition of $[0,T]$, we get

$$\begin{aligned}
&\int_0^T \left[\int_s^T \left(f(\varphi^{(n)}(t)) - f(\varphi^{(n)}(s)) - (f(B_t) - f(B_s)) \right) |K|(dt,s) \right]^2 ds \\
&\le 4 \sum_{i=0}^{2^n-1} \int_{t_i^n}^{\bar{t}_i^n} \left[\int_{\bar{t}_i^n}^{t_{i+1}^n} \left(f(B_{t_{i+1}^n}) - f(B_{t_i^n}) \right) |K|(dt,s) \right]^2 ds \\
&\quad + 4 \int_0^T \left[\int_s^T \sum_{i=0}^{2^n-1} (f(B_t) - f(B_s)) \, 1_{(t_i^n, t_{i+1}^n]^2}(s,t) |K|(dt,s) \right]^2 ds \\
&\quad + 2 \int_0^T \left[\int_s^T \sum_{i=0}^{2^n-2} \left(f(\varphi^{(n)}(t)) - f(\varphi^{(n)}(s)) - (f(B_t) - f(B_s)) \right) \right. \\
&\qquad \left. \times 1_{(t_i^n, t_{i+1}^n]}(s) 1_{(t_{i+1}^n, T]}(t) |K|(dt,s) \right]^2 ds.
\end{aligned} \tag{1.139}$$

Proceeding as in (1.136) and (1.137), we can find a constant $C > 0$ such that

$$E\left(\sum_{i=0}^{2^n-1}\int_{t_i^n}^{\bar{t}_i^n}\left[\int_{\bar{t}_i^n}^{t_{i+1}^n}\left(f(B_{t_{i+1}^n})-f(B_{t_i^n})\right)|K|(dt,s)\right]^2 ds\right)$$

$$\leq C\sum_{i=0}^{2^n-1}\int_{t_i^n}^{\bar{t}_i^n}\left[\int_{\bar{t}_i^n}^{t_{i+1}^n}\left\|B_{t_{i+1}^n}-B_{t_i^n}\right\|_{L^2(\Omega)}|K|(dt,s)\right]^2 ds \to 0, \text{ as } n\to\infty, \quad (1.140)$$

where last convergence follows by hypothesis. Similarly, we can obtain

$$E\left(\int_0^T\left[\int_s^T\sum_{i=0}^{2^n-1}(f(B_t)-f(B_s))\,1_{(t_i^n,t_{i+1}^n]^2}(s,t)|K|(dt,s)\right]^2 ds\right)$$

$$\leq C\int_0^T\left[\int_s^T\sum_{i=0}^{2^n-1}\|B_t-B_s\|_{L^2(\Omega)}\,1_{(t_i^n,t_{i+1}^n]^2}(s,t)|K|(dt,s)\right]^2 ds.$$

Hence, the facts that

$$\sum_{i=0}^{2^n-1}\|B_t-B_s\|_{L^2(\Omega)}\,1_{(t_i^n,t_{i+1}^n]^2}(s,t)\to 0, \quad \text{as } n\to\infty, \quad \text{for } s,t\in[0,T],$$

and

$$\sum_{i=0}^{2^n-1}\|B_t-B_s\|_{L^2(\Omega)}\,1_{(t_i^n,t_{i+1}^n]^2}(s,t)\leq\|B_t-B_s\|_{L^2(\Omega)}\leq g(|t-s|),$$

together with the dominated convergence theorem, yield

$$E\left(\int_0^T\left[\int_s^T\sum_{i=0}^{2^n-1}(f(B_t)-f(B_s))\,1_{(t_i^n,t_{i+1}^n]^2}(s,t)|K|(dt,s)\right]^2 ds\right)\to 0, \quad \text{as } n\to\infty. \quad (1.141)$$

Before studying the last term of (1.139), we observe

$$\|B_t-\varphi^{(n)}(t)\|_{L^2(\Omega)}+\|B_s-\varphi^{(n)}(s)\|_{L^2(\Omega)}\leq 2g(t-s), \quad \text{for } (s,t)\in(t_i^n,t_{i+1}^n]\times(t_{i+1}^n,T].$$

Indeed, for instance, we have that, for $s\in(t_i^n,t_{i+1}^n]$ and $t\in(t_{i+1}^n,\bar{t}_{i+1}^n]$,

$$\|B_t-\varphi^{(n)}(t)\|_{L^2(\Omega)}=\|B_t-B_{t_{i+1}^n}\|_{L^2(\Omega)}\leq g(t-t_{i+1}^n)\leq g(t-s)$$

and, for $s\in(t_i^n,t_{i+1}^n]$ and $t\in(\bar{t}_{i+1}^n,t_{i+2}^n]$,

$$\|B_t-\varphi^{(n)}(t)\|_{L^2(\Omega)}=\|B_t-B_{t_{i+2}^n}\|_{L^2(\Omega)}\leq g(t_{i+2}^n-t)\leq g(t-s).$$

With this property in mind, we finally establish

$$E\left(\int_0^T\left[\int_s^T\sum_{i=0}^{2^n-2}\left(f(\varphi^{(n)}(t))-f(\varphi^{(n)}(s))-(f(B_t)-f(B_s))\right)\right.\right.$$

$$\left.\left.\times 1_{(t_i^n,t_{i+1}^n]}(s)1_{(t_{i+1}^n,T]}(t)|K|(dt,s)\right]^2 ds\right)$$

$$\leq \int_0^T\left[\int_s^T\sum_{i=0}^{2^n-2}\left\|f(\varphi^{(n)}(t))-f(\varphi^{(n)}(s))-(f(B_t)-f(B_s))\right\|_{L^2(\Omega)}\right.$$

$$\left.\times 1_{(t_i^n,t_{i+1}^n]}(s)1_{(t_{i+1}^n,T]}(t)|K|(dt,s)\right]^2 ds.$$

Since we have already known that there is a constant $C > 0$ such that the continuity of g at zero and the fact that it is non-decreasing imply that, for $s, t \in [0, T]$,

$$\begin{aligned}
&\left\| f(\varphi^{(n)}(t)) - f(\varphi^{(n)}(s)) - (f(B_t) - f(B_s)) \right\|_{L^2(\Omega)} \\
\leq\ & \left\| f(\varphi^{(n)}(t)) - f(B_t) \right\|_{L^2(\Omega)} + \left\| f(\varphi^{(n)}(s)) - f(B_s) \right\|_{L^2(\Omega)} \\
\leq\ & C \left\| \varphi^{(n)}(t) - B_t \right\|_{L^2(\Omega)} + C \left\| \varphi^{(n)}(s) - B_s \right\|_{L^2(\Omega)} \leq 2Cg(T/2^{n+1}) \to 0 \quad \text{as } n \to \infty
\end{aligned}$$

and, for $(s, t) \in (t_i^n, t_{i+1}^n] \times (t_{i+1}^n, T]$,

$$\begin{aligned}
&\left\| f(\varphi^{(n)}(t)) - f(\varphi^{(n)}(s)) - (f(B_t) - f(B_s)) \right\|_{L^2(\Omega)} \\
\leq\ & C \left\| \varphi^{(n)}(t) - B_t \right\|_{L^2(\Omega)} + C \left\| \varphi^{(n)}(s) - B_s \right\|_{L^2(\Omega)} \leq 2Cg(t - s).
\end{aligned}$$

Thus, the dominated convergence theorem leads us to conclude

$$\begin{aligned}
E\Bigg(\int_0^T \Bigg[\int_s^T \sum_{i=0}^{2^n-2} & \Big(f(\varphi^{(n)}(t)) - f(\varphi^{(n)}(s)) - (f(B_t) - f(B_s)) \Big) \\
& \times 1_{(t_i^n, t_{i+1}^n]}(s) 1_{(t_{i+1}^n, T]}(t) |K|(dt, s) \Bigg]^2 ds \Bigg) \to 0, \quad \text{as } n \to \infty.
\end{aligned}$$

In consequence, (1.138)-(1.141) give that the sequence $\{f(\varphi^{(n)}) : n \in \mathbb{N}\}$ of simple processes converges to $f(B)$ in $L^2(\Omega; \mathcal{H}_K)$, as $n \to \infty$, which proves that the process $f(B)$ belongs to the space $L^2(\Omega; \mathcal{H}_K)$. □

We can apply the ideas of the last proof to see that Proposition 29 is still true for functions $f : [0, T] \times \mathbb{R} \to \mathbb{R}$ such that there is a constant $C > 0$ so that

$$|f(s, x)| \leq C \exp(\sigma x^2), \quad (s, x) \in [0, T] \times \mathbb{R}, \tag{1.142}$$

with $0 < \sigma < \left(\sup_{0 \leq t \leq T} R_B(t, t) \right)^{-1} / 4$.

Corollary 14 *Let assumptions of Proposition 29 be satisfied and $f \in C^1([0, T] \times \mathbb{R})$ such that (1.142) holds for $f, \partial_t f$ and $\partial_x f$. Also suppose that*

$$\sum_{i=0}^{2^n-1} \int_{t_i^n}^{\bar{t}_i^n} \left(\int_{\bar{t}_i^n}^{t_{i+1}^n} \left(t_{i+1}^n - t_i^n \right) |K|(dt, s) \right)^2 ds \to 0, \quad \text{as } n \to \infty,$$

and $\int_0^T \left[\int_s^T (t - s) |K|(dt, s) \right]^2 ds < \infty$. Then, the process $f(\cdot, B_\cdot)$ belongs to $L^2(\Omega; \mathcal{H}_K)$.

Proof. Here, for $n \in \mathbb{N}$, we consider the simple process

$$f(\varphi^{(n)}) = \sum_{i=0}^{2^n-1} \left[f(t_i^n, B_{t_i^n}) 1_{(t_i^n, \bar{t}_i^n]} + f(t_{i+1}^n, B_{t_{i+1}^n}) 1_{(\bar{t}_i^n, t_{i+1}^n]} \right].$$

Therefore, the proof of this result follows the ideas of that of Proposition 29 by noting that, for instance, the mean value theorem and (1.142) imply that, for $s, t \in [0, T]$,

$$|f(s, B_s) - f(t, B_t)| \leq C \exp\left(\sigma \sup_{u \in [0, T]} B_u^2 \right) (|t - s) + |B_t - B_s|).$$

□

As examples of processes that satisfy the assumptions of Proposition 29, we state the following two results.

Proposition 30 *Let $B = \{B_t : t \in [0,T]\}$ be fractional Brownian motion with Hurst parameter $H \in (1/4, 1/2)$. Then, B satisfies the hypotheses of Proposition 29 with $g(t) = t^H$, $t \in [0,T]$.*

Proof. From Theorem 9, the fractional Brownian motion B is a Volterra process with kernel K_H given by (1.67) (see also Definition 16 for the definition of Volterra processes). Moreover, (1.67) implies that

$$\frac{\partial K_H(t,s)}{\partial t} = \tilde{c}_H(H-\frac{1}{2})\left(\frac{t}{s}\right)^{H-\frac{1}{2}}(t-s)^{H-\frac{3}{2}}, \quad 0 < s < t \leq T. \tag{1.143}$$

Hence, the kernel K_H satisfies Condition (K1). Moreover, equality (1.44) establishes that $\|B_t - B_s\|_{L^2(\Omega)} = (t-s)^H = g(t-s)$, for $s < t$. For this function $g : [0,T] \to \mathbb{R}_+ \cup \{0\}$, we have

$$\begin{aligned}
\int_0^T \left[\int_s^T g(t-s)|K|(dt,s)\right]^2 ds &\leq \int_0^T \left[\int_s^T g(t-s)\left|\frac{\partial K_H(t,s)}{\partial t}\right| dt\right]^2 ds \\
&\leq C\int_0^T \left[\int_s^T g(t-s)(t-s)^{H-\frac{3}{2}}dt\right]^2 ds \\
&= C\int_0^T \left[\int_s^T (t-s)^{2H-\frac{3}{2}}dt\right]^2 ds < \infty
\end{aligned}$$

because $H > 1/4$.

Finally, in order to finish the proof, we see that the condition of Theorem 29 that remains to be verified is also fulfilled. That is,

$$\begin{aligned}
&\sum_{i=0}^{2^n-1}\int_{t_i^n}^{\bar{t}_i^n}\left(\int_{\bar{t}_i^n}^{t_{i+1}^n}\left\|B_{t_{i+1}^n} - B_{t_i^n}\right\|_{L^2(\Omega)}|K|(dt,s)\right)^2 ds \\
&\leq \sum_{i=0}^{2^n-1} g(t_{i+1}^n - t_i^n)^2\int_{t_i^n}^{\bar{t}_i^n}\left(\int_{\bar{t}_i^n}^{t_{i+1}^n}\left|\frac{\partial K_H(t,s)}{\partial t}\right| dt\right)^2 ds \\
&\leq C\sum_{i=0}^{2^n-1} g(t_{i+1}^n - t_i^n)^2\int_{t_i^n}^{\bar{t}_i^n}\left(\int_{\bar{t}_i^n}^{t_{i+1}^n}(t-s)^{H-\frac{3}{2}}dt\right)^2 ds \\
&= C\sum_{i=0}^{2^n-1} g(t_{i+1}^n - t_i^n)^2\int_{t_i^n}^{\bar{t}_i^n}\left((t_{i+1}^n - s)^{H-\frac{1}{2}} - (\bar{t}_i^n - s)^{H-\frac{1}{2}}\right)^2 ds \\
&\leq 2\sum_{i=0}^{2^n-1} g(t_{i+1}^n - t_i^n)^2\int_{t_i^n}^{\bar{t}_i^n}\left((t_{i+1}^n - s)^{2H-1} + (\bar{t}_i^n - s)^{2H-1}\right) ds \\
&\leq 4C\sum_{i=0}^{2^n-1} g(t_{i+1}^n - t_i^n)^2(t_{i+1}^n - t_i^n)^{2H} = C\sum_{i=0}^{2^n-1}(t_{i+1}^n - t_i^n)^{4H} \\
&= C(T2^{-n})^{4H-1}\sum_{i=0}^{2^n-1}(t_{i+1}^n - t_i^n) = CT(T2^{-n})^{4H-1} \to 0, \quad \text{as } n \to \infty.
\end{aligned}$$

Thus, the proof is complete now. □

A Volterra process that is used in quantitative finance is the Riemann-Liouville fractional Brownian process because it has similar properties to those of fractional Brownian motion and is simple to handle (see Alòs and García Lorite [4], and Alòs et al. [8] for an application on the construction of fractional volatility models).

Definition 25 *A Volterra process $B = \{B_t : t \in [0,T]\}$ is called a Riemann-Liouville fractional Brownian motion with Hurst parameter $H \in (0,1)$ if it has the kernel*

$$K(t,s) = (t-s)^{H-\frac{1}{2}}, \quad 0 \leq s < t \leq T.$$

That is, there is a Brownian motion $W = \{W_t : t \in [0,T]\}$ such that

$$B_t = \int_0^t (t-s)^{H-\frac{1}{2}} dW_s, \quad t \in [0,T].$$

Note that $K(t,\cdot)1_{[0,t]}(\cdot)$ belongs to $L^2([0,t])$, for all $t \in [0,T]$. Furthermore, observe that Riemann-Liouville fractional Brownian process appears in the Mandelbrot and Van Ness integral representation (1.45) of fractional Brownian motion. This process is the second example of a process that also enjoys the assumptions of Proposition 29, as the following result establishes.

Proposition 31 *Let $B = \{B_t : t \in [0,T]\}$ be Riemann-Liouville fractional Brownian process with Hurst parameter $H \in (1/4, 1/2)$. Then, B satisfies the assumptions of Proposition 29 with $g(t) = t^H$, $t \in [0,T]$.*

Remark 29 *Note that fractional Brownian motion and Riemann-Liouville fractional Brownian process fulfill the conditions of Proposition 29 with the same function g (see Proposition 30).*

Proof. Let $s,t \in [0,T]$, $s < t$, then Corollary 4.iii) yields

$$\begin{aligned}
E\left((B_t - B_s)^2\right) &= E\left(B_t^2 - 2B_tB_s + B_s^2\right) \\
&= (2H)^{-1}\left(t^{2H} + s^{2H}\right) - 2\int_0^s K(t,r)K(s,r)dr \\
&= (2H)^{-1}\left(t^{2H} + s^{2H}\right) - 2\int_0^s (t-r)^{H-\frac{1}{2}}(s-r)^{H-\frac{1}{2}}dr \\
&\leq (2H)^{-1}\left(t^{2H} + s^{2H}\right) - 2\int_0^s (t-r)^{2H-1}dr \\
&= (2H)^{-1}\left[t^{2H} + s^{2H} + 2(t-s)^{2H} - 2t^{2H}\right] \\
&= (2H)^{-1}\left[s^{2H} - t^{2H} + 2(t-s)^{2H}\right] \leq \frac{2}{2H}(t-s)^{2H}. \qquad (1.144)
\end{aligned}$$

Also, we have $\left|\frac{\partial K(t,s)}{\partial t}\right| \leq (\frac{1}{2} - H)(t-s)^{H-\frac{3}{2}}$, for $0 < s < t \leq T$.

Finally, the last two inequalities are the main ingredients to show that Proposition 30 holds. So, now we only have to follow the same lines of the proof of Proposition 30 to show that the result is satisfied. □

Chapter 2

Riemann-Stieltjes Integral

Young integration can be studied in several ways. For instance, we are able to mention the convergence of Riemann sums using p-variation [216], the fractional calculus setting [218], the algebraic approach introduced in [70] and developed e.g. in [71], among others. That is, for two functions $f, g : [a,b] \to \mathbb{R}$, $a, b \in \mathbb{R}$, we are interested in analyzing the convergence of the Riemann-Stieltjes sum

$$\mathrm{RS}(f,g,\pi,\tilde{\pi}) := \sum_{i=1}^{n} f(s_i)\left[g(t_i) - g(t_{i-1})\right], \tag{2.1}$$

as $|\pi| \to 0$, where $\pi = \{a = t_0 < t_1 < \ldots < t_n = b\}$ is a partition of $[a,b]$ and $\tilde{\pi} = \{a \le s_1 \le \ldots \le s_n \le b\}$ with $s_i \in [t_{i-1}, t_i]$. Note that, by convention, this limit has to be independent of the choice of the points of the set $\tilde{\pi}$. Also observe that, in order to establish this convergence, we need to show that for $\varepsilon > 0$ there is $\delta > 0$ such that

$$\left|\mathrm{RS}(f,g,\pi^{'},\tilde{\pi}^{'}) - \mathrm{RS}(f,g,\pi^{''},\tilde{\pi}^{''})\right| \le \varepsilon \quad \text{if } |\pi^{'}|, |\pi^{''}| \le \delta.$$

Indeed, in this case, we can find a sequence $\{\pi_n : n \in \mathbb{N}\}$ of partitions of $[a,b]$ such that $|\pi_n| < 2^{-n}$ and

$$|\mathrm{RS}(f,g,\pi_n,\tilde{\pi}_n) - \mathrm{RS}(f,g,\pi_{n+1},\tilde{\pi}_{n+1})| \le 2^{-n}.$$

So, we have that there is $\lambda \in \mathbb{R}$ such that $\mathrm{RS}(f,g,\pi_n,\tilde{\pi}_n)$ goes to λ, as $n \to \infty$, due to $\{\mathrm{RS}(f,g,\pi_n,\tilde{\pi}_n) : n \in \mathbb{N}\}$ being a Cauchy sequence in $\mathbb{R}$. Now it is easy to prove that our claim is satisfied since

$$\left|\mathrm{RS}(f,g,\pi^{'},\tilde{\pi}^{'}) - \lambda\right| \le \left|\mathrm{RS}(f,g,\pi^{'},\tilde{\pi}^{'}) - \mathrm{RS}(f,g,\pi_n,\tilde{\pi}_n)\right| + |\mathrm{RS}(f,g,\pi_n,\tilde{\pi}_n) - \lambda|.$$

In the remaining of this book, we use the notation

$$\int_a^b f(s)d^Y g(s) = \lim_{|\pi|\to 0} RS(f,g,\pi,\tilde{\pi}) \tag{2.2}$$

if this limit exists. In this case, we say either that f is Riemann-Stieltjes, or Young integrable with respect to g, or that f belongs to the domain of the integral $\int_a^b \cdot d^Y g$. This is denoted by $f \in \mathrm{Dom}(\int_a^b \cdot d^Y g)$. It is not difficult to see that $\int_a^b \cdot d^Y g(s)$ is a linear operator on its domain. Also, we will see that this operator satisfies some types of dominated convergence results later on. In other words, the limit $\int_a^b f(s)d^Y g(s)$ deserves the name integral.

Now let f and g be two bounded functions. It is well-known that if f is Riemann-Stieltjes integrable with respect to g (i.e., (2.1) converges as $|\pi| \to 0$), then f and g do not have discontinuities at a same point. To see that this is true, consider $x \in (a,b)$ and a partition $\pi = \{a = t_0 < t_1 < \ldots < t_n = b\}$ such that there is $i > 2$ satisfying $x \in (t_{i-1}, t_i)$. Define

$$\tilde{\pi}^{'} = \begin{cases} t_j, & \text{if either } j \le i-1, \text{ or } j > i+1 \\ s_i \in (t_{i-1}, x), & j = i \end{cases}$$

DOI: 10.1201/9781003484912-2

and

$$\tilde{\pi}^{''} = \begin{cases} t_j, & \text{if either } j \leq i-1, \text{ or } j > i+1 \\ \tilde{s}_i \in (x, t_i), & j = i. \end{cases}$$

Hence,

$$\mathrm{RS}(f, g, \pi, \tilde{\pi}^{'}) - \mathrm{RS}(f, g, \pi, \tilde{\pi}^{''}) = [f(s_i) - f(\tilde{s}_i)]\,[g(t_i) - g(t_{i-1})]\,. \tag{2.3}$$

Consequently, f cannot be integrable with respect to g if both f and g are discontinuous at x. The case, $x \in \{a, b\}$ is proven similarly.

As an example, we have the following result, where we use the notation

$$\mathrm{var}_1(g; [a, b]) := \sup\{\sum_{i=1}^{n} |g(t_i) - g(t_{i-1})| : \pi \text{ partition of } [a, b]\}. \tag{2.4}$$

Proposition 32 *Let $f : [a, b] \to \mathbb{R}$ be a continuous function and $g : [a, b] \to \mathbb{R}$ a function of bounded variation (i.e., $\mathrm{var}_1(g; [a, b]) < \infty$). Then, the Riemann-Stieltjes sum (2.1) converges as $|\pi| \to 0$.*

Proof. Let $\pi^{'} = \{a = t_0^{'} < t_1^{'} < \ldots < t_m^{'} = b\}$ and $\pi^{''} = \{a = t_0^{''} < t_1^{''} < \ldots < t_k^{''} = b\}$ be two partitions of the interval $[a, b]$. Set $\pi = \pi^{'} \cup \pi^{''} \cup \tilde{\pi}^{'} \cup \tilde{\pi}^{''} = \{a = t_0 < t_1 < \ldots < t_n = b\}$. For each index $j \in \{0, \ldots, m\}$, let $i(j) \in \{0, \ldots, n\}$ be such that $t_{i(j)} = t_j^{'}$, then

$$\begin{aligned}
&\left|\mathrm{RS}(f, g, \pi, \{a \leq t_1 \leq \ldots \leq t_n = b\}) - \mathrm{RS}(f, g, \pi^{'}, \tilde{\pi}^{'})\right| \\
&\leq \sum_{j=1}^{m} \left| f(s_j^{'}) \left[g(t_j^{'}) - g(t_{j-1}^{'})\right] - \sum_{\ell=i(j-1)+1}^{i(j)} f(t_\ell)\,[g(t_\ell) - g(t_{\ell-1})] \right| \qquad (2.5) \\
&= \sum_{j=1}^{m} \left| \sum_{\ell=i(j-1)+1}^{i(j)} \left[f(s_j^{'}) - f(t_\ell)\right] [g(t_\ell) - g(t_{\ell-1})] \right| \\
&\leq \left(\sup_{|t-s| \leq |\pi^{'}|} |f(t) - f(s)| \right) \mathrm{var}_1(g; [a, b]). \qquad (2.6)
\end{aligned}$$

Similarly, we can establish

$$\begin{aligned}
&\left|\mathrm{RS}(f, g, \pi, \{a \leq t_1 \leq \ldots \leq t_n = b\}) - \mathrm{RS}(f, g, \pi^{''}, \tilde{\pi}^{''})\right| \\
&\leq \left(\sup_{|t-s| \leq |\pi^{''}|} |f(t) - f(s)| \right) \mathrm{var}_1(g; [a, b]).
\end{aligned}$$

Hence, (2.6) and the continuity of f imply that, for $\varepsilon > 0$, we have

$$\left|\mathrm{RS}(f, g, \pi^{'}, \tilde{\pi}^{'}) - \mathrm{RS}(f, g, \pi^{''}, \tilde{\pi}^{''})\right| \leq \varepsilon$$

whenever $|\pi^{'}|$ and $|\pi^{''}|$ are small enough. Therefore the proof is complete. □

Examples of functions of bounded variation are continuous functions $g : [a, b] \to \mathbb{R}$ with a derivative that is continuous in (a, b) and Lebesgue integrable on $[a, b]$ due to

$$\sum_{i=1}^{n} |g(t_i) - g(t_{i-1})| = \sum_{i=1}^{n} \left| \int_{t_{i-1}}^{t_i} g^{'}(s) ds \right| \leq \int_a^b \left| g^{'}(s) \right| ds < \infty,$$

which gives that $\mathrm{var}_1(g, [a, b]) \leq \int_a^b \left| g^{'}(s) \right| ds$. Note that such a function g is an absolutely continuous function and that the last inequality is also satisfied for any absolutely continuous function g on $[a,b]$ (see Royden [187]). Moreover, if g' is continuous on $[a,b]$, then the mean value theorem, implies that, for any continuous function f,

$$\int_a^b f(s)d^Y g(s) = \lim_{|\pi|\to 0} RS(f,g,\pi,\tilde{\pi}) = \int_a^b f(s)g'(s)ds. \tag{2.7}$$

Indeed, for $\pi = \{a = t_0 < t_1 < \ldots < t_n = b\}$ and $i = 1,\ldots,n$, there is $\theta_i \in (t_{i-1}, t_i)$ satisfying

$$\sum_{i=1}^n f(s_i)\left[g(t_i) - g(t_{i-1})\right] = \sum_{i=1}^n f(s_i)g'(\theta_i)\left(t_i - t_{i-1}\right),$$

which, together with Proposition 32, the continuity of g' and

$$\begin{aligned}
&\left|\sum_{i=1}^n f(s_i)g'(\theta_i)\left(t_i - t_{i-1}\right) - \sum_{i=1}^n f(s_i)g'(s_i)\left(t_i - t_{i-1}\right)\right| \\
&\quad \leq \sum_{i=1}^n |f(s_i)| \left|g'(\theta_i) - g'(s_i)\right| \left(t_i - t_{i-1}\right) \leq CT\left(\sup_{|t-s|\leq|\pi|} \left|g'(t) - g'(s)\right|\right),
\end{aligned}$$

implies

$$\begin{aligned}
\int_a^b f(s)d^Y g(s) &= \lim_{|\pi|\to 0} RS(f,g,\pi,\tilde{\pi}) \\
&= \lim_{|\pi|\to 0} \sum_{i=1}^n f(s_i)g'(\theta_i)\left(t_i - t_{i-1}\right) = \lim_{|\pi|\to 0} \sum_{i=1}^n f(s_i)g'(s_i)\left(t_i - t_{i-1}\right).
\end{aligned}$$

That is, our claim holds.

The domain of the integral $\int_a^b \cdot d^Y g$ contains all the continuous functions on $[a,b]$ if $\mathrm{var}_1(g,[a,b]) < \infty$ because of Proposition 32. So, a natural question is: does the domain of $\int_a^b \cdot d^Y g$ contain all continuous functions if $\mathrm{var}_1(g,[a,b]) = \infty$? The answer to this question is given by Proposition 33, whose proof utilizes the following auxiliary result.

Lemma 24 *Let $var_1(g,[a,b]) < \infty$ and $a < c < b$. Then,*

$$var_1(g,[a,b]) = var_1(g,[a,c]) + var_1(g,[c,b]).$$

In particular, we obtain that $x \mapsto var_1(g,[a,x])$ is a non-decreasing function on $[a,b]$.

The proof of this result is straightforward and is left to the reader as an exercise.

Now we are ready to answer the above question.

Proposition 33 *Suppose that g is a bounded function such that $var_1(g,[a,b]) = \infty$. Then, there is a continuous function $f : [a,b] \to \mathbb{R}$ such that $\int_a^b f(s)d^Y g(s)$ does not exist (i.e., (2.1) does not converge as $|\pi| \to 0$).*

Proof. We will divide the proof into two steps.
Step 1: Here we assume that there is $\eta \in (a,b)$ such that $\mathrm{var}_1(g,[a,t]) = \infty$, for any $t \in (a,\eta]$.

Under the assumptions of this step, we can find two sequences $\{t_n \in (0,T] : n \in \mathbb{N}\}$ and $\{n_k \in \mathbb{N} : k \in \mathbb{N}\}$ such that $t_1 = b$, $n_1 = 1$, $t_k \downarrow a$, $n_k \uparrow \infty$ as $k \to \infty$, and $\sum_{j=n_k+1}^{n_{k+1}} (g(t_j) - g(t_{j-1}))\xi_j > 1$, for all $k \in \mathbb{N}$. Here,

$$\xi_j = \begin{cases} 1, & \text{if } g(t_j) - g(t_{j-1}) \geq 0, \\ -1, & \text{otherwise.} \end{cases}$$

A mentioned continuous function is given as $f(a) = 0$, $f(t_j) = \xi_j/k$ for $j \in [n_k, n_{k+1})$, and f is lineal on the interval $[t_{j+1}, t_j]$ for $j \in \mathbb{N}$. It is clear that this function f is continuous on $[a, b]$. Note that we have

$$\sum_{j=n_k+1}^{n_{k+1}} f(t_j)(g(t_j) - g(t_{j-1})) > \frac{1}{k}, \quad k \in \mathbb{N}.$$

We claim that $\int_a^b f(s)d^Y g(s)$ does not exist. Otherwise, there is $\mu > 0$ such that, for $t_{n_q} < \mu$ and $p > q$,

$$|RS(f, g, \pi, \pi) - RS(f, g, \tilde{\pi}, \tilde{\pi})| = \left| f(a)\left[g(t_{n_q}) - g(a)\right] - \left(f(a)\left[g(t_{n_p}) - g(a)\right] + \sum_{j=n_q+1}^{n_p} f(t_j)\left[g(t_j) - g(t_{j-1})\right] \right) \right| < 1, \tag{2.8}$$

with $\pi = \{a = t_0 < t_{n_q} < s_1 < \ldots < s_\ell = b\}$ and $\tilde{\pi} = \{a = t_0 < t_{n_p} < \ldots < t_{n_q} < s_1 < \ldots < s_\ell = b\}$. Remember that we need that $s_1 - t_{n_q}, s_i - s_{i-1} < \mu$, $i = 2, \ldots, \ell$. Using $f(a) = 0$ and (2.8), we get that $\sum_{\ell=q}^{p-1} \frac{1}{\ell} < 1$ for all $p > q$, which is not possible. Thus, $\int_a^b f(s)d^Y g(s)$ does not exist.

Step 2: Now we suppose that there exists $t \in (a, b)$ so that $\mathrm{var}_1(g, [a, t]) < \infty$.

We consider the point $\tilde{t} = \sup\{s \in [a, b] : \mathrm{var}_1(g, [a, s]) < \infty\}$. Then, from Lemma 24, we have either $\mathrm{var}_1(g, [s, \tilde{t}]) = \infty$ for any $s \in (a, \tilde{t})$, or $\mathrm{var}_1(g, [\tilde{t}, r]) = \infty$ for any $r \in (\tilde{t}, b)$. In both cases, we are able to proceed as in Step 1 to build the continuous function f that we are looking for. The details are left to the reader as an exercise. □

The following is an immediate consequence of Propositions 32 and 33.

Theorem 16 *Let g be a real-valued and bounded function defined on $[a, b]$. Then, the domain of $\int_a^b \cdot d^Y g$ contains all real-valued continuous functions on $[a, b]$ if and only if $\mathrm{var}_1(g, [a, b]) < \infty$.*

Remark 30 *It is well-known that $\mathrm{var}_1(g, [a, b]) < \infty$ if and only if g is the difference of two non-decreasing functions $g_1, g_2 : [a, b] \to \mathbb{R}$ (see Lemma 106 below). So, this theorem is other characterization of the functions of bounded variation on a bounded interval $[a, b]$. In the case that g is a function of bounded variation that is continuous on the right, we have that g_1 and g_2 can be chosen to be also right-continuous due to Lemmas 106 and 107. Thus, we can associate g with a signed measure μ. This measure is defined on the Borel sets $\mathcal{B}([a, b])$ and satisfies, for $a \le c < d \le b$, $\mu((c, d]) = g(d) - g(c)$ and $\mu = \mu_1 - \mu_2$, where μ_i, $i = 1, 2$, is the measure generated by g_i through the Carathéodory extension theorem (Proposition 139), as it is stated in Rudin [189] and in Royden [187], as well as any book of measure theory. Consequently, in this last case, $\int_a^b f(s)d^Y g(s)$, with f continuous, is nothing else than the Lebesgue integral of f with respect to the involved signed measure μ.*

The following result is an immediate consequence of the definition of the Riemann-Stieltjes integral.

Theorem 17 *Let $g, g_i, f, f_i : [a, b] \to \mathbb{R}$ be bounded functions with $i = 1, 2$. Then,*

i) $\int_a^b c d^Y g(s) = c(g(b) - g(a))$*, for any constant* $c \in \mathbb{R}$.

ii) $\int_a^b f(s)d^Y g(s) = 0$ *if* g *is a constant function on* $[a,b]$. *Moreover,*

$$\int_a^b f(s)d^Y g(s) = f(a)(c - g(a)) + f(b)(g(b) - c)$$

if f *is continuous at* a *and* b, *and* g *is the constant* c *in the interval* (a,b).

iii) $(c_1 f_1 + c_2 f_2) \in Dom\left(\int_a^b \cdot d^Y g\right)$ *and*

$$\int_a^b (c_1 f_1 + c_2 f_2)(s)d^Y g(s) = c_1 \int_a^b f_1(s)d^Y g(s) + c_2 \int_a^b f_2(s)d^Y g(s)$$

if $f_1, f_2 \in Dom\left(\int_a^b \cdot d^Y g\right)$ *and* $c_1, c_2 \in \mathbb{R}$.

iv) $f \in Dom\left(\int_a^b \cdot d^Y (c_1 g_1 + c_2 g_2)\right)$ *and*

$$\int_a^b f(s)d^Y (c_1 g_1 + c_2 g_2)(s) = c_1 \int_a^b f(s)d^Y g_1(s) + c_2 \int_a^b f(s)d^Y g_2(s)$$

if $f \in Dom\left(\int_a^b \cdot d^Y g_1\right) \cap Dom\left(\int_a^b \cdot d^Y g_2\right)$ *and* $c_1, c_2 \in \mathbb{R}$.

v) $\left|\int_a^b f(s)d^Y g(s)\right| \leq \|f\|_{\infty,[a,b]} var_1(g;[a,b])$ *if* $f \in Dom\left(\int_a^b \cdot d^Y g\right)$.

vi) $\int_a^b f(s)d^Y g(s) = c(g(b) - g(a))$ *for some constant* c *between* $\min_{s\in[a,b]} f(s)$ *and* $\max_{s\in[a,b]} f(s)$ *if* g *is nondecreasing on* $[a,b]$ *and* $f \in Dom\left(\int_a^b \cdot d^Y g\right)$.

The proof of integration by parts formula for the Riemann-Stieltjes integral is quite simple and this formula could be used to analyze the existence of integrals.

Theorem 18 (Integration by parts formula) *Assume that* f *and* g *are bounded functions on* $[a,b]$, *and that* $f \in Dom\left(\int_a^b \cdot d^Y g\right)$. *Then,* $g \in Dom\left(\int_a^b \cdot d^Y f\right)$ *and*

$$\int_a^b f(s)d^Y g(s) = f(b)g(b) - f(a)g(a) - \int_a^b g(s)d^Y f(s).$$

Proof. For a partition $\pi = \{a = t_0 < t_1 < \ldots < t_n = b\}$ of $[a,b]$ and $\tilde{\pi} = \{a = s_0 \leq s_1 \leq \ldots \leq s_{n+1} = b\}$ such that $a = s_0 = t_0 \leq s_1 \leq t_1 \leq \ldots \leq s_n \leq t_n = b = s_{n+1}$, we have

$$\begin{aligned}
\sum_{i=1}^{n} g(s_i)\left(f(t_i) - f(t_{i-1})\right) &= \sum_{i=1}^{n} g(s_i)f(t_i) - \sum_{i=0}^{n-1} g(s_{i+1})f(t_i) \\
&= f(b)g(b) - f(a)g(a) - \sum_{i=0}^{n} f(t_i)\left(g(s_{i+1}) - g(s_i)\right) \\
&= f(b)g(b) - f(a)g(a) - \sum_{i=1}^{n+1} f(t_{i-1})\left(g(s_i) - g(s_{i-1})\right),
\end{aligned}$$

which implies that the result is satisfied. □

In order to motivate the following result, consider the functions $f, g : [0,2] \to \mathbb{R}$ given by

$$f(x) = \begin{cases} 0, & \text{if } x \in [0,1], \\ 1, & \text{if } x \in (1,2], \end{cases} \quad \text{and} \quad g(x) = \begin{cases} 0, & \text{if } x \in [0,1), \\ 1, & \text{if } x \in [1,2]. \end{cases}$$

Then, Theorem 17 yields that f belongs to the set Dom $\left(\int_0^1 \cdot d^Y g\right) \cap$ Dom $\left(\int_1^2 \cdot d^Y g\right)$, but $f \notin$ Dom $\left(\int_0^2 \cdot d^Y g\right)$ because f and g are discontinuous at 1 (see (2.3)).

Theorem 19 *Let $f, g : [a,b] \to \mathbb{R}$ be two bounded functions and $c \in (a,b)$.*

i) Assume that $f \in Dom\left(\int_a^b \cdot d^Y g\right)$. Then, $f \in Dom\left(\int_a^c \cdot d^Y g\right) \cap Dom\left(\int_c^b \cdot d^Y g\right)$.

ii) Suppose that f and g are not both discontinuous at c, and that $f \in Dom\left(\int_a^c \cdot d^Y g\right) \cap Dom\left(\int_c^b \cdot d^Y g\right)$. Then, $f \in Dom\left(\int_a^b \cdot d^Y g\right)$.

In both cases,

$$\int_a^b f(s) d^Y g(s) = \int_a^c f(s) d^Y g(s) + \int_c^b f(s) d^Y g(s). \tag{2.9}$$

Proof. We first assume that $f \in$ Dom $\left(\int_a^b \cdot d^Y g\right)$. Let π_1 and π_2 be partitions of $[a,c]$ and $[c,b]$, respectively. Then $\pi = \pi_1 \cup \pi_2$ is a partition of the interval $[a,b]$. Therefore, for $\varepsilon > 0$ there is $\delta > 0$ such that

$$\left| RS(f, g, \pi_1, \tilde{\pi}_1) + RS(f, g, \pi_2, \tilde{\pi}_2) - \int_a^b f(s) d^Y g(s) \right| < \frac{\varepsilon}{2} \tag{2.10}$$

if $|\pi| < \delta$. Hence, fixing π_2, we have $|RS(f, g, \pi_3, \tilde{\pi}_3) - RS(f, g, \pi_4, \tilde{\pi}_4)| < \varepsilon$ for any partition π_i of $[a,c]$ such that $|\pi_i| < \delta$, $i = 3, 4$. Consequently, we obtain that $f \in$ Dom $\left(\int_a^c \cdot d^Y g\right)$. Similarly, we get that f also belongs to Dom $\left(\int_c^b \cdot d^Y g\right)$. In this case, equality (2.9) follows from (2.10).

On the other hand, now let f and g be as in Statement *ii*), and $\pi = \{a = t_0 < t_1 < \ldots < t_n = b\}$ a partition of $[a,b]$. Then, for $|\pi|$ small, enough, there is $j \in \{2, \ldots, n\}$ such that $c \in [t_{j-1}, t_j]$. Thus,

$$\begin{aligned}
&\sum_{i=1}^n f(s_i)\left[g(t_i) - g(t_{i-1})\right] \\
&= f(s_j)\left(g(t_j) - g(t_{j-1})\right) - f(c)\left(g(c) - g(t_{j-1})\right) \\
&\quad + \left(\sum_{i=1}^{j-1} f(s_i)\left[g(t_i) - g(t_{i-1})\right] + f(c)\left[g(c) - g(t_{j-1})\right]\right) \\
&\quad + \left(f(c)\left(g(t_j) - g(c)\right) + \sum_{i=j+1}^n f(s_i)\left[g(t_i) - g(t_{i-1})\right]\right) - f(c)\left(g(t_j) - g(c)\right) \\
&= \left(f(s_j) - f(c)\right)\left(g(t_j) - g(t_{j-1})\right) + \left(\sum_{i=1}^{j-1} f(s_i)\left[g(t_i) - g(t_{i-1})\right] + f(c)\left[g(c) - g(t_{j-1})\right]\right) \\
&\quad + \left(f(c)\left(g(t_j) - g(c)\right) + \sum_{i=j+1}^n f(s_i)\left[g(t_i) - g(t_{i-1})\right]\right).
\end{aligned}$$

Hence, the hypotheses of Statement *ii*) allows us to conclude that $f \in$ Dom $\left(\int_a^b \cdot d^Y g\right)$ and that equality (2.9) is also satisfied in this case. That is, the proof is complete. □

The following result can be interpreted as a version of dominated convergence theorem for the Young integral. In the following sections of this chapter, we will obtain other versions of this fundamental theorem.

Theorem 20 *Let $f : [a,b] \to \mathbb{R}$ be a bounded function and $var_1(g;[a,b]) < \infty$.*

i) Suppose that $\{f_n : n \in \mathbb{N}\} \subset Dom\left(\int_a^b \cdot d^Y g\right)$ and that $\|f_n - f\|_{\infty,[a,b]} \to 0$. Then, $f \in Dom\left(\int_a^b \cdot d^Y g\right)$ and $\int_a^b f(s) d^Y g(s) = \lim_{n\to\infty} \int_a^b f_n(s) d^Y g(s)$.

ii) Assume $f \in \cap_{n=1}^{\infty} Dom\left(\int_a^b \cdot d^Y g_n\right)$ and $\lim_{n\to\infty} var_1(g - g_n;[a,b]) = 0$. Then, $f \in Dom\left(\int_a^b \cdot d^Y g\right)$ and $\int_a^b f(s) d^Y g(s) = \lim_{n\to\infty} \int_a^b f(s) d^Y g_n(s)$.

Proof. We first deal with Statement $i)$. So, Theorem 17.$v)$ implies that, for $n, m \in \mathbb{N}$,

$$\begin{aligned} & \left| \int_a^b f_n(s) d^Y g(s) - \int_a^b f_m(s) d^Y g(s) \right| \\ = & \left| \int_a^b (f_n(s) - f_m(s))\, d^Y g(s) \right| \le \|f_n - f_m\|_{\infty,[a,b]} \mathrm{var}_1(g;[a,b]). \end{aligned}$$

Hence, the fact that $\|f_n - f\|_{\infty,[a,b]} \to 0$ gives that $\{\int_a^b f_n(s) d^Y g(s) : n \in \mathbb{N}\}$ is a Cauchy sequence in $\mathbb{R}$. Consequently, the integral $\int_a^b f_n(s) d^Y g(s)$ goes to a real number I, as $n \to \infty$. Now, consider a partition π of $[a,b]$. Then, with $I_n = \int_a^b f_n(s) d^Y g(s)$, the triangle inequality leads us to write

$$\begin{aligned} |\mathrm{RS}(f,g,\pi,\tilde{\pi}) - I| & \le |\mathrm{RS}(f,g,\pi,\tilde{\pi}) - \mathrm{RS}(f_n,g,\pi,\tilde{\pi})| + |\mathrm{RS}(f_n,g,\pi,\tilde{\pi}) - I_n| + |I_n - I| \\ & \le \|f - f_n\|_{\infty,[a,b]} \mathrm{var}_1(g;[a,b]) + |\mathrm{RS}(f_n,g,\pi,\tilde{\pi}) - I_n| + |I_n - I|. \end{aligned}$$

Thus, for $\varepsilon > 0$, there is $n_0 \in \mathbb{N}$ such that

$$|\mathrm{RS}(f,g,\pi,\tilde{\pi}) - I| \le \frac{2\varepsilon}{3} + |\mathrm{RS}(f_{n_0},g,\pi,\tilde{\pi}) - I_{n_0}|.$$

Therefore, we are able to find $\delta > 0$ such that $|\mathrm{RS}(f,g,\pi,\tilde{\pi}) - I| < \varepsilon$ if $|\pi| < \delta$, which, together with the definition of I, establishes that Statement $i)$ holds.

Finally, to show that Statement $ii)$ is also true, we proceed as in the proof of Statement $i)$ via Theorem 17.$iv)$. □

An important fact is that we can consider indicator functions as integrands of some Riemann-Stieltjes integrals.

Theorem 21 *Let $g : [a,b] \to \mathbb{R}$ be a bounded function, which is continuous at $\tilde{a}$ and $\tilde{b}$ with $a \le \tilde{a} < \tilde{b} \le b$. Then, if f is either $1_{[\tilde{a},\tilde{b}]}$, $1_{(\tilde{a},\tilde{b}]}$, $1_{[\tilde{a},\tilde{b})}$ or $1_{(\tilde{a},\tilde{b})}$, we have that $f \in Dom\left(\int_a^b \cdot d^Y g\right)$ and*

$$\int_a^b f(s) d^Y g(s) = g(\tilde{b}) - g(\tilde{a}).$$

Proof. We only consider the case $f = 1_{[\tilde{a},\tilde{b}]}$, with $a < \tilde{a} < \tilde{b} < b$. For the proof of the other cases, we can proceed similarly.

Let $\pi = \{a = t_0 < t_1 < \ldots < t_n = b\}$ be a partition of $[a,b]$ and $\tilde{\pi} = \{a \le s_1 \le \ldots \le s_n \le b\}$ with $s_i \in [t_{i-1}, t_i]$. Then, for $|\pi|$ small enough, there are $1 \le j < k \le n$ such that

$\tilde{a} \in [t_{j-1}, t_j)$ and $\tilde{b} \in (t_{k-1}, t_k]$. This yields

$$\begin{aligned}
&\sum_{i=1}^{n} f(s_i)\left[g(t_i) - g(t_{i-1}\right] \\
&= 1_{\{s_k \le \tilde{b}\}} \left(g(t_k) - g(t_{k-1})\right) \\
&\quad +1_{\{\tilde{a}=s_{j-1}=t_{j-1}\}} \left(g(t_j) - g(t_{j-2}) + \sum_{i=j+1}^{k-1} f(s_i)\left[g(t_i) - g(t_{i-1}\right] \right) \\
&\quad +1_{\{s_{j-1}<t_{j-1}=\tilde{a}\}} \left(g(t_j) - g(t_{j-1}) + \sum_{i=j+1}^{k-1} f(s_i)\left[g(t_i) - g(t_{i-1}\right] \right) \\
&\quad +1_{\{t_{j-1}<\tilde{a}\le s_j\}} \left(g(t_j) - g(t_{j-1}) + \sum_{i=j+1}^{k-1} f(s_i)\left[g(t_i) - g(t_{i-1}\right] \right) \\
&\quad +1_{\{t_{j-1}\le s_j<\tilde{a}\}} \sum_{i=j+1}^{k-1} f(s_i)\left[g(t_i) - g(t_{i-1}\right] \\
&= 1_{\{s_k \le \tilde{b}\}} \left(g(t_k) - g(t_{k-1})\right) + 1_{\{\tilde{a}=s_{j-1}=t_{j-1}\}} \left(g(t_{k-1}) - g(t_{j-2})\right) \\
&\quad +1_{\{s_{j-1}<t_{j-1}=\tilde{a}\}} \left(g(t_{k-1}) - g(t_{j-1})\right) + 1_{\{t_{j-1}<\tilde{a}\le s_j\}} \left(g(t_{k-1}) - g(t_{j-1})\right) \\
&\quad +1_{\{t_{j-1}\le s_j<\tilde{a}\}} \left(g(t_{k-1}) - g(t_j)\right).
\end{aligned}$$

Now, it is easy to finish the proof. □

The following result deals with the continuity of the Young integral. Actually, we see that the regularity of the map $t \mapsto \int_a^t f(s)d^Y g(s)$ depends on that of g. Towards this end, remember that, for $x \in \mathbb{R}$, it is usual the notation

$$F(x-) = \lim_{h \downarrow 0} F(x-h) \quad \text{and} \quad F(x+) = \lim_{h \downarrow 0} F(x+h).$$

Theorem 22 *Let $f, g : [a,b] \to \mathbb{R}$ be two bounded functions such that $f \in Dom\left(\int_a^b \cdot d^Y g\right)$ and $F(t) = \int_a^t f(s)d^Y g(s)$, $t \in [a,b]$. Then,*

i) F is continuous at t_0 if g is so.

ii) $F(t_0+)$and $F(t_0-)$ are well-defined if the limits $g(t_0+)$and $g(t_0-)$ exist. Moreover, in this case,

$$F(t_0+) = \int_a^{t_0} f(s)d^Y g(s) + f(t_0)\left[g(t_0+) - g(t_0)\right]$$

and

$$F(t_0-) = \int_a^{t_0} f(s)d^Y g(s) - f(t_0)\left[g(t_0) - g(t_0-)\right].$$

Remark 31 *The function F is well-defined due to Theorem 19.i).*

Proof. For Statement *i)*, we first suppose that $t_0 \in (a,b)$. Let $h > 0$ be small enough (i.e. $a < t_0 + h < b$) and π a partition of the interval $[a, t_0]$. Then,

$$\begin{aligned}
&\left|\mathrm{RS}(f,g,\pi,\tilde{\pi}) - \int_a^{t_0+h} f(s)d^Y g(s)\right| \\
&\leq \left|f(t_0)\left[g(t_0+h)-g(t_0)\right] + \mathrm{RS}(f,g,\pi,\tilde{\pi}) - \int_a^{t_0+h} f(s)d^Y g(s)\right| \\
&\quad + \|f\|_{\infty,[a,b]} \left|g(t_0+h)-g(t_0)\right|. \qquad (2.11)
\end{aligned}$$

For $\varepsilon > 0$, there is $\delta > 0$ such that $\left|\mathrm{RS}(f,g,\pi,\tilde{\pi}) - \int_a^{t_0} f(s)d^Y g(s)\right| \leq \frac{\varepsilon}{3}$, if $|\pi| < \delta$. Remember that δ can be chosen independent of t_0 due to the proof of Theorem 19.i). That is, this inequality is still true if we change t_0 and π by $s \in [a,b]$ and a partition of $[0,s]$, respectively. Thus, (2.11) implies

$$\left|\int_a^{t_0} f(s)d^Y g(s) - \int_a^{t_0+h} f(s)d^Y g(s)\right| < \varepsilon,$$

whenever $h < \delta$ and $\|f\|_{\infty,[a,b]} |g(t_0+h) - g(t_0)| < \varepsilon/3$. Similarly, now let π be a partition of $[a, t_0 - h]$. In this case,

$$\begin{aligned}
&\left|\mathrm{RS}(f,g,\pi,\tilde{\pi}) - \int_a^{t_0} f(s)d^Y g(s)\right| \\
&\leq \left|f(t_0)\left[g(t_0)-g(t_0-h)\right] + \mathrm{RS}(f,g,\pi,\tilde{\pi}) - \int_a^{t_0} f(s)d^Y g(s)\right| \\
&\quad + \|f\|_{\infty,[a,b]} \left|g(t_0)-g(t_0-h)\right|.
\end{aligned}$$

Hence, we can also see that $F(t_0 - h)$ goes to $F(t_0)$ as $h \downarrow 0$ by proceeding as before. Also note that the previous calculations can be applied to the case $t_0 \in \{a, b\}$.

Now, we assume that $g(t_0+)$ and $g(t_0-)$ exist, and that they are not the same. Let π be a partition of $[a, t_0 - h]$. Then,

$$\begin{aligned}
&\left|\int_a^{t_0} f(s)d^Y g(s) - f(t_0)\left[g(t_0)-g(t_0-)\right] - \int_a^{t_0-h} f(s)d^Y g(s)\right| \\
&\leq \left|\int_a^{t_0} f(s)d^Y g(s) - f(t_0)\left[g(t_0)-g(t_0-h)\right] - \mathrm{RS}(f,g,\pi,\tilde{\pi})\right| \\
&\quad + \left|\mathrm{RS}(f,g,\pi,\tilde{\pi}) - \int_a^{t_0-h} f(s)d^Y g(s)\right| + \left|f(t_0)\left[g(t_0-)-g(t_0-h)\right]\right|.
\end{aligned}$$

Consequently, we can conclude that $F(t_0-) = \int_a^{t_0} f(s)d^Y g(s) - f(t_0)\left[g(t_0) - g(t_0-)\right]$.

Finally, take a partition of the interval $[a, t_0]$. So,

$$\begin{aligned}
&\left|\int_a^{t_0} f(s)d^Y g(s) + f(t_0)\left[g(t_0+)-g(t_0)\right] - \int_a^{t_0+h} f(s)d^Y g(s)\right| \\
&\leq \left|\int_a^{t_0} f(s)d^Y g(s) - \mathrm{RS}(f,g,\pi,\tilde{\pi})\right| \\
&\quad + \left|\mathrm{RS}(f,g,\pi,\tilde{\pi}) + f(t_0)\left[g(t_0+h)-g(t_0)\right] - \int_a^{t_0+h} f(s)d^Y g(s)\right| \\
&\quad + \|f\|_{\infty,[a,b]} \left|g(t_0+)-g(t_0+h)\right|.
\end{aligned}$$

Therefore, the proof is complete □.

The following result is a version of the fundamental theorem of calculus for Young integral. In this chapter, we will give other version of this important theorem.

Theorem 23 (A fundamental theorem of calculus) *Let $g : [a,b] \to \mathbb{R}$ be a continuous function such that $var_1(g;[a,b]) < \infty$ and $f \in C^1(\mathbb{R})$. Then,*

$$f(g(b)) - f(g(a)) = \int_a^b f^{'}(g(s))d^Y g(s). \tag{2.12}$$

Remark 32 *By Proposition 32, the integral on the right-hand side of the last equality is well-defined.*

Proof. Let $\pi = \{a = t_0 < t_1 < \ldots < t_n = b\}$ be a partition of $[a,b]$ and $\tilde{\pi} = \{a \le s_1 \le \ldots \le s_n \le b\}$ with $s_i \in [t_{i-1}, t_i]$. Then, the mean value theorem implies that there exists a real number θ_i between $g(t_i)$ and $g(t_{i-1})$, for each $i = 1, \ldots, n$, such that

$$f(g(b)) - f(g(a)) = \sum_{i=1}^{n} [f(g(t_i)) - f(g(t_{i-1}))] = \sum_{i=1}^{n} f^{'}(\theta_i)\,[g(t_i) - g(t_{i-1})]\,.$$

Hence, since the fact that g is a continuous function allows us to find $r_i \in (t_{i-1}, t_i)$ such that $\theta_i = g(r_i)$, for any $i = 1, \ldots, n$, we have

$$\begin{aligned} f(g(b)) - f(g(a)) &= \sum_{i=1}^{n} f^{'}(g(r_i))\,[g(t_i) - g(t_{i-1})] \\ &= \lim_{|\pi|\to 0} \sum_{i=1}^{n} f^{'}(g(r_i))\,[g(t_i) - g(t_{i-1})] \end{aligned} \tag{2.13}$$

We also have

$$\begin{aligned} &\left| \sum_{i=1}^{n} \left(f^{'}(g(s_i))\,[g(t_i) - g(t_{i-1})] - f^{'}(g(r_i))\,[g(t_i) - g(t_{i-1})] \right) \right| \\ &\le \left(\sup_{|t-s|\le|\pi|} \left| f^{'}(g(t)) - f^{'}(g(s)) \right| \right) \mathrm{var}_1(g;[a,b]), \end{aligned}$$

which, together with (2.13) and the continuity of the function $f^{'}(g)$, yields that (2.12) is satisfied. □

Now, we consider a right-continuous and non-decreasing function $g : [a,b] \to \mathbb{R}$. This function has bounded variation since $\mathrm{var}_1(g;[a,b]) = g(b) - g(a) < \infty$. Hence, by Proposition 139 below, there is a complete measure space $([a,b], \mathcal{F}, \mu)$ such that $\mathcal{F}$ contains the Borel σ-algebra $\mathcal{B}([a,b])$ and $\mu((c,d]) = g(d) - g(c)$, for $c, d \in [a,b]$, $c < d$. For details, the reader can consult Royden [187] (Section 12), or Rudin [189] (Chapter 11).

The following aim is to see that the Lebesgue integral with respect to the measure μ is an extension of the Riemann-Stieljes integral $\int_a^b \cdot\, d^Y g$. It means, if $f : [a,b] \to \mathbb{R}$ is Riemann-Stieltjes integrable with respect to g, then it is also Lebesgue integrable with respect to μ and

$$\int_a^b f(s)d^Y g(s) = \int_a^b f(s)\mu(ds).$$

This is done as follows. For a partition $\pi = \{a = x_0 < x_1 < \ldots < x_n = b\}$ of $[a,b]$ and a bounded function $f : [a,b] \to \mathbb{R}$, set

$$m_i = \inf\{f(x) : x \in [x_{i-1}, x_i]\} \quad \text{and} \quad M_i = \sup\{f(x) : x \in [x_{i-1}, x_i]\}, \tag{2.14}$$

for $i = 1, \ldots, n$. We define the lower and upper sums corresponding to the partition π by

$$L(f, g, \pi) = \sum_{i=1}^{n} m_i \left(g(x_i) - g(x_{i-1})\right) \quad \text{and} \quad U(f, g, \pi) = \sum_{i=1}^{n} M_i \left(g(x_i) - g(x_{i-1})\right),$$

respectively. Note that these definitions imply $L(f, g, \pi) \leq U(f, g, \pi)$ by taking into account that g is non-decreasing. Moreover, for two partitions π_1 and π_2 of $[a, b]$ such that π_2 is finer than π_1 (i.e., $\pi_1 \subset \pi_2$), we have that, using induction on the numbers of points of $\pi_2 \setminus \pi_1$,

$$L(f, g, \pi_1) \leq L(f, g, \pi_2) \leq U(f, g, \pi_2) \leq U(f, g, \pi_1).$$

Hence, for any two partitions $\tilde{\pi}_3$ and $\tilde{\pi}_4$ of $[a, b]$, we can establish

$$L(f, g, \tilde{\pi}_3) \leq U(f, g, \tilde{\pi}_4) \tag{2.15}$$

by using the partition $\pi = \tilde{\pi}_3 \cup \tilde{\pi}_4$, which is finer than $\tilde{\pi}_3$ and $\tilde{\pi}_4$. As a consequence, the quantities

$$\underline{\int_a^b} f(s) dg(s) := \sup\{L(f, g, \pi) : \pi \text{ a partition of } [a, b]\}$$

and

$$\overline{\int_a^b} f(s) dg(s) := \inf\{L(f, g, \pi) : \pi \text{ a partition of } [a, b]\}$$

are well-defined. Furthermore, $\underline{\int_a^b} f(s) dg(s) \leq \overline{\int_a^b} f(s) dg(s)$.

Theorem 24 *Let $g : [a, b] \to \mathbb{R}$ be a non-decreasing function and f a bounded real-valued function defined on $[a, b]$ such that $f \in Dom(\int_a^b \cdot d^Y g)$. Then,*

$$\underline{\int_a^b} f(s) dg(s) = \overline{\int_a^b} f(s) dg(s) = \int_a^b f(s) d^Y g(s).$$

Proof. Note that

$$\underline{\int_a^b} f(s) dg(s) \leq \int_a^b f(s) d^Y g(s) \leq \overline{\int_a^b} f(s) dg(s) \tag{2.16}$$

by considering (2.15), the definition of the three involved quantities and the fact that

$$L(f, g, \pi) \leq RS(f, g, \pi, \tilde{\pi}) \leq U(f, g, \pi),$$

for every partition π of $[a, b]$. Also note that the three sums in the last inequalities are zero if $g(a) = g(b)$ (i.e., g is a constant function), which implies that the result is true. So, we can assume that $g(a) < g(b)$.

For $\varepsilon > 0$, we can find a partition $\pi = \{a = x_0 < x_1 < \ldots < x_n = b\}$ of $[a, b]$ and $s_i \in [x_{i-1}, x_i]$ such that

$$f(s_i) - m_i \leq \frac{\varepsilon}{2(g(b) - g(a))} \quad \text{and} \quad \left| \int_a^b f(s) d^Y g(s) - RS(f, g, \pi, \tilde{\pi}) \right| \leq \varepsilon/2.$$

Then, we have

$$\begin{aligned} & \left| \int_a^b f(s) d^Y g(s) - L(f, g, \pi) \right| \\ \leq & \left| \int_a^b f(s) d^Y g(s) - RS(f, g, \pi, \tilde{\pi}) \right| + |RS(f, g, \pi, \tilde{\pi}) - L(f, g, \pi)| \leq \frac{\varepsilon}{2} + \frac{\varepsilon}{2} = \varepsilon. \end{aligned}$$

Since ε is arbitrary, we obtain that (2.16) leads us to conclude that

$$\underline{\int_a^b} f(s)dg(s) = \int_a^b f(s)d^Y g(s).$$

Similarly we can show that $\overline{\int_a^b} f(s)dg(s) = \int_a^b f(s)d^Y g(s)$, and, therefore, the proof is complete. $\square$

The last result motivates the following:

Definition 26 *Let $f, g : [a, b] \to \mathbb{R}$ be a bounded function and a non-decreasing function, respectively. We say that f is Darboux-Stieltjes integrable on $[a, b]$ with respect to g (represented by $f \in Dom(\int_a^b \cdot d^D g)$) if*

$$\underline{\int_a^b} f(s)dg(s) = \overline{\int_a^b} f(s)dg(s).$$

In this case, this common quantity is denoted by $\int_a^b f(s)d^D g(s)$.

Observe that Theorem 24 implies that the Darboux-Stieltjes integral is an extension of the Riemann-Stieltjes one, in the sense that $\text{Dom}(\int_a^b \cdot d^Y g) \subset \text{Dom}(\int_a^b \cdot d^D g)$ and

$$\int_a^b f(s)d^Y g(s) = \int_a^b f(s)d^D g(s), \quad \text{for } f \in \text{Dom}(\int_a^b \cdot d^Y g).$$

Now we study the relation between the Riemann-Stieltjes integral and the Lebesgue integral with respect to the measure μ, which are associated with the function g.

Theorem 25 *Let $g : [a, b] \to \mathbb{R}$ be a right-continuous and non-decreasing function, and μ the measure constructed using g via Proposition 139. Then, a bounded function $f : [a, b] \to \mathbb{R}$ is Lebesgue integrable on $[a, b]$ with respect to μ if $f \in Dom(\int_a^b \cdot d^Y g)$. Moreover, in this case,*

$$\int_a^b f(s)d^Y g(s) = \int_a^b f(s)\mu(ds).$$

Proof. Let $f \in \text{Dom}(\int_a^b \cdot d^Y g)$ be a bounded function. Then, by Theorem 24, and the definitions of $\underline{\int_a^b} f(s)dg(s)$ and $\overline{\int_a^b} f(s)dg(s)$, there exists a sequence $\{\pi_n : n \in \mathbb{N}\}$ of partitions of the interval $[a, b]$ such that

$$\underline{\int_a^b} f(s)dg(s) = \lim_{n\to\infty} L(f, g, \pi_n) \quad \text{and} \quad \overline{\int_a^b} f(s)dg(s) = \lim_{n\to\infty} U(f, g, \pi_n). \tag{2.17}$$

For $\pi_k = \{a = x_0 < x_1 < \ldots < x_n = b\}$, we use the convention

$$L_k(x) = \begin{cases} f(a), & \text{if } x = a, \\ m_i, & \text{if } x \in (x_{i-1}, x_i], \end{cases} \quad \text{and} \quad U_k(x) = \begin{cases} f(a), & \text{if } x = a, \\ M_i, & x \in (x_{i-1}, x_i], \end{cases}$$

where m_i and M_i are introduced in (2.14). Thus we have that $\{L_k(x) : k \in \mathbb{N}\}$ is a non-decreasing sequence and $\{U_k(x) : k \in \mathbb{N}\}$ is a non-increasing sequence, for every $x \in [a, b]$. Also, we have that, defining $g(a-) = g(a)$,

$$\int_a^b L_k(x)\mu(dx) = L(f, g, \pi_k) \quad \text{and} \quad \int_a^b U_k(x)\mu(dx) = U(f, g, \pi_k). \tag{2.18}$$

Now, define

$$L(x) = \lim_{k\to\infty} L_k(x) \quad \text{and} \quad U(x) = \lim_{k\to\infty} U_k(x), \quad \text{for } x \in [a,b].$$

Consequently, (2.14) leads to $L(x) \le f(x) \le U(x)$, for all $x \in [a,b]$. In this way, the fact that f is bounded, (2.17), (2.18), the dominated convergence theorem and Theorem 24 allow us to get

$$\int_a^b L(x)\mu(dx) = \underline{\int_a^b} f(s)dg(s) = \overline{\int_a^b} f(s)dg(s) = \int_a^b U(x)\mu(dx),$$

which gives that $L = U$ μ-almost surely due to $L \le U$. Therefore, $L = f = U$ μ-almost surely.

Finally, using Theorem 24 again, we obtain

$$\int_a^b f(s)d^Y g(s) = \int_a^b f(s)d^D g(s) = \int_a^b f(x)\mu(dx)$$

because the involved measure space is complete as Proposition 139 indicates. □

An immediate consequence of Theorem 25 is that the Lebesgue-Stieltjes integral is an extension of the Riemann-Stieltjes integral with respect to the right-continuous and nondecreasing function g. It means, a bounded function f belonging to $\mathrm{Dom}(\int_a^b \cdot d^Y g)$ is Lebesgue integrable with respect to μ and

$$\int_a^b f(s)d^Y g(s) = \int_a^b f(x)\mu(dx).$$

It is easy to find bounded and Lebesgue integrable functions that are not Riemann-Stieltjes integrable with respect to g. For instance, consider the function

$$f(x) = \begin{cases} 1, & \text{if } x \in \mathbb{I} \cap [0,1], \\ 0, & \text{if } x \in \mathbb{Q} \cap [0,1]. \end{cases}$$

Here $\mathbb{I}$ and $\mathbb{Q}$ are the irrational and the rational numbers, respectively. Then, $f = 1$ almost surely. Thus, f is integrable with respect to the Lebesgue measure on $[0,1]$ and

$$\int_0^1 f(x)dx = \int_0^1 1dx = 1.$$

But, $f \notin \mathrm{Dom}(\int_0^1 \cdot d^Y g) \cup \mathrm{Dom}(\int_0^1 \cdot d^D g)$ with $g(x) = x$, $x \in [0,1]$. Indeed, it is not difficult to see

$$\underline{\int_0^1} f(s)dg(s) = 0 \quad \text{and} \quad \overline{\int_0^1} f(s)dg(s) = 1.$$

Furthermore, for a partition $\{0 = x_0 < x_1 < \ldots < x_n = 1\}$, we can easily show

$$\sum_{i=1}^n f(s_i)(x_i - x_{i-1}) = 0, \quad \text{if } s_i \in \mathbb{Q}, \text{ for } i = 1, \ldots, n$$

and

$$\sum_{i=1}^n f(\tilde{s}_i)(x_i - x_{i-1}) = 1, \quad \text{if } \tilde{s}_i \in \mathbb{I}, \text{ for } i = 1, \ldots, n.$$

Thus, our claim is true. Actually, this can be explained utilizing the following result, which is established in Ter Horst [205]. In order to state it, we introduce the set

$$\mathcal{Z}_\mu([a,b]) = \{x \in [a,b] : \mu(\{x\}) = 0\}.$$

Proposition 34 *Let $f, g : [a,b] \to \mathbb{R}$ be two bounded functions. Also let g be non-decreasing on $[a,b]$. Then:*

i) If $f \in Dom(\int_a^b \cdot d^D g)$, we have that f is Lebesgue integrable with respect to μ and

$$\int_a^b f(s) d^D g(s) = \int_a^b f(s) \mu(ds).$$

ii) $f \in Dom(\int_a^b \cdot d^D g)$ if and only if f is continuous μ-almost surely on $\mathcal{Z}_\mu([a,b])$ and f and g do not have common discontinuous on either the left, or the right on $[a,b]$.

Remember that we have already pointed out that if f belongs to $\mathrm{Dom}(\int_a^b \cdot d^Y g)$, then f and g have no discontinuities at the same point on $[a,b]$. But, if $f \in \mathrm{Dom}(\int_a^b \cdot d^D g)$, f and g may have discontinuities at the same point, as it is indicated in Proposition 34.*ii*). Also, this proposition explains why $\mathrm{Dom}(\int_a^b \cdot d^Y g) \subset \mathrm{Dom}(\int_a^b \cdot d^D g)$ (see Theorem 24). In this way, it is quite simple to figure out a function f that belongs to $\mathrm{Dom}(\int_a^b \cdot d^D g) \setminus \mathrm{Dom}(\int_a^b \cdot d^Y g)$, which is left to the reader as an exercise.

2.1 Functions with finite p-variation

Now, we study the convergence of the Riemann-Stieltjes sum given in (2.1) for functions with finite p-variation. That is, we deal with the convergence of

$$\mathrm{RS}(f, g, \pi, \tilde{\pi}) := \sum_{i=1}^{n} f(s_i) \left[g(t_i) - g(t_{i-1}) \right], \quad \text{as } |\pi| \to 0.$$

Remember that $\pi = \{a = t_0 < t_1 < \ldots < t_n = b\}$ is a partition of $[a,b]$ and $\tilde{\pi} = \{a \le s_1 \le \ldots \le s_n \le b\}$ is a family of intermediate points. It means, $s_i \in [t_{i-1}, t_i]$, $i = 1, \ldots, n$.

The concept of p-variation was introduced by Wiener [213] and the idea of the convergence of the Riemann-Stieltjes sum $\mathrm{RS}(f, g, \pi, \tilde{\pi})$ via p-variation of the involved functions was developed by Young [216].

For $f : [a,b] \to \mathbb{R}$, $0 \le a < b < \infty$ and $p > 0$, in the remaining of this book, we use the convention

$$\mathrm{var}_p(f; [a,b]) := \sup\{\sum_{i=1}^{n} |f(t_i) - f(t_{i-1})|^p : \pi \text{ partition of } [a,b]\}, \tag{2.19}$$

which is called the p-variation of the function f on the interval $[a,b]$. In the literature, it is also called the total p-variation of f on $[a,b]$.

In this section, we work with the space of functions with finite p-variation. It means,

$$\mathcal{W}_p([a,b]) := \{f : [a,b] \to \mathbb{R} : \mathrm{var}_p(f; [a,b]) < \infty\}.$$

By Corollary 1, we know that the paths of Brownian motion belong to the space $\mathcal{W}_p([a,b])$, for all $0 \le a < b < \infty$ and $p > 2$, w.p.1.

Proposition 35 *Let f be a real-valued function defined on the interval $[a,b]$, $0 \le a < b < \infty$. Then, $p \mapsto (var_p(f; [a,b]))^{1/p}$ is non-increasing on $(0,\infty)$.*

Proof. Assume that $0 < p < q$ and $a_1, ..., a_n \geq 0$, with $n \in \mathbb{N}$. Then,

$$\begin{aligned}\sum_{j=1}^{n} a_j^q &\leq \left(\max_{j\in\{1,\dots,n\}} a_j^p\right)^{(q-p)/p} \sum_{j=1}^{n} a_j^p \leq \left(\sum_{j=1}^{n} a_j^p\right)^{(q-p)/p} \sum_{j=i}^{n} a_j^p \\ &= \left[\left(\sum_{j=1}^{n} a_j^p\right)^{q-p} \left(\sum_{j=i}^{n} a_j^p\right)^{p}\right]^{1/p} = \left(\sum_{j=1}^{n} a_j^p\right)^{q/p}. \end{aligned} \tag{2.20}$$

Now, let $\{a = t_0 < t_1 < \ldots < t_n = b\}$ be a partition of the interval $[a, b]$. Then, (2.20) leads us to conclude

$$\left(\sum_{j=1}^{n} |f(t_j) - f(t_{j-1})|^q\right)^{1/q} \leq \left(\sum_{j=1}^{n} |f(t_j) - f(t_{j-1})|^p\right)^{1/p} \leq (\mathrm{var}_p(f; [a, b]))^{1/p}.$$

Therefore, the result is true. □

The elements of $\mathcal{W}_p([a, b])$ are regulated functions as it is stated in the following result. That is, a function f belonging to $\mathcal{W}_p([a, b])$ has left and right-limits. It means, the limits

$$f(x+) = \lim_{h\downarrow 0} f(x + h) \quad \text{and} \quad f(x-) = \lim_{h\downarrow 0} f(x - h)$$

exist for every $x \in (a, b)$, as well as the limits $f(a+)$ and $f(b-)$.

Lemma 25 *Let $p > 0$ and $f \in \mathcal{W}_p([a, b])$. Then, f is a regulated function.*

Proof. Assume that, for example, $lim_{h\downarrow 0} f(x-h)$ does not exist for some $x \in (a, b]$. Then, there is an increasing sequence $\{x_n \in [a, x) : n \in \mathbb{N}\}$ and $\varepsilon_0 > 0$ such that $x_n \to x$ and, for $n \in \mathbb{N}$, $|f(x_{2n}) - f(x_{2n-1})| > \varepsilon_0$. Hence,

$$\mathrm{var}_p(f; [a, b]) \geq \sum_{i=1}^{n} |f(x_{2i}) - f(x_{2i-1})|^p > n\varepsilon_0^p, \quad \text{for all } n \in \mathbb{N},$$

which is impossible due to $\mathrm{var}_p(f; [a, b]) < \infty$ by hypothesis. Thus f has left-limits. Similarly, we can prove that f has also right-limits on $(a, b]$. □

The regulated functions are characterized by the following result, where we say that $f : [a, b] \to \mathbb{R}$ is a step-function, or simple function if it belongs to $\mathcal{E}([a, b])$. Note that a step-function is also regulated.

Proposition 36 *Let $f : [a, b] \to \mathbb{R}$. Then, f is a regulated function if and only if f is the uniform limit of a sequence of step functions.*

Proof. We first assume that there is a sequence $\{f_n : n \in \mathbb{N}\}$ of step functions such that $\|f - f_n\|_{\infty,[a,b]} \to 0$, as $n \to \infty$. Thus, for $\varepsilon > 0$, we can choose $n_0 \in \mathbb{N}$ such that $\|f - f_{n_0}\|_{\infty,[a,b]} \leq \varepsilon/3$. Also, for $x \in (a, b)$, the fact that f_{n_0} is a regulated function implies that there is an interval (x_1, x_2) containing x such that

$$|f_{n_0}(t) - f_{n_0}(s)| \leq \varepsilon/3 \quad \text{if both } s, t \in (x_1, x), \text{ or } s, t \in (x, x_2).$$

Consequently, under the same assumptions on t and s, we get $|f(t) - f(s)| \leq \varepsilon$. In this way, we can establish that the limits $f(x+)$ and $f(x-)$ exist. The cases $x = a$ and $x = b$ are proven similarly.

Conversely, assume that f is a regulated function. For $n \in \mathbb{N}$ and $x \in (a,b)$ (resp. $x \in \{a,b\}$), consider the intervals $V_x = (g_1(x), g_2(x))$ (resp. $V_a = [a, g_2(a))$ and $V_b = (g_1(b), b]$) containing x such that

$$|f(t) - f(s)| \leq 1/n \quad \text{if both } s,t \in (g_1(x), x), \text{ or } s,t \in (x, g_2(x)).$$

Since $[a,b]$ is a compact set of $\mathbb{R}$, there exists a finite family $\{x_i \in [a,b] : i = 1, \dots, m\}$ such that $[a,b] = \cup_{i=1}^m V_{x_i}$. Therefore, we can get a finite increasing sequence $\{c_j \in [a,b] : j = 1, \dots, 3m-2\}$ consisting of the points a, b, $g_1(x_i)$, x_i and $g_2(x_i)$, $1 \leq i \leq m$. Note that if $c_j \in V_{x_i}$, then either $c_{j+1} \in V_{x_i}$, or $c_{j+1} = g_2(x_i)$. In both cases, we obtain

$$|f(t) - f(s)| \leq 1/n \quad \text{for } s,t \in (c_j, c_{j+1}).$$

Now, we define the step-function

$$f_n(x) = \begin{cases} f(c_i), & \text{if } x = c_i, \\ f\left(\frac{c_j + c_{j+1}}{2}\right) & \text{if } x \in (c_j, c_{j+1}), \end{cases}$$

which satisfies $\|f - f_n\| \leq 1/n$. In this way the proof is complete. □

It is said that a function $f : [a,b] \to \mathbb{R}$ is α-Hölder continuous, or Hölder continuous with exponent $\alpha \in (0,1)$ and constant L if

$$L = \|f\|_{\alpha,[a,b]} := \sup_{\substack{s,t \in [a,b] \\ s \neq t}} \frac{|f(t) - f(s)|}{|t-s|^\alpha} < \infty.$$

For $\alpha \in (0,1)$, the function $t \mapsto |t|^\alpha$ is Hölder continuous with exponent α and constant $L = 1$, as the following result establishes.

Lemma 26 *Let* $\alpha \in (0,1)$. *Then,* $||x|^\alpha - |y|^\alpha| \leq |x-y|^\alpha$, $x, y \in \mathbb{R}$.

Proof. Since $\alpha \in (0,1)$, we have, for $|y| \leq |x|$,

$$|x|^\alpha - |y|^\alpha = \alpha \int_{|y|}^{|x|} z^{\alpha-1} dz \leq \alpha \int_{|y|}^{|x|} (z - |y|)^{\alpha-1} dz = ||x| - |y||^\alpha \leq |x-y|^\alpha.$$

□

Note that if f is α-Hölder continuous with $\alpha > 1$, then f is a constant function since, in this case, it has a derivative equals to zero.

Some properties of the elements of the space $\mathcal{W}_p([a,b])$ are given in the following result.

Proposition 37 *Let* $g : [a,b] \to \mathbb{R}$ *be a function,* $p > 0$ *and* $f \in \mathcal{W}_p([a,b])$. *Then,*

i) $|g(b) - g(a)|^p \leq \mathrm{var}_p(g; [a,b])$.

ii) f *is a bounded function. Furthermore,* $\|f\|_{\infty,[a,b]} \leq |f(a)| + (\mathrm{var}_p(f; [a,b]))^{1/p}$.

iii) $x \mapsto \mathrm{var}_p(f; [a,x])$ *is a nondecreasing function on* $[a,b]$.

iv) for $c \in (a,b)$,

$$\mathrm{var}_p(f; [a,c]) + \mathrm{var}_p(f; [c,b]) \leq \mathrm{var}_p(f; [a,b]) \leq 2^p \left(\mathrm{var}_p(f; [a,c]) + \mathrm{var}_p(f; [c,b])\right).$$

v) g *belongs to* $\mathcal{W}_p([a,b])$ *if it is* $\frac{1}{p}$*-Hölder continuous.*

vi) f has at most a countably infinite number of discontinuities.

Proof. i): This statement is satisfied since $\{a = x_0 < x_1 = b\}$ is a partition of the interval $[a, b]$.

ii): Let $x \in [a, b]$. Then, the triangle inequality implies

$$\begin{aligned}|f(x)| &\leq |f(a)| + |f(x) - f(a)| = |f(a)| + (|f(x) - f(a)|^p)^{1/p} \\ &\leq |f(a)| + (|f(x) - f(a)|^p + |f(b) - f(x)|^p)^{1/p},\end{aligned}$$

which yields $|f(x)| \leq |f(a)| + (\text{var}_p(f; [a, b]))^{1/p} < \infty$, for every $x \in [a, b]$.

iii): Consider $a \leq x < y \leq b$ and a partition $\{a = x_0 < x_1 < \ldots < x_n = x\}$ of the interval $[a, x]$. Thus, we obtain

$$\sum_{i=1}^{n} |f(x_i) - f(x_{i-1})|^p \leq \sum_{i=1}^{n} |f(x_i) - f(x_{i-1})|^p + |f(y) - f(x)|^p \leq \text{var}_p(f; [a, y]).$$

Now, it is easy to see that this part of the result is true.

iv): For the partitions $\pi_1 = \{a = x_0 < x_1 < \ldots < x_n = c\}$ and $\pi_2 = \{c = x_n < x_{n+1} < \ldots < x_m = b\}$ of $[a, c]$ and $[c, b]$, respectively, we have that $\pi_1 \cup \pi_2$ is a partition of the interval $[a, b]$. Hence,

$$\sum_{i=1}^{n} |f(x_i) - f(x_{i-1})|^p + \sum_{j=n+1}^{m} |f(x_j) - f(x_{j-1})|^p \leq \text{var}_p(f; [a, b]).$$

As a consequence, we can conclude that

$$\text{var}_p(f; [a, c]) + \text{var}_p(f; [c, b]) \leq \text{var}_p(f; [a, b]).$$

On the other hand, for a partition $\{a = x_0 < x_1 < \ldots < x_n = b\}$, there is $j \in \{1, \ldots, n\}$ such that $c \in (x_{j-1}, x_j]$. Consequently, the triangle inequality gives

$$\begin{aligned}&\sum_{i=1}^{n} |f(x_i) - f(x_{i-1})|^p \\ &= \sum_{i=1}^{j-1} |f(x_i) - f(x_{i-1})|^p + |f(x_j) - f(x_{j-1})|^p + \sum_{i=j+1}^{n} |f(x_i) - f(x_{i-1})|^p \\ &\leq \sum_{i=1}^{j-1} |f(x_i) - f(x_{i-1})|^p + 2^p |f(c) - f(x_{j-1})|^p + 2^p |f(x_j) - f(c)|^p \\ &\quad + \sum_{i=j+1}^{n} |f(x_i) - f(x_{i-1})|^p \leq 2^p \left(\text{var}_p(f; [a, c]) + \text{var}_p(f; [c, b])\right)\end{aligned}$$

Therefore the proof of this assertion is finished.

v): This fact is clearly true.

vi): By Lemma 25, we know that f is a regulated function on $[a, b]$. We claim that, for $n \in \mathbb{N}$,

$$\Delta_n = \left\{x \in [a, b] : |f(x+) - f(x)| > \frac{1}{n} \text{ or } |f(x) - f(x-)| > \frac{1}{n}\right\}$$

is a finite set. Otherwise, we can find an infinite sequence $\{\xi_m \in \Delta_n : m \in \mathbb{N}\}$. Note that for $k \in \mathbb{N}$, we have a permutation of the set $\{1, \ldots, k\}$ such that $\xi_{\sigma(1)} < \ldots < \xi_{\sigma(k)}$. Therefore, we can find $\eta_1 \in (a, \xi_{\sigma(2)})$ and $\eta_i \in (\eta_{i-1} \vee \xi_{\sigma(i-1)}, \xi_{\sigma(i+1)})$, with $i = 2, \ldots, k$ and $\xi_{\sigma(k+1)} = b$, such that $|f(\xi_{\sigma(i)}) - f(\eta_i)| > 1/(2n)$. Hence, we have

$$\mathrm{var}_p(f; [a,b]) \geq \frac{k}{(2n)^p}, \quad \text{for every } k \in \mathbb{N},$$

which is impossible due to $f \in \mathcal{W}_p([a,b])$. Thus, our claim holds. □

In general, we will work with the case that $p \geq 1$. In this case, we have:

Proposition 38 *Let $p \geq 1$. We then have*

i) $(\mathcal{W}_p([a,b]); \|\cdot\|_{[p]})$ *is a Banach space, where*

$$\|f\|_{[p]} = \|f\|_{\infty,[a,b]} + (\mathrm{var}_p(f; [a,b]))^{1/p}.$$

ii) The space $\mathcal{W}_p([a,b])$ equipped with the norm

$$\|f\|_{(p),a} = |f(a)| + (\mathrm{var}_p(f; [a,b]))^{1/p}$$

is a Banach space.

Proof. Let $f, g \in \mathcal{W}_p([a,b])$. Then, from Minkowski inequality (see [187], Section 6.2), we obtain

$$(\mathrm{var}_p(f+g; [a,b]))^{1/p} \leq (\mathrm{var}_p(f; [a,b]))^{1/p} + (\mathrm{var}_p(g; [a,b]))^{1/p}.$$

Indeed, for a partition $\{a = x_0 < x_1 < \ldots < x_n = b\}$, we get

$$\begin{aligned}
&\left(\sum_{i=1}^{n} |(f(x_i) + g(x_i)) - (f(x_{i-1}) + g(x_{i-1}))|^p \right)^{1/p} \\
\leq\ & \left(\sum_{i=1}^{n} |f(x_i) - f(x_{i-1})|^p \right)^{1/p} + \left(\sum_{i=1}^{n} |g(x_i) - g(x_{i-1})|^p \right)^{1/p} \\
\leq\ & (\mathrm{var}_p(f; [a,b]))^{1/p} + (\mathrm{var}_p(g; [a,b]))^{1/p}.
\end{aligned}$$

Also, for $\lambda \in \mathbb{R}$, it is not difficult to show $(\mathrm{var}_p(\lambda g; [a,b]))^{1/p} = |\lambda| (\mathrm{var}_p(g; [a,b]))^{1/p}$. Therefore $(\mathcal{W}_p([a,b]); \|\cdot\|_{[p]})$ and $(\mathcal{W}_p([a,b]); \|\cdot\|_{(p),a})$ are real normed vector spaces.

Now, we consider a Cauchy sequence $\{f_n \in \mathcal{W}_p([a,b]) : n \in \mathbb{N}\}$ with respect to $\|\cdot\|_{[p]}$. Then, this sequence is also a Cauchy one with respect to $\|\cdot\|_{\infty,[a,b]}$. Consequently, there is $f : [a,b] \to \mathbb{R}$ such that $\{f_n : n \in \mathbb{N}\}$ converges uniformly to f. Hence, for $\varepsilon > 0$ and a partition $\{a = x_0 < x_1 < \ldots, x_k = b\}$, we have, for $m > n$ large enough,

$$\begin{aligned}
&\sum_{i=1}^{k} |(f_n(x_i) - f(x_i)) - (f_n(x_{i-1}) - f(x_{i-1}))|^p \\
\leq\ & 2^p \sum_{i=1}^{k} |(f_n(x_i) - f_m(x_i)) - (f_n(x_{i-1}) - f_m(x_{i-1}))|^p \\
& +2^p \sum_{i=1}^{k} |(f_m(x_i) - f(x_i)) - (f_m(x_{i-1}) - f(x_{i-1}))|^p \\
\leq\ & 2^p \mathrm{var}_p(f_n - f_m; [a,b]) + 2^{2p} k \|f_m - f\|_{\infty,[a,b]} \leq \varepsilon,
\end{aligned}$$

where, in the last inequality, we have used that $\{f_n : n \in \mathbb{N}\}$ is a Cauchy sequence with respect to $\|\cdot\|_{[p]}$ and that it converges uniformly to f. In other words, we have proven that $\mathrm{var}_p(f - f_n; [a,b]) \le \varepsilon$, for n large enough, and therefore $(\mathcal{W}_p([a,b]); \|\cdot\|_{[p]})$ is complete.

Finally, Statement *ii)* is also true because Proposition 37.*ii)* establishes that a Cauchy sequence with respect to $\|\cdot\|_{(p),a}$ is also a Cauchy one with respect to $\|\cdot\|_{\infty,[a,b]}$. Thus, the proof is finished. □

In Remark 30, we indicate some characterizations of all the functions of bounded variation on $[a,b]$. The following result gives a characterization of functions in $\mathcal{W}_p([a,b])$ with $p \ge 1$. The case that $p \in (0,1)$ is stated in Theorem 27 below.

Theorem 26 *Let $p \ge 1$. Then, a function f belongs to $\mathcal{W}_p([a,b])$ if and only if it can be represented as the composition $f = \tau \circ g$, where $g : [a,b] \to [g(a), g(b)]$ is non-decreasing and $\tau : [g(a), g(b)] \to \mathbb{R}$ is a Hölder continuous function with exponent $1/p$ and constant $L = 1$.*

Proof. Let $f \in \mathcal{W}_p([a,b])$ and, for $x \in [a,b]$, $g(x) = \mathrm{var}_p(f; [a,x])$. We have that $g(x) < \infty$ since $g(x) \le \mathrm{var}_p(f; [a,b])$, $x \in [a,b]$. Now, for $g(x)$ in the range of g (denoted by $\mathcal{M}$), we define $\tau(g(x)) = f(x)$. Note that

$$\begin{aligned} |\tau(g(x)) - \tau(g(y))|^p &= |f(x) - f(y)|^p \le \mathrm{var}_p(f; [x \wedge y, x \vee y]) \\ &\le \mathrm{var}_p(f; [a, x \vee y]) - \mathrm{var}_p(f; [a, x \wedge y]) \\ &= g(x \vee y) - g(x \wedge y), \quad x, y \in [a,b], \end{aligned}$$

implies that $f(x) = f(y)$ if $g(x) = g(y)$, which gives that τ is well-defined, and that τ is $1/p$-Hölder continuous with constant $L = 1$ on the range of g. So, we need to extend τ to the whole interval $[g(a), g(b)]$ because g could be discontinuous. To do so, we use the McShane extension [146]. That is, define $\hat{\tau} : \mathbb{R} \to \mathbb{R}$ by $\hat{\tau}(x) := \sup\{\tau(y) - |x-y|^{1/p} : y \in \mathcal{M}\}$. We obtain $\hat{\tau}(x) = \tau(x)$ for any $x \in \mathcal{M}$. Indeed, in this case, the fact that $-|x-y|^{1/p} \le \tau(x) - \tau(y)$ implies $\hat{\tau}(x) \le \tau(x)$ and, clearly, $\tau(x) = \tau(x) - |x-x|^{1/p} \le \hat{\tau}(x)$. Moreover, Lemma 26 yields

$$\begin{aligned} |\hat{\tau}(x) - \hat{\tau}(y)|^{1/p} &= \left| \sup_{z \in \mathcal{M}} \left(\tau(z) - |x-z|^{1/p} \right) - \sup_{z \in \mathcal{M}} \left(\tau(z) - |y-z|^{1/p} \right) \right| \\ &\le \sup_{z \in \mathcal{M}} \left| |x-z|^{1/p} - |y-z|^{1/p} \right| \le |x-y|^{1/p}. \end{aligned}$$

In other words, $\hat{\tau}$ is an extension of τ that we are looking for.

Conversely, assume that $f = \tau \circ g$, where g is non-decreasing and $\tau : [g(a), g(b)] \to \mathbb{R}$ is a $1/p$-Hölder continuous. Then, for a partition $\pi = \{a = t_0 < t_1 < \ldots < t_n = b\}$ of $[a,b]$, we have

$$\sum_{i=1}^{n} |f(t_i) - f(t_{i-1})|^p = \sum_{i=1}^{n} |\tau(g(t_i)) - \tau(g(t_{i-1}))|^p \le \sum_{i=1}^{n} (g(t_i) - g(t_{i-1})) = g(b) - g(a).$$

Hence, f belongs to $\mathcal{W}_p([a,b])$ and the proof is complete. □

Our next aim is to give the characterization of the elements of $\mathcal{W}_p([a,b])$ with $p \in (0,1)$. Towards this end we use the notation

$$\Delta^+ f(x) = f(x+) - f(x), \quad \Delta_- f(x) = f(x) - f(x-) \quad \text{and} \quad \Delta_-^+ f(x) = f(x+) - f(x-),$$

and establish the following auxiliary result.

Lemma 27 *Let $p > 0$ and $f \in \mathcal{W}_p([a,b])$. Then,*

$$\sum_{a<x<b} \left(|\Delta_- f(x)|^p + |\Delta^+ f(x)|^p \right) \le var_p(f; [a,b]).$$

Proof. From Proposition 37.*vi*), f has at most a countably infinity number of discontinuities. So, there is a subset $\mathcal{N} \subset \mathbb{N}$ such that the set of all the discontinuities of f is represented by $\{\xi_i \in [a,b] : i \in \mathcal{N}\}$. Thus, for $m \in \mathbb{N}$, there is $n \in \mathbb{N}$ large enough such that

$$\sum_{\substack{i=1\\ i\in\mathcal{N}}}^{m} \left\{ \left| f(\xi_i) - f\left(\xi_i - \frac{1}{n}\right) \right|^p + \left| f\left(\xi_i + \frac{1}{n}\right) - f(\xi_i) \right|^p \right\} \leq \mathrm{var}_p(f;[a,b]).$$

Hence, taking the limit as $n \to \infty$, we have

$$\sum_{\substack{i=1\\ i\in\mathcal{N}}}^{m} \left(|\Delta_- f(\xi_i)|^p + |\Delta^+ f(\xi_i)|^p \right) \leq \mathrm{var}_p(f;[a,b]).$$

Therefore, the result is satisfied. □

Now, we are ready to state the missing characterization of the space $\mathcal{W}_p([a,b])$ with $p \in (0,1)$.

Theorem 27 *Let $p \in (0,1)$ and $f \in \mathcal{W}_p([a,b])$. Then, for $a < x \leq b$,*

$$f(x) = f(a) + \Delta^+ f(a) + \Delta_- f(x) + \sum_{a<y<x} \Delta_-^+ f(y), \tag{2.21}$$

where the series is absolutely convergent.

Remark 33 *Note that, for $p \in (0,1)$, the only continuous functions in $\mathcal{W}_p([a,b])$ are the constant functions. Also notice that (2.21) can be written as*

$$f(x) = f(a) + v_a^+ + v_x^- + \sum_{a<y<x} \left(v_y^+ + v_y^- \right),$$

with

$$\Delta^+ f(x) = v_x^+, \text{ for } x \in [a,b), \quad \text{and} \quad \Delta_- f(x) = v_x^-, \text{ for } x \in (a,b].$$

Proof. By inequality (2.20), $p \in (0,1)$ and Lemma 27, we obtain,

$$\begin{aligned} \sum_{a<y<b} |\Delta_-^+ f(y)| &\leq \left(\sum_{a<y<b} |\Delta_-^+ f(y)|^p \right)^{1/p} \\ &\leq \left(\sum_{a<y<b} \left(|\Delta_- f(y)|^p + |\Delta^+ f(y)|^p \right) \right)^{1/p} \leq (\mathrm{var}_p(f;[a,b]))^{1/p} < \infty. \end{aligned}$$

In other words, we have proven that the series in (2.21) is absolutely convergent.

On the other hand, set

$$g(x) = \begin{cases} f(a), & \text{if } x = a, \\ \text{right-hand side of (2.21)}, & \text{if } x \in (a,b]. \end{cases}$$

Hence, for a partition $\{a = x_0 < x_1 < \ldots < x_n = b\}$, the fact that $p \in (0,1)$ and Lemma 27 lead us to write

$$\begin{aligned} \sum_{i=1}^{n} |g(x_i) - g(x_{i-1})|^p &\leq \sum_{i=1}^{n} \left\{ |\Delta_- f(x_i)|^p + \sum_{x_{i-1}<y<x_i} |\Delta_-^+ f(y)|^p + |\Delta^+ f(x_{i-1})|^p \right\} \\ &\leq 3\mathrm{var}_p(f;[a,b]) < \infty. \end{aligned}$$

Thus, we have that $f-g$ belongs to the space $\mathcal{W}_p([a,b])$. Also, it is easy to see that $f-g:[a,b]\to\mathbb{R}$ is a continuous function, whose proof is left to the reader as an exercise. So, in order to finish the proof, we only need to show that the only continuous functions in $\mathcal{W}_p([a,b])$ are the constant functions due to $f(a)-g(a)=0$.

Let $h:[a,b]\to\mathbb{R}$ be a continuous function in $\mathcal{W}_p([a,b])$ such that there exists $a<\tilde{b}\le b$ satisfying $|h(a)-h(\tilde{b})|=C\neq 0$. Then, we can choose a point $x_{1/2}\in(a,\tilde{b})$ such that $|h(a)-h(x_{1/2})|=C/2$. Consequently, $|h(x_{1/2})-h(\tilde{b})|\ge C/2$. Similarly, we can find $x_{1/4}\in(a,x_{1/2})$ and $x_{3/4}\in(x_{1/2},\tilde{b})$ such that $|h(a)-h(x_{1/4})|, |h(x_{3/4})-h(x_{1/2})|=C/4$. As before, we obtain $|h(x_{1/4})-h(x_{1/2})|, |h(x_{3/4})-h(\tilde{b})|\ge C/4$. Using induction on n, we can define $x_t\in[a,b]$, for $t\in\{j/2^k : j=i,\dots,2^k\}$ such that

$$|h(x_{j/2^k})-h(x_{(j-1)/2^k})|\ge\frac{C}{2^k},\quad \text{for } j=1,\dots,2^k,$$

which implies that $\mathrm{var}_p(h;[a,\tilde{b}])\ge 2^k\frac{C^p}{2^{pk}}\to\infty$ as $k\to\infty$. But this is a contradiction and, therefore, h has to be a constant function. □

Conversely to Theorem 27, we can state the following result.

Theorem 28 *Assume that $p\in(0,1)$ and let $v^+:[a,b)\to\mathbb{R}$ and $v^-:(a,b]\to\mathbb{R}$ be two functions such that $\sum_{a<y<b}\left(|v_y^-|^p+|v_y^+|^p\right)<\infty$. Then, for $f(a)\in\mathbb{R}$, the function*

$$f(x)=f(a)+v_a^++v_x^-+\sum_{a<y<x}\left(v_y^-+v_y^+\right),\qquad x\in(a,b],$$

belongs to the space $\mathcal{W}_p([a,b])$, and

$$\Delta^+f(x)=v_x^+\quad\text{and}\quad\Delta_-f(x)=v_x^-.$$

Proof. Using the fact that $p\in(0,1)$ and (2.20), we have

$$\sum_{a<y<b}|v_y^-+v_y^+|\le\left(\sum_{a<y<b}\left(|v_y^-|^p+|v_y^+|^p\right)\right)^{1/p}<\infty,$$

which implies that the function f is well-defined.

On the other hand, let $\{a=x_0<x_1<\ldots<x_n=b\}$ be a partition of $[a,b]$. Note that the definition of f gives

$$f(x_1)-f(a)=v_a^++v_{x_1}^-+\sum_{a<y<x_1}\left(v_y^-+v_y^+\right)$$

and, for $i=2,\dots,n$,

$$\begin{aligned}f(x_i)-f(x_{i-1}) &= v_{x_i}^- - v_{x_{i-1}}^- + \sum_{x_{i-1}\le y<x_i}\left(v_y^-+v_y^+\right)\\ &= v_{x_{i-1}}^+ + v_{x_i}^- + \sum_{x_{i-1}<y<x_i}\left(v_y^-+v_y^+\right).\end{aligned}$$

Consequently, using that $p\in(0,1)$ again,

$$\sum_{i=1}^n|f(x_i)-f(x_{i-1})|^p\le|v_a^+|^p+|v_b^-|^p+\sum_{a<y<b}\left(|v_y^-|^p+|v_y^+|^p\right)<\infty$$

and, therefore, $f\in\mathcal{W}_p([a,b])$.

Finally, for $z > x$,

$$\begin{aligned} f(z) - f(x) &= v_z^- - v_x^- + \sum_{x \le y < z} (v_y^- + v_y^+) = v_z^- + v_x^+ + \sum_{x<y<z} (v_y^- + v_y^+) \\ &= v_x^+ + \sum_{x<y\le z} v_y^- + \sum_{x<y<z} v_y^+ \to v_x^+ \quad \text{as } z \downarrow x. \end{aligned}$$

It means, $\Delta^+ f(x) = v_x^+$. Similarly, we can see that $\Delta_- f(x) = v_x^-$. Thus, the proof is finished. □

Now, we deal with some inequalities needed to study the convergence of the Riemann-Stieltjes sum defined in (2.1) via the techniques of the p-variation theory. Towards this end, we observe that, in the proof of Proposition 32 (see inequality (2.5)), it is used an estimate of an expression similar to the quantity

$$\left| f(x_k) \left[g(b) - g(a) \right] - \sum_{i=1}^{n} f(x_i) \left[g(x_i) - g(x_{i-1}) \right] \right|,$$

for $k = 0, \ldots, n$, in order to study a case that guarantees the convergence of (2.1). Note that for $k = 0$, we have

$$\begin{aligned} &\sum_{1 \le i \le j \le n} (f(x_i) - f(x_{i-1}))(g(x_j) - g(x_{j-1})) \\ &= \sum_{j=1}^{n} (g(x_j) - g(x_{j-1})) \sum_{i=1}^{j} (f(x_i) - f(x_{i-1})) \\ &= \sum_{j=1}^{n} (g(x_j) - g(x_{j-1}))(f(x_j) - f(a)) \\ &= \sum_{i=1}^{n} f(x_i)(g(x_i) - g(x_{i-1})) - f(a)(g(b) - g(a)) \end{aligned} \tag{2.22}$$

and, for $k = n$, we obtain

$$\begin{aligned} &\sum_{1 \le j < i \le n} (f(x_i) - f(x_{i-1}))(g(x_j) - g(x_{j-1})) \\ &= \sum_{j=1}^{n-1} (g(x_j) - g(x_{j-1})) \sum_{i=j+1}^{n} (f(x_i) - f(x_{i-1})) \\ &= f(b)(g(x_{n-1}) - g(a)) - \sum_{i=1}^{n-1} f(x_i)(g(x_i) - g(x_{i-1})) \\ &= f(b)(g(b) - g(a)) - \sum_{i=1}^{n} f(x_i)(g(x_i) - g(x_{i-1})). \end{aligned} \tag{2.23}$$

Similarly, for $0 < k < n$, we can write

$$\begin{aligned} &\sum_{k < i \le j \le n} (f(x_i) - f(x_{i-1}))(g(x_j) - g(x_{j-1})) \\ &= \sum_{i=k+1}^{n} f(x_i)(g(x_i) - g(x_{i-1})) - f(x_k)(g(b) - g(x_k)) \end{aligned} \tag{2.24}$$

and

$$\sum_{1\le j<i\le k}(f(x_i)-f(x_{i-1}))(g(x_j)-g(x_{j-1}))$$
$$= f(x_k)(g(x_k)-g(a))-\sum_{i=1}^{k} f(x_i)(g(x_i)-g(x_{i-1})),$$

which, together with (2.24), implies

$$\begin{aligned}\sum_{i=1}^{n} f(x_i)(g(x_i)-g(x_{i-1}))-f(x_k)(g(b)-g(a))&\\ = \sum_{k<i\le j\le n}(f(x_i)-f(x_{i-1}))(g(x_j)-g(x_{j-1}))&\\ -\sum_{1\le j<i\le k}(f(x_i)-f(x_{i-1}))(g(x_j)-g(x_{j-1})).& \qquad (2.25)\end{aligned}$$

Hence, by (2.22) and (2.23), we must estimate the right-hand side of (2.25) by means of the p-variation theory and the ideas developed in the proof of Proposition 32 in order to analyze the convergence of the Riemann-Stieltjes sum introduced in (2.1). To achieve this goal, we proceed as follows.

Let $\chi=(\chi_1,\ldots,\chi_n)$ and $\theta=(\theta_1,\ldots,\theta_n)$ be two n-tuples of real numbers, and $p,q>0$. Set

$$V_p(\chi)=\max\left\{\left(\sum_{j=1}^{m}\left|\sum_{\ell=i_{j-1}+1}^{i_j}\chi_\ell\right|^p\right)^{1/p} : 0=i_0<i_1<\ldots<i_m=n\right\}.$$

Note that if, in (2.25), $\chi_i=f(x_i)-f(x_{i-1})$, $i=1,\ldots,n$, then $V_p(\chi)\le(\mathrm{var}_p(f;[a,b]))^{1/p}$. This is the way that we get our goal through the following result, where we use the fact that there is $1\le\ell\le n$ such that $|\chi_\ell\theta_\ell|=\min\{|\chi_i\theta_i| : i=1,\ldots,n\}$ and, therefore,

$$\begin{aligned}|\chi_\ell\theta_\ell| &\le (|\chi_1\theta_1|\cdots|\chi_n\theta_n|)^{1/n}=\left[(|\chi_1|^p\cdots|\chi_n|^p)^{1/n}\right]^{1/p}\left[(|\theta_1|^q\cdots|\theta_n|^q)^{1/n}\right]^{1/q}\\ &\le \left(\frac{1}{n}\sum_{i=1}^{n}|\chi_i|^p\right)^{1/p}\left(\frac{1}{n}\sum_{i=1}^{n}|\theta_i|^q\right)^{1/q}. \qquad (2.26)\end{aligned}$$

Here, the last inequality is a consequence of the well-known theorem of the arithmetic and geometric means.

Lemma 28 *Let $\chi=(\chi_1,\ldots,\chi_n)$ and $\theta=(\theta_1,\ldots,\theta_n)$ be in $\mathbb{R}^n$, and $p,q>0$. Then,*

$$\left|\sum_{k<i\le j\le n}\chi_i\theta_j-\sum_{1\le j<i\le k}\chi_i\theta_j\right|\le\left(1+\sum_{m=1}^{n}m^{-(p^{-1}+q^{-1})}\right)V_p(\chi)V_q(\theta), \qquad (2.27)$$

for any $k=0,1\ldots,n$.

Proof. Fix $k\in\{0,1,\ldots,n\}$. In order to simplify the notation, we set

$$S_L(\chi,\theta;k)=\sum_{0\le j<i\le k}\chi_i\theta_j \quad\text{and}\quad S_R(\chi,\theta;k,n)=\sum_{k<i\le j\le n}\chi_i\theta_j,$$

where $\theta_0 = 0$. We also use the convention, for $1 \le \ell \le n$,

$$\chi_j^{(\ell)} = \begin{cases} \chi_j, & \text{if } 1 \le j \le \ell - 1 \\ \chi_\ell + \chi_{\ell+1}, & \text{if } j = \ell \\ \chi_{j+1}, & \text{if } \ell < j < n, \end{cases}$$

where $\chi_{n+1} = 0$ if $\ell = n$, and

$$\theta_j^{(\ell)} = \begin{cases} \theta_j, & \text{if } 1 \le j < \ell - 1 \\ \theta_{\ell-1} + \theta_\ell, & \text{if } j = \ell - 1 \\ \theta_{j+1}, & \text{if } \ell \le j < n. \end{cases}$$

In this way, we now have two elements $\chi^{(\ell)} = (\chi_1^{(\ell)}, \ldots, \chi_{n-1}^{(\ell)})$ and $\theta^{(\ell)} = (\theta_1^{(\ell)}, \ldots, \theta_{n-1}^{(\ell)})$ of $\mathbb{R}^{n-1}$ such that, for $1 \le \ell < k$,

$$S_R(\chi^{(\ell)}, \theta^{(\ell)}; k-1, n-1) = S_R(\chi, \theta; k, n) \tag{2.28}$$

and

$$\begin{aligned} S_L(\chi^{(\ell)}, \theta^{(\ell)}; k-1) &= \sum_{0<i\le k-1} \chi_i^{(\ell)} \left(\theta_0^{(\ell)} + \cdots + \theta_{i-1}^{(\ell)}\right) = \sum_{1<i<\ell} \chi_i \left(\theta_0 + \cdots + \theta_{i-1}\right) \\ &\quad + \left(\chi_\ell + \chi_{\ell+1}\right)\left(\theta_1 + \cdots + \theta_\ell\right) + \sum_{\ell<i\le k-1} \chi_{i+1}\left(\theta_1 + \cdots + \theta_i\right) \\ &= \chi_\ell \theta_\ell + S_L(\chi, \theta; k). \end{aligned} \tag{2.29}$$

Moreover, for $k < \ell \le n$, we have

$$S_L(\chi^{(\ell)}, \theta^{(\ell)}; k) = S_L(\chi, \theta; k) \tag{2.30}$$

and

$$\begin{aligned} S_R(\chi^{(\ell)}, \theta^{(\ell)}; k, n-1) &= \sum_{k<i\le n-1} \chi_i^{(\ell)} \left(\theta_i^{(\ell)} + \cdots + \theta_{n-1}^{(\ell)}\right) = \sum_{k<i<\ell} \chi_i \left(\theta_i + \cdots + \theta_n\right) \\ &\quad + \left(\chi_\ell + \chi_{\ell+1}\right)\left(\theta_{\ell+1} + \cdots + \theta_n\right) + \sum_{\ell<i<n} \chi_{i+1}\left(\theta_{i+1} + \cdots + \theta_n\right) \\ &= -\chi_\ell \theta_\ell + S_R(\chi, \theta; k, n). \end{aligned} \tag{2.31}$$

Now, choose $\ell \in (\{1, \ldots, n\} \setminus \{k\})$ such that $|\chi_\ell \theta_\ell| = \min\{|\chi_i \theta_i| : i \in (\{1, \ldots, n\} \setminus \{k\})\}$, and use the notation $S_n = S_R(\chi, \theta; k, n) - S_L(\chi, \theta; k)$ and

$$S_{n-1} = \begin{cases} S_R(\chi^{(\ell)}, \theta^{(\ell)}; k-1, n-1) - S_L(\chi^{(\ell)}, \theta^{(\ell)}; k-1), & \text{if } \ell < k \\ S_R(\chi^{(\ell)}, \theta^{(\ell)}; k, n-1) - S_L(\chi^{(\ell)}, \theta^{(\ell)}; k), & \text{if } k < \ell \le n. \end{cases}$$

Then, (2.26) and (2.28)-(2.31) allow us to write

$$|S_n| \le |\chi_\ell \theta_\ell| + |S_{n-1}| \le C_{p,q}(n) V_p(\chi) V_q(\theta) + |S_{n-1}|,$$

with

$$C_{p,q}(n) = \begin{cases} n^{-(p^{-1}+q^{-1})}, & \text{if } k = 0 \\ (n-1)^{-(p^{-1}+q^{-1})}, & \text{otherwise.} \end{cases}$$

Hence, utilizing that $V_p(\chi^{(\ell)}) V_q(\theta^{(\ell)}) \le V_p(\chi) V_q(\theta)$ and proceeding as before, we can get

$$|S_n| \le \sum_{m=3}^{n} C_{p,q}(m) V_p(\chi) V_q(\theta) + |S_2|.$$

Thus, the fact that

$$|S_2| = \begin{cases} |S_R(\chi,\theta;0,2)|, & \text{if } k=0, \\ |S_R(\chi,\theta;1,2)|, & \text{if } k=1, \\ |S_L(\chi,\theta;1)|, & \text{if } k=2 \end{cases} = \begin{cases} \left|\chi_\ell\theta_\ell + S_R(\chi^{(\ell)},\theta^{(\ell)};0,1)\right|, & \text{if } k=0, \\ |\chi_2\theta_2|, & \text{if } k=1, \\ |\chi_2\theta_1|, & \text{if } k=2 \end{cases}$$

$$= \begin{cases} \left|\chi_\ell\theta_\ell + \chi_1^{(\ell)}\theta_1^{(\ell)}\right|, & \text{if } k=0, \\ |\chi_2\theta_2|, & \text{if } k=1, \\ |\chi_2\theta_1|, & \text{if } k=2 \end{cases} \le \begin{cases} \left(1+2^{-(p^{-1}+q^{-1})}\right) V_p(\chi)V_q\theta), & \text{if } k=0, \\ V_p(\chi)V_q\theta), & \text{if } k=1, \\ V_p(\chi)V_q(\theta), & \text{if } k=2 \end{cases}$$

implies that (2.27) holds and the proof is finished. □

As a consequence of Lemma 28, we have the following inequality.

Corollary 15 *Let $p,q>0$ be such that $p^{-1}+q^{-1}>1$, $f\in\mathcal{W}_p([a,b])$, $g\in\mathcal{W}_q([a,b])$, and $\pi^{'}=\{a=t_0^{'}<t_1^{'}<\ldots<t_m^{'}=b\}$ and $\pi^{''}=\{a=t_0^{''}<t_1^{''}<\ldots<t_k^{''}=b\}$ two partitions of the interval $[a,b]$. Then, for any sets $\tilde{\pi}^{'}=\{a\le s_1^{'}\le\ldots\le s_m^{'}\le b\}$ and $\tilde{\pi}^{''}=\{a\le s_1^{''}\le\ldots\le s_k^{''}\le b\}$ of intermediate points of $\pi^{'}$ and $\pi^{''}$, respectively, we have*

$$\begin{aligned} &\left|RS(f,g,\pi^{'},\tilde{\pi}^{'})-RS(f,g,\pi^{''},\tilde{\pi}^{''})\right| \\ &\le \left(1+\zeta\left(p^{-1}+q^{-1}\right)\right)\left\{\sum_{i=1}^{m}\left(var_p(f;[t_{i-1}^{'},t_i^{'}])\right)^{1/p}\left(var_q(g;[t_{i-1}^{'},t_i^{'}])\right)^{1/q}\right. \\ &\qquad \left.+\sum_{j=1}^{k}\left(var_p(f;[t_{j-1}^{''},t_j^{''}])\right)^{1/p}\left(var_q(g;[t_{j-1}^{''},t_j^{''}])\right)^{1/q}\right\}, \end{aligned}$$

where $\zeta(s)=\sum_{n=1}^{\infty}n^{-s}$ is the Riemann zeta function.

Proof. Set $\pi=\pi^{'}\cup\pi^{''}\cup\tilde{\pi}^{'}\cup\tilde{\pi}^{''}=\{a=t_0<t_1<\ldots<t_n=b\}$. For each index $j\in\{0,\ldots,m\}$ (resp. $j\in\{0,\ldots,k\}$), let $i(j)$ (resp. $\tilde{i}(j)$)$\in\{0,\ldots,n\}$ be such that $t_{i(j)}=t_j^{'}$ (resp. $t_{\tilde{i}(j)}=t_j^{''}$). Then, inequality (2.5) implies

$$\begin{aligned} &\left|\mathrm{RS}(f,g,\pi^{'},\tilde{\pi}^{'})-\mathrm{RS}(f,g,\pi^{''},\tilde{\pi}^{''})\right| \\ &\le \sum_{j=1}^{m}\left|f(s_j^{'})\left[g(t_j^{'})-g(t_{j-1}^{'})\right]-\sum_{\ell=i(j-1)+1}^{i(j)}f(t_\ell)\left[g(t_\ell)-g(t_{\ell-1})\right]\right| \\ &\quad +\sum_{j=1}^{k}\left|f(s_j^{''})\left[g(t_j^{''})-g(t_{j-1}^{''})\right]-\sum_{\ell=\tilde{i}(j-1)+1}^{\tilde{i}(j)}f(t_\ell)\left[g(t_\ell)-g(t_{\ell-1})\right]\right|. \end{aligned}$$

Hence, Lemma 28, (2.22), (2.23) and (2.25) yield that the result holds. □

Now, we are ready to state the convergence of the Riemann-Stieltjes sum defined in (2.1) of this section.

Theorem 29 *Let $p,q>0$ be such that $p^{-1}+q^{-1}>1$, $f\in\mathcal{W}_p([a,b])$ and $g\in\mathcal{W}_q([a,b])$. Then, $\int_a^b f(s)d^Y g(s)$ exists if f and g have no discontinuities at the same point.*

Remark 34 *Young [216] has shown through an example that the Riemann-Stieltjes sum defined in (2.1) may not converge if we only have $p^{-1}+q^{-1}=1$.*

Proof. Lemma 25 and the proof of Proposition 36 yield that, for a given $\varepsilon > 0$, there is a finite sequence $\{(c_{i-1}, c_i) : i = 1, \ldots, n\}$ of open intervals such that $a = c_0 < \ldots < c_n = b$ and

$$\sup_{x,y\in(c_{i-1},c_i)} |f(x) - f(y)| < \varepsilon \quad \text{and} \quad \sup_{x,y\in(c_{i-1},c_i)} |g(x) - g(y)| < \varepsilon,$$

for all $i = 1, \ldots, n$. Thus, we can choose $\delta > 0$ small enough such that any closed subinterval of $[a, b]$ with length less than δ has at most one of the points $\{c_0, \ldots, c_n\}$. Hence, for $p' > p\vee 1$ and $q' > q \vee 1$ such that $(p')^{-1} + (q')^{-1} > 1$, and any two partitions π' and π'' of $[a, b]$ as in Corollary 15 satisfying $|\pi'|, |\pi''| < \delta$, we have Corollary 15 and the fact that f and g has no common discontinuities imply

$$\begin{aligned}
&\left|\mathrm{RS}(f, g, \pi', \tilde{\pi}') - \mathrm{RS}(f, g, \pi'', \tilde{\pi}'')\right| \\
&\leq C_{\varepsilon,p,q,p',q'} \left\{ \sum_{i=1}^{m} \left(\mathrm{var}_p(f; [t'_{i-1}, t'_i])\right)^{1/p'} \left(\mathrm{var}_q(g; [t'_{i-1}, t'_i])\right)^{1/q'} \right. \\
&\qquad \left. + \sum_{j=1}^{k} \left(\mathrm{var}_p(f; [t''_{j-1}, t''_j])\right)^{1/p'} \left(\mathrm{var}_q(g; [t''_{j-1}, t''_j])\right)^{1/q'} \right\},
\end{aligned}$$

where

$$\begin{aligned}
C_{\varepsilon,p,q,p',q'} &= \left(1 + \zeta\left((p')^{-1} + (q')^{-1}\right)\right) \\
&\quad \times \left(\varepsilon^{1-\frac{p}{p'}} \vee \varepsilon^{1-\frac{q}{q'}}\right) \left((2\|f\|_{\infty,[a,b]})^{1-\frac{p}{p'}} \vee (2\|g\|_{\infty,[a,b]})^{1-\frac{q}{q'}}\right).
\end{aligned}$$

Therefore, for $\tilde{p} > 1$ such that $(p')^{-1} + (\tilde{p})^{-1} = 1$, the Hölder inequality and Proposition 37.*iv*) lead us to write

$$\begin{aligned}
&\left|\mathrm{RS}(f, g, \pi', \tilde{\pi}') - \mathrm{RS}(f, g, \pi'', \tilde{\pi}'')\right| \\
&\leq C_{\varepsilon,p,q,p',q'} \left(\mathrm{var}_p(f; [a, b])\right)^{1/p'} \left\{ \left(\sum_{i=1}^{m} \left(\mathrm{var}_q(g; [t'_{i-1}, t'_i])\right)^{\tilde{p}/q'} \right)^{1/\tilde{p}} \right. \\
&\qquad \left. + \left(\sum_{j=1}^{k} \left(\mathrm{var}_q(g; [t''_{j-1}, t''_j])\right)^{\tilde{p}/q'} \right)^{1/\tilde{p}} \right\}.
\end{aligned}$$

Consequently, we can use that $1 < \tilde{p}(q')^{-1}$, together with (2.20), to conclude

$$\left|\mathrm{RS}(f, g, \pi', \tilde{\pi}') - \mathrm{RS}(f, g, \pi'', \tilde{\pi}'')\right| \leq C_{\varepsilon,p,q,p',q'} \left(\mathrm{var}_p(f; [a, b])\right)^{1/p'} \left(\mathrm{var}_q(g; [a, b])\right)^{1/q'}.$$

Since ε is arbitrary, we finally obtain that the integral $\int_a^b f(s)d^Y g(s)$ exists. □

Other consequence of Lemma 28 is the following theorem.

Theorem 30 *Let p, q, f and g be as in Theorem 29. Then,*

$$\left|\int_a^b f(s)d^Y g(s)\right| \leq \left(1 + \zeta\left(p^{-1} + q^{-1}\right)\right) \|f\|_{[p]} \left(var_q(g; [a, b])\right)^{1/q},$$

where $\|\cdot\|_{[p]}$ is introduced in Proposition 38.

Proof. The result is an immediate consequence of Lemma 28, (2.22), (2.23), (2.25) and Theorem 29. □

Proceeding as in this section, Young [217] has also considered 2D integrals of the form $\int_a^b \int_c^d f(x,y) d_{x,y} g(x,y)$, where $f : [a,b] \times [c,d] \to \mathbb{R}$ has finite (p,q)-bivariation (see Definition 27 below) and $g : [a,b] \times [c,d] \to \mathbb{R}$ is a controlled path satisfying suitable (p,q)-variation conditions. In order to state the existence of these integrals, we introduce the following. The interested reader can consult Young [217] for details.

Let $\pi(\xi) = \{a = x_0 < \ldots < x_n = b\}$ and $\pi'(\eta) = \{c = y_0 < \ldots < y_m = d\}$ be partitions of the intervals $[a,b]$ and $[c,d]$ equipped with the sets of points $\xi = \{\xi_i : i = 1, \ldots, n\}$ and $\eta = \{\eta_j : j = 1, \ldots, m\}$, respectively, such that $x_{i-1} \le \xi_i \le x_i$ and $y_{j-1} \le \eta_j \le y_j$, with $i = 1, \ldots, n$ and $j = 1, \ldots, m$.

The step function of f based on the partitions $\pi(\xi)$ and $\pi'(\eta)$ is the function $f_{\pi(\xi),\pi'(\eta)} : [a,b] \times [c,d] \to \mathbb{R}$ given by

$$f_{\pi(\xi),\pi'(\eta)}(x,y) = \begin{cases} f(x_i, y_j), & x = x_i \text{ and } y = y_j, \\ f(x_i, \eta_j), & x = x_i \text{ and } y_{j-1} < y < y_j, \\ f(\xi_i, y_j), & x_{i-1} < x < x_i \text{ and } y = y_j, \\ f(\xi_i, \eta_j), & x_{i-1} < x < x_i \text{ and } y_{j-1} < y < y_j. \end{cases}$$

The definition of the integral of the step function $f_{\pi(\xi),\pi'(\eta)}$ with respect to g on $[a,b] \times [c,d]$ is defined as usual. That is,

$$\int_a^b \int_c^d f_{\pi(\xi),\pi'(\eta)} d_{x,y} g(x,y) = \sum_{j=1}^m \sum_{i=1}^n f(\xi_i, \eta_j) \Delta_i \Delta_j g(x_i, y_j),$$

where $\Delta_j g(x_i, y_j) = g(x_i, y_j) - g(x_i, y_{j-1})$.

We are ready to introduce the Fréchet-Stieltjes integral of f with respect to g: we say that the integral of f with respect to g on $[a,b] \times [c,d]$, denoted by $\int_a^b \int_c^d f(x,y) d_{x,y} g(x,y)$, exists if given $\varepsilon > 0$ there are two finite partitions Π and Π' of $[a,b]$ and $[c,d]$, respectively, such that for any tagged partitions $\pi(\xi)$ and $\pi'(\eta)$ that are refinement of Π and Π', respectively, we have

$$\left| \int_a^b \int_c^d f_{\pi(\xi),\pi'(\eta)} d_{x,y} g(x,y) - \int_a^b \int_c^d f(x,y) d_{x,y} g(x,y) \right| < \varepsilon.$$

Now, we deal with sufficient condition on f and g that guarantee the existence of this integral.

Definition 27 *Let $p, q \ge 0$. The function $f[a,b] \times [c,d] \to \mathbb{R}$ has finite (p,q)-bivariation if*

$$\sup_{y_1, y_2 \in [c,d]} \operatorname{var}_p (f(\cdot, y_1) - f(\cdot, y_2); [a,b]) < \infty$$

and

$$\sup_{x_1, x_2 \in [a,b]} \operatorname{var}_q (f(x_1, \cdot) - f(x_2, \cdot); [c,d]) < \infty.$$

We now state a particular case of Theorem 6.3 in [217] in order not to complicate the notation.

Theorem 31 *Let $f[a,b] \times [c,d] \to \mathbb{R}$ have finite (p,q)-bivariation with $p, q \ge 1$, which vanishes on the lines $x = a$ and $y = c$, and $g[a,b] \times [c,d] \to \mathbb{R}$ a function with points of discontinuity of first kind. Moreover, assume that there exist increasing functions $\rho, \sigma, \lambda, \mu : \mathbb{R}_+ \cup \{0\} \to \mathbb{R}_+ \cup \{0\}$ such that $\rho(u)\sigma(u) = u$,*

$$\sum_{n=1}^\infty \rho\left(\frac{1}{n^{1/p}}\right) \lambda\left(\frac{1}{n}\right), \quad \sum_{n=1}^\infty \sigma\left(\frac{1}{n^{1/q}}\right) \mu\left(\frac{1}{n}\right) < \infty$$

and $|\Delta_j \Delta_i G(x_i, y_j)| \leq \lambda(x_i - x_{i-1})\mu(y_j - y_{j-1})$*. Then, the integral* $\int_a^b \int_c^d f(x,y) d_{x,y} g(x,y)$ *exists.*

Note that there is no restriction on the common jump points of the functions f and g.

As an example of the last theorem, we can consider the functions $\rho(x) = x^\alpha$, $\sigma(x) = x^{1-\alpha}$, $\lambda(x) = x^{1/\tilde{p}}$ and $\mu(x) = x^{1/\tilde{q}}$, $x \geq 0$. Here, $\alpha \in (0,1)$ and $\tilde{p}, \tilde{q} > 1$ are such that

$$\frac{\alpha}{p} + \frac{1}{\tilde{p}} > 1 \quad \text{and} \quad \frac{1-\alpha}{q} + \frac{1}{\tilde{q}} > 1.$$

In order to give an upper bound for the integral $\int_a^b \int_c^d f(x,y) d_{x,y} g(x,y)$, we introduce the following.

Definition 28 *Let* $p \geq 1$*. We say that a function* $f : [a,b] \times [c,d] \to \mathbb{R}$ *has finite* p*-variation if*

$$V_p(f; [a,b] \times [c,d]) := \left(\sup \left\{ \sum_{i=1}^{n} \sum_{j=1}^{m} |\Delta_j \Delta_i f(x_i, y_j)|^p : \right. \right.$$
$$\left. \left. \pi \text{ and } \pi' \text{ are partitions of } [a,b] \text{ and } [c,d] \right\} \right)^{1/p} < \infty.$$

The proof of the following result is given in Towghi [208] (Theorem 1.2). This result also shows the existence of the integral $\int_a^b \int_c^d f(x,y) d_{x,y} g(x,y)$ and an upper bound for it in terms of the variations of f and g.

Theorem 32 (Towghi [208]) *Let* $p, q \geq 1$ *be such that* $\theta := \frac{1}{p} + \frac{1}{q} > 1$ *and consider two functions* $f, g : [a,b] \times [c,d] \to \mathbb{R}$ *of finite* p *and* q*-variation, respectively, which do not have common jumps and* $f(a, \cdot) = f(\cdot, c) = 0$*. Then, the 2D Young integral* $\int_a^b \int_c^d f(x,y) d_{x,y} g(x,y)$ *exists in the Riemann-Stieltjes sense and, for* $\alpha \in (1, \theta)$*, we have*

$$\begin{aligned} \left| \int_a^b \int_c^d f(x,y) d_{x,y} g(x,y) \right| &\leq \left[\left(1 + \zeta\left(\frac{\theta}{\alpha}\right)\right)^\alpha (1 + \zeta(\alpha)) + 2(1 + \zeta(\theta)) \right] \\ &\quad \times V_p(f; [a,b] \times [c,d]) V_q(g; [a,b] \times [c,d]), \end{aligned} \tag{2.32}$$

where $\zeta(s) = \sum_{n=1}^\infty 1/n^s$ *is the Riemann zeta function.*

Remember that the importance of inequality (2.32) is to be able to control the convergence of 2D Young integrals through the variations of the functions f and g. The procedure to prove that the last two theorems hold has already been introduced in this section.

2.2 Hölder continuous functions

Here, we follow the algebraic Young integration approach introduced in Gubinelli [70] and developed, for example, in Gubinelli and Tindel [71].

2.2.1 Increments

Let us begin with the basic algebraic structures, which allow us to define a pathwise integral (i.e., ω by ω) with respect to irregular processes.

For arbitrary real numbers $a < b$, a topological vector space V and an integer $k \geq 1$, we denote by $\mathcal{C}_k(V)$ (or by $\mathcal{C}_k([a,b];V)$) the set of all continuous functions $g : [a,b]^k \to V$ such that $g_{t_1 \cdots t_k} = 0$ whenever $t_i = t_{i+1}$ for some $1 \leq i \leq k-1$. Such a function will be called a $(k-1)$-*increment*. In the case that $k = 1$, $\mathcal{C}_1(V)$ is the family of all continuous functions from $[a,b]$ into V.

An important elementary operator is δ, which is defined on $\mathcal{C}_k(V)$ as follows: the operator $\delta : \mathcal{C}_k(V) \to \mathcal{C}_{k+1}(V)$ is defined by

$$(\delta g)_{t_1 \cdots t_{k+1}} = \sum_{i=1}^{k+1} (-1)^{k-i} g_{t_1 \cdots \hat{t}_i \cdots t_{k+1}}, \quad t_1, \ldots, t_{k+1} \in [a,b], \tag{2.33}$$

where $\hat{t}_i$ means that this particular argument is omitted. Note that if $t_i = t_{i+1}$ for some $1 \leq i \leq k$, then $(\delta g)_{t_1 \cdots t_{k+1}} = 0$. Some simple examples of actions of δ are obtained by letting $g \in \mathcal{C}_1(V)$ and $h \in \mathcal{C}_2(V)$. Then, for any $s, u, t \in [0,T]$, we have

$$(\delta g)_{st} = g_t - g_s \quad \text{and} \quad (\delta h)_{sut} = h_{st} - h_{su} - h_{ut}. \tag{2.34}$$

A fundamental property of δ is given by the following result.

Lemma 29 *Let $k \geq 1$ be an integer. Then, we have that $\delta\delta = 0$, where $\delta\delta$ is considered as an operator from $\mathcal{C}_k(V)$ to $\mathcal{C}_{k+2}(V)$.*

Proof. We first assume that $k = 1$. So, consider $g \in \mathcal{C}_1([a,b];V)$. Then, (2.34) implies that, for $t_1, t_2, t_3 \in [a,b]$,

$$\begin{aligned}
(\delta(\delta g))_{t_1 t_2 t_3} &= (\delta g)_{t_1 t_3} - (\delta g)_{t_1 t_2} - (\delta g)_{t_2 t_3} \\
&= (g_{t_3} - g_{t_1}) - (g_{t_2} - g_{t_1}) - (g_{t_3} - g_{t_2}) = 0.
\end{aligned}$$

Hence, for $k > 1$, we can suppose that the result is satisfied for $k-1$ by induction hypothesis on k. Thus, now, let g be in $\mathcal{C}_k(V)$. In this case, from (2.33), we have that, for $t_1, \ldots, t_{k+2} \in [a,b]$,

$$\begin{aligned}
(\delta(\delta g))_{t_1 \cdots, t_{k+2}} &= (-1)^{k+1-(k+2)} (\delta g)_{t_1 \cdots t_{k+1}} + \sum_{i=1}^{k+1} (-1)^{k+1-i} (\delta g)_{t_1 \cdots \hat{t}_i \cdots t_{k+2}} \\
&= -(\delta g)_{t_1 \cdots t_{k+1}} + \sum_{i=1}^{k+1} (-1)^{k+1-i} (-1)^{k-(k+1)} g_{t_1 \cdots \hat{t}_i \cdots t_{k+1}} \\
&\quad + \sum_{i=1}^{k+1} (-1)^{k+1-i} \left(\sum_{j=1}^{i-1} (-1)^{k-j} g_{t_1 \cdots \hat{t}_j \cdots \hat{t}_i \cdots t_{k+2}} \right. \\
&\quad \left. + \sum_{j=i+1}^{k+1} (-1)^{k-(j-1)} g_{t_1 \cdots \hat{t}_i \cdots \hat{t}_j \cdots t_{k+2}} \right) = \left(\delta \left(\delta g_{\cdot t_{k+2}}\right)\right)_{t_1 \cdots, t_{k+1}} = 0,
\end{aligned}$$

where the last equality follows from our induction hypothesis. In other words, the result is also true for k and therefore the proof is finished. □

We will denote $\mathcal{ZC}_k(V) = \mathcal{C}_k(V) \cap \mathrm{Ker}\delta$, for any $k \geq 1$, and $\mathcal{BC}_k(V) = \mathcal{C}_k(V) \cap \mathrm{Im}\delta$ for $k \geq 2$. That is

$$\mathcal{ZC}_k(V) = \{g \in \mathcal{C}_k(V) : \delta g = 0\} \text{ and } \mathcal{BC}_k(V) = \{g \in \mathcal{C}_k(V) : g = \delta f \text{ for } f \in \mathcal{C}_{k-1}(V)\}.$$

Other basic property of the operator δ is the following:

Lemma 30 *Let $k \geq 1$ and $h \in \mathcal{ZC}_{k+1}(V)$. Then there exists a (nonunique) $f \in \mathcal{C}_k(V)$ such that $h = \delta f$.*

Proof. Let $t \in [a,b]$ be an arbitrary point and $t_1, \ldots, t_{k+1} \in [a,b]$. Then, by hypothesis, we have

$$0 = (\delta h)_{t_1 \cdots t_{k+1} t} = -h_{t_1 \cdots t_{k+1}} + \sum_{i=1}^{k+1} (-1)^{k+1-i} h_{t_1 \cdots \hat{t}_i \cdots t_{k+1} t}.$$

It means, $h_{t_1 \cdots t_{k+1}} = \sum_{i=1}^{k+1} (-1)^{k+1-i} h_{t_1 \cdots \hat{t}_i \cdots t_{k+1} t} = (\delta(-h_{\cdot t}))_{t_1 \cdots t_{k+1}}$. Hence, taking $f = -h_{\cdot t}$, we obtain that the result is true. Finally, note that this also shows that such an f is not the only function such that $h = \delta f$ because, in general, $h_{\cdot t} \neq h_{\cdot s}$, for $s \neq t$. □

Observe that Lemmas 29 and 30 imply

$$\mathcal{ZC}_k(V) = \mathcal{BC}_k(V), \quad \text{for any } k \geq 2,$$

and that all the elements $h \in \mathcal{C}_2(V)$ such that $\delta h = 0$ can be written as $h = \delta f$ for some (non unique) $f \in \mathcal{C}_1(V)$. Thus, we get a heuristic interpretation of $\delta|_{\mathcal{C}_2(V)}$: it measures how much a given 1-increment is far from being an exact increment of a function, i.e., a finite difference.

Now, let f and g be two smooth $\mathbb{R}$-valued functions on $[a,b]$. Set $I \in \mathcal{C}_2(V;[a,b])$ by

$$I_{st} = \int_s^t df_v \int_s^v dg_w, \quad \text{for} \quad s,t \in [a,b].$$

Thus, some straightforward computations show

$$(\delta I)_{sut} = [g_u - g_s][f_t - f_u] = (\delta f)_{ut} (\delta g)_{su}. \tag{2.35}$$

This is a helpful property of the operator δ. That is, it transforms iterated integrals into products of increments, and we will be able to take advantage of both regularities of f and g in these products of the form $\delta f \, \delta g$. Therefore we obtain an important link between these algebraic structures and integration theory.

For sake of simplicity, let us specialize now our setting to the case $V = \mathbb{R}$. Sometimes, we still write V instead of $\mathbb{R}$ because the interested reader will be easily able to change V by either $\mathbb{R}^n$, or a Banach space. Our discussions mainly rely on k-increments with $k \leq 2$, for which we use some analytical assumptions. Namely, we measure the size of these increments by Hölder norms defined in the following way: for $a \leq a_1 < a_2 \leq b$ and $f \in \mathcal{C}_2([a_1,a_2];V)$, let

$$\|f\|_{\mu,[a_1,a_2]} = \sup_{\substack{r,t \in [a_1,a_2] \\ r \neq t}} \frac{|f_{rt}|}{|t-r|^\mu}$$

and

$$\mathcal{C}_2^\mu([a_1,a_2];V) = \left\{ f \in \mathcal{C}_2(V) : \|f\|_{\mu,[a_1,a_2]} < \infty \right\}.$$

Obviously, the usual Hölder spaces $\mathcal{C}_1^\mu([a_1,a_2];V)$ is determined in the following way: for a continuous function $g \in \mathcal{C}_1([a_1,a_2];V)$, we simply set

$$\|g\|_{\mu,[a_1,a_2]} = \|\delta g\|_{\mu,[a_1,a_2]}, \tag{2.36}$$

and we say that $g \in \mathcal{C}_1^\mu([a_1,a_2];V)$ if and only if $\|g\|_{\mu,[a_1,a_2]}$ is finite. Notice that $\|\cdot\|_{\mu,[a_1,a_2]}$ is only a semi-norm on $\mathcal{C}_1^\mu([a_1,a_2];V)$, but we generally work on spaces of the type

$$\mathcal{C}_{v,a_1,a_2}^\mu(V) = \left\{ g : [a_1,a_2] \to V : \, g_{a_1} = v, \, \|g\|_{\mu,[a_1,a_2]} < \infty \right\}, \tag{2.37}$$

for a given $v \in V$, or

$$\mathcal{C}^{\mu}_{\varrho,a_1,a_2}(V) := \{\zeta \in \mathcal{C}^{\mu}_1([a_1-h,a_2];V) : \zeta = \varrho \text{ on } [a_1-h,a_1]\}, \tag{2.38}$$

where $a \leq a_1 < a_2 \leq b$ and $\varrho \in \mathcal{C}^{\mu}_1([a_1-h,a_1];V)$. These last two spaces are complete metric spaces with the distance d_{μ,a_1,a_2}. Here, $d_{\mu,a_1,a_2}(f,g) = \|f-g\|_{\mu,[a_1,a_2]}$ on $\mathcal{C}^{\mu}_{v,a_1,a_2}(V)$; and $d_{\mu,a_1,a_2}(f,g) = \|f-g\|_{\mu,[a_1-h,a_2]}$ on the space $\mathcal{C}^{\mu}_{\varrho,a_1,a_2}(V)$.

In some cases we only write $\mathcal{C}^{\mu}_k(V)$ instead of $\mathcal{C}^{\mu}_k([a_1,a_2];V)$ when this does not lead to an ambiguity in the domain of the definition of the functions under consideration. For $h \in \mathcal{C}_3([a_1,a_2];V)$ set

$$\|h\|_{\gamma,\rho,[a_1,a_2]} = \sup_{\substack{s,u,t\in[a_1,a_2]\\ u\neq s,\ u\neq t}} \frac{|h_{sut}|}{|u-s|^{\gamma}|t-u|^{\rho}} \tag{2.39}$$

and

$$\|h\|_{\mu,[a_1,a_2]} = \inf\left\{\sum_i \|h_i\|_{\rho_i,\mu-\rho_i} : h = \sum_i h_i,\ 0<\rho_i<\mu\right\},$$

where the last infimum is taken over all sequences $\{h_i \in \mathcal{C}_3(V)\}$ such that $h = \sum_i h_i$ and for all choices of the numbers $\rho_i \in (0,\mu)$. Observe that $\|\cdot\|_{\mu,[a_1,a_2]}$ is easily seen to be a norm on $\mathcal{C}_3([a_1,a_2];V)$. We now introduce the families

$$\begin{aligned} \mathcal{C}^{\mu}_3([a_1,a_2];V) &:= \left\{h \in \mathcal{C}_3([a_1,a_2];V) : \|h\|_{\mu,[a_1,a_2]} < \infty\right\}, && (2.40)\\ \mathcal{C}^{1+}_3([a_1,a_2];V) &:= \cup_{\mu>1}\mathcal{C}^{\mu}_3([a_1,a_2];V) && (2.41)\end{aligned}$$

and

$$\mathcal{ZC}^{1+}_3([a_1,a_2];V) := \mathcal{C}^{1+}_3([a_1,a_2];V) \cap \ker\delta. \tag{2.42}$$

Remember that if $g \in \mathcal{C}^{\mu}_1(V)$ with $\mu > 1$, then g has a derivative equal to zero and, consequently, it is a constant function. But, if g belongs to $\mathcal{C}^{\mu}_3(V)$ with $\mu > 1$, then g is not necessarily a constant function as the following example shows: consider $f \in \mathcal{C}^{\alpha}_1(V)$ and $h \in \mathcal{C}^{\beta}_1(V)$, where $\alpha,\beta \in (0,1)$ are such that $\mu = \alpha+\beta > 1$. Consequently, the map

$$g_{sut} = (\delta f)_{ut}\,(\delta h)_{su}$$

belongs to the space $\mathcal{C}^{\mu}_3(V)$ and $\|g\|_{\mu,[a_1,a_2]} \leq \|f\|_{\alpha,[a_1,a_2]}\|h\|_{\beta,[a_1,a_2]}$.

Taking into account (2.40), (2.41) and (2.42), the crucial point in the approach to path-wise integration of irregular processes is that, under mild smoothness conditions, the operator δ can be inverted. This inverse is called Λ, and is defined in the following proposition.

Proposition 39 *Let $a \leq a_1 < a_2 \leq b$. Then there exists a unique linear map $\Lambda : \mathcal{ZC}^{1+}_3([a_1,a_2];V) \to \mathcal{C}^{1+}_2([a_1,a_2];V)$ such that*

$$\delta\Lambda = Id_{\mathcal{ZC}^{1+}_3([a_1,a_2];V)}.$$

Furthermore, for any $\mu > 1$, the map Λ is continuous from $\mathcal{ZC}^{\mu}_3([a_1,a_2];V)$ to $\mathcal{C}^{\mu}_2([a_1,a_2];V)$ with

$$\|\Lambda h\|_{\mu,[a_1,a_2]} \leq \frac{1}{1-2^{1-\mu}}\|h\|_{\mu,[a_1,a_2]}, \qquad h \in \mathcal{ZC}^{\mu}_3([a_1,a_2];V). \tag{2.43}$$

Remark 35 *Note that, for $h \in \mathcal{C}^{1+}_3([a_1,a_2];V)$ such that $\delta h = 0$, the result implies that there exists a unique $g = \Lambda(h) \in \mathcal{C}^{1+}_2([a_1,a_2];V)$ such that $\delta g = h$.*

Proof. Let $h \in \mathcal{C}_3^{1+}(V)$ be such that $\delta h = 0$. Then, Lemma 30 implies that there exists $\Xi \in \mathcal{C}_2$ so that $h = \delta\Xi$. Set $P_0 = \{s,t\}$, $(I^\circ\Xi)_{s,t} = \Xi_{s,t}$, with $s,t \in [a_1,a_2]$, and ρ_n the dyadic partition of order n of the interval $[s,t]$, which has 2^n intervals with length $2^{-n}(t-s)$. We also use the notation

$$(I^n\Xi)_{s,t} = \sum_{[u,v]\in\rho_n} \Xi_{u,v}.$$

Here $[u,v] \in \rho_n$ means that there is $0 \le k \le 2^n - 1$ such that $u = s + (t-s)k/2^n$ and $v = s + (t-s)(k+1)/2^n$. Therefore, the facts that $h \in \mathcal{C}_3^{1+}(V)$ and $h = \delta\Xi$ give

$$\left|\left(I^{n+1}\Xi\right)_{s,t} - (I^n\Xi)_{s,t}\right| = \left|\sum_{[u,v]\in\rho_n} (\delta\Xi)_{u\frac{u+v}{2}v}\right| \le 2^{n(1-\mu)} \left|\left|\delta\Xi\right|\right|_\mu |t-s|^\mu,$$

for some $\mu > 1$. Consequently, the limit $(I\Xi)_{s,t} = \lim_{n\to\infty} (I^n\Xi)_{s,t}$ is well-defined and

$$\begin{aligned}\left|(I\Xi)_{s,t} - \Xi_{s,t}\right| &\le \sum_{n\ge 0}\left|\left(I^{n+1}\Xi\right)_{s,t} - (I^n\Xi)_{s,t}\right| \le \|\delta\Xi\|_\mu |t-s|^\mu \sum_{n\ge 0} 2^{n(1-\mu)} \\ &= \frac{1}{1-2^{1-\mu}} \left|\left|\delta\Xi\right|\right|_\mu |t-s|^\mu .\end{aligned} \tag{2.44}$$

It is not difficult to see that the definition of $I\Xi$ is independent of the sequence ρ_n. It means, instead of ρ_n, we can use an arbitrary partition $\bar\rho_n$, $n \in \mathbb{N}$, such that $\bar\rho_{n+1}$ is finer than $\bar\rho_n$ and $\|\bar\rho_n\| \to 0$, as $n \to \infty$.

Now we use the convention $\Lambda h = \Xi - I\Xi$. Thus (2.44) yields that $\Lambda h \in \mathcal{C}_2^{1+}(V)$, (2.43) holds and

$$(\delta\Lambda h)_{s,u,t} = (\delta\Xi)_{s,u,t} - (\delta I\Xi)_{s,u,t} = h_{s,u,t} - I\Xi_{s,t} + I\Xi_{s,u} + I\Xi_{u,t} = h_{s,u,t}.$$

Finally, suppose that there is another operator $\tilde\Lambda$ satisfying the result. Then,

$$\delta\left(\Lambda h - \tilde\Lambda h\right) = h - h = 0.$$

Hence, Lemma 30 and (2.34) allow us to find $f \in \mathcal{C}_1^\mu(V)$ with

$$f(t) - f(s) = \Lambda_{s,t}h - \tilde\Lambda_{s,t}h,$$

which leads us to conclude that f is a constant since $\mu > 1$. In other words, $\Lambda_{s,t}h = \tilde\Lambda_{s,t}h$. That is, the proof is complete. □

Moreover, the operator Λ can be related to the limit of some Riemann sums, which gives a second link (see equality (2.35)) between the previous algebraic developments and some kind of generalized integration.

Corollary 16 *Let g be a 1-increment in $\mathcal{C}_2(V)$ such that $\delta g \in \mathcal{C}_3^{1+}(V)$ and $h = (Id - \Lambda\delta)g$. Then, for $a \le s < t \le b$,*

$$h_{st} = \lim_{|\pi_{st}|\to 0} \sum_{i=1}^n g_{t_{i-1}t_i},$$

where the limit is over any partition $\pi_{st} = \{s = t_0 < \cdots < t_n = t\}$ of $[s,t]$, whose mesh tends to zero (i.e., $|\pi_{st}| \to 0$).

Remark 36 *Note that the 1-increment h could be interpreted as the indefinite integral of the 1-increment g. The reader interested in this type of results can consult Matsuda and Perkowski [144], and references therein.*

Proof. Note that $\delta h = 0$ due to $\delta\Lambda = \mathrm{Id}_{\mathcal{Z}\mathcal{C}_3^{1+}(V)}$, as Proposition 39 establishes. Thus, from Lemma 30, there exists $f \in \mathcal{C}_1(V)$ such that $h = \delta f$. Hence,

$$\begin{aligned} h_{s,t} &= f_t - f_s = \sum_{i=1}^{n} (f_{t_1} - f_{t_{i-1}}) = \sum_{i=1}^{n} (\delta f)_{t_{i-1}t_i} \\ &= \sum_{i=1}^{n} ((\mathrm{Id} - \Lambda\delta)g)_{t_{i-1}t_i} = \sum_{i=1}^{n} g_{t_{i-1}t_i} + \sum_{i=1}^{n} (\Lambda\delta g)_{t_{i-1}t_i}. \end{aligned}$$

Therefore, the result is satisfied since Proposition 39 implies that there is $\mu > 1$ such that

$$\left| \sum_{i=1}^{n} (\Lambda\delta g)_{t_{i-1}t_i} \right| \le \|\Lambda\delta g\|_{\mu,[a,b]} \sum_{i=1}^{n} (t_i - t_{i-1})^{\mu} \le (b-a)\|\Lambda\delta g\|_{\mu,[a,b]} |\pi_{st}|^{\mu-1},$$

which converges to zero as $|\pi_{st}|$ goes to zero. □

2.2.2 Young integration

In this section, for $f \in \mathcal{C}_1^{\kappa}([a,b];\mathbb{R})$ and $g \in \mathcal{C}_1^{\gamma}([a,b];\mathbb{R})$ with $\kappa + \gamma > 1$, we define a generalized integral $\int_s^t f_u dg_u$ by means of the algebraic tools introduced in Section 2.2.1. Towards this end, we first assume that f and g are smooth functions, in which case the integral of f with respect to g can be defined in the Lebesgue-Stieltjes sense (see Remark 30), and then we express this integral in terms of the operator Λ introduced in Proposition 39. This leads us to a natural extension of the notion of integral, which coincides with the usual Young integral. In the sequel, we sometimes write $\mathcal{J}_{st}(f\,dg)$ instead of $\int_s^t f_u dg_u$ in order to simplify the notation.

So, for the moment, let us consider then two smooth functions f and g defined on $[a,b]$. Therefore, we can establish

$$\mathcal{J}_{st}(f\,dg) \equiv \int_s^t f_u\, dg_u = f_s(\delta g)_{st} + \int_s^t (\delta f)_{su}\, dg_u = f_s(\delta g)_{st} + \mathcal{J}_{st}(\delta f\, dg). \tag{2.45}$$

Let us analyze now the term $\mathcal{J}(\delta f\,dg)$. By (2.7), $\mathcal{J}(\delta f\,dg)$ belongs to the space $\mathcal{C}_2^{1+}([a,b];\mathbb{R})$ since f and g are smooth functions with a bounded derivative. Also, invoking equality (2.35), it is easily seen that, for $s,u,t \in [a,b]$,

$$h_{sut} := [\delta(\mathcal{J}(\delta f\,dg))]_{sut} = (\delta f)_{su}(\delta g)_{ut}.$$

The increment h is thus an element of $\mathcal{C}_3(\mathbb{R})$ satisfying $\delta h = 0$ (recall that $\delta\delta = 0$ due to Lemma 29). Let us estimate the regularity of h: if $f \in \mathcal{C}_1^{\kappa}([a,b];\mathbb{R})$ and $g \in \mathcal{C}_1^{\gamma}([0,T];\mathbb{R})$, from the Definition (2.39), it is readily checked that $h \in \mathcal{C}_3^{\gamma+\kappa}(\mathbb{R})$. Hence $h \in \mathcal{Z}\mathcal{C}_3^{\gamma+\kappa}(\mathbb{R})$, and if $\kappa + \gamma > 1$ (which is the case if f and g are regular), Proposition 39 yields that $\mathcal{J}(\delta f\,dg)$ can also be expressed as

$$\mathcal{J}(\delta f\,dg) = \Lambda(h) = \Lambda(\delta f\,\delta g),$$

and thus, plugging this identity into (2.45), we get:

$$\mathcal{J}_{st}(f\,dg) = f_s(\delta g)_{st} + \Lambda_{st}(\delta f\,\delta g). \tag{2.46}$$

Note that the right-hand side of the last equality is rigorously defined whenever $f \in \mathcal{C}_1^{\kappa}([a,b];\mathbb{R})$ and $g \in \mathcal{C}_1^{\gamma}([a,b];\mathbb{R})$ with $\kappa + \gamma > 1$. Therefore, this is the definition that we use in order to extend the notion of integral for Hölder continuous functions. Thus, we obtain the following:

Theorem 33 *Let $f \in \mathcal{C}_1^{\kappa}([a,b];\mathbb{R})$ and $g \in \mathcal{C}_1^{\gamma}([a,b];\mathbb{R})$ be two Hölder continuous functions with $\kappa+\gamma>1$. Set, for $s,t \in [a,b]$,*

$$\mathcal{J}_{st}(f\,dg) := f_s(\delta g)_{st} + \Lambda_{st}\left(\delta f\, \delta g\right). \tag{2.47}$$

Then,

(1) Whenever f and g are smooth functions, $\mathcal{J}_{st}(f\,dg)$ coincides with the usual Riemann-Stieltjes integral given by (2.2).

(2) The generalized integral $\mathcal{J}(f\,dg)$ satisfies:

$$|\mathcal{J}_{st}(f\,dg)| \leq \|f\|_{\infty}\|g\|_{\gamma}|t-s|^{\gamma} + c_{\gamma,\kappa}\|f\|_{\kappa}\|g\|_{\gamma}|t-s|^{\gamma+\kappa},$$

for a constant $c_{\gamma,\kappa}$ whose exact value is $2^{\gamma+\kappa}(2^{\gamma+\kappa}-2)^{-1}$.

(3) We have

$$\mathcal{J}_{st}(f\,dg) = \lim_{|\pi_{st}|\to 0} \sum_{i=1}^{n} f_{t_{i-1}}\delta g_{t_{i-1}\,t_i},$$

where the limit is over any partition $\pi_{st} = \{s = t_0 < \cdots < t_n = t\}$ of $[s,t]$ such that $|\pi_{st}|$ tends to zero. In particular, $\mathcal{J}_{st}(f\,dg)$ coincides with the Young integral as defined either in [216], or in (2.2).

Proof. The first claim is just (2.46). The second assertion follows directly from Definition (2.47) and inequality (2.43) concerning the operator Λ.

Now, we use the convention $(f\delta g)_{st} = f_s(\delta g)_{st}$. Thus, our third statement is a direct consequence of Corollary 16 and the fact that $\delta(f\,\delta g) = -\delta f \delta g$, which means that

$$\mathcal{J}(f\,dg) = [\mathrm{Id} - \Lambda\delta]\,(f\,\delta g).$$

Finally, we also have $\mathcal{J}_{st}(f\,dg) = \lim_{|\pi_{st}|\to 0}\sum_{i=1}^{n} f_{s_i}\,\delta g_{t_{i-1}\,t_i}$, where $\pi_{st} = \{s = t_0 < \cdots < t_n = t\}$ is a partition of $[s,t]$ and $s_i \in [t_{i-1},t_i]$. Indeed, the fact that $\kappa+\lambda>1$ implies

$$\begin{aligned}&\left|\sum_{i=1}^{n} f_{s_i}\,\delta g_{t_{i-1}\,t_i} - \sum_{i=1}^{n} f_{t_{i-1}}\,\delta g_{t_{i-1}\,t_i}\right| \\ &\quad \leq \sum_{i=1}^{n} \left|\delta f_{t_{i-1}\,s_i}\right|\,\left|\delta g_{t_{i-1}\,t_i}\right| \leq \|f\|_{\kappa,[a,b]}\|g\|_{\gamma,[a,b]}\sum_{i=1}^{n}(t_i-t_{i-1})^{\kappa+\lambda} \to 0, \quad \text{as } |\pi_{st}|\to 0,\end{aligned}$$

which says that $\mathcal{J}$ agrees with the Riemann-Stieltjes integral. □

The proofs of the following three results are good examples that show the importance of Proposition 39 and Theorem 33. We begin with a Fubini's type theorem for Young's integral.

Proposition 40 *Assume that $\gamma>\lambda>1/2$. Let f and g be two functions in $\mathcal{C}_1^{\gamma}([a,b]:\mathbb{R})$ and $h : \{(t,s)\in[a,b]^2; a\leq s\leq t\leq b\} \to \mathbb{R}$ a function such that $h(\cdot,t)$ (resp. $h(t,\cdot)$) belongs to $\mathcal{C}_1^{\lambda}([t,b];\mathbb{R})$ (resp. $\mathcal{C}_1^{\lambda}([a,t];\mathbb{R})$) uniformly in $t\in[a,b]$, and*

$$\|h(\cdot,r_1)-h(\cdot,r_2)\|_{\lambda,[r_1\vee r_2,b]} + \|h(r_1,\cdot)-h(r_2,\cdot)\|_{\lambda,[a,r_1\wedge r_2]} \leq C|r_1-r_2|^{\lambda}. \tag{2.48}$$

Then,

$$\int_s^t\left(\int_s^r h(r,u)dg_u\right)df_r = \int_s^t\left(\int_u^t h(r,u)df_r\right)dg_u, \quad a\leq s\leq t\leq b. \tag{2.49}$$

Proof. Fix $s, t \in [a, b]$, with $s < t$. Now, we divide the proof in several steps.
Step 1. Here we see that $\int_s^t \int_s^r h(r, u)dg_u df_r$ is well-defined. Note that we only need to show that $\int_s^{\cdot} h(\cdot, u)dg_u$ belongs to $\mathcal{C}_1^\lambda([s,t];\mathbb{R})$ due to Theorem 33.

Let $r_1, r_2 \in [s,t]$, $r_1 < r_2$, then Theorems 19 and 33.(2) give

$$\begin{aligned}
&\left|\int_s^{r_2} h(r_2,u)dg_u - \int_s^{r_1} h(r_1,u)dg_u\right| \\
&\leq \left|\int_s^{r_1} (h(r_2,u)-h(r_1,u))dg_u\right| + \left|\int_{r_1}^{r_2} h(r_2,u)dg_u\right| \\
&\leq \|g\|_\gamma \left(\|h(r_2,\cdot)-h(r_1,\cdot)\|_{\infty,[a,r_1]}(r_1-s)^\gamma\right. \\
&\quad \left.+c_{\gamma,\lambda}\|h(r_2,\cdot)-h(r_1,\cdot)\|_{\lambda,[a,r_1]}(r_1-s)^{\gamma+\lambda}\right) \\
&\quad +\|g\|_\gamma \left(\|h(r_2,\cdot)\|_{\infty,[a,r_2]}(r_2-r_1)^\gamma + c_{\gamma,\lambda}\|h(r_2,\cdot)\|_{\lambda,[a,r_2]}(r_2-r_1)^{\gamma+\lambda}\right).
\end{aligned}$$

Hence (2.48) and the inequality, for $u \in [a, r_1]$,

$$\begin{aligned}
|h(r_2,u)-h(r_1,u)| &\leq |h(r_2,u)-h(r_1,u)-[h(r_2,a)-h(r_1,a)]| + |h(r_2,a)-h(r_1,a)| \\
&\leq \|h(r_2,\cdot)-h(r_1,\cdot)\|_{\lambda,[a,r_1]}(u-a)^\lambda + \|h(\cdot,a)\|_{\lambda,[a,b]}(r_2-r_1)^\lambda
\end{aligned}$$

imply that our claim holds. The integral $\int_s^t \int_u^t h(r,u)df_r dg_u$ is also well-defined using the same lines.
Step 2. Consider the sequence $\{\pi_{st}^n : n \in \mathbb{N}\}$ of dyadic partitions of $[s,t]$. That is,

$$\pi_{st}^n = \{s = t_0^n < t_1^n < \cdots < t_{2^n}^n = t\}, \text{ with } t_i^n = s + \frac{i\,(t-s)}{2^n}.$$

So, observing that the limit in Theorem 33.(3) does not depend on the sequence of partitions under consideration, we have

$$\int_s^v h(v,u)dg_u = \lim_{|\pi_{st}^n|\to 0} \sum_{i=1}^{2^n} h(v,t_{i-1}^n)\,(\delta g)_{t_{i-1}^n\wedge v\, t_i^n\wedge v} := \lim_{|\pi_{st}^n|\to 0} H(v,n), \quad v \in [s,t].$$

The aim of this part of the proof is to show that the last limit also holds in the space $\mathcal{C}_1^\lambda([s,t];\mathbb{R})$. To do so, we choose $n, m \in \mathbb{N}$, $n < m$. Then, for $v \in [s,t]$,

$$\begin{aligned}
|H(v,m)-H(v,n)| &\leq \sum_{j=n}^{m-1} |H(v,j+1)-H(v,j)| \\
&\leq \sum_{j=n}^{m-1}\sum_{i=1}^{2^j} \left|\left(\delta\left(h(v,\cdot)(\delta g)_{\cdot\wedge v\, *\wedge v}\right)\right)_{t_{i-1}^j\wedge v\ \frac{t_{i-1}^j\wedge v + t_i^j\wedge v}{2}\ t_i^j\wedge v}\right|.
\end{aligned}$$

Thus, we can use that $\delta(f\delta g) = -\delta f \delta g$, for any $f \in \mathcal{C}_1(\mathbb{R})$, to get

$$\begin{aligned}
|H(v,m)-H(v,n)| &\leq \|h(v,\cdot)\|_{\lambda,[a,v]}\|g\|_{\gamma,[a,b]} \sum_{j=n}^{m-1}\sum_{i=1}^{2^j}(t_i^j - t_{i-1}^j)^{\lambda+\gamma} \\
&= (t-s)^{\lambda+\gamma}\|h(v,\cdot)\|_{\lambda,[a,v]}\|g\|_{\gamma,[a,b]} \sum_{j=n}^{m-1} 2^{-j(\lambda+\gamma-1)}.
\end{aligned}$$

Hence, taking the limit as m goes to infinity, we are able to establish

$$\begin{aligned}&\left\|\int_s^{\cdot} h(\cdot,u)dg_u - \sum_{i=1}^{2^n} h(\cdot, t_{i-1}^n)\,(\delta g)_{t_{i-1}^n\wedge\cdot\, t_i^n\wedge\cdot}\right\|_{\infty,[s,t]}\\ &\leq C(b-a)^{\lambda+\gamma}\|g\|_{\gamma,[a,b]}\sum_{j=n}^{\infty} 2^{-j(\lambda+\gamma-1)} \to 0, \quad \text{as } n\to\infty. \end{aligned} \tag{2.50}$$

Similarly, let $s \leq r_1 < r_2 \leq t$. Then,

$$\begin{aligned}&|H(r_2,m)-H(r_2,n)-(H(r_1,m)-H(r_1,n))|\\ &\leq \sum_{j=n}^{m-1} |H(r_2,j+1)-H(r_2,j)-(H(r_1,j+1)-H(r_1,j))|\\ &\leq \sum_{j=n}^{m-1}\sum_{t_i^j\leq r_1}\left|(\delta([h(r_2,\cdot)-h(r_1,\cdot)]\delta g))_{t_{i-1}^j\,\frac{t_{i-1}^j+t_i^j}{2}\,t_i^j}\right|\\ &\quad+\sum_{j=n}^{m-1}\sum_{r_1<t_{i-1}^j}\left|(\delta(h(r_2,\cdot)\delta g))_{t_{i-1}^j\wedge r_2\,\frac{t_{i-1}^j\wedge r_2+t_i^j\wedge r_2}{2}\,t_i^j\wedge r_2}\right|\\ &\quad+\sum_{j=n}^{m-1}\left(\left|(\delta h(r_1,\cdot)\delta g)_{t_{i_0}^j(t_{2i_0+1}^{j+1}\wedge r_1)r_1}\right|+\left|(\delta h(r_2,\cdot)\delta g)_{t_{i_0}^j t_{2i_0+1}^{j+1} t_{i_0+1}^j}\right|\right),\end{aligned}$$

where i_0 is such that $r_1 \in [t_{i_0}^j, t_{i_0+1}^j)$. Consequently, for n large enough, (2.48) leads us to write

$$\begin{aligned}&|H(r_2,m)-H(r_2,n)-(H(r_1,m)-H(r_1,n))|\\ &\leq C(r_2-r_1)^{\lambda}\|g\|_{\gamma,[a,b]}(t-s)^{\lambda+\gamma}\sum_{j=n}^{m-1}2^{-j(\lambda+\gamma-1)}\\ &\quad+C\|g\|_{\gamma,[a,b]}\sum_{j=n}^{m-1}\left[\sum_{r_1<t_{i-1}^j}(t_i\wedge r_2-t_{i-1}\wedge r_2)^{\lambda+\gamma}+\frac{(t-s)^{\lambda+\gamma}}{2^{j(\lambda+\gamma)}}\right]\\ &\leq C(r_2-r_1)^{\lambda}\|g\|_{\gamma,[a,b]}(t-s)^{\lambda+\gamma}\sum_{j=n}^{m-1}2^{-j(\lambda+\gamma-1)}\\ &\quad+C(r_2-r_1)\|g\|_{\gamma,[a,b]}(t-s)^{\lambda+\gamma-1}\sum_{j=n}^{m-1}2^{-j(\lambda+\gamma-1)}.\end{aligned}$$

Therefore, taking the limit as $m\to\infty$ and then using the continuity of the Young integral and $H(\cdot,n)$, we are able to deduce

$$\begin{aligned}&\left\|\int_s^{\cdot} h(\cdot,u)dg_u - \sum_{i=1}^{2^n} h(\cdot, t_{i-1}^n)\,(\delta g)_{t_{i-1}^n\wedge\cdot\, t_i^n\wedge\cdot}\right\|_{\lambda,[s,t]}\\ &\leq C\|g\|_{\gamma,[a,b]}\left((b-a)^{\lambda+\gamma}+(b-a)^{\gamma}\right)\sum_{j=n}^{\infty}2^{-j(\lambda+\gamma-1)}\to 0 \quad\text{as}\quad n\to\infty,\end{aligned}$$

which, together with (2.50), implies that the aim of this step is satisfied.

Step 3. From Theorem 33.(2) and Step 2, we have

$$\begin{aligned}\int_s^t \int_s^r h(r,u)dg_u df_r &= \lim_{|\pi_{st}|\to 0} \int_s^t \left(\sum_{i=1}^n h(r,t_{i-1})(g_{t_i\wedge r} - g_{t_{i-1}\wedge r}) \right) df_r \\ &= \lim_{|\pi_{st}|\to 0} \sum_{i=1}^n \int_{t_{i-1}}^t h(r,t_{i-1}) \left(g_{t_i\wedge r} - g_{t_{i-1}}\right) df_r \\ &= \lim_{|\pi_{st}|\to 0} \sum_{i=1}^n \left[\left(\int_{t_i}^t h(r,t_{i-1})df_r \right) \left(g_{t_i} - g_{t_{i-1}}\right) \right. \\ &\qquad \left. + \int_{t_{i-1}}^{t_i} h(r,t_{i-1}) \left(g_r - g_{t_{i-1}}\right) df_r \right].\end{aligned}$$

Moreover, thank the Hölder properties of f and g, we have

$$\sum_{i=1}^n \left| \int_{t_{i-1}}^{t_i} h(r,t_{i-1})(g_r - g_{t_{i-1}})df_r \right| \leq C \sum_{i=1}^n (t_i - t_{i-1})^{\gamma+\lambda} \to 0$$

as $|\pi_{st}| \to 0$, and thus

$$\int_s^t \int_s^r h(r,u)dg_u df_r = \lim_{|\pi_{st}|\to 0} \sum_{i=1}^n \left(\int_{t_i}^t h(r,t_{i-1})df_r \right) \left(g_{t_i} - g_{t_{i-1}}\right).$$

Consequently, Step 2 and Theorem 33 imply that (2.49) is satisfied and therefore the proof is complete. □

The following two results are integration by parts formulae.

Theorem 34 (Integration by parts formula) *Let f and g be two functions in $\mathcal{C}_1^\gamma([a,b];\mathbb{R})$ and $\mathcal{C}_1^\lambda([a,b];\mathbb{R})$, respectively, with $\gamma+\lambda>1$. Then,*

$$f_t g_t = f_a g_a + \int_a^t f_u dg_u + \int_a^t g_u df_u, \quad t\in[a,b].$$

Remark 37 *We have already proven that this result holds using Riemann-Stieltjes sums (see proof of Theorem 18).*

Proof. Set $q_t := f_t g_t - \int_a^t f_u dg_u - \int_a^t g_u df_u$, $t\in[a,b]$. It is easy to see that this function belongs to $\mathcal{C}_1^{\gamma+\lambda}([a,b];\mathbb{R})$ because of the equalities

$$f_t g_t - f_s g_s = f_s(\delta g)_{st} + g_s(\delta f)_{st} + (\delta g)_{st}(\delta f)_{st}$$

and

$$\int_s^t f_u dg_u + \int_s^t g_u df_u = f_s(\delta g)_{st} + g_s(\delta f)_{st} + \Lambda_{st}(\delta f \delta g) + \Lambda_{st}(\delta g \delta f),$$

which follows from (2.47). Now, since $q\in\mathcal{C}_1^{\gamma+\lambda}([a,b];\mathbb{R})$, with $\gamma+\lambda>1$, q is a constant function. Thus, $q_t = q_a = f_a g_a$. Therefore, the result is true. □

Theorem 35 *Let g and h be in $\mathcal{C}_1^\gamma([a,b],\mathbb{R})$ with $\gamma\geq 1/2$, $f\in\mathcal{C}_b^2(\mathbb{R})$ and $x_t = x_a + \int_a^t g_s dh_s$, $t\in[a,b]$. Then,*

$$f(x_t) = f(x_a) + \int_a^t f'(x_u) g_u dh_u, \quad t\in[a,b].$$

Remark 38 *Theorem 35 has been proven in [218] using Riemann-Stieltjes sums.*

Proof. Note that the mean value theorem and Theorem 33 imply that $t \mapsto \int_0^t f'(x_s)g_s dh_s$ is well-defined. Finally, we can proceed as in the proof of Theorem 34 to see that the mean value theorem yield that the function

$$q_t = f(x_t) - \int_0^t f'(x_s)g_s dh_s, \quad t \in [a,b],$$

is 2γ-Hölder continuous. Therefore the result holds. □

Now, we show how we can integrate with respect to a noise of the form $\tilde{\theta} = \int_0^{\cdot} g(s)d\theta_s$.

Proposition 41 *Let $f \in C_1^\tau([a,b];\mathbb{R})$, $g \in C_1^\lambda([a,b];\mathbb{R})$ and $\theta \in C_1^\gamma([a,b];\mathbb{R})$, where $\tau, \lambda, \gamma \in (0,1)$ and $\tau+\gamma, \lambda+\gamma \in (1,2)$. Also let $\tilde{\theta} = \int_0^{\cdot} g_s d\theta_s$ on $[a,b]$. Then,*

$$\int_a^b f_s d\tilde{\theta}_s = \int_a^b f_s g_s d\theta_s.$$

Proof. Let $\pi = \{a = t_0 < t_1 < \ldots < t_n = b\}$ be a partition of the interval $[a,b]$. Then, by Theorem 33.(2), there is a constant $C > 0$ such that

$$\begin{aligned}\left|\sum_{i=1}^n f_{t_{i-1}}(\tilde{\theta}_{t_i} - \tilde{\theta}_{t_{i-1}}) - \sum_{i=1}^n f_{t_{i-1}} g_{t_{i-1}}(\theta_{t_i} - \theta_{t_{i-1}})\right| &= \left|\sum_{i=1}^n f_{t_{i-1}} \int_{t_{i-1}}^{t_i} (g_s - g_{t_{i-1}})d\theta_s\right| \\ &\leq C\|f\|_\infty \sum_{i=1}^n |t_i - t_{i-1}|^{\lambda+\gamma} \to 0\end{aligned}$$

as $|\pi| \to 0$. Therefore, the result is an immediate consequence of Theorem 33.(3). □

2.2.3 Young integration through rough paths theory

In this section, we consider a γ-Hölder continuous function $x : [a,b] \to \mathbb{R}$ with $\gamma \in (1/3, 1/2)$. Now, suppose that we are interested in studying the existence of a unique solution to the equation

$$y_t = \xi_0 + \int_a^t \sigma(y_s)dx_s, \quad t \in [a,b], \tag{2.51}$$

where σ is smooth enough. But, according to Theorem 33, we have that $\sigma(y)$ is also a γ-Hölder continuous function. Hence, the main problem is that, in this case, $2\gamma < 1$ and the integral in the last equation cannot be defined through (2.47).

The theory of rough paths has been introduced by Lyons [133] (see also Lyons [134], and Lyons and Qian [132]) in order to give a meaning to the integral $\int_s^t \sigma(y_s)dx_s$, $a \leq s < t \leq b$, and to analyze equation (2.51). Lyons studied this equation in the case that x is a function with finite p-variation on $[a,b]$, with $p \geq 2$. In this way, the theory of rough paths provides an extension of the Young integral for Hölder continuous functions with exponents less than 1/2. Due to the importance of this theory, it has been improved in different works among which we can mention [57, 58, 70].

The main purpose of this section is to show the fundamental ideas of the theory of rough paths. To simplify the exposition, remember that x is a γ-Hölder continuous function, where γ belongs to the interval $(1/3, 1/2)$.

In order to give a meaning to the integral $\int_a^b y_s dx_s$, we assume that the function y has form

$$y_t = \xi_0 + \int_a^t \hat{\sigma}_s dx_s, \quad t \in [a,b].$$

Towards this end, we first suppose that $\hat{\sigma}$ and x are two smooth functions. Then, for $s,t \in [a,b]$, $s<t$, we get

$$(\delta y)_{st} = \hat{\sigma}_s(\delta x)_{st} + \int_s^t (\hat{\sigma}_u - \hat{\sigma}_s) dx_u = \hat{\sigma}_s(\delta x)_{st} + \rho_{st}. \tag{2.52}$$

Applying Theorem 33 again, we can assume that ρ is in the space $\mathcal{C}_2^{2\kappa}([a,b];\mathbb{R})$ with $1/3 < \kappa < \gamma$. Hence, last equality yields

$$\mathcal{J}_{st}(ydx) = y_s(\delta x)_{st} + \mathcal{J}_{st}\left((\delta y)_s . dx\right) = y_s(\delta x)_{st} + \mathcal{J}_{st}(\hat{\sigma}_s(\delta x)_s . dx) + \mathcal{J}_{st}(\rho_s . dx). \tag{2.53}$$

Note that

$$\mathcal{J}_{st}(\hat{\sigma}_s(\delta x)_s . dx) = \hat{\sigma}_s \int_s^t (\delta x)_{su} dx_u = \hat{\sigma}_s \int_{s \le u_1 < u_2 \le t} dx_{u_1} dx_{u_2} := \hat{\sigma}_s x^2_{st},$$

$(\delta x^2)_{sut} = (\delta x)_{su}(\delta x)_{ut}$ because of (2.35) and $\delta(\mathcal{J}(ydx)) = 0$ since

$$\left(\delta(\mathcal{J}(ydx))\right)_{sut} = \mathcal{J}(ydx)_{st} - \mathcal{J}(ydx)_{su} - \mathcal{J}(ydx)_{ut}.$$

Hence, (2.53) implies

$$[(\delta(\mathcal{J}(\rho dx))]_{sut} = (\delta y)_{su}(\delta x)_{ut} + (\delta\hat{\sigma})_{su} x^2_{ut} - \hat{\sigma}_s(\delta x^2)_{sut}.$$

Thus, by (2.52), we obtain

$$\begin{aligned}[(\delta(\mathcal{J}(\rho dx))]_{sut} &= \hat{\sigma}_s(\delta x)_{su}(\delta x)_{ut} + \rho_{su}(\delta x)_{ut} + (\delta\hat{\sigma})_{su} x^2_{ut} - \hat{\sigma}_s(\delta x)_{su}(\delta x)_{ut} \\ &= \rho_{su}(\delta x)_{ut} + (\delta\hat{\sigma})_{su} x^2_{ut}.\end{aligned}$$

In this way, we can make use of Proposition 39 and (2.53) again to introduce the following definition.

Definition 29 *Let $\gamma, \kappa \in (1/3, 1/2)$ be such that $\kappa < \gamma$. Assume that there exist $x^2 \in \mathcal{C}_2^{2\gamma}([a,b];\mathbb{R})$ with $(\delta x^2)_{sut} = (\delta x)_{su}(\delta x)_{ut}$, $\hat{\sigma} \in \mathcal{C}_1^{\gamma}([a,b];\mathbb{R})$ and $\rho \in \mathcal{C}_2^{2\kappa}([a,b];\mathbb{R})$ such that*

$$(\delta y)_{st} = \hat{\sigma}_s(\delta x)_{st} + \rho_{st}, \quad s,t \in [a,b]. \tag{2.54}$$

Then, we define the integral of y with resect to x as

$$\mathcal{J}_{st}(ydx) := y_s(\delta x)_{st} + \hat{\sigma}_s x^2_{s,t} + \Lambda_{st}(\rho\delta x + (\delta\hat{\sigma})x^2), \quad s,t \in [a,b].$$

Remarks 1 *i) If $\hat{\sigma}$ and x are smooth functions, $\mathcal{J}_{st}(\hat{\sigma}dx)$ coincides with the usual Riemann-Stieltjes integral given by (2.2) and $x^2_{st} = \int_{s \le u_1 < u_2 \le t} dx_{u_1} dx_{u_2}$.*

ii) Note that (2.54) yields that y belongs to $\mathcal{C}_1^{\gamma}([a,b];\mathbb{R})$.

iii) Choose $z \in \mathcal{C}_1^{2\gamma}([a,b];\mathbb{R})$ and set $\bar{x}^2 = x^2 + \delta z$. Then, by Lemma 29, we get

$$(\delta\bar{x}^2)_{sut} = (\delta x^2)_{sut} + (\delta^2 z)_{sut} = (\delta x^2)_{sut} = (\delta x)_{su}(\delta x)_{ut},$$

for $s,u,t \in [a.b]$. So, we can use $\bar{x}^2$ instead of x^2 in Definition 29. Therefore, we now have

$$\begin{aligned}\mathcal{J}_{st}(ydx) &= y_s(\delta x)_{st} + \hat{\sigma}_s \bar{x}^2_{st} + \Lambda_{st}(\rho\delta x + (\delta\hat{\sigma})\bar{x}^2) \\ &= y_s(\delta x)_{st} + \hat{\sigma}_s x^2_{st} + \Lambda_{st}(\rho\delta x + (\delta\hat{\sigma})x^2) \\ &\quad + \hat{\sigma}_s(\delta z)_{st} + \Lambda_{st}(\delta\hat{\sigma}\delta z) = I^1_{st} + I^2_{st}, \quad s,t \in [a,b].\end{aligned}$$

In consequence, we have that the integral $\mathcal{J}(ydx)$ depends on the function x^2 since I^1 is the integral given by Definition 29 when we consider x^2 and I^2 is the Young integral of $\hat{\sigma}$ with respect to z due to (2.47).

Reciprocally, assume that there are two functions x^2 and $\tilde{x}^2$ in $\mathcal{C}_2^{2\gamma}([a,b];\mathbb{R})$ such that $(\delta x^2)_{sut} = (\delta\tilde{x}^2)_{sut} = (\delta x)_{su}(\delta x)_{ut}$, for $s,u,t \in [a,b]$. Then, these equalities imply $(\delta[x^2 - \tilde{x}^2])_{sut} = 0$. Consequently, Lemma 30 leads us to find a function $z \in \mathcal{C}_1^{2\gamma}([a,b];\mathbb{R})$ such that $x^2 - \tilde{x} = \delta z$.

We have the following consequence of Definition 29.

Corollary 17 *Let $\pi = \{a = s_0 < s_1 < \ldots < s_n = b\}$ be a partition of the interval $[a,b]$. Then, under the conditions of Definition 29, we have*

$$\mathcal{J}_{st}(ydx) = \lim_{|\pi|\to 0} \sum_{i=0}^{n} \left(y_{s_i}(\delta x)_{s_i s_{i+1}} + \hat{\sigma}_{s_i} x^2_{s_i s_{i+1}} \right).$$

Proof. Let $s,u,t \in [a,b]$. Then, (2.52) implies

$$\begin{aligned}\delta(y\delta x)_{sut} &= y_s(x_t - x_s) - y_s(x_u - x_s) - y_u(x_t - x_u)\\ &= (y_s - y_u)(x_t - x_u) = -(\delta y)_{su}(\delta x)_{ut}\\ &= -(\hat{\sigma}\delta x + \rho)_{su}(\delta x)_{ut} = -\hat{\sigma}_s(\delta x^2)_{sut} - \rho_{su}(\delta x)_{ut}\end{aligned}$$

and

$$\delta(\hat{\sigma}x^2)_{sut} = (\hat{\sigma}_s - \hat{\sigma}_u)x^2_{ut} + \hat{\sigma}_s(\delta x^2)_{sut}.$$

Therefore, the result is a consequence of Corollary 16. □

We observe that in Definition 29 and the last corollary, the main ingredients are equality (2.54) (see also (2.52)) and the existence of the function x^2 satisfying the relation $(\delta x^2)_{sut} = (\delta x)_{su}(\delta x)_{ut}$, which is called the Chen's relation. Thus, according to the above corollary, we use a "Second order Taylor expansion" of the function y to build and approximate the integral $\int_a^b y_s dx_s$. In the general case (i.e., $\gamma < 1/3$), we must utilize an "Taylor expansion of y" of order bigger than 2 and a general Chen's relation, as it is explained in [57, 58, 132, 133, 134], which complicates the applications via this theory when γ is small since we must involve more terms.

Finally, we observe that if y and x are stochastic processes with paths satisfying the assumptions of this section, then we can define the integral $\int_s^t y_s dx_s$ pathwise. That is, ω by ω.

2.3 Extension of Young integral via fractional calculus

In this section, we deal with the integral defined by Zähle [218]. This integral is introduced by means of fractional calculus techniques and it is an extension of the Young integral presented in Section 2.2.2 that allows integration of either discontinuous integrands, or with respect to processes of unbounded variation. Thus, we can consider equations with Hölder-type and singular nonlinearities [118], or with discontinuous coefficients [61]. In some particular cases, we have that this integral agrees with the Young one, and the forward and Stratonovich integrals given by Russo and Vallois [190] (see Sections 5 and 6).

In order to motivate the definition of the Zähle's integral, for the moment, we suppose that f and g are two smooth functions in $C^1([a,b])$. We also suppose that $f \in I^1_{a+}(L^p)$ and $g \in I^1_{b-}(L^p)$. That is,

$$f(t) = \int_a^t f'(s)ds \quad \text{and} \quad g(t) = \int_t^b g'(s)ds, \quad t \in [a,b].$$

Note that, in this case $f', g' \in L^p([a,b])$, for all $p \geq 1$, since they are continuous on $[a,b]$. Set $g^{b-} := g(b-) - g$ whenever $g(b-)$ exists. Remember that, by definition, $g(b-) = \lim_{s\uparrow b} g(s)$. So, in this case, $g(b-) = g(b)$. Then, from (2.7), integration by parts formula and Remark 17.*iii*), the Riemann-Stieltjes integral of f with respect to g satisfies

$$\begin{aligned}\int_a^b f(s)d^Y g(s) &= \int_a^b f(s)g'(s)ds = f(b)g(b) - f(a)g(a) - \int_a^b f'(s)g(s)ds \\ &= -\int_a^b f'(s)g(s)ds = \int_a^b \left(D^1_{b-}f\right)(s)g^{b-}(s)ds.\end{aligned}$$

Hence, using the fact that $f \in I^1_{a+}(L^p)$ and $g^{b-} \in I^1_{b-}(L^p)$, for any $p > 1$, together with Propositions 18 and 19, we get

$$\begin{aligned}\int_a^b f(s)d^Y g(s) &= \int_a^b \left(D^{1-\alpha}_{a+}\left(D^{\alpha}_{a+}f\right)\right)(s)g^{b-}(s)ds \\ &= \int_a^b \left(D^{\alpha}_{a+}f\right)(s)\left(D^{1-\alpha}_{b-}g^{b-}\right)(s)ds, \quad \alpha \in (0,1).\end{aligned}$$

Now, it is natural to consider the following definition of integral.

Definition 30 *Let $\alpha \in [0,1]$ and $p, q > 1$ such that $p^{-1} + q^{-1} \leq 1$. Consider $f \in I^{\alpha}_{a+}(L^p)$ and $g^{b-} \in I^{1-\alpha}_{b-}(L^q)$. In this case, we define the fractional integral of f with respect to g as*

$$\int_a^b f(s)dg(s) := \int_a^b \left(D^{\alpha}_{a+}f\right)(s)\left(D^{1-\alpha}_{b-}g^{b-}\right)(s)ds. \tag{2.55}$$

Note that we assume that $g(b-)$ exists. Moreover, this definition of integral is independent of the choice of α, as the following result shows.

Proposition 42 *Assume that f and g satisfy the conditions of Definition 30 for (α, p, q) and $(\beta, \tilde{p}, \tilde{q})$. Then,*

$$\int_a^b \left(D^{\alpha}_{a+}f\right)(s)\left(D^{1-\alpha}_{b-}g^{b-}\right)(s)ds = \int_a^b \left(D^{\beta}_{a+}f\right)(s)\left(D^{1-\beta}_{b-}g^{b-}\right)(s)ds.$$

Proof. With loss of generality, we can suppose that $\beta = \alpha + \tilde{\alpha}$. Consequently, Propositions 18 and 19 imply

$$\begin{aligned}\int_a^b \left(D^{\beta}_{a+}f\right)(s)\left(D^{1-\beta}_{b-}g^{b-}\right)(s)ds &= \int_a^b \left(D^{\tilde{\alpha}}_{a+}\left(D^{\alpha}_{a+}f\right)\right)(s)\left(D^{1-\beta}_{b-}g^{b-}\right)(s)ds \\ &= \int_a^b \left(D^{\alpha}_{a+}f\right)(s)\left(D^{\tilde{\alpha}}_{b-}\left(D^{1-(\alpha+\tilde{\alpha})}_{b-}g^{b-}\right)\right)(s)ds \\ &= \int_a^b \left(D^{\alpha}_{a+}f\right)(s)\left(D^{1-\alpha}_{b-}g^{b-}\right)(s)ds.\end{aligned}$$

It means, the proof is complete. □

As an example consider a function $g : [a, b] \to \mathbb{R}$ such that $g^{b-} \in I_{b-}^{1-\alpha}(L^q)$ for some $\alpha \in (0, 1)$ and $q > \frac{1}{1-\alpha}$. Then, Corollary 7, Remark 18 and Theorem 98 yield

$$\begin{aligned}\int_a^b dg(r) &= \int_a^b \left(D_{a+}^{\alpha}1\right)(r) \left(D_{b-}^{1-\alpha}g^{b-}\right)(r)\, dr \\ &= \frac{1}{\Gamma(1-\alpha)} \int_a^b (r-a)^{-\alpha} \left(D_{b-}^{1-\alpha}g^{b-}\right)(r)\, dr = I_{b-}^{1-\alpha}(D_{b-}^{1-\alpha}g^{b-})(a) \\ &= g^{b-}(a) = g(b-) - g(a). \end{aligned} \tag{2.56}$$

Note that, in the last equality, we use Corollary 7, which establishes that, for $\alpha \in (0, 1)$, $p \geq 1$, $f \in I_{a+}^{\alpha}(L^p)$ and $g \in I_{b-}^{\alpha}(L^p)$,

$$I_{a+}^{\alpha}(D_{a+}^{\alpha}f) = f \quad \text{and} \quad I_{b-}^{\alpha}(D_{b-}^{\alpha}g) = g, \quad \text{in } L^p([a, b]).$$

But in (2.56), as in the remaining of this section, we assume that last two equalities are satisfied pointwise on $[a, b]$.

Equality (2.56) is extended by the following result.

Proposition 43 *Let $g : [a, b] \to \mathbb{R}$ be such that $g^{b-} \in I_{b-}^{1-\alpha}(L^q)$ for some $\alpha \in (0, 1)$ and $q > \frac{1}{1-\alpha}$. Then, for $a \leq c < d < b$, we have*

$$\int_a^b 1_{(c,d]}(s) dg(s) = g(d) - g(c) \quad \textit{and} \quad \int_a^b 1_{(c,b]}(s) dg(s) = g(b-) - g(c).$$

Remark 39 *Note that this result, together with (2.56), implies*

$$\int_a^t dg(r) = \int_a^b 1_{(a,t]}(r) dg(r), \quad t \in [a, b].$$

Proof. By (1.54), we have that, for $t \in [a, b]$,

$$\begin{aligned}\left(I_{a+}^{1-\alpha}1_{(c,d]}\right)(t) &= \frac{1}{\Gamma(1-\alpha)} \int_a^t (t-r)^{-\alpha} 1_{(c,d]}(r) dr = \frac{1}{\Gamma(1-\alpha)} \int_{c\wedge t}^{d\wedge t} (t-r)^{-\alpha} dr \\ &= \frac{1}{(1-\alpha)\Gamma(1-\alpha)} \left[(t - c\wedge t)^{1-\alpha} - (t - d\wedge t)^{1-\alpha}\right] \\ &= \frac{1}{\Gamma(1-\alpha)} \left(\int_a^t 1_{(c,b]}(y)(y-c)^{-\alpha} dy - \int_a^t 1_{(d,b]}(y)(y-d)^{-\alpha} dy\right).\end{aligned}$$

Hence, $I_{a+}^{1-\alpha}1_{(c,d]}$ is absolutely continuous on $[a, b]$ and $\left(I_{a+}^{1-\alpha}1_{(c,d]}\right)(a) = 0$. Therefore, Corollary 7 implies that $1_{(c,d]}$ belongs to $I_{a+}^{\alpha}(L^p)$, for any $p < 1/\alpha$, with

$$\left(D_{a+}^{\alpha}1_{(c,d]}\right)(t) = \frac{1}{\Gamma(1-\alpha)} \left(1_{(c,b]}(t)(t-c)^{-\alpha} - 1_{(d,b]}(t)(t-d)^{-\alpha}\right), \quad t \in [a, b].$$

Consequently, (2.55) leads us to write

$$\begin{aligned}&\int_a^b 1_{(c,d]}(s) dg(s) \\ &= \frac{1}{\Gamma(1-\alpha)} \int_c^b (s-c)^{-\alpha} \left(D_{b-}^{1-\alpha}g^{b-}\right)(s) ds - \frac{1}{\Gamma(1-\alpha)} \int_d^b (s-d)^{-\alpha} \left(D_{b-}^{1-\alpha}g^{b-}\right)(s) ds \\ &= \left(I_{b-}^{1-\alpha}D_{b-}^{1-\alpha}g^{b-}\right)(c) - \left(I_{b-}^{1-\alpha}D_{b-}^{1-\alpha}g^{b-}\right)(d) = g^{b-}(c) - g^{b-}(d) = g(d) - g(c).\end{aligned}$$

Finally, note that the last calculations with $d = b$ allow us to show that $\int_a^b 1_{(c,b]}(s)dg(s) = g(b-) - g(c)$. The proof is now complete. □

The Hölder property of a function $g : [a, b] \to \mathbb{R}$ establishes that it belongs to the space $I_{b-}^{1-\alpha}(L^q)$, for some $\alpha \in (0, 1)$ and all $q > 1$. So, we obtain the following consequence of Proposition 43.

Corollary 18 *Let $g : [a, b] \to \mathbb{R}$ be λ-Hölder continuous function for some $\lambda \in (0, 1)$ and $a \le c < d < b$. Then,*

$$\int_a^b 1_{(c,d]}(s)dg(s) = g(d) - g(c) \quad \text{and} \quad \int_a^b 1_{(c,b]}(s)dg(s) = g(b) - g(c).$$

Proof. Choose $\alpha > 1 - \lambda$. Then, $\frac{g^{b-}(\cdot)}{(b-\cdot)^{1-\alpha}}$ and $\int_\cdot^b \frac{g^{b-}(\cdot)-g_r^{b-}}{(r-\cdot)^{2-\alpha}}dr$ are two bounded and measurable functions on $[a, b]$ and, therefore, they belong to $L^q([a, b])$, for all $q > 1$. These guarantees that (1.65) is satisfied when we write $1 - \alpha$ and g^{b-} instead of α and f, respectively. Consequently, g^{b-} is in $I_{b-}^{1-\alpha}(L^q)$, for all $q > 1$. In this way, the result is now a consequence of Proposition 43. □

Now, we study the relation between the fractional and the Lebesgue-Stieltjes integrals with respect to a function of bounded variation on the interval $[a, b]$. Towards this end, remember that if $g : [a, b] \to \mathbb{R}$ is a right-continuous function of bounded variation, then, from Proposition 139, there exists a complete measure space $([a, b], \mathcal{F}, \mu)$ such that $\mathcal{F}$ contains the Borel σ-algebra $\mathcal{B}([a, b])$ and $\mu((c, d]) = g(d) - g(c)$, for $a \le c < d \le b$. In general, the right-continuous property of g can be omitted by assuming that $\mu((c, d]) = g(d+) - g(c+)$, for $a \le c < d \le b$. Also remember that, in this case, $g = g_1 - g_2$, where $g_1, g_2 : [a, b] \to \mathbb{R}$ are two increasing functions. Consequently, μ is a signed measure given by $\mu = \mu_1 - \mu_2$, where μ_1 and μ_2 are the measures associated with g_1 and g_2, respectively. For more details, see Remark 30. In the following result $|\mu|$ stands for the total variation of μ. In means, $|\mu| = \mu_1 + \mu_2$.

Theorem 36 *Assume that $g : [a, b] \to \mathbb{R}$ is a function of bounded variation with associated measure μ and that the fractional integral of the function $f : [a, b] \to \mathbb{R}$ with respect to g is well-defined. Also suppose that $\alpha \in (0, 1)$ is such that (2.55) holds and that $\int_{(a,b)} \left(I_{a+}^{\alpha} \left|D_{a+}^{\alpha} f\right|\right)(r)|\mu|(dr) < \infty$. Then,*

$$\int_a^b f(r)dg(r) = \int_{(a,b)} f(r)\mu(dr),$$

where the left-hand side is the fractional integral of f with respect to g (i.e., it is given by (2.55)). Moreover, if f is continuous and g is right-continuous, we have

$$\int_a^b f(s)d^Y g(s) = \int_{[a,b]} f(s)\mu(ds) = f(b)\left(g(b) - g(b-)\right) + \int_a^b f(r)dg(r),$$

where the left-hand side is given by (2.2).

Proof. Using that $I_{a+}^{\alpha}\left|D_{a+}^{\alpha} f\right|$ is integrable with respect to $|\mu|$, the set of discontinuities of g is at most countable and the Fubini theorem, we get

$$\begin{aligned}
\int_{(a,b)} f(r)\mu(dr) &= \int_{(a,b)} \left(I_{a+}^{\alpha} D_{a+}^{\alpha} f\right)(r)\mu(dr) = \frac{1}{\Gamma(\alpha)} \int_{(a,b)} \int_a^r \frac{\left(D_{a+}^{\alpha} f\right)(s)}{(r-s)^{1-\alpha}} ds\mu(dr) \\
&= \frac{1}{\Gamma(\alpha)} \int_a^b \left(D_{a+}^{\alpha} f\right)(s) \int_{(s,b)} (r-s)^{\alpha-1}\mu(dr)ds \\
&= \frac{1-\alpha}{\Gamma(\alpha)} \int_a^b \left(D_{a+}^{\alpha} f\right)(s) \int_{(s,b)} \int_r^{\infty} (y-s)^{\alpha-2} dy\mu(dr)ds. \qquad (2.57)
\end{aligned}$$

Hence, using that $\int_a^b \left(I_{a+}^{\alpha}\left|D_{a+}^{\alpha}f\right|\right)(r)|\mu|(dr)<\infty$ again, we can apply Fubini's theorem to write

$$\begin{aligned}
&\frac{1-\alpha}{\Gamma(\alpha)}\int_{(s,b)}\int_r^{\infty}(y-s)^{\alpha-2}dy\mu(dr)\\
&=\frac{1-\alpha}{\Gamma(\alpha)}\int_s^{\infty}\int_{(s,b\wedge y)}(y-s)^{\alpha-2}\mu(dr)dy=\frac{1-\alpha}{\Gamma(\alpha)}\int_s^{\infty}\frac{g(b\wedge y-)-g(s+)}{(y-s)^{2-\alpha}}dy\\
&=\frac{1-\alpha}{\Gamma(\alpha)}\int_s^b\frac{g(y)-g(s+)}{(y-s)^{2-\alpha}}dy+\frac{1}{\Gamma(\alpha)}\frac{g(b-)-g(s+)}{(b-s)^{1-\alpha}}\\
&=\frac{1-\alpha}{\Gamma(\alpha)}\int_s^b\frac{g(y)-g(s)}{(y-s)^{2-\alpha}}dy+\frac{1}{\Gamma(\alpha)}\frac{g(b-)-g(s)}{(b-s)^{1-\alpha}},\quad\text{for a.a. } s\in[a,b], \qquad (2.58)
\end{aligned}$$

where last equalities hold because the set of discontinuities of a function with bounded variation is at most countable.

On the other hand, Theorem 99 states that

$$s\mapsto\frac{1}{\Gamma(\alpha)}\frac{g(b-)-g(s)}{(b-s)^{1-\alpha}}+\frac{1-\alpha}{\Gamma(\alpha)}\int_{s+\varepsilon}^b\frac{g(y)-g(s)}{(y-s)^{2-\alpha}}dy$$

goes to $D_{b-}^{1-\alpha}g$ in $L^p([a,b])$, for some $p>1$, as ε goes to zero. But, for $s\in[a,b]$,

$$\int_{s+\varepsilon}^b\frac{g(y)-g(s)}{(y-s)^{2-\alpha}}dy\to\int_s^b\frac{g(y)-g(s)}{(y-s)^{2-\alpha}}dy,\quad\text{as } \varepsilon\to 0,$$

since g is the difference of two monotone functions on $[a,b]$. Consequently, (2.57) and (2.58) imply

$$\int_{(a,b)}f(r)\mu(dr)=\int_a^b\left(D_{a+}^{\alpha}f\right)(s)\left(D_{b-}^{1-\alpha}g\right)(s)ds.$$

Finally, the result follows from (2.55) and Remark 30. □

As an example, consider the function $g=1_{[1/2,1]}$ on $[0,1]$. Then, for a continuous function $f:[0,1]\to\mathbb{R}$ belonging to $I_{a+}^{\alpha}(L^p)$ for some $p>1/\alpha$, Theorem 36 implies that the fractional integral

$$\int_a^t f(r)dg(r)=\begin{cases}0, & \text{if } t\in[0,1/2),\\ f(1/2), & \text{if } t\in[1/2,1].\end{cases}$$

is not continuous as a function of t, in general. The following result establishes conditions that guarantee the continuity of fractional integrals when they are considered as functions.

Proposition 44 *Let $p,q>1$ be such that $p^{-1}+q^{-1}\le 1$, $\alpha\in(0,1)$, $f\in I_{a+}^{\alpha}(L^p)$ and $g:[a,b]\to\mathbb{R}$ a continuous function such that $g^{t-}\in I_{t-}^{1-\alpha}(L^q([a,t]))$, for all $t\in[a,b]$ and*

$$\sup_{a\le r\le t\le b}\left|(D_{t-}^{1-\alpha}g^{t-})(r)\right|<\infty. \qquad (2.59)$$

Then, the fractional integral $t\mapsto\int_a^t fdg$ is continuous on $[a,b]$.

Remark 40 *Here, we use (1.63) to deal with the right-sided Riemann-Liouville fractional derivative of g^{t-}. Also, note that f belongs to $I_{a+}^{\alpha}(L^p([a,t]))$, for all $t\in[a,b]$. So,*

$$\int_a^t fdg=\int_a^t\left(D_{a+}^{\alpha}f\right)(r)\left(D_{t-}^{1-\alpha}g^{t-}\right)(r)dr,\quad t\in[a,b].$$

Proof. Let $a \leq t_1 < t_2 \leq b$. Then, by previous remark, we have

$$\begin{aligned}
&\left| \int_a^{t_2} f(r)\,dg(r) - \int_a^{t_1} f(r)\,dg(r) \right| \\
&= \left| \int_a^{t_2} \left(D_{a+}^{\alpha} f\right)(r) \left(D_{t_2-}^{1-\alpha} g^{t_2-}\right)(r)\,dr - \int_a^{t_1} \left(D_{a+}^{\alpha} f\right)(r) \left(D_{t_1-}^{1-\alpha} g^{t_1-}\right)(r)\,dr \right| \\
&\leq \int_a^{t_1} \left|\left(D_{a+}^{\alpha} f\right)(r)\right| \left|\left(D_{t_2-}^{1-\alpha} g^{t_2-}\right)(r) - \left(D_{t_1-}^{1-\alpha} g^{t_1-}\right)(r)\right| dr \\
&\quad + \int_{t_1}^{t_2} \left|\left(D_{a+}^{\alpha} f\right)(r)\right| \left|\left(D_{t_2-}^{1-\alpha} g^{t_2-}\right)(r)\right| dr \\
&\leq C_{p,f} \left(\int_a^{t_1} \left|\left(D_{t_2-}^{1-\alpha} g^{t_2-}\right)(r) - \left(D_{t_1-}^{1-\alpha} g^{t_1-}\right)(r)\right|^q dr \right)^{1/q} \\
&\quad + \left(\int_{t_1}^{t_2} \left|\left(D_{0+}^{\alpha} f\right)(r)\right|^p dr \right)^{1/p} \left(\int_{t_1}^{t_2} \left|\left(D_{t_2-}^{1-\alpha} g^{t_2-}\right)(r)\right|^q dr \right)^{1/q} = C_{p,f} I_1 + I_{21} I_{22}.
\end{aligned}$$

Condition (2.59) implies that $I_{21}I_{22}$ converges to 0 as $t_2 - t_1 \to 0$. For I_1, we observe that (1.63) gives that there exist a subsequence $\{n_k \in \mathbb{N} : k \in \mathbb{N}\}$ of $\mathbb{N}$ such that, for almost all $r \in [a, b]$,

$$\begin{aligned}
&\left|\left(D_{t_2-}^{1-\alpha} g^{t_2-}\right)(r) - \left(D_{t_1-}^{1-\alpha} g^{t_1-}\right)(r)\right| 1_{(a,t_1)}(r) \\
&= \frac{1_{(a,t_1)}(r)}{\Gamma(\alpha)} \lim_{k\to\infty} \left| \frac{g(t_2) - g(r)}{(t_2 - r)^{1-\alpha}} + (1-\alpha) \int_{r+(1/n_k)}^{t_2} \frac{g(r) - g(u)}{(u-r)^{2-\alpha}}\,du \right. \\
&\qquad \left. - \frac{g(t_1) - g(r)}{(t_1 - r)^{1-\alpha}} - (1-\alpha) \int_{r+(1/n_k)}^{t_1} \frac{g(r) - g(u)}{(u-r)^{2-\alpha}}\,du \right| \\
&= \frac{1_{(a,t_1)}(r)}{\Gamma(\alpha)} \left| \frac{g(t_2) - g(r)}{(t_2 - r)^{1-\alpha}} - \frac{g(t_1) - g(r)}{(t_1 - r)^{1-\alpha}} + (1-\alpha) \int_{t_1}^{t_2} \frac{g(r) - g(u)}{(u-r)^{2-\alpha}}\,du \right| \\
&\to 0, \quad \text{as } t_2 - t_1 \to 0.
\end{aligned}$$

Hence, the continuity of g, Condition (2.59) and the dominated convergence theorem implies that I_1 also goes to zero as $t_2 - t_1 \to 0$. Thus, the proof is complete. □

We need to state the following auxiliary result developed by Zähle [218] in order to continue with our analysis of the fractional integral (2.55). Towards this end, we choose two nonnegative functions $\varphi, \psi \in C^\infty(\mathbb{R})$ with supports $[-1, 0]$ and $[0, 1]$, respectively, such that $\int_{\mathbb{R}} \varphi(v)dv, \int_{\mathbb{R}} \psi(v)dv = 1$. For $n \in \mathbb{N}$, we set $\varphi_n(x) = n\varphi(nx)$ and $\psi_n(x) = n\psi(nx)$, $x \in \mathbb{R}$.

Lemma 31 *Let $p \geq 1$, $\alpha \in (0, 1)$, $g \in I_{b-}^{\alpha}(L^p)$, $f \in I_{a+}^{\alpha}(L^p)$ and $n \in \mathbb{N}$. Then, $g * \varphi_n$ and $f * \psi_n$ also belong to $I_{b-}^{\alpha}(L^p)$ and $I_{a+}^{\alpha}(L^p)$, respectively. Moreover, on $[a, b]$, we have*

$$D_{b-}^{\alpha}(g * \varphi_n) = (D_{b-}^{\alpha} g) * \varphi_n \quad \text{and} \quad D_{a+}^{\alpha}(f * \psi_n) = (D_{a+}^{\alpha} f) * \psi_n.$$

Remark 41 *Remember that we are using the convention $g, f, D_{b-}^{\alpha} g, D_{a+}^{\alpha} f \equiv 0$ on $[a, b]^c$. Also, note that the result yields that $D_{b-}^{\alpha}(g * \varphi_n)$ and $D_{a+}^{\alpha}(f * \psi_n)$ converge to $D_{b-}^{\alpha} g$ and $D_{a+}^{\alpha} f$ in $L^p([a, b])$, respectively, as $n \to \infty$.*

Proof. Fix $u \in [a,b]$. From Definition 18 and the integration by parts formula, we have

$$\begin{aligned}\left((D^{\alpha}_{b-}g)*\varphi_n\right)(u) &= \int_{\mathbb{R}} \varphi_n(u-z)\left(D^{\alpha}_{b-}g\right)(z)dz \\ &= \frac{-1}{\Gamma(1-\alpha)}\int_a^b \varphi_n(u-z)\frac{d}{dz}\int_z^b \frac{g(v)}{(v-z)^{\alpha}}dvdz \\ &= \frac{-1}{\Gamma(1-\alpha)}\frac{d}{du}\int_a^b \varphi_n(u-z)\int_z^b \frac{g(v)}{(v-z)^{\alpha}}dvdz.\end{aligned}$$

Therefore, the changes of variables $\theta = u-z$ and $y = v+\theta$, together with the fact that $g(v)=0$, for $v \notin [a,b]$, imply

$$\begin{aligned}\left((D^{\alpha}_{b-}g)*\varphi_n\right)(u) &= \frac{-1}{\Gamma(1-\alpha)}\frac{d}{du}\int_{u-b}^{u-a} \varphi_n(\theta)\int_{u-\theta}^b \frac{g(v)}{(v+\theta-u)^{\alpha}}dvd\theta \\ &= \frac{-1}{\Gamma(1-\alpha)}\frac{d}{du}\int_{u-b}^{u-a} \varphi_n(\theta)\int_u^{b+\theta} \frac{g(y-\theta)}{(y-u)^{\alpha}}dyd\theta \\ &= \frac{-1}{\Gamma(1-\alpha)}\frac{d}{du}\int_{u-b}^{u-a} \varphi_n(\theta)\int_u^b \frac{g(y-\theta)}{(y-u)^{\alpha}}dyd\theta.\end{aligned}$$

So, Fubini's theorem and the change of variables $z = y-\theta$ allow us to get

$$\begin{aligned}\left((D^{\alpha}_{b-}g)*\varphi_n\right)(u) &= \frac{-1}{\Gamma(1-\alpha)}\frac{d}{du}\int_u^b (y-u)^{-\alpha}\int_{u-b}^{u-a} \varphi_n(\theta)g(y-\theta)d\theta dy \\ &= \frac{-1}{\Gamma(1-\alpha)}\frac{d}{du}\int_u^b (y-u)^{-\alpha}\int_{y-u+a}^{y-u+b} \varphi_n(y-z)g(z)dzdy \\ &= \frac{-1}{\Gamma(1-\alpha)}\frac{d}{du}\int_u^b (y-u)^{-\alpha}\left(g*\varphi_n\right)(y)dy,\end{aligned}$$

where last equality holds since $z < y-u+a$ if and only if $0 \le u-a < y-z$, which implies that $\varphi_n(y-z)=0$.

Now, proceeding as the last calculations (i.e., omitting $\frac{d}{du}$ and $\frac{d}{dz}$ in previous calculations), we can write

$$\int_u^b (y-u)^{-\alpha}\left(g*\varphi_n\right)(y)dy\Big|_{u=b} = \int_a^b \varphi_n(b-z)\int_z^b \frac{g(v)}{(v-z)^{\alpha}}dvdz = 0,$$

where last equality follows from the fact that $\varphi_n(b-z) \neq 0$ if and only if $-n^{-1} < b-z < 0$, which gives that $b < z$. Hence, the equality $D^{\alpha}_{b-}(g*\varphi_n) = (D^{\alpha}_{b-}g)*\varphi_n$ holds by applying Corollary 7.

Finally, using Corollary 7, Definition 18 and previous change of variable formulas again, we also get that $D^{\alpha}_{a+}(f*\psi_n) = (D^{\alpha}_{a+}f)*\psi_n$ is also satisfied. Thus, the proof is complete. □

We are ready to see that the fractional integral (2.55) is a linear operator.

Proposition 45 *Let $f_i, g_i : [a,b] \to \mathbb{R}$, $i = 1,2$, be measurable functions. Then,*

i) We have

$$\int_a^b f_1(s)d\left(g_1+g_2\right)(s) = \int_a^b f_1(s)dg_1(s) + \int_a^b f_1(s)dg_2(s)$$

if all the integrals in this equality are well-defined according to Definition 30.

ii) We can write

$$\int_a^b (f_1(s)+f_2(s))\,dg_1(s) = \int_a^b f_1(s)dg_1(s) + \int_a^b f_2(s)dg_1(s)$$

provided all the fractional integrals exist.

Remark 42 *Here, we are not assuming that f_1, f_2 (resp. g_1, g_2) belong to $I_{a+}^{\alpha}(L^p)$ (resp. $I_{b-}^{1-\alpha}(L^p)$) for some $\alpha \in [0,1]$.*

Proof. We first consider Statement *i*). By hypothesis, there is $\alpha \in (0,1)$ such that

$$\begin{aligned}\int_a^b f_1(s)d\,(g_1+g_2)\,(s) &= \int_a^b \left(D_{a+}^{\alpha} f_1\right)(s)\left(D_{b-}^{1-\alpha}\,(g_1+g_2)^{b-}\right)(s)ds\\ &= \int_a^b \left(D_{a+}^{\alpha} f_1\right)(s)\left(D_{b-}^{1-\alpha}\left(g_1^{b-}+g_2^{b-}\right)\right)(s)ds.\end{aligned}$$

Hence, Lemma 31 and Proposition 42 give

$$\begin{aligned}&\int_a^b f_1(s)d\,(g_1+g_2)\,(s)\\ &= \lim_{n\to\infty}\int_a^b \left(D_{a+}^{\alpha} f_1\right)(s)\left(D_{b-}^{1-\alpha}\left(\left(g_1^{b-}+g_2^{b-}\right)*\varphi_n\right)\right)(s)ds\\ &= \lim_{n\to\infty}\int_a^b f_1(s)\left(\left(g_1^{b-}+g_2^{b-}\right)*\varphi_n\right)'(s)ds\\ &= \lim_{n\to\infty}\int_a^b f_1(s)\left(g_1^{b-}*\varphi_n\right)'(s)ds + \lim_{n\to\infty}\int_a^b f_1(s)\left(g_2^{b-}*\varphi_n\right)'(s)ds.\end{aligned}$$

Therefore, utilizing Proposition 42 and Lemma 31 again, together with the fact that there exist $\alpha_1, \alpha_2 \in [0,1]$ such that

$$\int_a^b f_1(s)dg_i(s) = \int_a^b \left(D_{a+}^{\alpha_i} f_1\right)(s)\left(D_{b-}^{1-\alpha_i} g_i^{b-}\right)(s)ds, \quad i=1,2,$$

we get

$$\begin{aligned}\int_a^b f_1(s)d\,(g_1+g_2)\,(s) &= \lim_{n\to\infty}\int_a^b \left(D_{a+}^{\alpha_1} f_1\right)(s)\left(D_{b-}^{1-\alpha_1}\left(g_1^{b-}*\varphi_n\right)\right)(s)ds\\ &\quad + \lim_{n\to\infty}\int_a^b \left(D_{a+}^{\alpha_2} f_1\right)(s)\left(D_{b-}^{1-\alpha_2}\left(g_2^{b-}*\varphi_n\right)\right)(s)ds\\ &= \int_a^b f_1(s)dg_1(s) + \int_a^b f_1(s)dg_2(s).\end{aligned}$$

Thus, Statement *i*) holds.

Finally, we have that affirmation *ii*) is likewise satisfied, which implies that the proof is finished. □

We observe that last proof is based on φ utilized in Lemma 31. But instead, we could make use of the function ψ also considered in Lemma 31. That is, we could apply the property

$$\begin{aligned}\int_a^b f_1(s)d\,(g_1+g_2)\,(s) &= \lim_{n\to\infty}\int_a^b \left(D_{a+}^{\alpha_1}(f*\psi_n)\right)(s)\left(D_{b-}^{1-\alpha_1} g^{b-}\right)(s)ds\\ &= \lim_{n\to\infty}\int_a^b \left((f*\psi_n)'(s)g^{b-}(s)ds.\right.\end{aligned}$$

Now, we study Theorem 19 for the fractional integral.

Proposition 46 *Let $f, g : [a,b] \to \mathbb{R}$ be measurable functions and $x \in (a,b)$ such that $f(x-)$, $g(x-)$ and $g(x+)$ exist. Then,*

$$\int_a^b f(s)dg(s) = f(x-)\left(g(x-) - g(x+)\right) + \int_a^x f(s)dg(s) + \int_x^b f(s)dg(s)$$

provided all the fractional integrals are well-defined according to Definition 30.

Proof. As in the proof of Proposition 45, we can make use of Lemma 31 to establish

$$\begin{aligned}
\int_a^b f(s)dg(s) &= \lim_{n\to\infty} \int_a^b f(s)\left(g^{b-} * \varphi_n\right)'(s)ds \\
&= \lim_{n\to\infty} \left(\int_a^x f(s)\left(g^{b-} * \varphi_n\right)'(s)ds + \int_x^b f(s)\left(g^{b-} * \varphi_n\right)'(s)ds \right) \\
&= \lim_{n\to\infty} \left(I_1 + I_2\right). \qquad (2.60)
\end{aligned}$$

Note that Proposition 42, Lemma 31 and the existence of the integral $\int_x^b f(s)dg(s)$ yield that there is $\beta \in [0,1]$ such that

$$\begin{aligned}
\lim_{n\to\infty} I_2 &= \lim_{n\to\infty} \int_x^b \left(D^\beta_{x+} f\right)(s)\left(\left(D^{1-\beta}_{b-} g^{b-}\right) * \varphi_n\right)(s)ds \\
&= \int_x^b \left(D^\beta_{x+} f\right)(s)\left(D^{1-\beta}_{b-} g^{b-}\right)(s)ds = \int_x^b f(s)dg(s). \qquad (2.61)
\end{aligned}$$

On the other hand, we observe that we cannot apply Lemma 31 to I_1 in (2.60) because g could be different from zero on the interval $(x,b]$. So, we have to proceed as follows: we observe

$$\begin{aligned}
\int_a^x f(s)\int_a^b (\varphi_n)'(s-y)dyds &= \int_a^x f(s)\left(\varphi_n(s-a) - \varphi_n(s-b)\right)ds \\
&= -\int_a^x f(s)\varphi_n(s-b)ds \to 0, \quad \text{as } n\to\infty, \qquad (2.62)
\end{aligned}$$

due to $s - b < x - b < 0$, for any $s \in (a,x)$. In consequence, the Fubini's theorem leads us to have

$$\begin{aligned}
\lim_{n\to\infty} I_1 &= \lim_{n\to\infty} \int_a^x f(s)\left(g^{x-} * \varphi_n\right)'(s)ds = \lim_{n\to\infty} \int_{\mathbb{R}} g^{x-}(y) \int_a^x (\varphi_n)'(s-y)f(s)dsdy \\
&= \lim_{n\to\infty} \int_a^{x+\frac{1}{n}} g^{x-}(y) \int_a^x (\varphi_n)'(s-y)f(s)dsdy,
\end{aligned}$$

where we utilize the definition of φ_n and the fact that $-1/n < s - y < 0$ gives that $s < y < s + (1/n)$. Thus, using Fubini's theorem again,

$$\begin{aligned}
\lim_{n\to\infty} I_1 &= \lim_{n\to\infty} \int_a^x g^{x-}(y) \int_a^x (\varphi_n)'(s-y)f(s)dsdy \\
&\quad + \lim_{n\to\infty} \int_x^{x+\frac{1}{n}} g^{x-}(y) \int_a^x (\varphi_n)'(s-y)f(s)dsdy \\
&= \lim_{n\to\infty} \int_a^x f(s) \int_a^x (\varphi_n)'(s-y)g^{x-}(y)dyds \\
&\quad + \lim_{n\to\infty} \int_x^{x+\frac{1}{n}} g^{x-}(y) \int_a^x (\varphi_n)'(s-y)f(s)dsdy = \lim_{n\to\infty} \left(J_1 + J_2\right). \qquad (2.63)
\end{aligned}$$

For J_1, we can proceed as in (2.61) to see

$$\lim_{n\to\infty} J_1 = \int_a^x f(s)dg(s). \tag{2.64}$$

Finally, we deal with J_2. The definition of φ_n implies

$$\begin{aligned} J_2 &= \int_x^{x+\frac{1}{n}} g^{x-}(y) \int_{x-\frac{1}{n}}^x (\varphi_n)'(s-y)f(s)dsdy \\ &= \int_{x-\frac{1}{n}}^x f(s) \int_x^{x+\frac{1}{n}} (g(x+)-g(y))\,(\varphi_n)'(s-y)dyds \\ &\quad + (g(x-)-g(x+)) \int_{x-\frac{1}{n}}^x f(s) \int_x^{x+\frac{1}{n}} (\varphi_n)'(s-y)dyds = J_{2,1}+J_{2,2}. \end{aligned} \tag{2.65}$$

Note that, given $\varepsilon > 0$, there exists $N \in \mathbb{N}$ such that, for $n > N$, we get

$$\begin{aligned} J_{2,1} &\leq \varepsilon \int_{x-\frac{1}{n}}^x |f(s)| \int_x^{x+\frac{1}{n}} |(\varphi_n)'(s-y)|dyds \\ &= n^2\varepsilon \int_{x-\frac{1}{n}}^x |f(s)| \int_x^{x+\frac{1}{n}} |\varphi'(n(s-y))|dyds \leq n\varepsilon\|\varphi'\|_{\infty,\mathbb{R}} \int_{x-\frac{1}{n}}^x |f(s)|ds. \end{aligned}$$

Hence, we obtain

$$\limsup_{n\to\infty} J_{2,1} \leq \varepsilon\|\varphi'\|_{\infty,\mathbb{R}}|f(x-)|, \quad \text{for any } \varepsilon > 0. \tag{2.66}$$

The term $J_{2,2}$ in (2.65) is analyzed as follows.

$$\begin{aligned} J_{2,2} &= (g(x-)-g(x+)) \int_{x-\frac{1}{n}}^x f(s)\left(\varphi_n(s-x)-\varphi_n(s-x-\frac{1}{n})\right)ds \\ &= (g(x-)-g(x+)) \int_{x-\frac{1}{n}}^x f(s)\varphi_n(s-x)ds \\ &= (g(x-)-g(x+)) \int_{\mathbb{R}} f(s)\varphi_n(s-x)ds \to f(x-)\,(g(x-)-g(x+))\,, \quad \text{as } n\to\infty. \end{aligned}$$

Therefore, we have proven that the result is satisfied by considering (2.60)-(2.66). □

Imitating last proof, we also have the following result.

Proposition 47 *Let $f, g : [a,b] \to \mathbb{R}$ be measurable functions and $x, y \in [a,b]$ such that $x < y$, and $f(y-)$, $g(x-)$ and $g(y+)$ exist. Then,*

$$\int_a^b 1_{(x,y)}(s)f(s)dg(s) = \begin{cases} \int_x^b f(s)dg(s), & \text{if } y = b, \\ \int_x^y f(s)dg(s) - f(y-)(g(y+)-g(y-)), & \text{otherwise}, \end{cases}$$

provided all the fractional integrals are well-defined.

Proof. From Lemma 31 and Proposition 42, we can write

$$\begin{aligned} \int_a^b 1_{(x,y)}(s)f(s)dg(s) &= \lim_{n\to\infty} \int_a^b 1_{(x,y)}(s)f(s)\left(g^{b-}*\varphi_n\right)'(s)ds \\ &= \lim_{n\to\infty} \int_x^y f(s)\left(g^{b-}*\varphi_n\right)'(s)ds. \end{aligned}$$

Hence, using Lemma 31 and Proposition 42 again, the result is true in the case that $y = b$. Moreover, for $y < b$ and proceeding as in (2.62), we obtain

$$\begin{aligned}
&\int_a^b 1_{(x,y)}(s)f(s)dg(s) \\
&= \lim_{n\to\infty}\int_x^y f(s)\left(g^{y-} * \varphi_n\right)'(s)ds = \lim_{n\to\infty}\int_{\mathbb{R}} g^{y-}(z)\int_x^y (\varphi_n)'(s-z)f(s)dsdz \\
&= \lim_{n\to\infty}\int_x^{y+\frac{1}{n}} g^{y-}(z)\int_x^y (\varphi_n)'(s-z)f(s)dsdz = \lim_{n\to\infty}\int_x^y g^{y-}(z) \\
&\quad\times\int_x^y (\varphi_n)'(s-z)f(s)dsdz + \lim_{n\to\infty}\int_y^{y+\frac{1}{n}} g^{y-}(z)\int_x^y (\varphi_n)'(s-z)f(s)dsdz.
\end{aligned}$$

Thus, the analysis developed in (2.63) yields that the result is satisfied. □

The fractional integral $\int_a^b f(s)dg(s)$ introduced in Definition 30 is an extension of the Young integral $\int_a^b f(s)dg^Y(s)$, which is given in (2.2) through Riemann-Stieltjes sums and the analysis throughout this section (i.e., Section 2). It means, on the space of Hölder continuous functions, the integral defined in (2.55) agrees with the generalized integral considered in Section 2.2.2 by means of the algebraic tools introduced in Section 2.2.1 as the following result shows.

Theorem 37 *Let λ and μ in $(0,1)$ be such that $\lambda + \mu > 1$, $f \in C_1^\lambda([a,b])$ and $g \in C_1^\mu([a,b])$. Then, the fractional integral $\int_a^b f(s)dg(s)$ exists and agrees with the Young integral $\int_a^b f(s)dg^Y(s)$, which are defined in (2.55) and (2.2), respectively.*

Remark 43 *In Theorem 33.(3), we have already proven that the Young integral $\int_a^b f(s)dg^Y(s)$ is well-defined. Moreover, under the assumptions of Theorem 36, both integrals are the same if g is continuous.*

Proof. Let $\pi = \{a = t_0 < t_1 < \ldots < t_n = b\}$ be a partition of the interval $[a,b]$ and $f_\pi = \sum_{i=1}^n f(t_{i-1})1_{(t_{i-1},t_i]}$. Then, from the proof of Theorem 33.(3), it is enough to show that $\int_a^b f(s)dg(s) = \lim_{|\pi|\to 0}\sum_{i=1}^n f(t_{i-1})\left(g(t_i) - g(t_{i-1})\right)$.

Now, choose $\alpha \in (1-\mu, \lambda)$. So, by (1.65), we have that g^{b-} belongs to $I_{b-}^{1-\alpha}(L^q)$, for any $q > 1$, and $D_{b-}^{1-\alpha}g^{b-}$ is a bounded function. Moreover, the proofs of Proposition 43 and Corollary 18 imply that $f_\pi \in I_{a+}^\alpha(L^p)$, for $p < 1/\alpha$, and, likewise (1.65), we can see that f is in $I_{a+}^\alpha(L^p)$, for $p \in (1, 1/\alpha)$. In consequence, the fractional integral $\int_a^b f(s)dg(s)$ exists. Hence, we only need to show that

$$\lim_{|\pi|\to 0}\left\|D_{a+}^\alpha f_\pi - D_{a+}^\alpha f\right\|_{L^1([a,b])} = 0 \tag{2.67}$$

due to (2.55) and Corollary 18.

On the other hand, in order to calculate $D_{a+}^\alpha f_\pi - D_{a+}^\alpha f$, we first consider the integral

$$\int_a^u \frac{|f_\pi(u) - f(u) - (f_\pi(y) - f(y))|}{(u-y)^{\alpha+1}}dy,$$

which is equal to

$$\sum_{i=1}^{n} 1_{(t_{i-1},t_i]}(u) \left(\sum_{k=1}^{i-1} \int_{t_{k-1}}^{t_k} \frac{|f_\pi(u) - f(u) - (f_\pi(y) - f(y))|}{(u-y)^{\alpha+1}} dy \right.$$
$$\left. + \int_{t_{i-1}}^{u} \frac{|f_\pi(u) - f(u) - (f_\pi(y) - f(y))|}{(u-y)^{\alpha+1}} dy \right)$$
$$= \sum_{i=1}^{n} 1_{(t_{i-1},t_i]}(u) \left(\sum_{k=1}^{i-1} \int_{t_{k-1}}^{t_k} \frac{|f(t_{i-1}) - f(u) - (f(t_{k-1}) - f(y))|}{(u-y)^{\alpha+1}} dy \right.$$
$$\left. + \int_{t_{i-1}}^{u} \frac{|f(t_{i-1}) - f(u) - (f(t_{i-1}) - f(y))|}{(u-y)^{\alpha+1}} dy \right).$$

Consequently, the Hölder continuity of f allows us to conclude

$$\int_a^u \frac{|f_\pi(u) - f(u) - (f_\pi(y) - f(y))|}{(u-y)^{\alpha+1}} dy$$
$$\leq \|f\|_\lambda \sum_{i=1}^{n} 1_{(t_{i-1},t_i]}(u) \left((u-t_{i-1})^\lambda \int_a^{t_{i-1}} \frac{dy}{(u-y)^{\alpha+1}} \right.$$
$$\left. + \sum_{k=1}^{i-1} (t_k - t_{k-1})^\lambda \int_{t_{k-1}}^{t_k} \frac{dy}{(u-y)^{\alpha+1}} + \int_{t_{i-1}}^{u} \frac{dy}{(u-y)^{\alpha-\lambda+1}} \right) \tag{2.68}$$
$$\leq \|f\|_\lambda \sum_{i=1}^{n} 1_{(t_{i-1},t_i]}(u) \left(\frac{(u-t_{i-1})^{\lambda-\alpha}}{\alpha} + \frac{|\pi|^\lambda}{\alpha} \sum_{k=1}^{i-1} (u-t_{i-1})^{-\alpha} + \frac{1}{\lambda-\alpha}(u-t_{i-1})^{\lambda-\alpha} \right),$$

where the last expression belongs to $L^p([a,b])$, $p < 1/\alpha$. Thus, proceeding as in (1.64), we have

$$\begin{aligned} &\left| \left(D_{a+}^\alpha f_\pi\right)(u) - \left(D_{a+}^\alpha f\right)(u) \right| \\ &= \frac{1}{\Gamma(1-\alpha)} \left| \frac{f_\pi(u) - f(u)}{(u-a)^\alpha} + \alpha \int_a^u \frac{f_\pi(u) - f(u) - (f_\pi(y) - f(y))}{(u-y)^{\alpha+1}} \right| \\ &\leq \frac{1}{\Gamma(1-\alpha)} \left(\frac{|f_\pi(u) - f(u)|}{(u-a)^\alpha} + \alpha \int_a^u \frac{|f_\pi(u) - f(u) - (f_\pi(y) - f(y))|}{(u-y)^{\alpha+1}} \right) \\ &= I_1 + I_2. \end{aligned} \tag{2.69}$$

Note that

$$\|I_1\|_{L^1([a,b])} \leq \frac{\|f\|_{\lambda,[a,b]}}{(1-\alpha)\Gamma(1-\alpha)} (b-a)^{1-\alpha} |\pi|^\lambda. \tag{2.70}$$

For I_2, we apply (2.68) to obtain

$$\begin{aligned} \|I_2\|_{L^1([a,b])} &\leq \frac{\alpha \|f\|_{\lambda,[a,b]}}{\Gamma(1-\alpha)} \left(\left[\frac{1}{\alpha} + \frac{1}{\lambda-\alpha} \right] \sum_{i=1}^{n} \int_{t_{i-1}}^{t_i} (u-t_{i-1})^{\lambda-\alpha} du \right. \\ &\quad \left. + \sum_{i=1}^{n} \sum_{k=1}^{i-1} (t_k - t_{k-1})^\lambda \int_{t_{i-1}}^{t_i} \int_{t_{k-1}}^{t_k} \frac{1}{(u-y)^{\alpha+1}} dy du \right) \\ &\leq \frac{\alpha \|f\|_{\lambda,[a,b]}}{\Gamma(1-\alpha)} \left((1+\lambda-\alpha)^{-1} \left[\frac{1}{\alpha} + \frac{1}{\lambda-\alpha} \right] \sum_{i=1}^{n} (t_i - t_{i-1})^{1+\lambda-\alpha} \right. \\ &\quad \left. + \sum_{k=1}^{n-1} (t_k - t_{k-1})^\lambda \int_{t_k}^{b} \int_{t_{k-1}}^{t_k} \frac{1}{(u-y)^{\alpha+1}} dy du \right) \leq C_{\alpha,\lambda} \|f\|_{\lambda]} (b-a) |\pi|^{\lambda-\alpha}. \end{aligned}$$

Therefore, (2.69) and (2.70) give that (2.67) is satisfied. Thus, the proof is complete. □

Immediate consequences of Theorem 37 are the following four results. We first deal with the integration by parts formula for the fractional integral.

Proposition 48 *Let $f \in \mathcal{C}_1^{\lambda}([a,b])$ and $g \in \mathcal{C}_1^{\mu}([a,b])$, with $\lambda + \mu > 1$. Then,*

$$f(t)g(t) = f(a)g(a) + \int_a^t f(s)dg(s) + \int_a^t g(s)df(s), \quad t \in [a,b],$$

where the integrals are introduced in Definition 30.

Proof. The result follows from Theorems 34 and 37. □

Now, we consider the change of variables formula for the fractional integral given by (2.55).

Proposition 49 *Let $\lambda, \mu \in (0,1)$ be such that $\lambda + \mu > 1$, $f \in \mathcal{C}_1^{\lambda}([a,b])$, $g \in \mathcal{C}_1^{\mu}([a,b])$ and $F \in \mathcal{C}_b^2(\mathbb{R})$. Also let $x : [a,b] \to \mathbb{R}$ be given by $x(t) = x(a) + \int_a^t f(s)dg(s)$, $t \in [a,b]$. Then, the fractional integral satisfies*

$$F(x(t)) = F(x(a)) + \int_a^b F'(x(s))f(s)dg(s), \quad t \in [a,b].$$

Remark 44 *In particular, if $\lambda > 1/2$, we obtain*

$$F(f(t)) = F(f(a)) + \int_a^t F'(f(s))df(s), \quad t \in [a,b].$$

Proof. We only need to apply Theorem 37 and proceed as in the proof of Theorem 35. □

We also have the following Fubini's type theorem for the fractional integral (2.55).

Proposition 50 *Under the assumption of Proposition 40, the fractional integral satisfies*

$$\int_s^t \left(\int_s^r h(r,u)dg(u) \right) df(r) = \int_s^t \left(\int_u^t h(r,u)df(r) \right) dg(u), \quad a \le s \le t \le b.$$

Remark 45 *In Proposition 40, we have already stated that the iterated Young integrals*

$$\int_s^t \left(\int_s^r h(r,u)d^Y g(u) \right) d^Y f(r) \quad \text{and} \quad \int_s^t \left(\int_u^t h(r,u)d^Y f(r) \right) d^Y g(u)$$

are well-defined.

Proof. In order to see that the result holds, we only need to apply Theorem 37, together with Proposition 40. □

Proposition 51 *Let $f \in C_1^{\tau}([a,b] : \mathbb{R})$, $g \in C_1^{\lambda}([a,b] : \mathbb{R})$ and $\theta \in C_1^{\gamma}([a,b] : \mathbb{R})$, where $\tau, \lambda, \gamma \in (0,1)$ and $\tau + \gamma, \lambda + \gamma > 1$. Also let $\tilde{\theta} = \int_0^{\cdot} g_s d\theta_s$ on $[a,b]$. Then*

$$\int_a^t f_s d\tilde{\theta}_s = \int_a^t f_s g_s d\theta_s, \quad t \in [a,b].$$

Proof. The result is an immediate consequence of Proposition 41 and Theorem 37. □

Fubini's theorem for the fractional integral (i.e., Proposition 50) can be stated for functions not necessarily Hölder continuous, as follows:

Theorem 38 (Fubini's theorem) *Let $f, g : [a,b] \to \mathbb{R}$ be two measurable functions such that $f \in I_{b-}^{1-\alpha}(L^p)$ and $g \in I_{b-}^{1-\beta}(L^{\tilde{p}})$ for some $\alpha, \beta \in (0,1)$ and $p, \tilde{p} > 1$. Also, let $q, \tilde{q} > 1$ and $h : [a,b]^2 \to \mathbb{R}$ be a measurable function such that $p^{-1} + q^{-1} = 1$, $\tilde{p}^{-1} + \tilde{q}^{-1} = 1$ and*

i) $h(\cdot, u)$ is in $I_{a+}^{\alpha}(L^q)$, for all $u \in [a,b]$.

ii) $h(r, \cdot)$ belongs to $I_{a+}^{\beta}(L^{\tilde{q}})$, for all $r \in [a,b]$.

iii) $\int_a^b h(\cdot, u)dg(u)$ and $\int_a^b h(r, \cdot)df(r)$ belong to $I_{a+}^{\alpha}(L^q)$ and $I_{a+}^{\beta}(L^{\tilde{q}})$, respectively.

iv) $z \mapsto \left\| D_{a+}^{\beta}(h(z, \cdot)) \right\|_{L^{\tilde{q}}([a,b])}$ belongs to $L^q([a,b])$.

v) For $m \in \mathbb{N}$ and the function ψ_m used in Lemma 31, we have

$$\lim_{m\to\infty} \int_a^b \left(\int_a^b (h(\cdot, r) * \psi_m)(u) df(u) \right) dg(r) = \int_a^b \left(\int_a^b h(r,u) df(r) \right) dg(u).$$

Then,

$$\int_a^b \left(\int_a^b h(r,u) dg(u) \right) df(r) = \int_a^b \left(\int_a^b h(r,u) df(r) \right) dg(u).$$

Proof. For $n \in \mathbb{N}$, ψ_n is the function introduced in Lemma 31. Now, we divide the proof into two steps.
Step 1: Here we prove

$$\lim_{m\to\infty} \lim_{n\to\infty} \int_a^b \left(\int_a^b (h(r, \cdot) * \psi_n)(u) dg(u) \right) * \psi_m(r) df(r) = \int_a^b \left(\int_a^b h(r,u) dg(u) \right) df(r).$$

Lemma 31 and Proposition 42 yield

$$\begin{aligned}
&\int_a^b \left(\int_a^b (h(r, \cdot) * \psi_n)(u) dg(u) \right) * \psi_m(r) df(r) - \int_a^b \left(\int_a^b h(r,u) dg(u) \right) df(r) \\
&= \int_a^b \int_{\mathbb{R}} \left(\int_a^b \left(D_{a+}^{\beta}(h(z, \cdot) * \psi_n \right)(u) \left(D_{b-}^{1-\beta} g^{b-} \right)(u) du \right) \psi_m'(r-z) f^{b-}(r) dz dr \\
&\quad - \int_a^b \int_{\mathbb{R}} \left(\int_a^b \left(D_{a+}^{\beta} h(z, \cdot) \right)(u) \left(D_{b-}^{1-\beta} g^{b-} \right)(u) du \right) \psi_m'(r-z) f^{b-}(r) dz dr \\
&\quad + \int_a^b \left(\int_a^b h(r, \cdot)(u) dg(u) \right) * \psi_m(r) df(r) - \int_a^b \left(\int_a^b h(r,u) dg(u) \right) df(r) \\
&= I_1 - I_2 + I_3 - I_4.
\end{aligned} \tag{2.71}$$

Note that

$$\begin{aligned}
&\int_a^b \left| \left(D_{a+}^{\beta}(h(z, \cdot) * \psi_n \right)(u) - \left(D_{a+}^{\beta} h(z, \cdot) \right)(u) \right| \left| \left(D_{b-}^{1-\beta} g^{b-} \right)(u) \right| du \\
&\leq \left\| D_{a+}^{\beta}(h(z, \cdot) * \psi_n - D_{a+}^{\beta} h(z, \cdot) \right\|_{L^{\tilde{q}}([a,b])} \left\| D_{b-}^{1-\beta} g^{b-} \right\|_{L^{\tilde{p}}([a,b])} \to 0, \quad \text{as } n \to \infty,
\end{aligned}$$

and that

$$\left\| D_{a+}^{\beta}(h(z, \cdot) * \psi_n \right\|_{L^{\tilde{q}}([a,b])} \leq \left\| D_{a+}^{\beta}(h(z, \cdot)) \right\|_{L^{\tilde{q}}([a,b])}.$$

Hence, the dominated convergence theorem and (2.71) imply that $\lim_{n\to\infty}(I_1 - I_2) = 0$. Indeed, by Hypothesis *iv*), we have

$$\begin{aligned}
&\int_a^b \int_{\mathbb{R}} \left\| D_{a+}^{\beta}(h(z,\cdot) * \psi_n\right\|_{L^{\tilde{q}}([a,b])} |\psi'_m(r-z)|\,|f^{b-}(r)|\,dzdr \\
&= \int_{\mathbb{R}} |\psi'_m(z)| \int_a^b \left\| D_{a+}^{\beta}(h(r-z,\cdot) * \psi_n\right\|_{L^{\tilde{q}}([a,b])} \left|f^{b-}(r)\right| drdz \\
&\leq \left\|f^{b-}\right\|_{L^p([a,b])} \left(\int_a^b \left\| D_{a+}^{\beta}(h(r,\cdot) * \psi_n\right\|^q_{L^{\tilde{q}}([a,b])} dr \right)^{1/q} \int_{\mathbb{R}} |\psi'_m(z)|\,dz < \infty.
\end{aligned}$$

Also, it is clear that (2.71) also gives that $lim_{m\to\infty}(I_3 - I_4) = 0$. Thus, the claim of this step is true.

Step 2: Now, we see that we can change the order of the integrals in I_1 given by (2.71).

Fubini's theorem, Lemma 31 and Proposition 42 lead to get

$$\begin{aligned}
I_1 &= \int_a^b \int_{\mathbb{R}} \left(\int_a^b (h(z,\cdot) * \psi_n)'(u) g^{b-}(u) du \right) \psi'_m(r-z) f^{b-}(r) dzdr \\
&= \int_a^b \int_{\mathbb{R}} \left(\int_a^b \int_{\mathbb{R}} h(z,\tilde{z}) \psi'_n(u-\tilde{z}) g^{b-}(u) d\tilde{z}du \right) \psi'_m(r-z) f^{b-}(r) dzdr \\
&= \int_a^b \int_{\mathbb{R}} \left(\int_a^b \int_{\mathbb{R}} h(z,\tilde{z}) \psi'_m(r-z) f^{b-}(r) dzdr \right) \psi'_n(u-\tilde{z}) g^{b-}(u) d\tilde{z}du \\
&= \int_a^b \left(\int_a^b (h(\cdot,r) * \psi_m)(u) df(u) \right) * \psi_n(r) dg(r).
\end{aligned}$$

In consequence, Hypothesis *v*) and (2.71) allow us to write

$$\begin{aligned}
\lim_{m\to\infty} \lim_{n\to\infty} I_1 &= \lim_{m\to\infty} \int_a^b \left(\int_a^b (h(\cdot,r) * \psi_m)(u) df(u) \right) dg(r) \\
&= \int_a^b \left(\int_a^b h(r,u) df(r) \right) dg(u).
\end{aligned}$$

Therefore, Step 1 leads us to conclude that the result is satisfied. □

The change of variables formula stated in Remark 44 is also true if F' is a function of locally finite variation, which could have jumped. The proof that we provide here is due to Chen et al. [26]. This is based on some auxiliary results, whose proofs are long. So, we include this auxiliary tool in Section 8.6 to avoid a tedious presentation.

Proposition 52 *Let $F : \mathbb{R} \to \mathbb{R}$ be an absolutely continuous function of the form*

$$F(x) = F(0) + \int_0^x F'(s)ds, \quad x \in \mathbb{R},$$

where $F' : \mathbb{R} \to \mathbb{R}$ has locally bounded variation. Also let $g : \Omega \times [a,b] \to \mathbb{R}$ be a process with Hölder continuous paths in $\mathcal{C}_1^{\alpha}(\mathbb{R})$, where $\alpha > 1/2$, such that, for almost $t \in [a,b]$, the random variable $g(t)$ is absolutely continuous with density $p_t : \mathbb{R} \to \mathbb{R}_+ \cup \{0\}$ satisfying

$$\sup_{x\in\mathbb{R}} p_t(x) \leq \hat{p}_t, \quad \textit{where} \quad \int_a^b \hat{p}_t dt < \infty. \tag{2.72}$$

Then,

$$F(g(t)) - F(g(a)) = \int_a^t F'(g(s))dg(s), \quad t \in [a,b].$$

Here, the integral is the fractional one given in Definition 30.

Remarks 2 *i) Note that a convex function F satisfies the assumption of this result since F' is a non-decreasing function in this case.*

ii) The last equality holds ω by ω (i.e., pathwise).

iii) As it is pointed out in [26], if g is a Gaussian process with covariance R such that $\int_0^T R(s,s)^{-1/2}ds < \infty$, then the condition of this result is satisfied with $\hat{p}_t = R(t,t)^{-1/2}$. In particular, the fractional Brownian motion with Hurst parameter $H \in (1/2, 1)$ fulfills Condition (2.72).

iv) The fractional integral $\int_a^t F'(g(s))dg(s)$, $t \in [a,b]$, exists due to Proposition 136.

v) Condition (2.72) guarantees that, for $\alpha \in (0,1)$, $\sup_{x\in\mathbb{R}} E\left(\int_0^T |y_t - x|^{-\alpha}\, dt\right) < \infty$. Indeed, the Fubini's theorem allows us to see

$$\begin{aligned}
&E\left(\int_0^T |y_t - x|^{-\alpha}\, dt\right) \\
&= \int_0^T \int_{\mathbb{R}} |z-x|^{-\alpha} p_t(z)dzdt \\
&= \int_0^T \int_{|z-x|\le 1} |z-x|^{-\alpha} p_t(z)dzdt + \int_0^T \int_{|z-x|>1} |z-x|^{-\alpha} p_t(z)dzdt \\
&\le \int_0^T \hat{p}_t \int_{|u|\le 1} |u|^{-\alpha} dudt + \int_0^T \int_{\mathbb{R}} p_t(z)dzdt < \infty.
\end{aligned}$$

Proof. By Condition (2.72), we can assume that F' is right-continuous without loss of generality (see the proof of Proposition 136 for details). Let $\varphi \in C^\infty(\mathbb{R})$ be a nonnegative function with support $[-1,0]$ and $\int_{\mathbb{R}} \varphi(s)ds = 1$, and $F_n = F * \varphi_n$, where $n \in \mathbb{N}$ and $\varphi_n(x) = n\varphi(nx)$. Then, Proposition 49 and the fact that $\alpha > 1/2$ imply

$$F_n(g(t)) - F_n(g(a)) = \int_a^t F_n'(g(s))dg(s), \quad t \in [a,b] \text{ and } n \in [a,b]. \tag{2.73}$$

Now, fix $t \in [a,b]$ and $\gamma \in (1-\alpha, \alpha)$. Then, g^{t-} belongs to $I_{t-}^{1-\gamma}(L^q)$ with probability 1, for any $q > 1$, since g has α-Hölder continuous paths (see (1.65)). Furthermore, from the proof of Proposition 136, we have that

$$\left(u \mapsto \frac{|F'(g(u))|}{(u-a)^\gamma} + \gamma \int_a^u \frac{|F'(g(u)+) - F'(g(r)+)|}{(u-r)^{1+\gamma}} dr\right) \in L^p([a,t]),$$

for some $p > 1$ and, consequently, it is equal to $D_{a+}^\gamma(F'(g))$, which is shown proceeding as in (1.65). Also, note that

$$|F_n'(g(u)) - F'(g(u))| \le \sup_{|x|\le \|g\|_{\infty,[a,b]}+1} |F'(x)| < \infty,$$

with probability 1, due to F' being bounded on bounded intervals of $\mathbb{R}$. Thus, the fact that F'_n converges pointwise to F' on account of the right-continuity of F', the dominated convergence theorem and Lemma 91 yield

$$\int_a^t \left(\frac{|F'_n(g(u)) - F'(g(u))|}{(u-a)^\gamma} + \gamma \int_a^u \frac{|F'_n(g(u)) - F'_n(g(r)) - (F'(g(u)) - F'(g(r)))|}{(u-r)^{1+\gamma}} dr \right) du$$

goes to zero in probability, as $n \to \infty$. Therefore, the right-hand side of (2.73) converges to $\int_a^t F'(g(s))dg(s)$ in probability since $D_{t-}^{1-\gamma}(g^{t-})$ is bounded with probability 1 because of (1.65). Finally, using that $F_n \to F$ pointwise again, we also get that the left-hand side of (2.73) converges to $F(g(t)) - F(g(a))$, with probability 1, as $n \to \infty$. So, the result holds.□

On the other hand, Zähle [218] has pointed out that the fractional integral (2.55) is related to the Wiener integral analyzed in Section 1.2.2. Namely, we have the following result.

Proposition 53 *Let $f : [a,b] \to \mathbb{R}$ be a square-integrable function in $I_{a+}^{\alpha}(L^2)$ for some $\alpha \in (1/2, 1)$ and $B = \{B_t : t \in [a,b]\}$ a Brownian motion defined on a complete probability space $(\Omega, \mathcal{F}, P)$. Then,*

$$\int_a^b f(t)dB_t = \int_a^b f(s)dB(s),$$

where the left-hand side is the Wiener integral of f with respect to B and the right-hand side is the fractional integral.

Remark 46 *In Section 1.2, we have already seen that B has Hölder continuous paths for any exponent less than 1/2. Thus, B^{b-} belongs to $I_{b-}^{1-\alpha}(L^q)$, for all $q > 1$, with probability 1 because of (1.65). So, $\int_a^b f(s)dB(s)$ is defined pathwise. That is,*

$$\left(\int_a^b f(s)dB(s) \right)(\omega) = \int_a^b \left(D_{a+}^{\alpha} f\right)(s) \left(D_{b-}^{1-\alpha} B^{b-}(\omega)\right)(s)ds.$$

Proof. For $n \in \mathbb{N}$, ψ_n is the smooth function considered in Lemma 31. Therefore, $f * \psi_n$ is in $C^\infty(\mathbb{R})$ and, consequently, it is a Lipschitz function. Thus, Theorem 37 gives that the fractional integral satisfies

$$\int_a^b (f * \psi_n)(s)dB(s) = \lim_{|\pi| \to 0} \sum_{i=1}^m (f * \psi_n)(t_{i-1}) \left(B_{t_i} - B_{t_{i-1}}\right), \quad \text{w.p.1},$$

where $\pi = \{a = t_0 < t_1 < \ldots < t_m = b\}$ is a partition of the interval $[a,b]$. Hence,

$$\int_a^b (f * \psi_n)(s)dB(s) = \int_a^b (f * \psi_n)(s)dB_s$$

because $\sum_{i=1}^m (f * \psi_n)(t_{i-1}) 1_{]t_{i-1}, t_i]}$ goes to $(f * \psi_n)$ in $L^2([a,b])$, as $m \to \infty$, due to the continuity of $(f * \psi_n)$, and

$$\int_a^b \sum_{i=1}^m (f * \psi_n)(t_{i-1}) 1_{]t_{i-1}, t_i]}(s)dB_s = \sum_{i=1}^m (f * \psi_n)(t_{i-1}) \left(B_{t_i} - B_{t_{i-1}}\right).$$

Finally, by Lemma 31 and the facts that $f * \psi_n$ and $\left[D_{a+}^{\alpha} f\right] * \psi_n$ converge to f and $D_{a+}^{\alpha} f$ in $L^2([a,b])$, respectively, as $n \to \infty$, we have

$$\begin{aligned} \int_a^b (f * \psi_n)(s)dB(s) &= \int_a^b \left(\left[D_{a+}^{\alpha} f\right] * \psi_n\right)(s) \left(D_{b-}^{1-\alpha} B^{b-}\right)(s)ds \\ &\to \int_a^b \left(D_{a+}^{\alpha} f\right)(s) \left(D_{b-}^{1-\alpha} B^{b-}\right)(s)ds = \int_a^b f(s)dB(s), \quad \text{w.p.1}, \end{aligned}$$

and

$$\int_a^b (f * \psi_n)(s)dB_s \to \int_a^b f(s)dB_s \quad \text{in probability, as } n \to \infty.$$

For this reason, the proof is finished. □

Concerning the relation between the fractional integrals and the integrals defined by means of the rough paths theory (see Section 2.2.3), Hu and Nualart [82] have combined the techniques of the fractional calculus and rough paths theory to introduce integrals of the form $\int_a^b f(y_s)dx_s$. Here, $x, y \in C_1^\alpha([a,b];\mathbb{R}^d)$, where $\alpha \in (1/3, 1/2)$, and f is a suitable smooth function such that the derivative f' is β-Hölder continuous for some $\beta > \frac{1}{\alpha} - 2$. Note that $\frac{1}{\alpha} - 2 < 1$ due to $1/3 < \alpha$. Thus, x and $f(y)$ are two Hölder continuous functions with exponents less than $1/2$. Moreover, Hu and Nualart [82] have also proved that this integral agrees with the Riemann-Stieltjes integral, and therefore with the fractional integral, when x is a smooth function (see Theorem 37). In particular, these authors obtain an estimation of this integral that allows them to apply the fixed point theorem to study the existence of a unique solution to differential equation driven by x. It is worth mentioning that they give an explicit expression for the integral $\int_a^b f(y_s)dx_s$ in terms of x, y and $y \otimes x$, which plays the role of x^2 in Definition 29. Consequently, it is not necessary provide approximation arguments to know this integral, as it is done in Corollary 17.

2.4 Fractional Brownian motion

In this section, $B^H = \{B_t^H : t \in [0;T]\}$ stands for a fractional Brownian motion with Hurst parameter $H \in (0,1) \setminus \{1/2\}$. In this case, by Proposition 14, B^H is not a semimartingale no matter the filtration that we are considering on the underlying probability space $(\Omega, \mathcal{F}, P)$. So, we cannot use the techniques of the classical Itô's calculus analyzed in Section 1.2.2 (see also Chapter 3 for a more detailed exposition on this topic) to introduce a stochastic integral with respect to B^H. Therefore, we need another definition of stochastic integral to deal with applications of stochastic calculus based on the process B^H via fractional stochastic differential equations. In order to see that we are able to use the Young integral to resolve this problem, we take into account the following facts.

From (1.49) and (1.47), which gives that B^H has stationary increments, we have that, for $p > 0$,

$$\sum_{j=0}^{n-1} \left| B_{\frac{(j+1)T}{n}} - B_{\frac{jT}{n}} \right|^p \to \begin{cases} 0, & \text{if } pH - 1 > 0, \\ T^{pH}, & \text{if } p = 1/H, \\ \infty, & \text{elsewhere}, \end{cases} \qquad \text{w.p.1, as } n \to \infty.$$

Furthermore, (1.47) and Proposition 5 yield that the fractional Brownian motion has Hölder continuous paths for any exponent less than the Hurst parameter H. Hence, we can utilize the approaches studied in this section to analyze the existence of a unique pathwise solution (i.e., figured out ω by ω) to stochastic differential equations driven by B^H in the case that $H > 1/2$ because the different interpretations of stochastic integral with respect to it inherit the properties of the integrator in general. For instance, the Wiener integral agrees with the Young integral for any continuous integrand with finite q-variation, for $q < 2$ (See Corollary 5), the Itô's integral has continuous paths as the Brownian motion has (see Proposition 59) and the Young integral given by (2.47) is Hölder continuous with the same exponent of the integrator as Theorem 33.(2) establishes, among another examples.

Concerning the case that $H < 1/2$, we can apply the theory of rought paths given by Lyons [133] due to Nualart and Tindel [161]. Remember that in Section 2.2.3, we show the fundamental ideas on the construction of the theory of rough paths in the case that $H \in (1/3, 1/2)$ and we comment on the difficulties to utilize this theory in the general case (i.e., $H < 1/2$). In Section 6, we study the Stratonovich integral with respect to Gaussian Volterra processes, which include fractional Brownian motion and Riemann-Liouville fractional process, both with Hurst parameter less than one half. In consequence, we are also able to work with fractional differential equations of Stratonovich type. However, it is not easy to deal with this integral since it is a limit in probability. Thus, we need to apply the Malliavin calculus via the divergence operator to handle the moments of the Stratonovich integral.

Chapter 3

Classical Stochastic Integration

Now, we study the stochastic integral with respect to Brownian motion B in the Itô sense. That is, the integrands are measurable and adapted processes to the underlying filtration satisfying suitable integrability conditions. The main difficulty to define a stochastic integral with respect to B is that Brownian motion does not have paths of bounded variation and, therefore, we cannot interpret it as a Riemman-Stieltjes integral, in general. Thus, we must proceed as in Section 1.2.2 to introduce this stochastic integral in the Itô sense. It means, we have to consider a family of adapted step, or simple processes that includes the simple functions (see Definition 13), for which the definition of stochastic integral is inspired by the right-hand side of (2.2), and then we must analyze the convergence of integrals of simple processes in some suitable spaces of random variables. This is the approach that we use in Section 1.2.2 to construct the Wiener integral (i.e., the Itô's integral of square-integrable functions with respect to Brownian motion).

The Itô stochastic integral is one of the basic tools of the classical stochastic analysis, which allows us to consider differential equation involving stochastic processes. Moreover, it is a linear operator satisfying versions of the dominated convergence theorem and an integration by parts formula (known as Itô's formula). Consequently, this linear operator deserves the name "stochastic integral". In general, the solution of a stochastic differential equation is a stochastic process X that models a random phenomenon (i.e., an event with different possible results) that evolves in time. Hence, it is natural to suppose that X is adapted to the information that we have on the problem that we are analyzing, which is represented by the underlying filtration.

Due to above, it is important to investigate some properties of the stochastic integral. So, the main aim of this Chapter is to analyze some basic concepts that are used to deal with Itô type stochastic differential equations driven by Brownian motion and their applications.

3.1 Definition and properties of the stochastic integral in the Itô's sense

In this section, we extend the Wiener integral to the space of measurable and adapted processes to the underlying filtration, which either are square-integrable (as functions of $(\omega, t) \in \Omega \times [0,T]$), or have square-integrable paths. Actually, the family of square-integrable processes is included in the space of all the measurable processes with square-integrable paths by Fubini's theorem. But, we will see that the stochastic integral of a process $X \in L^2(\Omega \times [0,T])$ is a martingale (see Definition 57), while it is a local martingale if X only has square-integrable paths. Moreover, the stochastic integrals appearing in the Itô's formula are local martingales in general.

For $0 < T < \infty$, we consider a filtration $(\mathcal{F}_t)_{t\in[0,T]}$ on the complete probability space $(\Omega, \mathcal{F}, P)$ and a Brownian motion, or Wiener process, $B = \{B_t : t \in [0,T]\}$ with respect to

DOI: 10.1201/9781003484912-3

$(\mathcal{F}_t)_{t\in[0,T]}$ (see Definition 11). We observe that the fact that we only consider $\mathcal{F}_t$-adapted integrands allows us to have a stochastic integral satisfying estimates similar to the isometry relation introduced in Corollary 4.*iii*), which is a way to control the convergence of stochastic integrals in $L^2(\Omega)$.

As we have already pointed out in Section 1.2.2, the main problem to introduce a stochastic integral with respect B via Riemman-Stieltjes sums is that Brownian motion does not have bounded variation paths (see Remark 5). But, we now have an extra problem: If f is a stochastic process, then the limit on the right-hand side of (1.25) could also depend on the intermediate points $s_1, \ldots, s_n$, as the following calculations show.

Let $\pi = \{0 = t_0 < t_1 < \ldots < t_n = t\}$ be a partition of the interval $[0,t]$ with $t \in [0,T]$, $\alpha \in [0,1]$ and $s_i = (1-\alpha)t_{i-1} + \alpha t_i$, which belongs to $[t_{i-1}, t_i]$, $i = 1, \ldots, n$. Then, straightforward calculations imply

$$\begin{aligned}\sum_{i=1}^{n} B_{s_i}\left(B_{t_i} - B_{t_{i-1}}\right) &= \frac{B_t^2}{2} - \frac{1}{2}\sum_{i=1}^{n}\left(B_{t_i} - B_{t_{i-1}}\right)^2 + \sum_{i=1}^{n}\left(B_{s_i} - B_{t_{i-1}}\right)^2 \\ &\quad + \sum_{i=1}^{n}\left(B_{t_i} - B_{s_i}\right)\left(B_{s_i} - B_{t_{i-1}}\right) \\ &= \frac{B_t^2}{2} - I_t^{(1)} + I_t^{(2)} + I_t^{(3)}. \end{aligned} \tag{3.1}$$

Note that we can get, by proceeding as in the proof of Proposition 7,

$$I_t^{(2)} - \sum_{i=1}^{n}\left(s_i - t_{i_{i-1}}\right) \to 0 \quad \text{and} \quad I_t^{(3)} \to 0, \quad \text{as } |\pi| \to 0, \text{ in } L^2(\Omega). \tag{3.2}$$

Now, imitating the approach used in Section 1.2.2, where we change the pathwise convergence by $L^2(\Omega)$-one, we are able to define a stochastic integral $(\alpha)\int_0^t B_s dB_s$ as the $L^2(\Omega)$-limit of the Riemman-Stieltjes sum $\sum_{i=1}^{n} B_{s_i}\left(B_{t_i} - B_{t_{i-1}}\right)$. Thus, using that $\sum_{i=1}^{n}\left(s_i - t_{i_{i-1}}\right) = \alpha\sum_{i=1}^{n}\left(t_i - t_{i_{i-1}}\right)$, Proposition 7, (3.1) and (3.2), we obtain

$$(\alpha)\int_0^t B_s dB_s = \frac{B_t^2}{2} + \left(\alpha - \frac{1}{2}\right)t, \quad t \in [0,T]. \tag{3.3}$$

In other words, unlike the Riemman-Stieltjes integral given in Chapter 2, this stochastic integral depends on the intermediate points $s_1, \ldots, s_n$.

The stochastic integral in (3.3) is a martingale if we choose the left-endpoint of each interval of the partition π (i.e., $\alpha = 0$), due to (1.12), and it satisfies the fundamental theorem of calculus if we choose the middle point of the intervals in π (i.e., $\alpha = 1/2$) taking into account $\frac{dx^2}{dx} = 2x$. In this chapter we consider the case $\alpha = 0$, and we study the case $\alpha = 1/2$ in Chapter 6. Namely, the so-called Itô and Stratonovich integrals, respectively.

Remember that we are considering a Brownian motion $B = \{B_t : t \in [0,T]\}$ with respect to $(\mathcal{F}_t)_{t\in[0,T]}$ in this section. Henceforth, we fix $a, b \in [0,T]$, $a < b$. As in Section 1.2.2, we first introduce the family of simple $\mathcal{F}_t$-adapted processes and their integral with respect to B.

Definition 31 *We say that $X : \Omega \times [a,b] \to \mathbb{R}$ is a simple and $\mathcal{F}_t$-adapted process, or a nonanticipating simple process if there exist a partition $\pi = \{a = t_0 < t_1 < \ldots < t_n = b\}$ of $[a,b]$ and random variables $F_1, \ldots, F_n$ such that F_i is $\mathcal{F}_{t_{i-1}}$-measurable, $i = 1, \ldots, n$, and*

$$X_t = \sum_{i=1}^{n} F_i 1_{]t_{i-1}, t_i]}(t), \quad t \in [a,b]. \tag{3.4}$$

In this case, we define the stochastic integral of X with respect to Brownian motion B in the Itô sense as the random variable

$$\int_a^b X_s dB_s := \sum_{i=1}^n F_i \left(B_{t_i} - B_{t_{i-1}}\right). \tag{3.5}$$

Note that if $F_1, \ldots, F_n \in \mathbb{R}$ (i.e., they are deterministic), then X is also a simple function introduced in Definition 13 and that its Itô's integral with respect to B can be given by both (1.27) and (3.5). Moreover, we can show that the integral (3.5) is independent of the representation of X as a nonanticipating simple process by following the proof of Proposition 9. Also note that, as in Section 1.2.2, the equality

$$\int_0^T Y_s 1_{[a,b]}(s) dB_s = \int_a^b Y_s dB_s$$

holds for any simple and $\mathcal{F}_t$-adapted process either $Y : \Omega \times [0,T] \to \mathbb{R}$, or $Y : \Omega \times [a,b] \to \mathbb{R}$.

Now, we see that Corollary 3 is still true for nonanticipating simple processes.

Lemma 32 *Consider two simple and $\mathcal{F}_t$-adapted processes $X^{(1)}, X^{(2)} : \Omega \times [a,b] \to \mathbb{R}$.*

i) Let $c_1, c_2 \in \mathbb{R}$. Then, $c_1 X^{(1)} + c_2 X^{(2)}$ is also a nonanticipating simple process and

$$\int_a^b \left(c_1 X_s^{(1)} + c_2 X_s^{(2)}\right) dB_s = c_1 \int_a^b X_s^{(1)} dB_s + c_2 \int_a^b X_s^{(1)} dB_s.$$

ii) Assume that $X^{(1)} \in L^1(\Omega \times [a,b])$. We then have

$$E\left(\int_a^b X_s^{(1)} dB_s\right) = 0.$$

iii) If $X^{(1)}, X^{(2)} \in L^2(\Omega \times [a,b])$, we get

$$E\left\{\left(\int_a^b X_s^{(1)} dB_s\right)\left(\int_a^b X_s^{(2)} dB_s\right)\right\} = E\left(\int_a^b X_s^{(1)} X_s^{(2)} ds\right). \tag{3.6}$$

In particular,

$$E\left\{\left(\int_a^b X_s^{(1)} dB_s\right)^2\right\} = E\left(\int_a^b \left(X_s^{(1)}\right)^2 ds\right).$$

iv) Let F be an $\mathcal{F}_a$-measurable random variable. Then,

$$\int_a^b F X_s^{(1)} dB_s = F \int_a^b X_s^{(1)} dB_s.$$

Proof. Following the proof of Proposition 9, we can choose a partition $\pi = \{a = t_0 < t_1 < \ldots < t_n = b\}$ of $[a,b]$ such that

$$X^{(1)} = \sum_{i=1}^n F_i^{(1)} 1_{]t_{i-1},t_i]} \quad \text{and} \quad X^{(2)} = \sum_{i=1}^n F_i^{(2)} 1_{]t_{i-1},t_i]}, \tag{3.7}$$

where $F_i^{(1)}$ and $F_i^{(2)}$ are two $\mathcal{F}_{t_{i-1}}$-measurable random variables. Hence, it is easy to see that the stochastic integral is a linear operator on the space of all the nonanticipating simple processes. That is, Statement i) is satisfied.

Now we deal with Assertion ii). Note that the fact that $X^{(1)} \in L^1(\Omega \times [a,b])$ implies that $F_1^{(1)}, \ldots, F_n^{(1)}$ in (3.7) are in $L^1(\Omega)$. Thus, Definition 11.ii) yields

$$\begin{aligned} E\left(\left|\int_a^b X_s^{(1)} dB_s\right|\right) &\leq \sum_{i=1}^n E\left(\left|F_i^{(1)}\right| \left|B_{t_i} - B_{t_{i-1}}\right|\right) = \sum_{i=1}^n E\left(\left|F_i^{(1)}\right|\right) E\left(\left|B_{t_i} - B_{t_{i-1}}\right|\right) \\ &\leq \sum_{i=1}^n E\left(\left|F_i^{(1)}\right|\right) \sqrt{t_i - t_{i-1}} < \infty. \end{aligned}$$

Similarly, applying Definition 11.ii) again, we have

$$E\left(\int_a^b X_s^{(1)} dB_s\right) = \sum_{i=1}^n E\left(|F_i^{(1)}\right) E\left(B_{t_i} - B_{t_{i-1}}\right) = 0,$$

as we wanted to show.

Concerning iii), using that $B_t - B_s$ is a Gaussian random variable with zero mean and variance $t-s$, and independent of $\mathcal{F}_s$, for $a \leq s < t \leq b$, together with (3.7), we get

$$\begin{aligned} & E\left\{\left(\int_a^b X_s^{(1)} dB_s\right)\left(\int_a^b X_s^{(2)} dB_s\right)\right\} \\ &= E\left(\sum_{i=1}^n F_i^{(1)} F_i^{(2)} \left(B_{t_i} - B_{t_{i-1}}\right)^2\right) + E\left(\sum_{i \neq j} F_i^{(1)} F_j^{(2)} \left(B_{t_i} - B_{t_{i-1}}\right)\left(B_{t_j} - B_{t_{j-1}}\right)\right) \\ &= \sum_{i=1}^n E\left(F_i^{(1)} F_i^{(2)}\right) E\left(\left(B_{t_i} - B_{t_{i-1}}\right)^2\right) \\ &\quad + \sum_{i<j} E\left(F_i^{(1)} F_j^{(2)} \left(B_{t_i} - B_{t_{i-1}}\right)\right) E\left(B_{t_j} - B_{t_{j-1}}\right) \\ &\quad + \sum_{j<i} E\left(F_i^{(1)} F_j^{(2)} \left(B_{t_j} - B_{t_{j-1}}\right)\right) E\left(B_{t_i} - B_{t_{i-1}}\right) \\ &= \sum_{i=1}^n E\left(F_i^{(1)} F_i^{(2)}\right) (t_i - t_{i-1}) = E\left(\int_a^b X_s^{(1)} X_s^{(2)} ds\right). \end{aligned} \tag{3.8}$$

Therefore, we obtain that $\int_a^b X_s^{(1)} dB_s \in L^2(\Omega)$ by writing $X^{(1)}$ instead of $X^{(2)}$ in previous calculations. Consequently, from (3.8), Statement iii) holds.

Finally, Affirmation iv) is an immediate consequence of the definition of the stochastic integral for simple and $\mathcal{F}_t$-adapted processes. Thus, the proof is finished. □

In (3.8), we can apply that the processes B and $t \mapsto B_t^2 - t$ are two $\mathcal{F}_t$-martingales instead of the fact that $B_t - B_s \sim \mathcal{N}(0, t-s)$ is independent of the σ-algebra $\mathcal{F}_s$ in order to see that (3.6) holds. Indeed, we now show a more general result to show that this claim is true. Towards this end, we suppose that we are considering a càdlàg and square-integrable $\mathcal{F}_t$-martingale $M = \{M_t : t \in [0,T]\}$ such that $M_0 = 0$. Remember that this means that $E[M_t|\mathcal{F}_s] = M_s$ and $E(M_t^2) < \infty$, for any $0 \leq s < t \leq T$. We assume that the filtration $(\mathcal{F}_t)_{t\in[0,T]}$ satisfies the usual conditions. From the Doob-Meyer decomposition (see Bojdecki [22], Dellacherie and Meyer [38]. Ikeda and Watanabe [86] or Protter [180]), there exists a unique right-continuous, increasing and $\mathcal{F}_t$- predictable process $\langle M\rangle$ such that $\langle M\rangle_0 = 0$,

$E(\langle M\rangle_t) < \infty$ and $M^2 - \langle M\rangle$ is an $\mathcal{F}_t$-martingale. The fact that $\langle M\rangle$ is an $\mathcal{F}_t$- predictable process means that

$$\langle M\rangle : (\Omega \times [0,T], \mathcal{P}_{[0,T]}) \to \mathbb{R},$$

where $\mathcal{P}_{[a,b]}$ is the predictable σ-algebra, which is generated by the predictable cylinders of the form

$$\{F\times]s,t] : a \le s < t \le b \text{ and } F \in \mathcal{F}_s\} \cup \{F \times [0,a] : F \in \mathcal{F}_0\}.$$

Note that $\langle B\rangle_t = t$. Now, the Itô's stochastic integral $\int_a^b X_s dM_s$ of the nonanticipating simple process X in (3.4) with respect to M is given by the right-hand side of (3.5) when we change B by M. Thus, if $X^{(1)}$ and $X^{(2)}$ are the processes in (3.7) with $F_i^{(1)}, F_i^{(2)} \in L^2(\Omega)$, then the calculations in (3.8) become

$$\begin{aligned}
&E\left\{\left(\int_a^b X_s^{(1)} dM_s\right)\left(\int_a^b X_s^{(2)} dM_s\right)\right\} \\
&= \sum_{i=1}^n E\left(F_i^{(1)}F_i^{(2)}\left(M_{t_i} - M_{t_{i-1}}\right)^2\right) \\
&\quad + \sum_{i<j} E\left(F_i^{(1)}F_j^{(2)}\left(M_{t_i} - M_{t_{i-1}}\right)\left(M_{t_j} - M_{t_{j-1}}\right)\right) \\
&\quad + \sum_{j<i} E\left(F_i^{(1)}F_j^{(2)}\left(M_{t_j} - M_{t_{j-1}}\right)\left(M_{t_i} - M_{t_{i-1}}\right)\right) \\
&= \sum_{i=1}^n E\left(F_i^{(1)}F_i^{(2)}\left(M_{t_i}^2 - 2M_{t_i}M_{t_{i-1}} + M_{t_{i-1}}^2\right)\right) \\
&\quad + \sum_{i<j} E\left(F_i^{(1)}F_j^{(2)}\left(M_{t_i} - M_{t_{i-1}}\right) E\left[M_{t_j} - M_{t_{j-1}} | \mathcal{F}_{t_{j-1}}\right]\right) \\
&\quad + \sum_{j<i} E\left(F_i^{(1)}F_j^{(2)}\left(M_{t_j} - M_{t_{j-1}}\right) E\left[M_{t_i} - M_{t_{i-1}} | \mathcal{F}_{t_{i-1}}\right]\right).
\end{aligned}$$

Hence, the martingale property of M leads us to write

$$\begin{aligned}
&E\left\{\left(\int_a^b X_s^{(1)} dM_s\right)\left(\int_a^b X_s^{(2)} dM_s\right)\right\} \\
&= \sum_{i=1}^n E\left(F_i^{(1)}F_i^{(2)} E\left[M_{t_i}^2 - 2M_{t_i}M_{t_{i-1}} + M_{t_{i-1}}^2 | \mathcal{F}_{t_{i-1}}\right]\right) \\
&= \sum_{i=1}^n E\left(F_i^{(1)}F_i^{(2)} E\left[M_{t_i}^2 - M_{t_{i-1}}^2 | \mathcal{F}_{t_{i-1}}\right]\right) = \sum_{i=1}^n E\left(F_i^{(1)}F_i^{(2)}\left(\langle M\rangle_{t_i} - \langle M\rangle_{t_{i-1}}\right)\right) \\
&= E\left(\int_a^b X_s^{(1)} X_s^{(2)} d\langle M\rangle_s\right) = \int_a^b X^{(1)} X^{(2)} d\lambda_{M^2},
\end{aligned} \tag{3.9}$$

where λ_{M^2} is the measure associated with the increasing process $\langle M\rangle$. That is, λ_{M^2} is the Doléans measure defined on $(\Omega \times [0,T], \mathcal{P}_{[a,b]})$, which satisfies

$$\lambda_{M^2}(F\times]s,t]) = E\left(\left(\langle M\rangle_t - \langle M\rangle_s\right) 1_F\right) \quad \text{if } a \le s < t \le b \text{ and } F \in \mathcal{F}_s$$

and $\lambda_{M^2}(F \times [0,a]) = 0$, if $F \in \mathcal{F}_0$. Observe that (3.9) is satisfied when $X^{(1)}$ and $X^{(2)}$ belong to the space $L^2(\Omega \times [0,T]; \lambda_{M^2})$.

Henceforth, the space of all the measurable and $\mathcal{F}_t$-adapted processes in $L^2(\Omega \times [a,b])$ is denoted by $\mathcal{L}^2_{\mathrm{a}}(\Omega \times [a,b])$. In order to extend the definition of the Itô's integral to last space, we proceed as in Proposition 10. That is, we see that the family of nonanticipating simple process of the form (3.4) is a dense subset of $\mathcal{L}^2_{\mathrm{a}}(\Omega \times [a,b])$.

Lemma 33 *Let X be a process in $\mathcal{L}^2_a(\Omega \times [a,b])$ and $\mathcal{F}_a$ contain all the P-null sets. Then, there exists a sequence $\{X^{(n)} \in L^2(\Omega \times [a,b]) : n \in \mathbb{N}\}$ of nonanticipating simple processes of the form (3.4) such that*

$$\left\| X - X^{(n)} \right\|_{L^2(\Omega\times[a,b])} \to 0, \quad \text{as } n \to \infty. \tag{3.10}$$

Proof. Let $\{a = t_0^n < t_1^n < \ldots < t_n^n = b\}$ be the partition of $[a,b]$ with $t_i^n = a + (i(b-a))/n$. In the remaining of this proof, we write t_i instead of t_i^n in order to simplify the notation. From Theorem 2, we can choose a progressively measurable modification $\bar{X}$ of the process X. Hence, we have that, for any $i = 2, \ldots, n$, the integral $\int_{t_{i-2}}^{t_{i-1}} \bar{X}_s ds$ is $\mathcal{F}_{t_{i-1}}$-measurable due to Fubini's theorem. Thus, the fact that $(\Omega, \mathcal{F}_{t_{i-1}}, P)$ is a complete probability space and

$$E\left(\left| \int_{t_{i-2}}^{t_{i-1}} \bar{X}_s ds - \int_{t_{i-2}}^{t_{i-1}} X_s ds \right| \right) \le \int_{t_{i-2}}^{t_{i-1}} E\left(\left| \bar{X}_s - X_s \right| \right) ds = 0,$$

which follows from Fubini's theorem again, imply that

$$X^{(n)} = \sum_{i=2}^{n} \left(\frac{1}{t_{i-1} - t_{i-2}} \int_{t_{i-2}}^{t_{i-1}} X_s ds \right) 1_{]t_{i-1}, t_i]}$$

is a nonanticipating simple process such that

$$\left| X_r^{(n)} - X_r \right|^2 \le 2X_r^2 1_{[a,t_1]}(r) + 2\sum_{i=2}^{n} \left(\frac{1}{t_{i-1} - t_{i-2}} \int_{t_{i-2}}^{t_{i-1}} X_s ds - X_r \right)^2 1_{]t_{i-1}, t_i]}(r). \tag{3.11}$$

Let $\omega \in \Omega$ such that $X(\omega) \in L^2([a,b])$ and $\{f_m : [a,b] \to \mathbb{R} : m \in \mathbb{N}\}$ a sequence of continuous functions that goes to $X(\omega)$ in $L^2([a,b])$. Then,

$$\begin{aligned}
&\sum_{i=2}^{n} \int_{t_{i-1}}^{t_i} \left(\frac{1}{t_{i-1} - t_{i-2}} \int_{t_{i-2}}^{t_{i-1}} X_s(\omega) ds - X_r \right)^2 dr \\
&\le C \sum_{i=2}^{n} \int_{t_{i-1}}^{t_i} \left(\frac{1}{t_{i-1} - t_{i-2}} \int_{t_{i-2}}^{t_{i-1}} \left(X_s(\omega) - f_m(s) \right) ds \right)^2 dr \\
&\quad + C \sum_{i=2}^{n} \left[\int_{t_{i-1}}^{t_i} \left(\frac{1}{t_{i-1} - t_{i-2}} \int_{t_{i-2}}^{t_{i-1}} f_m(s) ds - f_m(r) \right)^2 dr + \int_{t_{i-1}}^{t_i} \left(f_m(r) - X_r(\omega) \right)^2 dr \right],
\end{aligned}$$

which, from Hölder inequality, is bounded by

$$\begin{aligned}
&C \sum_{i=2}^{n} \frac{1}{t_{i-1} - t_{i-2}} \int_{t_{i-1}}^{t_i} \int_{t_{i-2}}^{t_{i-1}} \left(X_s(\omega) - f_m(s) \right)^2 ds dr \\
&\quad + C \sum_{i=2}^{n} \left[\frac{1}{t_{i-1} - t_{i-2}} \int_{t_{i-1}}^{t_i} \int_{t_{i-2}}^{t_{i-1}} \left(f_m(s) - f_m(r) \right)^2 ds dr + \int_{t_{i-1}}^{t_i} \left(f_m(r) - X_r(\omega) \right)^2 dr \right] \\
&\le 2C \int_a^b \left(f_m(r) - X_r(\omega) \right)^2 dr + C \sum_{i=2}^{n} \frac{1}{t_{i-1} - t_{i-2}} \int_{t_{i-1}}^{t_i} \int_{t_{i-2}}^{t_{i-1}} \left(f_m(s) - f_m(r) \right)^2 ds dr,
\end{aligned}$$

which, together with (3.11) and Fubini theorem, gives that $\int_a^b \left|X_r^{(n)} - X_r\right|^2 dr \to 0$, as $n \to \infty$, w.p.1. Consequently, (3.10) is true if the process X is bounded by a constant because of the dominated convergence theorem. So, for $m \in \mathbb{N}$ and $t \in [a,b]$, we introduce the process $X_{m,t} = (-m) \vee (X_t \wedge m)$, in order to finish the proof. Therefore, proceeding as in the last calculations, we have

$$E\left(\int_a^b \left|X_r^{(n)} - X_r\right|^2 dr\right) \leq CE\left(\int_a^{t_1} X_r^2 dr\right) + CE\left(\int_a^b |X_{m,r} - X_r|^2 dr\right) + CE\left(\int_a^b \left|X_{m,r}^{(n)} - X_{m,r}\right|^2 dr\right).$$

Finally, we obtain that (3.10) holds by taking first the limit as $n \to \infty$ and then the limit as $m \to \infty$. That means, the proof is complete. □

An immediate consequence of the last proof is the following result.

Corollary 19 *Let $X : \Omega \times [a,b] \to \mathbb{R}$ be a measurable and $\mathcal{F}_t$-adapted process with square-integrable paths, $\mathcal{F}_a$ contain all the P-null sets and $\{a = t_0^n < t_1^n < \ldots < t_n^n = b\}$ the partition of $[a,b]$ such that $t_i^n = a + (i(b-a))/n$. Then, for $n \in \mathbb{N}$,*

$$X_t^{(n)} := \sum_{i=2}^{n} \left(\frac{1}{t_{i-1} - t_{i-2}} \int_{t_{i-2}}^{t_{i-1}} X_s ds \right) 1_{]t_{i-1},t_i]}(t), \quad t \in [a,b],$$

is a nonanticipating simple process such that

$$\int_a^b \left|X_r^{(n)} - X_r\right|^2 dr \to 0, \quad \text{as } n \to \infty, \text{ w.p.1.}$$

Moreover, if $X \in \mathcal{L}_a^2(\Omega \times [a,b])$, then we also have

$$E\left(\int_a^b \left|X_r^{(n)} - X_r\right|^2 dr\right) \to 0, \quad \text{as } n \to \infty.$$

As in Lemma 5, for a process X in $\mathcal{L}_a^2(\Omega \times [a,b])$ and a sequence of nonanticipating simple processes $\{X^{(n)} \in L^2(\Omega \times [a,b]) : n \in \mathbb{N}\}$ such that (3.10) holds, Lemma 32.*iii*) implies that $\{\int_a^b X_s^{(n)} dB_s : n \in \mathbb{N}\}$ converges in $L^2(\Omega)$ to a random variable $\int_a^b X_s dB_s$. Moreover, the definition of this square-integrable random variable is independent of the choice of the sequence of simple processes $\{X^{(n)} \in L^2(\Omega \times [a,b]) : n \in \mathbb{N}\}$. This allows us to extend the domain of the stochastic integral with respect to B as follows (compare it with Definition 14).

Definition 32 *Assume that $\mathcal{F}_a$ contains all the P-null sets and let $X \in \mathcal{L}_a^2(\Omega \times [a,b])$. The stochastic integral of X with respect to B on $[a,b]$ in the Itô sense, denoted by $\int_a^b X_s dB_s$, is the square-integrable random variable given as*

$$\int_a^b X_s dB_s := L^2(\Omega) - \lim_{n\to\infty} \int_a^b X_s^{(n)} dB_s,$$

where $\{X^{(n)} \in L^2(\Omega \times [a,b]) : n \in \mathbb{N}\}$ is a sequence of $\mathcal{F}_t$-adapted simple processes such that $\left\|X - X^{(n)}\right\|_{L^2(\Omega\times[a,b])} \to 0$, as $n \to \infty$.

As a consequence of this definition and Lemma 32, we have the following properties of the Itô's integral with respect to B:

Proposition 54 *Let X and Y be two processes in $\mathcal{L}^2_a(\Omega \times [a,b])$ and $\mathcal{F}_a$ contains all the P-null sets. Then,*

i) For $c_1, c_2 \in \mathbb{R}$, the process $c_1X + c_2Y$ also belongs to the space $\mathcal{L}^2_a(\Omega \times [a,b])$ and

$$\int_a^b (c_1 X_s + c_2 Y_s)\, dB_s = c_1 \int_a^b X_s dB_s + c_2 \int_a^b Y_s dB_s.$$

ii) We have that $E\left(\int_a^b X_s dB_s\right) = 0$ and

$$E\left\{\left(\int_a^b X_s dB_s\right)\int_a^b Y_s dB_s\right\} = E\left(\int_a^b X_s Y_s ds\right).$$

iii) The process $\int_a^\cdot X_s dB_s = \{\int_a^t X_s dB_s : t \in [a,b]\}$ is a square-integralble $\mathcal{F}_t$-martingale.

Proof. Statements $i)$ and $ii)$ are an immediate consequence of Definition 32 and Lemma 32.

In order to see that Assertion $iii)$ holds, we first assume that X is a nonanticipating simple process of the form (3.4). Then the martingale property of B yields that, for $a \leq s < t \leq T$,

$$\begin{aligned} E\left[\int_a^t X_r dB_r \middle| \mathcal{F}_s\right] &= \sum_{i=1}^n E\left[F_i\left(B_{t_i \wedge t} - B_{t_{i-1}\wedge t}\right) \middle| \mathcal{F}_s\right] \\ &= \sum_{i=1}^n F_i\left(B_{t_i\wedge s} - B_{t_{i-1}\wedge s}\right) = \int_a^s X_r dB_r. \end{aligned}$$

Finally the Jensen's inequality for conditional expectation implies that, for a sequence of nonanticipating simple processes $\{X^{(n)} \in L^2(\Omega \times [0,T]) : n \in \mathbb{N}\}$ such that $\|X - X^{(n)}\|_{L^2(\Omega\times[a,b])} \to 0$, we obtain

$$\begin{aligned} E\left[\int_a^t X_r dB_r \middle| \mathcal{F}_s\right] &= L^2(\Omega) - \lim_{n\to\infty} E\left[\int_a^t X_r^{(n)} dB_r \middle| \mathcal{F}_s\right] \\ &= L^2(\Omega) - \lim_{n\to\infty} \int_a^s X_r^{(n)} dB_r = \int_a^s X_r dB_r. \end{aligned}$$

Thus, the proof is finished. □

Other important fact of Itô's integral with respect to B on the space $\mathcal{L}^2_a(\Omega \times [a,b])$ is the local property for it, which is established in the following result.

Proposition 55 *Suppose that $\mathcal{F}_a$ contains all the P-null sets. Let $X, Y \in \mathcal{L}^2_a(\Omega \times [a,b])$ and $A \in \mathcal{F}_b$ such that*

$$X = Y \quad \textit{almost surely on } A \times [a,b]$$

(i.e., $1_{A\times[a,b]}X = 1_{A\times[a,b]}Y$ almost surely). Then,

$$1_A \int_a^b X_s dB_s = 1_A \int_a^b Y_s dB_s \quad \textit{w.p.1.}$$

That is, the stochastic integrals $\int_a^b X_s dB_s$ and $\int_a^b Y_s dB_s$ are the same on the set A.

Proof. Let $X^{(n)}$ be the nonanticipating simple process introduced in Corollary 19. Then, by hypothesis,

$$\begin{aligned} 1_A \int_a^b X_s^{(n)} dB_s &= \sum_{i=2}^n 1_A \left(\frac{1}{t_{i-1} - t_{i-2}} \int_{t_{i-2}}^{t_{i-1}} X_s ds \right) (B_{t_i} - B_{t_{i-1}}) \\ &= \sum_{i=2}^n 1_A \left(\frac{1}{t_{i-1} - t_{i-2}} \int_{t_{i-2}}^{t_{i-1}} Y_s ds \right) (B_{t_i} - B_{t_{i-1}}). \end{aligned}$$

Hence, by Corollary 19, we can conclude

$$\begin{aligned} 1_A \int_a^b X_s dB_s &= L^2(\Omega) - \lim_{n\to\infty} 1_A \int_a^b X_s^{(n)} dB_s \\ &= L^2(\Omega) - \lim_{n\to\infty} 1_A \int_a^b Y_s^{(n)} dB_s = 1_A \int_a^b Y_s dB_s, \end{aligned}$$

where

$$Y_t^{(n)} := \sum_{i=2}^n \left(\frac{1}{t_{i-1} - t_{i-2}} \int_{t_{i-2}}^{t_{i-1}} Y_s ds \right) 1_{]t_{i-1}, t_i]}(t), \quad t \in [a, b].$$

Therefore, the result is satisfied. □

The local property of the Itô's integral with respect to B allows us to extend its domain as follows.

Definition 33 *A measurable process X belongs to $\mathcal{L}_a^2(\Omega\times[a,b])_{loc}$ if there exists a sequence $\{(\Omega_n, X^{(n)}) \in \mathcal{F}_b \times \mathcal{L}_a^2(\Omega \times [a,b]) : n \in \mathbb{N}\}$ such that*

i) $\Omega_n \uparrow \Omega$ w.p.1, as $n \to \infty$ (i.e., $\Omega_n \subset \Omega_{n+1}$ and $P(\Omega \setminus \cup_m \Omega_m) = 0$).

ii) $X_t = X_t^{(n)}$ for all $t \in [a,b]$, w.p.1 on Ω_n.

In this case, we say that the sequence $\{(\Omega_n, X^{(n)}) : n \in \mathbb{N}\}$ localizes the process X in $\mathcal{L}_a^2(\Omega \times [a,b])$.

Observe that if $X \in \mathcal{L}_a^2(\Omega\times[a,b])_{loc}$ and all the P-null sets of Ω are in $\mathcal{F}_a$, then Statement *ii)* implies that X is also an $\mathcal{F}_t$-adapted process. Moreover, in this case, we can extend the domain of the stochastic Itô's integral to the space $\mathcal{L}_a^2(\Omega \times [a,b])_{loc}$ by defining $\int_a^b X_s dB_s$ as the random variable given by

$$\int_a^b X_s dB_s = \int_a^b X_s^{(n)} dB_s \quad \text{on } \Omega_n. \tag{3.12}$$

Notice that this random variable $\int_a^b X_s dB_s$ is well-defined because, for $m < n$, Proposition 55 establishes

$$\int_a^b X_s^{(m)} dB_s = \int_a^b X_s^{(n)} dB_s \quad \text{w.p.1 on } \Omega_m.$$

Also note that, using Proposition 55 again, the definition of $\int_a^b X_s dB_s$ via (3.12) is independent of the sequence that localizes the process X in $\mathcal{L}_a^2(\Omega \times [a,b])$. Indeed, suppose that $\{(\tilde{\Omega}_n, Y^{(n)}) : n \in \mathbb{N}\}$ is another localizing sequence of X. Therefore, Proposition 55 yields

$$1_{\Omega_n \cap \tilde{\Omega}_m} \int_a^b Y_s^{(m)} dB_s = 1_{\Omega_n \cap \tilde{\Omega}_m} \int_a^b X_s^{(n)} dB_s = 1_{\Omega_n \cap \tilde{\Omega}_m} \int_a^b X_s dB_s,$$

w.p.1, for $m, n \in \mathbb{N}$, which allows us to have $1_{\tilde{\Omega}_m} \int_a^b Y_s^{(m)} dB_s = 1_{\tilde{\Omega}_m} \int_a^b X_s dB_s$ w.p.1, by letting n goes to ∞. But, unlike Proposition 54.*ii*), now we cannot guarantee that $E\left(\int_a^b X_s dB_s\right) = 0$, as we will see later on. Furthermore, it is not difficult to see that $\mathcal{L}_a^2(\Omega \times [a,b]) \subset \mathcal{L}_a^2(\Omega \times [a,b])_{loc}$. Indeed, concerning Definition 33, we can choose any sequence $\{\Omega_n \in \mathcal{F} : n \in \mathbb{N}\}$ such that $\Omega_n \uparrow \Omega$ and $X^{(n)} = X$, for all $n \in \mathbb{N}$. Thus, for $X \in \mathcal{L}_a^2(\Omega \times [a,b])$, the stochastic integrals of X with respect to B defined via Definition 32 and (3.12) are the same. In other words, the integral given in (3.12) is an extension of the Itô's integral introduced in Definition 32.

Other consequence of Definition 33 is the following:

Corollary 20 *Let $X \in \mathcal{L}_a^2(\Omega \times [a,b])_{loc}$ and $\mathcal{F}_a$ have all the P-null sets. Then, X belongs to the space $\mathcal{L}_a(\Omega; L^2([a,b]))$ of all the measurable and $\mathcal{F}_t$-adapted processes with square-integrable paths.*

Proof. Let $\{(\Omega_n, X^{(n)}) : n \in \mathbb{N}\}$ be a sequence that localizes the process X in $\mathcal{L}_a^2(\Omega \times [a,b])$. Then,

$$1_{\Omega_n} \int_a^b X_s^2 ds = 1_{\Omega_n} \int_a^b (X_s^{(n)})^2 ds < \infty \quad \text{w.p.1},$$

where the equality follows from Definition 33.*ii*) and Fubini theorem. Finally, the fact that $\Omega_n \uparrow \Omega$ implies that the result is true. □

Now, we will see that $\mathcal{L}_a(\Omega; L^2([a,b])) \subset \mathcal{L}_a^2(\Omega \times [a,b])_{loc}$. That is, the domain of the stochastic integral in the Itô sense introduced in (3.12) is $\mathcal{L}_a^2(\Omega; L^2([a,b]))$. Towards this end, we recall that $B = \{B_t : t \in [0,T]\}$ is a Brownian motion with respect to the filtration $(\mathcal{F}_t)_{t\in[0,T]}$.

Definition 34 *Let $(\mathcal{G}_t)_{t\geq 0}$ be a filtration on $(\Omega, \mathcal{F}, P)$. A function $S : \Omega \to \mathbb{R}_+ \cup \{\infty\}$ is called a $\mathcal{G}_t$-stopping time if $[S \leq t] \in \mathcal{G}_t$, for any $t \geq 0$.*

In order to answer our claim, now we introduce a stopping time in the following result.

Lemma 34 *Let $X : \Omega \times [a,b] \to \mathbb{R}$ be a continuous and $\mathcal{F}_t$-adapted process and $x \in \mathbb{R}$. Then,*

$$S = \inf\{t \in [a,b] : X_t \geq x\} \wedge b, \quad \textit{with } \inf \emptyset = \infty,$$

is an $\mathcal{F}_t$-stopping time.

Remark 47 *Remember that $\mathcal{F}_t$ is only defined for $t \in [0,T]$. In this case, we define $\mathcal{F}_t = \mathcal{F}_T$, for $t \geq T$. But, if $B = \{B_t : t \geq 0\}$ is a Brownian motion with respect to the filtration $(\mathcal{F}_t)_{t\geq 0}$ and $X : \Omega \times [a,\infty) \to \mathbb{R}$, then S is given by*

$$S = \inf\{t \geq a : X_t \geq x\}, \quad \textit{with } \inf \emptyset = \infty.$$

As the reader can see, the proof of this fact is easier than the one of the lemma.

Proof. Notice that the continuity of the process X and the definition of S imply

$$\begin{aligned} [S \leq t] &= \begin{cases} \emptyset, & \text{if } t \in [0,a), \\ [S = a], & \text{if } t = a, \end{cases} = \begin{cases} \emptyset, & \text{if } t \in [0,a), \\ [X_a \geq x], & \text{if } t = a, \end{cases} \\ &\in \begin{cases} \mathcal{F}_t, & \text{if } t \in [0,a), \\ \mathcal{F}_a, & \text{if } t = a. \end{cases} \end{aligned}$$

In order to continue with the proof, for $n \in \mathbb{N}$, set $S_n = \inf\{s \in [a,b] : X_s > x - \frac{1}{n}\} \wedge b$. Thus, the fact that X is a continuous process gives

$$[S_n < t] = \bigcup_{\substack{a \leq s < t \\ s \in \mathbb{Q}}} \left[X_s > x - \frac{1}{n}\right] \in \mathcal{F}_t, \quad \text{for } n \in \mathbb{N} \text{ and } t \in (a,b]. \tag{3.13}$$

It is easy to see that $a \leq S_n \leq S_m \leq S$, for any $n < m$. Hence, if $S(\omega) = a$, we obtain that $S_n(\omega) = a$, for all $n \in \mathbb{N}$. If $\sup_{t\in[a,b]} |X_t(\omega)| \geq x$ and $X_a(\omega) < x$, then, using the continuity of X again, $X_{S_n(\omega)}(\omega) = x - \frac{1}{n}$ for n large enough and $S(\omega) = \lim_{n\to\infty} S_n(\omega)$ since $x - \frac{1}{n}$ goes to x as $n \to \infty$. And if $\sup_{t\in[a,b]} |X_t(\omega)| < x$, then $S(\omega) = S_n(\omega) = b$, for n large enough. Thus, (3.13) leads to conclude

$$[S \leq t] = \begin{cases} \cap_{n=1}^{\infty} [S_n < t], & \text{if } t \in (a,b), \\ (\cap_{n=1}^{\infty} [S_n < b]) \cup \left[\sup_{s\in[a,b]\cap\mathbb{Q}} |X_s| < x\right], & \text{if } t \geq b. \end{cases}$$

Now, the proof is complete. □

In general, we have the following result, whose proof can be found in Dellacherie an Meyer [37] (Section IV, no. 50).

Theorem 39 *Let $(\mathcal{F}_t)_{t\geq 0}$ be a filtration defined on a complete probability space $(\Omega, \mathcal{F}, P)$ satisfying the usual conditions and $X : \Omega \times [0,\infty) \to \mathbb{R}$ an $\mathcal{F}_t$-progressively measurable process. Then, for any $B \in \mathcal{B}(\mathbb{R})$,*

$$S = \inf\{t \geq 0 : X_t \in B\}, \quad \textit{with } \inf \emptyset = \infty,$$

is an $\mathcal{F}_t$-stopping time.

Now, we are ready to show that our claim is true:

Proposition 56 *Assume that $\mathcal{F}_a$ contains all the P-null sets. Let $X \in \mathcal{L}_a(\Omega; L^2([a,b]))$. Then, X also belongs to the space $\mathcal{L}_a^2(\Omega \times [a,b])_{loc}$.*

Proof. By definition of $\mathcal{L}_a(\Omega; L^2([a,b]))$, X is a measurable and $\mathcal{F}_t$-adapted process. So, Theorem 2 implies that X has a progressively measurable version $\tilde{X}$. Thus, the process $t \mapsto \int_a^t X_s^2 ds$ is a continuous and $\mathcal{F}_t$-adapted process because the Fubini's theorem allows to get

$$\int_a^t X_s^2 ds = \int_a^t \tilde{X}_s^2 ds \quad \text{w.p.1.}$$

On the other hand, for $n \in \mathbb{N}$, Lemma 34 yields that

$$S_n = \inf\{t \in [a,b] : \int_0^t X_s^2 ds \geq n\} \wedge b$$

is an $\mathcal{F}_t$-stopping time. Consequently, the fact that

$$[S_n < t] = \cup_{m=1}^{\infty} [S_n \leq (t - 1/m) \vee 0] \in \mathcal{F}_t$$

gives that the process $t \mapsto 1_{(0,S_n]}(t)$ is left-continuous and $\mathcal{F}_t$-adapted. Moreover,

$$E\left(\int_a^b 1_{(0,S_n]}(t) X_t^2 ds\right) = E\left(\int_a^{S_n} X_t^2 ds\right) \leq n.$$

That is, $X^{(n)} = 1_{(0,S_n]}(\cdot)X_\cdot \in \mathcal{L}_a^2(\Omega \times [a,b])$. Also note that if $\Omega_n = \left[\int_a^b X_s^2 ds < n\right]$, then $\Omega_n \uparrow \Omega$ and $X^{(n)} = X_\cdot$ on $\Omega_n \times [a,b]$. In other words, the sequence $\left\{(\Omega_n, X^{(n)}) : n \in \mathbb{N}\right\}$ localizes the process X in $\mathcal{L}_a^2(\Omega \times [a,b])$. □

Now, we see that (3.5) is still true for a simple and $\mathcal{F}_t$-adapted process in $\mathcal{L}_a(\Omega; L^2([a,b]))$ of the form

$$X_s = \sum_{i=1}^{n} F_i 1_{]t_{i-1},t_i]}(s), \quad s \in [a,b], \tag{3.14}$$

where $\{a = t_0 < t_1 < \ldots < t_n = b\}$ is a partition of the interval $[a,b]$. Note that, in this case, F_i is an $\mathcal{F}_{t_{i-1}}$-measurable random variable such that $P([|F_i| < \infty]) = 1$, for $i = 1, \ldots, n$.

Corollary 21 *Let $X \in \mathcal{L}_a(\Omega; L^2([a,b]))$ be the simple and $\mathcal{F}_t$-adapted process given in (3.14). Then*

$$\int_a^b X_s dB_s := \sum_{i=1}^{n} F_i \left(B_{t_i} - B_{t_{i-1}}\right). \tag{3.15}$$

Proof. Let

$$\Omega_m = [|F_1| \le m, \dots, |F_n| \le m] \quad \text{and} \quad X_\cdot^{(m)} = \sum_{i=1}^{n} (-m \vee (F_i \wedge m))\, 1_{]t_{i-1},t_i]}(\cdot).$$

Then, we have that $\{(\Omega_m, X_\cdot^{(m)}) : m \in \mathbb{N}\}$ localizes X in $\mathcal{L}_a^2(\Omega \times [a,b])$. Thus, Definition 33 allows us to conclude

$$\begin{aligned} 1_{\Omega_m} \int_a^b X_s dB_s &= 1_{\Omega_m} \int_a^b X_s^{(m)} dB_s \\ &= 1_{\Omega_m} \sum_{i=1}^{n} (-m \vee (F_i \wedge m)) \left(B_{t_i} - B_{t_{i-1}}\right) = 1_{\Omega_m} \sum_{i=1}^{n} F_i \left(B_{t_i} - B_{t_{i-1}}\right), \end{aligned}$$

which implies the result. □

Note that an important application of the isometry property of the Itô stochastic integral given by Proposition 54.ii) is the analysis of the $L^2(\Omega)$-convergence of stochastic integrals. Indeed, suppose that $\{X^{(n)} : n \in \mathbb{N} \cup \{0\}\}$ is a sequence in $\mathcal{L}_a^2(\Omega \times [a,b])$ such that

$$E\left(\int_a^b \left|X_s^{(n)} - X_s^{(0)}\right|^2 dt\right) \to 0, \quad \text{as } n \to \infty,$$

then

$$\int_a^b X_s^{(0)} dB_s = L^2(\Omega) - \lim_{n\to\infty} \int_a^b X_s^{(n)} dB_s. \tag{3.16}$$

In particular, when $\mathcal{F}_a$ has all the P-null sets, the isometry property of Itô's integral allows us to get versions of the dominated convergence theorem. For instance, assume now that $\{X^{(n)} : n \in \mathbb{N}\}$ is a sequence of measurable and $\mathcal{F}_t$-adapted process that goes to a measurable process $X^{(0)}$, pointwise, and that there exists $Y \in \mathcal{L}^2(\Omega \times [a,b])$ such, that $|X^{(n)}| \le Y$ almost surely. Then, $X^{(0)}$ is also Itô integrable with respect to B and (3.16) holds due to the dominated convergence theorem applied in $\mathcal{L}^2(\Omega \times [a,b])$. However, we do not have that (3.16) is true if X only belongs to the space $\mathcal{L}_a^2(\Omega \times [a,b])_{loc}$ because, in general, X only has square-integrable paths as Corollary 20 and Proposition 56 state. But, in this case, (3.16) is satisfied if we change the $L^2(\Omega)$-convergence by convergence in probability, which is obtained by means of the following result.

Proposition 57 *Let $\mathcal{F}_a$ contain all the P-null sets and $X \in \mathcal{L}_a^2(\Omega \times [a,b])_{loc} = \mathcal{L}_a(\Omega; L^2([a,b]))$. Then, for $N, c > 0$, we have*

$$P\left[\left|\int_a^b X_s dB_s\right| > c\right] \le \frac{N}{c^2} + P\left[\int_a^b X_s^2 ds \ge N\right]. \tag{3.17}$$

Proof. Let $\Omega_N = \left[\int_a^b X_s^2 ds < N\right]$, $S_N = \inf\{t \in [a,b] : \int_0^t X_s^2 ds \ge N\} \wedge b$ and $\tilde{X} =$

$1_{[0,S_N]}(\cdot)X_\cdot$. Then, Definition 33 (see also (3.12)) and Chevyshev's inequality imply

$$\begin{aligned} P\left[\left|\int_a^b X_s dB_s\right| > c\right] &= P\left[\left|\int_a^b X_s dB_s\right| > c, \Omega_N\right] + P\left[\left|\int_a^b X_s dB_s\right| > c, \Omega_N^c\right] \\ &\leq P\left[\left|\int_a^b \tilde{X}_s dB_s\right| > c, \Omega_N\right] + P\left[\Omega_N^c\right] \\ &\leq \frac{E\left(\left(\int_a^b \tilde{X}_s dB_s\right)^2\right)}{c^2} + P\left[\int_a^b X_s^2 ds \geq N\right] \\ &\leq \frac{N}{c^2} + P\left[\int_a^b X_s^2 ds \geq N\right], \end{aligned}$$

as we want to establish. So, the proof is complete. □

Note that (3.17) plays the role of the isometry relation for the Itô's stochastic integral when we are dealing with the convergence of stochastic integrals with integrands in the space $\mathcal{L}_a(\Omega; L^2([a,b]))$. That is, if $\{X^{(n)} \in \mathcal{L}_a(\Omega; L^2([a,b])) : n \in \mathbb{N}\}$ is such that $\int_a^b \left(X_s^{(n)} - X_s\right)^2 ds$ converges to zero in probability, as $n \to \infty$, then (3.17) yields

$$\int_a^b X_s dB_s = P - \lim_{n\to\infty} \int_a^b X_s^{(n)} dB_s,$$

where the last limit is in probability. In particular, this convergence in probability is satisfied if $X^{(n)}$ is as in Corollary 19. Hence, an immediate consequence of Corollary 21 and Lemma 32.*iv*) is the following:

Proposition 58 *Suppose that all the P-null sets are included in $\mathcal{F}_a$. Let X, Y be two processes in $\mathcal{L}_a(\Omega; L^2([a,b]))$. Then,*

i) For $c_1, c_2 \in \mathbb{R}$, we have

$$\int_a^b (c_1 X_s + c_2 Y_s)\, dB_s = c_1 \int_a^b X_s dB_s + c_2 \int_a^b Y_s B_s.$$

ii) Proposition 55 holds. That is, if there is $A \in \mathcal{F}_b$ such that

$$X = Y \quad \textit{almost surely on } A \times [a,b],$$

we get

$$1_A \int_a^b X_s dB_s = 1_A \int_a^b Y_s dB_s \quad \textit{w.p.1.}$$

iii) For an $\mathcal{F}_a$-measurable random variable F, we obtain

$$\int_a^b F X_s dB_s = F \int_a^b X_s dB_s.$$

We remark that, for $X \in \mathcal{L}_a(\Omega; L^2([a,b]))$, $\int_a^{\cdot} X_s dB_s$ is not an $\mathcal{F}_t$- martingale in general and $E(\int_a^b X_s dB_s)$ could be different than zero unlike Statements *ii*) and *iii*) of Proposition 54. Indeed, let F be a finite and $\mathcal{F}_a$-measurable random variable that is not in $L^1(\Omega)$. Then,

$$\int_a^t F dB_s = F\,(B_t - B_a) \notin L^1(\Omega), \quad t \in [a,b],$$

since the independence of the increments of Brownian motion yields

$$E(|F|\,|B_t - B_a|) = E(|F|)\,E(|B_t - B_a|) = \infty.$$

So, in this case, $t \mapsto \int_a^t FdB_s$ is not an $\mathcal{F}_t$-martingale. For the second claim, the reader can read the paper by Dudley [44] to see that there exist stochastic integrals in $L^1(\Omega)$ with expectation different than 0. However, in the following two results, we will show that in general $t \mapsto \int_a^t X_s dB_s$ is a continuous and an $\mathcal{F}_t$-local martingale, for $X \in \mathcal{L}_a(\Omega; L^2([a,b]))$. That is, there is a sequence $\{\tau_n : n \in \mathbb{N}\}$ of stopping times such that $t \mapsto \int_a^{t\wedge\tau_n} X_s dB_s$ is a martingale, for all $n \in \mathbb{N}$, and $\tau_n \uparrow b$.

Proposition 59 *Let X be a processes in $\mathcal{L}_a(\Omega; L^2([a,b]))$ and $(\mathcal{F}_t)_{t\in[0,T]}$ satisfy the usual conditions. Then, the process $t \mapsto \int_a^t X_s dB_s$ has continuous paths.*

Proof. We first assume that X is the simple and $\mathcal{F}_t$-adapted process given in (3.14). Then,

$$\int_a^t X_s dB_s = \sum_{i=1}^n F_i\left(B_{t_i\wedge t} - B_{t_{i-1}\wedge t}\right), \quad t \in [a,b],$$

which allows us to show that the result is satisfied in this case.

Now assume that $X \in \mathcal{L}_a^2(\Omega \times [a,b])$. Then, Proposition 54.*iii*), the fact that $(\mathcal{F}_t)_{t\in[0,T]}$ satisfies the usual conditions and [38] (Theorem VI.4) allow us to assume that the integral $\int_a^{\cdot} X_s dB_s$ is càdlàg without loss of generality. Moreover, Lemma 33 implies that there is a sequence $\{X^{(n)} : n \in \mathbb{N}\}$ of simple and $\mathcal{F}_t$-adapted processes satisfying (3.10). Consequently, the Doob's inequality (see Theorem 100 below) and Proposition 54 lead us to conclude

$$\begin{aligned} &E\left(\left(\sup_{t\in[a,b]}\left|\int_a^t X_s dB_s - \int_a^t X_s^{(n)} dB_s\right|\right)^2\right) \\ &= E\left(\left(\sup_{t\in[a,b]}\left|\int_a^t \left[X_s - X_s^{(n)}\right] dB_s\right|\right)^2\right) \le 4E\left(\int_a^b \left[X_s - X_s^{(n)}\right]^2 ds\right). \end{aligned}$$

Hence, using (3.10), we have seen that the result is also true for $X \in \mathcal{L}_a^2(\Omega \times [a,b])$ since convergence in $L^2(\Omega)$ of a sequence of random variables implies the convergence w.p.1 for some subsequence of it.

Finally, for $X \in \mathcal{L}_a(\Omega; L^2([a,b]))$, we have that Definition 33, the definition of the Itô's integral through (3.12), Corollary 20 and Proposition 56 yield the result is true by using that $\cup_{n\ge1}\Omega_n = \Omega$ w.p.1. □

In Proposition 56, we use a sequence $\{S_n : n \in \mathbb{N}\}$ of stopping times satisfying $S_n \uparrow \infty$ w.p.1 to show that $\mathcal{L}_a(\Omega; L^2([a,b])) = \mathcal{L}_a^2(\Omega \times [a,b])_{loc}$. As a consequence of the following result, we have that this sequence is also useful to see that the Itô's integral of any process $X \in \mathcal{L}_a(\Omega; L^2([a,b]))$ is a local martingale. That is, in this case, $t \mapsto \int_a^{t\wedge S_n} X_s dB_s$ is an $\mathcal{F}_t$-martingale. Towards this end, we first state the following auxiliary result.

Lemma 35 *Let τ be an $\mathcal{F}_t$-stopping time. Moreover suppose that $\mathcal{F}_a$ contains all the P-null sets. Then,*

$$\int_0^{t\wedge\tau} dB_s = \int_0^t 1_{(0,\tau]}(s) dB_s = B_{t\wedge\tau}, \quad t \in [0,T].$$

Proof. Note that $1_{(0,\tau]}(\cdot)$ is a measurable and $\mathcal{F}_t$-adapted process because it is left-continuous and $1_{(0,\tau(\omega)]}(\cdot) = 1_{[\tau<\cdot]^c}(\omega)1_{[\tau>0]}(\omega)$, $\omega \in \Omega$ (see the proof of Proposition 56 for details). Also note that we only need to show that the last equality holds because, for $t \in [0,T]$,

$$\int_0^{t\wedge\tau} dB_s = \left(\int_0^u dB_s\right)\Big|_{u=t\wedge\tau} = (B_u)\,|_{u=t\wedge\tau}.$$

Now fix $t \in [0,T]$. Consider a partition $\pi = \{0 = t_0 < t_1 < \ldots < t_n = t\}$ of $[0,t]$, and the measurable and $\mathcal{F}_t$-adapted simple process

$$X_s^\pi = \sum_{i=1}^n 1_{(0,\tau]}(t_{i-1})1_{(t_{i-1},t_i]}(s), \quad s \in [0,t].$$

It is easy to see that $\lim_{|\pi|\to 0} X_s^\pi = 1_{(0,\tau]}(s)$ almost all $(\omega,s) \in \Omega \times [0,t]$. Hence, since $|X^\pi| \le 1$ almost surely, the dominated convergence theorem and (3.16) imply

$$\begin{aligned}
\int_0^t 1_{(0,\tau]}(s)dB_s &= L^2(\Omega) - \lim_{|\pi|\to 0}\int_0^t X_s^\pi dB_s \\
&= L^2(\Omega) - \lim_{|\pi|\to 0}\sum_{i=1}^n 1_{(0,\tau]}(t_{i-1})\left(B_{t_i} - B_{t_{i-1}}\right) \\
&= L^2(\Omega) - \lim_{|\pi|\to 0}\sum_{i=1}^n B_{t_i}1_{(t_{i-1},t_i]}(\tau\wedge t) = B_{\tau\wedge t},
\end{aligned}$$

where last equality follows from the continuity of B, the Doob's $L^2(\Omega)$ inequality (i.e., Theorem 100) and the dominated convergence theorem. Thus, the proof is complete. □

Now, we are ready to see what the stopped Itô's integral is.

Proposition 60 *Assume that $(\mathcal{F}_t)_{t\in[0,T]}$ satisfies the usual conditions. Let τ be an $\mathcal{F}_t$-stopping time and X a process in $\mathcal{L}_a(\Omega; L^2([a,b]))$. Then,*

$$\int_0^{t\wedge\tau} 1_{[a,b]}(s)X_s dB_s = \int_a^t 1_{(0,\tau]}(s)X_s dB_s, \quad t \in [a,b].$$

Remark 48 *For a process M, in the literature, it is defined the process M^τ by $M_t^\tau := M_{t\wedge\tau}$, where $M_{t\wedge\tau}(\omega) = M_{t\wedge\tau(\omega)}(\omega)$, $\omega \in \Omega$. M^τ is called the stopped process associated with M and τ. These processes are considered in order to obtain properties that the original process does not have. For instance, we can see Corollary 22 below.*

Proof. Fix $t \in [a,b]$. We divide the proof into three steps.
Step 1: Here, we assume that X is a simple process in $\mathcal{L}_a(\Omega; L^2([a,b]))$ of the form (3.4). Hence,

$$\begin{aligned}
&\int_0^{t\wedge\tau} 1_{[a,b]}(s)X_s dB_s \\
&= \int_0^{t\wedge\tau} X_s dB_s - \int_0^{\tau\wedge a} X_s dB_s = \left(\int_0^u X_s dB_s\right)\Big|_{u=t\wedge\tau} - \left(\int_0^u X_s dB_s\right)\Big|_{u=\tau\wedge a} \\
&= \sum_{i=1}^n F_i\left(B_{t\wedge t_i}^\tau - B_{t\wedge t_{i-1}}^\tau\right) - \sum_{i=1}^n F_i\left(B_{a\wedge t_i}^\tau - B_{a\wedge t_{i-1}}^\tau\right). \qquad (3.18)
\end{aligned}$$

On the other hand, using Lemma 35, Proposition 58.*iii)* and (3.4), we can obtain

$$\begin{aligned}
&\int_a^t 1_{(0,\tau]}(s)X_s dB_s \\
&= \sum_{i=1}^n F_i \int_a^t 1_{(t_{i-1},t_i]}(s)1_{(0,\tau]}(s)dB_s = \sum_{i=1}^n F_i \int_a^t 1_{(\tau\wedge t_{i-1},\tau\wedge t_i]}(s)dB_s \\
&= \sum_{i=1}^n F_i \left(\int_0^t 1_{(\tau\wedge t_{i-1},\tau\wedge t_i]}(s)dB_s - \int_0^a 1_{(\tau\wedge t_{i-1},\tau\wedge t_i]}(s)dB_s \right) \\
&= \sum_{i=1}^n F_i \left(B^\tau_{t\wedge t_i} - B^\tau_{t\wedge t_{i-1}} \right) - \sum_{i=1}^n F_i \left(B^\tau_{a\wedge t_i} - B^\tau_{a\wedge t_{i-1}} \right),
\end{aligned}$$

which, together with (3.18), implies that the result holds when X is an $\mathcal{F}_t$-adapted simple process.

Step 2: Now, we assume that X belongs to $\mathcal{L}^2_{\mathrm{a}}(\Omega \times [a,b])$.

In this case, from Lemma 33, we can choose a sequence $\{X^{(n)} \in \mathcal{L}^2_{\mathrm{a}}(\Omega \times [a,b]) : n \in \mathbb{N}\}$ of $\mathcal{F}_t$-adapted simple processes satisfying (3.10). Then, by the $L^2(\Omega)$ Doob's inequality (i.e., Theorem 100), Proposition 59 and Step 1, we have

$$\begin{aligned}
&E\left(\left(\int_0^{t\wedge\tau} 1_{[a,b]}(s)X_s dB_s - \int_a^t 1_{(0,\tau]}(s)X_s dB_s \right)^2 \right) \\
&\le 2E\left(\left(\int_0^{t\wedge\tau} 1_{[a,b]}(s)X_s dB_s - \int_0^{t\wedge\tau} 1_{[a,b]}(s)X^{(n)}_s dB_s \right)^2 \right) \\
&\quad +2E\left(\left(\int_a^t 1_{(0,\tau]}(s)X_s dB_s - \int_a^t 1_{(0,\tau]}(s)X^{(n)}_s dB_s \right)^2 \right) \\
&\le 2E\left(\sup_{u\in[a,b]} \left| \int_0^u 1_{[a,b]}(s) \left(X_s - X^{(n)}_s \right) dB_s \right|^2 \right) \\
&\quad +2E\left(\left(\int_a^t 1_{(0,\tau]}(s)X_s dB_s - \int_a^t 1_{(0,\tau]}(s)X^{(n)}_s dB_s \right)^2 \right) \\
&\le 10E\left(\int_a^b \left(X_s - X^{(n)}_s \right)^2 ds \right) \to 0, \quad \text{as } n \to \infty.
\end{aligned}$$

In other words, the result is also true for $X \in \mathcal{L}^2_{\mathrm{a}}(\Omega \times [a,b])$.

Step 3: Finally, we suppose that X is in $\mathcal{L}_a(\Omega; L^2([a,b]))$.

Let S_n and Ω_n be the $\mathcal{F}_t$-stopping time and the set introduced in the proof of Proposition 56, respectively. Then, (3.12) and Step 2 yield

$$\begin{aligned}
1_{\Omega_n} \int_0^{t\wedge\tau} 1_{[a,b]}(s)X_s dB_s &= 1_{\Omega_n} \int_0^{t\wedge\tau} 1_{[a,b]}(s)1_{(0,S_n]}(s)X_s dB_s \\
&= 1_{\Omega_n} \int_a^t 1_{(0,\tau]}(s)1_{(0,S_n]}(s)X_s dB_s = 1_{\Omega_n} \int_a^t 1_{(0,\tau]}(s)dB_s.
\end{aligned}$$

Consequently, the fact that $\Omega_n \uparrow \Omega$ w.p.1 allows us to conclude that the proposition is satisfied. □

Corollary 22 *Let $(\mathcal{F}_t)_{t\in[0,T]}$ satisfy the usual conditions and $X \in \mathcal{L}_a(\Omega; L^2([a,b]))$. Then, the Itô stochastic integral $\int_a^\cdot X_s dB_s$ is a square-integrable $\mathcal{F}_t$-local martingale.*

Proof. Let $\{S_n : n \in \mathbb{N}\}$ be the sequence of stopping times introduced in the proof of Proposition 56. Then, Proposition 60 gives

$$\int_a^{t\wedge S_n} X_s dB_s = \int_a^t 1_{]0,S_n]}(s) X_s dB_s, \quad t \in [a,b]. \tag{3.19}$$

Moreover, we also see in the proof of Proposition 56 that

$$E\left(\int_a^t 1_{]0,S_n]}(s) X_s^2 ds\right) < \infty.$$

Hence, the integral on the right-hand side of (3.19) is a square-integrable martingale by Proposition 54.*iii*). Thus, the proof is complete. □

Using the proof of Proposition 56 again, we know that, for $X \in L^2(\Omega; L^2([a,b]))$, the sequence

$$\left\{\left(\left[\int_a^b X_s^2 ds < n\right], 1_{]0,S_n]}(\cdot)X_\cdot\right) : n \in \mathbb{N}\right\}$$

localizes the process X in $\mathcal{L}_a^2(\Omega \times [a,b])$, which together with (3.19), implies

$$\int_a^t X_s dB_s = \int_a^t 1_{]0,S_n]}(s) X_s dB_s = \int_a^{t\wedge S_n} X_s dB_s, \quad \text{on } \left[\int_a^b X_s^2 ds < n\right].$$

This establishes the relation between the localization given in Definition 33 and the localization of the stochastic integral as a local square-integrable $\mathcal{F}_t$-martingale by means of stopping times.

The Itô's stochastic integral is also associated with the fractional integral (2.55) in the case that both integrals exist, as the following result establishes.

Proposition 61 *Let $f : [a,b] \to \mathbb{R}$ be a process belonging to $\mathcal{L}_a(\Omega; L^2([a,b])) \cap I_{a+}^{\alpha}(L^2)$, for some $\alpha \in (1/2, 1)$, with probability 1. Moreover, suppose that $\mathcal{F}_a$ contains all the P-null sets. Then,*

$$\int_a^b f(t) dB_t = \int_a^b f(s) dB(s),$$

where the left-hand side is the Itô's stochastic integral of f with respect to B and the right-hand side is the fractional integral given by (2.55).

Remark 49 *We have that B^{b-} belongs to $I_{b-}^{1-\alpha}(L^q)$, for all $q > 1$, with probability 1 (for details see Remark 46), and that the fractional integral $\int_a^b f(s)dB(s)$ is defined pathwise (i.e., ω by ω). It means,*

$$\left(\int_a^b f(s)dB(s)\right)(\omega) = \int_a^b \left(D_{a+}^{\alpha} f(\omega)\right)(s) \left(D_{b-}^{1-\alpha} B^{b-}(\omega)\right)(s) ds, \quad \textit{w.p.1.}.$$

Also, we remember that, in the proof, we use the convention $f \equiv 0$ in $[a,b]^c$.

Proof. Let ψ_n be the smooth function introduced in Lemma 31. Then, the process

$$(f * \psi_n)(t) = \int_{\mathbb{R}} f(t-z)\psi_n(z) dz, \quad t \in [a,b],$$

is $\mathcal{F}_t$-adapted due to Theorem 2 and that $\psi_n = 0$ on $[0, 1/n]^c$. Hence, in order to finish the proof, we only have to follow the steps developed in the proof of Proposition 53. □

Now we introduce one of the main tools of the stochastic calculus based on the Itô's integral. Namely, the Itô's formula, which is the chain rule for the Itô's stochastic integral with respect to Brownian motion. Towards this end, we introduce the following definition, where $\mathcal{L}_a(\Omega; L^1([a,b]))$ stands for all the measurable and $\mathcal{F}_t$-adapted processes X with integrable paths. That is,

$$\int_a^b |X_s|\, ds < \infty \quad \text{w.p.1.}$$

Definition 35 *Assume that $\mathcal{F}_a$ contains all the P-null sets. We say that a measurable and $\mathcal{F}_t$-adapted processes X is an Itô process if it has the form*

$$X_t = \xi + \int_a^t \beta_s ds + \int_a^t \sigma_s dB_s, \quad t \in [a,b], \tag{3.20}$$

where ξ is a $\mathcal{F}_a$-measurable random variable, $\beta \in \mathcal{L}_a(\Omega; L^1([a,b]))$ and $\sigma \in \mathcal{L}_a(\Omega; L^2([a,b]))$.

Theorem 40 (Itô's formula) *Let X be the Itô process given in (3.20) and $f : [a,b] \times \mathbb{R} \to \mathbb{R}$ a function in $C^{1,2}([a,b] \times \mathbb{R})$ (i.e., it is a continuous function with continuous partial derivatives ∂_t, $\partial_x f$ and $\partial_{xx} f$). Also let $(\mathcal{F}_t)_{t\in[0,T]}$ satisfy the usual conditions. Then,*

$$\begin{aligned} f(t, X_t) \quad &= \quad f(a,\xi) + \int_a^t \partial_t f(s, X_s) ds + \int_a^t \partial_x f(s, X_s)\sigma_s dB_s \\ &\quad + \int_a^t \partial_x f(s, X_s)\beta_s ds + \frac{1}{2}\int_a^t \partial_{xx} f(s, X_s)\sigma_s^2 ds \quad w.p.1, \end{aligned} \tag{3.21}$$

for each $t \in [a,b]$.

Remark 50 *Note that both sides of equality (3.21) are continuous processes due to Proposition 59. So, from Proposition 1, the equality is also true for all $t \in [a,b]$ w.p.1. It means, the set of probability 1 where equality (3.21) holds is independent of t. Also, the last stochastic integral is defined via (3.12), Corollary 20 and Proposition 56 since we can only guarantee that $\partial_x f(\cdot, X_\cdot)\sigma_\cdot \in \mathcal{L}_a(\Omega; L^2([a,b]))$, which follows from the continuity of f and X.*

Proof. We fix $t \in [a,b]$ and divide the proof into three steps.
Step 1: Here we show that (3.21) holds for $f \in C_b^{1,2}([a,b] \times \mathbb{R})$ (i.e., f and its partial derivatives are bounded) if it is satisfied for every simple and $\mathcal{F}_t$-adapted process σ, and $f \in C_b^{1,2}([a,b] \times \mathbb{R})$.

First, we consider the case that $f \in C_b^{1,2}([a,b] \times \mathbb{R})$, $\beta \in \mathcal{L}^1_{\mathrm{a}}(\Omega \times [a,b])$ and $\sigma \in \mathcal{L}^2_{\mathrm{a}}(\Omega \times [a,b])$. So, from Lemma 33, there is a sequence $\{\sigma^{(n)} : n \in \mathbb{N}\}$ of simple and $\mathcal{F}_t$-adapted processes such that

$$E\left(\int_a^b \left|\sigma_r^{(n)} - \sigma_r\right|^2 dr\right) \to 0, \quad \text{as } n \to \infty.$$

We set

$$X_r^{(n)} = \xi + \int_a^r \beta_s ds + \int_a^r \sigma_s^{(n)} dB_s, \quad r \in [a,b].$$

Hence, for $r \in [a,b]$, $X_r^{(n)}$ goes in probability to X_r due to (3.17), as $n \to \infty$, and

$$\begin{aligned} f(t, X_t^{(n)}) \quad &= \quad f(a,\xi) + \int_a^t \partial_t f(s, X_s^{(n)}) ds + \int_a^t \partial_x f(s, X_s^{(n)})\sigma_s^{(n)} dB_s \\ &\quad + \int_a^t \partial_x f(s, X_s^{(n)})\beta_s ds + \frac{1}{2}\int_a^t \partial_{xx} f(s, X_s^{(n)}) \left(\sigma_s^{(n)}\right)^2 ds \quad w.p.1. \end{aligned}$$

Consequently, letting $n \to \infty$, we have that our claim is true. That is, (3.21) is valid when $f \in C_b^{1,2}([a,b]\times\mathbb{R})$, and $\sigma \in \mathcal{L}_a^2(\Omega\times[a,b])$. Indeed, for instance, the dominated convergence theorem implies

$$
\begin{aligned}
&E\left(\int_0^t \left|\partial_{xx}f(s,X_s^{(n)})\left(\sigma_s^{(n)}\right)^2 - \partial_{xx}f(s,X_s)\sigma_s^2\right| ds\right)\\
&= E\left(\int_0^t \left|\partial_{xx}f(s,X_s)\left(\sigma_s^2-\left(\sigma_s^{(n)}\right)^2\right)+\left(\sigma_s^{(n)}\right)^2\left(\partial_{xx}f(s,X_s)-\partial_{xx}f(s,X_s^{(n)})\right)\right| ds\right)\\
&\leq 3\|\partial_{xx}f\|_\infty E\left(\int_0^t\left(\sigma_s^2-\left(\sigma_s^{(n)}\right)^2\right)ds\right)\\
&\quad +E\left(\int_0^t \sigma_s^2\left|\partial_{xx}f(s,X_s)-\partial_{xx}f(s,X_s^{(n)})\right| ds\right). \qquad (3.22)
\end{aligned}
$$

Now, consider $\beta \in \mathcal{L}_a(\Omega;L^1([a,b]))$ and $\sigma \in \mathcal{L}_a(\Omega;L^2([a,b]))$. For $\in \mathbb{N}$, introduce the stopping time

$$
S_n = \inf\{r \geq 0 : \int_0^r 1_{[a,b]}(s)\left(|\beta_s|+\sigma_s^2\right)ds \geq n\} \wedge b,
$$

the set $\Omega_n := \left[\int_a^b \left(|\beta_s|+\sigma_s^2\right)ds < n\right]$ and the Itô's process

$$
X_r^{(n)} = \xi + \int_a^r \beta_s 1_{(a,S_n](s)}ds + \int_a^r \sigma_s 1_{(a,S_n](s)}dB_s, \quad r\in[a,b].
$$

Then, the first part of the proof of this step and the local property (3.12) yield

$$
\begin{aligned}
f(t,X_t)1_{\Omega_n} &= f(t,X_t^{(n)})1_{\Omega_n} = 1_{\Omega_n}\left(f(a,\xi)+\int_a^t \partial_t f(s,X_s^{(n)})ds\right.\\
&\quad +\int_a^t \partial_x f(s,X_s^{(n)})\sigma_s 1_{(a,S_n](s)}dB_s + \int_a^t \partial_x f(s,X_s^{(n)})\beta_s 1_{(a,S_n](s)}ds\\
&\quad \left.+\frac{1}{2}\int_a^t \partial_{xx}f(s,X_s^{(n)})\sigma_s^2 1_{(a,S_n](s)}ds\right)\\
&= 1_{\Omega_n}\left(f(a,\xi)+\int_a^t \partial_t f(s,X_s)ds + \int_a^t \partial_x f(s,X_s)\sigma_s dB_s\right.\\
&\quad \left.+\int_a^t \partial_x f(s,X_s)\beta_s ds + \frac{1}{2}\int_a^t \partial_{xx}f(s,X_s)\sigma_s^2 ds\right) \quad w.p.1.
\end{aligned}
$$

In other words, the proof of this step is finished due to $\Omega_n \uparrow \Omega$ w.p.1.
Step 2: Now, we see that the result is true if it is satisfied for every simple process σ and $f \in C_b^{1,2}([a,b]\times\mathbb{R})$.

For $n \in \mathbb{N}$, consider the function $f^{(n)}(s,x) = f(s,x)\varphi(x/n)$, $(s,x)\in[a,b]\times\mathbb{R}$, where $\varphi \in C^\infty(\mathbb{R})$ is such that

$$
\varphi(x) = \begin{cases} 1, & \text{if } |x|\leq 1\\ 0, & \text{if } |x|\geq 2,\end{cases}
$$

which gives that $f^{(n)} \in C_b^{1,2}([a,b]\times\mathbb{R})$. Thus Step 1 leads us to write

$$
\begin{aligned}
f^{(n)}(t,X_t) &= f^{(n)}(a,\xi) + \int_a^t \partial_t f^{(n)}(s,X_s)ds + \int_a^t \partial_x f^{(n)}(s,X_s)\sigma_s dB_s\\
&\quad + \int_a^t \partial_x f^{(n)}(s,X_s)\beta_s ds + \frac{1}{2}\int_a^t \partial_{xx}f^{(n)}(s,X_s)\sigma_s^2 ds \quad w.p.1.
\end{aligned}
$$

Therefore, the result is valid because of the dominated convergence theorem applied together with the facts that X is a continuous process,

$$\left|\partial_t f^{(n)}(t,x)\right| \leq |\partial_t f(t,x)|, \quad \left|\partial_x f^{(n)}(t,x)\right| \leq |\partial_x f(t,x)| + C\,|f(t,x)| \tag{3.23}$$

and

$$\left|\partial_{xx} f^{(n)}(t,x)\right| \leq |\partial_{xx} f(t,x)| + C\left(|f(t,x)| + |\partial_x f(t,x)|\right), \tag{3.24}$$

for some constant $C > 0$.

Step 3: Finally, from Steps 1 and 2, we only need to state the result for a simple and $\mathcal{F}_t$-adapted process σ, and $f \in C_b^{1,2}([a,b] \times \mathbb{R})$.

Note that it is enough to analyze this case for $f \in C_b^{1,2}([a,b] \times \mathbb{R})$ and $\sigma_s = \sigma 1_{[a,t]}(s)$, $s \in [a,b]$, where σ is a $\mathcal{F}_a$-measurable random variable, since the involved integrals in the Itô's formula are linear operators. Let $\pi = \{a = t_0 < t_1 < \ldots < t_n = t\}$ be a partition of the interval $[a,t]$. Then,

$$\begin{aligned} f(t,X_t) - f(a,X_a) &= \sum_{i=1}^{n} \left[f(t_i, X_{t_i}) - f(t_{i-1}, X_{t_{i-1}})\right] = \sum_{i=1}^{n} \left[f(t_i, X_{t_i}) - f(t_{i-1}, X_{t_i})\right] \\ &\quad + \sum_{i=1}^{n} \left[f(t_{i-1}, X_{t_i}) - f(t_{i-1}, X_{t_{i-1}})\right] := I_1 + I_2. \end{aligned} \tag{3.25}$$

By the mean value theorem, we have that there is $\theta_i \in (t_{i-1}, t_i)$, $i = 1, \ldots, n$, such that

$$\begin{aligned} I_1 &= \sum_{i=1}^{n} \partial_t f(\theta_i, X_{t_i})\,(t_i - t_{i-1}) \\ &= \sum_{i=1}^{n} \left(\partial_t f(\theta_i, X_{t_i}) - \partial_t f(t_i, X_{t_i})\right)(t_i - t_{i-1}) + \sum_{i=1}^{n} \partial_t f(t_i, X_{t_i})\,(t_i - t_{i-1}). \end{aligned}$$

Hence, using that $f \in C_b^{1,2}([a,b] \times \mathbb{R})$, we obtain

$$\lim_{|\pi| \to 0} I_1 = \int_a^t \partial_t f(s, X_s)ds, \quad \text{w.p.1.} \tag{3.26}$$

From Taylor's formula, it follows that there exist $\eta_i \in (0,1)$, $i \in \{1, \ldots, n\}$ such that

$$\begin{aligned} I_2 &= \sum_{i=1}^{n} \partial_x f(t_{i-1}, X_{t_{i-1}})\left(X_{t_i} - X_{t_{i-1}}\right) \\ &\quad + \frac{1}{2}\sum_{i=1}^{n} \partial_{xx} f\left(t_{i-1}, X_{t_{i-1}} + \eta_i\left(X_{t_i} - X_{t_{i-1}}\right)\right)\left(X_{t_i} - X_{t_{i-1}}\right)^2 \\ &= \sum_{i=1}^{n} \partial_x f(t_{i-1}, X_{t_{i-1}})\left(X_{t_i} - X_{t_{i-1}}\right) + \frac{1}{2}\sum_{i=1}^{n} \partial_{xx} f\left(t_{i-1}, X_{t_{i-1}}\right)\left(X_{t_i} - X_{t_{i-1}}\right)^2 \\ &\quad + \frac{1}{2}\sum_{i=1}^{n} \left\{\partial_{xx} f\left(t_{i-1}, X_{t_{i-1}} + \eta_i\left(X_{t_i} - X_{t_{i-1}}\right)\right) - \partial_{xx} f\left(t_{i-1}, X_{t_{i-1}}\right)\right\}\left(X_{t_i} - X_{t_{i-1}}\right)^2 \\ &:= I_{2,1} + I_{2,2} + I_{2,3}. \end{aligned} \tag{3.27}$$

Now, we deal with $I_{2,3}$. Note that (3.20) and the hypotheses of this Step imply

$$\sum_{i=1}^{n}\left(X_{t_i}-X_{t_{i-1}}\right)^2 = \sum_{i=1}^{n}\left(\int_{t_{i-1}}^{t_i}\beta_s ds\right)^2+2\sigma\sum_{i=1}^{n}\left(\int_{t_{i-1}}^{t_i}\beta_s ds\right)\left(B_{t_i}-B_{t_{i-1}}\right) +\sigma^2\sum_{i=1}^{n}\left(B_{t_i}-B_{t_{i-1}}\right)^2.$$

Therefore, (1.14) allows us to conclude

$$\sum_{i=1}^{n}\left(X_{t_i}-X_{t_{i-1}}\right)^2\to\sigma^2(t-a),\quad \text{as}\quad |\pi|\to 0,\quad \text{in probability.} \tag{3.28}$$

Consequently, utilizing that $f\in C_b^{1,2}([a,b]\times\mathbb{R})$ again, we can state

$$\lim_{|\pi|\to 0} I_{2,3}=0 \quad \text{in probability} \tag{3.29}$$

Concerning $I_{2,2}$, we are able to write

$$I_{2,2} = \frac{1}{2}\sum_{i=1}^{n}\partial_{xx}f\left(t_{i-1},X_{t_{i-1}}\right)\left\{\left(X_{t_i}-X_{t_{i-1}}\right)^2-\sigma^2\left(t_i-t_{i-1}\right)\right\} +\frac{\sigma^2}{2}\sum_{i=1}^{n}\partial_{xx}f\left(t_{i-1},X_{t_{i-1}}\right)\left(t_i-t_{i-1}\right).$$

Thus, (3.28) and $f\in C_b^{1,2}([a,b]\times\mathbb{R})$ give

$$\lim_{|\pi|\to 0} I_{2,2}=\frac{\sigma^2}{2}\int_a^t\partial_{xx}f\left(s,X_s\right)ds \quad \text{in probability.} \tag{3.30}$$

In order to finish the proof, we must analyze the term $I_{2,1}$. Note that (3.20) and (3.27) yield

$$I_{2,1}=\sum_{i=1}^{n}\partial_x f(t_{i-1},X_{t_{i-1}})\left[\int_{t_{i-1}}^{t_i}\beta_s ds+\sigma\left(B_{t_i}-B_{t_{i-1}}\right)\right],$$

which together with (3.17) leads to

$$\lim_{|\pi|\to 0} I_{2,1}=\int_a^t\partial_x f(s,X_s)\beta_s ds+\int_a^t\partial_x f(s,X_s)\sigma dB_s \quad \text{in probability.}$$

Hence, from (3.25)-(3.27), (3.29) and (3.30), we have that (3.21) is satisfied for every simple $\mathcal{F}_t$-adapted process σ and $f\in C_b^{1,2}([a,b]\times\mathbb{R})$. Finally, the result is true due to Steps 1 and 2. □

As a consequence of the proof of Theorem 40, we have that the integral $\frac{1}{2}\int_0^t\partial_{xx}f(s,X_s)\sigma_s^2ds$ appears on the right-hand side of (3.21) since the quadratic variation of B on an interval $[a,t]$ agrees with $t-a$, which is established in Proposition 7. Also, if σ is equal to zero in (3.20), then (3.21) becomes

$$f(t,X_t)=f(a,\xi)+\int_0^t\partial_t f(s,X_s)ds+\int_0^t\partial_x f(s,X_s)\beta_s ds,$$

which follows from the fundamental theorem of calculus. In this case, $\partial_{xx}f$ is not involve in last equality because the quadratic variation of the map $t \mapsto t$ is zero. It means, for a partition $\pi = \{a = t_0 < t_1 < \ldots < t_n\}$ of $[a, b]$, we get

$$\lim_{|\pi|\to 0} \sum_{i=1}^{n} (t_i - t_{i-1})^2 \leq |\pi|(b-a) = 0.$$

In the same way, we can establish

$$\lim_{|\pi|\to 0} \sum_{i=1}^{n} (t_i - t_{i-1}) \left(B_{t_i} - B_{t_{i-1}}\right) \leq \lim_{|\pi|\to 0} \left(\sup_{|t-s|<|\pi|} |B_t - B_s| \right) (b-a) = 0. \tag{3.31}$$

Thus, we can write (3.21) by means of a shorter expression, which is easy to remember. It means,

$$df(t, X_t) = \partial_t f(t, X_t)dt + \partial_x f(t, X_t)dX_t + \frac{1}{2}\partial_{xx} f(t, X_t)(dX_t)^2$$

with

$$dX_t = \beta_t dt + \sigma_t dB_t \tag{3.32}$$

and $(dX_t)^2$ is calculated using the equalities $(dW_t)^2 = dt$ and $dt dW_t = dW_t dt = 0$, which are motivated by (3.31) and Proposition 7.

Now, the aim is to give a multidimensional Itô's formula. Towards this end we need to introduce a d-dimensional Brownian motion.

Definition 36 *We say that an $\mathbb{R}^d$-valued stochastic process $W = \{(W_t^{(1)}, \ldots, W_t^{(d)}) : t \in [0,T]\}$ is a d-dimensional Brownian motion with respect to the filtration $(\mathcal{F}_t)_{t\in[0,T]}$ if $W_t^{(1)}, \ldots, W_t^{(d)}$ are d independent real-valued $\mathcal{F}_t$-Brownian motions.*

According to Section 1.1.1, for $s, t \in [0, T]$, $s < t$, $W_t - W_s$ is independent of the σ-algebra $\mathcal{F}_s$ and $W_t - W_s$ is a Gaussian random vector with mean 0 and covariance $(t-s)I_d$, where I_d is the identity matrix on $\mathbb{R}^d$ (compare it with Definition 11).

Now, we are ready to state the multidimensional Itô's formula, whose demonstration is base on Taylor's theorem for functions on $\mathbb{R}^d$ to $\mathbb{R}$. We omit the proof because it is quite similar to those of Theorems 40 and 53 with a more cumbersome notation, as the reader can easily realize.

Theorem 41 *Let $(\mathcal{F}_t)_{t\in[0,T]}$ be a filtration satisfying the usual conditions, W a d-dimensional $\mathcal{F}_t$-Brownian motion, $i, j \in \{1, \ldots, d\}$, $\beta^i \in \mathcal{L}_a(\Omega; L^1([a,b]))$, $\sigma^{i,j} \in \mathcal{L}_a(\Omega; L^2([a,b]))$ and*

$$X_t^{(i)} = \xi^{(i)} + \int_a^t \beta_s^i ds + \sum_{j=1}^{d} \int_a^t \sigma_s^{i,j} dW_s^{(j)}, \quad t \in [a,b], \tag{3.33}$$

where $\xi^{(i)}$ is an $\mathcal{F}_a$-measurable random variable. Also let $f \in C^{1,2}([a,b] \times \mathbb{R}^d)$. Then,

$$\begin{aligned} f(t, X_t) &= f(a, \xi) + \int_a^t \partial_t f(s, X_s)ds + \sum_{j=1}^{d} \int_a^t \partial_{x_j} f(s, X_s)dX_s^{(j)} \\ &\quad + \frac{1}{2} \sum_{i,j=1}^{d} \int_a^t \partial_{x_i x_j}^2 f(s, X_s)dX_s^{(i)} dX_s^{(j)}, \quad t \in [a,b]. \end{aligned} \tag{3.34}$$

Here, the product $dX_s^{(i)}dX_s^{(j)}$ is obtained by via the Itô multiplication table

$\times$	$dW_t^{(i)}$	dt
$dW_t^{(j)}$	$\delta_{ij}dt$	0
dt	0	0

We remark that, in the last table, it is applied the multiplication rule $dW_t^{(i)}dW_t^{(j)} = 0$, for $i \neq j$, which is a symbolic expression motivated by the fact that, for a partition $\pi = \{a = t_0 < t_1 < \ldots < t_n\}$ of $[a, b]$,

$$L^2(\Omega) - \lim_{|\pi|\to 0} \sum_{\ell=1}^{n} \left(W_{t_\ell}^{(i)} - W_{t_{\ell-1}}^{(i)}\right)\left(W_{t_\ell}^{(j)} - W_{t_{\ell-1}}^{(j)}\right) = 0.$$

Note that this can be proven by using that $W^{(i)}$ and $W^{(j)}$ are independent, and by proceeding similarly as in the proof of Proposition 7 (see also (3.31)).

Examples 2 *Here we give two applications of Itô's formula.*

i) Let X and Y be two Itô processes of the form (3.20) (or (3.32)). Then, Theorem 41 applied to $f(x, y) = xy$ implies that the integration by parts formula

$$d(XY)_t = Y_t dX_t + X_t dY_t + dX_t dY_t$$

holds. That is,

$$X_t Y_t = X_a Y_a + \int_a^t Y_s dX_s + \int_a^t X_s dY_s + \int_a^t dY_s dX_s, \quad t \in [a, b]. \tag{3.35}$$

ii) Consider the Itô process X introduced in (3.20) and $f(x) = e^x$, $x \in \mathbb{R}$. Then, Theorem 40 gives that the process

$$Z_t = \exp\left(X_t - \frac{1}{2}\int_a^t \sigma_s^2 ds\right), \quad t \in [a, b]$$

is a solution of the linear stochastic differential equation

$$\begin{aligned} dZ_t &= \beta_t Z_t ds + \sigma_t Z_t dB_t, \quad t \in (a, b], \\ Z_a &= \exp(\xi). \end{aligned} \tag{3.36}$$

Indeed, using (3.21) and $\frac{de^x}{dx} = e^x$, $x \in \mathbb{R}$, we have

$$\begin{aligned} Z_t &= \exp(\xi) + \int_a^t \sigma_s Z_s dB_s + \int_a^t \left(\beta_s - \frac{\sigma_s^2}{2}\right) Z_s ds + \frac{1}{2}\int_a^t \sigma_s^2 Z_s ds \\ &= \exp(\xi) + \int_a^t \sigma_s Z_s dB_s \int_a^t \beta_s Z_s ds, \quad t \in [a, b], \end{aligned}$$

which implies that our claim is satisfied.

Now, we see that equation (3.36) has a unique solution through Theorem 41. To do so, consider a solution X of equation (3.36) and $Y_t = \frac{1}{2}\int_a^t \sigma_s^2 ds - \int_a^t \sigma_s dB_s$, $t \in [a, b]$.

Then, if $f(x,y) = x\exp(y)$, we can write from Theorem 41

$$\begin{aligned}
X_t \exp(Y_t) &= \exp(\xi) + \int_a^t X_s \exp(Y_s) dY_s + \int_a^t \exp(Y_s) dX_s \\
&\quad + \frac{1}{2}\int_a^t X_s \exp(Y_s) dY_s dY_s + \int_a^t \exp(Y_s) dY_s dX_s \\
&= \exp(\xi) + \int_a^t X_s \exp(Y_s)\left(-\sigma_s dB_s + \frac{1}{2}\sigma_s^2 ds\right) \\
&\quad + \int_a^t \exp(Y_s)\left(\sigma_s X_s dB_s + \beta_s X_s ds\right) \\
&\quad + \frac{1}{2}\int_a^t X_s \exp(Y_s)\sigma_s^2 ds - \int_a^t \exp(Y_s)\sigma_s^2 X_s ds \\
&= \exp(\xi) + \int_a^t \beta_s X_s \exp(Y_s) ds, \quad t \in [a,b].
\end{aligned}$$

In other words, we have that $X_\cdot \exp(Y_\cdot)$ is the pathwise solution (i.e., ω by ω) of the equation

$$g_t = \exp(\xi) + \int_a^t \beta_s g_s ds, \quad t \in [a,b],$$

whose solution is $g_t = \exp(\xi + \int_a^t \beta_s ds)$. Consequently,

$$X_t = \exp\left(\xi + \int_a^t \beta_s ds + \int_a^t \sigma_s dB_s - \frac{1}{2}\int_a^t \sigma_s^2 ds\right), \quad t \in [a,b].$$

Thus, the uniqueness for the solution to equation (3.36) is true.

In order to give an Itô's formula for $\mathcal{F}_t$-adapted random fields, we analyze the following Fubini's theorem for the Itô's integral with respect to the Brownian motion.

Theorem 42 *Let $\mathcal{F}_a$ contain all the P-null-sets and $f : \Omega \times [a,b] \times \mathbb{R} \to \mathbb{R}$ a random field such that:*

i) There exists $g \in L^1(\mathbb{R}, \lambda)$, where λ is the Lebesgue measure, so that $|f(t,x)| \le g(x)$ for all $(\omega, t, x) \in \Omega \times [a,b] \times \mathbb{R}$.

ii) $f(t,y)$ is an $\mathcal{F}_t$-adapted process, for every $y \in \mathbb{R}$.

iii) $f(\cdot, y)$ is left-continuous and $f(r,\cdot)$ is continuous for all $(\omega, r, y) \in \Omega \times [a,b] \times \mathbb{R}$.

Then, for each $t \in [a,b]$ there exists a measurable function $Y^t : \Omega \times \mathbb{R} \to \mathbb{R}$, belonging to $L^1(\mathbb{R}, \lambda)$ w.p.1, such that

$$\int_a^t f(r,x) dB_r = Y^t(x) \quad \text{w.p.1 for almost all } x \in \mathbb{R}, \tag{3.37}$$

and

$$\int_a^t \int_{\mathbb{R}} f(r,x) dx dB_r = \int_{\mathbb{R}} Y^t(x) dx \quad \text{w.p.1.} \tag{3.38}$$

Remarks 3 *i) The set of probability 1 where equality (3.37) (resp. (3.38)) holds depends on t and x (resp. on t).*

ii) If $x \mapsto \int_a^t f(r,x)dB_r$ is $\mathcal{F}\otimes\mathcal{B}(\mathbb{R})$-measurable, then

$$\int_a^t \int_{\mathbb{R}} f(r,x)dxdB_r = \int_{\mathbb{R}}\int_a^t f(r,x)dB_r dx \quad w.p.1.$$

iii) Let $M>0$ and $\pi = \{-M = y_0 < y_1 < \ldots < y_n = M\}$ a partition of the interval $[-M,M]$. Then, the process

$$r \mapsto \int_{\mathbb{R}} \sum_{j=1}^{n} f(r,y_j)1_{(y_{j-1},y_j]}(x)dx = \sum_{j=1}^{n} f(r,y_j)(y_j - y_{j-1})$$

is measurable and $\mathcal{F}_t$-adapted due to Hypotheses ii) and iii), and Proposition 2. Hence, using Hypothesis iii) again and letting $|\pi|$ go to zero, we have that $r\mapsto \int_{-M}^{M} f(r,x)dx$ is also a measurable and $\mathcal{F}_t$-adapted process. In this way, by Condition i), we have that the left-hand sides of (3.37) and (3.38) are well-defined.

Proof. Fix $t\in[a,b]$ and let $\pi_n = \{a = t_0 < t_1 < \ldots < t_n = t\}$ be a partition of $[a,t]$ such that $|\pi_n| < 1/n$. Set $f_n(r,x) = \sum_{i=1}^n f(t_{i-1},x)1_{(t_{i-1},t_i]}(r)$, $(r,x)\in[a,t]\times\mathbb{R}$ and $n\in\mathbb{N}$. Thus, Hypotheses *ii)* and *iii)* imply that

$$Y_n^t(x) := \int_a^t f_n(r,x)dB_r = \sum_{i=1}^{n} f_n(t_{i-1},x)\left(B_{t_i} - B_{t_{i-1}}\right)$$

is an $\mathcal{F}\otimes\mathcal{B}(\mathbb{R})$-measurable function such that

$$\int_a^t \int_{\mathbb{R}} f_n(r,x)dxdB_r = \int_{\mathbb{R}} Y_n^t(x)dx \quad \text{for } \omega\in\Omega. \tag{3.39}$$

Consequently, the dominated convergence theorem, together with Condition *i)*, leads us to get

$$\begin{aligned}\int_{\mathbb{R}} E\left(|Y_n^t(x) - Y_m^t(x)|\right)dx &= \int_{\mathbb{R}} E\left(\left|\int_a^t [f_n(r,x) - f_m(r,x)]\, dB_r\right|\right)dx \\ &\le \int_{\mathbb{R}} \left(E\int_a^t [f_n(r,x) - f_m(r,x)]^2\, dr\right)^{1/2} dx \to 0\end{aligned}$$

as $m,n\to\infty$. Therefore, there exists $Y^t\in L^1(\Omega\times\mathbb{R})$ and a subsequence $\{n_k : k\in\mathbb{N}\}\subset\mathbb{N}$ such that

$$\begin{aligned}Y^t(x) &= \lim_{k\to\infty} Y_{n_k}^t(x) = \lim_{k\to\infty}\int_a^t f_{n_k}(r,x)dB_r \\ &= \int_a^t f(r,x)dB_r \quad \text{w.p.1 for almost all } x\in\mathbb{R}\end{aligned}$$

and, from (3.39),

$$\begin{aligned}\int_{\mathbb{R}} Y^t(x)dx &= L^1(\Omega) - \lim_{n\to\infty}\int_{\mathbb{R}} Y_n^t(x)dx \\ &= L^1(\Omega) - \lim_{n\to\infty}\int_a^t\int_{\mathbb{R}} f_n(r,x)dxdB_r = \int_a^t\int_{\mathbb{R}} f(r,x)dxdB_r,\end{aligned}$$

where last equality follows by using the dominated convergence theorem and Hypothesis *i)* again. So, we have finished the proof and the result holds □

In order to avoid the continuity of f in Theorem 42, we need to deal with the predictable σ-algebra, which is introduced as follows:

Definition 37 *We introduce the predictable* σ*-algebra on* $\Omega \times [a,b]$ *(denoted by* $\mathcal{P}([a,b])$*) as the* σ*-algebra generated by all the set of the form* $F \times (u,v]$*, with* $F \in \mathcal{F}_u$ *and* $a \le u < v \le b$*, together with the sets* $F \times \{a\}$*,* $F \in \mathcal{F}_a$*. Moreover, we say that a process* $X : \Omega \times [a,b] \to \mathbb{R}$ *is predictable if it is* $\mathcal{P}([a,b])$*-measurable.*

Note that the class monotone lemma (i.e., Theorem 97) implies that a predictable process is also an $\mathcal{F}_t$-adapted and measurable process.

Theorem 43 *Let* $(\Omega, \mathcal{F}_a, P)$ *be complete,* $(X, \mathcal{X}, \mu)$ *a measure space and* $f : \Omega \times [a,b] \times X \to \mathbb{R}$ *a* $\mathcal{P}([a,b]) \otimes \mathcal{X}$*-measurable random field such that there exists a non-negative function* $g \in L^1(X,\mu)$ *such that* $|f(t,x)| \le g(x)$*, for all* $(\omega,t,x) \in \Omega \times [a,b] \times X$*. Then, for each* $t \in [a,b]$ *there exists a measurable function* $Y^t : \Omega \times X \to \mathbb{R}$ *belonging to* $L^1(X,\mu)$ *w.p.1 such that*

$$\int_a^t f(r,x)dB_r = Y^t(x) \quad \text{w.p.1 for almost all } x \in X, \tag{3.40}$$

and

$$\int_a^t \int_X f(r,x)\mu(dx)dB_r = \int_X Y^t(x)\mu(dx) \quad \text{w.p.1.} \tag{3.41}$$

Remarks 4 *i) In this case, we have that the right-hand side of (3.41) is well-defined since the Fubini's theorem establishes that the process* $\int_X f(\cdot,x)\mu(dx)$ *is* $\mathcal{P}([a,b])$*-measurable.*

ii) When the map $x \mapsto \int_a^t f(r,x)dB_r$ *is* $\mathcal{F} \otimes \mathcal{X}$*-measurable, we have*

$$\int_a^t \int_X f(r,x)\mu(dx)dB_r = \int_X \int_a^t f(r,x)dB_r\mu(dx) \quad \text{w.p.1.}$$

iii) The Fubini's theorem for stochastic integrals with respect to semimartingales in Hilbert space is proven in León [109] and next proof follows the ideas given in this paper. The existence of the measurable function Y^t *satisfying (3.40) and (3.41) has been realized by Jacod [89]. The reader interested in Fubini's theorems for another stochastic integrals can consult the references in [109].*

Proof. Fix $t \in (a,b]$. So we divide the proof into two steps.
Step 1: Here we assume that there is a sequence $\{f_n : n \in \mathbb{N}\}$ of $\mathcal{P}([a,b]) \otimes \mathcal{X}$-measurable random fields for which the result is true and, for every $(\omega,s,x) \in \Omega \times [a,b] \times X$, $|f_n(t,x)| \le g(x)$ and $|f_n(t,x) - f(t,x)| \to 0$ as $n \to \infty$. Under this assumption, we have a sequence $\{Y_n^t : n \in \mathbb{N}\}$ of $\mathcal{F} \otimes \mathcal{X}$-measurable functions such that (3.40) and (3.41) hold when we write f_n and Y_n^t instead of f and Y^t. Hence,

$$\begin{aligned}
&\int_{\mathbb{R}} E\left(\left|Y_n^t(x) - Y_m^t(x)\right|\right)\mu(dx) \\
&= \int_{\mathbb{R}} E\left(\left|\int_a^t [f_n(r,x) - f_m(r,x)]\,dB_r\right|\right)\mu(dx) \\
&\le \int_{\mathbb{R}} \left(E\int_a^t [f_n(r,x) - f_m(r,x)]^2\,dr\right)^{1/2} \mu(dx) \to 0 \quad \text{as } m,n \to \infty.
\end{aligned}$$

So, we have that the result is satisfied by proceeding as in the last part of Theorem 42.
Step 2: Let $c \in \mathbb{R}$ and $n \in \mathbb{N}$. Consider the family

$$\mathcal{L} := \{A \in \mathcal{P}([a,b]) \otimes \mathcal{X} : \text{the theorem holds for } c\,(1_A(\omega,r,x) \wedge ng(x))\}.$$

The set $\mathcal{H} = \{F \times B \subset \Omega \times [a,b] \times X : F \in \mathcal{P}([a,b]) \text{ and } B \in \mathcal{X}\}$ is closed under intersection and included in $\mathcal{L}$ with

$$Y^t(x) = c\,(1_B(x) \wedge ng(x)) \int_a^t 1_F(r)dB_r.$$

Consequently, Step 1 and the class monotone lemma (i.e., Theorem 97) yield that $\mathcal{L}$ agrees with $\mathcal{P}([a,b]) \otimes \mathcal{X}$. Thus, for $c \in \mathbb{R}$ and $A \in \mathcal{P}([a,b]) \otimes \mathcal{X}$ such that $|c1_A(\omega, r, x)| \le g(x)$, we also have $|c(1_A(\omega, r, x) \wedge ng(x))| \le g(x)$, which allows us to conclude that the result is also true for the process $c1_A$.

Finally, the result follows from Proposition 11.7 in [187], which states that any nonnegative measurable function is the limit of an increasing sequence of simple functions. □

Now, we are ready to state the Itô's formula for $\mathcal{F}_t$-adapted random fields. The proof of the following result follows the ideas developed by Ocone and Pardoux [164] in order to obtain a generalized Itô-Ventzell formula, which includes random fields and processes not necessarily adapted to the underlying filtration.

Theorem 44 *Suppose that $(\mathcal{F}_t)_{t\in[0,T]}$ satisfies the usual conditions. Let X be an Itô process of the form (3.20) and $f : \Omega \times [a,b] \times \mathbb{R} \to \mathbb{R}$ a random field such that $f(\cdot, y)$ is an $\mathcal{F}_t$-adapted process, for every $y \in \mathbb{R}$, and $f \in C^{1,2}([a,b] \times \mathbb{R})$, for all $\omega \in \Omega$. Then, $f(\cdot, X_\cdot)$ is an Itô process such that*

$$\begin{aligned} f(t, X_t) &= f(a, \xi) + \int_a^t \partial_t f(s, X_s)ds + \int_a^t \sigma_s \partial_x f(s, X_s)dB_s \\ &\quad + \int_a^t \left(\beta_s \partial_x f(s, X_s) + \frac{\sigma_s^2}{2}\partial_{xx}^2 f(s, X_s)\right) ds, \quad t \in [a,b]. \end{aligned} \tag{3.42}$$

Remark 51 *Let M and π be as in Remarks 3. Then,*

$$\sum_{j=1}^n f(s, y_j)1_{(y_{j-1}, y_j]}(X_s) \to 1_{(-M,M]}(X_s)f(s, X_s) \quad as\ |\pi| \to 0.$$

Hence, the fact that $f(\cdot, y_j)$ is a continuous and $\mathcal{F}_t$-adapted process implies that the process $s \mapsto f(s, X_s)$ is a measurable and $\mathcal{F}_t$-adapted process. Actually, it is a continuous and $\mathcal{F}_t$-adapted process. Similarly, we also have that $\partial_s f(s, X_s)$, $\partial_x f(s, X_s)$ and $\partial_{xx}^2 f(s, X_s)$ are continuous and $\mathcal{F}_t$-adapted processes. Thus, the right-hand side of the Itô's formula (3.42) is well-defined.

Proof. Let $\phi, \psi \in C^\infty(\mathbb{R})$ be two nonnegative functions with compact support such that $\int_{\mathbb{R}} \phi(x)dx = 1$ and $\psi(x) = 1$ for $|x| \le 1$, and $\psi(x) = 0$ for $|x| \ge 2$. Set

$$\phi_n(x) = n\phi(nx) \quad \text{and} \quad g_n(t,x) = f(t,x)\psi\left(\frac{x}{n}\right), \quad \text{for } (n,t,x) \in \mathbb{N} \times [a,b] \times \mathbb{R}.$$

Then, (3.35) allows us to get

$$\begin{aligned} g_m(t,y)\phi_n(X_t - y) &= g_m(a,y)\phi_n(\xi - y) + \int_a^t \sigma_s g_m(s,y)\phi_n'(X_s - y)dB_s \\ &\quad + \int_a^t \Big(\phi_n(X_s - y)\partial_t g_m(s,y) + \beta_s g_m(s,y)\phi_n'(X_s - y) \\ &\quad + \frac{\sigma_s^2}{2} g_m(s,y)\phi_n''(X_s - y)\Big) ds, \quad t \in [a,b] \text{ and } y \in \mathbb{R}. \end{aligned} \tag{3.43}$$

Fix $t \in [a,b]$. Now, we divide the proof into three steps.
Step 1: Here we assume that f is bounded by a nonnegative constant and $\sigma = F1_{(c,d]}$, where F is a bounded and $\mathcal{F}_c$-measurable random variable and $c, d \in [a,b]$.

In this step, we have that there is a constant $C > 0$ such that

$$|\sigma_s g_m(s,y)\phi_n'(X_s - y)| \leq Cn^2 1_{[-2m,2m]}(y), \quad \text{for all } (\omega, s, y) \in \Omega \times [a,b] \times \mathbb{R}.$$

Therefore, by Theorem 42, we can find a measurable function $Y^t : \Omega \times \mathbb{R} \to \mathbb{R}$ belonging to $L^1(\mathbb{R})$ w.p.1 such that

$$Y^t(y) = \int_a^t \sigma_s g_m(s,y)\phi_n'(X_s - y)dB_s \quad \text{w.p.1 for almost all } y \in \mathbb{R}$$

and

$$\int_a^t \int_{\mathbb{R}} \sigma_s g_m(s,y)\phi_n'(X_s - y)dydB_s = \int_{\mathbb{R}} Y^t(y)dy \quad \text{w.p.1},$$

which, together with (3.43), yield

$$\begin{aligned}
&\int_{\mathbb{R}} g_m(t,y)\phi_n(X_t - y)dy \\
&= \int_{\mathbb{R}} g_m(a,y)\phi_n(\xi - y)dy + \int_a^t \int_{\mathbb{R}} \sigma_s g_m(s,y)\phi_n'(X_s - y)dydB_s \\
&\quad + \int_a^t \int_{\mathbb{R}} \phi_n(X_s - y)\partial_t g_m(s,y)dyds + \int_a^t \int_{\mathbb{R}} \beta_s g_m(s,y)\phi_n'(X_s - y)dyds \\
&\quad + \frac{1}{2}\int_a^t \int_{\mathbb{R}} \sigma_s^2 g_m(s,y)\phi_n''(X_s - y)dyds, \quad \text{w.p.1.}
\end{aligned}$$

Hence, letting $m \to \infty$, and using that ϕ a ψ have compact support, together with the dominated convergence theorem, we conclude

$$\begin{aligned}
&\int_{\mathbb{R}} f(t,y)\phi_n(X_t - y)dy \\
&= \int_{\mathbb{R}} f(a,y)\phi_n(\xi - y)dy + \int_a^t \int_{\mathbb{R}} \sigma_s f(s,y)\phi_n'(X_s - y)dydB_s \\
&\quad + \int_a^t \int_{\mathbb{R}} \phi_n(X_s - y)\partial_t f(s,y)dyds + \int_a^t \int_{\mathbb{R}} \beta_s f(s,y)\phi_n'(X_s - y)dyds \\
&\quad + \frac{1}{2}\int_a^t \int_{\mathbb{R}} \sigma_s^2 f(s,y)\phi_n''(X_s - y)dyds, \quad \text{w.p.1.}
\end{aligned}$$

Consequently, the integration by parts formula becomes the last equality in

$$\begin{aligned}
&\int_{\mathbb{R}} f(t,y)\phi_n(X_t - y)dy \\
&= \int_{\mathbb{R}} f(a,y)\phi_n(\xi - y)dy + \int_a^t \int_{\mathbb{R}} \sigma_s \partial_x f(s,y)\phi_n(X_s - y)dydB_s \\
&\quad + \int_a^t \int_{\mathbb{R}} \phi_n(X_s - y)\partial_t f(s,y)dyds + \int_a^t \int_{\mathbb{R}} \beta_s \partial_x f(s,y)\phi_n(X_s - y)dyds \\
&\quad + \frac{1}{2}\int_a^t \int_{\mathbb{R}} \sigma_s^2 \partial_{xx}^2 f(s,y)\phi_n(X_s - y)dyds, \quad \text{w.p.1.}
\end{aligned}$$

Thus, the facts that, for $\omega \in \Omega$ and a continuous function $h : [a,b]\times\mathbb{R} \to \mathbb{R}$, $\int_{\mathbb{R}} h(s,y)\phi_n(X_s - y)dy \to h(s,X_s)$ and that h is bounded on compact sets of $[a,b]\times\mathbb{R}$ imply that (3.42) hold under the assumptions of this step.

Step 2: In this part of the proof, we assume that $f \in C_b^{1,2}([a,b]\times\mathbb{R})$ and that σ is in the space $\mathcal{L}_a(\Omega; L^2([a,b]))$.

By Corollary 19, there is a sequence $\{\sigma^{(n)} : n \in \mathbb{N}\}$ of simple and $\mathcal{F}_t$-adapted processes of the form

$$\sigma_s^{(n)} = \sum_{i=1}^{k_n} F_i^{(n)} 1_{]t_{i-1}^n, t_i^n]}(s), \quad s \in [a,b],$$

such that

$$\int_a^b \left(\sigma_s^{(n)} - \sigma_s\right)^2 ds \to 0, \quad \text{as } n \to \infty, \text{ w.p.1.}$$

Let $m \in \mathbb{N}$ and

$$\sigma_s^{(n,m)} = \sum_{i=1}^{k_n} \left[(F_i^{(n)} \wedge m) \vee (-m)\right] 1_{]t_{i-1}^n, t_i^n]}(s), \quad s \in [a,b],$$

Therefore, from Step 1, we have that (3.42) is satisfied when we change σ by $\sigma^{(n,m)}$. In this way, (3.42) is true in this case by using (3.22) and letting first $m \to \infty$, and then taking the limit as $n \to \infty$.

Step 3: Finally, we deal with the general case. For $m \in \mathbb{N}$, we consider the random field $f^{(m)}(t,x) = f(t,x)\psi(x/m)$.

From Step 2, we get that (3.42) holds when we write $f^{(m)}$ instead of f. This gives that the proof is finished because we can let $m \to \infty$, and use the dominated convergence theorem, together with (3.23) and (3.24). □

An immediate consequence of Theorem 44 is the following result.

Corollary 23 *Assume that $(\mathcal{F}_t)_{t\in[0,T]}$ satisfies the usual conditions and that Ω_0 belongs to $\mathcal{F}$. Let X be an Itô process of the form (3.20), $P(\Omega_0) = 1$ and $f : \Omega\times[a,b]\times\mathbb{R} \to \mathbb{R}$ a random field such that, for every $y \in \mathbb{R}$, $f(\cdot,y)$ is an $\mathcal{F}_t$-adapted process and $f \in C^{1,2}([a,b]\times\mathbb{R})$ for $\omega \in \Omega_0$. Then, $f(\cdot,X_\cdot)$ is an Itô process such that*

$$\begin{aligned} f(t,X_t) &= f(0,\xi) + \int_a^t \partial_s f(s,X_s)ds + \int_a^t \sigma_s \partial_x f(s,X_s) 1_{\Omega_0} dB_s \\ &\quad + \int_a^t \left(\beta_s \partial_x f(s,X_s) + \frac{\sigma_s^2}{2}\partial_{xx}^2 f(s,X_s)\right) ds, \quad t \in [a,b]. \end{aligned}$$

Using Theorem 41 and proceeding as in the proof of Theorem 44, we have the following Itô's type formula.

Theorem 45 *Assume that $(\mathcal{F}_t)_{t\in[0,T]}$ satisfies the usual conditions. Let W be a d-dimensional $\mathcal{F}_t$-Brownian motion and $X^{(i)}$, $i = 1,\dots,d$, the Itô's processes defined in (3.33). Moreover, assume that $\{f(t,x) : t \in [a,b]$ and $x \in \mathbb{R}^d\}$ is an $\mathcal{F}_t$-adapted and continuous (in (t,x)) random field such that*

i) For $(\omega,t) \in \Omega\times[a,b]$, $f(t,\cdot)$ belongs to $C^2(\mathbb{R}^d)$.

ii) For $(\omega,t,x) \in \Omega\times[a,b]\times\mathbb{R}^d$,

$$f(t,x) = f(a,x) + \int_a^t f^{(0)}(s,x)ds + \sum_{j=1}^d \int_a^t f^{(j)}(s,x)dW_s^{(j)},$$

where, for $j = 0, 1, \dots, d$, $f^{(j)}(\omega, \cdot)$ *is continuous,* $f^{(j)}(\omega, t, \cdot)$ *is in* $C^2(\mathbb{R}^d)$ *and* $f^{(j)}(\cdot, x)$ *is an* $\mathcal{F}_t$*-adapted process.*

Then,

$$\begin{aligned} f(s, X_t) &= f(a, \xi) + \int_a^t f^{(0)}(s, X_s) ds + \sum_{j=1}^d \int_a^t f^{(j)}(s, X_s) dW_s^{(j)} \\ &\quad + \sum_{j=1}^d \int_a^t \partial_{x_i} f(s, X_s) dX_s^{(i)} + \frac{1}{2} \sum_{i,j=1}^d \int_a^t \partial_{x_i x_j} f(s, X_s) dX_s^{(i)} dX_s^{(j)} \\ &\quad + \sum_{i,j=1}^d \int_a^t \partial_{x_i} f^{(j)}(s, X_s) dW_s^{(j)} dX_s^{(i)}, \quad t \in [a, b]. \end{aligned}$$

This Itô's type formula has been analyzed by Kunita [105] (Theorem 8.1) in the case that $W^{(i)}$ and $X^{(i)}$ are continuous semimartingales, for $i = 1, \dots, d$. In this case, $dX^{(i)} dX^{(j)}$ and $dW^{(j)} dX^{(i)}$ are changed by $d[X^{(i)}, X^{(j)}]$ and $d[W^{(j)}, X^{(i)}]$, respectively. Here, the joint quadratic variation of two continuous semimartingales Z and Y is represented by $[Z, Y]$. Moreover, in [105], the reader can find other proof of Theorem 45. The process $[Z, Y]$ is defined as

$$[Z, Y]_t = P - \lim_{|\pi| \to 0} \sum_{i=0}^{n-1} \left(Z_{t_{i+1} \wedge t} - Z_{t_i \wedge t} \right) \left(Y_{t_{i+1} \wedge t} - Y_{t_i \wedge t} \right), \quad t \in [a, b], \tag{3.44}$$

where $\pi = \{a = t_0 < t_1 < \dots < t_n = b\}$ is a partition of $[a, b]$.

Chapter 4

Divergence Operator

In this section, we consider two complete orthonormal systems $\{e_i : i \in \mathbb{N}\}$ and $\{g_i : i \in \mathbb{N}\}$ of the real separable Hilbert spaces $\mathcal{H}$ and $\mathcal{K}$, respectively, and an isonormal Gaussian process $W = \{W(h) : h \in \mathcal{H}\}$ defined on $\mathcal{H}$.

For $p \geq 1$, the divergence operator δ_p with respect to W is defined as the adjoint of the derivative operator $D : \mathbb{D}^{1,q}(\mathcal{K}) \subset L^q(\Omega;\mathcal{K}) \to L^q(\Omega;\mathcal{H}\otimes\mathcal{K})$, where $p^{-1}+q^{-1}=1$. The operator D is introduced in Section 1.5.4. Moreover, if $q=\infty$, the domain of D is

$$\mathbb{D}^{1,\infty}(\mathcal{K}) = \{F \in \mathbb{D}^{1,2}(\mathcal{K}) : \|F\|_{\mathcal{K}}, \|DF\|_{\mathcal{H}\otimes\mathcal{K}} \in L^\infty(\Omega)\}. \tag{4.1}$$

So, δ_p is a closed operator from $L^p(\Omega;\mathcal{H}\otimes\mathcal{K})$ into $L^p(\Omega;\mathcal{K})$, whose domain $\operatorname{Dom}\delta_p(\mathcal{K})$ is a dense subset of $L^p(\Omega;\mathcal{H}\otimes\mathcal{K})$ since it is the adjoint of an unbounded closed operator with dense domain (see Reed and Simon [182]).

As we have already mention in Section 1.5, in the case that $\mathcal{H} = L^2([0,T])$ and the underlying probability space $(\Omega,\mathcal{F},P)$ is the Canonical Wiener space defined in Section 1.2.1, Gaveau and Trauber [63] have obtained that the divergence operator on $L^2(\Omega\times[0,T])$ is the Skorohod integral [198], which is defined using the chaos decomposition and is an extension of the Itô's integral with respect to Brownian motion in the sense that δ_2 and the Itô's integral are the same on the space $\mathcal{L}^2_{\mathrm{a}}(\Omega\times[0,T])$ (given in Section 3.1).

Note that, by Hytönen et al. [85] (Corollary 1.3.22), $u \in L^p(\Omega;\mathcal{H}\otimes\mathcal{K})$ belongs to $\operatorname{Dom}\delta_p(\mathcal{K})$ if and only if there exists a constant $C_u > 0$ depending on u such that

$$\left|E\left(\langle u, DF\rangle_{\mathcal{H}\otimes\mathcal{K}}\right)\right| \leq C_u \|F\|_{L^q(\Omega;\mathcal{K})}, \quad \text{for all } F \in \mathbb{D}^{1,q}(\mathcal{K}).$$

For $u \in \operatorname{Dom}\delta_p(\mathcal{K})$, $\delta_p(u)$ is the only element in $L^p(\Omega;\mathcal{K})$ satisfying

$$E\left(\langle F, \delta_p(u)\rangle_{\mathcal{K}}\right) = E\left(\langle u, DF\rangle_{\mathcal{H}\otimes\mathcal{K}}\right), \quad \text{for all } F \in \mathbb{D}^{1,q}(\mathcal{K}). \tag{4.2}$$

Hence, for $1 < p < r$, the fact that $\|\cdot\|_{1,p,\mathcal{K}} \leq \|\cdot\|_{1,r,\mathcal{K}}$ implies that $\operatorname{Dom}\delta_r(\mathcal{K}) \subset \operatorname{Dom}\delta_p(\mathcal{K})$ and $\delta_p(u) = \delta_r(u)$, for all $u \in \operatorname{Dom}\delta_r(\mathcal{K})$. Also, Remark 24 yields that, for $h \in \mathcal{H}$, $k \in \mathcal{K}$ and $p > 1$, $h\otimes k$ is in $\operatorname{Dom}\delta_p(\mathcal{K})$ and $\delta_p(h\otimes k) = W(h)k$.

The operator δ_p is a continuous linear operator from $\mathbb{D}^{1,p}(\mathcal{H}\otimes\mathcal{K})$ into $L^p(\Omega;\mathcal{K})$, which is an important consequence of Meyer's inequality. The proof of this inequality is quite technical and uses tools that are not the goal of this book. The reader interested in this result can consult Maas [135] and Nualart [158], where the authors follow the approach given by Pisier [178]. Moreover, the proof of the following result is also in these references.

Proposition 62 *The operator δ_p is continuous from $\mathbb{D}^{1,p}(\mathcal{H}\otimes\mathcal{K})$ into $L^p(\Omega;\mathcal{K})$ for all $p > 1$. That is, there is a constant $c_p > 0$ such that*

$$E\left(|\delta_p(u)|^p_{\mathcal{K}}\right) \leq c_p \|u\|^p_{1,p,\mathcal{H}\otimes\mathcal{K}}, \quad \text{for all } u \in \mathbb{D}^{1,p}(\mathcal{H}\otimes\mathcal{K}).$$

We will need the following two results later on.

Lemma 36 *Let $p, q > 1$, $F \in \mathbb{D}^{1,q}$ and $u \in \operatorname{Dom}\delta_p(\mathcal{K})$ such that $p^{-1}+q^{-1}=1$, $Fu \in L^p(\Omega;\mathcal{H}\otimes\mathcal{K})$ and $(F\delta(u) - \langle DF, u\rangle_{\mathcal{H}}) \in L^p(\Omega;\mathcal{K})$. Then, Fu belongs to Dom $\delta_p(\mathcal{K})$ and*

$$F\delta_p(u) = \delta_p(Fu) + \langle u, DF\rangle_{\mathcal{H}}. \tag{4.3}$$

DOI: 10.1201/9781003484912-4

Remark 52 *We observe that on the right-hand side of (4.3), $\langle u, \cdot \rangle_{\mathcal{H}}$ stands for the operator Ξ related to u through Proposition 144.*

Proof. Let G be a smooth random variable as in (1.102). Thus, $|G|_{\mathcal{K}}$ is bounded and consequently FG belongs to the space $\mathbb{D}^{1,q}(\mathcal{K})$. Therefore, the duality relation (4.2) implies

$$E\left(\langle GF, \delta_p(u)\rangle_{\mathcal{K}}\right) = E\left(\langle u, D(FG)\rangle_{\mathcal{H}\otimes\mathcal{K}}\right).$$

Hence,

$$\begin{aligned} E\left(\langle Fu, DG\rangle_{\mathcal{H}\otimes\mathcal{K}}\right) &= E\left(\langle G, F\delta_p(u)\rangle_{\mathcal{K}} - \langle u, DF\otimes G\rangle_{\mathcal{H}\otimes\mathcal{K}}\right) \\ &= E\left(\langle G, F\delta_p(u)\rangle_{\mathcal{K}} - \langle G, \langle u, DF\rangle_{\mathcal{H}}\rangle_{\mathcal{K}}\right). \end{aligned}$$

Thus, we can utilize the duality relation (4.2) again to see that the result is satisfied. □

Lemma 37 *Let $p > 1$ and $z \in \mathcal{H}\otimes\mathcal{K}$. Then, z belongs to Dom $\delta_p(\mathcal{K})$ and*

$$\delta_p(z) = I_1(z).$$

Proof. By hypothesis, we have that $z = \sum_{i,j=1}^{\infty} c_{i,j} e_i \otimes g_j$, with $\sum_{i,j=1}^{\infty} c_{i,j}^2 < \infty$. Then, Remark 24 leads us to have that $z_n = \sum_{i,j=1}^{n} c_{i,j} e_i \otimes g_j$ is in Dom $\delta_p(\mathcal{K})$ and

$$\delta_p(z_n) = \sum_{i,j=1}^{n} c_{i,j} W(e_i) g_j = I_1(z_n), \quad \text{for every } n \in \mathbb{N}.$$

Also note that

$$\|z - z_n\|_{\mathcal{H}\otimes\mathcal{K}}^p = \left(\sum_{i>n \text{ or } j>n} c_{i,j}^2\right)^{p/2} \to 0, \quad \text{as } n \to \infty.$$

Consequently, Proposition 62 gives that $\{\delta_p(z_n) : n \in \mathbb{N}\}$ is a Cauchy sequence in $L^p(\Omega; \mathcal{K})$. Since δ_p : Dom $\delta_p(\mathcal{K}) \subset L^p(\Omega; \mathcal{H}\otimes\mathcal{K}) \to L^p(\Omega; \mathcal{K})$ is a closed operator, we have that $z \in \text{Dom}\delta_p(\mathcal{K})$ and

$$\delta_p(z) = L^p(\Omega; \mathcal{K}) - \lim_{n\to\infty} I_1(z_n).$$

Finally, the result follows from Remark 22.*iii*), which establishes that $I_1(z_n)$ goes to $I_1(z)$ in $L^2(\Omega; \mathcal{K})$. □

Lemma 38 *Let u be an $\mathcal{H}\otimes\mathcal{K}$-valued random variables in $\mathcal{S}(\mathcal{H}\otimes\mathcal{K})$ and $h\otimes k \in \mathcal{H}\otimes\mathcal{K}$. Then, for every $p > 1$, $\delta_p(u)$ is in $\mathbb{D}^{1,p}(\mathcal{K})$ and*

$$\langle D\left(\delta_p(u)\right), h\otimes k\rangle_{\mathcal{H}\otimes\mathcal{K}} = \langle u, h\otimes k\rangle_{\mathcal{H}\otimes\mathcal{K}} + \delta_p\left(\langle Du^*(k), h\rangle_{\mathcal{H}}\right).$$

Remark 53 *u^* is the adjoint of u and, therefore, u^* is an $\mathcal{K}\otimes\mathcal{H}$-valued random variable.*

Proof. The definition of $\mathcal{S}(\mathcal{H}\otimes\mathcal{K})$ gives that u has the form $u = \sum_{i=1}^{n} F_i z_i$, with $F_i \in \mathcal{S}$ and $z_i \in \mathcal{H}\otimes\mathcal{K}$, $i = 1, \ldots, n$. Then, Lemmas 36 and 37 imply

$$\delta_p(u) = \sum_{i=1}^{n} \left(F_i \delta_p(z_i) - \langle z_i, DF_i\rangle_{\mathcal{H}}\right) = \sum_{i=1}^{n} \left(F_i I_1(z_i) - \langle z_i, DF_i\rangle_{\mathcal{H}}\right).$$

Consequently, from Corollary 11 and Proposition 21, we can deduce

$$D\left(\delta_p(u)\right) = \sum_{i=1}^{n} \left(F_i z_i + (DF_i)\otimes\delta_p(z_i) - D\left(\langle z_i, DF_i\rangle_{\mathcal{H}}\right)\right). \tag{4.4}$$

Thus Lemma 36, Proposition 144 and the fact that $D^2F_i = (D^2F_i)^*$ yield that, for $h \otimes k \in \mathcal{H} \otimes \mathcal{K}$,

$$\begin{aligned}
&\langle D(\delta_p(u)), h \otimes k\rangle_{\mathcal{H}\otimes\mathcal{K}} - \langle u, h\otimes k\rangle_{\mathcal{H}\otimes\mathcal{K}} \\
&= \sum_{i=1}^{n}\Big(\langle DF_i, h\rangle_{\mathcal{H}}\langle \delta_p(z_i), k\rangle_{\mathcal{K}} - \langle D(\langle z_i, DF_i\rangle_{\mathcal{H}}), h\otimes k\rangle_{\mathcal{H}\otimes\mathcal{K}}\Big) \\
&= \sum_{i=1}^{n}\Big(\langle DF_i, h\rangle_{\mathcal{H}}\,\delta_p(z_i^*(k)) - \langle D(\langle z_i, DF_i\rangle_{\mathcal{H}}), h\otimes k\rangle_{\mathcal{H}\otimes\mathcal{K}}\Big) \\
&= \sum_{i=1}^{n}\delta_p\left(\langle DF_i, h\rangle_{\mathcal{H}}\, z_i^*(k)\right) = \delta_p\left(\langle Du^*(k), h\rangle_{\mathcal{H}}\right).
\end{aligned}$$

In this way, the proof is complete. □

Two consequences of the last lemma are the following two results.

Proposition 63 *Let $p, q > 1$ be such that $p^{-1} + q^{-1} = 1$, and u and v two $\mathcal{H}\otimes\mathcal{K}$-valued random variables in $\mathbb{D}^{1,p}(\mathcal{H}\otimes\mathcal{K})$ and $\mathbb{D}^{1,q}(\mathcal{H}\otimes\mathcal{K})$, respectively. Then,*

$$E\left(\langle \delta_p(u), \delta_q(v)\rangle_{\mathcal{K}}\right) = E\left(\langle u, v\rangle_{\mathcal{H}\otimes\mathcal{K}} + \langle (Du)^*, Dv^*\rangle_{\mathcal{H}\otimes\mathcal{K}\otimes\mathcal{H}}\right). \tag{4.5}$$

Remark 54 *If $\mathcal{K} = \mathbb{R}$, then (4.5) has the form*

$$E\left(\delta_p(u)\delta_q(v)\right) = E\left(\langle u, v\rangle_{\mathcal{H}} + \langle (Du)^*, Dv\rangle_{\mathcal{H}^{\otimes 2}}\right).$$

Proof. We first assume that u and v are in $\mathcal{S}(\mathcal{H}\otimes\mathcal{K})$. In this case, the duality relation (4.2) and Lemma 38 imply

$$\begin{aligned}
&E\left(\langle \delta_p(u), \delta_q(v)\rangle_{\mathcal{K}}\right) \\
&= E\Big(\langle u, D(\delta_q(v))\rangle_{\mathcal{H}\otimes\mathcal{K}}\Big) = E\left(\sum_{i,j=1}^{\infty}\langle u, e_i\otimes g_j\rangle_{\mathcal{H}\otimes\mathcal{K}}\langle D(\delta_q(v)), e_i\otimes g_j\rangle_{\mathcal{H}\otimes\mathcal{K}}\right) \\
&= E\left(\sum_{i,j=1}^{\infty}\langle u, e_i\otimes g_j\rangle_{\mathcal{H}\otimes\mathcal{K}}\Big(\langle v, e_i\otimes g_j\rangle_{\mathcal{H}\otimes\mathcal{K}} + \delta_q\left(\langle Dv^*(g_j), e_i\rangle_{\mathcal{H}}\right)\Big)\right) \\
&= E\left(\langle u, v\rangle_{\mathcal{H}\otimes\mathcal{K}} + \sum_{i,j=1}^{\infty}\Big\langle D\langle u, e_i\otimes g_j\rangle_{\mathcal{H}\otimes\mathcal{K}}, \langle Dv^*(g_j), e_i\rangle_{\mathcal{H}}\Big\rangle_{\mathcal{H}}\right) \\
&= E\left(\langle u, v\rangle_{\mathcal{H}\otimes\mathcal{K}} + \sum_{i,j=1}^{\infty}\langle D\langle u^*(g_j), e_i\rangle_{\mathcal{H}}, \langle Dv^*(g_j), e_i\rangle_{\mathcal{H}}\rangle_{\mathcal{H}}\right) \\
&= E\left(\langle u, v\rangle_{\mathcal{H}\otimes\mathcal{K}} + \sum_{i,j=1}^{\infty}\langle\langle (Du^*(g_j))^*, e_i\rangle_{\mathcal{H}}, \langle Dv^*(g_j), e_i\rangle_{\mathcal{H}}\rangle_{\mathcal{H}}\right) \\
&= E\left(\langle u, v\rangle_{\mathcal{H}\otimes\mathcal{K}} + \sum_{i,j=1}^{\infty}\Big\langle\langle (Du)^*, e_i\otimes g_j\rangle_{\mathcal{H}\otimes\mathcal{K}}, \langle Dv^*, e_i\otimes g_j\rangle_{\mathcal{H}\otimes\mathcal{K}}\Big\rangle_{\mathcal{H}}\right) \\
&= E\left(\langle u, v\rangle_{\mathcal{H}\otimes\mathcal{K}} + \sum_{i,j,\ell=1}^{\infty}\Big\langle\langle (Du)^*, e_i\otimes g_j\rangle_{\mathcal{H}\otimes\mathcal{K}}, e_\ell\Big\rangle_{\mathcal{H}}\Big\langle\langle Dv^*, e_i\otimes g_j\rangle_{\mathcal{H}\otimes\mathcal{K}}, e_\ell\Big\rangle_{\mathcal{H}}\right).
\end{aligned}$$

Hence, Proposition 144 allows to conclude

$$\begin{aligned}
&E\left(\langle \delta_p(u), \delta_q(v)\rangle_{\mathcal{K}}\right)\\
&= E\left(\langle u, v\rangle_{\mathcal{H}\otimes\mathcal{K}} + \sum_{i,j,\ell=1}^{\infty} \left\langle (Du)^*, e_i\otimes g_j\otimes e_\ell\right\rangle_{\mathcal{H}\otimes K\otimes\mathcal{H}} \langle Dv^*, e_i\otimes g_j\otimes e_\ell\rangle_{\mathcal{H}\otimes\mathcal{K}\otimes\mathcal{H}}\right)\\
&= E\left(\langle u, v\rangle_{\mathcal{H}\otimes\mathcal{K}} + \left\langle (Du)^*, Dv^*\right\rangle_{\mathcal{H}\otimes\mathcal{K}\otimes\mathcal{H}}\right).
\end{aligned}$$

Finally, from this equality turns out that the result is an immediate consequence of Proposition 62. □

Corollary 24 *Let $p > 1$ and $u \in \mathbb{D}^{2,p}(\mathcal{H}\otimes\mathcal{K})$. Then, $\delta_p(u) \in \mathbb{D}^{1,p}(\mathcal{K})$ and*

$$D\left(\delta_p(u)\right) = u + \delta_p\left((Du)^*\right)^*. \tag{4.6}$$

Remark 55 *If $\mathcal{K} = \mathbb{R}$, then the last equality becomes*

$$D\left(\delta_p(u)\right) = u + \delta_p\left((Du)^*\right).$$

Proof. First suppose that u belongs to $\mathcal{S}(\mathcal{H}\otimes\mathcal{K})$. In this case, we can assume that u is as in the proof of Lemma 38 with $F_i = f(W(h_1), \dots, W(h_m))$. Therefore,

$$\begin{aligned}
&(DF_i)\otimes\delta_p(z_i) - D\left(\langle z_i, DF_i\rangle_{\mathcal{H}}\right)\\
&= \sum_{\ell=1}^{m} \{\partial_{x_\ell} f(W(h_1), \dots, W(h_m))h_\ell\otimes\delta_p(z_i) - D\left(\langle z_i, h_\ell\rangle_{\mathcal{H}}\,\partial_{x_\ell} f(W(h_1), \dots, W(h_m))\right)\}\\
&= \sum_{\ell=1}^{m} \{\partial_{x_\ell} f(W(h_1), \dots, W(h_m))h_\ell\otimes\delta_p(z_i)\\
&\quad -\sum_{j=1}^{m} \partial_{x_j}\partial_{x_\ell} f(W(h_1), \dots, W(h_m))h_j\otimes\langle z_i, h_\ell\rangle_{\mathcal{H}}\}\\
&= \sum_{\ell=1}^{m} h_\ell\otimes\{\partial_{x_\ell} f(W(h_1), \dots, W(h_m))\delta_p(z_i)\\
&\quad -\sum_{j=1}^{m} \partial_{x_j}\partial_{x_\ell} f(W(h_1), \dots, W(h_m))\,\langle z_i, h_j\rangle_{\mathcal{H}}\}.
\end{aligned}$$

Hence, from Lemma 36, we have

$$\begin{aligned}
(DF_i)\otimes\delta_p(z_i) - D\left(\langle z_i, DF_i\rangle_{\mathcal{H}}\right) &= \sum_{\ell=1}^{m} h_\ell\otimes\delta_p(\partial_{x_\ell} f(W(h_1), \dots, W(h_m))z_i)\\
&= \delta_p\left((DF_i\otimes z_i)^*\right)^*.
\end{aligned}$$

So, (4.4) gives that (4.6) holds for u in $\mathcal{S}(\mathcal{H}\otimes\mathcal{K})$.

Finally, there is a sequence $\{u_n \in \mathcal{S}(\mathcal{H}\otimes\mathcal{K})\}$ such that $\|u_n - u\|_{2,p,\mathcal{K}} \to 0$ as $n\to\infty$. Thus, Proposition 62 yields that $\delta_p(u_n) \to \delta_p(u)$ in $L^p(\Omega;\mathcal{K})$, and $u_n \to u$ and $\delta_p\left((Du_n)^*\right)^* \to \delta_p\left((Du)^*\right)^*$ in $L^p(\Omega;\mathcal{H}\otimes\mathcal{K})$, as $n\to\infty$. This completes the proof because D is a closed operator and (4.6) is true for $u\in\mathcal{S}(\mathcal{H}\otimes\mathcal{K})$. □

Notice that Corollary 24 establishes that δ_p is a continuous linear operator from $\mathbb{D}^{2,p}(\mathcal{K})$ into $\mathbb{D}^{1,p}(\mathcal{K})$) because of Proposition 62 (see Nualart [158], Proposition 1.5.7, for a general result).

Now, our aim is to characterize the domain of δ_2 via the chaos decomposition introduced in Theorem 12. Towards this end, we recall you that, for $f \in \mathcal{H}^{\otimes n} \otimes \mathcal{K}$, the function $\tilde{f}$ is defined in Section 1.5.3.

Lemma 39 *Let u be a $\mathcal{H} \otimes \mathcal{K}$-valued random variable in $\mathrm{Dom}\,\delta_2$ having the chaos decomposition*

$$u = \sum_{n=0}^{\infty} I_n(f_n), \quad \textit{in } L^2(\Omega; \mathcal{H} \otimes \mathcal{K}).$$

Then, $\delta_2(u)$ has the chaos decomposition

$$\delta_2(u) = \sum_{n=1}^{\infty} I_n(u_n), \quad \textit{in } L^2(\Omega; \mathcal{K}),$$

with $u_n = \tilde{f}_{n-1}$, for $n \geq 1$.

Proof. Let $n, \ell \in \mathbb{N}$ and $\alpha \in \Sigma_{2,n}$. Then, the duality relation (4.2), Remark 22.v) and Lemma 17 imply

$$\begin{aligned}
n!\,\langle u_n, e(\alpha) \otimes g_\ell \rangle_{\mathcal{H}^{\otimes n} \otimes \mathcal{K}} &= E\left(\langle \delta_2(u), I_n\left(e(\alpha) \otimes g_\ell\right)\rangle_{\mathcal{K}}\right) \\
&= E\left\{\left\langle u, nI_{n-1}\left(\sum_{i=1}^{\infty} \langle e(\alpha), e_i \rangle_{\mathcal{H}} \otimes (e_i \otimes g_\ell)\right)\right\rangle_{\mathcal{H}\otimes\mathcal{K}}\right\} \\
&= n!\left\langle f_{n-1}, \sum_{i=1}^{\infty} \langle e(\alpha), e_i \rangle_{\mathcal{H}} \otimes (e_i \otimes g_\ell)\right\rangle_{\mathcal{H}^{\otimes n}\otimes\mathcal{K}} \\
&= n!\,\langle f_{n-1}, e(\alpha) \otimes g_\ell \rangle_{\mathcal{H}^{\otimes n}\otimes\mathcal{K}} = n!\left\langle \tilde{f}_{n-1}, e(\alpha) \otimes g_\ell \right\rangle_{\mathcal{H}^{\otimes n}\otimes\mathcal{K}},
\end{aligned}$$

where last equality follows from the fact that operator $\mathrm{Symm} \otimes I_{\mathcal{K}}$ is an orthogonal projection. Hence, using that $\{e(\alpha) \otimes g_\ell : \alpha \in \Sigma_{2,n} \text{ and } \ell \in \mathbb{N}\}$ is a complete orthogonal system of $\mathcal{H}^{\odot n} \otimes \mathcal{K}$, we have that the proof is complete. □

In order to consider the reciprocal of Lemma 39, we see that some multiple integrals belong to $\mathrm{Dom}\,\delta_2$.

Lemma 40 *Let $\alpha = ((\alpha_1, j_1), \ldots, (\alpha_{\hat{\alpha}}, j_{\hat{\alpha}}))$ be in $\Sigma_{2,n}$ and $j, \ell \in \mathbb{N}$. Then, $I_n(e(\alpha) \otimes (e_j \otimes g_\ell))$ is in $\mathrm{Dom}\,\delta_2$.*

Remark 56 *Note that Lemma 39 gives*

$$\delta_2\left(I_n(e(\alpha) \otimes (e_j \otimes g_\ell))\right) = I_{n+1}\left(e(\alpha(j)) \otimes g_\ell\right),$$

with

$$\alpha(j) = \begin{cases} ((\alpha_1, j_1), \ldots, (\alpha_k + 1, j_k), \ldots, (\alpha_{\hat{\alpha}}, j_{\hat{\alpha}})), & \text{if } j = k \in \{j_1, \ldots, j_{\hat{\alpha}}\} \\ ((\alpha_1, j_1), \ldots, (\alpha_k, j_k), (1, j), (\alpha_{k+1}, j_{k+1}), \ldots, (\alpha_{\hat{\alpha}}, j_{\hat{\alpha}})), & \text{if } j \notin \{j_1, \ldots, j_{\hat{\alpha}}\}. \end{cases}$$

Proof. The result is an immediate consequence of Remark 24, (8.65), (8.66) and Lemma 36. □

An immediate consequence of the last lemma is the reciprocal of Corollary 12.

Theorem 46 *Let X be in $L^2(\Omega; \mathcal{K})$ with chaos decomposition*

$$X = E(X) + \sum_{n=1}^{\infty} I_n(h_n),$$

where $h_n \in \mathcal{H}^{\odot n} \otimes \mathcal{K}$. *Then,* X *belongs to* $\mathbb{D}^{1,2}(\mathcal{K})$ *if and only if*

$$\sum_{n=1}^{\infty} nn!\, |h_n|^2_{\mathcal{H}^{\otimes n}\otimes\mathcal{K}} < \infty. \tag{4.7}$$

In this case, we have that the chaos decomposition

$$DX = \sum_{n=1}^{\infty} nI_{n-1}\left(\sum_{i,\ell=1}^{\infty} \langle h_n^*(g_\ell), e_i\rangle_{\mathcal{H}} \otimes (e_i \otimes g_\ell)\right), \quad \text{in } L^2(\Omega : \mathcal{H}\otimes\mathcal{K}), \tag{4.8}$$

holds.

Remark 57 *Note that Remark 22.v) and (4.8) imply that* $E(DX) = h_1$.

Proof. We first assume that $X \in \mathbb{D}^{1,2}(\mathcal{K})$ and that DX has the chaos decomposition

$$DX = \sum_{n=0}^{\infty} I_n(f_n), \quad \text{in } L^2(\Omega; \mathcal{H}\otimes\mathcal{K}).$$

Here, $f_m = \tilde{f}_m$, $m \geq 1$. Choose $n, j, \ell \in \mathbb{N}$ and $\alpha \in \Sigma_{2,n}$. Then, from Remark 56, we obtain

$$\begin{aligned} &n!\, \langle e(\alpha)\otimes(e_j\otimes g_\ell), f_n\rangle_{\mathcal{H}^{\otimes(n+1)}\otimes\mathcal{K}} \\ &= \langle I_n\left(e(\alpha)\otimes(e_j\otimes g_\ell)\right), DX\rangle_{L^2(\Omega;\mathcal{H}\otimes\mathcal{K})} = \langle I_{n+1}\left(e(\alpha(j))\otimes g_\ell\right), X\rangle_{L^2(\Omega;\mathcal{K})} \\ &= (n+1)!\, \langle h_{n+1}, e(\alpha(j))\otimes g_\ell\rangle_{\mathcal{H}^{\otimes(n+1)}\otimes\mathcal{K}}. \end{aligned}$$

Therefore, using Proposition 144 and $h_n \in \mathcal{H}^{\odot n}\otimes\mathcal{K}$, we get

$$\begin{aligned} &\langle e(\alpha)\otimes(e_j\otimes g_\ell), f_n\rangle_{\mathcal{H}^{\otimes(n+1)}\otimes\mathcal{K}} \\ &= (n+1)\, \langle h_{n+1}, e(\alpha)\otimes(e_j\otimes g_\ell)\rangle_{\mathcal{H}^{\otimes(n+1)}\otimes\mathcal{K}} \\ &= (n+1)\, \langle \langle h_{n+1}^*(g_\ell), e_j\rangle_{\mathcal{H}}, e(\alpha)\rangle_{\mathcal{H}^{\otimes n}} \\ &= (n+1)\left\langle \sum_{i,k=1}^{\infty} \langle h_{n+1}^*(g_k), e_i\rangle_{\mathcal{H}} \otimes (e_i\otimes g_k), e(\alpha)\otimes(e_j\otimes g_\ell)\right\rangle_{\mathcal{H}^{\otimes(n+1)}\otimes\mathcal{K}}. \end{aligned}$$

Consequently, we have that (4.7) and (4.8) are satisfied because

$$\{e(\alpha)\otimes(e_i\otimes g_\ell) : \alpha\in\Sigma_{2,n} \text{ and } i,\ell\in\mathbb{N}\}$$

is a complete orthogonal system of $\mathcal{H}^{\odot n}\otimes\mathcal{H}\otimes\mathcal{K}$.

Finally, it is easy to finish the proof by applying Corollary 12. □

Note that Theorem 46 allows us to find a random variable

$$X = E(X) + \sum_{n=1}^{\infty} I_n(h_n)$$

in $L^2(\Omega;\mathcal{K}) \setminus \mathbb{D}^{1,2}(\mathcal{K})$ by choosing the family $\{h_n : n\in\mathbb{N}\}$ such that (4.7) is not satisfied. Other example is the following result.

Proposition 64 *Let* $A \in \mathcal{F}$. *Then,* 1_A *belongs to* $\mathbb{D}^{1,2}$ *if and only if* $P(A) \in \{0, 1\}$.

Proof. We first assume that $1_A \in \mathbb{D}^{1,2}$. Then, Proposition 21 implies

$$D1_A = D(1_A)^2 = 21_A D1_A.$$

Therefore, Proposition 22 (i.e., the local property for D) implies that $D1_A = 0$ with probability 1. Thus, from Theorem 46, we can conclude that $1_A = E(1_A) = P(A)$. In other words, $P(A) \in \{0,1\}$.

Conversely, 1_A is a constant if $P(A) \in \{0,1\}$. Consequently, 1_A is in $\mathbb{D}^{1,2}$ because it is a random variable in $\mathcal{S}$ in this case. □

The last proposition also holds if we write $\mathbb{D}^{1,p}$, $p \geq 1$, instead of $\mathbb{D}^{1,2}$, which is established in Nualart [158] (Proposition 1.2.6). Thus, $\left(L^p(\Omega) \setminus \mathbb{D}^{1,p}\right) \neq \emptyset$.

Lemma 41 *Let $u \in L^2(\Omega; \mathcal{H} \otimes \mathcal{K})$ have the chaos decomposition $u = \sum_{n=0}^{\infty} I_n(f_n)$ such that*

$$\sum_{n=1}^{\infty} n! \left| \tilde{f}_{n-1} \right|^2_{\mathcal{H}^{\otimes n} \otimes \mathcal{K}} < \infty. \tag{4.9}$$

Then, u belongs to $\operatorname{Dom} \delta_2$ and

$$\delta_2(u) = \sum_{n=1}^{\infty} I_n(\tilde{f}_{n-1}), \quad \text{in } L^2(\Omega; \mathcal{K}).$$

Proof. We divide the proof into two steps.
Step 1: Here we suppose that $u = I_n(f_n)$, for some $n \in \mathbb{N}$. Without loss of generality, we can assume that $f_n \in \mathcal{H}^{\odot n} \otimes (\mathcal{H} \otimes \mathcal{K})$ because of (1.97) and, therefore, it has the form

$$f_n = \sum_{\substack{\alpha \in \Sigma_{2,n} \\ j,\ell \in \mathbb{N}}} c_{\alpha,\ell}\, e(\alpha) \otimes (e_j \otimes g_\ell), \quad c_{\alpha,\ell} \in \mathbb{R}.$$

Hence, by Lemma 39 and Remark 55, the result is satisfied if $c_{\alpha,\ell} = 0$ except for finitely many terms since δ_2 is a linear operator. So, in this case, the theorem is a consequence of (4.9), and the facts that $\text{Symm} : \mathcal{H}^{\otimes n} \to \mathcal{H}^{\otimes n}$ is a linear bounded operator with norm 1 and that δ_2 is a closed operator from $L^2(\Omega; \mathcal{H} \otimes \mathcal{K})$ into $L^2(\Omega; \mathcal{K})$.
Step 2: Finally, we deal with the general case. From Step 1 and $N \in \mathbb{N}$, the $\mathcal{H} \otimes \mathcal{K}$-valued random variable $u_N = \sum_{n=0}^{N} I_n(f_n)$ belongs to $\operatorname{Dom} \delta_2$ and $\delta_2(u_N) = \sum_{n=1}^{N+1} I_n(\tilde{f}_{n-1})$ because δ_2 is a linear operator. Thus, the proof is complete using (4.9) and the property that δ_2 is a closed operator again. □

An immediate consequence of Lemmas 39 and 41, and Remark 22 is the characterization of $\operatorname{Dom} \delta_2$ through the chaos decomposition:

Theorem 47 *Let u be a random variable in $L^2(\Omega; \mathcal{H} \otimes \mathcal{K})$ with the chaos decomposition*

$$u = \sum_{n=0}^{\infty} I_n(f_n).$$

Then, u belongs to $\operatorname{Dom} \delta_2$ if and only if $\sum_{n=1}^{\infty} n! \left| \tilde{f}_{n-1} \right|^2_{\mathcal{H}^{\otimes n} \otimes \mathcal{K}} < \infty$. In this case, we have

$$\delta_2(u) = \sum_{n=1}^{\infty} I_n(\tilde{f}_{n-1}), \quad \text{in } L^2(\Omega; \mathcal{K}).$$

The following result could be interpreted as a Clark-Ocone-type result (see Proposition 72 below), which is an integral representation of square-integrable random variables measurable with respect to the σ-algebra generate by a d-dimensional Brownian motion.

Proposition 65 *Let $F \in L^2(\Omega;\mathcal{K})$. Then, there is a $\mathcal{K}$-valued random variable u in $\mathbb{D}^{1,2}(\mathcal{K})$ such that*

$$F = E(F) + \delta_2(Du). \tag{4.10}$$

Moreover, if u_1 and u_2 are such that $F = E(F) + \delta_2(Du_1) = E(F) + \delta_2(Du_2)$, Then, $u_1 - E(u_1) = u_2 - E(u_2)$.

Proof. Consider the chaos decomposition $F - E(F) = \sum_{n=1}^{\infty} I_n(f_n)$ and define $u = \sum_{n=1}^{\infty} n^{-1} I_n(f_n)$, with $f_n \in \mathcal{H}^{\odot n} \otimes \mathcal{K}$. Therefore, Theorem 46 gives that u is in $\mathbb{D}^{1,2}(\mathcal{K})$ and

$$Du = \sum_{n=1}^{\infty} I_{n-1}\left(\sum_{i,\ell=1}^{\infty} \langle f_n^*(g_\ell), e_i\rangle_{\mathcal{H}} \otimes (e_i \otimes g_\ell)\right), \quad \text{in } L^2(\Omega : \mathcal{H}\otimes\mathcal{K}).$$

Hence, it is easy to see that this u is the $\mathcal{K}$-valued random variable that we are looking for. That is, we have that (4.10) is satisfied for u.

Finally, assume that there exist u_1 and u_2 in $\mathbb{D}^{1,2}(\mathcal{K})$ such that

$$F = E(F) + \delta_2(Du_1) = E(F) + \delta_2(Du_2).$$

Thus, $v = u_2 - u_1$ is an element in $\mathbb{D}^{1,2}(\mathcal{K})$ such that $0 = \delta_2(Dv)$. Hence, the result holds because if v has the chaos decomposition

$$v = \sum_{n=0}^{\infty} I_n(v_n), \quad v_n \in \mathcal{H}^{\odot n} \otimes \mathcal{K},$$

then $0 = \sum_{n=1}^{\infty} n I_n(v_n)$. It means that $v - E(v) = 0$ because of the uniqueness of the chaos decomposition. □

In the last proof, we have applied that $\delta_2(DF) = -LF$, where L is the infinitesimal generator of the Ornstein-Uhlenbeck semigroup (for details see Nualart [158]). That is, if $F \in L^2(\Omega;\mathcal{K})$ is such that

$$F = \sum_{n=0}^{\infty} I_n(f_n), \quad \text{with} \quad \sum_{n=1}^{\infty} n^2 n! \|\tilde{f}_n\|^2_{\mathcal{H}^{\odot n}\otimes\mathcal{K}} < \infty,$$

then $LF = -\sum_{n=0}^{\infty} n I_n(f_n)$. Consequently, in the last proof, $u = L^{-1}(E(F) - F)$. Moreover, Maas [135] (Proposition 6.11) has studied Proposition 65 in the case that F belongs to the space $L^p(\Omega;\mathcal{K})$, $p > 1$, using the Meyer's inequality again.

Now, we deal with the local property of the divergence operator. To do so, we recall that $\mathbb{D}^{1,\infty}(\mathcal{H}\otimes\mathcal{K})$ is introduced in (4.1).

Proposition 66 *Let $p \in [1,\infty)$, $u \in Dom\, \mathbb{D}^{1,p}(\mathcal{H}\otimes\mathcal{K}) \cap \mathbb{D}^{1,q}(\mathcal{H}\otimes\mathcal{K})$ and $A \in \mathcal{F}$ be such that $u = 0$ on A and $p^{-1} + q^{-1} = 1$. Then, $\delta_p(u) = 0$ on A w.p.1.*

Proof. Set $\varepsilon > 0$, $\psi : \mathbb{R} \to [0,1]$ a function in $C^\infty(\mathbb{R})$ that satisfies

$$\psi(x) = \begin{cases} 1, & \text{if } |x| \le 1, \\ 0, & \text{if } |x| \ge 2 \end{cases}$$

and $\varphi_\varepsilon(x) = \psi(x/\varepsilon)$. Remember that we can find such a function ψ due to Spivak [200]. Now, we divide the proof into some steps.

Step 1: Here, we show that $\varphi_\varepsilon(\|u\|^2_{\mathcal{H}})F$ belongs to $\mathbb{D}^{1,q}$ if $\mathcal{K}=\mathbb{R}$ and $F\in\mathcal{S}$. Towards this end, we apply Proposition 21. So, Proposition 144 leads us to write

$$\begin{aligned}
&\left\|\varphi_\varepsilon(\|u\|^2_{\mathcal{H}})DF + F\varphi_\varepsilon'(\|u\|^2_{\mathcal{H}})D\langle u,u\rangle_{\mathcal{H}}\right\|_{\mathcal{H}}\\
&\quad\le \varphi_\varepsilon(\|u\|^2_{\mathcal{H}})\,\|DF\|_{\mathcal{H}} + \left|F\varphi_\varepsilon'(\|u\|^2_{\mathcal{H}})\right|\,\|D\langle u,u\rangle_{\mathcal{H}}\|_{\mathcal{H}}\\
&\quad\le \|DF\|_{\mathcal{H}} + 2\left|F\varphi_\varepsilon'(\|u\|^2_{\mathcal{H}})\right|\,\|\langle (Du)^*,u\rangle_{\mathcal{H}}\|_{\mathcal{H}} \le \|DF\|_{\mathcal{H}} + \frac{2C}{\sqrt{\varepsilon}}\|Du\|_{\mathcal{H}^{\otimes 2}},
\end{aligned}$$

which is in $L^q(\Omega)$. Thus, Proposition 21 implies that the claim of this step holds.
Step 2: Now, we see that the result is true for $\mathcal{K}=\mathbb{R}$. From Step 1, we have, for $F\in\mathcal{S}$,

$$\left|E\left(\delta_p(u)\varphi_\varepsilon(\|u\|^2_{\mathcal{H}})F\right)\right| \le E\left(\left|\left\langle u,\varphi_\varepsilon(\|u\|^2_{\mathcal{H}})DF\right\rangle_{\mathcal{H}}\right|\right) + 2E\left(\left|\left\langle u, F\varphi_\varepsilon'(\|u\|^2_{\mathcal{H}})\,\langle (Du)^*,u\rangle_{\mathcal{H}}\right\rangle_{\mathcal{H}}\right|\right).$$

Hence, the fact that

$$\left|\left\langle u, F\varphi_\varepsilon'(\|u\|^2_{\mathcal{H}})\,\langle (Du)^*,u\rangle_{\mathcal{H}}\right\rangle_{\mathcal{H}}\right| \le 2\|\psi'\|_{\infty,\mathbb{R}}|F|\|Du\|_{\mathcal{H}^{\otimes 2}},$$

together with the dominated convergence theorem, allows us to take the limit as $\varepsilon\to 0$ to get

$$E\left(\delta_p(u)1_{\{u=0\}}F\right)=0.$$

Since $F\in\mathcal{S}$ is arbitrary, we have that the result is satisfied under the assumptions of this step.
Step 3: Finally, Step 2 implies

$$\langle \delta_p(u)1_A, g_\ell\rangle_{\mathcal{K}} = \delta_p\left(u^*(g_\ell)\right)1_A, \quad \text{w.p.1, for each } \ell\in\mathbb{N},$$

which finishes the proof of the proposition. □

Proceeding as in Definition 23, we are able to extend the domain of the divergence operator δ_p via Proposition 66.

Definition 38 *Let $p,q\ge 1$ be as in Proposition 66, u a random variable in $\left(\mathbb{D}^{1,p}(\mathcal{H}\otimes\mathcal{K})\cap\mathbb{D}^{1,q}(\mathcal{H}\otimes\mathcal{K})\right)_{loc}$ and $\{(\Omega_n,u^{(n)}):n\in\mathbb{N}\}$ a localizing sequence for u in $\mathbb{D}^{1,p}(\mathcal{H}\otimes\mathcal{K})\cap\mathbb{D}^{1,q}(\mathcal{H}\otimes\mathcal{K})$. Then, we define the divergence δ_p of u as the $\mathcal{K}$-valued random variable*

$$\delta_p(u)=\delta_p(u^{(n)}) \quad \text{on } \Omega_n.$$

Note that, in this definition, $u^{(n)}$ belongs to the space $\mathbb{D}^{1,p}(\mathcal{H}\otimes\mathcal{K})\cap\mathbb{D}^{1,q}(\mathcal{H}\otimes\mathcal{K})$ and that $\delta_p(u)$ is well-defined due to Proposition 66. Also, note that

$$u^{(n)}=\left\{\exp\left(\psi\left(g\left(W(h_1)\right)/n\right)g\left(W(h_1)\right)\right)-1\right\}h_2\otimes k$$

belongs to $\mathcal{S}(\mathcal{H}\otimes\mathcal{K})$, where $h_1,h_2\in\mathcal{H}$, $k\in\mathcal{K}$, g is a polynomial function of degree bigger than 2 and ψ is defined in (1.105). Thus, we have that

$$u=\left\{\exp\left(g\left(W(h_1)\right)\right)-1\right\}h_2\otimes k$$

is in $\left(\mathbb{D}^{1,p}(\mathcal{H}\otimes\mathcal{K})\cap\mathbb{D}^{1,q}(\mathcal{H}\otimes\mathcal{K})\right)_{loc}\setminus\left(\mathbb{D}^{1,p}(\mathcal{H}\otimes\mathcal{K})\cap\mathbb{D}^{1,q}(\mathcal{H}\otimes\mathcal{K})\right)$. To show this fact, the reader can see the example of Definition 23.

On the other hand, the divergence operator δ_p also satisfies Fubini's type theorems. For the definition of the Bochner integral, the reader can see Section 8.3 in the appendix. The following is an immediate consequence of the duality relation (4.2).

Theorem 48 *Let μ be a σ-finite measure on the measurable space $(X,\mathcal{C})$ and $p>1$. Assume that $\varphi:\Omega\times X\to\mathcal{H}\otimes\mathcal{K}$ is an $\mathcal{F}\otimes\mathcal{C}$-measurable function satisfying the following conditions:*

i) $\varphi(x)\in$Dom $\delta_p(\mathcal{K})$, for almost all $x\in X$.

ii) $(\omega,x)\mapsto\delta_p(\varphi(x))$ is $\mathcal{F}\otimes\mathcal{C}$-measurable.

iii) The Bochner integrals $\int_X\varphi(x)\mu(dx)$ and $\int_X\delta_p(\varphi(x))\mu(dx)$ are well-defined w.p.1 and belong to $L^p(\Omega;\mathcal{H}\otimes\mathcal{K})$ and $L^p(\Omega;\mathcal{K})$, respectively.

Then, $\int_X\varphi(x)\mu(dx)$ is in Dom $\delta_p(\mathcal{K})$ and

$$\int_X\delta_p(\varphi(x))\mu(dx)=\delta_p\left(\int_X\varphi(x)\mu(dx)\right)\quad w.p.1.$$

Proof. Let $q>1$ be such that $q^{-1}+p^{-1}=1$ and $F\in\mathbb{D}^{1,q}(\mathcal{K})$. Then, Lemma 85 and Fubini's theorem imply

$$\begin{aligned}
&E\left(\left\langle F,\int_X\delta_p(\varphi(x))\mu(dx)\right\rangle_{\mathcal{K}}\right)\\
&=E\left(\int_X\langle F,\delta_p(\varphi(x))\rangle_{\mathcal{K}}\mu(dx)\right)=\int_X E\left(\langle F,\delta_p(\varphi(x))\rangle_{\mathcal{K}}\right)\mu(dx)\\
&=\int_X E\left(\langle DF,\varphi(x)\rangle_{\mathcal{H}\otimes\mathcal{K}}\right)\mu(dx)=E\left(\int_X\langle DF,\varphi(x)\rangle_{\mathcal{H}\otimes\mathcal{K}}\mu(dx)\right)\\
&=E\left(\left\langle DF,\int_X\varphi(x)\mu(dx)\right\rangle_{\mathcal{H}\otimes\mathcal{K}}\right).
\end{aligned}$$

Therefore, the result is satisfied by using the definition δ_p via the duality relation (4.2). □

In Theorem 48, we suppose that $x\mapsto\delta_p(\varphi(x))$ is $\mathcal{C}$-measurable w.p.1 in order to have that $\int_X\delta_p(\varphi(x))\mu(dx)$ is well-defined w.p.1. But, it is not so, we can state the following Fubini's type result.

Theorem 49 *Let μ be a σ-finite measure on the measurable space $(X,\mathcal{C})$, $p>1$ and $\varphi:\Omega\times X\to\mathcal{H}\otimes\mathcal{K}$ a measurable function such that $\varphi(x)\in\mathbb{D}^{1,p}(\mathcal{H}\otimes\mathcal{K})$, for almost all $x\in X$, and the $\mathbb{D}^{1,p}(\mathcal{H}\otimes\mathcal{K})$-valued Bochner integral $\int_X\varphi(x)\mu(dx)$ is well-defined. Then, there exists a measurable function $Y:\Omega\times X\to\mathcal{K}$ belonging to $L^1(X;\mathcal{K})$ w.p.1 such that*

$$\delta_p(\varphi(x))=Y_x\quad w.p.1,\ for\ almost\ all\ x\in X,\tag{4.11}$$

and

$$\delta_p\left(\int_X\varphi(x)\mu(dx)\right)=\int_X Y_x\mu(dx)\quad w.p.1.\tag{4.12}$$

Remark 58 *Note that that $\int_X\varphi(x)\mu(dx)$ belongs to Dom $\delta_p(\mathcal{K})$ since it also belongs to $\mathbb{D}^{1,p}(\mathcal{H}\otimes\mathcal{K})$.*

Proof. By hypothesis (see Definition 56), there exists a sequence $\{\varphi_n:n\in\mathbb{N}\}$ of elementary $\mathbb{D}^{1,p}(\mathcal{H}\otimes\mathcal{K})$-valued processes defined on X such that

$$\int_X\|\varphi(x)-\varphi_n(x)\|_{1,p,\mathcal{H}\otimes\mathcal{K}}\mu(dx)\to 0,\quad \text{as } n\to\infty.\tag{4.13}$$

We remark that φ_n has the form $\varphi_n = \sum_{i=1}^N \psi_{i,n} 1_{\mathcal{C}_{i,n}}$, with $\psi_{i,n} \in \mathbb{D}^{1,p}(\mathcal{H}\otimes\mathcal{K})$ and $\mathcal{C}_{i,n} \in \mathcal{C}$. Hence, we have

$$Y_x^{(n)} := \delta_p(\varphi_n(x)) = \sum_{i=1}^N \delta_p(\psi_{i,n}) 1_{\mathcal{C}_{i,n}}(x) \tag{4.14}$$

and

$$\begin{aligned}\int_X Y_x^{(n)} \mu(dx) &= \sum_{i=1}^N \delta_p(\psi_{i,n}) \mu(\mathcal{C}_{i,n}) = \delta_p\left(\sum_{i=1}^N \psi_{i,n}\mu(\mathcal{C}_{i,n})\right) \\ &= \delta_p\left(\int_X \varphi_n(x)\mu(dx)\right), \quad \text{w.p.1.}\end{aligned} \tag{4.15}$$

It means, the result is satisfied for φ_n.

Now, we see that $\{Y^{(n)} : n \in \mathbb{N}\}$ is a Cauchy sequence in $L^1(\Omega\times X;\mathcal{K})$. Towards this end, we observe that Fubini's theorem, (4.14) and Proposition 62 lead us to write, for $n, m \in \mathbb{N}$,

$$\begin{aligned}&\int_{\Omega\times X} \left|Y_x^{(n)} - Y_x^{(m)}\right|_{\mathcal{K}} P\times\mu(d\omega, dx) \\ &= \int_X E\left(\left|Y_x^{(n)} - Y_x^{(m)}\right|_{\mathcal{K}}\right)\mu(dx) = \int_X E\left(\left|\delta_p(\varphi_n(x)) - \delta_p(\varphi_m(x))\right|_{\mathcal{K}}\right)\mu(dx) \\ &\le \int_X \left[E\left(\left|\delta_p(\varphi_n(x)) - \delta_p(\varphi_m(x))\right|_{\mathcal{K}}^p\right)\right]^{1/p}\mu(dx) \\ &\le c_p\int_X \left\|\varphi_n(x) - \varphi_m(x)\right\|_{1,p,\mathcal{H}\otimes\mathcal{K}}\mu(dx).\end{aligned}$$

Consequently, (4.13) gives that our claim is true. Therefore, there exists a measurable function $Y : \Omega\times X \to \mathcal{K}$ in $L^1(\Omega\times X;\mathcal{K})$ such that

$$E\left(\int_X \left|Y_x - Y_x^{(n)}\right|_{\mathcal{K}}\right) \to 0, \quad \text{as } n\to\infty. \tag{4.16}$$

Thus, we are able to find a subsequence $\{n_k : k\in\mathbb{N}\}$ of $\mathbb{N}$ such that

$$E\left(\left|Y_x - Y_x^{(n_k)}\right|_{\mathcal{K}}\right) \to 0, \quad \text{for almost all } x\in X, \text{ as } k\to\infty,$$

which, together with Proposition 62, (4.13) and (4.14), allows us to establish

$$\begin{aligned}E\left(\left|\delta_p(\varphi(x)) - Y_x\right|\right) &\le E\left(\left|\delta_p(\varphi(x)) - \delta_p\left(\varphi_{n_{k'}}(x)\right)\right|\right) + E\left(\left|Y_x^{(n_{k'})} - Y_x\right|\right) \\ &\le c_p\left\|\varphi(x) - \varphi_{n_{k'}}(x)\right\|_{1,p,\mathcal{H}\otimes\mathcal{K}} + E\left(\left|Y_x^{(n_{k'})} - Y_x\right|\right) \\ &\to 0 \quad \text{for almost all } x\in X, \text{ as } k\to\infty,\end{aligned}$$

for some subsequence $\{n_{k'} : k\in\mathbb{N}\}$ of $\{n_k : k\in\mathbb{N}\}$. In this way, we have that Y satisfies (4.11).

Finally, from (4.15), we get

$$\begin{aligned}&E\left(\left|\int_X Y_x\mu(dx) - \delta_p\left(\int_X \varphi(x)\mu(dx)\right)\right|_{\mathcal{K}}\right) \\ &\le E\left(\int_X \left|Y_x - Y_x^{(n)}\right|_{\mathcal{K}}\mu(dx)\right) + E\left(\left|\delta_p\left(\int_X [\varphi_n(x) - \varphi(x)]\mu(dx)\right)\right|_{\mathcal{K}}\right).\end{aligned}$$

Hence, Proposition 62, (4.13) and (4.16) imply

$$\begin{aligned} E&\left(\left|\int_X Y_x\mu(dx) - \delta_p\left(\int_X \varphi(x)\mu(dx)\right)\right|_{\mathcal{K}}\right) \\ &\leq E\left(\int_X \left|Y_x - Y_x^{(n)}\right|_{\mathcal{K}}\mu(dx)\right) + c_p\left\|\int_X [\varphi_n(x) - \varphi(x)]\,\mu(dx)\right\|_{1,p,\mathcal{H}\otimes\mathcal{K}} \\ &\leq E\left(\int_X \left|Y_x - Y_x^{(n)}\right|_{\mathcal{K}}\mu(dx)\right) + c_p\int_X \|\varphi_n(x) - \varphi(x)\|_{1,p,\mathcal{H}\otimes\mathcal{K}}\,\mu(dx) \to 0, \text{ as } n \to \infty. \end{aligned}$$

Therefore, the proof is complete. □

As an example of the utility of the localization argument to weaken the hypotheses of a result, we have the following:

Theorem 50 *Let $(X, \mathcal{C}, \mu)$ be a σ-finite measure space and $\varphi : \Omega\times X \to \mathcal{H}\otimes\mathcal{K}$. Assume that $\{\Omega_n : n \in \mathbb{N}\}$ is a sequence of measurable subsets of Ω and $\{\varphi_n : \Omega \times X \to \mathcal{H}\otimes\mathcal{K} : n \in \mathbb{N}\}$ is a sequence of measurable functions satisfying the assumptions of Theorem 49 such that $\Omega_n \uparrow \Omega$ w.p.1 and, for each $n \in \mathbb{N}$,*

$$\varphi(x) = \varphi_n(x), \quad \textit{for all } x \in X, \ \textit{w.p.1 on } \Omega_n. \tag{4.17}$$

Then, there exists a measurable function $Y : \Omega \times X \to \mathcal{K}$ belonging to $L^1(X;\mathcal{K})$ w.p.1 such that (4.11) and (4.12) are satisfied.

Remarks 5 *i) The set of probability 1 in (4.17) is independent of x.*

ii) In the proof of this result, we show that the $\mathcal{H}\otimes\mathcal{K}$-valued Bochner integral $\int_X \varphi(\omega, x)\mu(dx)$ is well-defined w.p.1 and

$$\int_X \varphi(\omega, x)\mu(dx) = \left(\int_X \varphi_n(x)\mu(dx)\right)(\omega), \quad \textit{on } \Omega_n \textit{ w.p.1.} \tag{4.18}$$

Thus, the left-hand side belongs to $\mathbb{D}^{1,p}_{loc}(\mathcal{H}\otimes\mathcal{K})$ and the one of (4.12) makes sense. Therefore, if φ is as in Theorem 49, then the $\mathbb{D}^{1,p}(\mathcal{H}\otimes\mathcal{K})$-valued Bochner integral $\int_X \varphi(x)\mu(dx)$ satisfies (4.18) w.p.1, for each $n \in \mathbb{N}$, due to Proposition 22.

iii) We have already known that we can find a random variable in $\mathbb{D}^{1,p}_{loc}(\mathcal{H}\otimes\mathcal{K})\backslash\mathbb{D}^{1,p}(\mathcal{H}\otimes\mathcal{K})$ (see the example of Definition 23). So, the conditions of this theorem do not imply that $\varphi(x)$, $x \in X$, is in $\mathbb{D}^{1,p}(\mathcal{H}\otimes\mathcal{K})$. In other words, the hypotheses of this theorem are weaker than those of Theorem 49.

Proof. We first establish that the $\mathcal{H}\otimes\mathcal{K}$-valued Bochner integral $\int_X \varphi(x)\mu(dx)$ is well-defined and that satisfied (4.18) w.p.1. To do so, fix $n \in \mathbb{N}$. Then, there is a sequence $\{\varphi_{n,m} : m \in \mathbb{N}\}$ of elementary $\mathbb{D}^{1,p}(\mathcal{H}\otimes\mathcal{K})$-valued processes of the form

$$\varphi_{n,m} = \sum_{i=1}^{M} \psi_{i,n,m} 1_{C_{i,n,m}},$$

where $\psi_{i,n,m} \in \mathbb{D}^{1,p}(\mathcal{H}\otimes\mathcal{K})$ and $C_{i,n,m} \in \mathcal{C}$, such that

$$\int_X \|\varphi_n(x) - \varphi_{n,m}(x)\|_{1,p,\mathcal{H}\otimes\mathcal{K}}\mu(dx) \to 0, \quad \text{as } m \to \infty.$$

Hence, we have

$$\begin{aligned} E & \left(\int_X \|\varphi_n(x) - \varphi_{n,m}(x)\|_{\mathcal{H}\otimes\mathcal{K}}\, \mu(dx) \right) \\ & \leq \int_X \|\varphi_n(x) - \varphi_{n,m}(x)\|_{L^p(\Omega;\mathcal{H}\otimes\mathcal{K})}\, \mu(dx) \\ & \leq \int_X \|\varphi_n(x) - \varphi_{n,m}(x)\|_{1,p,\mathcal{H}\otimes\mathcal{K}}\, \mu(dx) \to 0, \quad \text{as } m \to \infty. \end{aligned} \tag{4.19}$$

and

$$\begin{aligned} E & \left(\left\| \int_X \varphi_n(x)\mu(dx) - \int_X \varphi_{n,m}(x)\mu(dx) \right\|_{\mathcal{H}\otimes\mathcal{K}} \right) \\ & \leq \left\| \int_X \varphi_n(x)\mu(dx) - \int_X \varphi_{n,m}(x)\mu(dx) \right\|_{L^p(\Omega;\mathcal{H}\otimes\mathcal{K})} \\ & \leq \left\| \int_X \varphi_n(x)\mu(dx) - \int_X \varphi_{n,m}(x)\mu(dx) \right\|_{1,p,\mathcal{H}\otimes\mathcal{K}} \\ & \leq \int_X \|\varphi_n(x) - \varphi_{n,m}(x)\|_{1,p,\mathcal{H}\otimes\mathcal{K}}\mu(dx) \to 0, \quad \text{as } m \to \infty, \end{aligned}$$

which, together with (4.19), yields

$$\int_X \varphi_n(\omega, x)\mu(dx) = \left(\int_X \varphi_n(x)\mu(dx) \right)(\omega), \quad \text{w.p.1.}$$

Here, the integral on the left-hand side is the $\mathcal{H} \otimes \mathcal{K}$-valued Bochner integral of $\varphi_n(\omega, \cdot)$ and the integral on the right-hand side is the $\mathbb{D}^{1,p}(\mathcal{H} \otimes \mathcal{K})$-valued Bochner integral of φ_n, both with respect to μ. In particular, we obtain that the $\mathcal{H} \otimes \mathcal{K}$-valued function$\varphi(\omega, \cdot)$ is Bochner integrable with respect to μ w.p.1 and

$$\int_X \varphi(\omega, x)\mu(dx) = \left(\int_X \varphi_n(x)\mu(dx) \right)(\omega), \quad \text{on } \Omega_n \text{ w.p.1.}$$

It means, (4.18) holds.

On the other hand, (4.11) gives that there exists a measurable function $Y^{(n)} : \Omega \times X \to \mathcal{K}$ belonging to $L^1(X; \mathcal{K})$ w.p.1 such that

$$\delta_p\left(\varphi_n(x)\right) = Y_x^{(n)} \quad \text{w.p.1, for almost all } x \in X.$$

Finally, it is not difficult to see that Proposition 66 leads us to conclude that (4.11) and (4.12) hold if we introduce $Y : \Omega \times X \to \mathcal{K}$ as

$$Y_x = Y_x^{(n)} \quad \text{on } \Omega_n.$$

Now, the proof is complete. □

4.1 Brownian motion

In this section, we suppose that $p > 1$ and $B = \{B_t : t \in [0, T]\}$ is a d-dimensional Brownian motion defined on a complete probability space $(\Omega, \mathcal{F}, P)$. We also suppose that

$(\mathcal{F}_t)_{t\in[0,T]}$ is the filtration given by $\mathcal{F}_t = \sigma\{B_s : 0 \le s \le t\} \vee \mathcal{N}$, where $\mathcal{N}$ is the class of all the P-negligible sets (i.e., sets of probability zero). From Theorem 3, this filtration satisfies the usual conditions. Throughout this section, we assume that $\mathcal{F} = \mathcal{F}_T$.

In this section, the space of all square-integrable functions in $L^2([0,T];\mathbb{R}^d)$ is represented by $\mathcal{H}$ and we denote by $W = \{W(h) : h \in \mathcal{H}\}$ the isonormal Gaussian process introduced in Examples 1.i). That is,

$$W(h) = \sum_{i=1}^{d} \int_0^T h^{(i)}(s) dB_s^{(i)}, \quad \text{for } h \in \mathcal{H}. \tag{4.20}$$

For a random variable $F \in \mathcal{S}$ (i.e., it has the form (1.100)), the definition given by (1.101) implies that its derivative, in the Malliavin calculus sense, is the d-dimensional stochastic process

$$D_t F = \sum_{j=1}^{n} \frac{\partial f}{\partial x_j}(W(h_1), \ldots, W(h_n)) h_j(t), \quad t \in [0,T].$$

We have already known that DF belongs to the space $L^p(\Omega;\mathcal{H})$. Observe that its ℓ-th component is given by

$$D_t^{(\ell)} F = \sum_{j=1}^{n} \frac{\partial f}{\partial x_j}(W(h_1), \ldots, W(h_n)) h_j^{(\ell)}(t), \quad t \in [0,T].$$

Hence $D^{(\ell)}F = 0$ if $h_1^{(\ell)}, \ldots, h_n^{(\ell)} = 0$, and $D^{(\ell)}F = D^{B^{(\ell)}}F$ if $h_i^{(j)} = 0$ for $i = 1, \ldots, n$ and $j \neq \ell$, where $D^{B^{(\ell)}}$ stands for the derivative of F with respect to the Brownian motion $B^{(\ell)}$ (see Remark 25, which establishes that $D^{(\ell)}B^{(\ell)}(h_j^{(\ell)}) = h_j^{(\ell)}$, for $j = 1, \ldots, n$). It means, Definition (1.101) leads us to have that, for $f \in L^2([0,T])$, $D^{(j)}B^{(\ell)}(f) = 0$ if $j \neq \ell$, and $D^{(\ell)}B^{(\ell)}(f) = f$. In conclusion, $D^{(\ell)}$ is an extension of $D^{B^{\ell}}$. Moreover, $D^{(\ell)}F = 0$ if F is in $L^p(\Omega)$ and is measurable with respect to the σ-algebra generated by the random variables $\{B^{(1)}, \ldots, B^{(\ell-1)}, B^{(\ell+1)}, \ldots, B^{(d)}\}$ due to the monotone class lemma (i.e., Theorem 96).

By (8.58), we have

$$\begin{aligned} & L^2([0,T],\lambda;\mathbb{R}^d) \otimes L^2[0,T],\lambda;\mathbb{R}^d) \\ & = L^2\left(([0,T] \times \{1,\ldots,d\}) \times ([0,T] \times \{1,\ldots,d\}), (\lambda \times \delta) \times (\lambda \times \delta)\right), \end{aligned}$$

with $\delta(\{\ell\}) = 1$, for $\ell = 1, \ldots, d$. Therefore, still with F as in (1.100),

$$D^2 F = \sum_{i,j=1}^{n} \frac{\partial^2 f}{\partial x_i \partial x_j}(W(h_1), \ldots, W(h_n)) h_i \otimes h_j$$

and (8.59) allows us to conclude that $D^2F \in L^2\left(([0,T] \times \{i, \ldots, d\})^2, (\lambda \times \delta)^2\right)$ and

$$\begin{aligned} \left(D^2F\right)_{t_1,t_2}^{\ell_1,\ell_2} &= \sum_{i,j=1}^{n} \frac{\partial^2 f}{\partial x_i \partial x_j}(W(h_1), \ldots, W(h_n)) h_i^{(\ell_1)}(t_1) h_j^{(\ell_2)}(t_2) \\ &= D_{t_1}^{(\ell_1)}\left[D_{t_2}^{(\ell_2)}F\right], \quad t_1, t_2 \in [0,T] \text{ and } \ell_1, \ell_2 \in \{1, \ldots, d\}. \end{aligned}$$

In general, for $m > 1$ and F as in (1.100), (8.58) yields that D^mF belongs to $L^2\left(([0,T] \times \{i, \ldots, d\})^m, (\lambda \times \delta)^m\right)$ and, for $t_1, \ldots, t_m \in [0,T]$ and $\ell_1, \ldots, \ell_m \in \{1, \ldots, d\}$,

$$\begin{aligned} & (D^mF)_{t_1,\ldots,t_m}^{\ell_1,\ldots\ell_m} \\ & = \sum_{i_i,\ldots,i_m=1}^{n} \frac{\partial^m f}{\partial x_{i_1} \cdots \partial x_{i_m}}(W(h_1), \ldots, W(h_n)) h_{i_1}^{(\ell_1)}(t_1) \cdots h_{i_m}^{(\ell_m)}(t_m). \end{aligned} \tag{4.21}$$

Thus, the random variable $(D^m F)^{\ell_1,\dots\ell_m}_{t_1,\dots,t_m}$ and the iterated derivative

$$D^{(\ell_1)}_{t_1}\left[D^{(\ell_2)}_{t_2}\cdots D^{(\ell_{m-1})}_{t_{m-1}}\left[D^{(\ell_m)}_{t_m}F\right]\right]$$

are the same. Consequently, using that, for $p>1$, D is a closable operator from $L^p(\Omega)$ into $L^p(\Omega; L^2(([0,T]\times\{i,\dots,d\})^m, (\lambda\times\delta)^m))$ (see, Theorem 14), $\mathbb{D}^{m,p}$ is the closure of the family $\mathcal{S}$ of all smooth random variables of the form (1.100) with respect to the norm

$$\|F\|^p_{m,p} = \|F\|^p_{L^p(\Omega)} + \sum_{j=1}^{n} \left\|\|D^j F\|_{\mathcal{H}^{\otimes j}}\right\|^p_{L^p(\Omega)},$$

where

$$\left\|D^j F\right\|^2_{\mathcal{H}^{\otimes j}} = \sum_{\ell_1,\dots,\ell_j=1}^{d} \int_{[0,T]^j} \left[(D^j F)^{\ell_1,\dots\ell_j}_{s_1,\dots,s_j}\right]^2 ds_1\cdots ds_j.$$

The duality relation (4.2) between δ_p and D has now the form

$$E\left(F\delta_p(u)\right) = E\left(\langle u, DF\rangle_{\mathcal{H}}\right) = E\left(\sum_{\ell=1}^{d}\left\langle u^{(\ell)}, D^{(\ell)}F\right\rangle_{L^2([0,T])}\right). \tag{4.22}$$

As a consequence of this expression, we can extend equality (4.20), where $\delta_p^{B^{(\ell)}}$ stands for the adjoint of the operator $D^{(\ell)}$ (see Remark 24).

Proposition 67 *Let $p>1$ and $u\in L^p(\Omega;\mathcal{H})$ a process such that $u^{(\ell)}$ is in Dom $\delta_p^{B^{(\ell)}}$, for all $\ell\in\{1,\dots,d\}$. Then, u belongs to Dom δ_p and*

$$\delta_p(u) = \sum_{\ell=1}^{d} \delta_p^{B^{(\ell)}}(u^{(\ell)})$$

Proof. Let $F\in\mathcal{S}$. Then, our hypotheses yield

$$E\left(F\sum_{\ell=1}^{d}\delta_p^{B^{(\ell)}}(u^{(\ell)})\right) = \sum_{\ell=1}^{d} E\left(F\delta_p^{B^{(\ell)}}(u^{(\ell)})\right) = \sum_{\ell=1}^{d} E\left(\left\langle u^{(\ell)}, D^{(\ell)}F\right\rangle_{L^2([0,T])}\right),$$

which, together with the duality relation (4.22), implies that the result holds. □

Sometimes, we use the notation

$$\delta_p(u1_{[0,t]}) = \int_0^t u_s\delta_p B_s, \quad \text{for } u1_{[0,t]}\in \text{Dom}\,\delta_p.$$

On the other hand, Lemma 36 is a useful tool to calculate some Skorohod integrals as the following example shows.

Consider $d=1$ and the process

$$u_\cdot = 1_{\{B_T>0\}}\left(1_{[0,T/2]}(\cdot)h(\cdot) - 1_{[T/2,T]}(\cdot)h(\cdot)\right),$$

where $h\in L^2([0,T];\mathbb{R})$ is such that $\int_0^{T/2}h(s)ds = \int_{T/2}^{T}h(s)ds$. We claim that this process is in Dom δ_p. Indeed, let $\psi:\mathbb{R}\to[0,1]$ be a function in $C^\infty(\mathbb{R})$ such that

$$\psi(x) = \begin{cases} 1, & \text{if } x\geq 1,\\ 0, & \text{if } x\leq 0\end{cases}$$

and $\psi_k(x) = \psi(kx)$, for $x \in \mathbb{R}$. Then, Remark 24 and Lemma 36 imply that the process

$$u_k(\cdot) = \psi_k(B_T)\left(1_{[0,T/2]}(\cdot)h(\cdot) - 1_{[T/2,T]}h(\cdot)\right)$$

is in $\operatorname{Dom}\delta_p$ and

$$\int_0^T u_k(s)\delta_p B_s = \psi_k(B_T)\left(\int_0^{T/2} h(s)\delta_p B_s - \int_{T/2}^T h(s)\delta_p B_s\right),$$

which goes to $1_{[B_T>0]}\left(\int_0^{T/2} h(s)\delta_p B_s - \int_{T/2}^T h(s)\delta_p B_s\right)$ in $L^p(\Omega)$, as $k \to \infty$. Hence, the facts that $\delta_p : L^p(\Omega; L^2([0,T])) \to L^p(\Omega)$ is a closed operator and $u_k \to u$ in $L^p(\Omega; L^2([0,T]))$, as $k \to \infty$, give that $u \in \operatorname{Dom}\delta_p$ and

$$\int_0^T u(s)\delta_p B_s = 1_{[B_T>0]}\left(\int_0^{T/2} h(s)\delta_p B_s - \int_{T/2}^T h(s)dB_s\right).$$

In Nualart [156] is pointed out that $u1_{[0,T/2]}$ is not in $\operatorname{Dom}\delta_p$, unlike of the processes v in $\mathbb{D}^{1,p}(\mathcal{H})$, which satisfy that, for every $t \in [0,T]$, $v1_{[0,t]}$ are still in $\mathbb{D}^{1,p}(\mathcal{H})$ and, therefore, they are in $\operatorname{Dom}\delta_p$ due to Proposition 62.

Concerning multiple Wiener-Itô's integrals, we have the following results.

Proposition 68 *Let $n \geq 1$ and $h \in \mathcal{H}^{\otimes n}$. Then,*

$$D_t^{(j)} I_n(h) = nI_{n-1}\left(\tilde{h}_{t,\cdot}^{j,\cdot}\right), \quad (t,j) \in [0,T] \times \{1,\ldots,d\}.$$

Remark 59 *By Corollary 11, we have already known that $I_n(h) \in \mathbb{D}^{1,2}$. Also, we have that $DI_0(h) = 0$.*

Proof. From Corollary 11, we have $D(I_n(h)) = nI_{n-1}\left(\sum_{\ell=1}^{\infty}\left\langle \tilde{h}, e_\ell\right\rangle_{\mathcal{H}} \otimes e_\ell\right)$, where $\{e_\ell : \ell \in \mathbb{N}\}$ is a complete orthonormal system of $\mathcal{H}$. Now, it is easy to finish the proof using (1.109). Indeed, for $m \in \mathbb{N}$,

$$I_{n-1}\left(\sum_{\ell=1}^{m}\left\langle \tilde{h}, e_\ell\right\rangle_{\mathcal{H}} \otimes e_\ell\right) = \sum_{\ell=1}^{m} I_{n-1}\left(\left\langle \tilde{h}, e_\ell\right\rangle_{\mathcal{H}}\right) e_\ell.$$

Hence the result is satisfied due to Remark 22.ii).

□

In the remaining of this section, for $n \in \mathbb{N}$, $\mathcal{E}_n$ stands for the family of all the step function in $L^2\left(([0,T] \times \{i,\ldots,d\})^n, (\lambda \times \delta)^n\right)$ of the form

$$\phi = \sum_{i_1,\ldots,i_n=1}^{m} a_{i_1,\ldots,i_n} 1_{A_{i_1}} \cdots 1_{A_{i_n}}. \tag{4.23}$$

Here, $A_1,\ldots,A_m$ are pairwise disjoint sets of $\mathcal{B}([0,T]) \otimes 2^{\{1,\ldots,d\}}$ and $a_{i_1,\ldots,i_n} \in \mathbb{R}$ with $a_{i_1,\ldots,i_n} = 0$ if at least two of the indices $i_1,\ldots,i_n$ are the same. Note that $A_i \cap A_j = \emptyset$ if and only if $A_i^\ell \cap A_j^\ell = \emptyset$, for all $\ell = 1,\ldots,d$.

The proof of the following result is in Nualart [158] (Section 1.1.2) for $d = 1$, which can be applied in the general case.

Lemma 42 *The space $\mathcal{E}_n$ is dense in $L^2\left(([0,T] \times \{i,\ldots,d\})^n, (\lambda \times \delta)^n\right)$.*

Proof. We claim that $([0,T] \times \{1,\dots,d\}, \lambda \times \delta)$ is a measure space without atoms. Indeed, let $A \in \mathcal{B}([0,T]) \otimes 2^{\{1,\dots,d\}}$ such that $\lambda \times \delta(A) > 0$. Then, Fubini's theorem implies

$$\lambda \times \delta(A) = \int_{\{1,\dots,d\}} \int_0^T 1_A(t,i) dx \delta(di) = \int_{\{1,\dots,d\}} \int_0^T 1_{A^i}(t) dx \delta(di).$$

Thus, there is $i_0 \in \{1,\dots,d\}$ such that $\lambda(A^{i_0}) > 0$. Therefore, there exists a Borel subset C of $[0,T]$ such that $C \subset A^{i_0}$ and $0 < \lambda(C) < \lambda(A^{i_0})$ because $([0,T], \lambda)$ is a measure space without atoms. Consequently, $C \times \{i_0\} \subset A$ and

$$0 < \lambda(C) = \lambda \times \delta(C \times \{i_0\}) < \lambda(A^{i_0}) \leq \lambda \times \delta(A).$$

It means, our claim holds.

On the other hand, note that the lemma is satisfied if, for $B_1,\dots,B_n \in \mathcal{B}([0,T]) \otimes 2^{\{1,\dots,d\}}$, the function $1_{B_1 \times \dots \times B_n}$ can be approximated by step functions in $L^2\left(([0,T] \times \{i,\dots,d\})^n, (\lambda \times \delta)^n\right)$ since the span of this indicator functions is dense in the last space of square-integrable functions.

Let $\varepsilon > 0$, the fact that $\lambda \times \delta$ has no atoms leads us to write

$$1_{B_1 \times \dots \times B_n} = \sum_{i_1,\dots,i_n=1}^{m} a_{i_1,\dots,i_n} 1_{A_{i_1}} \cdots 1_{A_{i_n}},$$

for some some family $\{A_1,\dots,A_m\} \subset \mathcal{B}([0,T]) \otimes 2^{\{1,\dots,d\}}$ of pairwise disjoint sets such that $\lambda \times \delta(A_{i_j}) < \epsilon$ and $a_{i_1,\dots,i_n} \in \{0,1\}$, $i_j = 1,\dots,m$. We observe that the last equality can be written as

$$1_{B_1 \times \dots \times B_n} = \sum_{(i_1,\dots,i_n)\in \mathbb{I}}^{m} a_{i_1,\dots,i_n} 1_{A_{i_1}} \cdots 1_{A_{i_n}} + \sum_{(i_1,\dots,i_n)\in \mathbb{J}}^{m} a_{i_1,\dots,i_n} 1_{A_{i_1}} \cdots 1_{A_{i_n}},$$

where $\mathbb{I} = \{(i_1,\dots,i_n) \in \{1,\dots,m\}^n : \#\{i_1,\dots,i_n\} = n\}$ and

$$\begin{aligned} &\left\| \sum_{(i_1,\dots,i_n)\in \mathbb{J}}^{m} a_{i_1,\dots,i_n} 1_{A_{i_1}} \cdots 1_{A_{i_n}} \right\|^2_{L^2(([0,T]\times\{i,\dots,d\})^n,(\lambda\times\delta)^n)} \\ &\leq \binom{n}{2} \sum_{i=1}^{m} \lambda \times \delta(A_i)^2 \left(\sum_{i=1}^{m} \lambda \times \delta(A_i) \right)^{n-2} \leq \binom{n}{2} \varepsilon T^{n-1}. \end{aligned}$$

So, the proof is complete. That is, we are able to find a sequence that we are looking for using $\sum_{(i_1,\dots,i_n)\in \mathbb{I}}^{m} a_{i_1,\dots,i_n} 1_{A_{i_1}} \cdots 1_{A_{i_n}}$. □

In order to see that the divergence operator δ_2 is an extension of the Itô's integral with respect to the Brownian motion B, we deal with the σ-algebra $\mathcal{F}_A$ associated with the set $A \in \mathcal{B}([0,T]) \otimes 2^{\{1,\dots,d\}}$, which is generated by the family $\{W(1_B) : B \subset A, B \in \mathcal{B}([0,T]) \otimes 2^{\{1,\dots,d\}}\}$ of square-integrable random variables.

Lemma 43 *Let $X \in L^2(\Omega)$ have the chaos decomposition*

$$X = E(X) + \sum_{n=1}^{\infty} I_n(h_n) \tag{4.24}$$

and $A \in \mathcal{B}([0,T]) \otimes 2^{\{1,\dots,d\}}$. Then,

$$E\left(X \middle| \mathcal{F}_A\right) = E(X) + \sum_{n=1}^{\infty} I_n(h_n 1_{A^n}).$$

Remark 60 *Remember that in this section, we are assuming that $\mathcal{F} = \mathcal{F}_T$.*

Proof. By properties of the conditional expectation, together with the fact that the equality in (4.24) is in $L^2(\Omega)$, it is enough to show the result holds for a multiple integral $I_n(h_n)$. Thus, from Lemma 42, we only need to show that the result is satisfied for $h = 1_B$, with $B = B_1 \times \cdots \times B_n$, where $B_1, \ldots, B_n \in \mathcal{B}([0,T]) \otimes 2^{\{1,\ldots,d\}}$ are pairwise disjoint sets.

The multiplication formula for multiple integrals given by Lemma 12 implies

$$I_n(1_B) = W(1_{B_1}) \cdots W(1_{B_n}). \tag{4.25}$$

Then, utilizing the independence of $W(1_{B_1 \cap A^c})$ and $\mathcal{F}_A \vee \mathcal{F}_{B_2} \vee \cdots \vee \mathcal{F}_{B_n}$, we have

$$\begin{aligned} E\left(I_n(1_B)\middle|\mathcal{F}_A\right) &= E\left(W(1_{B_2}) \cdots W(1_{B_n}) E\left(W(1_{B_1})\middle|\mathcal{F}_A \vee \mathcal{F}_{B_2} \vee \cdots \vee \mathcal{F}_{B_n}\right)\middle|\mathcal{F}_A\right) \\ &= E\left(W(1_{B_2}) \cdots W(1_{B_n}) E\left(W(1_{B_1 \cap A})\middle|\mathcal{F}_A \vee \mathcal{F}_{B_2} \vee \cdots \vee \mathcal{F}_{B_n}\right)\middle|\mathcal{F}_A\right) \\ &\quad + E\left(W(1_{B_2}) \cdots W(1_{B_n}) E\left(W(1_{B_1 \cap A^c})\middle|\mathcal{F}_A \vee \mathcal{F}_{B_2} \vee \cdots \vee \mathcal{F}_{B_n}\right)\middle|\mathcal{F}_A\right) \\ &= W(1_{B_1 \cap A}) E\left(W(1_{B_2}) \cdots W(1_{B_n})\middle|\mathcal{F}_A\right). \end{aligned}$$

Finally, we can use induction on n to finish the proof. □

An immediate consequence of the last lemma is the following:

Corollary 25 *Let X and A be as in Lemma 43. Moreover, assume that $X \in \mathbb{D}^{1,2}$. Then, $E\left(X\middle|\mathcal{F}_A\right)$ is also in $\mathbb{D}^{1,2}$ and, for $i = 1, \ldots, d$ and almost all $(\omega, t) \in \Omega \times [0,T]$,*

$$D_t^i\left(E\left(X\middle|\mathcal{F}_A\right)\right) = E\left(D_t^i X\middle|\mathcal{F}_A\right) 1_A(t,i).$$

Proof. From Proposition 68 and Lemma 43, we can obtain

$$D_t^i\left(E\left(X\middle|\mathcal{F}_A\right)\right) = \sum_{n=1}^{\infty} I_{n-1}\left((\tilde{h}_n)_{t,\cdot}^{i,\cdot} 1_{A^{n-1}}\right) 1_A(t,i) = E\left(D_t^i X\middle|\mathcal{F}_A\right) 1_A(t,i),$$

as we wanted to establish. □

Now, we interpret the multiple integrals with respect to the Brownian motion B as iterated Itô's integrals

Proposition 69 *Suppose that $n \geq 1$ and $h \in \mathcal{H}^{\otimes n}$. Then,*

$$\begin{aligned} I_n(h) &= n! \int_0^T \int_0^{t_n} \cdots \int_0^{t_2} \tilde{h}(t_1, \ldots, t_n) dB_{t_1} \cdots dB_{t_n} \\ &= n! \sum_{i_1, \ldots, i_n = 1}^{d} \int_0^T \int_0^{t_n} \cdots \int_0^{t_2} \tilde{h}_{t_1, \ldots, t_n}^{i_1, \ldots, i_n} dB_{t_1}^{(i_1)} \cdots dB_{t_n}^{(i_n)}. \end{aligned}$$

Remark 61 *Remember that $\tilde{h}$ is a symmetric function in the variables (t_ℓ, i_ℓ). That is,*

$$\tilde{h}_{t_1, \ldots, t_n}^{i_1, \ldots, i_n} = \frac{1}{n!} \sum_{\sigma \in S_n} \tilde{h}_{t_{\sigma(1)}, \ldots, t_{\sigma(n)}}^{i_{\sigma(1)}, \ldots, i_{\sigma(n)}},$$

which implies that $\tilde{h}_{t_1, \ldots, t_n}^{i_1, \ldots, i_n} = \tilde{h}_{t_{\sigma(1)}, \ldots, t_{\sigma(n)}}^{i_{\sigma(1)}, \ldots, i_{\sigma(n)}}$, for any $\sigma \in S_n$.

Proof. By Proposition 54, we have

$$\begin{aligned} & E\left(\left(n! \int_0^T \int_0^{t_n} \cdots \int_0^{t_2} \tilde{h}(t_1, \ldots, t_n) dB_{t_1} \cdots dB_{t_n}\right)^2\right) \\ &= (n!)^2 \sum_{i_n=1}^{d} \int_0^T E\left(\left(\int_0^{t_n} \cdots \int_0^{t_2} \tilde{h}(t_1, \cdots, t_{n-1}, (t_n, i_n)) dB_{t_1} \ldots dB_{t_{n-1}}\right)^2\right) dt_{t_n}. \end{aligned}$$

Thus, using induction on $n \in \mathbb{N}$ and Remark 22, we get

$$
\begin{aligned}
&E\left(\left(n!\int_0^T\int_0^{t_n}\cdots\int_0^{t_2}\tilde{h}(t_1,\ldots,t_n)dB_{t_1}\cdots dB_{t_n}\right)^2\right)\\
&= (n!)^2\sum_{i_1,\ldots,i_n=1}^{d}\int_0^T\int_0^{t_n}\cdots\int_0^{t_2}\left(\tilde{h}^{i_1,\ldots,i_n}_{t_1,\ldots,t_n}\right)^2 dt_1\cdots dt_n\\
&= n!\sum_{i_1,\ldots,i_n=1}^{d}\int_0^T\cdots\int_0^T\left(\tilde{h}^{i_1,\ldots,i_n}_{t_1,\ldots,t_n}\right)^2 dt_1\cdots dt_n = E\left(I_n(h)\right)^2. \qquad (4.26)
\end{aligned}
$$

Therefore, from Lemma 42, we only need to show that the result holds for $h \in \mathcal{E}_n$. So, we can assume that $h = 1_{A_1\times\cdots A_n}$ without loss of generality, where $A_1,\ldots,A_n$ are pairwise disjoint sets of $\mathcal{B}([0,T])\otimes 2^{\{1,\ldots,d\}}$.

Now, let $0 \le a_1 < b_1 \le a_2 < b_2 \le \cdots \le a_n < b_n \le T$, S_n the family of all the permutations of the set $\{1,\ldots,n\}$ and $C_i = 1_{(a_i,b_i]\times\{1,\ldots,d\}}$, $i = 1,\ldots,n$. Then, we have

$$
\begin{aligned}
&n!\int_0^T\int_0^{t_n}\cdots\int_0^{t_2}\left(\frac{1}{n!}\sum_{\eta\in S_n}1_{A_{\eta(1)}}(t_1)\cdots 1_{A_{\eta(n)}}(t_n)\right)\\
&\qquad\times\left(\sum_{\sigma\in S_n}1_{C_{\sigma(1)}}(t_1)\cdots 1_{C_{\sigma(n)}}(t_n)\right)dB_{t_1}\cdots dB_{t_n}\\
&= \sum_{\eta\in S_n}\int_0^T\int_0^{t_n}\cdots\int_0^{t_2}1_{A_{\eta(1)}\cap C_1}(t_1)\cdots 1_{A_{\eta(n)}\cap C_n}(t_n)dB_{t_1}\cdots dB_{t_n}\\
&= n!\sum_{\eta\in S_n}\int_0^T\cdots\int_0^T 1_{A_{\eta(1)}\cap C_1}(t_1)\cdots 1_{A_{\eta(n)}\cap C_n}(t_n)dB_{t_1}\cdots dB_{t_n}\\
&= n!\sum_{\eta\in S_n}\prod_{i=1}^{d}\int_0^T 1_{A_{\eta(i)}\cap C_i}(t_i)dB_{t_i} = \sum_{\eta\in S_n}\prod_{i=1}^{d}W\left(1_{A_{\eta(i)}\cap C_n}\right)\\
&= \sum_{\eta\in S_n}I_n\left(1_{A_{\eta(1)}\cap C_1}\cdots 1_{A_{\eta(n)}\cap C_n}\right) = I_n\left(\sum_{\eta\in S_n}1_{A_{\eta(1)}\cap C_1}\cdots 1_{A_{\eta(n)}\cap C_n}\right)\\
&= I_n\left(1_{A_1}\cdots 1_{A_n}\sum_{\eta\in S_n}1_{C_{\eta^{-1}(1)}}\cdots 1_{C_{\eta^{-1}(n)}}\right), \qquad (4.27)
\end{aligned}
$$

where the antepenultimate equality follows from (4.25).

On the other hand, from either measure theory, or the monotone class lemma (see Theorem 96), there exists a sequence $\{\phi_k : k \in \mathbb{N}\}$, with $\phi_k = \sum_{\ell=1}^{m_k} b_{\ell,k}1_{B_{\ell,k}}$, where $B_{1,k},\ldots,B_{m_k,k}$ are rectangles included in $\{0 \le t_1 < \cdots < t_n \le T\}$, and $\|\phi_k - 1_{\{0\le t_1<\cdots<t_n\le T\}}\|_{L^2([0,T]^n)} \to 0$, as $k\to\infty$. Hence, the sequence

$$
\bar{\phi}_k = \sum_{\ell=1}^{m_k} b_{\ell,k}1_{\{((t_1,i_1),\ldots,(t_n,i_n)):(t_1,\ldots,t_n)\in B_{\ell,k}\}}
$$

converges in $L^2(([0,T]\times\{1,\ldots,d\})^n)$ to $1_{\{((t_1,i_1),\ldots,(t_n,i_n)):t_1<t_2<\ldots<t_n)\}}$, as $k\to\infty$. Therefore, we can applied (4.26), Remark 22.*iii*) and (4.27) to deduce

$$
n!\int_0^T\int_0^{t_n}\cdots\int_0^{t_2}\tilde{1}_{A_1\times\cdots A_n}(t_1,\ldots,t_n)dB_{t_1}\cdots dB_{t_n} = I_n\left(1_{A_1\times\cdots A_n}\right).
$$

Consequently, the proof is finished. □

The divergence operator δ_2 with respect to W (see (4.20)) is an extension of the Itô's integral with respect to B in the sense that the set $\mathcal{L}^2_{\rm a}(\Omega \times [0,T];\mathbb{R}^d)$ of the $\mathbb{R}^d$-valued, square-integrable and $\mathcal{F}_t$-adapted processes is included in $\operatorname{Dom}\delta_2$, and the operator δ_2 restricted to $\mathcal{L}^2_{\rm a}(\Omega \times [0,T];\mathbb{R}^d)$ coincides with the Itô's stochastic integral with respect to the Brownian motion B. Note that $u \in \mathcal{L}^2_{\rm a}(\Omega\times[0,T];\mathbb{R}^d)$ if and only if $u^{(\ell)} \in \mathcal{L}^2_{\rm a}(\Omega\times[0,T])$, for all $\ell = 1,\dots,d$. In order to see that our claim holds, we choose an $\mathbb{R}^d$-valued process u in $\mathcal{L}^2_{\rm a}(\Omega\times[0,T];\mathbb{R}^d)$. Then, Corollary 19 applied to each component of u implies that there exists a sequence $\{u_n \in \mathcal{L}^2_{\rm a}(\Omega \times [0,T];\mathbb{R}^d) : n \in \mathbb{N}\}$ of $\mathbb{R}^d$-valued, simple and $\mathcal{F}_t$-adapted processes such that $\|u - u_n\|^2_{\mathcal{L}^2_{\rm a}(\Omega\times[0,T];\mathbb{R}^d)} \to 0$, as $n \to \infty$. Remember that the $\mathbb{R}^d$-valued stochastic process u_n has the form

$$u_n(t) = \sum_{i=1}^{m} F_i 1_{]t_{i-1},t_i]}(t), \quad t \in [0,T], \tag{4.28}$$

where $\{0 = t_0 < t_1 < \ldots < t_m = T\}$ is a partition of $[0,T]$ and F_i is an $\mathcal{F}_{t_{i-1}}$-measurable and square-integrable random vector taking values in $\mathbb{R}^d$, for all $i = 1,\dots,m$. We observe that we can also assume that $F_1^{(\ell)},\dots,F_m^{(\ell)} \in \mathcal{S}$ without loss of generality due to $\mathcal{S}$ being dense in $L^2(\Omega)$. Thus, u_n is in $\mathcal{S}(\mathcal{H})$ because $F_i 1_{]t_{i-1},t_i]} = \sum_{j=1}^d F_i^{(j)} e_j 1_{]t_{i-1},t_i]}$, where $\{e_1,\dots,e_d\}$ is the canonical basis of $\mathbb{R}^d$.

Now, we can prove that our claim is satisfied.

Proposition 70 *Suppose that $u \in \mathcal{L}^2_a(\Omega \times [0,T];\mathbb{R}^d)$. Then, u belongs to $\operatorname{Dom}\delta_2$ and*

$$\delta_2(u) = \int_0^T u_s dB_s, \tag{4.29}$$

where the right-hand side is the Itô's integral of u with respect to the Brownian motion B.

Proof. Choose $\ell \in \{1,\dots,d\}$. Thus, applying (4.28), Lemma 36, Remark 24 and Corollary 25, we can state

$$\begin{aligned} \delta_2^{B^{(\ell)}}(u^{(\ell)}) &= \sum_{i=1}^m \delta_2^{B^{(\ell)}}\left(F_i^{(\ell)} 1_{]t_{i-1},t_i]}\right) = \sum_{i=1}^m \left\{ F_i^{(\ell)} \delta_2^{B^{(\ell)}}\left(1_{]t_{i-1},t_i]}\right) - \int_{t_{i-1}}^{t_i} D_s^{(\ell)} F_i^{(\ell)} ds \right\} \\ &= \sum_{i=1}^m F_i^{(\ell)} \delta_2^{B^{(\ell)}}\left(1_{]t_{i-1},t_i]}\right) = \sum_{i=1}^m F_i^{(\ell)} \int_{t_{i-1}}^{t_i} dB_s^{(\ell)} = \int_0^T u_s^{(\ell)} dB_s^{(\ell)}. \end{aligned}$$

Hence, by Proposition 67, the process u introduced in (4.28) belongs to $\operatorname{Dom}\delta_2$ and (4.29) holds.

Finally, the result follows using that, for $u \in \mathcal{L}^2_{\rm a}(\Omega \times [0,T];\mathbb{R}^d)$, there is a sequence $\{u_n \in \mathcal{S}(\mathcal{H}) : n \in \mathbb{N}\}$ of $\mathbb{R}^d$-valued and $\mathcal{F}_t$-adapted processes that goes to u in the space $L^2(\Omega \times [0,T];\mathbb{R}^d)$, as $n \to \infty$ (see Corollary 19). Thus, the fact that δ_2 is a closed operator from $L^2(\Omega \times [0,T];\mathbb{R}^d)$ into $L^2(\Omega)$ and the isometry relation

$$E\left(\left(\int_0^T v_s dB_s\right)^2\right) = E\left(\int_0^T \|v_s\|^2_{\mathbb{R}^d} ds\right), \quad \text{for } v \in \mathcal{L}^2_{\rm a}(\Omega \times [0,T];\mathbb{R}^d),$$

imply that the result is satisfied. □

We can also establish the following local version of Proposition 70.

Proposition 71 *Suppose that* $u \in \mathcal{L}^2_{\mathrm{a}}(\Omega\times[0,T];\mathbb{R}^d)_{loc}$. *Then,* u *also belongs to* $(Dom\,\delta_2)_{loc}$ *and*

$$\delta_2(u) = \int_0^T u_s dB_s,$$

where the divergence $\delta_2(u)$ *is given by Definition 38, while the right-hand side is defined as in (3.12).*

Remark 62 *We have that* u *is an* $\mathbb{R}^d$*-valued, measurable and* $\mathcal{F}_t$*-adapted process with square-integrable paths because of Corollary 20 and Proposition 56.*

Proof. The result is an immediate consequence of Propositions 55, 56 and 70, and Corollary 20. □

The Brownian motion has the integral representation (see, for instance, Bojdecki [22], or Dellacherie and Meyer [38]). That is, let $F \in L^2(\Omega, \mathcal{F}_T, P)$, then there exists a unique $\phi \in \mathcal{L}^2_{\mathrm{a}}(\Omega \times [0,T]; \mathbb{R}^d)$ such that

$$F = E(F) + \int_0^T \phi(s) dB_s = E(F) + \sum_{i=1}^d \int_0^T \phi^{(i)}(s) dB_s^{(i)}.$$

However, it is not easy to figure out the $\mathbb{R}^d$-valued process ϕ in general. The first author who calculates this ϕ is Clark [30] in the case that F has a derivative in the Fréchet sense. Later, Haussmann [74] deals with this problem when F is a function of a solution of a stochastic differential equation driven by B. The process ϕ can be written in terms of the derivative DF, as it was established by Ocone [163]. Namely, we have the following:

Proposition 72 (Clark-Haussmann-Ocone formula) *Let* $F \in \mathbb{D}^{1,2}$ *be an* $\mathcal{F}_{[0,T]\times\{1,\dots,d\}}$*-measurable random variable. Then,*

$$\begin{aligned} F &= E(F) + \int_0^T E\left[D_s F \middle| \mathcal{F}_{[0,s]\times\{1,\dots,d\}}\right] dB_s \\ &= E(F) + \sum_{i=1}^d \int_0^T E\left[D_s^i F \middle| \mathcal{F}_{[0,s]\otimes\{1,\dots,d\}}\right] dB_s^{(i)}, \end{aligned}$$

where the stochastic integral with respect to B *is in the Itô's sense.*

Proof. Assume that F has the chaos decomposition

$$F = E(F) + \sum_{n=1}^\infty I_n\left(\tilde{h}_n\right), \tag{4.30}$$

with $h_n \in \mathcal{H}^{\otimes n}$, for $n \geq 1$. From Corollary 12 and Proposition 68, we have

$$D_t^j F = \sum_{n=1}^\infty n I_{n-1}\left((\tilde{h}_n)_{t,\cdot}^{j,\cdot}\right), \quad (t,j) \in [0,T] \times \{1,\dots,d\}.$$

Thus, Lemma 43 allows us to establish

$$E\left[D_t^j F \middle| \mathcal{F}_{[0,t]\otimes\{1,\dots,d\}}\right] = \sum_{n=1}^\infty n I_{n-1}\left((\tilde{h}_n)_{t,\cdot}^{j,\cdot} 1_{([0,t]\times\{1,\dots,d\})^{n-1}}(\cdot)\right), \quad t \in [0,T].$$

Consequently, Lemma 39 implies

$$\delta_2\left(E\left[D_t^j F|\mathcal{F}_{[0,t]\otimes\{1,\ldots,d\}}\right]\right)=\sum_{n=1}^{\infty} I_n\left(\sum_{i=1}^{n}\left[(\tilde{h}_n)_{t_i,\cdot}^{j_i,\cdot}1_{([0,t_i]\times\{1,\ldots,d\})^{n-1}}\right]((t_i,\hat{j}_i))\right),$$

where $(t_i,\hat{j}_i))$ stands for the vector $((t_1,j_1),\ldots,(t_{i-1},j_{i-1}),(t_{i+1},j_{i+1}),\ldots,(t_n.j_n))$, (i.e.,the i-th component is omitted). Hence, the result is a consequence of (4.30), and Propositions 69 and 70. □

The Clark-Haussmann-Ocone formula has been extended to random variables in $\mathbb{D}^{1,1}$ by Karatzas et al. [95], and this extension was used by Ocone and Karatzas [165] to deal with hedging strategies in complete financial markets driven by B.

On the other hand, note that an $\mathcal{F}_t$-adapted and simple stochastic process u as in (4.28) does not necessarily belong to $\mathbb{D}^{1,2}(\mathcal{H})$, but it is in $\operatorname{Dom}\delta_2$, at least locally, according to Propositions 70 and 71. In order to state this fact, in the proof of Proposition 70, we approximate the $\mathbb{R}^d$-valued process $F_i 1_{]t_{i-1},t_i]}$ by a sequence $\{F_{i,n}1_{]t_{i-1},t_i]} : n\in\mathbb{N}\}$, with $F_{i,n}^{(\ell_1)}\in\mathbb{D}^{1,2}$, such that $D_t^{(\ell_2)}\left(F_{i,n}^{(\ell_1)}1_{]t_{i-1},t_i]}(s)\right)=0$ for $0\le s<t\le T$ and $\ell_1,\ell_2=1,\ldots,d$. This idea is used by Alòs and Nualart [12] to introduce the space $\mathbb{L}^F$ of the $\mathbb{R}^d$-valued stochastic processes having two derivatives in the future sense, which is contained in $\operatorname{Dom}\delta_2$. Moreover, $\mathbb{L}^F$ contains the space $\mathbb{D}^{2,2}(\mathcal{H})$ and the square-integrable $\mathcal{F}_t$-adapted processes taking values in $\mathbb{R}^d$. This space is built as follows. Consider the sets

$$\Delta_1^T=\{(s,t)\in[0,T]^2 : s\ge t\}\quad\text{and}\quad \Delta_2^T=\{(r,s,t)\in[0,T]^3 : r\vee s\ge t\}. \tag{4.31}$$

The spaces $\mathbb{L}^{1,2,f}$ and $\mathbb{L}^F$ are the completion of $\mathcal{S}(\mathcal{H})$ with respect to the seminorms

$$\begin{aligned}\|u\|_{1,2,f}^2 &= E\left(\|u\|_{\mathcal{H}}^2\right)+E\left(\|1_{\Delta_1^T}Du\|_{\mathcal{H}\otimes\mathcal{H}}^2\right)\\ &= E\left(\int_0^T\|u_t\|_{\mathbb{R}^d}^2dt\right)+\sum_{\ell_1,\ell_2=1}^{d}E\left(\int_{\Delta_1^T}\left[(Du_t^{\ell_2})_s^{\ell_1}\right]^2dsdt\right)\end{aligned} \tag{4.32}$$

and

$$\begin{aligned}\|u\|_{\mathbb{L}^F}^2 &= \|u\|_{1,2,f}^2+E\left(\|1_{\Delta_2^T}D^2u\|_{\mathcal{H}^{\otimes 2}\otimes\mathcal{H}}^2\right)\\ &= \|u\|_{1,2,f}^2+\sum_{\ell_1,\ell_2,\ell_3=1}^{d}E\left(\int_{\Delta_2^T}\left[\left(D^2u_t^{\ell_3}\right)_{r,s}^{\ell_1,\ell_2}\right]^2drdsdt\right),\end{aligned} \tag{4.33}$$

respectively.

In order to show that these two spaces are well-defined, we state the following result.

Lemma 44 *The operator* $D:\mathcal{S}(\mathcal{H})\to L^2\left(\Omega\times\tilde{\Delta}_1^T;P\times(\lambda\times\delta)^2\right)$ *is closable from* $L^2(\Omega\times[0,T];\mathbb{R}^d)$ *into* $L^2\left(\Omega\times\tilde{\Delta}_1^T;P\times(\lambda\times\delta)^2\right)$, *where*

$$\tilde{\Delta}_1^T=\left\{((s,i),(t,j))\in([0,T]\times\{1,\ldots,d\})^2 : (s,t)\in\Delta_1^T\right\}.$$

Proof. The proof is similar to that of Theorem 14. So, consider a sequence $\{F_n\in\mathcal{S}(\mathcal{H}) : n\in\mathbb{N}\}$ such that it converges to zero in $L^2\left(\Omega\times[0,T];\mathbb{R}^d\right)$ and DF_n goes to Y in $L^2\left(\Omega\times\tilde{\Delta}_1^T;P\times(\lambda\times\delta)^2\right)$. Then, for $G\in\mathcal{S}$ and $h\in L^2\left(\tilde{\Delta}_1^T;(\lambda\times\delta)^2\right)$, we have that

Remark 24 yields

$$\begin{aligned}
\langle Y, hG\rangle_{L^2(\Omega\times\tilde{\Delta}_1^T;P\times(\lambda\times\delta)^2)} &= \lim_{n\to\infty} E\left(\langle DF_n, h\rangle_{L^2(\tilde{\Delta}_1^T;(\lambda\times\delta)^2)}\, G\right) \\
&= \lim_{n\to\infty} E\left(\langle D(GF_n), h\rangle_{L^2(\tilde{\Delta}_1^T;(\lambda\times\delta)^2)}\right) \\
&\quad - \lim_{n\to\infty} E\left(\langle (DG)\otimes F_n, h\rangle_{L^2(\tilde{\Delta}_1^T;(\lambda\times\delta)^2)}\right) \\
&= \lim_{n\to\infty} E\left(\left\langle D(GF_n), 1_{\tilde{\Delta}_1^T} h\right\rangle_{L^2(([0,T]\times\{1,\ldots,d\})^2;(\lambda\times\delta)^2)}\right) \\
&= \lim_{n\to\infty} E\left(G\left\langle F_n, W\left(1_{\tilde{\Delta}_1^T} h\right)\right\rangle_{\mathcal{H}}\right) = 0,
\end{aligned}$$

which gives that D is a closable operator because

$$\operatorname{Span}\left(\left\{Gh : G\in\mathcal{S} \text{ and } h\in L^2(\tilde{\Delta}_1^T;(\lambda\times\delta)^2)\right\}\right)$$

is a dense set of $L^2\left(\Omega\times\tilde{\Delta}_1^T; P\times(\lambda\times\delta)^2\right)$. □

We observe that last lemma implies that the space $\mathbb{L}^{1,2,f}$ is well-defined and that an d-dimensional stochastic process u belongs to this space if and only if there exists a kernel $Du = \{D_s u_t : (s,t)\in\Delta_1^T\} \in L^2(\Omega\times\tilde{\Delta}_1^T; P\times(\lambda\times\delta)^2)$ and a sequence $\{u_n\in\mathcal{S}(\mathcal{H}) : n\in\mathbb{N}\}$ such that

$$E\left(\|u_n-u\|_{\mathcal{H}}^2 + \|1_{\Delta_1^T}(D_s u_n(t) - D_s u_t)\|_{\mathcal{H}\otimes\mathcal{H}}^2\right)\to 0 \quad \text{as } n\to\infty.$$

Hence, as it is pointed out by Alòs and Nualart [12], $\mathbb{L}^{1,2,f}$ can be interpreted as the space of the d-dimensional stochastic processes with a derivative in the future sense, in the Malliavin calculus sense. Also observe that a d-dimensional stochastic process in $\mathbb{D}^{1,2}(\mathcal{H})$ is a process in the space $\mathbb{L}^{1,2,f}$, but a process in $\mathbb{L}^{1,2,f}$ does not necessarily belong to $\mathbb{D}^{1,2}(\mathcal{H})$, as it is the case of the $\mathcal{F}_t$-adapted process given in (4.28), when $F_i^{(\ell)}\notin\mathbb{D}^{1,2}$, for some $i\in\{1,\ldots,m\}$ and $\ell\in\{1,\ldots,d\}$. Moreover, by choosing a sequence $\{u_n : n\in\mathbb{N}\}$ as in the proof of Proposition 70, we have that the space $\mathbb{L}^{1,2,f}$ contains $\mathcal{L}_{\mathrm{a}}^2(\Omega\times[0,T];\mathbb{R}^d)$ because the $\mathcal{H}$-valued smooth random variables in the last space have a derivative in the future sense equal to zero, which is used in the proof of Proposition 70.

The kernel Du, for $u\in\mathbb{L}^{1,2,f}$, also enjoys the local property. In order to prove this claim, we first analyze the following result.

Proposition 73 *Let $\varphi\in C^1(\mathbb{R}^{nd};\mathbb{R}^d)$ be a bounded continuously differentiable function with bounded partial derivatives of order* 1 *and $u_1,\ldots,u_n$ processes in $\mathbb{L}^{1,2,f}$. Then, $\varphi(u_1,\ldots,u_n)\in\mathbb{L}^{1,2,f}$ and*

$$D_s^{\cdot}\left(\varphi^{(\ell_1)}(u_1(t),\ldots,u_n(t))\right) = \sum_{i=1}^{n}\sum_{\ell=1}^{d}\partial_{x_{i\ell}}\varphi^{(\ell_1)}(u_1(t),\ldots,u_n(t))D_s^{\cdot}(u_i)_t^{\ell},$$

for almost all $(\omega,(s,t))\in\Omega\times\Delta_1^T$ and all $\ell_1\in\{1,\ldots,d\}$.

Proof. The result is true for $u\in\mathcal{S}(\mathcal{H})$ due to Corollary 10. Thus, the facts that φ and its partial derivatives of order 1 are bounded, together with the dominated convergence theorem, yield that the result is satisfied. □

Proposition 74 *Let $u\in\mathbb{L}^{1,2,f}$ and $A\in\mathcal{F}$ such that $u_t=0$ for almost all $(\omega,t)\in A\times[0,T]$. Then, $D_s u_t = 0$ for almost all $(\omega,(s,t))$ in $A\times\Delta_1^T$.*

Proof. The proof is quite similar to that of Proposition 22. So, we only sketch it and use the notation introduced in the proof of Proposition 22.

Let $\bar{\phi}_\varepsilon : \mathbb{R}^d \rightarrow: \mathbb{R}^d$ given by $\bar{\phi}_\varepsilon(x) = (\phi_\varepsilon(x_1), \ldots, \phi_\varepsilon(x_d))$. Also, let $v = \sum_{i=1}^m G_i h_i$, with $m \in \mathbb{N}$, $G_i \in \mathcal{S}$ and $h_i \in \mathcal{H}^{\otimes 2}$. Then, Proposition 73 gives

$$E\left(\sum_{\ell_1,\ell_2=1}^{d} \int_{\Delta_1^T} \psi_\varepsilon(u_t^{\ell_2})(Du_t^{\ell_2})_s^{\ell_1} v_{s,t}^{\ell_1,\ell_2} dsdt\right) = E\left(\sum_{\ell_1,\ell_2=1}^{d} \int_{\Delta_1^T} (D\bar{\phi}^{(\ell_2)}(u_t))_s^{\ell_1} v_{s,t}^{\ell_1,\ell_2} dsdt\right).$$

Therefore, proceeding as in the proof of Proposition 22, we can write

$$\begin{aligned}
&\left| E\left(\sum_{\ell_1,\ell_2=1}^{d} \int_{\Delta_1^T} \psi_\varepsilon(u_t^{\ell_2})(Du_t^{\ell_2})_s^{\ell_1} v_{s,t}^{\ell_1,\ell_2} dsdt\right)\right| \\
&= \left| \sum_{\ell_1,\ell_2=1}^{d} \sum_{i=1}^{m} E\left\{ G_i \int_0^T \phi_\varepsilon(u_t^{\ell_2}) \delta_2^{B^{(\ell_1)}}((h_i)_{\cdot,t}^{\ell_1,\ell_2} 1_{\Delta_1^T}(\cdot,t))dt \right\} \right. \\
&\quad \left. - \sum_{\ell_1,\ell_2=1}^{d} \sum_{i=1}^{m} E\left\{ \int_{\Delta_1^T} \phi_\varepsilon(u_t^{\ell_2})(DG_i)_s^{\ell_1} (h_i)_{s,t}^{\ell_1,\ell_2}) dsdt \right\} \right| \\
&\leq 4\varepsilon \sum_{\ell_1\ell_2=1}^{d} \sum_{i=1}^{m} E\left\{ |G_i| \int_0^T \left| \delta_2^{B^{(\ell_1)}}((h_i)_{\cdot,t}^{\ell_1,\ell_2} 1_{\Delta_1^T}(\cdot,t))\right| dt + \int_{\Delta_1^T} \left| (DG_i)_s^{\ell_1} (h_i)_{s,t}^{\ell_1,\ell_2}\right| dsdt \right\}.
\end{aligned}$$

Hence, letting $\varepsilon \downarrow 0$, and applying the dominated convergence theorem, we can get

$$\sum_{\ell_1,\ell_2=1}^{d} E\left(\int_{\Delta_1^T} 1_{\{u_t^{\ell_2}=0\}} (Du_t^{\ell})_s^{\ell_1} v_{s,t}^{\ell_1,\ell_2} dsdt\right) = 0.$$

Thus, $1_{\{u_t^{\ell_2}=0\}}(Du_t^{\ell_2})_s^{\ell_1} = 0$, for almost all $(\omega,(s,t)) \in \Omega \times \Delta_1^T$ and for all $\ell_1, \ell_2 \in \{1, \ldots, d\}$. It means, the proof is complete. □

As in Definition 23, Proposition 74 allows to consider the operator D on a bigger family of process than $\mathbb{L}^{1,2,f}$.

Definition 39 *A stochastic process u belongs to $\mathbb{L}_{loc}^{1,2,f}$ if there exists a sequence $\{(\Omega_n, u_n) \in \mathcal{F} \times \mathbb{L}^{1,2,f} : n \in \mathbb{N}\}$ such that*

i) $\Omega_n \uparrow \Omega$ w.p.1, as $n \to \infty$.

ii) $u = u_n$, for almost all $(\omega,(s,t)) \in \Omega_n \times \Delta_1^T$.

In this case, we define the kernel $\{D_s u(t) : (s,t) \in \Delta_1^T\}$ as

$$D_s u(t) = D_s u_n(t) \quad \text{on } \Omega_n \times \Delta_1^T.$$

Note that, by Corollary 20 and Proposition 56, we have the space $\mathcal{L}_a(\Omega; L^2([0,T];\mathbb{R}^d))$ of all the d-dimensional $\mathcal{F}_t$-adapted and measurable process u such that $\int_0^T \|u_s\|_{\mathbb{R}^d}^2 ds < \infty$, w.p.1, is a subset of $\mathbb{L}_{loc}^{1,2,f}$ since

$$\mathcal{L}_a^2(\Omega \times [0,T];\mathbb{R}^d)_{loc} = \mathcal{L}_a(\Omega; L^2([0,T];\mathbb{R}^d)).$$

Also observe that Propositions 73 and 74, and Definition 39, together with the proofs of Propositions 21 and 23, allow us to state the following result. So, the proof of this result is left as an exercise for the reader.

Proposition 75 *Let $\varphi \in C^1(\mathbb{R}^{nd};\mathbb{R}^d)$ be a continuously differentiable function and $u_1,\ldots,u_n$ stochastice processes in $\mathbb{L}^{1,2,f}$. Then, $\varphi(u_1,\ldots,u_n) \in \mathbb{L}^{1,2,f}$ if and only if $\varphi(u_1,\ldots,u_n) \in L^2(\Omega \times [0,T];\mathbb{R}^d)$ and*

$$\sum_{i=1}^{n}\sum_{\ell=1}^{d} \partial_{x_{i\ell}}\varphi^{(\ell_1)}(u_1(t),\ldots,u_n(t))(D(u_i)_t^{\ell})_s^{\ell_2} \in L^2\left(\Omega \times \tilde{\Delta}_1^T; P\times(\lambda\times\delta)^2\right),$$

for all $\ell_1,\ell_2 \in \{1,\ldots,d\}$. Moreover, in this case,

$$D_s^{\cdot}\left(\varphi^{(\ell_1)}(u_1(t),\ldots,u_n(t))\right) = \sum_{i=1}^{n}\sum_{\ell=1}^{d} \partial_{x_{i\ell}}\varphi^{(\ell_1)}(u_1(t),\ldots,u_n(t))D_s^{\cdot}((u_i)_t^{\ell}),$$

for almost all $(\omega,(s,t)) \in \Omega\times\Delta_1^T$ and all $\ell_1 \in \{1,\ldots,d\}$.

Now, we deal with the space $\mathbb{L}^F$. In the following result, we utilize the notation

$$\tilde{\Delta}_2^T = \left\{((r,\ell_1),(s,\ell_2),(t,\ell_3)) \in ([0,T]\times\{1,\ldots,d\})^3 : (r,s,t)\in\Delta_2^T\right\}.$$

Lemma 45 *The operator*

$$D : \mathcal{S}(\mathcal{H}) \subset L^2\left(\Omega\times[0,T];\mathbb{R}^d\right) \to L^2\left(\Omega\times\tilde{\Delta}_1^T; P\times(\lambda\times\delta)^2\right) \times L^2\left(\Omega\times\tilde{\Delta}_2^T; P\times(\lambda\times\delta)^3\right)$$

is closable.

Proof. Let $\{F_n \in \mathcal{S}(\mathcal{H}) : n \in \mathbb{N}\}$ be a sequence such that it goes to zero in $L^2\left(\Omega\times[0,T];\mathbb{R}^d\right)$, and there exist $Y_1 \in L^2(\Omega\times\tilde{\Delta}_1^T; P\times(\lambda\times\delta)^2)$ and $Y_2 \in L^2(\Omega\times\tilde{\Delta}_2^T; P\times(\lambda\times\delta)^3)$ such that

$$E\left(\|1_{\tilde{\Delta}_1^T}(DF_n - Y_1)\|^2_{\mathcal{H}^{\otimes 2}} + \|1_{\tilde{\Delta}_2^T}(D^2F_n - Y_2)\|^2_{\mathcal{H}^{\otimes 3}}\right) \to 0, \quad \text{as } n\to 0.$$

Therefore, the proof of Lemma 44 yields that $Y_1 = 0$. Thus, in order to finish the proof, we only need to show that $Y_2 = 0$. Towards this end, we have that, for $h \in L^2(([0,T]\times\{1,\ldots,d\})^3;(\lambda\times\delta)^3)$,

$$\begin{aligned}
&\sum_{\ell_1,\ell_2,\ell_3=1}^{d}\int_0^T\int_0^T\int_0^{r\vee s}\left[D^2((F_n)_t^{\ell_3})\right]_{r,s}^{\ell_1,\ell_2}(h)_{r,s,t}^{\ell_1,\ell_2,\ell_3}dtdrds\\
&= \sum_{\ell_1,\ell_2,\ell_3=1}^{d}\int_0^T\int_0^s\int_0^s\left[D^2((F_n)_t^{\ell_3})\right]_{r,s}^{\ell_1,\ell_2}(h)_{r,s,t}^{\ell_1,\ell_2,\ell_3}dtdrds\\
&\quad+\sum_{\ell_1,\ell_2,\ell_3=1}^{d}\int_0^T\int_s^T\int_0^r\left[D^2((F_n)_t^{\ell_3})\right]_{r,s}^{\ell_1,\ell_2}(h)_{r,s,t}^{\ell_1,\ell_2,\ell_3}dtdrds\\
&= \sum_{\ell_1,\ell_2,\ell_3=1}^{d}\int_0^T\int_0^s\int_0^s\left[D^2((F_n)_t^{\ell_3})\right]_{r,s}^{\ell_1,\ell_2}(h)_{r,s,t}^{\ell_1,\ell_2,\ell_3}dtdrds\\
&\quad+\sum_{\ell_1,\ell_2,\ell_3=1}^{d}\int_0^T\int_0^r\int_0^r\left[D^2((F_n)_t^{\ell_3})\right]_{r,s}^{\ell_1,\ell_2}(h)_{r,s,t}^{\ell_1,\ell_2,\ell_3}dtdsdr\\
&= \sum_{\ell_1,\ell_2,\ell_3=1}^{d}\int_0^T\int_0^s\int_0^s\left[D^2((F_n)_t^{\ell_3})\right]_{r,s}^{\ell_1,\ell_2}\left((h)_{r,s,t}^{\ell_1,\ell_2,\ell_3}+(h)_{s,r,t}^{\ell_2,\ell_1,\ell_3}\right)dtdrds.
\end{aligned}$$

Hence, for $G \in \mathcal{S}$ and $h = h_1 \otimes h_2 \otimes h_3 \in L^2(([0,T] \times \{1,\dots,d\})^3; (\lambda \times \delta)^3)$, we obtain

$$\begin{aligned}
&\sum_{\ell_1,\ell_2,\ell_3=1}^{d} E\left(G \int_0^T \int_0^T \int_0^{r\vee s} \left[D^2((F_n)_t^{\ell_3})\right]_{r,s}^{\ell_1,\ell_2} (h)_{r,s,t}^{\ell_1,\ell_2,\ell_3} dtdrds\right) \\
&= \sum_{\ell_1,\ell_2,\ell_3=1}^{d} E\left(\int_0^T \int_0^s \int_0^s \left[D\left[GD(F_n)_t^{\ell_3})\right]_s^{\ell_2}\right]_r^{\ell_1} \left((h)_{r,s,t}^{\ell_1,\ell_2,\ell_3} + (h)_{s,r,t}^{\ell_2,\ell_1,\ell_3}\right) dtdrds\right) \\
&\quad - \sum_{\ell_1,\ell_2,\ell_3=1}^{d} E\left(\int_0^T \int_0^s \int_0^s [DG]_r^{\ell_1} \left[D((F_n)_t^{\ell_3})\right]_s^{\ell_2} \left((h)_{r,s,t}^{\ell_1,\ell_2,\ell_3} + (h)_{s,r,t}^{\ell_2,\ell_1,\ell_3}\right) dtdrds\right) \\
&= \sum_{\ell_1,\ell_2,\ell_3=1}^{d} E\Bigg(G \int_0^T \int_0^s \left[D(F_n)_t^{\ell_3})\right]_s^{\ell_2} \\
&\qquad \times \left(\delta_2^{B^{(\ell_1)}}\left((h_1)_\cdot^{\ell_1} 1_{[0,s]}(\cdot)\right)(h_2)_s^{\ell_2}(h_3)_t^{\ell_3} + \delta_2^{B^{(\ell_1)}}\left((h_2)_\cdot^{\ell_1} 1_{[0,s]}(\cdot)\right)(h_1)_s^{\ell_2}(h_3)_t^{\ell_3}\right) dtds\Bigg) \\
&\quad - \sum_{\ell_1,\ell_2,\ell_3=1}^{d} E\left(\int_0^T \int_0^s \int_0^s [DG]_r^{\ell_1} \left[D((F_n)_t^{\ell_3})\right]_s^{\ell_2} \left((h)_{r,s,t}^{\ell_1,\ell_2,\ell_3} + (h)_{s,r,t}^{\ell_2,\ell_1,\ell_3}\right) dtdrds\right) \\
&\to 0, \quad \text{as } n \to \infty,
\end{aligned}$$

which follows from equality (4.5) and the fact that $Y_1 = 0$. Consequently,

$$\begin{aligned}
&E\left(G \int_0^T \int_0^T \int_0^{r\vee s} (Y_2)_{r,s,t}^{\ell_1,\ell_2,\ell_3})(h)_{r,s,t}^{\ell_1,\ell_2,\ell_3} dtdrds\right) \\
&= \lim_{n\to\infty} \sum_{\ell_1,\ell_2,\ell_3=1}^{d} E\left(G \int_0^T \int_0^T \int_0^{r\vee s} \left[D^2((F_n)_t^{\ell_3})\right]_{r,s}^{\ell_1,\ell_2} (h)_{r,s,t}^{\ell_1,\ell_2,\ell_3} dtdrds\right) \\
&= 0.
\end{aligned}$$

Thus $Y_2 = 0$ and the proof is complete. □

Note that $\mathbb{D}^{2,2}(\mathcal{H})$ is contained in $\mathbb{L}^F$, but, as we have already pointed out, $\mathbb{L}^F \not\subset \mathbb{D}^{2,2}(\mathcal{H})$ because we can find $\mathbb{R}^d$-valued simple and $\mathcal{F}_t$-adapted processes that are not in $\mathbb{L}^{1,2,f}$. Furthermore, a d-dimensional stochastic process u belongs to $\mathbb{L}^F$ if and only if there exist two kernels $Du = \{D_s u_t : (s,t) \in \Delta_1^T\} \in L^2(\Omega \times \tilde{\Delta}_1^T; P \times (\lambda \times \delta)^2)$ and $DDu = \{D_r D_s u_t : (r,s,t) \in \Delta_2^T\} \in L^2(\Omega \times \tilde{\Delta}_2^T; P \times (\lambda \times \delta)^3)$, and a sequence $\{u_n \in \mathcal{S}(\mathcal{H}) : n \in \mathbb{N}\}$ such that

$$E\left(\|u_n - u\|_{\mathcal{H}}^2 + \|1_{\tilde{\Delta}_1^T}(Du_n - Du)\|_{\mathcal{H}^{\otimes 2}} + \|1_{\tilde{\Delta}_2^T}(DDu_n - DDu)\|_{\mathcal{H}^{\otimes 3}}\right) \to 0$$

as $n \to \infty$. Note that the proof of Lemma 45 implies that we can find a subsequence $\{n_k : k \in \mathbb{N}\}$ of $\mathbb{N}$ such that, for almost all $s \in [0,T]$ and all $\ell_1, \ell_2, \ell_3 \in \{1,\dots,d\}$, we have

$$E\left\{\int_0^s \int_0^s \left(\left[D^2((u_{n_k})_t^{\ell_3})\right]_{r,s}^{\ell_1,\ell_2} - \left[DD((u)_t^{\ell_3})\right]_{r,s}^{\ell_1,\ell_2}\right)^2 dtdr\right\} \to \infty$$

as $k \to \infty$. Consequently, the kernels Du_n and $D^2 u_n$ are two types of derivatives of an $\mathcal{H}$-valued smooth functional in the Malliavin calculus sense (see (4.21)). In particular, an $\mathbb{R}^d$-valued simple and $\mathcal{F}_t$-adapted process v belongs to $\mathbb{L}^F$ with kernels $\{D_s v_t : (s,t) \in \Delta_1^T\}$ and $\{D_r D_s v_t : (r,s,t) \in \Delta_2^T\}$ equal to zero. Thus, we have established the following result.

Proposition 76 *Let $u \in \mathcal{L}^2_a(\Omega \times [0,T]; \mathbb{R}^d)$. Then, u also belongs to $\mathbb{L}^F$ with $Du, DDu \equiv 0$ and*

$$\|u\|^2_{\mathbb{L}^F} = \|u\|^2_{L^2(\Omega\times[0,T];\mathbb{R}^d)}.$$

The following Lemma is an isometry property for the Skorohod integral of some processes in $\mathbb{L}^{1,2,f}$ (compare it with equality (4.5)).

Lemma 46 *Consider a process u in $\mathbb{D}^{2,2}(\mathcal{H})$. Then,*

$$E\left(\left(\int_0^T u_s\delta_2 B_s\right)^2\right) = E\left(\int_0^T \|u_s\|^2_{\mathbb{R}^d} ds\right) + 2\sum_{\ell_1,\ell_2=1}^d E\left(\int_0^T u_s^{\ell_2}\delta_2^{(B^{\ell_1})}\left(1_{[0,s]}(\cdot)(Du_\cdot^{\ell_1})_s^{\ell_2}\right)ds\right). \quad (4.34)$$

Proof. Proposition 62 and equality (4.5) lead us to write

$$\begin{aligned}
&E\left(\left(\int_0^T u_s\delta_2 B_s\right)^2\right)\\
&= E\left(\int_0^T \|u_s\|^2_{\mathbb{R}^d} ds\right) + \sum_{\ell_1,\ell_2=1}^d E\left(\int_0^T\int_0^T (Du_t^{\ell_1})_s^{\ell_2}(Du_s^{\ell_2})_t^{\ell_1} dtds\right)\\
&= E\left(\int_0^T \|u_s\|^2_{\mathbb{R}^d} ds\right) + 2\sum_{\ell_1,\ell_2=1}^d E\left(\int_0^T\int_0^s (Du_t^{\ell_1})_s^{\ell_2}(Du_s^{\ell_2})_t^{\ell_1} dtds\right).
\end{aligned}$$

Thus, the fact that $u \in \mathbb{D}^{2,2}(\mathcal{H})$ and (4.2) yield that the result is satisfied. □

The derivative $(Du_t^{\ell_1})_s^{\ell_2}$ that appears in (4.34) is only evaluated in $\tilde{\Delta}_1^T$. This inspires us to consider the following theorem.

Theorem 51 *We have that $\mathbb{L}^F \subset Dom\,\delta_2$ and*

$$E\left((\delta_2(u))^2\right) \leq 2\|u\|^2_{\mathbb{L}^F}. \quad (4.35)$$

Proof. We first choose u in $\mathcal{S}(\mathcal{H})$. Then, Lemma 46, Proposition 67 and Hölder inequality lead us to get

$$\begin{aligned}
&E\left(\left(\int_0^T u_s\delta_2 B_s\right)^2\right)\\
&\leq E\left(\int_0^T \|u_s\|^2_{\mathbb{R}^d} ds\right) + 2\sum_{\ell_2=1}^d E\left\{\left(\int_0^T (u_s^{\ell_2})^2 ds\right)^{1/2}\right.\\
&\qquad \left.\times\left(\int_0^T\left[\sum_{\ell_1=1}^d \delta_2^{(B^{\ell_1})}\left(1_{[0,s]}(\cdot)(Du_\cdot^{\ell_1})_s^{\ell_2}\right)\right]^2 ds\right)^{1/2}\right\}\\
&\leq 2E\left(\int_0^T \|u_s\|^2_{\mathbb{R}^d} ds\right) + \sum_{\ell_2=1}^d E\left(\int_0^T\left[\sum_{\ell_1=1}^d \delta_2^{(B^{\ell_1})}\left(1_{[0,s]}(\cdot)(Du_\cdot^{\ell_1})_s^{\ell_2}\right)\right]^2 ds\right).
\end{aligned}$$

Consequently, equality (4.5) implies

$$E\left(\left(\int_0^T u_s\delta_2 B_s\right)^2\right) \le 2E\left(\int_0^T \|u_s\|^2_{\mathbb{R}^d} ds\right) + \sum_{\ell_1,\ell_2=1}^d E\left(\int_0^T\int_0^s \left((Du_r^{\ell_1})_s^{\ell_2}\right)^2 drds\right)$$
$$+ \sum_{\ell_1,\ell_2,\ell_3=1}^d E\left(\int_0^T\int_0^T\int_0^s \left((D^2u_t^{\ell_1})_{s,r}^{\ell_2,\ell_3}\right)^2 dtdrds\right),$$

which gives that (4.35) is true for any u in $\mathcal{S}(\mathcal{H})$.

Finally, observe that if the sequence $\{u_n \in \mathcal{S}(\mathcal{H}) : n \in \mathbb{N}\}$ goes to u in $\mathbb{L}^F$, then, from (4.35) , we have that $\{\delta_2(u_n) : n \in \mathbb{N}\}$ is a Cauchy sequence in $L^2(\Omega)$. Therefore, $u \in \text{Dom}\,\delta_2$ because δ_2 is a closed operator from $L^2(\Omega\times[0,T];\mathbb{R}^d)$ into $L^2(\Omega)$ and inequality (4.35) holds for u. □

A consequence of Proposition 67 and Theorem 51 is the following:

Proposition 77 *Let $u \in \mathbb{L}^F$. Then, $u^{(\ell)}$ is in $Dom\,\delta_2^{B^{(\ell)}}$, for all $\ell \in \{1,\dots,d\}$, and*

$$\delta_2(u) = \sum_{\ell=1}^d \delta_2^{B^{(\ell)}}(u^{(\ell)}).$$

Proof. Let $\{u_n \in \mathcal{S}(\mathcal{H}) : n \in \mathbb{N}\}$ be a sequence that converges to u in $\mathbb{L}^F$, as $n \to \infty$. Then, for $\ell \in \{1,\dots,d\}$, we have that (4.32) and (4.33) yield that the processes $u_n^{(\ell)}e_\ell$ and $u^{(\ell)}e_\ell$ belong to $\mathbb{L}^F$, where $\{e_1,\dots,e_d\}$ is the canonical basis of $\mathbb{R}^d$. Moreover, $u_n^{(\ell)}$ is also in $\mathcal{S}(L^2([0,T]))$ and $\|u_n^{(\ell)}e_\ell - u^{(\ell)}e_\ell\|_{\mathbb{L}^F}$ goes to zero as $n \to \infty$. Hence, Propositions 62 and 67 give

$$\delta_2(u_n^{(\ell)}e_\ell) = \int_0^T u_n^{(\ell)}(s)\delta_2 B_s^{(\ell)}$$

and Theorem 51 yields

$$E\left(\left(\delta_2(u_n^{(\ell)}e_\ell) - \delta_2(u^{(\ell)}e_\ell)\right)^2\right) \to 0, \quad \text{as } n \to \infty.$$

Consequently, using that $\delta_2^{B^{(\ell)}}$ is a closed operator from $L^2(\Omega\times[0.T])$ to $L^2(\Omega)$ and Proposition 67, we obtain that the result is true and, therefore, the proof is finished. □

Now, we study the commutative relation between the operators δ_2 and D for processes in $\mathbb{L}^F$.

Lemma 47 *Let $u \in \mathbb{L}^F$. Then, the process $\{\delta_2(u1_{[0,t]}) : t \in [0,T]\}$ belongs to $\mathbb{L}^{1,2,f}$ and, for almost all $(s,t) \in \Delta_1^T$ and $\ell \in \{1,\dots,d\}$, we have*

$$\left(D\delta_2(u1_{[0,t]})\right)_s^\ell = \delta_2\left((Du)_s^\ell 1_{[0,t]}\right). \tag{4.36}$$

Remark 63 *Observe that $u1_{[0,t]}$ also belongs to $\mathbb{L}^F$ by using the definition of this space. Thus, Theorem 51 implies that the process $u.1_{[0,t]}(\cdot)$ is in $Dom\,\delta_2$. In the proof, we state that $(Du)_s^\ell 1_{[0,t]}$ is in $\mathbb{D}^{1,2}(\mathcal{H})$, for almost all $s > t$. Therefore, (4.36) makes-sense. Moreover, note that $\delta_2(u1_{[0,t]})$ is an $\mathbb{R}$-valued random variable. So in this case, we mean that, in the definition of the space $\mathbb{L}^{1,2,f}$, we omit ℓ_2 in (4.33). We leave as an exercise to see that the operator D is still a closed operator.*

Proof. Let $\ell \in \{1, \dots, d\}$ and $t \in [0,T]$. We first claim that the process $r \mapsto (Du_r)^\ell_s 1_{[0,t]}(r)$ belongs to $\mathbb{D}^{1,2}(\mathcal{H})$, for almost all $s > t$, in order to show that the right-hand side of (4.36) is well-defined. Indeed, by hypothesis, there exists a sequence $\{u_n \in \mathcal{S}(\mathcal{H}) : n \in \mathbb{N}\}$ such that $\|u_n - u\|_{\mathbb{L}^F} \to 0$ as $n \to \infty$. Hence, we obtain

$$\sum_{\ell_1=1}^d E\left(\int_0^T \int_0^s \left|(Du^{\ell_1}_{n,r} 1_{[0,t]}(r))^\ell_s - (Du^{\ell_1}_r)^\ell_s 1_{[0,t]}(r)\right|^2 drds\right)$$
$$= \sum_{\ell_1=1}^d E\left(\int_0^T \int_0^s \left|\left\{(Du^{\ell_1}_{n,r})^\ell_s - (Du^{\ell_1}_r)^\ell_s\right\} 1_{[0,t]}(r)\right|^2 drds\right) \le \|u_n - u\|_{\mathbb{L}^F} \to 0$$

and

$$\sum_{\ell_1,\ell_2=1}^d \int_0^T \int_0^T E\left(\int_0^{r\vee s} \left|(D^2 u^{\ell_1}_{n,r} 1_{[0,t]}(r))^{\ell,\ell_2}_{s,\theta} - (D^2 u^{\ell_1}_r)^{\ell,\ell_2}_{s,\theta} 1_{[0,t]}(r)\right|^2 drd\theta ds\right)$$
$$\le \|u_n - u\|_{\mathbb{L}^F} \to 0, \quad \text{as } n \to \infty.$$

Consequently, we can find a subsequence $\{n_k : k \in \mathbb{N}\}$ of $\mathbb{N}$ such that

$$\sum_{\ell_1=1}^d E\left(\int_0^T \left|(Du^{\ell_1}_{n_k,r} 1_{[0,t]}(r))^\ell_s - (Du^{\ell_1}_r)^\ell_s 1_{[0,t]}(r)\right|^2 dr\right) \to 0$$

and

$$\sum_{\ell_1,\ell_2=1}^d E\left(\int_0^T \int_0^T \left|(D^2 u^{\ell_1}_{n_k,r} 1_{[0,t]}(r))^{\ell,\ell_2}_{s,\theta} - (D^2 u^{\ell_1}_r)^{\ell,\ell_2}_{s,\theta} 1_{[0,t]}(r)\right|^2 drd\theta\right) \to 0,$$

as $k \to \infty$, for almost all $s > t$. Therefore, our claim is true. Additionally, we have also seen that $\|(Du_{n_k,\cdot} 1_{[0,t]}(\cdot))^\ell_s - (Du_\cdot)^\ell_s 1_{[0,t]}(\cdot)\|_{1,2,\mathcal{H}} \to 0$, as $k \to \infty$, for almost all $s > t$.

On the other hand, Lemma 36 yields that $t \mapsto \delta_2(u_n 1_{[0,t]})$ is in $\mathbb{L}^F$ and Corollary 24 (see also Remark 55) gives that (4.36) holds for $u \in \mathcal{S}(\mathcal{H})$. Thus, applying Proposition 63 and Theorem 51, we can conclude

$$\begin{aligned}
E&\left(\int_0^T \left|\delta_2(u_{n_k} 1_{[0,t]}) - \delta_2(u 1_{[0,t]})\right|^2 dt\right) \\
&\quad + E\left(\sum_{\ell=1}^d \int_0^T \int_0^s \left|(D\delta_2(u_{n_k} 1_{[0,t]}))^\ell_s - \delta_2((Du)^\ell_s 1_{[0,t]})\right|^2 dtds\right) \\
&= E\left(\int_0^T \left|\delta_2(u_{n_k} 1_{[0,t]}) - \delta_2(u 1_{[0,t]})\right|^2 dt\right) \\
&\quad + \sum_{\ell=1}^d \int_0^T \int_0^s E\left(\left|\delta_2((Du_{n_k})^\ell_s 1_{[0,t]}) - \delta_2((Du)^\ell_s 1_{[0,t]})\right|^2\right) dtds \\
&\le 2\int_0^T \|(u_{n_k} - u) 1_{[0,t]}\|^2_{\mathbb{L}^F} dt + \sum_{\ell=1}^d \int_0^T \int_0^s \|(Du_{n_k,\cdot})^\ell_s 1_{[0,t]}(\cdot) - (Du_\cdot)^\ell_s 1_{[0,t]}(\cdot)\|_{1,2,\mathcal{H}} dtds \\
&=: 2I_1 + I_2.
\end{aligned} \tag{4.37}$$

Using the definition of the norm $\|\cdot\|_{\mathbb{L}^F}$, we have

$$\begin{aligned} I_1 &\leq E\left(\int_0^T \|u_{n_k,t}-u_t\|_{\mathbb{R}^d}^2\,dt\right) \\ &+\sum_{\ell,\ell_1=1}^{d} E\left(\int_0^T\int_0^T\int_0^{s\wedge t}\left|[(Du^{\ell_1}_{n_k,\theta})^{\ell}_s-(Du^{\ell_1}_{\theta})^{\ell}_s\right|^2 d\theta ds dt\right) \\ &+\sum_{\ell,\ell_1,\ell_2=1}^{d} E\left(\int_0^T\int_0^T\int_0^T\int_0^{r\vee s}\left|\left[(D^2u^{\ell_1}_{n_k,\theta})^{\ell,\ell_2}_{s,r}-(D^2u^{\ell_1}_{\theta})^{\ell,\ell_2}_{s,r}\right]1_{[0,t]}(\theta)\right|^2 d\theta ds dt\right) \\ &\leq (T+1)\|u_{n_k}-u\|_{\mathbb{L}^F} \end{aligned}$$

and applying the definition of the norm $\|\cdot\|_{1,2,\mathcal{H}}$, we get

$$\begin{aligned} I_2 &= \sum_{\ell=1}^{d} E\left(\int_0^T\int_0^s\int_0^T \left|(Du_{n_k,\theta})^{\ell}_s 1_{[0,t]}(\theta)-(Du_{\theta})^{\ell}_s 1_{[0,t]}(\theta)\right|_{\mathbb{R}^d}^2 d\theta dt ds\right. \\ &\left.+\sum_{\ell_1=1}^{d}\int_0^T\int_0^s\int_0^T\int_0^T \left|(D^2u_{n_k,\theta})^{\ell,\ell_1}_{s,r}1_{[0,t]}(\theta)-(D^2u_{\theta})^{\ell,\ell_1}_{s,r}1_{[0,t]}(\theta)\right|_{\mathbb{R}^d}^2 d\theta dr dt ds\right) \\ &\leq (T+1)\|u_{n_k}-u\|_{\mathbb{L}^F}, \end{aligned}$$

which, together with (4.37), implies that the proof is finished. □

Note that if $u\in\mathbb{L}^F$, then, proceeding as in the last proof, we can see that the process $t\mapsto\int_0^t u_r dr$ also belongs to $\mathbb{L}^F$ with

$$D^{\ell_1}_s\left(\int_0^t u^{\ell_2}_r dr\right)=\int_0^t \left(Du^{\ell_2}_r\right)^{\ell_1}_s dr, \quad (s,t)\in\Delta^T_1 \text{ and } \ell_1,\ell_2\in\{1,\dots,d\}, \tag{4.38}$$

and, for $\ell_1,\ell_2,\ell_3\in\{1,\dots,d\}$,

$$D^{\ell_2}_u D^{\ell_1}_s\left(\int_0^t u^{\ell_3}_r dr\right)=\int_0^t \left(D^2u^{\ell_3}_r\right)^{\ell_2\ell_1}_{u,s} dr, \quad (u,s,t)\in\Delta^T_2. \tag{4.39}$$

The divergence operator δ_2 also satisfies the following local property on the space $\mathbb{L}^F$:

Proposition 78 *Let $u\in\mathbb{L}^F$ and $A\in\mathcal{F}$ such that $u=0$ on $A\times[0,T]$, almost surely. Then $\delta(u)=0$ on A, with probability 1.*

Proof. Consider the sequence $(\pi^m)_{m\in\mathbb{N}}$ of dyadic partitions of $[0,T]$. That is,

$$\pi^m=\{0=t^m_0<t^m_1<\cdots<t^m_{2^m}=T\}, \text{ with } t^m_i=\frac{iT}{2^m}.$$

For $m\in\mathbb{N}$, set

$$u_{m,t}=\sum_{j=1}^{2^m-1}\frac{2^m}{T}\left(\int_{t^m_{j-1}}^{t^m_j} u_s ds\right)1_{(t^m_j,t^m_{j+1}]}(t), \quad t\in[0,T].$$

Now, we divide the proof into two steps.

Step 1: Here, for a fixed $m\in\mathbb{N}$, we show that the process u_m belongs to $\mathbb{L}^F$.

Choose a sequence $\{v_n \in \mathcal{S}(\mathcal{H}) : n \in \mathbb{N}\}$ such that $\|v_n - u\|_F \to 0$ as $n \to \infty$. It is clear that the process

$$v_{m,n,t} = \sum_{j=1}^{2^m-1} \frac{2^m}{T} \left(\int_{t_{j-1}^m}^{t_j^m} v_{n,s} ds \right) 1_{(t_j^m, t_{j+1}^m]}(t), \quad t \in [0,T],$$

belongs to the space $\mathbb{L}^F$ with

$$\left[D\left(v_{m,n,t}^{\ell_1} \right) \right]_r^{\ell_2} = \sum_{j=1}^{2^m-1} \frac{2^m}{T} \left(\int_{t_{j-1}^m}^{t_j^m} \left[D\left(v_{n,s}^{\ell_1} \right) \right]_r^{\ell_2} ds \right) 1_{(t_j^m, t_{j+1}^m]}(t), \quad (r,t) \in \Delta_1^T,$$

and

$$\left[D^2\left(v_{m,n,t}^{\ell_1} \right) \right]_{r,u}^{\ell_2,\ell_3} = \sum_{j=1}^{2^m-1} \frac{2^m}{T} \left(\int_{t_{j-1}^m}^{t_j^m} \left[D^2\left(v_{n,s}^{\ell_1} \right) \right]_{r,u}^{\ell_2,\ell_3} ds \right) 1_{(t_j^m, t_{j+1}^m]}(t), \quad (r,u,t) \in \Delta_2^T,$$

for $\ell, \ell_2, \ell_3 \in \{1, \dots, d\}$. This suggests that we can define $D\left(u_{m,t}^{\ell_1} \right)$ and $D^2\left(u_{m,t}^{\ell_1} \right)$ using the last two equalities when we write $u_{m,t}^{\ell_1}$ and $u_s^{\ell_1}$ instead of $v_{m,n,t}$ and $v_{n,s}^{\ell_1}$, respectively. Hence, we observe that $u_m \in \mathbb{L}^F$ if $\|u_m - v_{m,n}\|_F \le \|u - v_n\|_F$. This inequality is a consequence of Fubini's theorem. For instance, we have

$$\begin{aligned}
&\sum_{j=1}^{2^m-1} \frac{2^{2m}}{T^2} \int_0^T \int_0^T \int_0^{r \vee u} \left(\int_{t_{j-1}^m}^{t_j^m} \left[D^2\left(v_{n,s}^{\ell_1} - u_s^{\ell_1} \right) \right]_{r,u}^{\ell_2,\ell_3} ds \right)^2 1_{(t_j^m, t_{j+1}^m]}(t) dt dr du \\
&= \sum_{j=1}^{2^m-1} \frac{2^{2m}}{T^2} \int_0^T \int_0^u \int_{u \wedge t_j^m}^{u \wedge t_{j+1}^m} \left(\int_{t_{j-1}^m}^{t_j^m} \left[D^2\left(v_{n,s}^{\ell_1} - u_s^{\ell_1} \right) \right]_{r,u}^{\ell_2,\ell_3} ds \right)^2 dt dr du \\
&\quad + \sum_{j=1}^{2^m-1} \frac{2^{2m}}{T^2} \int_0^T \int_u^T \int_{r \wedge t_j^m}^{r \wedge t_{j+1}^m} \left(\int_{t_{j-1}^m}^{t_j^m} \left[D^2\left(v_{n,s}^{\ell_1} - u_s^{\ell_1} \right) \right]_{r,u}^{\ell_2,\ell_3} ds \right)^2 dt dr du \\
&\le \sum_{j=1}^{2^m-1} \frac{2^m}{T} \int_0^T \int_0^u \int_{u \wedge t_j^m}^{u \wedge t_{j+1}^m} \int_{t_{j-1}^m}^{t_j^m} \left(\left[D^2\left(v_{n,s}^{\ell_1} - u_s^{\ell_1} \right) \right]_{r,u}^{\ell_2,\ell_3} \right)^2 1_{\Delta_2^T}(u,r,s) ds dt dr du \\
&\quad + \sum_{j=1}^{2^m-1} \frac{2^m}{T} \int_0^T \int_u^T \int_{r \wedge t_j^m}^{r \wedge t_{j+1}^m} \int_{t_{j-1}^m}^{t_j^m} \left(\left[D^2\left(v_{n,s}^{\ell_1} - u_s^{\ell_1} \right) \right]_{r,u}^{\ell_2,\ell_3} \right)^2 1_{\Delta_2^T}(u,r,s) ds dt dr du \\
&\le \int_0^T \int_0^T \int_0^T \left(\left[D^2\left(v_{n,s}^{\ell_1} - u_s^{\ell_1} \right) \right]_{r,u}^{\ell_2,\ell_3} \right)^2 1_{\Delta_2^T}(u,r,s) ds dt dr du.
\end{aligned}$$

Thus, the claim of this step is true.

Step 2: Finally, we deal with the local property of δ_2 on $\mathbb{L}^F$.

Note that we can proceed as in the proof of Lemma 33 to prove that u_m converges to u in $\mathbb{L}^F$, as $m \to \infty$. Consequently, Theorem 51 yields that $\delta_2(u_m)$ goes to $\delta_2(u)$ in $L^2(\Omega)$, as $m \to \infty$, and Lemma 36 and Proposition 67 imply

$$\begin{aligned}
\delta_2(u_m) &= \sum_{\ell=1}^{d} \delta_2^{B^{(\ell)}}(u_m^\ell) = \sum_{\ell=1}^{d} \sum_{j=1}^{2^m-1} \frac{2^m}{T} \left(\int_{t_{j-1}^m}^{t_j^m} u_s^\ell ds \right) \left(B_{t_{j+1}}^{(\ell)} - B_{t_j}^{(\ell)} \right) \\
&\quad - \sum_{\ell=1}^{d} \sum_{j=1}^{2^m-1} \frac{2^m}{T} \int_{t_j^m}^{t_{j+1}^m} \int_{t_{j-1}^m}^{t_j^m} \left(D u_s^\ell \right)_r^\ell ds dr.
\end{aligned}$$

Hence, Proposition 74 allows us to conclude

$$\delta_2(u)1_A = L^2(\Omega) - \lim_{m\to\infty} \delta_2(u_m)1_A = 0$$

and, therefore, the proof is complete. □

As in Definition 38, we can extend the domain of δ_2 again: let $\{(\Omega_n, u_n) \in \mathcal{F} \times \mathbb{L}^F : n \in \mathbb{N}\}$ be a sequence that localizes u in $\mathbb{L}^F$. Then, we are able to define $\delta_2(u)$ as

$$\delta(u) = \delta(u_n) \text{ on } \Omega_n, \quad n \in \mathbb{N}. \tag{4.40}$$

Other consequence of Theorem 51 is the following proposition.

Proposition 79 *Let u be an $\mathbb{R}^d$-valued process in $\mathbb{L}^F_{loc}$, $t \in [0,T]$ and $\pi = \{0 = t_0 < t_1 < \ldots < t_n = t\}$ a partition of the interval $[0,t]$. Then,*

$$\sum_{i=1}^{n} \left(\int_{t_{i-1}}^{t_i} u_s \delta_2 B_s \right)^2 \to \int_0^t |u_s|^2_{\mathbb{R}^d} \, ds \quad \textit{in probability as } |\pi| \to 0.$$

Furthermore, this convergence is also in $L^1(\Omega)$ whenever $u \in \mathbb{L}^F$.

Remarks 6 *i) If $u \in \mathcal{L}^2_a(\Omega \times [0,T]; \mathbb{R}^d)$, then we can change the divergence operator δ_2 by the Itô's integral due to Propositions 70 and 76.*

ii) Let u and v be two processes in $\mathbb{L}^F_{loc}$ (resp. in $\mathbb{L}^F$). Then,

$$\sum_{i=1}^{n} \left(\int_{t_{i-1}}^{t_i} u_s \delta_2 B_s \right) \left(\int_{t_{i-1}}^{t_i} v_s \delta_2 B_s \right) \to \int_0^t \langle u_s, v_s \rangle_{\mathbb{R}^d} \, ds$$

in probability (resp. in $L^1(\Omega)$), as $|\pi| \to 0$. In order to show this, we only need to use that $ab = \frac{1}{2}[(a+b)^2 - a^2 - b^2]$, $a, b \in \mathbb{R}$ and properties of the inner product $\langle \cdot, \cdot \rangle_{\mathbb{R}^d}$.

iii) Let u and v be as in Statement ii) and $\ell_1, \ell_2 \in \{1, \ldots, d\}$. Then, the proof of Proposition 77 gives

$$\sum_{i=1}^{n} \left(\int_{t_{i-1}}^{t_i} u_s^{(\ell_1)} \delta_2 B_s^{(\ell_1)} \right) \left(\int_{t_{i-1}}^{t_i} v_s^{(\ell_2)} \delta_2 B_s^{(\ell_2)} \right) \to 1_{\{\ell_1\}}(\ell_2) \int_0^t u_s^{(\ell_1)} v_s^{(\ell_2)} ds$$

in probability (resp. in $L^1(\Omega)$), as $|\pi| \to 0$.

Proof. Here, we use the notation $V^\pi(u) := \sum_{i=1}^n \left(\int_{t_{i-1}}^{t_i} u_s \delta_2 B_s \right)^2$.

We first see that it is enough to show the result for any process u in $\mathbb{L}^F$. So, let $\{(\Omega_n, u_n) \in \mathcal{F} \times \mathbb{L}^F : n \in \mathbb{N}\}$ be a sequence that localizes u in $\mathbb{L}^F$. Then, for $n \in \mathbb{N}$ and $\varepsilon > 0$,

$$\begin{aligned}
&P\left(\left[\left| V^\pi(u) - \|u 1_{[0,t]}\|^2_{L^2(\Omega;\mathbb{R}^d)} \right| > \varepsilon \right] \right) \\
&\quad \le P\left(\left[\Omega_n, \left| V^\pi(u) - \|u 1_{[0,t]}\|^2_{L^2(\Omega;\mathbb{R}^d)} \right| > \varepsilon \right] \right) + P\left(\Omega_n^c\right) \\
&\quad \le P\left(\left[\left| V^\pi(u_n) - \|u_n 1_{[0,t]}\|^2_{L^2(\Omega;\mathbb{R}^d)} \right| > \varepsilon \right] \right) + P\left(\Omega_n^c\right) \to P\left(\Omega_n^c\right), \quad \text{as } |\pi| \to 0,
\end{aligned}$$

where the last convergence holds because we are assuming that the result is satisfied for processes in $\mathbb{L}^F$. Thus, we have shown that

$$\limsup_{|\pi|\to 0} P\left(\left[\left|V^\pi(u) - \|u1_{[0,t]}\|^2_{L^2(\Omega;\mathcal{R}^d)}\right| > \varepsilon\right]\right) = 0.$$

Consequently, we are now able to suppose that $u \in \mathbb{L}^F$ and we divide the proof into several steps.

Step 1: Here, we prove that we can assume that u is a process in $\mathcal{S}(\mathcal{H})$ of the form

$$u = \sum_{i=1}^m c_i F_i 1_{]s_i, s_{i+1}]}, \tag{4.41}$$

with $c_i \in \mathbb{R}^d$, $F_i \in \mathcal{S}$ and $0 \le s_i < s_{i+1} \le t$, for $i = 1, \ldots, m$.

By definition of the space $\mathbb{L}^F$, there exists a sequence $\{u_n \in \mathcal{S}(\mathcal{H}) : n \in \mathbb{N}\}$ of simple processes as on the right-hand side of (4.41) converging to u in $\mathbb{L}^F$, as $n \to \infty$. Hence, for $n \in \mathbb{N}$, we have

$$\begin{aligned}
&E\left(\left|V^\pi(u) - \|u1_{[0,t]}\|^2_{L^2(\Omega;\mathbb{R}^d)}\right|\right)\\
&\le E\left(|V^\pi(u) - V^\pi(u_n)|\right) + E\left(\left|V^\pi(u_n) - \|u_n 1_{[0,t]}\|^2_{L^2(\Omega;\mathbb{R}^d)}\right|\right)\\
&\quad + E\left(\left|\|u1_{[0,t]}\|^2_{L^2(\Omega;\mathbb{R}^d)} - \|u_n 1_{[0,t]}\|^2_{L^2(\Omega;\mathbb{R}^d)}\right|\right)\\
&\le \left[E\left(V^\pi(u+u_n)\right)\right]^{1/2}\left[E\left(|V^\pi(u-u_n)|\right)\right]^{1/2} + E\left(\left|V^\pi(u_n) - \|u_n 1_{[0,t]}\|^2_{L^2(\Omega;\mathbb{R}^d)}\right|\right)\\
&\quad + E\left(\left|\|u1_{[0,t]}\|^2_{L^2(\Omega;\mathbb{R}^d)} - \|u_n 1_{[0,t]}\|^2_{L^2(\Omega;\mathbb{R}^d)}\right|\right).
\end{aligned}$$

Therefore, Theorem 51 leads us to obtain that there is a constant C depending on u such that

$$E\left(\left|V^\pi(u) - \|u1_{[0,t]}\|^2_{L^2(\Omega;\mathbb{R}^d)}\right|\right) \le C\|u-u_n\|_{\mathbb{L}^F} + E\left(V^\pi(u_n) - \|u_n 1_{[0,t]}\|^2_{L^2(\Omega;\mathbb{R}^d)}\right),$$

which implies that

$$\limsup_{|\pi|\to 0} E\left(\left|V^\pi(u) - \|u1_{[0,t]}\|^2_{L^2(\Omega;\mathbb{R}^d)}\right|\right) \le C\|u-u_n\|_{\mathbb{L}^F}$$

if the result is true for u_n, as we wish to establish in this part of the proof.

Step 2: Let u be as in (4.41). Then, without loss of generality, we can assume that $\{s_1, \ldots, s_m\} \subset \{t_0, \ldots, t_n\}$.

Let π be such that $|\pi| \le |\{s_1 < \ldots < s_m\}|$, then there is $t_{\tilde{j}} \in \pi$ such that $s_j \in [t_{\tilde{j}}, t_{\tilde{j}+1}]$ since π is a partition of the interval $[0,t]$. Thus, the claim of this step is satisfied due to Theorem 51 giving

$$E\left(\sum_{j=1}^m \left\{\left(\int_{t_{\tilde{j}}}^{s_j} u_s \delta_2 B_s\right)^2 + \left(\int_{t_{\tilde{j}+1}}^{s_{j+1}} u_s \delta_2 B_s\right)^2\right\}\right) \to 0, \quad \text{as } |\pi| \to 0.$$

Step 3: Now we assume that u is as in (4.41).

From Step 2, we can assume that $\{s_1, \dots, s_m\} \subset \{t_0, \dots, t_n\}$. Consequently, Lemma 36 allows us to obtain

$$
\begin{aligned}
V^{\pi}(u) &= \sum_{j=1}^{m} \sum_{\{s_j<t_i\le s_{j+1}\}} \left(F_j \langle c_j, B_{t_{i+1}} - B_{t_i}\rangle_{\mathbb{R}^d} - \int_{t_i}^{t_{i+1}} \langle D_s F_j, c_j\rangle_{\mathbb{R}^d} ds \right)^2 \\
&= \sum_{j=1}^{m} \sum_{\{s_j<t_i\le s_{j+1}\}}^{m} \Bigg\{ (F_j^2 \langle c_j, B_{t_{i+1}} - B_{t_i}\rangle_{\mathbb{R}^d}^2 + \left(\int_{t_i}^{t_{i+1}} \langle D_s F_j, c_j\rangle_{\mathbb{R}^d} ds \right)^2 \\
&\qquad -2F_j \langle c_j, B_{t_{i+1}} - B_{t_i}\rangle_{\mathbb{R}^d} \int_{t_i}^{t_{i+1}} \langle D_s F_j, c_j\rangle_{\mathbb{R}^d} ds \Bigg\} \\
&= \sum_{j=1}^{m} \sum_{\{s_j<t_i\le s_{j+1}\}} (I_{1,i,j} + I_{2,i,j} - 2I_{3,i,j}) .
\end{aligned}
$$

Now, fix $j \in \{1, \dots, m\}$. Note that for $I_{2,i,j}$ we have

$$
\begin{aligned}
\sum_{\{s_j<t_i\le s_{j+1}\}} I_{2,i,j} &\le |\pi| \sum_{\{s_j<t_i\le s_{j+1}\}} \int_{t_i}^{t_{i+1}} \langle D_s F_j, c_j\rangle_{\mathbb{R}^d}^2 ds = |\pi| \int_{s_j}^{s_{j+1}} \langle D_s F_j, c_j\rangle_{\mathbb{R}^d}^2 ds \\
&\to 0 \quad \text{in } L^1(\Omega) \text{ as } |\pi| \to 0.
\end{aligned}
$$

Similarly, for $I_{3,i,j}$, Theorem 100 and the dominated convergence theorem yield

$$
\begin{aligned}
\sum_{\{s_j<t_i\le s_{j+1}\}} |I_{3,i,j}| &\le C \left(\sup_{|r-s|\le|\pi|} \|B_s - B_r\|_{\mathbb{R}^d} \right) \int_{s_j}^{s_{j+1}} |\langle D_s F_j, c_j\rangle_{\mathbb{R}^d}| \, ds \\
&\to 0 \quad \text{in } L^1(\Omega) \text{ as } |\pi| \to 0.
\end{aligned}
$$

Now, for $I_{1,i,j}$

$$
\begin{aligned}
&\sum_{\{s_j<t_i\le s_{j+1}\}}^{m} \left(\langle c_j, B_{t_{i+1}} - B_{t_i}\rangle_{\mathbb{R}^d}^2 - \sum_{\ell=1}^{d} \left(c_j^{(\ell)}\right)^2 (t_{i+1} - t_i) \right) \\
&= \sum_{\ell=1}^{d} \sum_{\{s_j<t_i\le s_{j+1}\}}^{m} (c_j^{(\ell)})^2 \left(\left(B_{t_{i+1}}^{(\ell)} - B_{t_i}^{(\ell)} \right)^2 - (t_{i+1} - t_i) \right) \\
&\quad +2 \sum_{1\le\ell<\ell_1\le d} \sum_{\{s_j<t_i\le s_{j+1}\}}^{m} c_j^{(\ell)} c_j^{(\ell_1)} \left(B_{t_{i+1}}^{(\ell)} - B_{t_i}^{(\ell)} \right) \left(B_{t_{i+1}}^{(\ell_1)} - B_{t_i}^{(\ell_1)} \right) \\
&= \sum_{\ell=1}^{d} (c_j^{(\ell)})^2 I_{1,1,\ell} + 2 \sum_{1\le\ell<\ell_1\le d} c_j^{(\ell)} c_j^{(\ell_1)} I_{1,2,\ell,\ell_1}.
\end{aligned}
$$

From Proposition 7, $I_{1,1,\ell}$ converges to zero in $L^2(\Omega)$, for all $\ell \in \{1, \dots, d\}$ and the independence of the components of B give

$$
\begin{aligned}
E\left((I_{1,2,\ell,\ell_1})^2 \right) &= E\left(\sum_{\{s_j<t_i\le s_{j+1}\}}^{m} \left(B_{t_{i+1}}^{(\ell)} - B_{t_i}^{(\ell)} \right)^2 \left(B_{t_{i+1}}^{(\ell_1)} - B_{t_i}^{(\ell_1)} \right)^2 \right) \\
&= \sum_{\{s_j<t_i\le s_{j+1}\}}^{m} (t_{i+1} - t_i)^2 \le |\pi| (s_{j+1} - s_j) \to 0 \quad \text{as } |\pi| \to 0,
\end{aligned}
$$

which, together with the analysis of $I_{2,i,j}$, $I_{3,i,j}$ and $I_{1,1,\ell}$, implies that the result holds for simple processes of the form (4.41). □

Finally, the proposition follows from Steps 1 and 2. □

On the other hand, Proposition 60 is still true for the divergence operator δ_2. Before giving the proof of this fact, we need to state some auxiliary results. Remember that the proof of Proposition 60 is based on the Doob's maximal inequality (i.e., Theorem 100 below). For the divergence operator, we use the following result.

Theorem 52 *Let u be a process in $\mathbb{L}^F$, $p \in (2,4)$ and $q = 2p/(4-p)$ such that $u \in L^q(\Omega \times [0,T]; \mathbb{R}^d)$. Then, there exists a constant C depending only on p and T such that*

$$E\left(\sup_{t\in[0,T]}\left|\int_0^t u_s\delta_2 B_s\right|^p\right) \leq CE\left[\int_0^T |u_s|^q_{\mathbb{R}^d} ds + \sum_{\ell_1,\ell_2=1}^d \int_{\Delta_1^T} \left\{(Du_t^{\ell_2})_s^{\ell_1}\right\}^2 dsdt + \sum_{\ell_1,\ell_2,\ell_3=1}^d \int_{\Delta_2^T} \left\{\left(D^2 u_t^{\ell_3}\right)_{r,s}^{\ell_1,\ell_2}\right\}^2 drdsdt\right]. \quad (4.42)$$

Proof. The proof is divided into several steps.
Step 1: Here we suppose that the result is true for all the processes in $\mathcal{S}(\mathcal{H})$ and that there is a sequence $\{u_n \in \mathcal{S}(\mathcal{H}) : n \in \mathbb{N}\}$ such that

$$\|u_n - u\|_{L^q(\Omega\times[0,T];\mathbb{R}^d)} + \|u_n - u\|_{\mathbb{L}^F} \to 0, \quad \text{as } n \to \infty. \quad (4.43)$$

It is well-known that the process $t \mapsto \int_0^t u_s\delta_2 B_s$ has a separable version as it is established in Todorovic [206] (see also Doob [41], Gikhman and Skorokhod [64], Meyer [147] and Neveu [153], among others). Hence, we can assume without loss of generality that the process $t \mapsto \int_0^t u_s\delta_2 B_s$ is separable. Let $\Lambda = \{t_n \in [0,T] : n \in \mathbb{N}\}$ be a separant set. Then, for $n \in \mathbb{N}$, we have that (4.43) and Theorem 51, together with the fact that (4.42) holds for $u \in \mathcal{S}(\mathcal{H})$, imply

$$E\left(\sup_{i\in\{1,\dots,n\}}\left|\int_0^{t_i} u_s\delta_2 B_s\right|^p\right) \leq CE\left[\int_0^T |u_s|^q_{\mathbb{R}^d} ds + \sum_{\ell_1,\ell_2=1}^d \int_{\Delta_1^T} \left\{(Du_t^{\ell_2})_s^{\ell_1}\right\}^2 dsdt + \sum_{\ell_1,\ell_2,\ell_3=1}^d \int_{\Delta_2^T} \left\{\left(D^2 u_t^{\ell_3}\right)_{r,s}^{\ell_1,\ell_2}\right\}^2 drdsdt\right].$$

Thus, the monotone convergence theorem gives that (4.42) is satisfied under the assumptions of this case.
Step 2: Now, we show that there exists a sequence $\{u_n \in \mathcal{S}(\mathcal{H}) : n \in \mathbb{N}\}$ satisfying (4.43), for $u \in \mathbb{L}^F \cap L^q(\Omega \times [0,T]; \mathbb{R}^d)$.

Choose a sequence $\{\tilde{u}_n \in \mathcal{S}(\mathcal{H}) : n \in \mathbb{N}\}$ that converges to u in $L^q(\Omega \times [0,T]; \mathbb{R}^d)$. Then, for $t > 0$, we have that $T_t(\tilde{u}_n)$ also belongs to $\mathcal{S}(\mathcal{H})$, where $\{T_t : t \geq 0\}$ is the Ornstein-Uhlenbeck semigroup (see Section 8.14) and $T_t(Fh) = (T_t(F))h$, for $F \in \mathcal{S}$ and $h \in \mathcal{H}$. In order to find the sequence that satisfies (4.43), we deal with the sequence

$\{T_{1/k}(\tilde{u}_n) : k, n \in \mathbb{N}\}$. Therefore,

$$E\left(\int_0^T \left\{\sum_{\ell=1}^d \left|u_s^\ell - T_{1/k}((\tilde{u}_n(s)^\ell)\right|^2\right\}^{q/2} ds\right)$$
$$\leq 2^q E\left(\int_0^T \left\{\sum_{\ell=1}^d \left|u_s^\ell - T_{1/k}(u_s^\ell)\right|^2\right\}^{q/2} ds\right.$$
$$\left. + \int_0^T \left\{\sum_{\ell=1}^d \left|T_{1/k}(u_s^\ell) - T_{1/k}((\tilde{u}_n(s)^\ell)\right|^2\right\}^{q/2} ds\right).$$

So, the fact that $T_{1/k}$ is a contraction on $L^q(\Omega)$ and Hölder inequality yield

$$E\left(\int_0^T \left\{\sum_{\ell=1}^d \left|u_s^\ell - T_{1/k}((\tilde{u}_n(s)^\ell)\right|^2\right\}^{q/2} ds\right) \leq 2^q E\left(\int_0^T \left\{\sum_{\ell=1}^d \left|u_s^\ell - T_{1/k}(u_s^\ell)\right|^2\right\}^{q/2} ds\right.$$
$$\left. + C_d \int_0^T \sum_{\ell=1}^d \left|u_s^\ell - (\tilde{u}_n(s)^\ell\right|^q ds\right). \quad (4.44)$$

Now, consider a sequence $\{v_n \in \mathcal{S}(\mathcal{H}) : n \in \mathbb{N}\}$ that goes to u in $\mathbb{L}^F$. Thus, Corollary 34 leads us to establish that for $k, n, m \in \mathbb{N}$,

$$\sum_{\ell_1,\ell_2=1}^d E\left(\int_{\Delta_1^T} \left[(DT_{1/k}((\tilde{u}_n(t))^{\ell_2}))_s^{\ell_1} - (Du_t^{\ell_2})_s^{\ell_1}\right]^2 dsdt\right)$$
$$\leq 4 \sum_{\ell_1,\ell_2=1}^d E\left(\int_{\Delta_1^T} \left[(DT_{1/k}((\tilde{u}_n(t))^{\ell_2}))_s^{\ell_1} - \left(DT_{1/k}((v_m)_t^{\ell_2})\right)_s^{\ell_1}\right]^2 dsdt\right)$$
$$+4 \sum_{\ell_1,\ell_2=1}^d E\left(\int_{\Delta_1^T} \left[e^{-1/k} T_{1/k}\left((D(v_m)_t^{\ell_2})_s^{\ell_1} - (Du_t^{\ell_2})_s^{\ell_1}\right)\right]^2 dsdt\right)$$
$$+4 \sum_{\ell_1,\ell_2=1}^d E\left(\int_{\Delta_1^T} \left[e^{-1/k} T_{1/k}\left((Du_t^{\ell_2})_s^{\ell_1}\right) - (Du_t^{\ell_2})_s^{\ell_1}\right]^2 dsdt\right).$$

Hence, we can use that T_t is a contraction on $L^2(\Omega)$ again, (1.109) and Corollary 34 to write

$$\sum_{\ell_1,\ell_2=1}^d E\left(\int_{\Delta_1^T} \left[(DT_{1/k}((\tilde{u}_n(t))^{\ell_2}))_s^{\ell_1} - (Du_t^{\ell_2})_s^{\ell_1}\right]^2 dsdt\right)$$
$$\leq Ck \sum_{\ell=1}^d E\left(\int_0^T \left[(\tilde{u}_n(t))^\ell - (v_m)_t^\ell\right]^2 dt\right)$$
$$+C \sum_{\ell_1,\ell_2=1}^d E\left(\int_{\Delta_1^T} \left[(D(v_m)_t^{\ell_2})_s^{\ell_1} - (Du_t^{\ell_2})_s^{\ell_1}\right]^2 dsdt\right)$$
$$+C \sum_{\ell_1,\ell_2=1}^d E\left(\int_{\Delta_1^T} \left[e^{-1/k} T_{1/k}\left((Du_t^{\ell_2})_s^{\ell_1}\right) - (Du_t^{\ell_2})_s^{\ell_1}\right]^2 dsdt\right). \quad (4.45)$$

Note that we are able to proceed as in the last two inequalities in order to get

$$\begin{aligned}
&\sum_{\ell_1,\ell_2,\ell_3=1}^{d} E\left(\int_{\Delta_2^T}\left[\left(D^2T_{1/k}((\tilde{u}_n(t))^{\ell_3})\right)_{r,s}^{\ell_1,\ell_2}-(D^2u_t^{\ell_3})_{r,s}^{\ell_1,\ell_2}\right]^2 drdsdt\right)\\
&\leq Ck^2\sum_{\ell=1}^{d}E\left(\int_0^T\left[(\tilde{u}_n(t))^{\ell}-(v_m)_t^{\ell}\right]^2dt\right)\\
&\quad+\sum_{\ell_1,\ell_2,\ell_3=1}^{d}E\left(\int_{\Delta_2^T}\left[(D(v_m)_t^{\ell_3})_{r,s}^{\ell_1,\ell_2}-(Du_t^{\ell_3})_{r,s}^{\ell_1,\ell_2}\right]^2 drdsdt\right.\\
&\quad+\left.\int_{\Delta_2^T}\left[e^{-1/k}T_{1/k}\left((Du_t^{\ell_3})_{r,s}^{\ell_1,\ell_2}\right)-(Du_t^{\ell_3})_{r,s}^{\ell_1,\ell_2}\right]^2 drdsdt\right). \qquad (4.46)
\end{aligned}$$

We can obtain the sequence that we are looking for by means of inequalities (4.44)-(4.46). Indeed, for instance, in (4.45), given $m\in\mathbb{N}$, we can find $k_m\in\mathbb{N}$ such that

$$\begin{aligned}
&\sum_{\ell_1,\ell_2=1}^{d}E\left(\int_{\Delta_1^T}\left[\left(DT_{1/k_m}((\tilde{u}_n(t))^{\ell_2})\right)_s^{\ell_1}-(Du_t^{\ell_2})_s^{\ell_1}\right]^2dsdt\right)\\
&\leq Ck_m\sum_{\ell=1}^{d}E\left(\int_0^T\left[(\tilde{u}_n(t))^{\ell}-(u)_t^{\ell}\right]^2dt\right)+\frac{1}{m}
\end{aligned}$$

due to Proposition 147. Consequently, now we can choose $n_m\in\mathbb{N}$ such that

$$\sum_{\ell_1,\ell_2=1}^{d}E\left(\int_{\Delta_1^T}\left[\left(DT_{1/k_m}((\tilde{u}_{n_m}(t))^{\ell_2})\right)_s^{\ell_1}-(Du_t^{\ell_2})_s^{\ell_1}\right]^2dsdt\right)\leq\frac{2}{m}.$$

Therefore, our claim is true. That means, Step 2 is satisfied.
Step 3: Finally, we sketch the proof of the fact that the result is satisfied for $u\in\mathcal{S}(\mathcal{H})$.

The idea of the proof of this step is as follows: from Fubini's theorem for δ_2 (see Theorem 49) and the fact that $u\in\mathcal{S}(\mathcal{H})$, we have that, for $\alpha\in(1/p,1/2)$, there is a measurable process Y, which is a version of the processes $\{\int_0^t(t-s)^{-\alpha}u_s\delta_2B_s : t\in[0,T]\}$, such that

$$\begin{aligned}
\int_0^t u_s\delta_2B_s &= \frac{\Gamma(1)}{\Gamma(\alpha)\Gamma(1-\alpha)}\int_0^t\left(\int_s^t(t-r)^{\alpha-1}(r-s)^{-\alpha}dr\right)u_s\delta_2B_s\\
&= \frac{\Gamma(1)}{\Gamma(\alpha)\Gamma(1-\alpha)}\int_0^t(t-r)^{\alpha-1}Y_rdr,\quad t\in[0,T],
\end{aligned}$$

where Γ is the gamma function given by (8.10). Consequently,

$$\begin{aligned}
E\left(\sup_{t\in[0,T]}\left|\int_0^t u_s\delta_2B_s\right|^p\right) &\leq \frac{\Gamma(1)}{\Gamma(\alpha)\Gamma(1-\alpha)}E\left(\int_0^t|Y_s|^pds\left(\int_0^t(t-r)^{p(\alpha-1)/(p-1)}dr\right)^{p-1}\right)\\
&= \frac{\Gamma(1)}{\Gamma(\alpha)\Gamma(1-\alpha)}\left(\frac{p-1}{p\alpha-1}\right)^{p-1}T^{p\alpha-1}E\left(\int_0^T|Y_s|^pds\right).
\end{aligned}$$

In order to finish the proof of this theorem, we can follow that of Theorem 3.1 in Alòs and Nualart [11], which implies that this step holds for $d=1$. For $d>1$, we leave the details to the reader as an exercise. □

A consequence of the last theorem, we obtain the following result.

Proposition 80 *Let $p \in (2,4)$, $q = 2p/(4-p)$ and $u \in \mathbb{L}^F \cap L^q(\Omega \times [0,T])$. Then, the process $\{\int_0^t u_s \delta_2 B_s : t \in [0,T]\}$ has continuous paths.*

Proof. We first assume that u belongs to $\mathcal{S}(\mathcal{H})$. That is, $u = \sum_{i=1}^n F_i h_i$ with $F_i \in \mathcal{S}$ and $h_i \in \mathcal{H}$, $i \in \{1, \dots, n\}$. Hence, Lemma 36 yields

$$\int_0^t u_s \delta_2 B_s = \sum_{i=1}^n \left(F_i \int_0^t h_i(s) dB_s - \sum_{\ell=1}^d \int_0^t h_i^\ell(s) D_s^\ell F_i ds \right), \quad t \in [0,T].$$

Thus, the result is true for $u \in \mathcal{S}(\mathcal{H})$ due to Propositions 59 and 70.

Finally, there exists a sequence $\{u_n \in \mathcal{S}(\mathcal{H}) : n \in \mathbb{N}\}$ satisfying (4.43), which is established in Step 2 of Theorem 52. Thus, the result is satisfied because of inequality (4.42). □

We are ready to see that Proposition 60 is also true for the divergence operator δ_2.

Lemma 48 *Let u be as in Proposition 80 and τ an $\mathcal{F}_t$-stopping time. Then,*

$$\int_0^s u_r \delta_2 B_r \Big|_{s=t\wedge\tau} = \int_0^t 1_{[0,\tau]}(r) u_r \delta_2 B_r. \quad t \in [0,T].$$

Remark 64 *Remember that, in this section, $(\mathcal{F}_t)_{t\in[0,T]}$ is the filtration generated by B and all the P-negligible sets. On the other hand, $\int_0^{\cdot} u_r \delta_2 B_r$ has continuous paths in view of Proposition 80. Thus, $\int_0^s u_r \delta_2 B_r \big|_{s=t\wedge\tau}$ is well-defined.*

Proof. We first assume that process u has the form $u = Fh$, with $F \in \mathcal{S}$ and $h \in \mathcal{H}$. Therefore, Lemma 36 allows us to get

$$\int_0^s u_r \delta_2 B_r = F \int_0^s h(r) dB_r - \sum_{\ell=1}^d \int_0^s h(r)^\ell D_r^\ell F dr, \quad s \in [0,T].$$

Hence, the fact that $\int_0^s h(r) dB_r$ is an Itô's integral, together with Lemma 36 and Proposition 60, implies

$$\begin{aligned} \int_0^s u_r \delta_2 B_r|_{s=t\wedge\tau} &= F \int_0^t 1_{[0,\tau]}(r) h(r) dB_r - \sum_{\ell=1}^d \int_0^t h^\ell(r) 1_{[0,\tau]}(r) D_r^\ell F dr \\ &= \int_0^t u_r 1_{[0,\tau]}(r) \delta_2 B_r, \quad t \in [0,T]. \end{aligned}$$

Now, we know that there is a sequence $\{u_n : n \in \mathbb{N}\}$ of processes in $\mathcal{S}(\mathcal{H})$ satisfying Step 2 of the proof of Theorem 52. As a consequence, we have that, for $t \in [0,T]$,

$$\begin{aligned} & E\left(\left| \int_0^s u_r \delta_2 B_r|_{s=t\wedge\tau} - \int_0^t u_n(r) 1_{[0,\tau]}(r) \delta_2 B_r \right|^p \right) \\ &= E\left(\left| \int_0^s u_r \delta_2 B_r|_{s=t\wedge\tau} - \int_0^s u_n(r) \delta_2 B_r|_{s=t\wedge\tau} \right|^p \right) \\ &\le E\left(\sup_{s\in[0,T]} \left| \int_0^s u_r \delta_2 B_r - \int_0^s u_n(r) \delta_2 B_r \right|^p \right) \to 0, \quad \text{as } n \to \infty. \end{aligned}$$

In particular, from this, we deduce that, for $t \in [0,T]$, the integral $\int_0^t u_n(r) 1_{[0,\tau]}(r) \delta_2 B_r$ goes to the random variable $\int_0^s u_r \delta_2 B_r|_{s=t\wedge\tau}$ in $L^2(\Omega)$, as $n \to \infty$. Finally, since the divergence δ_2

is a closed operator from $L^2(\Omega\times[0,T])$ into $L^2(\Omega)$ and $u_n 1_{[0,\tau]}$ goes to $u1_{[0,\tau]}$ in $L^2(\Omega\times[0,T])$, we obtain that the result holds. □

Now, our aim is an Itô's type formula for the operator δ_2. Towards this end, we proceed as follows.

Let X be a process in $\mathbb{L}^{1,2,f}$ and $q \in \{1,2\}$. We say that X belongs to the space $\mathbb{L}^{1,2,f}_{q-}$ if there exists process $D^-X \in L^q(\Omega\times[0,T];\mathbb{R}^{d\times d})$ such that

$$\lim_{n\to\infty}\int_0^T \sup_{0\vee(s-(1/n))\leq t\leq s} E\left(\left|D_sX_t-(D^-X)_s\right|^q_{\mathbb{R}^{d\times d}}\right)ds$$
$$= \lim_{n\to\infty}\int_0^T \sup_{0\vee(s-(1/n))\leq t\leq s} E\left[\left(\sum_{\ell_1,\ell_2=1}^{d}\left|(DX_t^{\ell_2})_s^{\ell_1}-(D^-X^{\ell_2})_s^{\ell_1}\right|^2\right)^{q/2}\right]ds$$
$$=0. \tag{4.47}$$

Remember that, in Remark 63, we introduce the space $\mathbb{L}^{1,2,f}$ for $\mathbb{R}$-valued stochastic processes. In this case, (4.47) has the form

$$\lim_{n\to\infty}\int_0^T \sup_{0\vee(s-(1/n))\leq t\leq s} E\left[\left(\sum_{\ell=1}^{d}\left|(DX_t)_s^{\ell}-(D^-X)_s^{\ell}\right|^2\right)^{q/2}\right]ds=0.$$

Note that if X belongs to $\left(\mathbb{L}^{1,2,f}_{q-}\right)_{loc}$ and $\{(\Omega_n,X_n): n\in\mathbb{N}\}$ localizes X in $\mathbb{L}^{1,2,f}_{q-}$, we can define the process D^-X as

$$D^-X = D^-X_n \quad \text{on } \Omega_n\times[0,T]. \tag{4.48}$$

Indeed, for $m,n,k\in\mathbb{N}$, $n\leq m$, we have

$$\int_0^T E\left(1_{\Omega_n}\left|D^-X_{m,s}-D^-X_{m+1,s}\right|^q_{\mathbb{R}^{d\times d}}\right)ds$$
$$= \int_0^T \sup_{0\vee(s-(1/k))\leq t\leq s} E\left(1_{\Omega_n}\left|D^-X_{m,s}-D^-X_{m+1,s}\right|^q_{\mathbb{R}^{d\times d}}\right)ds$$
$$\leq C\int_0^T \sup_{0\vee(s-(1/k))\leq t\leq s} E\left(1_{\Omega_n}\left|D^-X_{m,s}-D_sX_t\right|^q_{\mathbb{R}^{d\times d}}\right)ds$$
$$+C\int_0^T \sup_{0\vee(s-(1/k))\leq t\leq s} E\left(1_{\Omega_n}\left|D^-X_{m+1,s}-D_sX_t\right|^q_{\mathbb{R}^{d\times d}}\right)ds$$
$$= C\int_0^T \sup_{0\vee(s-(1/k))\leq t\leq s} E\left(1_{\Omega_n}\left|D^-X_{m,s}-D_sX_{m,t}\right|^q_{\mathbb{R}^{d\times d}}\right)ds$$
$$+C\int_0^T \sup_{0\vee(s-(1/k))\leq t\leq s} E\left(1_{\Omega_n}\left|D^-X_{m+1,s}-D_sX_{m+1,t}\right|^q_{\mathbb{R}^{d\times d}}\right)ds \to 0, \text{ as } k\to\infty,$$

which implies that the process D^-X is well-defined.

As an example, consider the $\mathbb{R}$-valued process

$$X_t = X_0+\int_0^t u_s\delta_2B_s+\int_0^t v_sds, \quad t\in[0,T],$$

with $X_0 \in \mathbb{D}^{1,2}$, $u \in \mathbb{L}^F$ and $v \in \mathbb{L}^{1,2,f}$. Here, u and v are $\mathbb{R}^d$-valued and $\mathbb{R}$-valued processes, respectively. Then, X is a process in $\mathbb{L}^{1,2,f}_{2-}$ with

$$(D^-X)^\ell_t = (DX_0)^\ell_t + \int_0^t (Du_s)^\ell_t \delta_2 B_s + \int_0^t (Dv_s)^\ell_t ds, \quad t \in [0,T]. \tag{4.49}$$

Indeed, in Lemma 47 (see also Remark 63), we show that the process $(Du.)^\ell_s 1_{[0,t]}(\cdot)$ belongs to $\mathbb{D}^{1,2}(\mathcal{H})$ for almost all $s > t$, for each $t \in [0,T]$. Hence, we can find a Borel subset $\mathcal{N}$ of $[0,T]$ such that $[0,T] \setminus \mathcal{N}$ has measure zero with respect to Lebesgue measure and $(Du.)^\ell_s 1_{[0,t]}(\cdot) \in \mathbb{D}^{1,2}(\mathcal{H})$ for all $s \in \mathcal{N}$ and $t \in (0,s) \cap \mathbb{Q}$. Therefore, using the definitions of the norms $\|\cdot\|_{\mathbb{L}^F}$ and $\|\cdot\|_{\mathbb{L}^{1,2,f}}$, it is now easy to prove that, for almost all $s \in [0,T]$, the process $(Du.)^\ell_s 1_{[0,t_n]}(\cdot)$ goes in to $(Du.)^\ell_s 1_{[0,s]}(\cdot)$ in $\mathbb{D}^{1,2}(\mathcal{H})$, for any sequence $\{t_n \in \mathbb{Q} : n \in \mathbb{N}\}$ that converges to s. Finally, (4.49) holds because Proposition 63 and Lemma 47 allow us to establish

$$\begin{aligned}
&\int_0^T \sup_{0\vee(s-(1/n))\le t\le s} E\left(\left|(DX_t)^\ell_s - (D^-X)^\ell_s\right|^2\right) ds \\
&\le E\left(\frac{2}{n}\int_0^T \int_{(s-1/n)\vee 0}^s \left|(Dv_r)^\ell_s\right|^2 drds + \sum_{\ell_1=1}^d \int_0^T \int_{(s-1/n)\vee 0}^s \left|(Du^{\ell_1}_r)^\ell_s\right|^2 drds\right) \\
&\quad + \sum_{\ell_1,\ell_2=1}^d \int_0^T \int_0^T \int_{(s-1/n)\vee 0}^s \left|(D^2 u^{\ell_1}_r)^{\ell,\ell_2}_{s,\theta}\right|^2 drd\theta ds \to 0, \quad \text{as } n \to \infty.
\end{aligned}$$

We are ready to state the Itô's formula for δ_2.

Theorem 53 *Let $m \in \mathbb{N}$, $X_{i,0} \in \mathbb{D}^{1,2}$, u_i a process in $\mathbb{L}^F$ and v_i an $\mathbb{R}$-valued process in $\mathbb{L}^{1,2,f}$ such that $\int_0^\cdot u_{i,s}\delta_2 B_s$ has continuous paths and $\| \int_0^T \|u_{i,s}\|^2_{\mathbb{R}^d} ds\|_{L^\infty(\Omega)}, \| \int_0^T |v_{i,s}|^2 ds\|_{L^\infty(\Omega)} < \infty$, for $i \in \{1,\dots,m\}$. Consider, for $i \in \{1,\dots,m\}$, the $\mathbb{R}$-valued processes*

$$X_{i,t} = X_{i,0} + \int_0^t u_{i,s}\delta_2 B_s + \int_0^t v_{i,s} ds, \quad t \in [0,T], \tag{4.50}$$

and a bounded function $f : \mathbb{R}^m \to \mathbb{R}$ that is twice continuously differentiable with bounded derivatives. Then,

$$\begin{aligned}
f(X_t) &= f(X_0) + \sum_{i=1}^m \left\{ \sum_{\ell=1}^d \int_0^t \partial_{x_i} f(X_s) u^{(\ell)}_{i,s} \delta_2 dB^{(\ell)}_s \right. \\
&\quad + \int_0^t \partial_{x_i} f(X_s) v_{i,s} ds + \frac{1}{2}\sum_{\ell=1}^d \sum_{j=1}^m \int_0^t \partial^2_{x_i,x_j} f(X_s) u^{(\ell)}_{i,s} u^{(\ell)}_{j,s} ds \\
&\quad \left. + \sum_{\ell=1}^d \sum_{j=1}^m \int_0^t \partial^2_{x_i,x_j} f(X_s) \left(D^-X_j\right)^\ell_s u^{(\ell)}_{i,s} ds \right\}, \quad t \in [0,T].
\end{aligned} \tag{4.51}$$

Remarks 7 *i) Note that the last summand in (4.51) does not appear in (3.34). The reason we have this additional term is, as we will see in the proof, because of Property (4.3), which is not used in the proof of Theorem 41 (see also Proposition 58.iii)).*

ii) Assume that $u^{(\ell)}_i, v_i \in \mathcal{L}_a(\Omega; L^2([a,b]))$ and that $X_{i,0}$ is a constant, for $i \in \{1,\dots,m\}$ and $\ell \in \{1,\dots,d\}$. Then, $u^{(\ell)}_i, v_i \in (\mathbb{L}^F)_{loc}$ and (4.51) holds with $D^-X_i \equiv 0$ (see

Corollary 20, Proposition 56 and Theorem 41). In this case, from the local property of Itô's integral (i.e., Proposition 55) and a localization argument (see the proof of Theorem 54), we only need to have that f is a twice continuously differentiable function and the proof of Proposition 56 allows us to omit the conditions $\| \int_0^T \|u_{i,s}\|_{\mathbb{R}^d}^2 ds\|_{L^\infty(\Omega)}, \| \int_0^T |v_{i,s}|^2 ds\|_{L^\infty(\Omega)} < \infty$ *since it is satisfied locally.*

iii) *Proposition 80 provides us a sufficient condition for the continuity of* $\int_0^\cdot u_{i,s}\delta_2 B_s$, $i = 1, \ldots, m$.

iv) *We need to show that* $\partial_{x_i} f(X.)u_{i,\cdot}^{(\ell)} \in Dom\,\delta_2^{B^{(\ell)}}$ *because this process is not in* $(\mathbb{L}^F)_{loc}$ *in general, for* $i = 1\ldots, m$, *since the product of two* $\mathbb{R}$*-valued processes* $\tilde{u}, \tilde{v}$ *in* $\mathbb{L}^F$ *is not necessarily in this space. Indeed, note that we could have that* $(D\tilde{u}_t\tilde{v}_t)_s^\ell = \tilde{u}_t(D\tilde{v}_t)_s^\ell + \tilde{v}_t(D\tilde{u}_t)_s^\ell$, *for* $s > t$ *and* $\ell \in \{1, \ldots, d\}$ *due to Proposition 21. But* $\tilde{u}$ *and* $\tilde{v}$ *are not necessarily in* $\mathbb{D}^{1,2}$.

Proof of Theorem 53. Fix $t \in [0,T]$ and let $\{\pi_n = \{0 = t_0^n < \ldots < t_{k_n}^n = t\} : n \in \mathbb{N}\}$ be a sequence of partitions of $[0,t]$ such that $\pi_n \subset \pi_{n+1}$ and $\lim_{n\to\infty} |\pi_n| = 0$. Then, the Taylor's theorem for functions on $\mathbb{R}^m$ to $\mathbb{R}$ yields

$$\begin{aligned} f(X_t) - f(X_0) &= \sum_{j=1}^{k_n} \left(f(X_{t_j^n}) - f(X_{t_{j-1}^n}) \right) = \sum_{j=1}^{k_n} \sum_{i=1}^{m} \partial_{x_i} f(X_{t_{j-1}^n}) \left(X_{i,t_j^n} - X_{i,t_{j-1}^n} \right) \\ &+ \frac{1}{2} \sum_{j=1}^{k_n} \sum_{i,\ell=1}^{m} \partial_{x_i,x_\ell}^2 f(\bar{X}_{j,n}) \left(X_{i,t_j^n} - X_{i,t_{j-1}^n} \right) \left(X_{\ell,t_j^n} - X_{\ell,t_{j-1}^n} \right), \quad (4.52) \end{aligned}$$

where $\bar{X}_{j,n}$ is a $\mathbb{R}^m$-valued random vector on the line segment joining the random points $X_{t_{j-1}^n}$ and $X_{t_j^n}$.

Now, in order to analyze the convergence of the last two terms on the right-hand side of (4.52), we divide the proof into three steps.

Step 1: Here we prove that the last term in (4.52) goes to

$$\frac{1}{2} \sum_{\ell=1}^{d} \sum_{i,j=1}^{m} \int_0^t \partial_{x_i,x_j}^2 f(X_s) u_{i,s}^{(\ell)} u_{j,s}^{(\ell)} ds, \quad \text{in } L^1(\Omega), \text{ as } |\pi| \to 0.$$

We first observe that the fact that $\left\| \int_0^T \|v_s\|_{\mathbb{R}^m}^2 ds \right\|_{L^\infty(\Omega)} < \infty$, Hölder inequality, Theorem 51 and Proposition 79 imply that, for $i, \ell \in \{1, \ldots, m\}$,

$$\begin{aligned} \sum_{j=1}^{k_n} \Bigg[& \left| \int_{t_{j-1}^n}^{t_j^n} v_{i,s} ds \right| \left| \int_{t_{j-1}^n}^{t_j^n} v_{\ell,s} ds \right| + \left| \int_{t_{j-1}^n}^{t_j^n} v_{i,s} ds \right| \left| \int_{t_{j-1}^n}^{t_j^n} u_{\ell,s} \delta_2 B_s \right| \\ & + \left| \int_{t_{j-1}^n}^{t_j^n} v_{\ell,s} ds \right| \left| \int_{t_{j-1}^n}^{t_j^n} u_{i,s} \delta_2 B_s \right| \Bigg] \to 0, \quad \text{in } L^1(\Omega) \text{ as } |\pi| \to 0. \end{aligned}$$

For example,

$$\begin{aligned} \sum_{j=1}^{k_n} \left| \int_{t_{j-1}^n}^{t_j^n} v_{i,s} ds \right| \left| \int_{t_{j-1}^n}^{t_j^n} v_{\ell,s} ds \right| &\leq \left[\sum_{j=1}^{k_n} \left| \int_{t_{j-1}^n}^{t_j^n} v_{i,s} ds \right|^2 \right]^{1/2} \left[\sum_{j=1}^{k_n} \left| \int_{t_{j-1}^n}^{t_j^n} v_{\ell,s} ds \right|^2 \right]^{1/2} \\ &\leq |\pi_n| \left[\int_0^t |v_{i,s}|^2 ds \right]^{1/2} \left[\int_0^t |v_{\ell,s}|^2 ds \right]^{1/2}. \end{aligned}$$

Therefore, in order to show that the claim of this step holds , we only need to establish

$$\sum_{j=1}^{k_n}\sum_{i,\ell=1}^{m} \partial^2_{x_i,x_\ell} f(\bar{X}_{j,n}) \left(\int_{t_{j-1}^n}^{t_j^n} u_{i,s}\delta_2 B_s\right)\left(\int_{t_{j-1}^n}^{t_j^n} u_{\ell,s}\delta_2 B_s\right)$$
$$\to \sum_{\ell=1}^{d}\sum_{i,j=1}^{m}\int_0^t \partial^2_{x_i,x_j} f(X_s) u_{i,s}^{(\ell)} u_{j,s}^{(\ell)} ds, \quad \text{in } L^1(\Omega), \text{ as } |\pi|\to 0.$$

Towards this end, we take $\theta < n$ and denote by $t_i^{(\theta)} \in \pi_\theta$ the point that is closer to t_i^n from the left. Consequently, we get, for $i,\ell \in \{1,\dots,m\}$,

$$\begin{aligned}
I := E&\left|\sum_{j=1}^{k_n} \partial^2_{x_i,x_\ell} f(\bar{X}_{j,n}) \left(\int_{t_{j-1}^n}^{t_j^n} u_{i,s}\delta_2 B_s\right)\left(\int_{t_{j-1}^n}^{t_j^n} u_{\ell,s}\delta_2 B_s\right)\right. \\
&\qquad \left. - \int_0^t \partial^2_{x_i,x_\ell} f(X_s)\langle u_{i,s}, u_{\ell,s}\rangle_{\mathbb{R}^d} ds\right| \\
\le\ & E\left|\sum_{j=1}^{k_n} \partial^2_{x_i,x_\ell}\left(f(\bar{X}_{j,n}) - f(X_{t_j^{(\theta)}})\right)\left(\int_{t_{j-1}^n}^{t_j^n} u_{i,s}\delta_2 B_s\right)\left(\int_{t_{j-1}^n}^{t_j^n} u_{\ell,s}\delta_2 B_s\right)\right| \\
&+E\left|\sum_{j=1}^{k_\theta} \partial^2_{x_i,x_\ell} f(X_{t_j^{(\theta)}}) \sum_{t_j^n\in[t_j^{(\theta)}, t_{j+1}^{(\theta)}]}\left[\left(\int_{t_{j-1}^n}^{t_j^n} u_{i,s}\delta_2 B_s\right)\left(\int_{t_{j-1}^n}^{t_j^n} u_{\ell,s}\delta_2 B_s\right)\right.\right. \\
&\qquad \left.\left. - \int_{t_j^n}^{t_{j+1}^n} \langle u_{i,s}, u_{\ell,s}\rangle_{\mathbb{R}^d}\, ds\right]\right| \\
&+E\left|\sum_{j=1}^{k_\theta} \partial^2_{x_i,x_\ell} f(X_{t_j^{(\theta)}}) \sum_{t_j^n\in[t_j^{(\theta)}, t_{j+1}^{(\theta)}]} \int_{t_j^n}^{t_{j+1}^n} \langle u_{i,s}, u_{\ell,s}\rangle_{\mathbb{R}^d}\, ds\right. \\
&\qquad \left. - \int_0^t \partial^2_{x_i,x_\ell} f(X_s)\,\langle u_{i,s}, u_{\ell,s}\rangle_{\mathbb{R}^d}\, ds\right| := I_1 + I_2 + I_3. \qquad (4.53)
\end{aligned}$$

The terms I_1 and I_3 are bounded as follows:

$$\begin{aligned}
|I_1| \ &\le\ E\left\{\left[\sup_{|s-r|\le|\pi_\theta|}\left|\partial^2_{x_i,x_\ell}\left(f(X_r)-f(X_s)\right)\right|\right]\sum_{j=1}^{k_n}\left|\int_{t_{j-1}^n}^{t_j^n} u_{i,s}\delta_2 B_s\right|\left|\int_{t_{j-1}^n}^{t_j^n} u_{\ell,s}\delta_2 B_s\right|\right\} \\
&=\ E\left\{\left[\sup_{|s-r|\le|\pi_\theta|}\left|\partial^2_{x_i,x_\ell}\left(f(X_r)-f(X_s)\right)\right|\right] I_{1,n}\right\}
\end{aligned}$$

and

$$|I_3| \le E\left\{\left[\sup_{|s-r|\le|\pi_m|}\left|\partial^2_{x_i,x_\ell}\left(f(X_r)-f(X_s)\right)\right|\right]\int_0^t \left|\langle u_{i,s}, u_{\ell,s}\rangle_{\mathbb{R}^d}\right| ds\right\}.$$

We also have that, for $M>0$,

$$|I_1| \le ME\left\{\left[\sup_{|s-r|\le|\pi_m|}\left|\partial^2_{x_i,x_\ell}\left(f(X_r)-f(X_s)\right)\right|\right]\right\} + 2\|\partial^2_{x_i,x_\ell} f\|_{\infty,\mathbb{R}^m} E\left(I_{1,n}1_{\{I_{1,n}>M\}}\right).$$

Hence, the claim of this step is satisfied due to the continuity of X, the fact that $\left\|\int_0^T \|u_{i,s}\|^2_{\mathbb{R}^d} ds\right\|_{L^\infty(\Omega)} < \infty$, the dominated convergence theorem and Proposition 79.

Step 2: Proceeding as in the analysis of I_3 in Step 1, we have

$$\sum_{j=1}^{k_n}\sum_{i=1}^{m} \partial_{x_i} f(X_{t^n_{j-1}}) \left(\int_{t^n_{j-1}}^{t^n_j} v_{i,s} ds\right) \to \sum_{i=1}^{m} \int_0^t \partial_{x_i} f(X_s) v_{i,s} ds, \quad \text{in } L^1(\Omega), \text{ as } n \to \infty.$$

Step 3: In order to finish the proof, we deal with the convergence of

$$\sum_{j=1}^{k_n}\sum_{i=1}^{m} \partial_{x_i} f(X_{t^n_{j-1}}) \left(\int_{t^n_{j-1}}^{t^n_j} u_{i,s} \delta_2 B_s\right).$$

Note that, by Corollary 10, Lemma 36 and Proposition 77, we have that, for $i \in \{1, \dots, m\}$ and $j \in \{1, \dots, k_n\}$,

$$\begin{aligned}
&\partial_{x_i} f(X_{t^n_{j-1}}) \left(\int_{t^n_{j-1}}^{t^n_j} u_{i,s} \delta_2 B_s\right) \\
&= \sum_{\ell=1}^{d} \partial_{x_i} f(X_{t^n_{j-1}}) \left(\int_{t^n_{j-1}}^{t^n_j} u^{(\ell)}_{i,s} \delta_2 B^{(\ell)}_s\right) \\
&= \sum_{\ell=1}^{d} \left(\int_{t^n_{j-1}}^{t^n_j} \partial_{x_i} f(X_{t^n_{j-1}}) u^{(\ell)}_{i,s} \delta_2 B^{(\ell)}_s + \int_{t^n_{j-1}}^{t^n_j} (D\partial_{x_i} f(X_{t^n_{j-1}}))^\ell_s u^{(\ell)}_{i,s} ds\right) \\
&= \sum_{\ell=1}^{d} \left(\int_{t^n_{j-1}}^{t^n_j} \partial_{x_i} f(X_{t^n_{j-1}}) u^{(\ell)}_{i,s} \delta_2 B^{(\ell)}_s + \sum_{\tilde{i}=1}^{m} \partial^2_{x_i, x_{\tilde{i}}} f(X_{t^n_{j-1}}) \int_{t^n_{j-1}}^{t^n_j} (DX_{\tilde{i}, t^n_{j-1}})^\ell_s u^{(\ell)}_{i,s} ds\right). \quad (4.54)
\end{aligned}$$

We claim that

$$\begin{aligned}
&\sum_{j=1}^{k_n}\sum_{i,\tilde{i}=1}^{m}\sum_{\ell=1}^{d} \partial^2_{x_i, x_{\tilde{i}}} f(X_{t^n_{j-1}}) \int_{t^n_{j-1}}^{t^n_j} (DX_{\tilde{i}, t^n_{j-1}})^\ell_s u^{(\ell)}_{i,s} ds \\
&\quad \to \sum_{\ell=1}^{d}\sum_{i,j=1}^{m} \int_0^t \partial^2_{x_i, x_j} f(X_s) \left(D^- X_j\right)^\ell_s u^{(\ell)}_{i,s} ds \text{ in } L^1(\Omega), \text{ as } n \to \infty. \quad (4.55)
\end{aligned}$$

In effect, for $i, \tilde{i} \in \{1, \dots, m\}$ and $\ell \in \{1, \dots, d\}$, we can use the boundedness of f to get

$$\begin{aligned}
&\left|\sum_{j=1}^{k_n} \partial^2_{x_i, x_{\tilde{i}}} f(X_{t^n_{j-1}}) \int_{t^n_{j-1}}^{t^n_j} (DX_{\tilde{i}, t^n_{j-1}})^\ell_s u^{(\ell)}_{i,s} ds - \int_0^t \partial^2_{x_i, x_{\tilde{i}}} f(X_s) \left(D^- X_{\tilde{i}}\right)^\ell_s u^\ell_{i,s} ds\right| \\
&\leq \left|\sum_{j=1}^{k_n} \partial^2_{x_i, x_{\tilde{i}}} f(X_{t^n_{j-1}}) \int_{t^n_{j-1}}^{t^n_j} \left[(DX_{\tilde{i}, t^n_{j-1}})^\ell_s - \left(D^- X_{\tilde{i}}\right)^\ell_s\right] u^{(\ell)}_{i,s} ds\right| \\
&\quad + \left|\sum_{j=1}^{k_n} \int_{t^n_{j-1}}^{t^n_j} \left[\partial^2_{x_i, x_{\tilde{i}}} f(X_{t^n_{j-1}}) - \partial^2_{x_i, x_{\tilde{i}}} f(X_s)\right] \left(D^- X_{\tilde{i}}\right)^\ell_s u^\ell_{i,s} ds\right| \\
&\leq C \sum_{j=1}^{k_n} \int_{t^n_{j-1}}^{t^n_j} \left|\left[(DX_{\tilde{i}, t^n_{j-1}})^\ell_s - \left(D^- X_{\tilde{i}}\right)^\ell_s\right] u^{(\ell)}_{i,s}\right| ds \\
&\quad + \left(\sup_{|r-s| \leq |\pi_n|} \left|\partial^2_{x_i, x_{\tilde{i}}} f(X_r) - \partial^2_{x_i, x_{\tilde{i}}} f(X_s)\right|\right) \int_0^t \left|\left(D^- X_{\tilde{i}}\right)^\ell_s\right| ds.
\end{aligned}$$

Now, we assume that $|\pi_n| \le 1/n$, for $n \in \mathbb{N}$. Consequently, (4.55) holds since X is an $\mathbb{R}^m$-valued continuous process by hypothesis and satisfies (4.49), $\partial^2_{x_i, x_{\bar{i}}} f$ is a bounded function on $\mathbb{R}^m$ and the inequality

$$\begin{aligned} & E\left(\sum_{j=1}^{k_n} \int_{t_{j-1}^n}^{t_j^n} \left| \left[(DX_{\bar{i}, t_{j-1}^n})_s^\ell - \left(D^- X_{\bar{i}}\right)_s^\ell \right] u_{i,s}^{(\ell)} \right| ds \right) \\ & \le \left(\sum_{j=1}^{k_n} \int_{t_{j-1}^n}^{t_j^n} E\left(\left[(DX_{\bar{i}, t_{j-1}^n})_s^\ell - \left(D^- X_{\bar{i}}\right)_s^\ell \right]^2 \right) ds \right)^{1/2} \left(\int_0^t E\left((u_{i,s}^{(\ell)})^2 \right) ds \right)^{1/2} \\ & \le C \left(\int_0^t \sup_{0 \vee (s-(1/n)) \le t < s} E\left(\left[(DX_{\bar{i}, t})_s^\ell - \left(D^- X_{\bar{i}}\right)_s^\ell \right]^2 \right) ds \right)^{1/2}. \end{aligned}$$

Finally, observe that, in (4.54), we have

$$\sum_{j=1}^{k_n} \partial_{x_i} f(X_{t_{j-1}^n}) u_{i,\cdot}^{(\ell)} 1_{]t_{j-1}^n, t_j^n]}(\cdot) \to \partial_{x_i} f(X_\cdot) u_{i,\cdot}^{(\ell)}, \quad \text{in } L^2(\Omega \times [0,T]) \text{ as } n \to \infty,$$

and that Proposition 67, (4.54), (4.55), and Steps 1 and 2 yield that

$$\delta_2 \left(\sum_{j=1}^{k_n} \sum_{i=1}^{m} \partial_{x_i} f(X_{t_{j-1}^n}) u_{i,\cdot} 1_{]t_{j-1}^n, t_j^n]}(\cdot) \right)$$

converges in $L^1(\Omega)$, as $n \to \infty$ to the random variable

$$\begin{aligned} f(X_t) - f(X_0) - \sum_{i=1}^{m} \Bigg\{ & \frac{1}{2} \sum_{\ell=1}^{d} \sum_{j=1}^{m} \int_0^t \partial^2_{x_i, x_j} f(X_s) u_{i,s}^{(\ell)} u_{j,s}^{(\ell)} ds \\ & + \int_0^t \partial_{x_i} f(X_s) v_{i,s} ds + \sum_{\ell=1}^{d} \sum_{j=1}^{m} \int_0^t \partial^2_{x_i, x_j} f(X_s) \left(D^- X_j\right)_s^\ell u_{i,s}^\ell ds \Bigg\}, \end{aligned}$$

which belongs to $L^2(\Omega)$. Therefore, (4.51) is true because of the duality relation (4.2), Proposition 67 and the fact that, for $i \in \{1, \ldots, m\}$, $\partial_{x_i} f(X_\cdot) u_{i,\cdot}$ is in $L^2(\Omega \times [0,T]; \mathbb{R}^d)$. □

Theorem 54 *Consider the $\mathbb{R}$-valued stochastic processes given by (4.50) with $X_{i,0} \in \mathbb{D}^{1,2}_{loc}$, $u_i \in (\mathbb{D}^{2,2}(\mathcal{H}) \cap \mathbb{D}^{1,4}(\mathcal{H}))_{loc}$ such that $\int_0^\cdot u_{i,s} \delta_2 B_s$ has continuous paths and $v_i \in (\mathbb{D}^{1,2}(L^2([0,T])))_{loc}$, for $i \in \{1, \ldots, m\}$. Let $f : \mathbb{R}^m \to \mathbb{R}$ be a twice continuously differentiable function. Then, the Itô's type formula (4.51) is also satisfied.*

Remark 65 *We will see in the proof that the process $\sum_{i=1}^m \partial_{x_i} f(X_\cdot) u_{i,\cdot}$ is in $(\mathbb{D}^{1,2}(\mathcal{H}))_{loc}$. Therefore, the divergence operator in (4.51) is defined locally through Proposition 66.*

Proof. By hypothesis, for $i = 1, \ldots, m$, there exist localizing sequences $\{(\Omega_{i,1,n}, X_{i,0,n}) : n \in \mathbb{N}\}$, $\{(\Omega_{i,2,n}, u_{i,n}) : n \in \mathbb{N}\}$ and $\{(\Omega_{i,3,n}, v_{i,n}) : n \in \mathbb{N}\}$ of $X_{i,0}$, u_i and v_i in $\mathbb{D}^{1,2}$, $\mathbb{D}^{2,2}(\mathcal{H}) \cap \mathbb{D}^{1,4}(\mathcal{H})$ and $\mathbb{D}^{1,2}(L^2([0,T]))$. respectively. Consider a $C^\infty(\mathbb{R})$-function $\varphi : \mathbb{R} \to [0,1]$ such that

$$\varphi(x) = \begin{cases} 1, & \text{if } |x| \le 1 \\ 0, & \text{if } |x| \ge 2, \end{cases}$$

and set

$$\Omega_{n,k} := \left(\bigcap_{\substack{1\le i\le m,\\ 1\le j\le 3}} \Omega_{i,j,n} \right) \cap \left[\sup_{\substack{t\in[0,T],\\ 1\le i\le m}} |X_{i.t}| \le k \right] \cap \left[\sup_{1\le i\le m} \int_0^T \left(|v_{i,n,s}|^2 + \|u_{i,n,s}\|_{\mathbb{R}^d}^2 \right) ds \le k \right],$$

$f_n(x) := f(x)\prod_{i=1}^m \varphi(x^{(i)}/n)$, for $x \in \mathbb{R}^m$,

$$\tilde{v}_{i,n,k} := v_{i,n}\varphi\left(\int_0^T |v_{i,n,s}|^2 ds/k \right) \quad \text{and} \quad \tilde{u}_{i,n,k} := u_{i,n}\varphi\left(\int_0^T \|u_{i,n,s}\|_{\mathbb{R}^d}^2 ds/k \right).$$

Note that $\tilde{v}_{i,n,k} \in \mathbb{D}^{1,2}(L^2([0,T]))$ and $\tilde{u}_{i,n,k} \in \mathbb{D}^{2,2}(\mathcal{H})$ due to Proposition 21, and the facts that $\int_0^T |\tilde{v}_{i,n,,k,s}|^2 ds \le 2k$, $\int_0^T \|\tilde{u}_{i,n,k,s}\|_{\mathbb{R}^d}^2 ds \le 2k$, $u_{i,n} \in \mathbb{D}^{2,2}(\mathcal{H}) \cap \mathbb{D}^{1,4}(\mathcal{H})$ and $v_{i,n} \in \mathbb{D}^{1,2}(L^2([0,T]))$. In order to show that this claim holds, we fix $\ell_1, \ell_2 \in \{1, \dots, d\}$. Then, Proposition 21 yields

$$\begin{aligned}
& E\left[\int_0^T \int_0^T \left\{ \left| (Du_{i,n,t}^{(\ell_1)})_s^{\ell_2} \varphi\left(\int_0^T \|u_{i,n,r}\|_{\mathbb{R}^d}^2 dr/k \right) \right|^2 \right.\right. \\
& \qquad \left.\left. + \frac{2}{k} \left| u_{i,n,t}^{(\ell_1)} \varphi'\left(\int_0^T \|u_{i,n,r}\|_{\mathbb{R}^d}^2 dr/k \right) \right|^2 \left| \sum_{\ell=1}^d \int_0^T u_{i,n,r}^{(\ell)} (Du_{i,n,r}^{(\ell)})_s^{\ell_2} dr \right|^2 \right\} dsdt \right] \\
\le\ & E\left[\int_0^T \int_0^T \left\{ \left| (Du_{i,n,t}^{(\ell_1)})_s^{\ell_2} \right|^2 + \left| \varphi'\left(\int_0^T \|u_{i,n,r}\|_{\mathbb{R}^d}^2 dr/k \right) \right|^2 \right.\right. \\
& \qquad \left.\left. \times \left| u_{i,n,t}^{(\ell_1)} \sum_{\ell=1}^d \int_0^T u_{i,n,r}^{(\ell)} (Du_{i,n,r}^{(\ell)})_s^{\ell_2} dr \right|^2 \right\} dsdt \right] \\
\le\ & C + E\left[\left| \varphi'\left(\int_0^T \|u_{i,n,r}\|_{\mathbb{R}^d}^2 dr/k \right) \right|^2 \right. \\
& \qquad \left. \times \int_0^T \int_0^T \left(u_{i,n,t}^{(\ell_1)} \right)^2 \left(\sum_{\ell=1}^d \int_0^T u_{i,n,r}^{(\ell)} (Du_{i,n,r}^{(\ell)})_s^{\ell_2} dr \right)^2 dsdt \right] \\
\le\ & C + CE\left[\left| \varphi'\left(\int_0^T \|u_{i,n,r}\|_{\mathbb{R}^d}^2 dr/k \right) \right|^2 \int_0^T \left(\sum_{\ell=1}^d \int_0^T u_{i,n,r}^{(\ell)} (Du_{i,n,r}^{(\ell)})_s^{\ell_2} dr \right)^2 ds \right] \\
\le\ & C + C\sum_{\ell=1}^d E\left[\int_0^T \int_0^T \left((Du_{i,n,r}^{(\ell)})_s^{\ell_2} \right)^2 drds \right] < \infty,
\end{aligned}$$

which implies that u_i belongs to $\mathbb{D}^{1,2}(\mathcal{H})$.

Similarly, we can see that u_i is also in $\mathbb{D}^{2,2}(\mathcal{H})$ through Proposition 21. In effect, we have, for $\ell_1, \ell_2, \ell_3 \in \{1, \dots, d\}$,

$$\begin{aligned}
& E\left[\int_0^T \int_0^T \int_0^T \left| (D^2 u_{i,n,t}^{(\ell_1)})_{s_1,s_2}^{\ell_2,\ell_3} \varphi\left(\int_0^T \|u_{i,n,r}\|_{\mathbb{R}^d}^2 dr/k \right) \right|^2 ds_1 ds_2 dt \right] \\
& \qquad \le E\left[\int_0^T \int_0^T \int_0^T \left| (D^2 u_{i,n,t}^{(\ell_1)})_{s_1,s_2}^{\ell_2,\ell_3} \right|^2 ds_1 ds_2 dt \right] < \infty,
\end{aligned}$$

and

$$E\left\{\int_0^T\int_0^T\int_0^T\left|(Du_{i,n,t}^{(\ell_1)})_{s_1}^{\ell_2}\varphi'\left(\int_0^T\|u_{i,n,r}\|_{\mathbb{R}^d}^2dr/k\right)\right|^2\right.$$
$$\left.\times\frac{2}{k}\left|\sum_{\ell=1}^d\int_0^T u_{i,n,r}^{(\ell)}(Du_{i,n,r}^{(\ell)})_{s_2}^{\ell_3}dr\right|^2 ds_1ds_2dt\right\}$$
$$\leq C\sum_{\ell=1}^d E\left\{\int_0^T\int_0^T\left|(Du_{i,n,t}^{(\ell_1)})_{s_1}^{\ell_2}\right|^2\left(\int_0^T\int_0^T\left|(Du_{i,n,r}^{(\ell)})_{s_2}^{\ell_3}\right|^2drds_2\right)ds_1dt\right\}$$
$$\leq C\sum_{\ell=1}^d\left[E\left(\int_0^T\int_0^T\left|(Du_{i,n,t}^{(\ell_1)})_{s_1}^{\ell_2}\right|^2ds_1dt\right)^2\right]^{1/2}$$
$$\times\left[E\left(\int_0^T\int_0^T\left|(Du_{i,n,r}^{(\ell)})_{s_2}^{\ell_3}\right|^2drds_2\right)^2\right]^{1/2}<\infty,$$

where, in the last inequality, we apply that $u_{i,n}\in\mathbb{D}^{1,4}(\mathcal{H})$. Moreover. proceeding as in the last two calculations, we can obtain

$$E\left\{\int_0^T\int_0^T\int_0^T\left|\left(D\left(u_{i,n,t}^{(\ell_1)}\varphi'\left(\int_0^T\|u_{i,n,r}\|_{\mathbb{R}^d}^2dr/k\right)\right.\right.\right.\right.$$
$$\left.\left.\left.\left.\times\sum_{\ell=1}^d\int_0^T u_{i,r}^{(\ell)}(Du_{i,n,r}^{(\ell)})_{s_1}^{\ell_2}dr\right)\right)_{s_2}^{\ell_3}\right|^2 ds_1ds_2dt\right\}<\infty.$$

In this way, we have proven that $\tilde{u}_{i,n,k}\in\mathbb{D}^{2,2}(\mathcal{H})$. Consequently, if we can show that the process $\sum_{i=1}^m\partial_{x_i}f_n(X_{n,\cdot})u_{i,n,\cdot}$ belongs to the space $\mathbb{D}^{1,2}(\mathcal{H})$, then the local property of the operator δ_2 (see Proposition 66) and Theorem 53 allow us to conclude that the Itô's type formula (4.51) is true when we consider the involved divergence operator with respect to B defined locally. Thus, by Proposition 21, we only need to show that, for $i,j\in\{1,\ldots,m\}$ and $\ell\in\{1,\ldots,d\}$,

$$(s,t)\mapsto\partial_{x_i,x_j}^2f_n(X_{n,k,t})\tilde{u}_{i,n,k,t}^{(\ell)}D_sX_{j,n,k,t}+\partial_{x_i}f_n(X_{n,k,t})D_s\tilde{u}_{i,n,k,t}^{(\ell)}$$

belongs to $L^2(\Omega\times[0,T]^2;\mathbb{R}^d)$, where $X_{i,n,k}$ is given by (4.50) when we write $X_{i,0,n}$, $\tilde{v}_{i,n,k}$ and $\tilde{u}_{i,n,k}$ instead of $X_{i,0}$, v_i and u_i, respectively. Since $\partial_{x_i}f_n$ is a bounded function, we have

$$\left\|\partial_{x_i}f_n(X_{n,k,t})D.\tilde{u}_{i,n,k,t}^{(\ell)}\right\|_{\mathbb{R}^d}\leq C\left\|D.\tilde{u}_{i,n,t}^{(\ell)}\right\|_{\mathbb{R}^d}\in L^2(\Omega\times[0,T]^2;\mathbb{R}^d).$$

For the remaining term, we make use of the duality relation (4.2) as follows:

$$E\left\{\int_0^T\int_0^T\left\|\partial_{x_i,x_j}^2f_n(X_{n,k,t})\tilde{u}_{i,n,k,t}^{(\ell)}DX_{j,n,k,t}\right\|_{\mathbb{R}^d}^2dsdt\right\}$$
$$\leq C\sum_{\ell_1=1}^d E\left\{\int_0^T\int_0^T\left(\tilde{u}_{i,n,k,t}^{(\ell)}(DX_{j,n,k,t})_s^{\ell_1}\right)^2dsdt\right\}.$$

In consequence,equality (4.6) implies

$$
\begin{aligned}
&E\left\{\int_0^T\int_0^T \left\|\partial^2_{x_i,x_j} f_n(X_{n,k,t})\tilde{u}^{(\ell)}_{i,n,k,t} DX_{j,n,k,t}\right\|^2_{\mathbb{R}^d} dsdt\right\}\\
&\le\ C\sum_{\ell_1=1}^d E\left\{\int_0^T\int_0^T \left(\tilde{u}^{(\ell)}_{i,n,k,t}\left[\tilde{u}^{(\ell_1)}_{j,n,k,s} + (DX_{j,0,n})^{\ell_1}_s\right.\right.\right.\\
&\qquad\left.\left.\left. + \int_0^t (D\tilde{u}_{j,n,k,r})^{\ell_1}_s \delta_2 B_r + \int_0^t (D\tilde{v}_{j,n,k,r})^{\ell_1}_s dr\right]\right)^2 dsdt\right\}\\
&\le\ C\sum_{\ell_1=1}^d E\left\{\int_0^T \left(\tilde{u}^{(\ell)}_{i,n,k,t}\right)^2 \left[\int_0^T \left(\tilde{u}^{(\ell_1)}_{j,n,k,s}\right)^2 ds + \int_0^T \left((DX_{j,0,n})^{\ell_1}_s\right)^2 ds\right.\right.\\
&\qquad\left. + \int_0^T\int_0^T \left((D\tilde{v}_{j,n,k,r})^{\ell_1}_s\right)^2 drds\right] dt\\
&\quad\left. + \int_0^T\int_0^T \left(\tilde{u}^{(\ell)}_{i,n,k,t}\int_0^t (D\tilde{u}_{j,n,k,r})^{\ell_1}_s \delta_2 B_r\right)^2 dsdt\right\}\\
&\le\ C + C\sum_{\ell_1=1}^d E\left\{\int_0^T\int_0^T \left(\tilde{u}^{(\ell)}_{i,n,k,t}\int_0^t (D\tilde{u}_{j,n,k,r})^{\ell_1}_s \delta_2 B_r\right)^2 dsdt\right\}. \qquad (4.56)
\end{aligned}
$$

Now, we apply the duality relation (4.2) to see that the last expectation is bounded by

$$
\begin{aligned}
&\sum_{\tilde{\ell}=1}^d E\left\{\int_0^T\int_0^T\int_0^t \left(2\tilde{u}^{(\ell)}_{i,n,k,t}(D\tilde{u}^{(\ell)}_{i,n,k,t})^{\tilde{\ell}}_r (D\tilde{u}^{\tilde{\ell}}_{j,n,k,r})^{\ell_1}_s \int_0^t (D\tilde{u}_{j,n,k,\theta})^{\ell_1}_s \delta_2 B_\theta\right.\right.\\
&\qquad\left.\left. + \left(\tilde{u}^{(\ell)}_{i,n,k,t}\right)^2 (D\tilde{u}^{\tilde{\ell}}_{j,n,k,r})^{\ell_1}_s \left[(D\tilde{u}^{\tilde{\ell}}_{j,n,k,r})^{\ell_1}_s + \int_0^t (D^2\tilde{u}_{j,n,k,\theta})^{\ell_1,\tilde{\ell}}_{s,r} \delta_2 B_\theta\right]\right) drdsdt\right\}\\
&:=\ \sum_{\tilde{\ell}=1}^d E\left\{\int_0^T\int_0^T\int_0^t (I_1 + I_2 + I_3)\, drdsdt\right\}. \qquad (4.57)
\end{aligned}
$$

Hence, from the duality relation again, we obtain

$$
\begin{aligned}
&E\left\{\int_0^T\int_0^T\int_0^t I_1 drdsdt\right\}\\
&=\ 2\sum_{\ell_2=1}^d E\left\{\int_0^T\int_0^T\int_0^t\int_0^t (D\tilde{u}^{(\ell_2)}_{j,n,k,\theta})^{\ell_1}_s \left[(D\tilde{u}^{(\ell)}_{i,n,k,t})^{\ell_2}_\theta (D\tilde{u}^{(\ell)}_{i,n,k,t})^{\tilde{\ell}}_r (D\tilde{u}^{\tilde{\ell}}_{j,n,k,r})^{\ell_1}_s\right.\right.\\
&\quad\left. + \tilde{u}^{(\ell)}_{i,n,k,t}(D^2\tilde{u}^{(\ell)}_{i,n,k,t})^{\tilde{\ell},\ell_2}_{r,\theta}(D\tilde{u}^{\tilde{\ell}}_{j,n,k,r})^{\ell_1}_s + \tilde{u}^{(\ell)}_{i,n,k,t}(D\tilde{u}^{(\ell)}_{i,n,k,t})^{\tilde{\ell}}_r (D^2\tilde{u}^{\tilde{\ell}}_{j,n,k,r})^{\ell_1,\ell_2}_{s,\theta}\right]\\
&\qquad\left. \times\, d\theta drdsdt\right\}.
\end{aligned}
$$

Therefore, Hölder inequality and the fact that $\int_0^T \|\tilde{u}_{i,n,k,s}\|^2_{\mathbb{R}^d} ds \leq 2k$ give

$$\begin{aligned}
& E\left\{\int_0^T\int_0^T\int_0^t I_1 drdsdt\right\} \\
& \leq \sum_{\ell_2=1}^d E\left\{\left[\int_0^T\int_0^T\left((D\tilde{u}^{(\ell_2)}_{j,n,k,\theta})^{\ell_1}_s\right)^2 dsd\theta\right]^{1/2}\left[\int_0^T\int_0^T\left((D\tilde{u}^{(\ell)}_{i,n,k,t})^{\ell_2}_\theta\right)^2 dtd\theta\right]^{1/2}\right. \\
& \quad\times\left.\left[\int_0^T\int_0^T\left((D\tilde{u}^{(\ell)}_{i,n,k,t})^{\tilde{\ell}}_r\right)^2 dtdr\right]^{1/2}\left[\int_0^T\int_0^T\left((D\tilde{u}^{\tilde{\ell}}_{j,n,k,r})^{\ell_1}_s\right)^2 drds\right]^{1/2}\right\} \\
& \quad +C\sum_{\ell_2=1}^d E\left\{\left[\int_0^T\int_0^T\left((D\tilde{u}^{(\ell_2)}_{j,n,k,\theta})^{\ell_1}_s\right)^2 dsd\theta\right]^{1/2}\left[\int_0^T\int_0^T\left((D\tilde{u}^{\tilde{\ell}}_{j,n,k,r})^{\ell_1}_s\right)^2 drds\right]^{1/2}\right. \\
& \quad\times\left.\left[\int_0^T\int_0^T\int_0^T\left((D^2\tilde{u}^{(\ell)}_{i,n,k,t})^{\tilde{\ell},\ell_2}_{r,\theta}\right)^2 dtdrd\theta\right]^{1/2}\right\} \\
& \quad +C\sum_{\ell_2=1}^d E\left\{\left[\int_0^T\int_0^T\left((D\tilde{u}^{(\ell_2)}_{j,n,k,\theta})^{\ell_1}_s\right)^2 dsd\theta\right]^{1/2}\left[\int_0^T\int_0^T\left((D\tilde{u}^{(\ell)}_{i,n,k,t})^{\tilde{\ell}}_r\right)^2 drdt\right]^{1/2}\right. \\
& \quad\times\left.\left[\int_0^T\int_0^T\int_0^T\left((D^2\tilde{u}^{\tilde{\ell}}_{j,n,k,r})^{\ell_1,\ell_2}_{s,\theta}\right)^2 drdsd\theta\right]^{1/2}\right\} < \infty. \qquad (4.58)
\end{aligned}$$

We observe that we are able to proceed as these last calculations to have

$$E\left\{\int_0^T\int_0^T\int_0^t I_3 drdsdt\right\} < \infty. \qquad (4.59)$$

For the integral of I_2, we employ that $\int_0^T \|\tilde{u}_{i,n,k,s}\|^2_{\mathbb{R}^d} ds \leq 2k$ again to get

$$\begin{aligned}
E\left\{\int_0^T\int_0^T\int_0^t I_2 drdsdt\right\} &= E\left\{\left(\int_0^T\left[\tilde{u}^{(\ell)}_{i,n,k,t}\right]^2 dt\right)\int_0^T\int_0^T\left[(D\tilde{u}^{(\tilde{\ell})}_{j,n,k,r})^{\ell_1}_s\right]^2 drds\right\} \\
&\leq 2kE\left\{\int_0^T\int_0^T\left[(D\tilde{u}^{(\tilde{\ell})}_{j,n,k,r})^{\ell_1}_s\right]^2 drds\right\} < \infty. \qquad (4.60)
\end{aligned}$$

Finally, (4.56)-(4.60) establish that the process $\sum_{i=1}^m \partial_{x_i} f_n(X_{n,\cdot})u_{i,n,\cdot}$ belongs to the space $\mathbb{D}^{1,2}(\mathcal{H})$. Thus, the proof is complete. □

4.2 An extension of the divergence operator for some class of Gaussian processes

The purpose of this section is to construct an extension of the divergence operator for Gaussian processes. It means, we extend the domain of the stochastic integral of divergence type with respect to suitable Gaussian processes (i.e., processes satisfying the assumptions required in Section 1.5.5). The drawback of the divergence operator is that, for instance, the fractional Brownian motion is not integrable with respect to itself when the Hurst parameter H is less than $1/4$. That is, B does not belong to the domain of δ_p, for all $p \geq 1$, in this case. This fact was proven by Cheridito and Nualart [27] (See also Proposition 85 below).

As in Section 1.5.5, we suppose that we have two real-separable Hilbert spaces $\mathcal{H}$ and $\mathcal{H}_0$ with inner products $\langle \cdot, \cdot \rangle_{\mathcal{H}}$ and $\langle \cdot, \cdot \rangle_{\mathcal{H}_0}$, respectively, and an isonormal Gaussian process $\{W(h) : h \in \mathcal{H}\}$ on $\mathcal{H}$ defined on a complete probability space $(\Omega, \mathcal{F}, P)$ (see Definition 19). Since we will use the results stated in Section 1.5.5, we also assume that $\mathcal{H}$ is densely and continuously embedded in $\mathcal{H}_0$, and that there exists a linear operator $\mathcal{T} : \mathcal{H} \subset \mathcal{H}_0 \to \mathcal{H}_0$ (whose domain $\mathcal{D}(T)$ is $\mathcal{H}$) satisfying the following two conditions:

(H1) $|Th|_{\mathcal{H}_0} = |h|_{\mathcal{H}}$, for all $h \in \mathcal{H}$.

(H2) $\mathcal{T}_{\mathcal{H}} := \{h \in \mathcal{H} : Th \in \mathcal{D}(T^*)\}$ is a dense subset of $\mathcal{H}$.

Additionally, we ask that the linear operator $\mathcal{T}$ fulfills the condition

(H3) $\mathcal{T}_{\mathcal{H}_0} = \{T^*Th : h \in \mathcal{T}_{\mathcal{H}}\}$ is dense in $\mathcal{H}_0$.

Now, we are ready to introduce our extended divergence operator. Remember that the spaces $\mathcal{S}_{\mathcal{T}}(\mathbb{R})$ and $\mathbb{D}_{\mathcal{T}}^{1,2}(\mathbb{R})$ are given in Section 1.5.5 (see (1.110) and Lemma 18).

Definition 40 *Let $u \in L^2(\Omega, \mathcal{F}, P; \mathcal{H}_0)$. We say that u belongs to Dom δ^* if and only if there exists $\delta(u) \in L^2(\Omega)$ such that*

$$E\langle \mathcal{T}^*\mathcal{T}DF, u\rangle_{\mathcal{H}_0} = E(F\delta(u)), \quad \text{for every} \quad F \in \mathcal{S}_{\mathcal{T}}(\mathbb{R}). \tag{4.61}$$

In this case, the random variable $\delta(u)$ is called the extended divergence *of u.*

Remarks 8 *i) Hypothesis (H2) gives that there is at most one square-integrable random variable $\delta(u)$ such that (4.61) holds.*

ii) If $\mathcal{H}_0 = \mathcal{H}$ and $\mathcal{T} = I_{\mathcal{H}}$, then δ is equal to the usual divergence operator δ_2 presented in (4.2).

iii) Observe that the duality relation (4.61) also holds for $F \in \mathbb{D}_{\mathcal{T}}^{1,2}(\mathbb{R})$ due to Lemmas 19 and 20.

iv) We have that Dom $\delta_2 \subset$ Dom δ^*. Indeed, for $u \in$ Dom δ_2 and $F \in \mathbb{D}_{\mathcal{T}}^{1,2}(\mathbb{R})$, we have that the duality relations (4.2), (4.61), (1.111), *the fact that $u. \in \mathcal{H}$ with probability 1 and (H1) imply*

$$E\left(F\delta_2(u)\right) = E\left(\langle DF, u\rangle_{\mathcal{H}}\right) = E\left(\langle \mathcal{T}DF, \mathcal{T}u\rangle_{\mathcal{H}_0}\right) = E\left(\langle D_{\mathcal{T}}F, u\rangle_{\mathcal{H}_0}\right).$$

That is, $E\left(F\delta_2(u)\right) = E\left(\langle D_{\mathcal{T}}F, u\rangle_{\mathcal{H}_0}\right)$, which gives that our claim is true and

$$\delta(u) = \delta_2(u).$$

In other words, δ is an extension of the usual divergence operator δ_2. In Section 4.3, we will give an example where Dom $\delta_2 \subsetneq$ Dom δ^*.

The following lemma is needed to identify the domain of the operator δ in terms of the chaos decomposition of the integrand (see Theorem 12).

Lemma 49 *Let $n \geq 1$, $f \in \mathcal{H}^{\otimes n} \otimes \mathcal{H}_0$, $h \in \mathcal{H}_0$ and $\{g_1, \ldots, g_n\} \subset \mathcal{T}_{\mathcal{H}}$. Then, (H1) implies*

$$\langle f, g_1 \otimes \cdots \otimes g_n \otimes h\rangle_{\mathcal{H}^{\otimes n}\otimes\mathcal{H}_0} = \langle f, \mathcal{T}\mathcal{T}^*g_1 \otimes \cdots \otimes \mathcal{T}\mathcal{T}^*g_n \otimes h\rangle_{\mathcal{H}_0^{\otimes(n+1)}}.$$

Proof. Let $\{h_i\}$ and $\{k_i\}$ be orthonormal bases of $\mathcal{H}^{\otimes n}$ and $\mathcal{H}_0$, respectively. Then

$$f = \sum_{i,j=1}^{\infty} a_{ij} h_i \otimes k_j \quad \text{in} \quad \mathcal{H}^{\otimes n} \otimes \mathcal{H}_0, \tag{4.62}$$

with $a_{ij} = \langle f, h_i \otimes k_j \rangle_{\mathcal{H}^{\otimes n} \otimes \mathcal{H}_0}$, $i, j \in \mathbb{N}$. Thus, Hypothesis (H1), Proposition 143 and (8.56) imply that, for $m \in \mathbb{N}$,

$$\begin{aligned}
&\langle \sum_{i,j=1}^{m} a_{ij} h_i \otimes k_j, g_1 \otimes \cdots \otimes g_n \otimes h \rangle_{\mathcal{H}^{\otimes n} \otimes \mathcal{H}_0} \\
&= \sum_{i,j=1}^{m} a_{ij} \langle h_i, g_1 \otimes \cdots \otimes g_n \rangle_{\mathcal{H}^{\otimes n}} \langle k_j, h \rangle_{\mathcal{H}_0} \\
&= \sum_{i,j=1}^{m} a_{ij} \langle \mathcal{T}^{\otimes n}(h_i), \mathcal{T}^{\otimes n}(g_1 \otimes \cdots \otimes g_n) \rangle_{\mathcal{H}_0^{\otimes n}} \langle k_j, h \rangle_{\mathcal{H}_0} \\
&= \langle \mathcal{T}^{\otimes n} \otimes I_{\mathcal{H}_0} (\sum_{i,j=1}^{m} a_{ij} h_i \otimes k_j), \mathcal{T} g_1 \otimes \cdots \otimes \mathcal{T} g_n \otimes h \rangle_{\mathcal{H}_0^{\otimes (n+1)}},
\end{aligned}$$

where $I_{\mathcal{H}_0}$ is the identity operator on $\mathcal{H}_0$. Therefore, using Proposition 143 again and (4.62), we get

$$\langle f, g_1 \otimes \cdots \otimes g_n \otimes h \rangle_{\mathcal{H}^{\otimes n} \otimes \mathcal{H}_0} = \langle \mathcal{T}^{\otimes n} \otimes I_{\mathcal{H}_0}(f), \mathcal{T} g_1 \otimes \cdots \otimes \mathcal{T} g_n \otimes h \rangle_{\mathcal{H}_0^{\otimes (n+1)}}.$$

Consequently, the proof is complete. □

Now, we are ready to characterized the set Dom δ^*. Towards this end, we apply Definition 21 and Theorem 12.

Theorem 55 *Assume that (H1)–(H3) hold and that $u \in L^2(\Omega; \mathcal{H}_0)$ has the chaos representation*

$$u = \sum_{n=0}^{\infty} I_n(f_n), \quad f_n \in \mathcal{H}^{\odot n} \otimes \mathcal{H}_0.$$

Then $u \in$ Dom δ^* if and only if $\widetilde{f}_n$ (symmetrization of f_n as an element of $\mathcal{H}_0^{\otimes (n+1)}$) belongs to $\mathcal{H}^{\odot (n+1)}$ for all $n \geq 0$, and

$$\sum_{n=1}^{\infty} n! |\widetilde{f}_{n-1}|^2_{\mathcal{H}^{\otimes n}} < \infty. \tag{4.63}$$

In this case $\delta(u) = \sum_{n=1}^{\infty} I_n(\widetilde{f}_{n-1})$.

Remark 66 *Note that a consequence of this result is that the random variable $\delta(u)$ and the set Dom δ^* are independent of the operator $\mathcal{T}$.*

Proof. Fix $n \in \mathbb{N}$. Let $\{n_1, \ldots, n_k\}$ be a finite sequence of positive integers such that $n_1 + \cdots + n_k = n$ and $\{g_1, \ldots, g_k\} \subset \mathcal{T}_{\mathcal{H}}$ an orthonormal system on $\mathcal{H}$.

Necessity: From Lemmas 20 and 49, and (1.97), we have

$$\begin{aligned}
&E\left\langle u, D_{\mathcal{T}}\Big(H_{n_1}(W(g_1)) \cdots H_{n_k}(W(g_k))\Big)\right\rangle_{\mathcal{H}_0} \\
&= \sum_{j=1}^{k} n_j (n-1)! \, \langle f_{n-1}, \\
&\quad (\mathcal{T}^* \mathcal{T})^{\otimes (n-1)} (g_1^{\otimes n_1} \otimes \cdots \otimes g_{j-1}^{\otimes n_{j-1}} \otimes g_j^{\otimes (n_j - 1)} \otimes \cdots \otimes g_{n_k}^{\otimes n_k}) \otimes \mathcal{T}^* \mathcal{T} g_j \Big\rangle_{\mathcal{H}_0^{\otimes n}}.
\end{aligned}$$

In consequence, using that Symm is an orthonormal projection (see (1.95)) and that f_{n-1} belongs to $\mathcal{H}^{\odot(n-1)} \otimes \mathcal{H}_0$, we get

$$\begin{aligned} E\left\langle u, D_{\mathcal{T}}\Big(H_{n_1}(W(g_1))\cdots H_{n_k}(W(g_k))\Big)\right\rangle_{\mathcal{H}_0} &= \sum_{j=1}^{k} n_j(n-1)!\left\langle f_{n-1}, \text{Symm}\left((\mathcal{T}^*\mathcal{T})^{\otimes(n-1)}(g_1^{\otimes n_1}\right.\right. \\ &\quad \left.\left.\otimes\cdots\otimes g_{j-1}^{\otimes n_{j-1}} \otimes g_j^{\otimes(n_j-1)} \otimes\cdots\otimes g_{n_k}^{\otimes n_k})\right)\otimes \mathcal{T}^*\mathcal{T}g_j\right\rangle_{\mathcal{H}_0^{\otimes n}} \\ &= \left\langle f_{n-1}, \text{Symm}\left((\mathcal{T}^*\mathcal{T})^{\otimes n}(g_1^{\otimes n_1}\otimes\cdots\otimes g_{n_k}^{\otimes n_k})\right)\right\rangle_{\mathcal{H}_0^{\otimes n}} \\ &= \left\langle \tilde{f}_{n-1}, (\mathcal{T}^*\mathcal{T})^{\otimes n}(g_1^{\otimes n_1}\otimes\cdots\otimes g_{n_k}^{\otimes n_k})\right\rangle_{\mathcal{H}_0^{\otimes n}}. \end{aligned}$$

Hence, if $\delta(u)$ has the chaos representation

$$\delta(u) = \sum_{n=0}^{\infty} I_n(v_n), \quad v_n \in \mathcal{H}^{\odot n},$$

then the duality relation (4.61) and (H3) yield that $v_n = \widetilde{f}_{n-1}$, and therefore (4.63) is true.

Sufficiency: Let $F = f(W(g_1), \ldots, W(g_k))$ be a random variable in $\mathcal{S}_{\mathcal{T}}(\mathbb{R})$ and $\mathcal{K}$ the linear subspace of $\mathcal{H}$ generated by $\{g_1, \ldots, g_k\}$. Then, by Theorem 12, F has the chaos decomposition given by

$$F = \sum_{n=0}^{\infty} I_n(k_n), \quad k_n \in \mathcal{K}^{\odot n}.$$

Consequently, using Lemmas 20 and 49 again, and the fact that Symm is an orthonormal projection again, we obtain

$$\begin{aligned} E\langle u, D_{\mathcal{T}}F\rangle_{\mathcal{H}_0} &= \sum_{n=0}^{\infty}(n+1)!\langle f_n, (\mathcal{T}^*\mathcal{T})^{\otimes(n+1)}(k_{n+1})\rangle_{\mathcal{H}_0^{\otimes(n+1)}} \\ &= \sum_{n=0}^{\infty}(n+1)!\langle \widetilde{f}_n, (\mathcal{T}^*\mathcal{T})^{\otimes(n+1)}(k_{n+1})\rangle_{\mathcal{H}_0^{\otimes(n+1)}} \\ &= \sum_{n=0}^{\infty}(n+1)!\langle \widetilde{f}_n, k_{n+1}\rangle_{\mathcal{H}^{\otimes(n+1)}} = E(F\sum_{n=1}^{\infty} I_n(\widetilde{f}_{n-1})). \end{aligned}$$

That is, the duality relation (4.61) is satisfied for u and $\delta(u) := \sum_{n=1}^{\infty} I_n(\widetilde{f}_{n-1})$, in this case. So, the proof is complete. □

The following result shows that the operator δ has the local property in $\mathbb{D}_{\mathcal{T}}^{1,2}(\mathcal{H}_0)$ (see Definition 38).

Proposition 81 *Let $u \in$ Dom $\delta^* \cap \mathbb{D}_{\mathcal{T}}^{1,2}(\mathcal{H}_0)$ and $A \in \mathcal{F}$ such that $u = 0$ on A. Then $\delta(u) = 0$ on A w.p.1.*

Proof. As in the proof of Proposition 66, let $\phi : \mathbb{R} \to \mathbb{R}_+$ be a C^∞-function such that $\phi(0) = 1$ and its support is included in $[-2, 2]$. For $\varepsilon > 0$, set $\phi_\varepsilon(x) = \phi(\frac{x}{\varepsilon})$. Now, let $F = f(W(g_1), \ldots, W(g_n))$ be in $\mathcal{S}_{\mathcal{T}}(\mathbb{R})$ with $f \in C_0^\infty(\mathbb{R}^n)$ (i.e., f has compact support).

Then, Lemma 21 and the duality relation (4.61) gives

$$\begin{aligned} E[\delta(u)\phi_\varepsilon(|u|^2_{\mathcal{H}_0})F] &= E\langle D_{\mathcal{T}}\phi_\varepsilon(|u|^2_{\mathcal{H}_0}), Fu\rangle_{\mathcal{H}_0} + E\langle D_{\mathcal{T}}F, \phi_\varepsilon(|u|^2_{\mathcal{H}_0})u\rangle_{\mathcal{H}_0} \\ &= E\{2F\langle((\mathcal{T}\otimes I_{\mathcal{H}_0})^*(\mathcal{T}\otimes I_{\mathcal{H}_0})Du)^*(u), \phi'_\varepsilon(|u|^2_{\mathcal{H}_0})u\rangle_{\mathcal{H}_0} \\ &\quad +\langle D_{\mathcal{T}}F, \phi_\varepsilon(|u|^2_{\mathcal{H}_0})u\rangle_{\mathcal{H}_0}. \end{aligned}$$

Finally, proceeding as in the proof of Proposition 66, we obtain

$$\delta(u)1_{\{|u|^2_{\mathcal{H}_0}=0\}} = 0 \quad \text{w.p.1}$$

and therefore the proof is finished. □

As an immediate consequence of Proposition 81, we can localize the domain of δ as follows. We say that $u \in (\mathrm{Dom}\,\delta^*)_{loc}$ if there exists a sequence $\{(\Omega_n, u^{(n)}) : n \geq 1\} \subset \mathcal{F} \times (\mathbb{D}^{1,2}_{\mathcal{T}}(\mathcal{H}_0) \cap \mathrm{Dom}\,\delta^*)$ such that $\Omega_n \uparrow \Omega$ and $u = u^{(n)}$ on Ω_n w.p.1. In this case we define

$$\delta(u) = \delta(u^{(n)}) \quad \text{on } \Omega_n, \ n \geq 1.$$

4.3 Fractional Brownian motion

In this section, we assume that $B = \{B_t : t \in [0,T]\}$ is fractional Brownian motion with Hurst parameter H in $(0,1)$ and $W = \{W_t : t \in [0,T]\}$ is the Brownian motion related to B through Theorem 9 and (1.90). Also, we assume that K^* is given by (1.73) and (1.87). For $p \geq 1$, we represent the divergence operators (see (4.2)) with respect to B and W by δ_p and δ_p^W, respectively.

Property (1.43) allows us to write

$$E\left(\|u\|^p_{\mathcal{H}}\right) = E\left(\|K^*(u)\|^p_{L^2([0,T])}\right).$$

It means, u belongs to $L^p(\Omega; \mathcal{H})$ if and only if $K^*(u)$ is in $L^p(\Omega; L^2([0,T]))$. Therefore, Proposition 24 leads us to establish that, for $u \in L^p(\Omega; \mathcal{H})$,

$$E\left(\langle u, DF\rangle_{\mathcal{H}}\right) = E\left(\langle K^*(u), K^*(DF)\rangle_{L^2([0,T])}\right) = E\left(\langle K^*(u), D^W F\rangle_{L^2([0,T])}\right), \tag{4.64}$$

for all $\mathbb{R}$-valued smooth functional F of B.

Now it is easy to state the relation between δ_p and δ_p^W.

Proposition 82 *Let $H \in (0,1)$, $p \geq 1$ and $u \in L^p(\Omega; \mathcal{H})$. Then, u belongs to Dom$\,\delta_p(\mathbb{R})$ if and only if $K^*(u)$ is in Dom$\,\delta_p^W(\mathbb{R})$. In this case, we have*

$$\delta_p(u) = \delta_p^W\left(K^*(u)\right).$$

Proof. The result is an immediate consequence of (4.2) and (4.64). □

Propositions 24 and 82 are known as the transfer principle.

Now, our next aim is to analyze a maximal inequality for the divergence operator with respect to fBm with Hurts parameter $H > 1/2$, similar to that in Theorem 52. In Section 4.3.1, we analyze some extensions of the divergence operator with respect to fBm with $H < 1/2$. The proof of the following two results have been developed by Alòs and Nualart [13]. Remember that the space $\mathbb{L}^{1,p}_{H-\varepsilon}$ is introduced in (1.129).

Before stating our maximal inequality, we establish the continuity of the divergence operator with integrands in $\mathbb{L}^{1,p}_H$.

Proposition 83 *Let $u = \{u_t : t \in [0,T]\}$ be a process in $\mathbb{L}_H^{1,p}$ with $pH > 2$ and $H > 1/2$. Also assume*

$$E\left\{\int_0^T |u_r|^p\, dr + \int_0^T \left(\int_0^T |D_t u_r|^{\frac{1}{H}}\, dt\right)^{pH} dr\right\} < \infty.$$

Then, the process $X_t = \int_0^t u_s \delta_p B_s$, $t \in [0,T]$, has a continuous modification, which is also denoted by X. Moreover, for $\gamma < H - (1/p)$ and the continuous modification X, there exist a random variable C_γ such that it is finite with probability 1 and

$$|X_t - X_s| \le C_\gamma |t-s|^\gamma, \quad w.p.1.$$

Proof. By Proposition 62, (1.129), (1.130) and Hölder inequality, we have that, for $0 \le s \le t \le T$,

$$\begin{aligned}
&E\left(|X_t - X_s|^p\right)\\
&\le C_{p,H} E\left(\left[\int_s^t |u_r|^{\frac{1}{H}} dr\right]^{pH} + \left[\int_s^t \int_0^T |D_\theta u_r|^{\frac{1}{H}} d\theta dr\right]^{pH}\right)\\
&\le C_{p,H}(t-s)^{pH-1} E\left(\int_s^t |u_r|^p dr + \int_s^t \left[\int_0^T |D_\theta u_r|^{\frac{1}{H}} d\theta\right]^{pH} dr\right) \qquad (4.65)\\
&\le C_{p,H,T}(t-s)^{pH-1}.
\end{aligned}$$

Hence, the fact that $PH - 1 > 1$ and the Kolmogorov-Čentsov continuity theorem (i.e., Theorem 1) imply that the process X has a continuous modification. In the remaining of this proof, X represents this continuous modification. Then, (4.65) yields that there is an integrable function $A : [0,T] \to \mathbb{R}_+$ such that, for $\alpha \in (2, PH+1)$, we obtain

$$E\left(\int_0^T \int_0^T \frac{|X_t - X_s|^p}{|t-s|^\alpha} dsdt\right) \le 2C_{p,H,T} \int_0^T \int_0^t (t-s)^{pH-1-\alpha} \int_s^t A(r) drdsdt.$$

In consequence, Fubini's theorem and the fact that $\alpha < PH + 1$

$$\begin{aligned}
&E\left(\int_0^T \int_0^T \frac{|X_t - X_s|^p}{|t-s|^\alpha} dsdt\right)\\
&\le C \int_0^T A(r) \left(\int_0^r \int_r^T (t-s)^{pH-1-\alpha} dtds\right) dr \le C \int_0^T A(r) dr < \infty.
\end{aligned}$$

In other words, the random variable $\Gamma := \int_0^T \int_0^T \frac{|X_t - X_s|^p}{|t-s|^\alpha} dsdt$ is finite with probability 1. In particular, for $\gamma = (\alpha - 2)$, the Corollary of Garsia-Rodemish-Ramsey lemma (i.e., Corollary 32) allows us to conclude the proof. □

Now, we are ready to state the maximal inequality for the divergence operator with respect to fBm with Hurst parameter $H > 1/2$.

Theorem 56 *Let B be fBm with $H > 1/2$, $p > 2/H$ and $u = \{u_t : t \in [0,T]\}$ a process in $\mathbb{L}_{H-\varepsilon}^{1,p}$, for some $\varepsilon < H - (1/p)$. Moreover assume*

$$E\left\{\int_0^T |u_r|^p\, dr + \int_0^T \left(\int_0^T |D_t u_r|^{\frac{1}{H}}\, dt\right)^{pH} dr\right\} < \infty.$$

Then,

$$E\left(\sup_{t\in[0,T]}\left|\int_0^t u_s\delta_p dB_s\right|^p\right) \le C_{H,p,\varepsilon,T}E\left[\left(\int_0^T |u_s|^{\frac{1}{H-\varepsilon}}ds\right)^{p(H-\varepsilon)} + \left(\int_0^T\left(\int_0^T |D_su_r|^{\frac{1}{H}}dr\right)^{\frac{H}{H-\varepsilon}}ds\right)^{p(H-\varepsilon)}\right].$$

Remark 67 *From Proposition 83, we know that* $t\mapsto\int_0^t u_s\delta_p dB_s$ *has a continuous modification. So, in this result we are considering this continuous version.*

Proof. We use the notation $\alpha = 1-\frac{1}{p}-\varepsilon$. Then, for $s,t\in[0,T]$, $s<t$, Lemma 87 in the appendix gives that $C_\alpha := \int_s^t (t-r)^{-\alpha}(r-s)^{\alpha-1}dr$, where C_α is a constant that only depends on α. Moreover, in the remaining of this proof, we show that $s\mapsto(r-s)^{\alpha-1}u_s 1_{[0,r]}(s)$ belongs to $\mathbb{L}_H^{1,p}$, for almost all $r\in[0,T]$. Therefore, for $t\in[0,T]$, Fubini's theorem for the divergence operator (i.e., Theorem 49) implies that there exists a measurable process $Y^t:\Omega\times[0,t]\to\mathbb{R}$ such that

$$Y_r^t = (t-r)^{-\alpha}\int_0^r (r-s)^{\alpha-1}u_s\delta_p B_s, \quad \text{w.p.1, for almost all } r\in[0,t],$$

and

$$\int_0^t u_s\delta_p B_s = C_\alpha^{-1}\int_0^t Y_r^t dr, \quad \text{w.p.1.}$$

Set $\tilde{Y}_r = (T-r)^\alpha Y_r^T$. Then,

$$\tilde{Y}_r = \int_0^r (r-s)^{\alpha-1}u_s\delta_p B_s, \quad \text{w.p.1, for almost all } r\in[0,T].$$

Thus, we have proven that Fubini's theorem establishes that $(t-r)^\alpha Y_r^t = \tilde{Y}_r$ w.p.1, for almost all $r\in[0,t]$, and Hölder inequality leads us to write

$$\left|\int_0^t u_s\delta_p B_s\right|^p = C_\alpha^{-p}\left|\int_0^t (t-r)^{-\alpha}\tilde{Y}_r dr\right|^p \le C\int_0^t \left|\tilde{Y}_r\right|^p dr, \quad \text{w.p.1.}$$

Hence, the continuity of $\int_0^\cdot u_s\delta_p B_s$, Proposition 62 and (1.129) yield

$$\begin{aligned} E\left(\sup_{t\in[0,T]}\left|\int_0^t u_s\delta_p dB_s\right|^p\right) &\le C\int_0^T E\left(|\tilde{Y}_r|^p\right)dr \\ &\le CE\left[\int_0^T\left(\int_0^r (r-s)^{\frac{\alpha-1}{H}}|u_s|^{\frac{1}{H}}ds\right)^{pH}dr \right.\\ &\quad \left.+\int_0^T\left(\int_0^r\int_0^T (r-s)^{\frac{\alpha-1}{H}}|D_\theta u_s|^{\frac{1}{H}}d\theta ds\right)^{pH}dr\right] \\ &\le C_{H,p,\varepsilon,T}E\left[\left(\int_0^T |u_s|^{\frac{1}{H-\varepsilon}}ds\right)^{p(H-\varepsilon)}\right.\\ &\quad \left.+\left(\int_0^T\left(\int_0^T |D_su_r|^{\frac{1}{H}}dr\right)^{\frac{H}{H-\varepsilon}}ds\right)^{p(H-\varepsilon)}\right]. \end{aligned}$$

where, in last inequality, we use Theorem 7 with $(H-p^{-1}-\varepsilon)/H$, $H/(H-\varepsilon)$ and $H^2/(H-\varepsilon)$ instead of α, p and q, respectively. Consequently, the proof is finished. □

4.3.1 The extended divergence operator with respect to fBm

Here, we deal with the extension of the divergence operator with respect to fractional Brownian motion B when the Hurst parameter is less than $1/2$. As we will see, the divergence of fBm with respect itself is not well-defined while it is in the domain of the extended divergence operator analyzed in Section 4.2.

In Section 1.5.6.1, we point out that fBm is an isonormal Gaussian process on its reproducing kernel Hilbert space $\mathcal{H}$, which is defined by (1.83) and (1.118). That is, a function f in $L^2([0,T])$ belongs to $\mathcal{H}$ if there is $\varphi_f \in L^2([0,T])$ such that

$$f(u) = u^{\alpha}\left(I^{\alpha}_{T-}(s^{-\alpha}\phi_f(s))\right)(u), \quad u \in [0,T],$$

where I^{α}_{T-} is the right-sided Riemann-Liouville fractional integral of order $\alpha = \frac{1}{2} - H$, which is defined in (1.56). Moreover, Proposition 25 establishes that the Hilbert space $\mathcal{H}$ is densely and continuously embedded in $L^2([0,T])$, which is a condition needed to deal with the extended divergence operator studied in Section 4.2 (see Definition 40)

In Section 1.5.6.1, in order to consider the extended divergence with respect to B, we deal with the operator $\mathcal{T} : \mathcal{H} \subset L^2([0,T]) \to L^2([0,T])$ introduced in (1.119). It means,

$$\mathcal{T}f = \tilde{c}_H \Gamma(H + \frac{1}{2})\varphi_f.$$

Here, $\tilde{c}_H$ is defined in (1.118). Also, we know that the operator $\mathcal{T}$ satisfies Hypotheses (H1) and (H2) contemplated in Section 4.2 because of Theorem 15. Concerning Condition (H3), we have the following:

Theorem 57 *The operator $\mathcal{T}$ given by (1.119) satisfies Condition (H3).*

Proof. In Theorem 15, we deal with the set

$$\mathcal{H}_* = \{f \in \mathcal{H} : \exists f^* \in L^{\infty}([0,T]) \text{ s.t. } \phi_f(u) = u^{-\alpha} I^{\alpha}_{0+}(s^{\alpha} f^*(s))(u)\}.$$

Therefore, (1.119) and (1.121) yield that $\mathcal{H}_* \subset \mathrm{Dom}(\mathcal{T}^*\mathcal{T})$ and that $L^{\infty}([0,T]) \subset \mathcal{T}_{L^2([0,T])}$, which implies that (H3) holds since the space $L^{\infty}([0,T])$ is a dense set of $L^2([0,T])$. Thus, the proof is complete. □

Note that Theorems 15 and 57 allow us to consider the extended divergence operator given in Definition 40. So, now, our next aim is to show that B is not in the domain of the usual divergence operator δ_2, but it is in Dom δ^*, for $H < 1/4$. That is, Dom $\delta_2 \subsetneq$ Dom δ^* in this case (see Remarks 8.iv)).

Proposition 84 *Fractional Brownian motion B with $H \in (0,1/2)$ belongs to Dom δ^*.*

Proof. Note that the fact that $E\left(\int_0^T B_s^2 ds\right) = \int_0^T s^{2H} ds < \infty$, Remark 22.i) and Theorem 12 imply $B_t = I_1(1_{[0,t]}) \in L^2(\Omega; L^2([0,T]))$. Since $\widetilde{1_{[0,t]}}(\cdot) = \frac{1}{2}(1 \otimes 1)$ (symmetrization as an element of $(L^2([0,T]))^{\otimes 2}$), we get the result is true due to Theorem 55 and the step functions being in $\mathcal{H}$ (see equality (1.74)). □

In order to continue with our analysis, we need to state the following auxiliary result.

Lemma 50 *Let $H < 1/2$. Then, $\mathcal{H} \subset I^{\alpha}_{T-}(L^p([0,T]))$, with $\alpha = \frac{1}{2} - H$ and $p \in (1,2)$.*

Proof. Fix φ in $\mathcal{H}$ and $\varepsilon > 0$ small enough. Then, using the convention $\varphi \equiv 0$ on $[0,T]^c$, we have, for $s \in [0,T]$ an $\tilde{\varepsilon} > 0$,

$$\begin{aligned}
&\frac{\varphi(s)}{(T-s)^\alpha} + \alpha\int_{s+\tilde{\varepsilon}}^T \frac{\varphi(s)-\varphi(u)}{(u-s)^{\alpha+1}}du + \alpha\int_{s+\tilde{\varepsilon}}^T \frac{\varphi(u)}{(u-s)^{\alpha+1}}\left(1-\left(\frac{u}{s}\right)^{-\alpha}\right)du \\
&= \frac{\varphi(s)}{(T-s)^\alpha} + \alpha\int_{s+\tilde{\varepsilon}}^T \frac{\varphi(s)}{(u-s)^{\alpha+1}}du - s^\alpha\alpha\int_{s+\tilde{\varepsilon}}^T \frac{u^{-\alpha}\varphi(u)}{(u-s)^{\alpha+1}}du \\
&= s^\alpha\left(\frac{s^{-\alpha}\varphi(s)}{(T-s)^\alpha} + \alpha\int_{s+\tilde{\varepsilon}}^T \frac{s^{-\alpha}\varphi(s)-u^{-\alpha}\varphi(u)}{(u-s)^{\alpha+1}}du\right).
\end{aligned}$$

In other words, using the notation introduced in (8.13), we have proven

$$\begin{aligned}
&\left(D_{T-}^{\alpha,\tilde{\varepsilon}}\varphi\right)(s) + \frac{\alpha}{\Gamma(1-\alpha)}\int_{s+\tilde{\varepsilon}}^T \frac{\varphi(u)}{(u-s)^{\alpha+1}}\left(1-\left(\frac{u}{s}\right)^{-\alpha}\right)du \\
&\quad = s^\alpha\left(D_{T-}^{\alpha,\tilde{\varepsilon}}u^{-\alpha}\varphi(u)\right)(s), \quad s\in[0,T].
\end{aligned} \tag{4.66}$$

In particular, for $\varphi \equiv 1$ and $\tilde{\varepsilon} = 0$, the last equality becomes

$$\frac{1}{(T-s)^\alpha} + \alpha\int_s^T \frac{1}{(u-s)^{\alpha+1}}\left(1-\left(\frac{u}{s}\right)^{-\alpha}\right)du = s^\alpha\left(\frac{s^{-\alpha}}{(T-s)^\alpha} + \alpha\int_s^T \frac{s^{-\alpha}-u^{-\alpha}}{(u-s)^{\alpha+1}}du\right).$$

Hence, Theorem 8 (see also (1.73)) implies

$$\frac{1}{(T-s)^\alpha} + \alpha\int_s^T \frac{1}{(u-s)^{\alpha+1}}\left(1-\left(\frac{u}{s}\right)^{-\alpha}\right)du = \tilde{c}_H^{-1}K(T,s).$$

Then, Lemma 54 leads us to obtain

$$\int_s^T \frac{1}{(u-s)^{\alpha+1}}\left(1-\left(\frac{u}{s}\right)^{-\alpha}\right)du \le Cs^{-\alpha}. \tag{4.67}$$

On the other hand, from Theorem 8, there is $f \in L^2([0,T])$ such that

$$|\varphi(s)| = s^{\frac{1}{2}-H}\left|I_{T-}^{\frac{1}{2}-H}\left(u^{H-\frac{1}{2}}f(u)\right)(s)\right| \le I_{T-}^{\frac{1}{2}-H}(|f|)(s).$$

Therefore, φ belongs to $L^{1/H}([0,T])$ due to Remark 14. Moreover, we have the estimation

$$\begin{aligned}
0 &< \frac{1}{(u-s)^{\alpha+1}}\left(1-\left(\frac{u}{s}\right)^{-\alpha}\right) = u^{-\alpha}\sqrt{\frac{u^\alpha-s^\alpha}{(u-s)^{\alpha+1}}}\sqrt{\frac{u^\alpha-s^\alpha}{(u-s)^{\alpha+1}}} \\
&\le u^{-\alpha}\sqrt{\frac{1}{(u-s)}}\sqrt{\alpha\frac{s^{\alpha-1}}{(u-s)^\alpha}} = \sqrt{\alpha}u^{-\alpha}s^{(\alpha-1)/2}(u-s)^{-(\alpha+1)/2}.
\end{aligned} \tag{4.68}$$

Consequently, (4.67) and Hölder inequality allow us to estimate the integral in (4.66). That is,

$$\begin{aligned}
&\int_0^T\left(\int_s^T \frac{|\varphi(u)|}{(u-s)^{\alpha+1}}\left(1-\left(\frac{u}{s}\right)^{-\alpha}\right)du\right)^{2-\varepsilon}ds \\
&\le C\int_0^T s^{-\alpha(1-\varepsilon)}\left(\int_s^T \frac{|\varphi(r)|^{2-\varepsilon}}{(r-s)^{\alpha+1}}\left(1-\left(\frac{r}{s}\right)^{-\alpha}\right)dr\right)ds.
\end{aligned}$$

Now, taking into account (4.68), together with Fubini's theorem, we obtain

$$\begin{aligned}
&\int_0^T \left(\int_s^T \frac{|\varphi(u)|}{(u-s)^{\alpha+1}} \left(1 - \left(\frac{u}{s}\right)^{-\alpha}\right) du \right)^{2-\varepsilon} ds \\
&\leq C \int_0^T s^{-\alpha(1-\varepsilon)} \left(\int_s^T \frac{|\varphi(r)|^{2-\varepsilon} s^{\frac{\alpha}{2}-\frac{1}{2}}}{r^\alpha (r-s)^{\frac{\alpha}{2}+\frac{1}{2}}} dr \right) ds \\
&= C \int_0^T |\varphi(r)|^{2-\varepsilon} r^{-\alpha} \int_0^r s^{-\frac{\alpha}{2}-\frac{1}{2}+\alpha\varepsilon} (r-s)^{-\frac{\alpha}{2}-\frac{1}{2}} ds dr.
\end{aligned}$$

Hence, we have that Hölder inequality and Lemma 87 yield, for $\eta > 0$ small enough,

$$\begin{aligned}
&\int_0^T \left(\int_s^T \frac{|\varphi(u)|}{(u-s)^{\alpha+1}} \left(1 - \left(\frac{u}{s}\right)^{-\alpha}\right) du \right)^{2-\varepsilon} ds \\
&\leq C \int_0^T |\varphi(r)|^{2-\varepsilon} r^{-2\alpha+\varepsilon\alpha} dr \leq C \left(\int_0^T |\varphi(r)|^{\frac{1}{H+\eta}} dr \right)^{(2-\varepsilon)(H+\eta)} \\
&\leq C \int_0^T |\varphi(r)|^{2-\varepsilon} dr < \infty.
\end{aligned}$$

Thus, (4.66) gives that φ is in $I_{T-}^{\frac{\alpha}{2}}(L^{2-\varepsilon}([0,T]))$ and the proof is complete. □

We are ready to show that the paths of B do not belong to $\mathcal{H}$.

Proposition 85 *Let $H \in (0, 1/4)$. Then, the fractional Brownian motion B is not in $\mathcal{H}$, with probability 1.*

Proof. Let $p \in (1,2)$. We proceed as Cheridito and Nualart [27]. That is, using that B is a self-similar process with index H (see Proposition 13), we obtain that the processes

$$t^{-pH} \int_0^{T-t} |B_{s+t} - B_s|^p \, ds$$

and

$$\int_0^{T-t} |B_{\frac{s}{t}+1} - B_{\frac{s}{t}}|^p \, ds = \frac{T-t}{\frac{T}{t}-1} \int_0^{\frac{T}{t}-1} |B_{u+1} - B_u|^p \, du$$

have the same distribution. Thus, the ergodic theorem implies

$$L^1(\Omega) - \lim_{t \downarrow 0} t^{-pH} \int_0^{T-t} |B_{s+t} - B_s|^p \, ds = TE\left(|B_1|^p\right).$$

Remember that the reader can consult Lindgren [128] (Section 6) and Shiryayev [197] (Chapter 5) for details on the ergodic theorem. In consequence, we can find a sequence $\{t_n : n \in \mathbb{N}\}$ that decrease to zero, as $n \to \infty$ such that

$$t_n^{-pH} \int_0^{T-t_n} |B_{s+t_n} - B_s|^p \, ds \to T \quad \text{as } n \to \infty, \text{ with probability 1.} \tag{4.69}$$

Since the Hölder continuity of the fBm (see Section 1.4) implies that, for δ small enough,

$$t_n^{-pH} \int_{T-t_n}^T |B_{s+t_n} - B_s|^p \, ds \leq C t_n^{-p\delta+1} \to 0, \quad n \to \infty, \tag{4.70}$$

with probability 1.

Finally, if there is $w_0 \in \Omega$ such that $B_\cdot(\omega_0) \in \mathcal{H}$, then Lemmas 7 and 50 yield

$$\int_0^T |B_{t+s}(\omega_0) - B_s(\omega_0)|^p \, ds = o(t^{p\alpha}). \tag{4.71}$$

But (4.69), (4.70) and (4.71) cannot be satisfied at the same time because $H \in (0, 1/4)$. Thus, the proof is finished. □

Note that Proposition 85 implies that $B \notin \text{Dom } \delta_2$ due to (4.2). Hence, Proposition 84 allows us to conclude that our claim is true. It means, $\text{Dom } \delta_2 \subsetneq \text{Dom } \delta^*$.

4.4 Gaussian Volterra processes

Here, the purpose is to study some Itô's formulas for some Gaussian Volterra type processes. Towards this end, as in Section 1.6, we consider a Gaussian Volterra process $B = \{B_t : t \in [0, T]\}$ as in Definition 16. That is, this process has the form

$$B_t = \int_0^t K(t, s) dW_s, \quad t \in [0, T].$$

Here, $W = \{W_t : t \in [0, T]\}$ is a Brownian motion and K satisfies condition (1.34). It means,

$$\sup_{t \in [0,T]} \int_0^t K(t, s)^2 ds = \sup_{t \in [0,T]} R_B(t, t) < \infty.$$

Moreover, sometimes, we assume that the kernel K satisfies some of the following conditions:

(K1) The function $K(\cdot, s)$ has bounded variation on the interval $(r, T]$, for any $T \geq r > s \geq 0$.

(K2) Hypothesis (K1) holds and

$$\int_0^T \left[\int_s^T \|B_t - B_s\|_{L^2(\Omega)} |K|(dt, s) \right]^2 ds < \infty.$$

In Proposition 29, we use a stronger condition than (K2). Namely,

(K2') Condition (K1) is satisfied and there exists a measurable, continuous at zero and non-decreasing function $g : [0, T] \to \mathbb{R}_+ \cup \{0\}$ such that $g(0) = 0$,

$$\|B_t - B_s\|_{L^2(\Omega)} \leq g(t - s), \quad \text{for } 0 \leq s \leq t \leq T,$$

and

$$\int_0^T \left[\int_s^T g(t - s)|K|(dt, s) \right]^2 ds < \infty.$$

Moreover, suppose that, for $t \in [0, T]$,

$$\sum_{i=0}^{2^n - 1} \int_{t_i^n \wedge t}^{\bar{t}_i^n} \left(\int_{\bar{t}_i^n}^{t_{i+1}^n \wedge t} \left\| B_{t_{i+1}^n} - B_{t_i^n} \right\|_{L^2(\Omega)} |K|(dt, s) \right)^2 ds \to 0, \quad \text{as } n \to \infty,$$

holds. Here,

$$t_i^n = \frac{Ti}{2^n} \quad \text{and} \quad \bar{t}_i^n = t_i^n + \frac{T}{2^{n+1}}, \quad n \in \mathbb{N}.$$

(K3) The functions $s \mapsto R_B(s,s)$ and $s \mapsto \int_{s\wedge t}^{s} K(s,r)dr$, $t \in [0,T]$, have bounded variation on the interval $[0,T]$. Additionally, the image of K^* is the whole space $L^2([0,T])$.

Note that two processes that satisfy Hypothesis (K2') are fractional Brownian motion and Riemann-Liouville fractional Gaussian process, when the Hurst parameter $H \in (1/4, 1/2)$ (see the proofs of Propositions 30 and 89). Furthermore, Lemmas 9 and 10 establish that the kernel of fractional Brownian motion is such that the image of the linear operator $K^* : \mathcal{H} \to L^2([0,T])$ is the space $L^2([0,T])$. Concerning the RLfp, we have that Theorem 63 and (4.91) yield that $\mathcal{H} = I_{T-}^{\alpha}(L^2([0,T]))$ and $K^* = \Gamma(1-\alpha)D_{T-}^{\alpha}$. Therefore, for $f \in L^2([0,T])$, we obtain that $K^*\left(I_{T-}^{\alpha}\left(f/\Gamma(1-\alpha)\right)\right) = f$. Thus, we have that $K^*(\mathcal{H})$ is the whole space $L^2([0,T])$ again. Now, it is easy to see that (K3) holds for RLfp and fractional Brownian motion.

Our next aim is to study some Itô's formulae for Volterra Gaussian processes, which are established in Alòs et al. [10]. So, we first state some auxiliary results in order to study the first one.

In the remaining of this section, we consider the Function $\mathcal{K}_B : L^2([0,T]) \to L^2([0,T])$ given by

$$(\mathcal{K}_B f)(t) = \int_0^t K(t,s)f(s)ds, \quad \text{for } f \in L^2([0,T]) \text{ and } t \in [0,T].$$

Note that the image of $\mathcal{K}_B$ is indeed contained in $L^2([0,T])$ due to Hölder inequality. In fact,

$$\begin{aligned}\int_0^T \left(\int_0^t K(t,s)f(s)ds\right)^2 dt &\leq \int_0^T \left(\int_0^t K(t,s)^2 ds\right)\int_0^t f^2(r)drdt \\ &\leq T\left(\sup_{t\in[0,T]} R_B(t,t)\right)\int_0^T f^2(r)dr < \infty.\end{aligned}$$

The adjoint of the linear operator $\mathcal{K}_B$ is nothing else than K^*, as the following result establishes.

Lemma 51 *Let φ be a step function in $\mathcal{E}([0,T])$ and g in $L^2([0,T])$, Then,*

$$\int_0^T (K^*\varphi)(s)g(s)ds = \int_0^T \varphi(s)(\mathcal{K}_B g)(ds).$$

Remark 68 *Assume that (K2') and (K3) hold, $\psi \in \mathcal{E}([0,T])$, and that $f \in C^1([0,T]\times\mathbb{R})$ and K satisfy the conditions of Corollary 14. Then, in the proof of this corollary, we figure out a sequence $\{f(\varphi^{(n)}) : n \in \mathbb{N}\}$ of simple processes such that $f(\varphi^{(n)})(t) \to f(t,B_t)$ for all t and almost all $\omega \in \Omega$, and in $L^2(\Omega;\mathcal{H}_K)$, as $n \to \infty$, and $|f(\varphi^{(n)})| \leq C\exp\left(\sigma \sup_{t\in[0,T]}|B_t|\right)$, where σ and C are introduced in* (1.142). *Hence, the dominated convergence theorem, together with* (1.132), *leads us to*

$$\int_0^T (K^* f(\cdot, B_\cdot))(s)\psi(s)ds = \int_0^T f(s,B_s)(\mathcal{K}_B\psi)(ds) \quad w.p.1.$$

Proof. In order to see that this result holds, we only need to write $g(s)ds$ instead of dW_s in the proof of Proposition 11. □

Now, we deal with the second auxiliary result to establish our first Itô's formula.

Lemma 52 *Let the image of $\mathcal{H}$ under K^* be $L^2([0,T])$, $n \in \mathbb{N}$, $f \in C^{0,n}([0,T] \times \mathbb{R})$ such that the partial derivative $\partial_x^j f$ satisfies (1.142), for any $1 \leq j \leq n$, and $G = I_n^W(\varphi_1 \otimes \ldots \otimes \varphi_n)$, with $\varphi_1 \ldots \varphi_n \in \mathcal{E}([0,T])$. Then, we have*

$$E\left(Gf(s,B_s)\right) = \left[\prod_{j=1}^{n} \left(\mathcal{K}_B\varphi_j\right)(s)\right] E\left(\partial_x^n f(s,B_s)\right), \quad s \in [0,T].$$

Proof. We use induction on n to show that the result is true. So, we first assume that $n = 1$. Observe that, in this case, we have that $G = \delta_2\left((K^*)^{-1})\varphi_1\right)$ due to $K^*(\mathcal{H}) = L^2([0,T])$, by hypothesis. Then, (4.2), (1.39), (1.42) and Proposition 21 imply

$$\begin{aligned} E\left(Gf(s,B_s)\right) &= E\left(\partial_x f(s,B_s)\right)\left(\int_0^T \varphi_1(r)(K^* 1_{[0,s]})(r)dr\right) \\ &= E\left(\partial_x f(s,B_s)\right)\left(\mathcal{K}_B\varphi_1\right)(s). \end{aligned}$$

Thus, from the induction hypothesis, we can assume that the result is satisfied for $\ell \in \mathbb{N}$ such that $\ell \leq n-1$. Therefore, Lemma 39, (1.39) and (1.42) allow us to get that, for $G = I_{\ell+1}^W(\varphi_1 \otimes \ldots \otimes \varphi_{\ell+1})$,

$$E\left(Gf(s,B_s)\right) = E\left[\delta_2\left(I_\ell(g_{\ell+1})\right)f(s,B_s)\right],$$

with $g_{\ell+1} = \mathrm{Symm}\left((K^*)^{-1}\varphi_1 \otimes \cdots \otimes (K^*)^{-1}\varphi_{\ell+1}\right)$ (see (1.95)). Hence, using (4.2) and Proposition 21 again, we obtain

$$\begin{aligned} E\left(Gf(s,B_s)\right) &= E\left(\langle I_\ell(g_{\ell+1}), Df(s,B_s)\rangle_{\mathcal{H}}\right) \\ &= \frac{1}{\ell+1}\sum_{j=1}^{\ell+1}(\mathcal{K}_B\varphi_j)(s)E\left(\partial_x f(s,B_s)I_\ell^W(\hat{\varphi}_j)\right). \end{aligned}$$

Here, $\hat{\varphi}_j$ means $\varphi_1 \otimes \cdots \varphi_{j-1} \otimes \varphi_{j+1} \otimes \cdots \otimes \varphi_{\ell+1}$. Now, the result follows since it is true for ℓ by induction hypothesis. □

Finally, we state the third result that we use to establish Itô's type formulae for B.

Lemma 53 *Let $f : [0,T] \times \mathbb{R} \to \mathbb{R}$ be a function in $C^{1,2}([0,T] \times \mathbb{R})$ such that $f, \partial_t f, \partial_x f$ and $\partial_x^2 f$ satisfy (1.142). Then, there exists a sequence $\{f_n \in C^{1,\infty}([0,T] \times \mathbb{R}) : n \in \mathbb{N}\}$ such that, for each $\varepsilon > 0$, and $k, n \in \mathbb{N}$, $f_n, \partial_t f_n$ and $\partial_x^k f_n$ satisfies (1.142) with constants $C_{\varepsilon,n,k}$ and $\sigma + \varepsilon$. Moreover, $f_n, \partial_t f_n, \partial_x f_n$ and $\partial_x^2 f_n$ converge uniformly on compact sets to $f, \partial_t f, \partial_x f$ and $\partial_x^2 f$, respectively, as $n \to \infty$.*

Proof. Consider a function $\psi : \mathbb{R} \to [0,1]$ in $C^\infty(\mathbb{R})$ satisfying

$$\psi(x) = \begin{cases} 1, & \text{if } |x| \leq 1, \\ 0, & \text{if } |x| \geq 2 \end{cases}$$

Remember that we can find such a function due to Spivak [200]. Moreover, we have

$$\int_{\mathbb{R}} \psi(y)dy = c < \infty.$$

Set $\psi_n(x) = nc^{-1}\psi(nx)$, for $x \in \mathbb{R}$ and $n \in \mathbb{N}$. Now, we introduce the functions

$$f_n(t,x) = \int_{\mathbb{R}} \psi_n(x-y)f(t,y)dy, \quad x \in \mathbb{R},\ n \in \mathbb{N} \text{ and } t \in [0,T].$$

We claim that $\{f_n : n \in \mathbb{N}\}$ is the sequence that we are looking for. In effect, we observe that (1.142) and the change of variables $u = x - y$ yield that, for $n \in \mathbb{N}$,

$$\begin{aligned}|\partial_t f_n(x)| &\leq \int_{\mathbb{R}} |\psi_n(x-y)\partial_t f(t,y)|\, dy = \int_{\mathbb{R}} \psi_n(y)\, |\partial_t f(t,x-y)|\, dy \\ &\leq C\int_{\mathbb{R}} \psi_n(y)\exp(\sigma(x-y)^2) dy \leq Ce^{\sigma x^2} e^{2\sigma|x|+1} \int_{\mathbb{R}} \psi_n(y) dy.\end{aligned}$$

Hence, we can find a constant $C_\varepsilon > 0$ such that

$$|\partial_t f_n(x)| \leq C_\varepsilon e^{(\sigma+\varepsilon)x^2} \int_{\mathbb{R}} |\psi_n(y)|\, dy = C_\varepsilon e^{(\sigma+\varepsilon)x^2}.$$

Proceeding similarly, we have

$$\begin{aligned}\left|\partial_x^k f_n(t,x)\right| &\leq \int_{\mathbb{R}} \left|\partial_x^k \psi_n(x-y) f(t,y)\right| dy = \int_{\mathbb{R}} \left|\partial_x^k \psi_n(y) f(t,x-y)\right| dy \\ &\leq C\int_{\mathbb{R}} \left|\partial_x^k \psi_n(y)\right| \exp\left(\sigma(x-y)^2\right) dy \leq C_\varepsilon e^{2(\sigma+\varepsilon)x^2} \int_{\mathbb{R}} \left|\partial_x^k \psi_n(y)\right| dy \\ &= C_{\varepsilon,n,k} e^{2(\sigma+\varepsilon)x^2}.\end{aligned}$$

Finally, for $i \in \{1,2\}$, we can write

$$\partial_x^i f_n(t,x) = \int_{\mathbb{R}} \psi_n(y)\partial_x^i f(t,x-y) dy \quad \text{and} \quad \partial_t f_n(t,x) = \int_{\mathbb{R}} \psi_n(y)\partial_t f(t,x-y) dy,$$

which lead us to obtain that $f_n, \partial_t f_n, \partial_x f_n$ and $\partial_x^2 f_n$ go uniformly on compact sets of $[0,T] \times \mathbb{R}$ to $f, \partial_t f, \partial_x f$ and $\partial_x^2 f$, respectively. Thus, the proof is complete. □

Now, we are ready to state the first Itô's formula analyzed in this section.

Theorem 58 *Let $f : [0,T] \times \mathbb{R} \to \mathbb{R}$ be a function in $C^{1,2}(\mathbb{R})$ such that $f, \partial_t f, \partial_x f$ and $\partial_x^2 f$ satisfy inequality* (1.142). *Also, assume that Hypotheses (K1), (K2') and (K3) hold, and for $t \in [0,T]$,*

$$\sum_{i=0}^{2^n-1} \int_{t_i^n \wedge t}^{\bar{t}_i^n} \left(\int_{\bar{t}_i^n}^{t_{i+1}^n \wedge t} \left(t_{i+1}^n - t_i^n\right) |K|(dt,s) \right)^2 ds \to 0, \quad \text{as } n \to \infty,$$

and

$$\int_0^T \left[\int_s^T (t-s)|K|(dt,s) \right]^2 ds < \infty.$$

Then,

$$\begin{aligned} f(t,B_t) &= f(0,0) + \int_0^t \partial_t f(s,B_s) ds + \int_0^t \partial_x f(s,B_s)\delta_2 B_s \\ &\quad + \frac{1}{2}\int_0^t \partial_x^2 f(s,B_s) dR_s, \quad t \in [0,T]. \end{aligned} \tag{4.72}$$

Proof. Fix $t \in [0,T]$. Let $\varphi \in \mathcal{E}([0,T])$. Then, equality (1.37) implies

$$\begin{aligned} &(K^*\varphi 1_{[0,t]})(s) \\ &= 1_{[0,t]}(s) \left(\varphi(s)K(T,s) + \int_{(s,t]} (\varphi(r)-\varphi(s))\, K(dr,s) - \int_{(t,T]} \varphi(s)K(dr,s) \right) \\ &= 1_{[0,t]}(s) \left(\varphi(s)K(t,s) + \int_{(s,t]} (\varphi(r)-\varphi(s))\, K(dr,s) \right) = 1_{[0,t]}(s)(K_t^*\varphi)(s), \end{aligned} \tag{4.73}$$

where K_t^* is the operator given by (1.37) when we change the interval $[0,T]$ by $[0,t]$. Thus, the proof of Corollary 14, together with (1.132), allows us to conclude that $s \mapsto \partial_x f(s,B_s)1_{[0,t]}(s)$ belongs to $L^2(\Omega;\mathcal{H})$ with

$$\left(K^*\partial_x f(\cdot,B.)1_{[0,t]}(\cdot)\right) = 1_{[0,t]}\left(K_t^*\partial_x f(\cdot,B.)\right). \tag{4.74}$$

Therefore, in order to see that (4.72) is true, we only need to show that, for any $n \in \mathbb{N}\cup\{0\}$, $\varphi,\dots,\varphi_n \in \mathcal{E}([0,T])$, and $F = I_n^W(\varphi_1\otimes\cdots\otimes\varphi_n)$,

$$\begin{aligned} E\left(Ff(t,B_t)\right) &- E\left(Ff(0,B_0)\right) - E\left(F\int_0^t \partial_t f(s,B_s)ds\right) - \frac{1}{2}E\left(F\int_0^t \partial_x^2 f(s,B_s)dR_s\right) \\ &= E\left(\int_0^t \partial_x f(s,B_s)(\mathcal{K}_B D^W F)(ds)\right) \end{aligned} \tag{4.75}$$

because the set $\mathcal{E}([0,T])$ of all the step functions is dense in $L^2([0,T])$, $K^*(\mathcal{H}) = L^2([0,T])$, Span $\left(\{I_n^W(\varphi_1\otimes\cdots\otimes\varphi_n) : n\in\mathbb{N}\cup\{0\}\}\right)$ is dense in $L^2(\Omega)$ (see Theorem 12) and $I_n^W(\varphi_1\otimes\cdots\otimes\varphi_n) = I_n\left((K^*)^{-1}(\varphi_1)\otimes\cdots\otimes(K^*)^{-1}(\varphi_n)\right)$. Indeed, (4.2), (4.74) and Remark 68 give

$$\langle DF, 1_{[0,t]}(\cdot)f(\cdot .B.)\rangle_{\mathcal{H}} = \int_0^T \left(K^*f(\cdot,B.)1_{[0,t]}(\cdot)\right)(s)D_s^W F ds = \int_0^t f(s,B_s)(\mathcal{K}_B D^W F)(ds).$$

Moreover, Lemma 23 implies that

$$f(t,B_t) - f(0,0) - \int_0^t \partial_t f(s,B_s)ds - \frac{1}{2}\int_0^t \partial_x^2 f(s,B_s)dR_s$$

is a square-integrable random variable.

On the other hand, let $F\in\mathbb{R}$. That is, we now prove that (4.75) holds in the case that $n=0$. Thus, for $p(x,y) = (2x\pi)^{-1/2}\exp(-y^2/2x)$, we can use the equality $\partial_x p(x,y) = \frac{1}{2}\partial_y^2 p(x,y)$, the integration by parts formula and (1.142) to have

$$\begin{aligned} &E\left(Ff(t,B_t)\right) - E\left(Ff(0,0)\right) \\ &= F\int_0^t \frac{d}{ds}E\left(f(s,B_s)\right)ds = F\int_0^t \frac{d}{ds}\int_{\mathbb{R}} f(s,y)p(R_s,y)dyds \\ &= F\int_0^t\int_{\mathbb{R}}(\partial_s f(s,y))p(R_s,y)dyds + \frac{F}{2}\int_0^t\int_{\mathbb{R}} f(s,y)\partial_y^2 p(R_s,y)dydR_s \\ &= E\left(F\int_0^t \partial_s f(s,B_s)ds\right) + \frac{F}{2}\int_0^t\int_{\mathbb{R}}(\partial_y^2 f(s,y))p(R_s,y)dydR_s \\ &= E\left(F\int_0^t \partial_s f(s,B_s)ds\right) + \frac{1}{2}E\left(F\int_0^t \partial_y^2 f(s,B_s)dR_s\right). \end{aligned}$$

It means, (4.75) holds for $n=0$ due to $D^W F = 0$.

For $n\geq 1$, we apply Lemma 52 in order to finish the proof. Towards this end, for $k\in\mathbb{N}$, we change f by f_k, which is introduced in Lemma 53. Furthermore, in the following calculations, we utilize the notation introduced in the proof of Lemma 52. Note that Lemma 52 establishes

$$\begin{aligned} E\left(\int_0^t \partial_y f_k(s,B_s)(\mathcal{K}_B D^W F)(ds)\right) &= \sum_{j=1}^n\int_0^t E\left(\partial_y f_k(s,B_s)I_{n-1}(\hat{\varphi}_j)\right)(\mathcal{K}_B\varphi_j)(ds) \\ &= \int_0^t E\left(\partial_y^n f_k(s,B_s)\right)\left[\prod_{j=1}^n(\mathcal{K}_B\varphi_j)\right](ds). \end{aligned}$$

and that the integration by parts formula and Fubini's theorem yield

$$
\begin{aligned}
&E\left(\partial_y^n f_k(t,B_t)\right)\left[\prod_{j=1}^n (\mathcal{K}_B\varphi_j)(t)\right] \\
&= \int_{\mathbb{R}} \partial_y^n f_k(t,y)\left[\prod_{j=1}^n (\mathcal{K}_B\varphi_j)(t)\right] p(R_t,y)dy \\
&= \int_0^t\int_{\mathbb{R}} \partial_s\partial_y^n f_k(s,y)\left[\prod_{j=1}^n (\mathcal{K}_B\varphi_j)(s)\right] p(R_s,y)dyds \\
&\quad +\frac{1}{2}\int_0^t\int_{\mathbb{R}} \partial_y^n f_k(s,y)\left[\prod_{j=1}^n (\mathcal{K}_B\varphi_j)(s)\right] \partial_y^2 p(R_s,y)dydR_s \\
&\quad +\int_0^t\int_{\mathbb{R}} \partial_y^n f_k(s,y)p(R_s,y)dy\left[\prod_{j=1}^n (\mathcal{K}_B\varphi_j)\right](ds) \\
&= \int_0^t\int_{\mathbb{R}} \partial_s\partial_y^n f_k(s,y)\left[\prod_{j=1}^n (\mathcal{K}_B\varphi_j)(s)\right] p(R_s,y)dyds \\
&\quad +\frac{1}{2}\int_0^t\int_{\mathbb{R}} \partial_y^{n+2} f_k(s,y)\left[\prod_{j=1}^n (\mathcal{K}_B\varphi_j)(s)\right] p(R_s,y)dydR_s \\
&\quad +\int_0^t\int_{\mathbb{R}} \partial_y^n f_k(s,y)p(R_s,y)dy\left[\prod_{j=1}^n (\mathcal{K}_B\varphi_j)\right](ds).
\end{aligned}
$$

Therefore, the last two equalities, together with the fact that $E(F)=0$ and Lemma 52, imply that (4.75) is satisfied when we write f_k instead of f. Finally, Lemma 53 with $\varepsilon>0$ such that $\sigma+\varepsilon<\left(\sup_{0\le t\le T} R_B(t,t)\right)^{-1}/4$, (1.142) and the dominated convergence theorem allow us to see that the proof is complete. □

Now, we analyze the following result in order to show that the Itô's type formula (4.72) holds for either the fractional Brownian motion, or the Riemann-Liouville fractional process (see Definition 25) with Hurst parameter $H\in(0,1/2)$.

Lemma 54 *Let $K_H=\{K_H(t,s): 0\le s<t\le T\}$ be the square-integrable kernel given in* (1.67). *Then, K_H also have the expression*

$$
K_H(t,s)=\tilde{c}_H(t-s)^{H-\frac{1}{2}}+s^{H-\frac{1}{2}}F_1\left(\frac{t}{s}\right),\quad 0\le s<t\le T, \tag{4.76}
$$

with

$$
F_1(z)=\tilde{c}_H\left(\frac{1}{2}-H\right)\int_0^{z-1}\theta^{H-\frac{3}{2}}\left(1-(\theta+1)^{H-\frac{1}{2}}\right)d\theta.
$$

Here, $\tilde{c}_H$ is defined in (1.67).

Remark 69 *Note that, for $\varepsilon>0$ small enough and $\theta\in(0,\varepsilon)$, we have*

$$
\theta^{H-\frac{3}{2}}\left(1-(\theta+1)^{H-\frac{1}{2}}\right)=\theta^{H-\frac{1}{2}}\frac{\left(1-(\theta+1)^{H-\frac{1}{2}}\right)}{\theta}<2\left(\frac{1}{2}-H\right)\theta^{H-\frac{1}{2}}
$$

and, for $\theta > \varepsilon$, we also have that $\theta^{H-\frac{3}{2}}(1-(\theta+1)^{H-\frac{1}{2}}) \le \theta^{H-\frac{3}{2}}$. Thus,

$$\int_0^\infty \theta^{H-\frac{3}{2}}(1-(\theta+1)^{H-\frac{1}{2}})d\theta < \infty.$$

Proof. Let $\varepsilon > 0$ be small enough and $0 < s < t$. Then, the integration by parts formula implies

$$\begin{aligned}&\left(\frac{1}{2}-H\right)\int_\varepsilon^{\frac{t}{s}-1} \theta^{H-\frac{3}{2}}(1-(\theta+1)^{H-\frac{1}{2}})d\theta\\ &= \varepsilon^{H-\frac{1}{2}}\left(1-(\varepsilon+1)^{H-\frac{1}{2}}\right) - \left(\frac{t}{s}-1\right)^{H-\frac{1}{2}}\left(1-\left(\frac{t}{s}\right)^{H-\frac{1}{2}}\right)\\ &\quad -\left(H-\frac{1}{2}\right)\int_\varepsilon^{\frac{t}{s}-1}\theta^{H-\frac{1}{2}}(\theta+1)^{H-\frac{3}{2}}d\theta.\end{aligned}$$

Hence, letting ε goes to zero and using the change of variables $u = 1+\theta$, Remark 69 yields

$$\begin{aligned}F_1\left(\frac{t}{s}\right) &= -\tilde{c}_H\left(\frac{t}{s}-1\right)^{H-\frac{1}{2}}\left(1-\left(\frac{t}{s}\right)^{H-\frac{1}{2}}\right)\\ &\quad -\tilde{c}_H\left(H-\frac{1}{2}\right)\int_0^{\frac{t}{s}-1}\theta^{H-\frac{1}{2}}(\theta+1)^{H-\frac{3}{2}}d\theta\\ &= -\tilde{c}_H\left[s^{1-2H}(t-s)^{H-\frac{1}{2}}\left(s^{H-\frac{1}{2}}-t^{H-\frac{1}{2}}\right)+\left(H-\frac{1}{2}\right)\int_1^{\frac{t}{s}}(u-1)^{H-\frac{1}{2}}u^{H-\frac{3}{2}}du\right].\end{aligned}$$

Therefore, using now the change of variables formula $v = su$, we obtain

$$\begin{aligned}F_1\left(\frac{t}{s}\right) &= -\tilde{c}_H\Bigg[s^{1-2H}(t-s)^{H-\frac{1}{2}}\left(s^{H-\frac{1}{2}}-t^{H-\frac{1}{2}}\right)\\ &\quad +\left(H-\frac{1}{2}\right)s^{1-2H}\int_s^t (v-s)^{H-\frac{1}{2}}v^{H-\frac{3}{2}}dv\Bigg].\end{aligned}$$

Consequently, (1.67) allows us to conclude that (4.76) is true. □

Proposition 86 *Let $H \in (1/4, 1/2)$. Then, Hypotheses (K1), (K2') and (K3) are satisfied for either fractional Brownian motion, or Riemann-Liouville fractional process.*

Proof. By Propositions 30 and 31, we have already known that both fractional Brownian motion and Riemann-Liouville fractional process satisfy Conditions (K1) and (K2').

Now, we prove that Hypothesis (K3) holds. Towards this end, we divide the proof into two steps.

Step 1. Here we suppose that B is Riemann-Liouville fractional process with $H > 1/4$. So, we have that

$$R_B(s,s) = \frac{s^{2H}}{2H} = \int_0^s u^{2H-1}du, \quad s \in [0,T],$$

which gives that $s \mapsto R_B(s,s)$ has bounded variation on $[0,T]$. Additionally, we obtain

$$\int_t^s (s-r)^{H-\frac{1}{2}}dr = \frac{(s-t)^{H+\frac{1}{2}}}{H+\frac{1}{2}} = \int_t^s (u-t)^{H-\frac{1}{2}}du, \quad s \in [t,T].$$

Thus, the function $s \mapsto \int_t^s (s-r)^{H-\frac{1}{2}} dr$ also has bounded variation on $[t,T]$, for $t \in [0,T]$. In consequence, Condition (K3) is true for the Riemann-Liouville fractional process B.

Step 2. We now deal with the fractional Brownian motion case.

From Step 1, (1.44) and (4.76), we only need to show that, for $t \in [0,T]$,

$$s \mapsto \int_t^s r^{H-\frac{1}{2}} F_1(\frac{s}{r}) dr, \quad s \in [t,T],$$

is a function of bounded variation. To do so, chose $t \leq s_1 < s_2 \leq T$. Then,

$$\begin{aligned} & \left| \int_t^{s_2} r^{H-\frac{1}{2}} F_1(\frac{s_2}{r}) dr - \int_t^{s_1} r^{H-\frac{1}{2}} F_1(\frac{s_1}{r}) \right| \\ & \leq \left| \int_{s_1}^{s_2} r^{H-\frac{1}{2}} F_1(\frac{s_2}{r}) dr \right| + \left| \int_t^{s_1} r^{H-\frac{1}{2}} \left[F_1(\frac{s_2}{r}) - F_1(\frac{s_1}{r}) \right] dr \right| = I_1 + I_2. \quad (4.77) \end{aligned}$$

Using the convention $f(\theta) = \tilde{c}_H(\frac{1}{2} - H)\theta^{H-\frac{3}{2}}(1-(\theta+1)^{H-\frac{1}{2}})$, we have that Remark 68 gives there exists a constant $C > 0$ such that

$$\begin{aligned} I_1 &= \int_{s_1}^{s_2} r^{H-\frac{1}{2}} \int_0^{\frac{s_2}{r}-1} f(\theta) d\theta dr \\ &\leq \left(\int_0^\infty f(\theta) d\theta \right) \int_{s_1}^{s_2} r^{H-\frac{1}{2}} dr \leq C \int_{s_1}^{s_2} r^{H-\frac{1}{2}} dr. \end{aligned} \quad (4.78)$$

For I_2, we apply the Fubini's theorem to obtain

$$\begin{aligned} I_2 &= \int_t^{s_1} r^{H-\frac{1}{2}} \int_{\frac{s_1}{r}-1}^{\frac{s_2}{r}-1} f(\theta) d\theta dr = \int_0^{\frac{s_2}{t}-1} f(\theta) \int_{t \vee \frac{s_1}{\theta+1}}^{s_1 \wedge \frac{s_2}{\theta+1}} r^{H-\frac{1}{2}} dr d\theta \\ &\leq \int_0^{\frac{s_2}{t}-1} f(\theta) \int_{\frac{s_1}{\theta+1}}^{\frac{s_2}{\theta+1}} r^{H-\frac{1}{2}} dr d\theta. \end{aligned}$$

Therefore, the change of variables $u = (\theta+1)r$ implies

$$\begin{aligned} I_2 &\leq \int_0^{\frac{s_2}{t}-1} f(\theta)(\theta+1)^{-\frac{1}{2}-H} \int_{s_1}^{s_2} u^{H-\frac{1}{2}} du d\theta \\ &\leq \left(\int_0^\infty f(\theta) d\theta \right) \int_{s_1}^{s_2} u^{H-\frac{1}{2}} du \leq C \int_{s_1}^{s_2} u^{H-\frac{1}{2}} du. \end{aligned}$$

Hence, (4.77) and (4.78) allow us to conclude that the function $s \mapsto \int_t^s r^{H-\frac{1}{2}} F_1(\frac{s}{r}) dr$ has bounded variation on the interval $[t,T]$. Thus, the proof is complete. □

As an immediate consequence of Proposition 86, we obtain the following result.

Corollary 26 *Let $f : [0,T] \times \mathbb{R} \to \mathbb{R}$ be as in Theorem 58 and $H \in (1/4, 1/2)$. Then the Itô's formula (4.72) is satisfied for either the Riemann-Liouville fractional process, or the fractional Brownian motion.*

Concerning this corollary in the case that H is in $(1/2, 1)$, we have that, by Definition 25 and (1.66), the Kernel K for either the Riemann-Liouville fractional process, or the fractional Brownian motion satisfies the condition

(K4) For $s \in [0,T]$, $K(\cdot, s)$ has bounded variation on $(s,T]$ and

$$\int_0^T |K|((s,T],s)^2 ds < \infty.$$

Our next aim is to see that Condition (K4) implies Assumptions (K1)-(K3) are satisfied.

Lemma 55 *Let B be a Gaussian Volterra process such that its kernel $K = \{K(t,s) : 0 \le s < t \le T\}$ satisfies Hypothesis (K4). Then, Assumptions (K1) and (K2) are also satisfied. Moreover, (K3) holds if $K(\cdot, r)$ is right-continuous on $(r,T]$, for almost all $r \in [0,T]$, and the image of K^* is the space $L^2([0,T])$.*

Remark 70 *Hypothesis (K4) and Lemma 25 imply that, for almost all $s \in [0,T]$, $K(\cdot,s)$ is a regulated function on $(s,T]$. That is, the function $K(\cdot,s)$ has left- and right-limits on the interval $(s,T]$. Also, Lemma 108 gives that, for almost all $s \in [0,T]$, the function $K(\cdot,s)$ is continuous except at countable many points. However, in the following proof, we need that, given $r \in [0,T]$, $K(\cdot,s)$ is continuous at r for almost all $s < r$, which is not guaranteed by Hypothesis (K4) and Lemma 108. This is the reason that we assume the right-continuity of $K(\cdot,s)$. On the other hand, we do not need the assumption on the image of K^* in the proof of this lemma. We ask for it because it is part of Condition (K3).*

Proof. It is obvious that (K4) is stronger than Condition (K1). Concerning Hypothesis (K2), it is also true because

$$\begin{aligned}\int_0^T \left[\int_s^T \|B_t - B_s\|_{L^2(\Omega)} |K|(dt,s)\right]^2 ds &\le \int_0^T \left[\int_s^T 2\sqrt{\sup_{t\in[0,T]} R_B(t,t)}|K|(dt,s)\right]^2 ds \\ &= 4 \sup_{t\in[0,T]} R_B(t,t) \int_0^T |K|\left((s,T],s\right)^2 ds < \infty\end{aligned}$$

due to (K4).

On the other hand, note that $K(s+,s) = K(T,s) - K((s,T]),s)$, for all $s \in [0,T]$. Thus Assumption (K4) leads us to conclude that the function $s \mapsto K(s+,s)$ belongs to $L^2([0,T])$. Moreover, we also have the function

$$k(r) := |K(T,r)| + |K|((r,T],r), \quad r \in [0,T],$$

belongs to $L^2([0,T])$.

Now, in order to see that (K3) is true, we fix a partition $\{0 = t_0 < t_1 < \ldots < t_n = T\}$ of the interval $[0,T]$ and suppose that $K(\cdot,r)$ is right-continuous on $(r,T]$, for almost all $r \in [0,T]$. Thus, under this assumption, for almost all $r \in [0,T]$,

$$|K(s,r)|\, 1_{[0,s)}(r) = |K(T,r) - K((s,T],r)|\, 1_{[0,s)}(r) \le k(r), \quad \text{for all } s \in [0,T].$$

With this inequality in mind, we have

$$\begin{aligned}&\sum_{i=1}^n |R_B(t_i,t_i) - R_B(t_{i-1},t_{i-1})| \\ &= \sum_{i=1}^n \left| \int_{t_{i-1}}^{t_i} K(t_i,r)^2 dr + \int_0^{t_{i-1}} \left(K(t_i,r)^2 - K(t_{i-1},r)^2\right) dr \right| \\ &\le \int_0^T k(r)^2 dr + \sum_{i=1}^n \int_0^{t_{i-1}} |K((t_{i-1},t_i],r)\left(K(t_i,r) + K(t_{i-1},r)\right)| dr \\ &\le \int_0^T k(r)^2 dr + 2\int_0^T k(r)|K|((r,T],r) dr,\end{aligned}$$

which implies that $s \mapsto R_B(s,s)$ has bounded variation on the interval $[0,T]$.

Finally, proceeding similarly, we have

$$\sum_{i=1}^{n}\left|\int_{t_i\wedge t}^{t_i}K(t_i,r)dr-\int_{t_{i-1}\wedge t}^{t_{i-1}}K(t_{i-1},r)dr\right|$$
$$\leq \sum_{i=1}^{n}\left\{\int_{t_{i-1}}^{t_i}|K(t_i,r)|dr+\int_0^{t_{i-1}}|K(t_i,r)-K(t_{i-1},r)|\,dr\right\}$$
$$\leq \int_0^T k(r)dr+\int_0^T|K|((r,T],r)dr.$$

Therefore, (K3) is also satisfied because we are assuming that the image of K^* is the whole space $L^2([0,T])$. Now, the proof is complete □

Note that, under Hypothesis (K4), (1.37) leads us to write

$$(K^*\varphi)(s)=\varphi(s)K(s+,s)+\int_s^T\varphi(t)K(dt,s),\quad s\in[0,T]\text{ and }\varphi\in\mathcal{E}([0,T]).\tag{4.79}$$

Also, in Corollary 14, we apply Hypothesis (K2'), together with the assumptions of this corollary, to show that $f(\cdot,B.)$ belongs to $L^2(\Omega,\mathcal{H}_K)$ for a function $f\in C^1([0,T]\times\mathbb{R})$ such that f, $\partial_t f$ and $\partial_x f$ satisfy (1.142). Now, by (4.79), a similar result is stated as follows.

As in (1.131), we take advantage of the expression for K^*. In this case, we use (4.79) to introduce the seminorm on $\mathcal{E}([0,T])$

$$\|\varphi\|^2_{\mathcal{H}_{K_4}}=\int_0^T\varphi(s)^2K(s+,s)^2ds+\int_0^T\left(\int_s^T|\varphi(r)||K|(dr,s)\right)^2ds.$$

Therefore, now we can consider the completion $\mathcal{H}_{K_4}$ of the space $\mathcal{E}([0,T])$ with respect to the seminorm $\|\cdot\|_{\mathcal{H}_{K_4}}$. In consequence, as in (1.132), we get

$$\|\varphi\|_{L^2([0,T],K(s+,s)^2ds)}\leq\|\varphi\|_{\mathcal{H}_{K_4}}\quad\text{and}\quad\|\varphi\|_{\mathcal{H}}\leq\sqrt{2}\,\|\varphi\|_{\mathcal{H}_{K_4}}.$$

Thus, the elements of $\mathcal{H}_{K_4}$ are functions in $L^2([0,T],K(s+,s)^2ds)$ and $\mathcal{H}_{K_4}$ is included in $\mathcal{H}$. Hence, Proposition 62 yields that $\mathbb{D}^{1,p}(\mathcal{H}_{K_4})\subset\operatorname{Dom}\delta_p$, for $p>1$.

Now, let $f\in C^1([0,T]\times\mathbb{R})$ be such that f, $\partial_t f$ and $\partial_x f$ satisfy (1.142). We recall you that we use (K2'), together with the assumptions of Corollary 14, to see that the sequence $\{f(\varphi^{(n)}):n\in\mathbb{N}\}$ of simple processes introduced in the proof of Corollary 14 converges to $f(\cdot,B.)$ in $L^2(\Omega;\mathcal{H}_K)$. But, we do not need Hypothesis (K2') to obtain that $f(\varphi^{(n)})$ converges if Condition (K4) hods because, in this case, the dominated convergence theorem and (4.79) imply that $f(\varphi^{(n)})$ goes to $f(\cdot,B.)$ in $L^2(\Omega;\mathcal{H}_{K_4})$ without using a function g as in (K2'). In this way, this convergence allows us to state the following result.

Theorem 59 *Let $f:[0,T]\times\mathbb{R}\to\mathbb{R}$ be a function in $C^{1,2}(\mathbb{R})$ such that $f,\partial_t f,\partial_x f$ and $\partial_x^2 f$ satisfy inequality (1.142), and B a continuous Volterra process such that the image of K^* is the whole space $L^2([0,T])$ and Hypothesis (K4) holds. Then, for $t\in[0,T]$, we have that $\partial_x f(\cdot,B.)1_{[0,t]}(\cdot)$ belongs to $Dom\,\delta_2$ and*

$$f(t,B_t)=f(0,0)+\int_0^t\partial_t f(s,B_s)ds+\int_0^t\partial_x f(s,B_s)\delta_2 B_s+\frac{1}{2}\int_0^t\partial_x^2 f(s,B_s)dR_s.$$

Proof. The proof is an immediate consequence of Lemma 55, and the facts that $f(\cdot,B.)\in L^2(\Omega;\mathcal{H}_{K_4})$ and

$$(K^*\varphi 1_{[0,t]})(\cdot)=1_{[0,t]}(\cdot)(K_t^*\varphi)(\cdot),\quad t\in[0,T],\tag{4.80}$$

which follows from (4.79). Indeed, these permit to make use of the proof of Theorem 58 again to prove that the result is true. □

As in the fractional Brownian motion case, we can take advantage of (4.80) and the fact that we are assuming that the image of K^* is the space $L^2([0,T])$ to have the transfer principle

$$\begin{aligned}\delta_2\left(f(\cdot,B.)1_{[0,t]}(\cdot)\right) &= \int_0^T \left(K^*f(\cdot,B.)1_{[0,t]}(\cdot)\right)(s)\partial_2 W_s \\ &= \int_0^t \left(K_t^*f(\cdot,B.)\right)(s)\partial_2 W_s = \int_0^t f(s,B_s)K(s+,s)\partial_2 W_s \\ &+ \int_0^t \left(\int_s^t f(r,B_r)K(dr,s)\right)\partial_2 W_s, \quad t\in[0,T],\end{aligned}$$

where W is the Brownian motion in (1.32). Moreover, we have already pointed out that the kernel K of either fractional Brownian motion, or Riemann-Liouville fractional process with parameter $H>1/2$ satisfies condition (K4). Thus, we obtain that the Itô's type formula in Theorem 59 holds also for these two processes.

On the other hand, in order to deal with Itô's formulae for processes of the form $X_t = \int_0^t u_s\partial_2 B_s$, $t\in[0,T]$, where u is a suitable process, we first consider a stronger condition than Assumptions (K1) and (K2) again, which in particular implies that the process $B=\{B_t : t\in[0,T]\}$ has a continuous modification. So we first deal with the condition

(K5) There are constants $C>0$ and $\alpha>0$ such that, for $0\le s<t\le T$:

- *i)* The kernel K is differentiable in the variable t, and both K and $\partial_t K$ are continuous in the set $\{(t,s): 0\le s<t\le T\}$.
- *ii)* $|\partial_t K(t,s)|\le C(t-s)^{-1-\alpha}$.
- *iii)* $\int_s^t K(t,r)^2dr \le C(t-s)^{1-2\alpha}$.

For $H\in(0,1/2)$, Assumptions *i)*-*iii)* are trivially true for the Riemann-Liouville fractional process because the kernel of this process is $K(t,s)=(t-s)^{-\alpha}$, with $\alpha=1/2-H$. For fractional Brownian motion with $H<1/2$, either equality (1.67), or equality (4.76) implies that *i)* and *ii)* hold. Concerning *iii)*, we have

$$\begin{aligned}\int_s^t K_H(t,r)^2dr &\le \int_s^t K_H(t,r)^2dr + \int_0^s \left(K_H(t,r)-K_H(s,r)\right)^2dr \\ &= \int_0^t K_H(t,r)^2dr - 2\int_0^s K_H(t,r)K_H(s,r)dr + \int_0^s K_H(s,r)^2dr \\ &= E\left(B_t^2 - 2B_tB_s + B_s^2\right) \\ &= E\left([B_t-B_s]^2\right) = (t-s)^{1-2\alpha}, \quad s<t, \end{aligned} \tag{4.81}$$

where last equality follows from the fact that fractional Brownian motion has stationary increments (see (1.47)).

Now, we compare (K1) and (K2) with (K5). It is obvious that *i)* gives that $(K1)$ is satisfied. In order to deal with (K2), we first estimate the quantity $E\left((B_t-B_s)^2\right)$. Note that the change of variables formula $u=s-r$, and Hypotheses *i)*-*iii)* yield that, for

$\alpha \in (0, 1/2)$,

$$\begin{aligned} & E\left((B_t - B_s)^2\right) \\ = \;& E\left(B_t^2 - 2B_sB_t + B_s^2\right) = \int_s^t K(t,r)^2 dr + \int_0^s \left(K(t,r) - K(s,r)\right)^2 dr \\ \le \;& C(t-s)^{1-2\alpha} + \int_0^s \left(\int_s^t \partial_t K(u,r) du\right)^2 dr \\ \le \;& C(t-s)^{1-2\alpha} + C^2 \int_0^s \left(\int_s^t (u-r)^{-1-\alpha} du\right)^2 dr \\ = \;& C(t-s)^{1-2\alpha} + \frac{C^2}{\alpha^2} \int_0^s \left((s-r)^{-\alpha} - (t-r)^{-\alpha}\right)^2 dr \\ = \;& C(t-s)^{1-2\alpha} + \frac{C^2}{\alpha^2} \int_0^s \left(1 - \left[\frac{t-s}{u} + 1\right]^{-\alpha}\right)^2 u^{-2\alpha} du. \end{aligned}$$

Hence, we can apply the change of variables formula $v = (t-s)/u$ to get that there is a constant $\tilde{C} > 0$ such that

$$\begin{aligned} E\left((B_t - B_s)^2\right) &\le C(t-s)^{1-2\alpha} + \frac{C^2 (t-s)^{1-2\alpha}}{\alpha^2} \int_{(t-s)/s}^{\infty} \left(1 - (v+1)^{-\alpha}\right)^2 v^{2\alpha - 2} dv \\ &\le \tilde{C}(t-s)^{1-2\alpha}. \end{aligned} \tag{4.82}$$

Therefore, we have proven that (K2) holds if $\alpha < 1/4$ due to Condition *ii*). Moreover, we have also proven that the following result is true.

Lemma 56 *Suppose that Hypothesis (K5) is satisfied with $\alpha \in (0, 1/2)$. Then, the process B has a Hölder continuous modification for any exponent less than $1/2 - \alpha$.*

Proof. The result is an immediate consequence of Proposition 5 and (4.82). □

On the other hand, in the following result, we consider a process $u = \{u_t : t \in [0,T]\}$ that is λ-Hölder continuous in $\mathbb{D}_W^{1,p}$. That is, there is a constant $C > 0$ such that $\|u_t - u_s\|_{1,p} \le C|t-s|^{\lambda}$, $s, t \in [0,T]$, (see (1.104) with D^W, $L^2([0,T])$ and $\mathbb{R}$ instead of D, $\mathcal{H}$ and $\mathcal{K}$, respectively). Then, we will see that the process u satisfies Condition (K2') with $g(t) = t^{\lambda}$, when we write u instead of B. Indeed, we only need to use Condition *ii*) and proceed as in the proof of Proposition 30. In this way, we can utilize the ideas of the proof of Proposition 29 in order to establish that u belongs to $\mathbb{D}_W^{1,p}(\mathcal{H}_K)$, as we do in the following proposition. This last space is defined in the following remarks.

Proposition 87 *Let B be a Gaussian Volterra process whose kernel K satisfies Condition (K5) with $\alpha \in (0.1/2)$ and the image of K^* is $L^2([0,T])$), and $u = \{u_t : t \in [0,T]\}$ a λ-Hölder continuous process with respect to the norm of the space $\mathbb{D}_W^{1,p}$ with $\lambda > \alpha$ and $p \ge 2$. Then, u is in $\mathbb{D}_W^{1,p}(\mathcal{H}_K)$ and the process $X_t = \int_0^t u_s \delta_2 B_s$, $t \in [0,T]$, is such that*

$$E\left(|X_t - X_s|^p\right) \le C|t-s|^{p(1-2\alpha)/2}, \quad s, t \in [0,T].$$

Remarks 9 *i) The Gaussian Volterra process $B = \{B_t : t \in [0,T]\}$ has the form (1.32). So, the space $\mathbb{D}_W^{1,p}$ stands for the space $\mathbb{D}^{1,p}$ when the underlying Gaussian process is the involved Brownian motion W appearing in (1.32). So, in the following statements and in the proof, D^W and δ_p^W are the derivative and the divergence operators with respect to W, respectively.*

ii) The fact that K^*u belongs to the space $\mathbb{D}_W^{1,p}(L^2([0,T]))$ implies that $K^*u1_{[0,t]} \in Dom\,\delta_2^W$, for $t \in [0,T]$, due to (4.73) and Corollary 13. Consequently, the transfer principle allows us to conclude that $u1_{[0,t]}$ is in $Dom\,\delta_2$, for all $t \in [0,T]$. Thus, the process X is well-defined.

iii) A process u belongs to the space $\mathbb{D}_W^{1,p}(\mathcal{H}_K)$ if there exists a sequence $\{u^{(n)} : n \in \mathbb{N}\}$ of $\mathcal{E}([0,T])$-valued smooth random variables such that

$$E\left\{\left(\int_0^T (u_s - u_s^{(n)})^2 K(T,s)^2 ds\right)^{p/2} + \left(\int_0^T \left[\int_s^T \left|u_t - u_s - (u_t^{(n)} - u_s^{(n)})\right| |K|(dt,s)\right]^2 ds\right)^{p/2}\right\}$$

and

$$E\left\{\left(\int_0^T \int_0^T (D_r^W u_s - D_r^W u_s^{(n)})^2 K(T,s)^2 ds dr\right)^{p/2} + \left(\int_0^T \int_0^T \left[\int_s^T \left|D_r^W \left[u_t - u_s - (u_t^{(n)} - u_s^{(n)})\right]\right| |K|(dt,s)\right]^2 ds dr\right)^{p/2}\right\}$$

converge to zero, as $n \to \infty$.

Proof. We first show that u is in the space $\mathbb{D}_W^{1,p}(\mathcal{H}_K)$. Towards this end, consider an $\mathcal{E}([0,T])$-valued smooth random variable ψ. Then, (1.131) and (8.5) give

$$\begin{aligned} E\left(\|u-\psi\|_K^p\right) \leq\ & \left(\int_0^T \|u_s - \psi_s\|_{L^p(\Omega)}^2 K(T,s)^2 ds\right)^{p/2} \\ & + \left(\int_0^T \left[\int_s^T \|u_t - u_s - (\psi_t - \psi_s)\|_{L^p(\Omega)} |K|(dt,s)\right]^2 ds\right)^{p/2}. \end{aligned} \tag{4.83}$$

Furthermore, using (1.131) and Fubini's theorem again, we establish

$$\begin{aligned} & E\left[\left(\int_0^T \|D_r(u-\psi)\|_K^2 dr\right)^{p/2}\right] \\ & = E\left\{\left(\int_0^T K(T,s)^2 \int_0^T (D_r^W u_s - D_r^W \psi_s)^2 dr ds\right)^{p/2} \right. \\ & \left. + \left(\int_0^T \int_0^T \left[\int_s^T \left|D_r^W \left[u_t - u_s - (\psi_t - \psi_s)\right]\right| |K|(dt,s)\right]^2 dr ds\right)^{p/2}\right\}. \end{aligned}$$

Hence, from (1.104) and (8.5), we can conclude

$$
\begin{aligned}
E & \left[\left(\int_0^T \|D_r(u-\psi)\|_K^2 \, dr \right)^{p/2} \right] \\
& \leq \left(\int_0^T K(T,s)^2 \, \|(u_s-\psi_s)\|_{1,p}^2 \, ds \right)^{p/2} \\
& \quad + \left(\int_0^T \left[\int_s^T \|u_t-u_s-(\psi_t-\psi_s)\|_{1,p} \, |K|(dt,s) \right]^2 ds \right)^{p/2} . \qquad (4.84)
\end{aligned}
$$

We now introduce the sequence of simple processes $\{\varphi^{(n)} : n \in \mathbb{N}\}$ defined as

$$
\varphi^{(n)} = \sum_{i=0}^{2^n-1} \left[u_{t_i^n} 1_{(t_i^n, \bar{t}_i^n]} + u_{t_{i+1}^n} 1_{(\bar{t}_i^n, t_{i+1}^n]} \right],
$$

with

$$
t_i^n = \frac{Ti}{2^n} \quad \text{and} \quad \bar{t}_i^n = t_i^n + \frac{T}{2^{n+1}}, \qquad n \in \mathbb{N}.
$$

In Consequence, we can proceed as in the proofs of Propositions 30 and 29 to show that the right-hand sides of inequalities (4.83) and (4.84) go to zero, as $n \to \infty$, when we change ψ by $\varphi^{(n)}$. Therefore, we have that u belongs to $D_W^{1,p}(\mathcal{H}_K)$. In this way, (1.37), (1.104) and Corollary 13 lead us to obtain that K^*u belongs to the space $D_W^{1,p}(L^2([0,T]))$. Thus, Proposition 62 and Remarks 9.*ii*) imply that the process X is well-defined. Moreover, (1.37), (4.73) and the transfer principle yield that, for $s,t \in [0,T]$, $s<t$,

$$
\begin{aligned}
X_t - X_s &= \int_s^t u_r K(t,r) \delta_2 W_r + \int_0^s u_r \left(K(t,r)-K(s,r)\right) \delta_2 W_r \\
&\quad + \int_s^t \left(\int_v^t (u_r-u_v) \partial_r K(r,v) dr \right) \delta_2 W_v \\
&\quad + \int_0^s \left(\int_s^t (u_r-u_v) \partial_r K(r,v) dr \right) \delta_2 W_v \\
&= \int_s^t u_r K(t,r) \delta_2 W_r + \int_s^t \left(\int_v^t (u_r-u_v) \partial_r K(r,v) dr \right) \delta_2 W_v \\
&\quad + \int_0^s \left(\int_s^t u_r \partial_r K(r,v) dr \right) \delta_2 W_v .
\end{aligned}
$$

So, by applying Proposition 62 and (1.104), we can proceed as in (4.83) and (4.84) to get

$$
\begin{aligned}
E\left(|X_t-X_s|^p\right) &= C_p \left\{ \left(\int_s^t \|u_r\|_{1,p}^2 K(t,r)^2 dr \right)^{p/2} \right. \\
&\quad + \left(\int_s^t \left[\int_v^t \|u_r-u_v\|_{1,p} \, |\partial_r K(r,v)| \, dr \right]^2 dv \right)^{p/2} \\
&\quad \left. + \left(\int_0^s \left[\int_s^t \|u_r\|_{1,p} \, |\partial_r K(r,v)| \, dr \right]^2 dv \right)^{p/2} \right\} \\
&= I_{1,p} + I_{2,p} + I_{3,p}.
\end{aligned}
$$

Note that Condition (K5).*iii*) gives

$$I_{1,p} \leq C_p \left(\sup_{r\in[0,T]} \|u_r\|_{1,p}^p \right) (t-s)^{p(1-2\alpha)/2}.$$

Assumption (K5).*ii*) and the Hölder continuity of u allow us to write

$$I_{2,p} \leq C_p \left(\int_s^t \left[\int_v^t (r-v)^{\lambda-\alpha-1} dr \right]^2 dv \right)^{p/2} \leq C_{p,\alpha,\lambda} (t-s)^{p(1+2(\lambda-\alpha))/2}$$

and

$$\begin{aligned} I_{3,p} &\leq C_p \left(\sup_{r\in[0,T]} \|u_r\|_{1,p}^p \right) \left(\int_0^s \left[\int_s^t (r-v)^{-1-\alpha} dr \right]^2 dv \right)^{p/2} \\ &\leq C_{p,\alpha} \left(\sup_{r\in[0,T]} \|u_r\|_{1,p}^p \right) (t-s)^{p(1-2\alpha)/2}. \end{aligned}$$

Finally, the last three inequalities show that the proof is finished. □

Our next aim is to state an Itô's formula for a Gaussian Volterra process whose kernel K satisfies Hypothesis (K5). Towards this end, for $\varepsilon > 0$ small enough and $t \in [0, T-\varepsilon]$, we introduce the operator $K_t^{*,\varepsilon}$, which is defined as

$$\begin{aligned} \left(K_t^{*,\varepsilon}\varphi\right)(s) &= \varphi(s)K(t+\varepsilon,s) + \int_s^t (\varphi(r)-\varphi(s))\, \partial_r K(r+\varepsilon,s)dr \\ &= \varphi(s)K(s+\varepsilon,s) + \int_s^t \varphi(r)\partial_r K(r+\varepsilon,s)dr, \quad s\in[0,t], \end{aligned} \tag{4.85}$$

where $\varphi : [0,T] \to V$ is a suitable function taking values in a Banach space V. Observe that (8.5), (4.73), (K5) and the dominated convergence theorem give

$$\begin{aligned} &\int_0^t \left\| \left(K_t^{*,\varepsilon}\varphi\right)(s) - \left(K_t^{*}\varphi\right)(s) \right\|_V^2 ds \\ &\leq 2\int_0^t \|\varphi(s)\|_V^2 \left[K(t+\varepsilon,s) - K(t,s)\right]^2 ds \\ &\quad +2\int_0^t \left(\int_s^t \|\varphi(r)-\varphi(s)\|_V \left|\partial_r K(r+\varepsilon,s) - \partial_r K(r,s)\right| dr \right)^2 ds \\ &= 2\int_0^t \|\varphi(s)\|_V^2 \left[\int_t^{t+\varepsilon} \partial_r K(r,s)dr \right]^2 ds \\ &\quad +2\int_0^t \left(\int_s^t \|\varphi(r)-\varphi(s)\|_V \left|\partial_r K(r+\varepsilon,s) - \partial_r K(r,s)\right| dr \right)^2 ds \to 0, \end{aligned} \tag{4.86}$$

as $\varepsilon \to 0$, if φ is Hölder continuous with exponent λ bigger than α, which appears in Condition (K5).

The following Itô's type formula is stated in Alòs et al. [10] (Theorem 3) using the operator $K^{*,\varepsilon}$ and the last convergence. So, we only sketch the proof of this result because the reader can consult [10] for a detailed proof.

Theorem 60 *Let F be a function in $C_b^2(\mathbb{R})$, and $B = \{B_t : t \in [0,T]\}$ a Gaussian Volterra process such that Hypothesis (K5) for some $\alpha < 1/4$ holds and the image of K^* is the space $L^2([0,t])$. Also let $u = \{u_t : t \in [0,T]\} \in \mathbb{D}_W^{2,2}(L^2([0,T]))$ be an adapted process to the filtration generated by the Brownian motion W satisfying the following two conditions:*

a) The processes u and $D_r^W u$ are λ-Hölder continuous in $\mathbb{D}_W^{1,4}(L^2([0,T]))$ with $\lambda > \alpha$, for each $r \in [0,T]$, and the function

$$\gamma(r) = \sup_{0\leq s\leq T} \|D_r^W u_s\|_{1,4,\mathbb{R}} + \sup_{0\leq s\leq t\leq T} \frac{\|D_r^W u_t - D_r^W u_s\|_{1,4,\mathbb{R}}}{|t-s|^\lambda}, \quad r \in [0,T],$$

belongs to $L^p([0,T])$ for some $p > 2/(1-4\alpha)$.

b) The process u satisfies

$$\sup_{\varepsilon>0} E\left(\int_0^T \left|\partial_s \int_0^s (K_s^{*,\varepsilon} u)(r)dr\right|^2 ds\right) < \infty.$$

Set $X_t = \delta_2\left(u 1_{[0,t]}\right)$, $t \in [0,T]$. Then, for all $t \in [0,T)$, we have that $F'(X.)u.1_{[0,t]}(\cdot)$ is in $Dom\,\delta_2$ and that the following Itô's type formula for the divergence operator holds:

$$\begin{aligned} F(X_t) &= F(0) + \int_0^t F'(X_s)u_s \delta_2 B_s \\ &+ \int_0^t F''(X_s)u_s \left(\int_0^s (\partial_s K(s,r)) \left(\int_0^s D_r^W ((K_s^* u)(\theta))\delta_2 W_\theta\right) dr\right) ds \\ &+ \frac{1}{2}\int_0^t F''(X_s)\partial_s \left(\int_0^s ((K_s^* u)(r))^2\, dr\right) ds. \end{aligned}$$

Remarks 10 *i) The fact that u is an adapted process is utilized to apply the usual Itô's formula for the Brownian motion (i.e., Theorem 40).*

ii) The process X is well-defined because of Propositions 62 and 87, and the transfer principle. Indeed, we have that $K^*\left(\mathbb{D}^{1,2}(\mathcal{H})\right) = \mathbb{D}_W^{1,2}(L^2([0,T]))$ since the image of K^* is the whole space $L^2([0,T])$ by hypothesis.

iii) See (4.73) for the definition of the operator K_t^*.

iv) In Alòs et al. [10] (Proposition 2), It is provided a sufficient condition on u that guarantee Statement b) is satisfied.

v) We have already pointed out that the fractional Brownian motion and the Riemann-Liouville fractional process satisfies hypothesis (K5).

Proof. Let $\varepsilon > 0$ be small enough. From Lemma 36, we obtain that, for $s \in [0, T-\varepsilon]$,

$$\int_0^s u_s \partial_s K(s+\varepsilon, r)dW_r = u_s \int_0^s \partial_s K(s+\varepsilon,r)dW_r - \int_0^s (D_r^W u_s)\partial_s K(s+\varepsilon,r)dr.$$

Therefore, the process $s \mapsto \int_0^s u_s \partial_s K(s+\varepsilon, r)dW_r$ is measurable on $\Omega \times [0, T-\varepsilon]$ since, for any partition $\pi_n = \{0 = s_0^{(n)} < s_1^{(n)} < \ldots < s_n^{(n)} = T - \varepsilon\}$ of the interval $[0, T-\varepsilon]$ such that $|\pi_n| \to 0$, as $n \to \infty$, (K5) and the dominated convergence theorem imply

$$\int_0^T \left[\sum_{i=1}^n \partial_s K(s+\varepsilon, s_i^{(n)}) 1_{(s_i^{(n)}\wedge s, s_{i+1}^{(n)}\wedge s]}(r) - \partial_s K(s+\varepsilon, r) 1_{(0,s]}(r)\right]^2 dr \to 0,$$

as $n \to \infty$, and

$$\int_0^s \sum_{i=1}^n \partial_s K(s+\varepsilon, s_i^{(n)}) 1_{(s_i^{(n)}\wedge s, s_{i+1}^{(n)}\wedge s]}(r)dW_r = \sum_{i=1}^n \partial_s K(s+\varepsilon, s_i^{(n)}) \left(W_{s_{i+1}^{(n)}\wedge s} - W_{s_i^{(n)}\wedge s}\right).$$

With this measurability property in mind, set the process $X_t^\varepsilon = \int_0^t \left(K_t^{*,\varepsilon}u\right)(s)\delta_2 W_s$, $t \in [0, T-\varepsilon]$.

Now, fix $t \in [0,T)$. Then, (4.85) and the Fubini's theorem for the divergence operator (i.e., Theorem 48) allow us to write

$$\begin{aligned} X_t^\varepsilon &= \int_0^t u_s K(s+\varepsilon, s)dW_s + \int_0^t \left(\int_s^t u_r \partial_r K(r+\varepsilon, s)dr\right)\delta_2 W_s \\ &= \int_0^t u_s K(s+\varepsilon, s)dW_s + \int_0^t \left(\int_0^s u_s \partial_s K(s+\varepsilon, r)\delta_2 W_r\right)ds. \end{aligned}$$

Thus, we obtain that X^ε is an Itô's process (see (3.20)). Consequently, we can apply the Itô's formula for the Itô's integral with respect to W (i.e., Theorem 40) to get

$$\begin{aligned} F(X_t^\varepsilon) &= F(0) + \int_0^t F'(X_s^\varepsilon)u_s K(s+\varepsilon, s)dW_s \\ &\quad + \int_0^t F'(X_s^\varepsilon)\left(\int_0^s u_s\partial_s K(s+\varepsilon, r)\delta_2 W_r\right)ds + \frac{1}{2}\int_0^t F''(X_s^\varepsilon)u_s^2 K(s+\varepsilon, s)^2 ds. \end{aligned}$$

Hence, we can use property (4.3) of the divergence operator again to obtain

$$\begin{aligned} F(X_t^\varepsilon) &= F(0) + \int_0^t F'(X_s^\varepsilon)u_s K(s+\varepsilon, s)dW_s \\ &\quad + \int_0^t \left(\int_0^s F'(X_s^\varepsilon)u_s\partial_s K(s+\varepsilon, r)\delta_2 W_r\right)ds \\ &\quad + \int_0^t \left(\int_0^s \left[D_r^W F'(X_s^\varepsilon)\right] u_s\partial_s K(s+\varepsilon, r)dr\right)ds \\ &\quad + \frac{1}{2}\int_0^t F''(X_s^\varepsilon)u_s^2 K(s+\varepsilon, s)^2 ds. \end{aligned}$$

Therefore, Corollary 10 and Remark 55 yield

$$\begin{aligned} F(X_t^\varepsilon) &= F(0) + \int_0^t F'(X_s^\varepsilon)u_s K(s+\varepsilon, s)dW_s \\ &\quad + \int_0^t \left(\int_0^s F'(X_s^\varepsilon)u_s\partial_s K(s+\varepsilon, r)\delta_2 W_r\right)ds \\ &\quad + \int_0^t F''(X_s^\varepsilon)u_s \left(\int_0^s \left(K_s^{*,\varepsilon}u\right)(r)\partial_s K(s+\varepsilon, r)dr\right)ds \\ &\quad + \int_0^t F''(X_s^\varepsilon)u_s \left(\int_0^s \left[\int_0^s D_r^W \left(K_s^{*,\varepsilon}u\right)(\theta)\delta_2 W_\theta\right]\partial_s K(s+\varepsilon, r)dr\right)ds \\ &\quad + \frac{1}{2}\int_0^t F''(X_s^\varepsilon)u_s^2 K(s+\varepsilon, s)^2 ds. \end{aligned} \tag{4.87}$$

In consequence, we are able to apply Theorem 48 (i.e., Fubini's theorem) and (4.85) to get

$$F(X_t^\varepsilon) = I_{1,\varepsilon} + I_{2,\varepsilon} + I_{3,\varepsilon},$$

where

$$\begin{aligned} I_{1,\varepsilon} &= F(0) + \int_0^t \left(K_t^{*.\varepsilon}F'(X_\cdot^\varepsilon)u_\cdot\right)(s)\delta_2 W_s, \\ I_{2,\varepsilon} &= \int_0^t F''(X_s^\varepsilon)u_s \left(\int_0^s \left[\int_0^s D_r^W \left(K_s^{*,\varepsilon}u\right)(\theta)\delta_2 W_\theta\right]\partial_s K(s+\varepsilon, r)dr\right)ds \end{aligned}$$

and

$$\begin{aligned} I_{3,\varepsilon} &= \int_0^t F''(X_s^\varepsilon)u_s \left(\int_0^s \left(K_s^{*,\varepsilon}u\right)(r)\partial_s K(s+\varepsilon,r)dr \right) ds \\ &+\frac{1}{2}\int_0^t F''(X_s^\varepsilon)u_s^2 K(s+\varepsilon,s)^2 ds. \end{aligned}$$

By Proposition 62, transfer principle (see the observation after Proposition 92), (4.85) and (1.104), we have

$$\left[E\left((X_t - X_t^\varepsilon)^4\right)\right]^{1/4} \le c_4 \left[\int_0^t \left\|\left(K_t^* u\right)(s) - \left(K_t^{*,\varepsilon}u\right)(s)\right\|_{1,4,\mathbb{R}}^2 ds\right]^{1/2} \to 0, \quad \text{as } \varepsilon \downarrow 0,$$

because of the Hölder property assumed on u and (4.86). Also, we claim that, as $\varepsilon \downarrow 0$,

$$K_t^{*,\varepsilon}\left(F'(X_\cdot^\varepsilon)u_\cdot\right) \to K_t^*\left(F'(X_\cdot)u_\cdot\right) \quad \text{in } L^2(\Omega \times [0,t]),$$

$$I_{2,\varepsilon} \to \int_0^t F''(X_s)u_s \left(\int_0^s \left[\int_0^s D_r^W \left(K_s^* u\right)(\theta)\delta_2 W_\theta \right] \partial_s K(s,r)dr \right) ds$$

in $L^2(\Omega)$, and $I_{3,\varepsilon} \to \frac{1}{2}\int_0^t F''(X_s)\partial_s \left(\int_0^s \left(K_s^* u\right)^2(r)\right) dr$ in $L^1(\Omega)$.

Finally, for a smooth random variable G in $\mathcal{S}$ (see (1.101)), the duality relation (4.2), together with (4.87), and the last convergences imply

$$\begin{aligned} &E\left(\int_0^t \left[D_s^W G\right] K_t^*\left(F'(X_s)u_s\right) ds\right) \\ &= E\Bigg(G\Bigg\{F(X_t) - F(0) \\ &\quad -\int_0^t F''(X_s)u_s \left(\int_0^s (\partial_s K(s,r)) \left(\int_0^s D_r((K_s^* u)(\theta))\delta_2 W_\theta\right) dr\right) ds \\ &\quad -\frac{1}{2}\int_0^t F''(X_s)\partial_s\left(\int_0^s \left[\left(K_s^* u\right)(r)\right]^2 dr\right) ds\Bigg\}\Bigg). \end{aligned}$$

So, using the duality relation (4.2) again and the fact that the divergence operator δ_2^W is the adjoint of the derivative operator D^W, we prove that the result holds. □

As we have pointed out in Theorem 60, we have studied an Itô's type formula for Gaussian Volterra processes, which includes the fractional Brownian motion and the Riemann-Liouville fractional process with Hurst parameter H in $(1/4, 1/2)$. So, now, we are interested in the case that $H > 1/2$. To do so, note that (1.66) implies that, for $\alpha = H - \frac{1}{2}$, the Kernel of the fractional Brownian motion with $H > 1/2$ satisfies

$$\begin{aligned} 0 \le \partial_t K_H(t,s) &= c_H s^{\frac{1}{2}-H}(t-s)^{H-\frac{3}{2}} t^{H-\frac{1}{2}} \le C_{T,H} s^{\frac{1}{2}-H}(t-s)^{H-\frac{3}{2}} \\ &= C_{T,H} s^{-\alpha}(t-s)^{\alpha-1}, \quad 0 \le s \le t \le T, \end{aligned}$$

and the kernel of the Riemann-Liouvile fractional process K meets

$$\begin{aligned} \partial_t K(t,s) &= (H - \frac{1}{2})(t-s)^{H-\frac{3}{2}} \le (H - \frac{1}{2})T^{H-\frac{1}{2}}(t-s)^{H-\frac{3}{2}} s^{\frac{1}{2}-H} \\ &= C_{T,H} s^{-\alpha}(t-s)^{\alpha-1}, \quad 0 \le s \le t \le T. \end{aligned}$$

Moreover, proceeding as in (4.81), it is easy to see

$$\int_s^t K_H(t,r)^2 dr \le C(t-s)^{1+2\alpha}, \quad 0 \le s \le t \le T, \text{ and } H > 1/2,$$

and $K_H(s+,s) = 0$, which are also true for the kernel of Riemann-Liouville fractional process.

The last three inequalities motivate to introduce the following condition:

(K6) Assume that the kernel $K = \{K(t,s) : 0 \le s < t \le T\}$ satisfies the following conditions:

i) The kernel K is differentiable in the variable t, and both K and $\partial_t K$ are continuous in the set $\{(t,s) : 0 \le s < t \le T\}$.

ii) $K(s+,s) = 0$, for all $s \in [0,T]$.

iii) There is a constant $C > 0$ such that, for some $\alpha \in (0,1/2)$,

$$|\partial_t K(t,s)| \le C s^{-\alpha}(t-s)^{\alpha-1} \quad \text{and} \quad \int_s^t K(t,r)^2 dr \le C(t-s)^{1+2\alpha},$$

for $0 \le s < t \le T$.

Condition (K6) implies that Assumption (K4) holds, as the following result states.

Lemma 57 *Assume that the kernel $K = \{K(t,s) : 0 \le s < t \le T\}$ of a Gaussian Volterra process B satisfies Hypothesis (K6). Then, it also satisfies Condition (K4).*

Proof. Let $s \in [0,T)$. Then Hypotheses (K6).*i)* and (K6).*iii)* imply that $K(\cdot,s)$ has bounded variation on $(s,T]$. Furthermore, (K6)*ii)* gives

$$\begin{aligned}
&\int_0^T |K|((s,T],s)^2 ds \\
&\le \int_0^T \left(\int_s^T |\partial_u K(u,s)|\, du\right)^2 ds \le C \int_0^T \left(\int_s^T (u-s)^{\alpha-1} s^{-\alpha} du\right)^2 ds \\
&\le C \int_0^T s^{-2\alpha}(T-s)^{2\alpha} ds < \infty.
\end{aligned}$$

Therefore, the proof is complete. □

As a consequence of the last result, together with (4.79) and (4.80), we have that if Condition (K6) is satisfied, then

$$(K_t^* \varphi)(s) = \int_s^t \varphi(r)\partial_r K(r,s)dr, \quad 0 \le s < t \le T, \text{ and } \varphi \in \mathcal{E}([0,T]).$$

Proceeding as in Theorem 60, Alòs et al. [10] (Theorem 4) have also obtained the following Itô's type formula under Assumption (K6), where they now apply the operator

$$(K_t^{*,\varepsilon} \varphi)(s) = \int_s^t \varphi(r)\partial_r K(r+\varepsilon,s)dr, \quad \varepsilon > 0.$$

Theorem 61 *Let F be a function in $C_b^2(\mathbb{R})$ and $B = \{B_t : t \in [0,T]\}$ a Gaussian Volterra process whose kernel K satisfies Condition (K6) for some $\alpha < 1/2$. Also, assume that the image of K^* is the space $L^2([0,T])$ and that $u \in \mathbb{D}_W^{2,4}(L^2([0,T]))$ is an adapted process to the filtration generated by the Brownian motion W such that it is bounded with respect* *to the norm* $\|\cdot\|_{2,4,\mathbb{R}}$. *Set* $X_t = \int_0^t u_s \delta_2 B_s$, $t \in [0,T]$. *Then, the process* $F'(X)$ *belongs to* $\mathbb{D}^{1,2}(\mathcal{H}_K)$ *and, for each* $t \in [0,T]$, *we have that the Itô's type formula*

$$\begin{aligned} F(X_t) &= F(0) + \int_0^t F'(X_s) u_s \delta_2 B_s + \frac{1}{2} \int_0^t F''(X_s) \partial_s \left(\int_0^s \left[(K_s^* u)(r) \right]^2 dr \right) ds \\ &\quad + \int_0^t F''(X_s) u_s \left(\int_0^s \partial_s K(s,r) \left(\int_0^s D_r^W (K_s^* u)(\theta) \delta_2 W_\theta \right) dr \right) ds \end{aligned}$$

holds.

Remarks 11 *i) It is not difficult to use Theorem 1 to see that B is a continuous process. Indeed, for* $0 \le s \le t \le T$,

$$E\left((B_t - B_s)^2 \right) = \int_s^t K(t,r)^2 dr + \int_0^s \left[\int_s^t \partial_u K(u,r) du \right]^2 dr \le C(t-s)^{1+2\alpha}.$$

ii) The stochastic integrals are defined as divergence operators. Also, we only sketch the proof of this result. The interested reader can consult Alòs et al. [10], as in Theorem 60.

Proof. Let $\varepsilon \in (0,T)$, $t \in [0, T-\varepsilon]$ and

$$X_t^\varepsilon = \int_0^t \left(\int_s^t u_r \partial_r K(r+\varepsilon, s) dr \right) \delta_2 W_s = \int_0^t \left(\int_0^s u_s \partial_s K(s+\varepsilon, r) \delta_2 W_r \right) ds,$$

where last equality follows as in the proof of Theorem 60. That is, it is a consequence of Theorem 48 (i.e., Fubini's theorem for the divergence operator). Thus, the process X^ε has bounded variation paths. In consequence, we apply Lemma 36 to get

$$\begin{aligned} F(X_t^\varepsilon) &= F(0) + \int_0^t F'(X_s^\varepsilon) \left(\int_0^s u_s \partial_s K(s+\varepsilon, r) \delta_2 W_r \right) ds \\ &= F(0) + \int_0^t \left(\int_0^s F'(X_s^\varepsilon) u_s \partial_s K(s+\varepsilon, r) \delta_2 W_r \right) ds \\ &\quad + \int_0^t \left(\int_0^s \left[D_r^W F'(X_s^\varepsilon) \right] u_s \partial_s K(s+\varepsilon, r) dr \right) ds. \end{aligned}$$

Hence, Fubini's theorem for the divergence operator (i.e., Theorem 48) and Corollary 24 establish

$$\begin{aligned} F(X_t^\varepsilon) &= F(0) + \int_0^t \left(\int_r^t F'(X_s^\varepsilon) u_s \partial_s K(s+\varepsilon, r) ds \right) \delta_2 W_r \\ &\quad + \int_0^t F''(X_s^\varepsilon) u_s \left(\int_0^s \partial_s (K(s+\varepsilon, r) \int_0^s D_r^W \left[(K_s^{*,\varepsilon} u)(\theta) \right] \delta_2 W_\theta dr \right) ds \\ &\quad + \int_0^t F''(X_s^\varepsilon) u_s \left(\int_0^s (K_s^{*,\varepsilon} u)(r) \partial_s K(s+\varepsilon, r) dr \right) ds \\ &= F(0) + I_{1,\varepsilon} + I_{2,\varepsilon} + I_{3,\varepsilon}, \end{aligned}$$

where, the last summand is also given by

$$I_{3,\varepsilon} = \frac{1}{2} \int_0^t F''(X_s^\varepsilon) \partial_s \left(\int_0^s \left[(K_s^{*,\varepsilon} u)(r) \right]^2 dr \right) ds.$$

Finally, we only need to study the convergence of $F(X_t^\varepsilon)$,$I_{1,\varepsilon}$, $I_{2,\varepsilon}$ and $I_{3,\varepsilon}$, as $\varepsilon \to 0$, to finish the proof. □

In the case that the Gaussian process B is fractional Brownian motion with Hurst parameter $H > 1/2$, Alòs and Nualart [13] have used the Taylor expansion up to the second order and the techniques of the Malliavin calculus based on the divergence and derivative operators, as well as the kernel of the fractional Brownian motion (1.66), to state the following Itô type formula.

Theorem 62 *Let $B = \{B_t : t \in [0,T]\}$ be fractional Brownian motion with Hurst parameter $H > 1/2$, f a function of class $C^2(\mathbb{R})$ and $u = \{u_t : t \in [0,T]\}$ a process in the space $\mathbb{D}^{2,2}_{loc}(|\mathcal{H}|)$ such that $X_\cdot = \int_0^\cdot u_s \delta_2 B_s$ has continuous paths and $\|u\|_{L^2(\Omega)}$ is in $\mathcal{H}$. Then,*

$$\begin{aligned} f(X_t) &= f(0) + \int_0^t f'(X_s) u_s \delta_2 B_s \\ &\quad + (2H-1)H \int_0^t f''(X_s) u_s \left[\int_0^T |s-r|^{2H-2} \left(\int_0^s D_r u_\theta \delta_2 B_\theta \right) dr \right. \\ &\quad \left. + \int_0^s u_r (s-r)^{2H-2} dr \right] ds, \quad t \in [0,T]. \end{aligned}$$

4.5 Riemann-Liouville fractional Brownian process

In this section, we assume that $B = \{B_t : t \in [0,T]\}$ is Riemann-Liouville fractional process (RLfp) as introduced in Definition 25. That is, there exists a Brownian motion $W = \{W_t : t \in [0,T]\}$ such that

$$B_t = \int_0^t (t-s)^{-\alpha} dW_s, \quad t \in [0,T],$$

where, $\alpha = \frac{1}{2} - H$, with $H \in (0,1/2)$.

Proposition 88 *The RLfp B is a self-similar Gaussian process with index H.*

Remark 71 *Remember that we must prove that the processes $\{c^{-H} B_{ct} : t \in [0,T]\}$ and $\{B_t : t \in [0,T]\}$ have the same finite-dimensional distributions, for any constant $c > 0$, in order to show that B is self-similar.*

Proof. By definition, the RLfp is a Gaussian process (see Definition 25). Now, choose $s, t \geq 0$. Then, using the change of variables formula $u = cr$, we have

$$\begin{aligned} & E\left(\int_0^t (t-r)^{-\alpha} dW_r \left[\int_0^s (s-r)^{-\alpha} dW_r \right] \right) \\ &= \int_0^{t\wedge s} (t-r)^{-\alpha} (s-r)^{-\alpha} dr = c^{2\alpha} \int_0^{t\wedge s} (ct-cr)^{-\alpha} (cs-cr)^{-\alpha} dr \\ &= c^{2\alpha-1} \int_0^{(ct)\wedge(cs)} (ct-u)^{-\alpha} (cs-u)^{-\alpha} du = E\left\{ \left(c^{-H} B_{ct} \right) \left(c^{-H} B_{cs} \right) \right\}. \end{aligned}$$

Hence the proof is finished because the finite-dimensional distributions of a Gaussian process are characterized by their mean and covariance, as it is pointed out in Section 1.1.1. □

RLfp does not have stationary increments but the proof of this fact is not easy. The reader interested in this result can consult Lim and Sithi [126]. Thus RLfp is not fractional Brownian motion because fBm is the only self-similar Gaussian process with stationary increments (see (1.48) for details). However, the RLfp still has Hölder continuous paths for any exponent less than H, as the fractional Brownian motion.

Proposition 89 *RLfp has Hölder continuous paths for any exponent less than H.*

Proof. Let $0 \leq s \leq t$. Then,

$$\begin{aligned}
E\left((B_t - B_s)^2\right) &= E\left(B_t^2 + B_s^2 - 2B_tB_s\right) \\
&= \frac{1}{1-2\alpha}\left(t^{1-2\alpha} + s^{1-2\alpha}\right) - 2\int_0^s (t-r)^{-\alpha}(s-r)^{-\alpha}dr \\
&\leq \frac{1}{1-2\alpha}\left(t^{1-2\alpha} + s^{1-2\alpha}\right) - 2\int_0^s (t-r)^{-2\alpha}dr \\
&= \frac{1}{1-2\alpha}\left(s^{1-2\alpha} - t^{1-2\alpha}\right) + 2\int_s^t (t-r)^{-2\alpha}dr \\
&= \frac{1}{1-2\alpha}\left(s^{1-2\alpha} - t^{1-2\alpha} + 2(t-s)^{1-2\alpha}\right) \\
&\leq \frac{2}{1-2\alpha}(t-s)^{1-2\alpha} = \frac{2}{1-2\alpha}(t-s)^{2H},
\end{aligned}$$

where last inequality holds due to $t^{1-2\alpha} - s^{1-2\alpha} \leq (t-s)^{1-2\alpha}$. Hence, the result is a consequence of Proposition 5. □

RLfp appears in the Mandelbrot-van Ness integral representation of fractional Brownian motion (see (1.45)). It means, the RLfp can be written as $B = B^H + S^H$, where $B^H = \{B_t^H : t \in [0,T]\}$ is a fractional Brownian motion and

$$S_t^H = -C_H \int_{-\infty}^0 [(t-s)^{H-1/2} - (-s)^{H-1/2}]d\tilde{W}_s \quad t \in [0,T]. \tag{4.88}$$

Here C_H is a constant, which is given in (1.45), and $\{\tilde{W}_{-t} : t \geq 0\}$ is a Brownian motion independent of $W = \{W_t : t \geq 0\}$. It means

$$\bar{W}_t = \begin{cases} W_t, & \text{if } t \geq 0, \\ \tilde{W}_t, & \text{if } t \leq 0 \end{cases}$$

is a two-sided Wiener process (see Definition 15). As a consequence, we have that the RLfp is a fractional Brownian motion plus a process of bounded variation as the following result states.

Proposition 90 *The process $S^H = \{S_t^H : t \in [0,T]\}$ has absolutely continuous paths.*

Proof. Fix $t \in [0,T]$. In the proof of Proposition 12, we see that S_t^H belongs to $L^2(\Omega)$. Therefore, we obtain that

$$\begin{aligned}
S_t^H &= L^2(\Omega) - \lim_{\varepsilon \downarrow 0} C_H \int_{-\infty}^{-\varepsilon} [(-s)^{-\alpha} - (t-s)^{-\alpha}]d\tilde{W}_s \\
&= L^2(\Omega) - \lim_{\varepsilon \downarrow 0} C_H(\alpha)^{-1} \int_{-\infty}^{-\varepsilon}\int_0^t (r-s)^{-\alpha-1} dr d\tilde{W}_s.
\end{aligned} \tag{4.89}$$

On the other hand, for $\varepsilon > 0$, the facts that $(r-s)^{-\alpha-1} \leq (r+\varepsilon)^{-\alpha-1}$, for $s < -\varepsilon < 0 < r$, and $\int_0^t (r+\varepsilon)^{-\alpha-1} dr < \infty$, and Fubini's theorem for Itô's integral (i.e., Theorem 43) imply that there exists a measurable process $Y^\varepsilon : \Omega \times [0,T] \to \mathbb{R}$ such that

$$(\alpha)^{-1} \int_{-\infty}^{-\varepsilon} (r-s)^{-\alpha-1} d\tilde{W}_s = Y_r^\varepsilon, \quad \text{w.p.1 for almost all } r \in [0,t],$$

and

$$(\alpha)^{-1} \int_{-\infty}^{-\varepsilon} \int_0^t (r-s)^{-\alpha-1} dr d\tilde{W}_s = \int_0^t Y_r^\varepsilon dr, \quad \text{w.p.1.} \tag{4.90}$$

Moreover, we claim that $Y^\varepsilon \in L^p(\Omega \times [0,T])$, for $p \in (1, (1-H)^{-1})$. Indeed, Hölder inequality and Fubini's theorem give

$$\begin{aligned} \int_0^T E(|Y_r^\varepsilon|^p)\, dr &\leq \int_0^T \left[E(|Y_r^\varepsilon|^2)\right]^{p/2} dr \\ &= (\alpha)^{-p} \int_0^T \left[E\left(\left|\int_{-\infty}^{-\varepsilon} (r-s)^{-\alpha-1} d\tilde{W}_s\right|^2\right)\right]^{p/2} dr \\ &= (\alpha)^{-p} \int_0^T \left(\int_{-\infty}^{-\varepsilon} (r-s)^{-2\alpha-2} ds\right)^{p/2} dr \\ &= \left(\frac{1}{\alpha\sqrt{2\alpha+1}}\right)^p \int_0^T (r+\varepsilon)^{p(-\alpha-\frac{1}{2})} dr < \infty. \end{aligned}$$

Proceeding similarly, together with the mean value theorem, we also have that, for $0 < \varepsilon_1 < \varepsilon_2$ and $\eta > 0$ small enough,

$$\begin{aligned} \int_0^T E(|Y_r^{\varepsilon_2} - Y_r^{\varepsilon_1}|^p)\, dr &\leq (\alpha)^{-p} \int_0^T \left(\int_{-\varepsilon_2}^{-\varepsilon_1} (r-s)^{-2\alpha-2} ds\right)^{p/2} dr \\ &= \left(\frac{1}{\alpha\sqrt{2\alpha+1}}\right)^p \int_0^T \left[(r+\varepsilon_1)^{-2\alpha-1} - (r+\varepsilon_2)^{-2\alpha-1}\right]^{p/2} dr \\ &\leq \frac{(2\alpha+1)^\eta}{\alpha^p(2\alpha+1)^{p/2}} (\varepsilon_2-\varepsilon_1)^\eta \int_0^T r^{-\eta(2\alpha+2)} \left(r^{-2\alpha-1}\right)^{\frac{p}{2}-\eta} dr \\ &= \frac{(2\alpha+1)^\eta}{\alpha^p(2\alpha+1)^{p/2}} (\varepsilon_2-\varepsilon_1)^\eta \int_0^T r^{-p(\alpha+\frac{1}{2})-\eta} dr = C_{\alpha,p,\eta}(\varepsilon_2-\varepsilon_1)^\eta. \end{aligned}$$

Hence, $\{Y^\varepsilon : \varepsilon \in (0,1)\}$ is a Cauchy sequence in $L^p(\Omega \times [0,T])$, as ε goes to zero. Thus, (4.89) and (4.90) allow us to show that there is a process Y in $L^p(\Omega \times [0,T])$ such that

$$S_t^H = \int_0^t Y_r dr, \quad \text{w.p.1}$$

Consequently, the proof is complete. □

Our next aim is to identify the reproducing kernel Hilbert space $\mathcal{H}$ generated by the covariance of the process B.

Theorem 63 *The reproducing kernel Hilbert space $\mathcal{H}$ of the RLfp B is $I_{T-}^\alpha(L^2([0,T]))$, where I_{T-}^α is defined in* (1.56).

Proof. Note that (1.37) and the fact that B is a Volterra Gaussian process with kernel $K(t,s) = (t-s)^{-\alpha} 1_{[0,t]}(s)$ imply that, for $s,t \in [0,T]$,

$$(K^* 1_{[0,t]})(s) = \frac{1_{[0,t]}(s)}{(T-s)^\alpha} - \alpha \int_s^T \frac{1_{[0,t]}(r) - 1_{[0,t]}(s)}{(r-s)^{1+\alpha}} dr = \Gamma(1-\alpha)(D_{T-}^\alpha 1_{[0,t]})(s),$$

where last equality follows from Theorem 99 in the appendix. Hence, for a step function $\varphi \in \mathcal{E}([0,T])$ of the form $\varphi = \sum_{i=1}^n a_i 1_{[s_i,s_{i+1}]}$, we have

$$\varphi = \frac{1}{\Gamma(1-\alpha)} I_{T-}^{\alpha}(K^*\varphi) = I_{T-}^{\alpha}(D_{T-}^{\alpha}\varphi).$$

Thus, $\mathcal{E}([0,T]) \subset I_{T-}^{\alpha}(L^2([0,T]))$ and

$$\langle \varphi, \psi \rangle_{\mathcal{H}} = \Gamma(1-\alpha)^2 \left\langle D_{T-}^{\alpha}\varphi, D_{T-}^{\alpha}\psi \right\rangle_{L^2([0,T])}, \quad \text{for } \varphi, \psi \in \mathcal{E}([0,T]). \tag{4.91}$$

Hence, using that $I_{T-}^{\alpha} : L^2([0,T]) \to L^2([0,T])$ is a continuous linear operator (see the proof of Proposition 25), we have that $\mathcal{H} \subset I_{T-}^{\alpha}\left(L^2([0,T])\right)$.

Conversely, we now see that $I_{T-}^{\alpha}\left(L^2([0,T])\right) \subset \mathcal{H}$. To do so, we only need to show that the image of $\mathcal{E}([0,T])$ under K^* is dense in $L^2([0,T])$. So, assume that we have a function $f \in L^2([0,T])$ such that

$$0 = \int_0^T f(r)(K^* 1_{[0,t]})(r)dr = \int_0^t f(r)(t-r)^{-\alpha}dr, \quad \text{for all } t \in [0,T],$$

where last equality follows from (1.38). Thus, $f \equiv 0$ because of (1.58) and (1.59). Therefore, the proof is complete. □

A consequence of Theorem 63 is the transfer principle for RLfp. This is done as follows. In the following result, we use the definition given in (1.42).

Proposition 91 *Let $\varphi \in I_{T-}^{\alpha}(L^2([0,T]))$ and B the RLfp. Then,*

$$B(\varphi) = \int_0^T (K^*\varphi)(s)dW_s = \Gamma(1-\alpha)\int_0^T (D_{T-}^{\alpha}\varphi)(s)dW_s.$$

Remark 72 *It is worth mentioning that this result is related to Theorem 5 in Alòs et al. [9]. But in this theorem, the involved integral is in the Stratonovich sense. This integral will be studied in Section 6.*

Proof. From Theorem 63, we know that $\mathcal{H} = I_{T-}^{\alpha}(L^2([0,T]))$. Thus, the result is a consequence of Proposition 11 and (1.42). □

Other consequences of Theorem 63 are $K^*(\mathcal{H}) = L^2([0,T])$,

$$\mathbb{D}^{1,p} = \mathbb{D}_W^{1,p}, \quad \text{for } p \geq 1, \tag{4.92}$$

and

$$D^W F = K^* DF, \quad F \in \mathbb{D}^{1,p}, \tag{4.93}$$

as in Proposition 24. Here, $\mathbb{D}^{1,p}$ and $\mathbb{D}_W^{1,p}$ stand for the domains of the derivative operators D and D^W with respect to the RLfp B and the Brownian motion W, respectively.

Now, we can deal with the transfer principle for RLfp. As in the fractional Brownian motion case, for $p \geq 1$, we represent the divergence operators (see (4.2)) with respect to the RLfp B and the Brownian motion W by δ_p and δ_p^W, respectively.

Property (1.43) allows us to write

$$E\left(\|u\|_{\mathcal{H}}^p\right) = E\left(\|K^*(u)\|_{L^2([0,T])}^p\right) = \Gamma(1-\alpha)^p E\left(\left(\int_0^T \left[(D_{T-}^{\alpha}u)(s)\right]^2 ds\right)^{p/2}\right).$$

It means, u belongs to $L^p(\Omega;\mathcal{H})$ if and only if $K^*(u)$ is in $L^p(\Omega;L^2([0,T]))$. Therefore, (4.92) and (4.93) yield that, for $u\in L^p(\Omega;\mathcal{H})$,

$$E\left(\langle u,DF\rangle_{\mathcal{H}}\right)=E\left(\langle K^*(u),K^*(DF)\rangle_{L^2([0,T])}\right)=E\left(\langle K^*(u),D^WF\rangle_{L^2([0,T])}\right),\qquad(4.94)$$

for all $\mathbb{R}$-valued smooth functionals F of the RLfp B.

Now it is easy to state the relation between δ_p and δ_p^W.

Proposition 92 *Let B be a RLfp, $H\in(0,1)$, $p\geq 1$ and $u\in L^p(\Omega;\mathcal{H})$. Then, u belongs to $Dom\,\delta_p(\mathbb{R})$ if and only if $K^*(u)$ is in $Dom\,\delta_p^W(\mathbb{R})$. In this case, we have*

$$\delta_p(u)=\delta_p^W\left(K^*(u)\right).$$

Proof. The result is an immediate consequence of (4.2) and (4.94). □

As in the fractional Brownian case, (4.92), (4.93) and Proposition 92 are known as the transfer principle for the RLfp. Actually, this transfer principle is true for a Volterra Gaussian process with kernel $\tilde{K}$ such that the image of its reproducing kernel Hilbert space under the linear operator $\tilde{K}^*$ is the whole space $L^2([0,T])$.

4.5.1 The extended divergence operator with respect to Riemann-Liouville fractional Brownian process

The main purpose of this section is to show that it is possible to deal with an extension of the divergence operator δ_2 with respect to RLfp. We consider this possibility since the RLfp is an isonormal Gaussian process $\{B(\varphi):\varphi\in\mathcal{H}\}$ due to (1.41) and (1.42) (see also Proposition 91).

Theorem 7 establishes that $\mathcal{H}=I_{T-}^{\alpha}(L^2([0,T]))\subset L^2([0,T])$ and, by definition, $\mathcal{H}$ contains the space $\mathcal{E}([0,T])$ of all the step functions on $[0,T]$. In other words, Theorem 7 also states that the reproducing kernel Hilbert space $\mathcal{H}$ is densely and continuously embedded in $L^2([0,T])$. Now, by Section 4.2, we need to find a linear operator $\mathcal{T}_{RL}$ on $L^2([0,T])$ satisfying Hypotheses (H1)-(H3). Note that Proposition 63 and (4.91) suggest that this candidate is the linear operator $\mathcal{T}_{RL}:\mathcal{H}\subset L^2([0,T])\to L^2([0,T])$ given by

$$\mathcal{T}_{RL}(\varphi)=\Gamma(1-\alpha)D_{T-}^{\alpha}\varphi,\quad \varphi\in\mathcal{H}.$$

The following step is to see that this operator satisfies Hypotheses (H1)-(H3) with $\mathcal{H}_0=L^2([0,T])$.

Proposition 93 *The linear operator $\mathcal{T}_{RL}$ satisfies Conditions (H1), (H2) and (H3) introduced in Section 4.2.*

Proof. Condition (H1) is nothing else than Theorem 63 and equality (4.91).

Concerning Hypotheses (H2) and (H3), we must figure out a subset of the domain of the adjoin $\mathcal{T}_{RL}^*$ of the linear operator $\mathcal{T}_{RL}$. This is possible due to Proposition 18. In effect, this proposition implies that $I_{0+}^{\alpha}(L^2([0,T]))\subset \mathrm{Dom}\,\mathcal{T}_{RL}^*$ and

$$\mathcal{T}_{RL}^*(\psi)=\Gamma(1-\alpha)D_{0+}^{\alpha}\psi,\quad \psi\in I_{0+}^{\alpha}(L^2([0,T])).$$

Furthermore, by Theorem 7, we also have that $I_{0+}^{\alpha}(L^2([0,T]))\subset L^2([0,T])$. Hence, to show that Assumption (H2) holds, we only need to prove that $I_{0+}^{\alpha}(L^2([0,T]))$ is a dense set of $L^2([0,T])$ because, in this case, $I_{T-}^{\alpha}\left(I_{0+}^{\alpha}(L^2([0,T])\right)$ would be a dense set of $\mathcal{H}=I_{T-}^{\alpha}(L^2([0,T]))$. So, choose $f\in L^2([0,T])$ such that

$$\int_0^T f(s)\left(I_{0+}^{\alpha}g\right)(s)dds=0,\quad\text{for all } g\in L^2([0,T]).$$

Therefore, Corollary 6 gives

$$\int_0^T \left(I_{T-}^{\alpha} f\right)(s)g(s)ds = 0, \quad \text{for all } g \in L^2([0,T]).$$

In this way, we obtain that $f \equiv 0$ by applying (1.58) and (1.59). Resuming, we have proven that (H2) is true.

Finally, the set $I_{T-}^{\alpha}\left(I_{0+}^{\alpha}(L^2([0,T]))\right)$ is included in the domain of the operator $\mathcal{T}_{RL}^{*}\mathcal{T}_{RL}$ and the image of this set is $L^2([0,T])$. It means (H3) is also satisfied. □

Together with Definition 40, the importance of Proposition 93 is that we can consider the extension δ of the usual divergence operator δ_2 defined by means of (4.2). Also, from Remark 8.iv), we know that Dom $\delta_2 \subset$ Dom δ^*.

Now, as in the fractional Brownian motion case, we show that Dom $\delta_2 \subsetneq$ Dom δ^*.

Proposition 94 *Let B be RLfp. Then B is in Dom δ^*.*

Proof. The proof of this result follows the same lines as that of Proposition 84 because of the characterization of the extended divergence δ, which is stated in Theorem 55. □.

We now consider the problem of seeing whether the trajectories of B belong to the reproducing kernel Hilbert space $\mathcal{H}$.

Proposition 95 *Let $H \in (0,1/4)$. Then, the paths of RLfp $B = \{B_t : t \in [0,T]\}$ do not belong to $\mathcal{H}$ w.p.1.*

Remark 73 *Remember that $\mathcal{H} = I_{T-}^{\alpha}(L^2([0,T]))$ in this case. Also remember that B does not have stationary increments. So, apparently, we cannot use the ergodic theorem to show this result holds, as we do in the proof of Proposition 85.*

Proof. From the proof of Proposition 90, we have already known that $B = S^H + B^H$, where B^H is a fractional Brownian motion with Hurst parameter H and $S_t^H = \int_0^t Y_r dr$, for $t \in [0,T]$ and some process $Y \in L^p(\Omega \times [0,T])$, with $p \in (1,(1-H)^{-1})$. Therefore, Hölder inequality and Fubini's theorem lead us to write

$$\begin{aligned}
& t^{-pH} \int_0^{T-t} \left|S_{t+s}^H - S_s^H\right|^p ds \\
&= t^{-pH} \int_0^{T-t} \left|\int_s^{t+s} Y_u du\right|^p ds \leq t^{-pH+p-1} \int_0^{T-t} \int_s^{t+s} |Y_u|^p duds \\
&= t^{-pH+p-1} \int_0^T \int_{0\vee(u-t)}^{u\wedge(T-t)} |Y_u|^p dsdu \\
&\leq t^{p(1-H)} \int_0^T |Y_u|^p du \to 0, \quad \text{as } t \downarrow 0, \text{ w.p.1.}
\end{aligned}$$

In consequence, (4.69) implies that there is a sequence $\{t_{n_k} : k \in \mathbb{N}\}$ converging to zero, as $k \to \infty$, such that

$$\begin{aligned}
& t_{n_k}^{-pH} \int_0^{T-t_{n_k}} \left|B_{t_{n_k}+s} - B_s\right|^p ds \\
&= t_{n_k}^{-pH} \int_0^{T-t_{n_k}} \left|B_{t_{n_k}+s}^H - B_s^H + S_{t_{n_k}+s}^H - S_s^H\right|^p ds \to T \quad \text{w.p.1.}
\end{aligned}$$

Finally, as in the proof of Proposition 85, if there is $w_0 \in \Omega$ such that $B.(\omega_0) \in \mathcal{H} = I_T^{\alpha}(L^2([0,T]))$, then Lemmas 7 and 50 yield

$$\int_0^T \left|B_{t+s}^H(\omega_0) - B_s^H(\omega_0)\right|^p ds = o(t^{p\alpha})$$

since $L^2(\Omega) \subset L^p(\Omega)$. Thus, the proof is finished due to $H < 1/4$ (see the last part of the proof of Proposition 85). $\square$

We finish this section by observing that Propositions 93-95 lead us to conclude that δ is a real extension of δ_2 (i.e., Dom $\delta_2 \subsetneq$ Dom δ^*) because Dom δ_2 contains only $\mathcal{H}$-valued random variables as we can see in (4.2).

4.6 Another extension of the divergence operator using only the covariance function of the underlying Gaussian process

In Section 4.2, we study an extension of the divergence operator mainly because the paths of fractional Brownian motion B^H are not in the space $\mathcal{H}$ in the case that the Hurst parameter H is less than $1/4$, which has been stated in Proposition 85. In other words, in this case, the divergence of B^H with respect to itself is not defined. As we see in Section 4.3.1, Definition 40 provides a solution to this problem. It means, we obtain an extension of the divergence operator, whose domain contains the process B^H in the case that the integrator is also the fractional Brownian motion B^H with H less than $1/2$. However, Hypotheses (H1)-(H3) require knowledge of the space $\mathcal{H}$ in order to find a linear operator $\mathcal{T}$ satisfying these hypotheses. Thus, it is necessary to analyze another extensions of the divergence operator that do not involve the space $\mathcal{H}$ in their construction.

Currently, there has been interest in getting some extensions of the divergence operator using the Malliavin calculus. Between them, we can mention those established by Kruk and Russo [104], and Mocioalca and Viens [151]. Paper [104] includes the fractional Brownian motion and [151] considers Gaussian processes that can be more irregular than B^H even allows discontinuous processes.

In this section, we introduce the extension of the divergence operator given by Lei and Nualart [106]. The definition of this integral in the divergence sense only imposes suitable conditions on the covariance of the involved Gaussian process. Namely, the covariance is absolutely continuous in one variable such that its derivative satisfies a suitable integrability condition. This approach allows us to deal with the fractional Brownian motion and the bifractional Brownian motion introduced in Houdré and Villa [79]. This integral is defined as follows.

Now, in this chapter, the process $B = \{B_t : t \in [0, T]\}$ stands for a continuous centered Gaussian process with covariance

$$R(s,t) = E\left(B_s B_t\right), \quad s,t \in [0,T].$$

This covariance function satisfies the following condition:

(H4) Assume that, for all $t \in [0,T]$, the function $R(\cdot, t) : [0,T] \to \mathbb{R}$ is absolutely continuous and that there is a constant $\gamma > 1$ such that

$$\sup_{0 \le t \le T} \int_0^T |\partial_s R(s,t)|^\gamma \, ds < \infty.$$

The conjugate of γ is denoted by β (i.e., $\gamma^{-1} + \beta^{-1} = 1$).

On the other hand, in order to introduce the extended integral, remember that the reproducing kernel Hilbert space $\mathcal{H}$ is the completion of the step functions in $\mathcal{E}([0,T])$ with respect to the inner product

$$\left\langle 1_{[0,t]}, 1_{[0,s]} \right\rangle_{\mathcal{H}} = R(s,t), \quad s,t \in [0,T].$$

Therefore, for $\varphi \in \mathcal{E}([0,T])$ of the form $\varphi = \sum_{i=1}^n c_i 1_{[0,t_i]}$ and $t \in [0,T]$, Hypothesis (H4) leads us to write

$$\langle \varphi, 1_{[0,t]} \rangle_{\mathcal{H}} = \sum_{i=1}^n c_i R(t_i, t) = \int_0^T \varphi(s) \partial_s R(s,t) ds. \tag{4.95}$$

Hence, using (H4) again and the Hölder inequality, we obtain

$$\left| \langle \varphi, 1_{[0,t]} \rangle_{\mathcal{H}} \right| \leq \|\varphi\|_{L^\beta([0,T])} \sup_{0 \leq t \leq T} \int_0^T |\partial_s R(s,t)|^\gamma \, ds.$$

In consequence, (4.95) and this inequality imply that $\langle \cdot, 1_{[0,t]} \rangle_{\mathcal{H}}$ is an $\mathbb{R}$-valued linear and continuous operator on $\mathcal{E}([0,T])$. In this way, using that $\mathcal{E}([0,T])$ is dense in $L^\beta([0,T])$, we can extend the function $\langle \cdot, 1_{[0,t]} \rangle_{\mathcal{H}}$ to the space $L^\beta([0,T])$ as follows:

Definition 41 *For $\varphi \in L^\beta([0,T])$ and $\psi \in \mathcal{E}([0,T])$ of the form $\psi = \sum_{i=1}^n c_i 1[0, t_i]$, we use the convention*

$$\langle \varphi, \psi \rangle_{\mathcal{H}} = \sum_{i=1}^n c_i \int_0^T \varphi(s) \partial_s R(s, t_i) ds.$$

Remark 74 *Notice that the left-hand side of the above equality agrees with the one in (4.95) if φ is in $\mathcal{E}([0,T])$. Moreover, these left-hand sides are not the same in general because not all the elements of $L^\beta([0,T])$ belong to the reproducing kernel Hilbert space $\mathcal{H}$ as it happens with the fractional Brownian motion with parameter $H \in (0, 1/4)$. So, we are making an abuse of the notation.*

In order to introduce the extension of the divergence operator with respect to B, we consider the space $\mathcal{S}_\mathcal{E}$ of all the smooth functionals F in $\mathcal{S}$ (see (1.100)) of the form

$$F = f\left(B(\psi_1), \ldots, B(\psi_n)\right)$$

with $\psi_i \in \mathcal{E}([0,T])$, $i = 1, \ldots, n$, and $f \in C_b^\infty(\mathbb{R}^n)$.

Imitating the definition of the usual divergence operator δ_2 given by (4.2), we introduce the extended divergence operator considered by Lei and Nualart [106] in the following:

Definition 42 *Suppose that Hypothesis (H4) is satisfied. We say that a process $u \in L^1\left(\Omega; L^\beta([0,T])\right)$ belongs to the domain of the extended divergence, denoted by $Dom^E \delta$, if there exists a square-integrable random variable $\hat{\delta}(u)$ such that*

$$E\left(F \hat{\delta}(u)\right) = E\left(\langle u, DF \rangle_{\mathcal{H}}\right), \quad \textit{for all } F \in \mathcal{S}_\mathcal{E}. \tag{4.96}$$

Remarks 12 *i) There exists at most a square-integrable random variable $\hat{\delta}(u)$ satisfying (4.96) since the space $\mathcal{E}([0,T])$ is dense in $\mathcal{H}$ by definition and $\mathcal{S}$ is dense in $L^2(\Omega)$ due to Section 1.5.4.*

ii) The duality relation (4.96) yields that there is a constant $C_u \geq 0$ such that

$$\left| E\left(\langle u, DF \rangle_{\mathcal{H}}\right) \right| \leq C_u \|F\|_{L^2(\Omega)}.$$

Actually, $C_u = \|\hat{\delta}(u)\|_{L^2(\Omega)}$.

iii) Remember that the right-hand side of (4.96) is given through Definition 41.

iv) In general, we cannot guarantee that there exists a relation between $Dom\,\delta_2$ and $Dom^E\,\delta$ because the elements of the first set take values in $\mathcal{H}$ and the ones of the second set are $L^\beta([0,T])$-valued, with probability 1.

Unlike last Statement *iii)*, for the fractional Brownian motion, we have the following result, which establishes that $\hat{\delta}$ is an extension of the usual divergence operator δ_2.

Theorem 64 *Let B be fractional Brownian motion with parameter $H \in (0, 1/2)$ and $u \in Dom\,\delta_2$. Then, u also belongs to $Dom^E\,\delta$ and $\hat{\delta}(u) = \delta_2(u)$.*

Remark 75 *Note that fractional Brownian motion with Hurst parameter $H \in (0, 1/2)$ satisfies Condition (H4) if $\gamma < 1/(1-2H)$. Therefore, $\beta > 1/(2H)$.*

Proof. The proof is divided into two steps.
Step 1. Here we see that if f is in the Hilbert space $\mathcal{H}$ and $t \in [0,T]$, then the inner product $\langle f, 1_{[0,t]}\rangle_{\mathcal{H}}$ defined via the covariance function R agrees with the function introduced in Definition 41.

From the proof of Lemma 50, we have that $f = I^\alpha_{T-}(\varphi)$, where $\alpha = \frac{1}{2} - H$ and

$$\varphi(s) = \phi_f(s) - \frac{\alpha}{\Gamma(1-\alpha)} \int_s^T \frac{f(u)}{(u-s)^{\alpha+1}} \left(1 - \left(\frac{u}{s}\right)^{-\alpha}\right) du, \quad s \in [0,T].$$

Here, ϕ_f is defined in (1.83). Now, choose $\eta > 0$ such that $\eta + \frac{1}{2H} < 1/H$. Thus, using the proof of Lemma 50 again, together with Theorem 7, we have that there exist a constant $C > 0$ and $\varepsilon > 0$ such that

$$\|f\|_{L^{\frac{1}{2H}+\eta}([0,T])} \le C\|\varphi\|_{L^{2-\varepsilon}([0,T])} \le C\left(\|f\|_{L^{2-\varepsilon}([0,T])} + \|\phi_f\|_{L^{2-\varepsilon}([0,T])}\right).$$

Hence, the proof of Proposition 25 implies

$$\|f\|_{L^{\frac{1}{2H}+\eta}([0,T])} \le C\left(\|f\|_{L^2([0,T])} + \|\phi_f\|_{L^2([0,T])}\right) \le C\|\phi_f\|_{L^2([0,T])} = C\|f\|_{\mathcal{H}}. \quad (4.97)$$

Now, we are ready to show that the claim of this step is true.

By the definition of $\mathcal{H}$, we can find a sequence $\{f_n \in \mathcal{E}([0,T]) : n \in \mathbb{N}\}$ of the form $f_n = \sum_{i=1}^{N_n} c_{i,n} 1_{[0,t_i^n]}$ such that $\|f - f_n\|_{\mathcal{H}} \to 0$, as $\varepsilon \to 0$, and

$$\langle f_n, 1_{[0,t]}\rangle_{\mathcal{H}} = \int_0^T f_n(s)\partial_s R(s,t)ds.$$

Since Hypothesis (H4) holds with $\gamma < \frac{1}{1-2H}$, we get

$$\langle f, 1_{[0,t]}\rangle_{\mathcal{H}} = \int_0^T f(s)\partial_s R(s,t)ds.$$

due to (4.97). In this way, this step is satisfied.
Step 2. By Hypothesis, we have

$$E\left(\langle u, DF\rangle_{\mathcal{H}}\right) = E\left(F\delta_2(u)\right), \quad \text{for } F \in \mathcal{S}_{\mathcal{E}}.$$

In consequence, Step 1 allows us to write

$$E\left(\int_0^T u_s \partial_s \langle 1_{[0,s]}, DF\rangle_{\mathcal{H}} ds\right) = E\left(F\delta_2(u)\right), \quad \text{for } F \in \mathcal{S}_{\mathcal{E}},$$

which implies that the result is true. □

Henceforth, for $t \in [0,T]$, we use the convention

$$\int_0^t u_s \hat{\delta} B_s = \hat{\delta}(u 1_{[0,t]}).$$

whenever $u1_{[0,t]} \in \mathrm{Dom}^{\mathrm{E}}\,\delta$.

In order to establish an Itô's type formula for $\hat{\delta}$, we also take into account the following condition.

(H5) The function $t \mapsto R(t,t)$ has bounded variation on $[0,T]$.

Proceeding as in the proof of Theorem 58, we are able to state the following result, which can be compared with (4.72). So, we omit its proof. For details, the reader can consult Lei and Nualart [106] (Theorem 3.2).

Theorem 65 *Let $f : [0,T] \times \mathbb{R} \to \mathbb{R}$ be a function in $C^{1,2}([0,T] \times \mathbb{R})$ such that f, $\partial_t f$, $\partial_x f$ and $\partial_x^2 f$ satisfy* (1.142). *Also let B be a continuous Gaussian process satisfying $B_0 = 0$, and Hypotheses (H4) and (H5). Then, for all $t \in [0,T]$, $\partial_x f(\cdot, B_\cdot)1_{[0,t]}(\cdot) \in Dom^{E}\,\delta$ and*

$$f(t, B_t) = f(0,0) + \int_0^t \partial_s f(s, B_s) ds + \int_0^t \partial_x f(s, B_s) \hat{\delta} B_s + \frac{1}{2} \int_0^t \partial_x^2 f(s, B_s) dR_s.$$

Note that, in particular, for $f(x) = x^2$ and B the fractional Brownian motion with $H \in (0, 1/2)$, we obtain

$$B_t^2 = 2 \int_0^t B_s \hat{\delta} B_s + t^{2H}.$$

Thus, Lemma 12 implies that $\int_0^t \partial_x f(s, B_s) \hat{\delta} B_s = \frac{1}{2} I_2 \left(1_{[0,t]} \otimes 1_{[0,t]}\right)$. Therefore, this stochastic integral is equal to that introduced in Definition 40 because of Theorem 55 and the proof of Proposition 84. It means, $\int_0^t B_s \hat{\delta} B_s = \delta(B_s 1_{[0,t]})$, $t \in [0,T]$.

On the other hand, in order to see that we can apply Theorem 65 to Riemann-Liouville fractional process with $H \in (0, 1/4)$, we remember the following facts.

We say that B is a self-similar process with index H if, for any constant $c > 0$, the processes $\{c^{-H} B_{ct} : t \geq 0\}$ and $\{B_t : t \geq 0\}$ have the same finite-dimensional distributions. Without loss of generality, we are able to assume $E(B_1^2) = 1$. Then, for $s, t \in [0,T]$, we have

$$E\left(B_t^2\right) = t^{2H} E(B_1^2) = t^{2H} \quad \text{and} \quad R(cs, ct) = E\left(B_{cs} B_{ct}\right) = c^{2H} R(s,t).$$

In the following result, we utilize the notation $\varphi(s) := R(1, s)$, $s \in [0,T]$.

Lemma 58 *Let $B = \{B_t : t \in [0,T]\}$ be a zero mean continuous self-similar Gaussian process with index $H \in (0, 1/2)$, φ absolutely continuous on $[0,1]$ and $\gamma \in (1, \frac{1}{1-2H})$ such that $\int_0^1 |\varphi'(u)|^\gamma du < \infty$. Then, Condition (H4) is satisfied.*

Proof. Let $t \in (0,T]$. We observe that we can assume that $t > 0$ because $B_0 = 0$ in this case.

For $s \leq t$, we have that $R(s,t) = t^{2H} \varphi(\frac{s}{t})$, which implies that $\partial_s R(s,t) = t^{2H-1} \varphi'(\frac{s}{t})$. In consequence, the change of variables formula $u = s/t$ leads us to write

$$\int_0^t |\partial_s R(s,t)|^\gamma ds = t^{\gamma(2H-1)+1} \int_0^1 |\varphi'(u)|^\gamma \, du \leq T^{\gamma(2H-1)+1} \int_0^1 |\varphi'(u)|^\gamma \, du. \tag{4.98}$$

For $s > t$, we get $R(s,t) = s^{2H}\varphi(t/s)$ and

$$\int_t^T |\partial_s R(s,t)|^\gamma \, ds = \int_t^T \left|2Hs^{2H-1}\varphi(t/s) - s^{2H-2}t\varphi'(t/s)\right|^\gamma ds.$$

Since

$$\int_t^T \left|s^{2H-1}\varphi(t/s)\right|^\gamma ds \le \int_t^T s^{\gamma(2H-1)}\left(\frac{t}{s}\right)^{\gamma H} ds \le \int_0^T s^{\gamma(2H-1)} ds,$$

we only need to show that there is a constant $C > 0$ independent of t such that

$$\int_t^T s^{\gamma(2H-2)} t^\gamma \left|\varphi'(t/s)\right|^\gamma ds < C,$$

due to (4.98). Towards this end, notice

$$\begin{aligned}
&\int_t^T s^{\gamma(2H-2)} t^\gamma \left|\varphi'(t/s)\right|^\gamma ds \\
&= \int_t^T s^{\gamma(2H-1)-\gamma} t^\gamma \left|\varphi'(t/s)\right|^\gamma ds \le \int_t^T s^{\gamma(2H-1)-\gamma} t^{\gamma-1} \left|\varphi'(t/s)\right|^\gamma t ds \\
&\le \int_t^T s^{\gamma(2H-1)+1} \left|\varphi'(t/s)\right|^\gamma \frac{t}{s^2} ds \le T^{\gamma(2H-1)+1} \int_t^T \left|\varphi'(t/s)\right|^\gamma \frac{t}{s^2} ds.
\end{aligned}$$

Therefore, the change of variables formula $u = t/s$ allows us to conclude

$$\int_t^T s^{\gamma(2H-2)} t^\gamma \left|\varphi'(t/s)\right|^\gamma ds \le T^{\gamma(2H-1)+1} \int_{t/T}^1 |\varphi'(u)|^\gamma du.$$

Thus, the proof is complete. □

Now, we show that the Riemann-Liouville fractional process (RLfp) satisfies the conditions of Lemma 58. Remember that RLfp is introduced in Section 4.5.

Lemma 59 *Let $B = \{B_t : t \in [0,T]\}$ be Riemann-Liouville fractional process with $H \in (0, 1/4)$. Then, it is a zero mean continuous self-similar Gaussian process with index H and*

$$\int_0^1 |\varphi'(u)|^\gamma du < \infty, \quad \textit{for } 1 < \gamma < 1/(1-2H).$$

Proof. By Propositions 88 and 89, we know that RLfp is a continuous self-similar Gaussian process with index H.

Let $\alpha = \frac{1}{2} - H$. From Definition 25 and the change of variables formula $u = s - r$, we have

$$\varphi(s) = \int_0^s (s-r)^{-\alpha}(1-r)^{-\alpha} dr = \int_0^s u^{-\alpha}(1+u-s)^{-\alpha} du \quad s \in [0,1].$$

Hence, the dominated convergence theorem yields

$$\varphi'(s) = s^{-\alpha} + \alpha \int_0^s u^{-\alpha}(1+u-s)^{-\alpha-1} du.$$

Since it is easy to see $\int_0^1 s^{-\gamma\alpha} ds < \infty$, we only need to show

$$\int_0^1 \left(\int_0^s u^{-\alpha}(1+u-s)^{-\alpha-1} du\right)^\gamma ds < \infty.$$

So, we now prove it.

Observe that, for $\varepsilon > 0$ small enough, we can obtain the following calculations:

$$\begin{aligned}
&\int_0^1 \left(\int_0^s u^{-\alpha}(1+u-s)^{-\alpha-1} du \right)^\gamma ds \\
&\leq \int_0^1 \left| (1-s)^{-\alpha-\varepsilon} \int_0^s u^{-\alpha}(1+u-s)^{\varepsilon-1} du \right|^\gamma ds \\
&\leq C_\varepsilon \int_0^1 (1-s)^{-\gamma(\alpha+\varepsilon)} \left(\int_0^s u^{-\gamma\alpha}(1+u-s)^{\varepsilon-1} du \right) ds \\
&= C_\varepsilon \int_0^1 u^{-\gamma\alpha} \left(\int_u^1 (1-s)^{-\gamma(\alpha+\varepsilon)}(1+u-s)^{\varepsilon-1} ds \right) du \\
&\leq C_\varepsilon \int_0^1 u^{-2\gamma\alpha-\varepsilon} \left(\int_u^1 (1-s)^{-\gamma(\alpha+\varepsilon)}(1+u-s)^{2\varepsilon-1+\gamma\alpha} ds \right) du.
\end{aligned}$$

Consequently, the fact that $\gamma < 1/(1-2H) < 1/\alpha$ gives that, for $\varepsilon > 0$ small enough,

$$\begin{aligned}
&\int_0^1 \left(\int_0^s u^{-\alpha}(1+u-s)^{-\alpha-1} du \right)^\gamma ds \\
&\leq C_\varepsilon \int_0^1 u^{-2\gamma\alpha-\varepsilon} \left(\int_u^1 (1-s)^{-\gamma(\alpha+\varepsilon)+2\varepsilon+\gamma\alpha-1} ds \right) du \\
&= C_\varepsilon \int_0^1 u^{-2\gamma\alpha-\varepsilon} \left(\int_u^1 (1-s)^{\varepsilon(2-\gamma)-1} ds \right) du.
\end{aligned}$$

Thus, using that $H < 1/4$, we finally conclude

$$\int_0^1 \left(\int_0^s u^{-\alpha}(1+u-s)^{-\alpha-1} du \right)^\gamma ds \leq C_\varepsilon \int_0^1 u^{-2\gamma\alpha-\varepsilon} du.$$

Now, it is easy to finish the proof □

An immediate consequence of the last result, Lemma 58 and Theorem 65 is the following:

Corollary 27 *Let $f : [0,T] \times \mathbb{R} \to \mathbb{R}$ be as in Theorem 65 and $B = \{B_t : t \in [0,T]\}$ Riemann-Liouville fractional process with $H \in (0, 1/4)$. Then, for all $t \in [0,T]$, $\partial_x f(\cdot, B_\cdot)1_{[0,t]}(\cdot) \in Dom^E \delta$ and*

$$f(t, B_t) = f(0,0) + \int_0^t \partial_s f(s, B_s) ds + \int_0^t \partial_x f(s, B_s) \hat{\delta} B_s + \frac{1}{2} \int_0^t \partial_x^2 f(s, B_s) dR_s.$$

Remarks 13 *i) By Proposition 93, we have again*

$$\int_0^t B_s \hat{\delta} B_s = \delta(B_s 1_{[0,t]}), \quad t \in [0,T],$$

where δ is the divergence type operator introduced in Definition 40.

ii) The case $H > 1/4$ is considered in Theorem 58 and Proposition 86.

iii) Note that Theorem 63 implies that Step 1 in the proof of Theorem 64 is also satisfied for Riemann-Liouville fractional process. Therefore, we also have that $\hat{\delta}$ is an extension of the divergence operator δ_2 by proceeding as in the proof of Theorem 64 (see Step 2).

Chapter 5

Forward Integration

The forward integral is defined with respect to right-continuous with left-limits stochastic processes. However, in this chapter, we deal with continuous processes as integrators. This integral is an anticipating one when the integrator is a semimartingale. That is, in this case, it is an extension of the stochastic Itô's integral that allows us to integrate processes that are not necessarily adapted to the underlying filtration. Moreover, the forward integral allows integration with respect to processes that are not semimartingales, as the fractional Brownian motion. The forward integral has been introduced by Berger and Mizel [19]. But the approach that we use here has been given by Russo and Vallois [190, 191], where the forward integral is defined as a limit in probability. Under suitable conditions, the difference between the forward and the divergence operator is a trace term that depends on the derivative operator. In this way, we can obtain estimates for the $L^p(\Omega)$-norm of the forward integral via the Malliavin calculus. In general, it is difficult to deal with the forward integral because, roughly speaking, this integral is given as the limit (in probability) of Riemann sums defined taking values of the integrands at the left-point of each involved interval (see Propositions 96 and 98 below). Furthermore, as we have already pointed out, if we are dealing with a filtration for which Brownian motion B is a semimartingale, then the forward integral also agrees with the Itô's integral with respect to the semimartingale B. Hence, as a consequence, under suitable conditions, the Skorohod integral is the Itô's integral with respect to the semimartingale B plus an extra term. As will see, the forward integral leads us to establish if the integrator is a semimartingale with respect to a given filtration and, together with the Malliavin calculus, to calculate the decomposition of the integrator as a semimartingale.

The properties of forward integral become it a suitable tool to extend some results that are known using the classical Itô's calculus. For instance, the study of financial markets with an insider (see León et al. [116]), or the existence of a unique solution to evolution equations with random coefficients. In this case, the mild interpretation of the equation involves an integrand that is not adapted to the underlying filtration. We remark that this term is adapted when the coefficients of the equation are deterministic (see [6, 117]).

5.1 Some properties of forward integral

The forward integral in the sense of Russo and Vallois [190, 191] is introduced as follows:

Definition 43 *Let $u = \{u_t : t \in [0,T]\}$ and $x = \{x_t : t \in [0,T]\}$ be stochastic processes with integrable and continuous paths, respectively. We say that u is forward integrable with respect to x on the interval $[a,b]$, $0 \le a < b \le T$, (denoted by $u \in Dom\,\delta_x^-([a,b])$) if the integral*

$$\varepsilon^{-1}\int_a^b u_s\left(x_{(s+\varepsilon)\wedge T} - x_s\right) ds \tag{5.1}$$

DOI: 10.1201/9781003484912-5

converges in probability as $\varepsilon \downarrow 0$. *In this case, the limit is denoted by* $\int_a^b u_s dx_s^-$ *and it is called the forward integral of* u *with respect to* x.

Note that the integral in (5.1) is well-defined due to x having continuous paths and $u \in L^1([0,T])$ w.p.1. Also note that, by definition of the forward integral, we have

$$\int_0^T 1_{[a,b]}(s) u_s dx_s^- = \int_a^b u_s dx_s^-,$$

provided one of the integrals exists.

Observe that if $u \in \operatorname{Dom}\delta_x^-([a,b])$ and F is a random variable, then

$$\begin{aligned} \varepsilon^{-1}\int_a^b (Fu_s)(x_{(s+\varepsilon)\wedge T} - x_s)ds &= F\left(\varepsilon^{-1}\int_a^b u_s(x_{(s+\varepsilon)\wedge T} - x_s)ds\right) \\ &\to F\int_a^b u_s dx_s^- \quad \text{in probability as } \varepsilon \downarrow 0. \end{aligned}$$

In this way, we have shown that $Fu \in \operatorname{Dom}\delta_x^-([a,b])$ and

$$\int_a^b (Fu_s)dx_s^- = F\int_a^b u_s dx_s^- \tag{5.2}$$

(compare it with Proposition 58.*iii*) and (4.3)).

The following result states a relation between the forward integral and the Young one, which is introduced in (2.2).

Proposition 96 *Let* $\kappa + \nu > 1$, *and* u *and* x *two processes with paths in* $\mathcal{C}_1^\kappa([0,T];\mathbb{R})$ *and* $\mathcal{C}_1^\nu([0,T];\mathbb{R})$, *respectively. Then, for* $a, b \in [0,T]$, $a < b$, *we have that* $u \in \operatorname{Dom}\delta_x^-([a,b]) \cap \operatorname{Dom}(\int_a^b \cdot d^Y x)$ *and*

$$\int_a^b u_s dx_s^- = \int_a^b u_s d^Y x_s, \quad \textit{w.p.1},$$

where the right-hand side is the Young integral of u *with respect to* x.

Remark 76 *By Theorem 33, we have that* $u \in \operatorname{Dom}(\int_a^b \cdot d^Y x)$.

Proof. We fix $\omega \in \Omega$ such that $u(\omega)$ and $x(\omega)$ belong to $\mathcal{C}_1^\kappa([a,b];\mathbb{R})$ and $\mathcal{C}_1^\nu([a,b];\mathbb{R})$, respectively. In the remainder of this proof, in order to simplify the notation, we write u and x instead of $u(\omega)$ and $x(\omega)$, respectively. Consider $\varepsilon > 0$ and set

$$x_\varepsilon(t) = \frac{1}{\varepsilon}\int_a^t \left(x_{(s+\varepsilon)\wedge T} - x_s\right) ds, \quad t \in [a,b],$$

which is in $C^1([a,b])$. Then, Proposition 32 and (2.7) imply

$$\frac{1}{\varepsilon}\int_a^b u_s\left(x_{(s+\varepsilon)\wedge T} - x_s\right) ds = \int_0^T u_s dx_\varepsilon^Y(s).$$

Hence, choosing $\beta \in (0,\nu)$ such that $\kappa + \beta > 1$, Theorems 17 and 33 allow us to get

$$\begin{aligned} \sup_{t\in[a,b]}\left|\int_a^t u_s dx_s^Y - \int_a^t u_s dx_\varepsilon(s)\right| &= \sup_{t\in[a,b]}\left|\int_a^t u_s dx_s^Y - \int_a^t u_s dx_\varepsilon^Y(s)\right| \\ &\le C\left(||u||_\kappa + \|u\|_\infty\right)||x - x_\varepsilon||_\beta. \end{aligned} \tag{5.3}$$

We first consider $s, t \in [a, b]$ such that $a \le s < s + \varepsilon < t \le b$. For

$$\begin{aligned}\Delta_\varepsilon(t) := x_\varepsilon(t) - x_t &= \frac{1}{\varepsilon}\int_t^{t+\varepsilon} x_{r\wedge T}dr - \frac{1}{\varepsilon}\int_a^{a+\varepsilon} x_r dr - x_t \\ &= \frac{1}{\varepsilon}\int_t^{t+\varepsilon} (x_{r\wedge T} - x_t)\,dr - \frac{1}{\varepsilon}\int_a^{a+\varepsilon} x_r dr,\end{aligned}$$

we can write

$$\begin{aligned}|\Delta_\varepsilon(t) - \Delta_\varepsilon(s)| &\le \frac{1}{\varepsilon}\int_t^{t+\varepsilon} |x_{r\wedge T} - x_t|\,dr + \frac{1}{\varepsilon}\int_s^{s+\varepsilon} |x_{r\wedge T} - x_s|\,du \\ &\le ||x||_\nu \frac{1}{\varepsilon}\left(\int_t^{t+\varepsilon} (r-t)^\nu dr + \int_s^{s+\varepsilon} (r-s)^\nu dr\right) \\ &\le C_\nu \varepsilon^\nu \le C_\nu \varepsilon^{\nu-\beta}(t-s)^\beta. \end{aligned} \tag{5.4}$$

Finally, we deal with the case $a \le s < t < s + \varepsilon$. So, we have now

$$\begin{aligned}\Delta_\varepsilon(t) - \Delta_\varepsilon(s) &= \frac{1}{\varepsilon}\int_{s+\varepsilon}^{t+\varepsilon} (x_{r\wedge T} - x_{(s+\varepsilon)\wedge T})dr - \frac{1}{\varepsilon}\int_s^t (x_r - x_s)dr \\ &\quad + \frac{t-s}{\varepsilon}(x_{(s+\varepsilon)\wedge T} - x_s) + x_s - x_t.\end{aligned}$$

Thus, using $0 \le t - s < \varepsilon$,

$$\begin{aligned}|\Delta_\varepsilon(t) - \Delta_\varepsilon(s)| &\le \frac{2\|x\|_\nu}{(\nu+1)\varepsilon}(t-s)^{\nu+1} + \|x\|_\nu \left(\varepsilon^{\nu-1}(t-s) + (t-s)^\nu\right) \\ &\le C_\nu \left\{(t-s)^\nu + (t-s)^\beta \left(\varepsilon^{\nu-1}\varepsilon^{1-\beta} + \varepsilon^{\nu-\beta}\right)\right\} \le C_\nu \varepsilon^{\nu-\beta}(t-s)^\beta.\end{aligned}$$

Therefore, this inequality and (5.4) imply $\|\Delta_\varepsilon(t) - \Delta_\varepsilon(s)\|_\beta \le C_\nu \varepsilon^{\nu-\beta}$, which, together with (5.3), leads us to establish

$$\left\|\int_a^\cdot u_s dx_s^Y - \int_a^\cdot u_s dx_\varepsilon(s)\right\|_\infty \to 0, \quad \text{as } \varepsilon \downarrow 0.$$

Consequently, the proof is complete. □

The forward integral also satisfies the local property (compare it with (3.12), and Propositions 66 and 78). That is, we have the following result.

Lemma 60 *Let x be a continuous process, $A \in \mathcal{F}$, and u and v to measurable processes in $Dom\,\delta_x^-([0,T])$ such that*

$$u = v \quad \textit{almost surely on } \ A \times [0, T].$$

Then,

$$\int_0^t u_s dx_s^- = \int_0^t v_s dx_s^- \quad \textit{on the set } A \textit{ w.p.1, for each } t \in [0, T].$$

Proof. The result is an immediate consequence of the fact that

$$E\left|1_A \int_0^t (u_s - v_s)(x_{(s+\varepsilon)\wedge T} - x_s)ds\right| \le \int_0^T E\Big(|1_A(u_s - v_s)||x_{(s+\varepsilon)\wedge T} - x_s|\Big)ds = 0,$$

for all $\varepsilon > 0$ and $t \in [0, T]$. □

We could proceed as in Definition 33, Corollary 20 and Proposition 56 to extend the domain $\mathrm{Dom}\,\delta_x^-([0,T])$ of the forward integral with respect to x. However, we are able to state the following result.

Proposition 97 *Let u and x be two stochastic processes with integrable and continuous paths, respectively, for which there exists a sequence $\{(\Omega_n, u^{(n)}) : n \geq 1\} \subset \mathcal{F} \times$ $Dom\,\delta_x^-([0,T]))$ such that:*

i) $\Omega_n \nearrow \Omega$ w.p.1.

ii) $u = u^{(n)}$ almost surely on $\Omega_n \times [0,T]$.

iii) For $n \in \mathbb{N}$, u and $u^{(n)}$ are measurable processes.

Then, u belongs to $Dom\,\delta_x^-([0,T])$ and

$$\int_0^T u_s dx_s^- = \int_0^T u_s^{(n)} dx_s^- \quad \text{on } \Omega_n \text{ for all } n \in \mathbb{N} \text{ w.p.1.}$$

Proof. Set the random variable

$$Z = \int_0^T u_s^{(n)} dx_s^- \quad \text{on } \Omega_n \text{ for all } n \in \mathbb{N},$$

which is well-defined due to Lemma 60. Then, for $\eta, \varepsilon > 0$ an $m \in \mathbb{N}$, we get

$$\begin{aligned}
&P\left(\left[\left|\varepsilon^{-1}\int_0^T u_s\left(x_{(s+\varepsilon)\wedge T} - x_s\right) ds - Z\right| > \eta\right]\right) \\
&\leq P\left(\left[1_{\Omega_m^c}\left|\varepsilon^{-1}\int_0^T u_s\left(x_{(s+\varepsilon)\wedge T} - x_s\right) ds\right| > \eta/8\right]\right) \\
&\quad + P\left(\left[1_{\Omega_m}\left|\varepsilon^{-1}\int_0^T \left(u_s - u_s^{(m)}\right)\left(x_{(s+\varepsilon)\wedge T} - x_s\right) ds\right| > \eta/8\right]\right) \\
&\quad + P\left(\left[1_{\Omega_m}\left|\varepsilon^{-1}\int_0^T u_s^{(m)}\left(x_{(s+\varepsilon)\wedge T} - x_s\right) ds - \int_0^T u_s^{(m)} dx_s^-\right| > \eta/8\right]\right) \\
&\quad + P\left(\left[1_{\Omega_m^c}|Z| > \eta/8\right]\right) \\
&\leq 2P(\Omega_m^c) + P\left(\left[1_{\Omega_m}\left|\varepsilon^{-1}\int_0^T u_s^{(m)}\left(x_{(s+\varepsilon)\wedge T} - x_s\right) ds - \int_0^T u_s^{(m)} dx_s^-\right| > \eta/8\right]\right).
\end{aligned}$$

Hence,

$$\lim_{\varepsilon \downarrow 0} P\left(\left[\left|\varepsilon^{-1}\int_0^T u_s\left(x_{(s+\varepsilon)\wedge T} - x_s\right) ds - Z\right| > \eta\right]\right) \leq 2P(\Omega_m^c).$$

Consequently, the fact that $\Omega_n \nearrow \Omega$ w.p.1 yields that the result is satisfied. □

On the other hand, our next aim is to see that Proposition 52 is also true for the forward integral. Towards this end, as in the proof of this proposition, some auxiliary tool is established in the appendix (Section 8.7) in order to improve the presentation of the following results in the remaining of this section. So, we are interested in analyzing the forward integral $\int_a^t f(y_s)dy_s^-$, with $t \in [a,b]$. Actually, since $[a,b] \subset [0,T]$, we only need to consider the integral $\int_0^t f(y_s)dy_s^-$, with $t \in [0,T]$. Here, the function $f : \mathbb{R} \to \mathbb{R}$ and the process y satisfy the following hypotheses:

A1: (Hölder continuity): For some $0 < H < 1$ and each $\rho \in (0,H)$, there exists a nonnegative random variable G_ρ such that $E(G_\rho^p) < \infty$, for all $p \geq 1$, and

$$|y_r - y_s| \leq G_\rho |r-s|^{H-\rho}, \quad r,s \in [0,T]. \tag{5.5}$$

A2: For almost all $t \in [0,T)$ and $t = T$, the random variable y_t admits a probability density function p_t such that

$$\sup_{x\in\mathbb{R}} p_t(x) \leq \hat{p}_t, \tag{5.6}$$

where $\int_0^T \hat{p}_t dt < \infty$ and $\hat{p}_T < \infty$.

B: $f : \mathbb{R} \to \mathbb{R}$ is a function of bounded variation with finite total variation measure $|\mu_f|$, where $\mu_f((a,b]) = f(b+) - f(a+)$, $a < b$.

Remark 77 *Note that, for $\theta \in (0,1)$ and $x \in \mathbb{R}$, we have that A2 and Fubini's theorem imply*

$$\begin{aligned}
E\left(\int_0^T |Y_t - x|^{-\theta} dt\right) &= \int_0^T \int_{\mathbb{R}} |y-x|^{-\theta} p_t(y) dy dt \leq \int_0^T \int_{|y-x|\leq 1} |y-x|^{-\theta} p_t(y) dy dt \\
&\quad + \int_0^T \int_{|y-x|>1} p_t(y) dy dt \\
&\leq \int_0^T \hat{p}_t \int_{|y-x|\leq 1} |y-x|^{-\theta} dy dt + T = \int_0^T \hat{p}_t \int_{-1}^1 |u|^{-\theta} du dt + T.
\end{aligned}$$

In other words, Hypothesis A2 guarantees

$$\sup_{x\in\mathbb{R}} E\left(\int_0^T |Y_t - x|^{-\theta} dt\right) < \infty.$$

This inequality is an important tool in our analysis (see the proof of Lemma 94 in the appendix).

Now, we give some notation that we use to study the existence of the integral $\int_a^t f(y_s) dy_s^-$.

For $\varepsilon > 0$, a bounded and measurable function $f : \mathbb{R} \to \mathbb{R}$, and a process y, we use the convention

$$P_\varepsilon(x) = \frac{e^{-x^2/2\varepsilon}}{\sqrt{2\pi\varepsilon}}, \quad x \in \mathbb{R},$$

$$F'_\varepsilon(x) = \int_{\mathbb{R}} f(x-s) P_\varepsilon(s) ds, \quad x \in \mathbb{R}, \tag{5.7}$$

and

$$y_t^{(\varepsilon)} = \frac{1}{\varepsilon} \int_t^{t+\varepsilon} y_{s\wedge T} ds, \quad t \in [0,T]. \tag{5.8}$$

Note that if f satisfies Condition **B**, then it is a bounded and measurable function. As a matter of fact, by Royden [187], f is the difference of two bounded and increasing functions on $\mathbb{R}$. Thus, we can assume that f is a bounded and increasing function to see that f is measurable. In this case, for $c \in \mathbb{R}$ such that $\{x : f(x) \leq c\} \neq \emptyset$, we set $a = \sup\{x : f(x) \leq c\}$. Therefore, we have that $a \in \mathbb{R} \cup \{\infty\}$. Now, let $z < a$. In consequence, the definition of supremum implies that there is $x \in (z, \infty)$ such that $f(z) \leq f(x) \leq c$. In consequence, $f^{-1}((-\infty, c])$ is equal to either $(-\infty, a)$, or $(-\infty, a]$. In this way, our claim is true.

The mean value theorem yields $|P_\varepsilon(x-s) - P_\varepsilon(y-s)| \leq |x-y||P'_\varepsilon(\theta)|$, for some θ between $|x-s|$ and $|y-s|$. Since,

$$|P'_\varepsilon(\theta)| = \frac{1}{\sqrt{2\pi}\varepsilon^{3/2}} |\theta| e^{-\theta^2/2\varepsilon} \leq C\frac{1}{\varepsilon} e^{-\theta^2/4\varepsilon},$$

we get

$$\begin{aligned}|F'_\varepsilon(x)-F'_\varepsilon(y)| &\le \int_{\mathbb{R}}|f(s)||P_\varepsilon(x-s)-P_\varepsilon(y-s)|ds\\ &\le \frac{C}{\varepsilon}|x-y|\int_{\mathbb{R}}|f(s)|(e^{-(x-s)^2/4\varepsilon}+e^{-(y-s)^2/4\varepsilon})ds\\ &\le \frac{C}{\varepsilon^{1/2}}|x-y|,\quad x,y\in\mathbb{R}. \end{aligned}\tag{5.9}$$

In other words, we have shown that F'_ε is a Lipschitz function. So, the fundamental theorem of calculus gives

$$\int_{y_0^{(\varepsilon)}}^{y_t^{(\varepsilon)}} F'_\varepsilon(x)dx = F_\varepsilon(y_t^{(\varepsilon)})-F_\varepsilon(y_0^{(\varepsilon)}) = \varepsilon^{-1}\int_0^t F'_\varepsilon(y_s^{(\varepsilon)})(y_{(s+\varepsilon)\wedge T}-y_s)ds \tag{5.10}$$

if y either is continuous, or satisfies Assumption **A1**.

We will need the following auxiliary result later on.

Lemma 61 *Let f satisfy Hypothesis* **B**. *Then, there exists a constant $C>0$ such that*

$$|f(x+)-F'_\varepsilon(x)|\le C\int_{-\infty}^{\infty}e^{-(x-y)^2/4\varepsilon}|\mu_f|(dy),\quad \text{for all } \varepsilon>0 \text{ and } x\in\mathbb{R}. \tag{5.11}$$

Proof. Using that f is continuous almost surely because it is the difference of two increasing functions, we have

$$\begin{aligned}F'_\varepsilon(x)-f(x+) &= \int_{\mathbb{R}}[f((x-s)+)-f(x+)]P_\varepsilon(s)ds = \int_{-\infty}^{0}[f((x-s)+)-f(x+)]P_\varepsilon(s)ds\\ &\quad+\int_0^\infty [f((x-s)+)-f(x+)]P_\varepsilon(s)ds = I_1+I_2.\end{aligned}$$

Note that the definition of μ_f and Fubini's theorem imply

$$\begin{aligned}|I_1| &= \left|\int_{-\infty}^{0}\int_{(x,x-s]}d\mu_f(y)P_\varepsilon(s)ds\right|\\ &= \left|\int_{(x,\infty)}\int_{-\infty}^{x-y}P_\varepsilon(s)ds\mu_f(dy)\right| \le C\int_{(x,\infty)}e^{-(x-y)^2/4\varepsilon}|\mu_f|(dy)\end{aligned}$$

and

$$\begin{aligned}|I_2| &= \left|\int_0^\infty [f(x+)-f((x-s)+)]P_\varepsilon(s)ds\right|\\ &= \left|\int_0^\infty\int_{(x-s,x]}d\mu_f(y)P_\varepsilon(s)ds\right| = \int_{(-\infty,x]}\int_{x-y}^{\infty}P_\varepsilon(s)ds|\mu_f|(dy)\\ &\le C\int_{(-\infty,x]}e^{-(x-y)^2/4\varepsilon}|\mu_f|(dy),\end{aligned}$$

for all $\varepsilon>0$ and $x\in\mathbb{R}$. Hence (5.11) holds. □

Now, we are ready to state the a fundamental theorem of calculus for the forward integral.

Theorem 66 *Let $y = \{y_t : t \in [0,T]\}$ be a stochastic process satisfying Hypotheses* **A1** *and* **A2** *with $H > 1/2$, and f a function fulfilling Assumption* **B**. *Then,*

$$\int_0^t f(y_s)dy_s^- = \int_{y_0}^{y_t} f(x)dx, \quad t \in [0,T].$$

Remark 78 *Proposition 52 implies that $\int_0^t f(y_s)dy_s^-$ agrees with the generalized Stieltjes integral $\int_0^t f(y_s)dy_s$ given by Zähle [218] (see Section 2.3 for its definition). Indeed, we only need to observe that the function $x \mapsto \int_0^x f(z)dz$ is absolutely continuous having derivative equals to f almost surely. Thus, as in Proposition 52, we get that if $G : \mathbb{R} \to \mathbb{R}$ is an absolutely continuous function such that G' fulfills Hypothesis* **B**, *then the forward integral satisfies the change of variables formula*

$$G(y_t) - G(y_0) = \int_0^t G'(y_s)dy_s^-, \quad t \in [0,T].$$

Proof of Theorem 66. Let $t \in [0,T]$. In order to prove the result, we analyze the convergence of $F_\varepsilon(y_t^{(\varepsilon)}) - F_\varepsilon(y_0^{(\varepsilon)})$, as $\varepsilon \downarrow 0$, in two different ways. Here, $y^{(\varepsilon)}$ is given in (5.8). So, we first show that

$$F_\varepsilon(y_t^{(\varepsilon)}) - F_\varepsilon(y_0^{(\varepsilon)}) \to \int_{y_0}^{y_t} f(x)dx \text{ in } L^2(\Omega), \text{ as } \varepsilon \downarrow 0. \tag{5.12}$$

Towards this end, we recall that we have already known that F'_ε is continuous (see (5.9)). Thus, from (5.10), the fundamental theorem of calculus leads to write

$$\begin{aligned}
&\left| F_\varepsilon(y_t^{(\varepsilon)}) - F_\varepsilon(y_0^{(\varepsilon)}) - \int_{y_0}^{y_t} f(x)dx \right| \\
&= \left| \int_{y_0^{(\varepsilon)}}^{y_t^{(\varepsilon)}} F'_\varepsilon(x)dx - \int_{y_0}^{y_t} f(x)dx \right| \\
&\leq \left| \int_0^{y_t^{(\varepsilon)}} F'_\varepsilon(x)dx - \int_0^{y_t} f(x)dx \right| + \left| \int_0^{y_0^{(\varepsilon)}} F'_\varepsilon(x)dx - \int_0^{y_0} f(x)dx \right| \\
&\leq \left| \int_0^{y_t^{(\varepsilon)}} F'_\varepsilon(x)dx - \int_0^{y_t^{(\varepsilon)}} f(x)dx \right| + \left| \int_0^{y_t^{(\varepsilon)}} f(x)dx - \int_0^{y_t} f(x)dx \right| \\
&\quad + \left| \int_0^{y_0^{(\varepsilon)}} F'_\varepsilon(x)dx - \int_0^{y_0^{(\varepsilon)}} f(x)dx \right| + \left| \int_0^{y_0^{(\varepsilon)}} f(x)dx - \int_0^{y_0} f(x)dx \right| = I_1 + I_2 + I_3 + I_4.
\end{aligned}$$

For I_2, we apply that f is a bounded function because of Hypothesis **B**, together with (8.25) in the appendix, in order to get that, for $0 < \rho < H$,

$$I_2 \leq \left| \int_{y_t^{(\varepsilon)}}^{y_t} |f(x)|dx \right| \leq \left| \int_{y_t^{(\varepsilon)}}^{y_t} C dx \right| = C|y_t^{(\varepsilon)} - y_t| \leq CG_\rho \varepsilon^{H-\rho} \to 0, \quad \text{in } L^2(\Omega), \text{ as } \varepsilon \downarrow 0.$$

Now, we deal with the term I_1. To do so, we use Fubini's theorem, the fact that f is continuous almost surely and Lemma 61 to establish

$$\begin{aligned}
I_1 &\leq \left| \int_0^{y_t^{(\varepsilon)}} |F'_\varepsilon(x) - f(x+)|dx \right| \leq \left| \int_0^{y_t^{(\varepsilon)}} \int_{\mathbb{R}} Ce^{-(x-y)^2/4\varepsilon} |\mu_f|(dy)dx \right| \\
&\leq C\sqrt{4\pi\varepsilon} \int_{\mathbb{R}} \int_{\mathbb{R}} \frac{e^{-(x-y)^2/4\varepsilon}}{\sqrt{4\pi\varepsilon}} dx|\mu_f|(dy) = C\sqrt{4\pi\varepsilon} \int_{\mathbb{R}} |\mu_f|(dy) \to 0, \text{ in } L^2(\Omega), \text{ as } \varepsilon \downarrow 0.
\end{aligned}$$

We observe that it is easy to see that the terms I_3 and I_4 go to zero in $L^2(\Omega)$ by proceeding similarly. Consequently, (5.12) is true.

On the other hand, Lemma 94 in the appendix, together with the fundamental theorem of calculus, implies that the forward integral $\int_0^t f(y_s)dy_s^-$ exists and, for some $p > 1$, we have

$$\begin{aligned} F_\varepsilon(y_t^{(\varepsilon)}) - F_\varepsilon(y_0^{(\varepsilon)}) &= \varepsilon^{-1}\int_0^t F'_\varepsilon(y_s^{(\varepsilon)})(y_{(s+\varepsilon)\wedge T} - y_s)ds \\ &\to \int_0^t f(y_s)dy_s^-, \quad \text{in } L^p(\Omega), \text{ as } \varepsilon \downarrow 0. \end{aligned} \tag{5.13}$$

Finally, we observe that (5.12) and (5.13) lead us to conclude that, for some $p > 1$,

$$\begin{aligned} \int_0^t f(y_s)dy_s^- &= L^p(\Omega) - \lim_{\varepsilon\downarrow 0} \varepsilon^{-1}\int_0^t F'_\varepsilon(y_s^{(\varepsilon)})(y_{(s+\varepsilon)\wedge T} - y_s)ds \\ &= L^p(\Omega) - \lim_{\varepsilon\downarrow 0}\left(F_\varepsilon(y_t^{(\varepsilon)}) - F_\varepsilon(y_0^{(\varepsilon)})\right) = \int_0^{y_t} f(x)dx - \int_0^{y_0} f(x)dx. \end{aligned}$$

Therefore, the proof is complete. □

Let us mention two consequences of Theorem 66.

Corollary 28 *Let $y = \{y_t :\in [0,T]\}$ satisfy Hypotheses **A1** y **A2** with $H > 1/2$. Then,*

$$\int_0^t 1_{\{y_s>0\}}dy_s^- = y_t^+ - y_0^+, \quad t \in [0,T],$$

with $y_t^+ = y_t \vee 0$.

Proof. The result is an immediate consequence of Theorem 66. □

Corollary 29 *Under the assumptions of Theorem 66, we have*

$$\int_0^t sign(y_s)dy_s^- = |y_t| - |y_0|, \quad t \in [0,T].$$

Proof. Fix $t \in [0,T]$. By Hypothesis **A2** (see Remark 77) and the definition of the forward integral, we have that $\int_0^t 1_{\{y_s=0\}}dy_s^- = 0$ because Fubini's theorem yields that $1_{\{y_s=0\}} = 0$ for almost all $s \in [0,T]$, with probability 1. Thus, using that $sign(x) = 2\cdot 1_{\{x>0\}} - 1 + 1_{\{x=0\}}$ and Theorem 66, together with Hypothesis **A1** and **A2**, we obtain

$$\begin{aligned} \int_0^t sign(y_s)dy_s^- &= \int_0^t [2\cdot 1_{\{y_s>0\}} - 1 + 1_{\{y_s=0\}}]dy_s^- = 2\int_0^t 1_{\{y_s>0\}}dy_s^- - y_t + y_0 \\ &= 2y_t^+ - 2y_0^+ - y_t + y_0 = |y_t| - |y_0|. \end{aligned}$$

Hence, the proof is complete. □

5.2 Brownian motion case

Here, we study the forward integral with respect to Brownian motion. This integral has been given in Definition 43. So, let $B = \{B_t : t \in [0,T]\}$ be a $\mathcal{F}_t$-Brownian motion defined on a complete probability space $(\Omega, \mathcal{F}, P)$. We assume that $\mathcal{F}_0$ contains all the P-null sets. In order to understand the idea of the definition of the forward integral, consider a bounded, measurable and $\mathcal{F}_t$-adapted process $u = \{u_t : t \in [0,T]\}$. In other words, we have that u is Itô integrable with respect to B. For a partition $\pi = \{a = t_0 < t_1 < \ldots < t_n = b\}$ of $[a,b] \subset [0,T]$ such that $|\pi| < 1/n$ and $\varepsilon > 0$, we set

$$f_n(r,s) = u_s \sum_{i=1}^{n} 1_{(t_{i-1},t_i]}(s) 1_{(t_i,(t_i+\varepsilon)\wedge T]}(r), \quad \text{for every } (r,s) \in [0,T]^2.$$

Observe that $f(r,s) := u_s 1_{[a,b]}(s) 1_{(s,(s+\varepsilon)\wedge T]}(r) = \lim_{n\to\infty} f_n(r,s)$, which gives that f is $\mathcal{F} \otimes \mathcal{B}([0,T]^2)$-measurable. Also, we have

$$\int_{\mathbb{R}} \int_a^{(b+\varepsilon)\wedge T} f_n(r,s) dB_r ds = \int_a^{(b+\varepsilon)\wedge T} \int_{\mathbb{R}} f_n(r,s) ds dB_r.$$

Hence, noting that $f_n(r,s) \le C 1_{[a,b]}(s)$ and proceeding as in the proof of Theorem 42, we can conclude

$$\int_{\mathbb{R}} \int_a^{(b+\varepsilon)\wedge T} f(r,s) dB_r ds = \int_a^{(b+\varepsilon)\wedge T} \int_{\mathbb{R}} f(r,s) ds dB_r.$$

Consequently, Proposition 58.*iii*) implies

$$\begin{aligned} \varepsilon^{-1} \int_a^b u_s (B_{(s+\varepsilon)\wedge T} - B_s) ds &= \varepsilon^{-1} \int_a^b u_s \int_s^{(s+\varepsilon)\wedge T} dB_r ds = \varepsilon^{-1} \int_a^b \int_s^{(s+\varepsilon)\wedge T} u_s dB_r ds \\ &= \int_a^{(b+\varepsilon)\wedge T} \left(\varepsilon^{-1} \int_{a\vee(r-\varepsilon)}^{r\wedge b} u_s ds \right) dB_r. \end{aligned}$$

Remember that, in order to establish last equalities, we assume that u is a bounded process. But note that

$$\varepsilon^{-1} \int_a^b u_s (B_{(s+\varepsilon)\wedge T} - B_s) ds = \int_a^{(b+\varepsilon)\wedge T} \left(\varepsilon^{-1} \int_{a\vee(r-\varepsilon)}^{r\wedge b} u_s ds \right) dB_r \qquad (5.14)$$

is also true for any measurable and $\mathcal{F}_t$-adapted process u with square-integrable paths. Indeed, the last equality holds when we change u by $u^{(m)} = (-m) \vee (u \wedge m)$, which gives that our claim is satisfied by letting $m \to \infty$. Now, observe that the right-hand side of the last equality converges in probability, as $\varepsilon \downarrow 0$, to the Itô's integral $\int_a^b u_s dB_s$, as it is stated in Proposition 98 below. That is, the Itô and the forward integrals agree on the set of all the measurable and $\mathcal{F}_t$-adapted processes with square-integrable paths.

Proposition 98 *The forward integral with respect to B is an extension of Itô's integral and it allows us to integrate some stochastic processes with integrable paths that are not necessarily adapted to the underlying filtration.*

Proof. By (5.2), it is easy to see that $\operatorname{Dom} \delta_B^-([a,b])$ contains stochastic processes that are not necessarily adapted to the given filtration $(\mathcal{F}_t)_{t\in[a,b]}$. Hence, from (3.17) and (5.14), we only need to see that, as $\varepsilon \downarrow 0$,

$$\int_a^T \left(\varepsilon^{-1} 1_{(a,b+\varepsilon]}(r) \int_{a\vee(r-\varepsilon)}^{r} u_s ds - u_r 1_{(a,b]}(r) \right)^2 dr \to 0 \quad \text{in probability.} \qquad (5.15)$$

Fix $\omega \in \Omega$ such that $u(\omega) \in L^2([0,T])$. Then, for ε small enough, Hölder inequality yields

$$\begin{aligned}
&\int_b^{(b+\varepsilon)\wedge T} \left(\varepsilon^{-1} \int_{a\vee(r-\varepsilon)}^r u_s(\omega)ds \right)^2 dr \\
&\leq \varepsilon^{-1} \int_b^{(b+\varepsilon)\wedge T} \int_{(r-\varepsilon)}^r u_s^2(\omega)dsdr = \varepsilon^{-1} \int_{(b-\varepsilon)}^{(b+\varepsilon)\wedge T} u_s^2(\omega) \int_{b\vee s}^{(b+\varepsilon)\wedge(s+\varepsilon)\wedge T} drds \\
&\leq \int_{(b-\varepsilon)}^{(b+\varepsilon)\wedge T} u_s^2(\omega)ds \to 0 \quad as\ \varepsilon \downarrow 0. \qquad (5.16)
\end{aligned}$$

From measure theory, it is well-known that almost all $r \in [a,T]$ are Lebesgue-points of $u(\omega)$. It means,

$$\varepsilon^{-1} \int_{a\vee(r-\varepsilon)}^r u_s(\omega)ds \to u_r(\omega), \quad \text{for almost all } r \in [a,T], \text{ as } \varepsilon \downarrow 0.$$

Thus, we have that

$$\int_a^b \left(\varepsilon^{-1} \int_{a\vee(r-\varepsilon)}^r u_s(\omega)ds - u_r(\omega) \right)^2 dr \to 0 \quad \text{as } \varepsilon \downarrow 0 \qquad (5.17)$$

if u is bounded by a constant because of the dominated convergence theorem. So, for $n \in \mathbb{N}$ and $u^{(n)} = (-n) \vee (u \wedge n)$, we get

$$\begin{aligned}
&\int_a^b \left(\varepsilon^{-1} \int_{a\vee(r-\varepsilon)}^r u_s(\omega)ds - u_r(\omega) \right)^2 dr \\
&\leq C \int_a^b \left(\varepsilon^{-1} \int_{a\vee(r-\varepsilon)}^r \left(u_s(\omega) - u_s^{(n)}(\omega) \right) ds \right)^2 dr \\
&\quad + C \int_a^b \left(\varepsilon^{-1} \int_{a\vee(r-\varepsilon)}^r u_s^{(n)}(\omega)ds - u_r^{(n)}(\omega) \right)^2 dr + C \int_a^b \left(u_r(\omega) - u_r^{(n)}(\omega) \right)^2 dr.
\end{aligned}$$

Therefore, Fubini's theorem and Hölder inequality lead us to write

$$\begin{aligned}
&\int_a^b \left(\varepsilon^{-1} \int_{a\vee(r-\varepsilon)}^r u_s(\omega)ds - u_r(\omega) \right)^2 dr \\
&\leq C \int_a^b \varepsilon^{-1} \int_{a\vee(r-\varepsilon)}^r \left(u_s(\omega) - u_s^{(n)}(\omega) \right)^2 dsdr \\
&\quad + C \int_a^b \left(\varepsilon^{-1} \int_{a\vee(r-\varepsilon)}^r u_s^{(n)}(\omega)ds - u_r^{(n)}(\omega) \right)^2 dr + C \int_a^b \left(u_r(\omega) - u_r^{(n)}(\omega) \right)^2 dr \\
&\leq C \int_a^b \left(\varepsilon^{-1} \int_{a\vee(r-\varepsilon)}^r u_s^{(n)}(\omega)ds - u_r^{(n)}(\omega) \right)^2 dr \\
&\quad + 2C \int_a^b \left(u_r(\omega) - u_r^{(n)}(\omega) \right)^2 dr \to 2C \int_a^b \left(u_r(\omega) - u_r^{(n)}(\omega) \right)^2 dr,
\end{aligned}$$

as $\varepsilon \downarrow 0$ since $u^{(n)}$ is a bounded process. Consequently, (5.17) is also satisfied in this case due to the dominated convergence theorem.

Finally, the fact that u has square-integrable paths, (5.16) and (5.17) yield that (5.15) is satisfied. It means, the proof is complete □

Concerning Proposition 98, now assume that we have a filtration $(\mathcal{G}_t)_{t\in[0,T]}$ such that it contains the filtration generated by the Brownian motion B (i.e., $\mathcal{F}_t^B \subset \mathcal{G}_t$, $t \in [0,T]$), and B is a $\mathcal{G}_t$-semimartingale. That is,

$$B_t = M_t + A_t, \quad t \in [0,T].$$

Here M is a $\mathcal{G}_t$-martingale and A is a continuous and $\mathcal{G}_t$-adapted process with local variation. In this case, we obtain that the quadratic variation $[M]_t = [B]_t = t$, for $t \in [0,T]$. Thus the Lévy theorem (see, for example, Protter [180]) implies that the $\mathcal{G}_t$-martingale M is also a $\mathcal{G}_t$-Brownian motion. Hence, proceeding as in the proof of Proposition 98, we can show that, for any bounded and $\mathcal{G}_t$-adapted process u,

$$\int_0^T u_s dB_s^- = \int_0^T u_s dB_s, \tag{5.18}$$

where the stochastic integral on the right-hand side is the Itô's integral with respect to the $\mathcal{G}_t$-semimartingale B (see Russo and Vallois [190] for details).

The theory of enlargement of a filtration was begun in 1976 by Itô [88]. This author has pointed out that one way to extend the domain of the stochastic integral (in the Itô sense) with respect to an $\mathcal{F}_t$-martingale Y is to enlarge the filtration $\mathcal{F} = (\mathcal{F}_t)_{t\in[0,T]}$ to another filtration $\mathcal{G}$ in such a way that Y remains a semimartingale with respect to the new and bigger filtration $\mathcal{G}$. In this way, we can now integrate processes that are $\mathcal{G}$-adapted, which include processes that are not necessarily adapted to the underlying filtration $\mathcal{F}$. In particular, Itô [88] shows that if $\mathcal{G}_1$ and $\mathcal{G}_2$ are two filtrations such that $\mathcal{G}_1 \subset \mathcal{G}_2$ and Y is a semimartingale with respect to both filtrations, then the stochastic integrals with respect to the $\mathcal{G}_1$- and $\mathcal{G}_2$-semimartingale Y are the same in the intersection of the domains of both integrals. Thus, the enlargement of a filtration produces an extension of Itô's integral. But, this problem has not been considered in [88] when $\mathcal{G}_1 \not\subset \mathcal{G}_2$ and $\mathcal{G}_2 \not\subset \mathcal{G}_1$. This trouble has been resolved by Russo and Vallois [190] by proving that (5.18) holds for both filtrations, as we did in the analysis of identity (5.18). Furthermore, using the Bichteler-Dellacherie theorem (see Proter [180]), Russo and Vallois [190] have also established that a càdlàg and $\mathcal{G}_t$-adapted process Y is a $\mathcal{G}_t$-semimartingale if the forward integral $\int_0^T u_s dY_s^-$ is well-defined for any bounded, càdlàg and $\mathcal{G}_t$-previsible process u.

On the other hand, the forward integral is also related to the divergence operator, as the following theorem shows. For the proof of this result, we need to state the following two auxiliary tools.

Lemma 62 *Let $t \in [0,T]$, $v \in L^2(\Omega \times [0,T])$ and $u \in \mathbb{L}^{1,2,f}$ be such that:*

i) $1_{[t,T]}(\cdot)v_\cdot \in \operatorname{Dom}\delta_2$.

ii) $u_s v_\cdot \in L^2(\Omega \times [0,T])$ *for almost all* $s \in [0,t]$.

iii) $\left(u_s \int_t^T v_r \delta_2 B_r - \int_t^T v_r (D_r u_s) dr\right) \in L^2(\Omega)$ *for almost all* $s \in [0,t]$.

Then, for a.a. $s \in [0,t]$, $1_{[t,T]}(\cdot)u_s v_\cdot \in \operatorname{Dom}\delta_2$ *and*

$$\int_t^T u_s v_r \delta_2 B_r = u_s \int_t^T v_r \delta_2 B_r - \int_t^T v_r (D_r u_s) dr.$$

Proof. By the definition of $\mathbb{L}^{1,2,f}$ and (4.32), we have that there exists a sequence $\{u^{(n)} \in \mathcal{S}(L^2([0,T])) : n \in \mathbb{N}\}$ such that

$$E\left(\left|u_s - u_s^{(n)}\right|^2 + \int_s^T \left|D_r\left(u_s - u_s^{(n)}\right)\right|^2 dr\right) \to 0, \quad \text{as } n \to \infty, \tag{5.19}$$

for almost all $s \in [0,T]$, and, for $G \in \mathcal{S}$,

$$E\left(Gu_s^{(n)} \int_t^T v_r \delta_2 B_r\right) = E\left(\int_t^T \left(D_r(Gu_s^{(n)})\right) v_r dr\right), \quad \text{for all } s \in [0,T].$$

Hence, taking limit as n goes to infinity and (5.19) imply that, for $G \in \mathcal{S}$,

$$E\left(u_s G \int_t^T v_r \delta_2 B_r\right) = E\left(\int_t^T \left(D_r(u_s G)\right) v_r dr\right), \quad \text{for almost all } s \in [0,t]. \tag{5.20}$$

Consequently, choosing $s \in [0,t]$ such that Assumptions i) and ii), and (5.20) hold, we obtain

$$\begin{aligned} E\left(\int_t^T (D_r G) u_s v_r dr\right) &= E\left(\int_t^T [D_r(Gu_s) - G(D_r u_s)] v_r dr\right) \\ &= E\left(G\left[u_s \int_t^T v_r \partial_2 B_r - \int_t^T (D_r u_s) v_r dr\right]\right). \end{aligned}$$

Thus, the definition of δ_2 (i.e., (4.2)) yields the result is satisfied. □

Lemma 63 *Let $u \in \mathbb{L}^F$ and $\varepsilon > 0$. Then, $r \mapsto \frac{1}{\varepsilon}\int_{(r-\varepsilon)\vee 0}^{r} u_s ds$ also belongs to $\mathbb{L}^F$ and*

$$\left\|\frac{1}{\varepsilon}\int_{(\cdot-\varepsilon)\vee 0}^{\cdot} u_s ds\right\|_{\mathbb{L}^F} \leq \|u\|_{\mathbb{L}^F}.$$

Proof. In the analysis of (4.38) and (4.39), we also establish that the integral $\frac{1}{\varepsilon}\int_{(\cdot-\varepsilon)\vee 0}^{\cdot} u_s ds$ belongs to the space $\mathbb{L}^F$. Moreover, using (4.38) and (4.39), together with Hölder inequality, we have, for $\varepsilon > 0$,

$$\begin{aligned} \left\|\frac{1}{\varepsilon}\int_{(\cdot-\varepsilon)\vee 0}^{\cdot} u_s ds\right\|_{\mathbb{L}^F}^2 &= \int_0^T \left(\frac{1}{\varepsilon}\int_{(r-\varepsilon)\vee 0}^{r} u_s ds\right)^2 dr + \int_0^T\int_0^\theta \left(\frac{1}{\varepsilon}\int_{(r-\varepsilon)\vee 0}^{r} D_\theta u_s ds\right)^2 dr d\theta \\ &\quad + \int_0^T\int_0^T\int_0^{\theta_1\vee\theta_2} \left(\frac{1}{\varepsilon}\int_{(r-\varepsilon)\vee 0}^{r} D^2_{\theta_1,\theta_2} u_s ds\right)^2 dr d\theta_1 d\theta_2 \\ &\leq \frac{1}{\varepsilon}\int_0^T\int_{(r-\varepsilon)\vee 0}^{r} u_s^2 ds dr + \frac{1}{\varepsilon}\int_0^T\int_0^\theta\int_{(r-\varepsilon)\vee 0}^{r} (D_\theta u_s)^2 ds dr d\theta \\ &\quad + \frac{1}{\varepsilon}\int_0^T\int_0^T\int_0^{\theta_1\vee\theta_2}\int_{(r-\varepsilon)\vee 0}^{r} \left(D^2_{\theta_1,\theta_2} u_s\right)^2 ds dr d\theta_1 d\theta_2. \end{aligned}$$

Hence, the result is a consequence of Fubini's theorem. □

We are ready to establish the relation between the forward and Skorohod integrals. In the following result, we use the convention $\mathbb{L}^F_{q-} = \mathbb{L}^F \cap \mathbb{L}^{1,2,f}_{q-}$, where $\mathbb{L}^{1,2,f}_{q-}$ is defined in (4.47).

Theorem 67 *Let* $u \in (\mathbb{L}^F_{1-})_{loc}$ *and* $0 \le a < b \le T$. *Then,* $u1_{[a,b]} \in (Dom\,\delta_2)_{loc} \cap Dom\,\delta^-_B([a,b])$ *and*

$$\int_a^b u_s dB_s^- = \delta_2(u1_{[a,b]}) + \int_a^b (D^-u)_s ds, \tag{5.21}$$

where D^-u *is introduced in (4.47) and (4.48).*

Proof. By Propositions 78 and 97, (4.40) and (4.48), we can assume that u belongs to $\mathbb{L}^F_{1-}$. Moreover, since $u1_{[a,b]}$ is also in $\mathbb{L}^F_{1-}$, we take the case that $a = 0$ and $b = T$. In order to simplify the notation, we use the convention $u \equiv 0$ on $[0,T]^c$. So, we can find a sequence $\{u^{(n)} \in \mathcal{S}(L^2([0,T])) : n \in \mathbb{N}\}$ that converges to u in $\mathbb{L}^F$. Furthermore, without loss of generality, we suppose that, for $n \in \mathbb{N}$, $u^{(n)}$, $Du^{(n)}$ and $D^2u^{(n)}$ are bounded. Now, let $\varepsilon > 0$. Then, Theorem 48 and Lemma 62 imply that, for $n \in \mathbb{N}$,

$$\begin{aligned}
&\frac{1}{\varepsilon}\int_0^T u_s^{(n)}\left(B_{(s+\varepsilon)\wedge T} - B_s\right) ds \\
&= \frac{1}{\varepsilon}\int_0^T \delta_2\left(u_s^{(n)} 1_{]s,(s+\varepsilon)\wedge T]}\right) ds + \frac{1}{\varepsilon}\int_0^T \int_s^{(s+\varepsilon)\wedge T} D_r u_s^{(n)} dr ds \\
&= \int_0^T \left(\frac{1}{\varepsilon}\int_{r-\varepsilon}^r u_s^{(n)} ds\right)\delta_2 B_r + \int_0^T \left(\frac{1}{\varepsilon}\int_{r-\varepsilon}^r D_r u_s^{(n)} ds\right) dr.
\end{aligned} \tag{5.22}$$

Note that

$$\begin{aligned}
&E\left(\int_0^T \left|u_s^{(n)} - u_s\right| \left|B_{(s+\varepsilon)\wedge T} - B_s\right| ds\right) \\
&\le 2E\left(\left[\sup_{r\in[0,T]} |B_r|\right] \int_0^T \left|u_s^{(n)} - u_s\right| ds\right) \\
&\le 2T^{1/2}\left(E\left[\sup_{r\in[0,T]} B_r^2\right]\right)^{1/2} \left(E\int_0^T \left|u_s^{(n)} - u_s\right|^2 ds\right)^{1/2},
\end{aligned}$$

$$\int_0^T \int_{r-\varepsilon}^r \left|D_r u_s^{(n)} - D_r u_s\right| ds dr \le \int_0^T \int_0^r \left|D_r u_s^{(n)} - D_r u_s\right| ds dr$$

and that Theorem 51 and Lemma 63 lead us to obtain

$$E\left(\left(\int_0^T \left(\frac{1}{\varepsilon}\int_{(r-\varepsilon)\vee 0}^r \left[u_s^{(n)} - u_s\right] ds\right)\delta_2 B_r\right)^2\right) \le 2\left\|u^{(n)} - u\right\|^2_{\mathbb{L}^F}.$$

Thus, the last three inequalities yield that (5.22) becomes

$$\begin{aligned}
&\frac{1}{\varepsilon}\int_0^T u_s\left(B_{(s+\varepsilon)\wedge T} - B_s\right) ds \\
&= \int_0^T \left(\frac{1}{\varepsilon}\int_{r-\varepsilon}^r u_s ds\right)\delta_2 B_r + \int_0^T \left(\frac{1}{\varepsilon}\int_{r-\varepsilon}^r D_r u_s ds\right) dr.
\end{aligned} \tag{5.23}$$

Now, we deal with the convergence of (5.23), as ε goes to zero. Towards this end, we observe that we can apply Theorem 51 and Lemma 63 again to get

$$
\begin{aligned}
&E\left(\left(\int_0^T\left[\frac{1}{\varepsilon}\int_{r-\varepsilon}^r u_s ds - u_r\right]\delta_2 B_r\right)^2\right)\\
&\leq 2\left\|\frac{1}{\varepsilon}\int_{\cdot-\varepsilon}^{\cdot} u_s ds - u_\cdot\right\|_{\mathbb{L}^F}^2 \leq 6\left\|\frac{1}{\varepsilon}\int_{\cdot-\varepsilon}^{\cdot}\left[u_s - u_s^{(n)}\right]ds\right\|_{\mathbb{L}^F}^2\\
&\quad +6\left\|\frac{1}{\varepsilon}\int_{\cdot-\varepsilon}^{\cdot} u_s^{(n)} ds - u_\cdot^{(n)}\right\|_{\mathbb{L}^F}^2 + 6\left\|u^{(n)} - u\right\|_{\mathbb{L}^F}^2\\
&\leq 6\left\|\frac{1}{\varepsilon}\int_{\cdot-\varepsilon}^{\cdot} u_s^{(n)} ds - u_\cdot^{(n)}\right\|_{\mathbb{L}^F}^2 + 12\left\|u^{(n)} - u\right\|_{\mathbb{L}^F}^2.
\end{aligned}
$$

Consequently, given $\eta > 0$, we can find $n_0 \in \mathbb{N}$ such that

$$
E\left(\left(\int_0^T\left[\frac{1}{\varepsilon}\int_{r-\varepsilon}^r u_s ds - u_r\right]\delta_2 B_r\right)^2\right) \leq \eta + 6\left\|\frac{1}{\varepsilon}\int_{\cdot-\varepsilon}^{\cdot} u_s^{(n_0)} ds - u^{(n_0)}\right\|_{\mathbb{L}^F}^2,
$$

which gives

$$
\limsup_{\varepsilon\to 0} E\left(\left(\int_0^T\left[\frac{1}{\varepsilon}\int_{(r-\varepsilon)\vee 0}^r u_s ds - u_r\right]\delta_2 B_r\right)^2\right) \leq \eta, \quad \text{for any } \eta > 0. \tag{5.24}
$$

Indeed, for instance, we have $\frac{1}{\varepsilon}\int_{r-\varepsilon}^r u_s^{(n_0)} ds$ converges to $u_r^{(n_0)}$, for almost all $(\omega, r) \in \Omega \times [0,T]$, as $\varepsilon \to 0$. Therefore, the dominated convergence theorem and the fact that $u^{(n_0)}$ is a bounded process by definition imply

$$
E\left(\int_0^T\left[\frac{1}{\varepsilon}\int_{r-\varepsilon}^r u_s^{(n_0)} ds - u_r^{(n_0)}\right]^2 dr\right) \to 0, \quad \text{as } \varepsilon \to 0.
$$

Finally, (4.47) allows us to conclude

$$
\begin{aligned}
&E\left(\left|\int_0^T\left(\frac{1}{\varepsilon}\int_{(r-\varepsilon)\vee 0}^r D_r u_s ds\right)dr - \int_0^T (D^- u)_r dr\right|\right)\\
&\leq E\left(\int_0^T\frac{1}{\varepsilon}\int_{r-\varepsilon}^r \left|D_r u_s - (D^- u)_r\right| ds dr\right)\\
&\leq \int_0^T \sup_{r-\varepsilon\leq s\leq r} E\left(\left|D_r u_s - (D^- u)_r\right|\right) dr \to 0, \quad \text{as } \varepsilon \to 0.
\end{aligned}
$$

Hence, Definition 43, (5.23) and (5.24) imply that the left-hand side of (5.23) converges to the right-hand side of (5.21) in probability, as ε goes to zero, and, in consequence, the result is satisfied. □

Remark 79 *Note that Theorem 67 is still true if* u *belongs to* $(\mathbb{D}^{1,2}(L^2([0,T])) \cap \mathbb{L}_{1-}^{1,2,f})_{loc}$. *Indeed, we only need to proceed as in the last proof and use Remark 54, Lemma 36 and the fact that*

$$
\left\|\frac{1}{\varepsilon}\int_{(r-\varepsilon)\vee 0}^{\cdot} u_s ds\right\|_{D^{1,2}(L^2([0,T]))} \leq \|u\|_{D^{1,2}(L^2([0,T]))}
$$

instead of Theorem 51, Lemma 62 and Lemma 63, respectively. We also observe that the proof of the last inequality is quite similar to that of Lemma 63. The details are left to the reader as an exercise.

By Definition 43, roughly speaking, the forward integral is the limit in probability of Riemann sums defined by taking the values of the integrands on the left-points of each interval (see Propositions 96 and 98). Thus, the main problem to handle the forward integral is to obtain suitable estimates for the $L^p(\Omega)$-norm of this integral. One way to archive this is to take advantage of the relation between this integral and the operator δ_2 stated in Theorem 67, together with properties that the operator δ_2 enjoys. For instance, we have that, for $u \in \mathbb{L}^F_{2-}$ and $0 \le a < b \le T$, Theorem 51 allows us to show

$$E\left(\left\{\int_a^b u_s dB_s^-\right\}^2\right) \le 4\|u.1_{[a,b]}(\cdot)\|^2_{\mathbb{L}^F} + 2(b-a)E\left(\int_a^b \left[(D^-u)_s\right]^2 ds\right). \tag{5.25}$$

Note that we are also able to apply Proposition 62 to deal with the $L^p(\Omega)$-norm of the forward integral.

Other example of the importance of relation (5.21) to estimate the p-moment of the forward integral is the following result.

Proposition 99 *Let u be a process in $\mathbb{L}^F_{1-}$, $p \in (2,4)$ and $q = 2p/(4-p)$ such that $u \in L^q(\Omega \times [0,T])$ and $D^-u \in L^p(\Omega \times [0,T])$. Then, there exists a constant C depending only on p and T such that*

$$\begin{aligned} E\left(\sup_{t\in[0,T]}\left|\int_0^t u_s dB_s^-\right|^p\right) \le\; & CE\Bigg[\int_0^T |u_s|^q ds + \int_{\Delta_1^T} \{(Du_t)_s\}^2 dsdt \\ & + \int_{\Delta_2^T} \left\{(D^2u_t)_{r,s}\right\}^2 drdsdt + \int_0^T \left|(D^-u)_s\right|^p ds\Bigg]. \end{aligned}$$

Proof. Since $u \in \mathbb{L}^F_{1-}$, we have that (5.21) implies

$$\int_0^t u_s dB_s^- = \delta_2(u1_{[0,t]}) + \int_0^t (D^-u)_s ds, \quad t \in [0,T].$$

Thus, the result is an immediate consequence of Theorem 52 and Hölder inequality. □

We observe that Theorem 67 is not only useful to deal with the p-moments of the forward integral, but also to study properties of this integral, as the following results show.

Proposition 100 *Let $p > 2$ and $u \in \mathbb{L}^F_{1-} \cap L^p(\Omega \times [0,T])$. Then, the process $\{\int_0^t u_s dB_s^- : t \in [0,T]\}$ has continuous paths.*

Proof. The result is an immediate consequence of (5.21) and Proposition 80. □

In Russo and Vallois [191], instead of Definition 43, the forward integral $\int_0^{\cdot} u_s dB_s^-$ is defined as the uniform limit on $[a,b]$ in probability of (5.1). Concerning this, we have the following result.

Proposition 101 *Let $u \in \mathbb{L}^F_{1-} \cap L^p(\Omega \times [0,T])$, for some $p > 2$. Then, for $0 \le a < b \le T$, $\varepsilon^{-1}\int_a^{\cdot} u_s \left(B_{(s+\varepsilon)\wedge T} - B_s\right) ds$ converges to $\int_a^{\cdot} u_s dB_s^-$ uniformly on $[a,b]$ in probability, as $\varepsilon \to 0$.*

Proof. From Theorem 67, without loss of generality, we can assume that $a = 0$ and $b = T$. Let $t \in [0, T]$. Then, proceeding as in the proof of Theorem 67, we have

$$\begin{aligned}
&\frac{1}{\varepsilon}\int_0^t u_s\left(B_{(s+\varepsilon)\wedge T} - B_s\right) ds \\
&= \int_0^{(t+\varepsilon)\wedge T}\left(\frac{1}{\varepsilon}\int_{(r-\varepsilon)\vee 0}^{r\wedge t} u_s ds\right)\delta_2 B_r + \int_0^t\left(\frac{1}{\varepsilon}\int_s^{(s+\varepsilon)\wedge T} D_r u_s dr\right) ds \\
&= \int_0^t\left(\frac{1}{\varepsilon}\int_{(r-\varepsilon)\vee 0}^{r} u_s ds\right)\delta_2 B_r + \int_0^t\left(\frac{1}{\varepsilon}\int_s^{(s+\varepsilon)\wedge T} D_r u_s dr\right) ds \\
&\quad + \frac{1}{\varepsilon}\int_t^{(t+\varepsilon)\wedge T}\left(\int_{(r-\varepsilon)\vee 0}^{t} u_s ds\right)\delta_2 B_r.
\end{aligned}$$

Hence, using Lemma 62 and Theorem 48 again, we get

$$\begin{aligned}
&\frac{1}{\varepsilon}\int_0^t u_s\left(B_{(s+\varepsilon)\wedge T} - B_s\right) ds \\
&= \int_0^t\left(\frac{1}{\varepsilon}\int_{(r-\varepsilon)\vee 0}^{r} u_s ds\right)\delta_2 B_r + \int_0^t\left(\frac{1}{\varepsilon}\int_s^{(s+\varepsilon)\wedge T} D_r u_s dr\right) ds \\
&\quad + \frac{1}{\varepsilon}\int_{(t-\varepsilon)\vee 0}^{t} u_s\left(B_{(s+\varepsilon)\wedge T} - B_t\right) ds - \frac{1}{\varepsilon}\int_{(t-\varepsilon)\vee 0}^{t}\int_t^{(s+\varepsilon)\wedge T} D_r u_s dr ds. \qquad (5.26)
\end{aligned}$$

Now, in order to analyze the uniform convergence on $[0, T]$ in probability of left-hand side of the last equality, we divide the proof into several steps.
Step 1: We first see that

$$\sup_{0\le t\le T}\left|\varepsilon^{-1}\int_{(t-\varepsilon)\vee 0}^{t} u_s\left(B_{(s+\varepsilon)\wedge T} - B_t\right) ds\right| \to 0, \quad \text{with probability 1, as } \varepsilon \to 0.$$

Note that Hölder inequality and the fact that Brownian motion has Hölder continuous paths for any exponent less than $1/2$ (see Section 1.2) lead us to write

$$\begin{aligned}
\left|\varepsilon^{-1}\int_{(t-\varepsilon)\vee 0}^{t} u_s\left(B_{(s+\varepsilon)\wedge T} - B_t\right) ds\right| &\le \varepsilon^{-1}\left(\sup_{|r-s|\le\varepsilon}|B_s - B_r|\right)\int_{(t-\varepsilon)\vee 0}^{t}|u_s| ds \\
&\le \varepsilon^{-1/p}\left(\sup_{|r-s|\le\varepsilon}|B_s - B_r|\right)\left(\int_0^T |u_s|^p ds\right)^{1/p} \\
&\le M_\rho \varepsilon^{\frac{1}{2}-\rho-\frac{1}{p}}\left(\int_0^T |u_s|^p ds\right)^{1/p},
\end{aligned}$$

where M_ρ is a random variable. Since $\frac{1}{2} - \rho - \frac{1}{p} > 0$ for ρ small enough, we have that the claim of this part of the proof is satisfied.
Step 2: Here, we show that $\sup_{t\in[0,T]}\left(\varepsilon^{-1}\int_{(t-\varepsilon)\vee 0}^{t}\int_t^{(s+\varepsilon)\wedge T} D_r u_s dr ds\right)$ goes to zero in probability.

For $M > 0$, set $\mathcal{D}u(r, s, M) = (-M) \vee (D_r u_s \wedge M)$, $0 \le s < r \le T$. Then, the triangle and Hölder inequalities yield

$$\begin{aligned}
&\left|\frac{1}{\varepsilon}\int_{(t-\varepsilon)\vee 0}^{t}\int_{t}^{(s+\varepsilon)\wedge T} D_r u_s drds\right| \\
&\leq \frac{1}{\varepsilon}\int_{(t-\varepsilon)\vee 0}^{t}\int_{t}^{(s+\varepsilon)\wedge T} |D_r u_s - \mathcal{D}u(r,s,M)|\, drds \\
&\quad + \frac{1}{\varepsilon}\int_{(t-\varepsilon)\vee 0}^{t}\int_{t}^{(s+\varepsilon)\wedge T} |\mathcal{D}u(r,s,M)|\, drds \\
&\leq \left(\int_{(t-\varepsilon)\vee 0}^{t}\int_{t}^{(s+\varepsilon)\wedge T} |D_r u_s - \mathcal{D}u(r,s,M)|^2\, drds\right)^{1/2} + M\varepsilon.
\end{aligned}$$

Therefore, given $\eta > 0$ there exists $M_\eta > 0$ such that

$$\begin{aligned}
&E\left(\sup_{t\in[0,T]}\left|\frac{1}{\varepsilon}\int_{(t-\varepsilon)\vee 0}^{t}\int_{t}^{(s+\varepsilon)\wedge T} D_r u_s drds\right|\right) \\
&\leq \left(\int_{0}^{T}\int_{s}^{T} E\left(|D_r u_s - \mathcal{D}u(r,s,M_\eta)|^2\right) drds\right)^{1/2} + M_\eta\varepsilon \leq \eta + M_\eta\varepsilon.
\end{aligned}$$

In order words, we have proven $\limsup_{\varepsilon\to 0} E\left(\sup_{t\in[0,T]}\left|\frac{1}{\varepsilon}\int_{(t-\varepsilon)\vee 0}^{t}\int_{t}^{(s+\varepsilon)\wedge T} D_r u_s drds\right|\right) \leq \eta$. Thus, the assertion of this step is also true.

Step 3: It is time to study the uniform convergence in probability on $[0,T]$ of $\varepsilon^{-1}\int_0^t\int_s^{(s+\varepsilon)\wedge T} D_r u_s drds$ to $\int_0^t (D^- u)_s ds$, as ε tends to zero.

By Fubini's theorem, we obtain

$$\begin{aligned}
&\left|\int_0^t\left(\frac{1}{\varepsilon}\int_s^{(s+\varepsilon)\wedge T} D_r u_s dr\right) ds - \int_0^t (D^- u)_s ds\right| \\
&\leq \frac{1}{\varepsilon}\int_0^T\int_{(r-\varepsilon)\vee 0}^{r} |D_r u_s - (D^- u)_r|\, dsdr + \frac{1}{\varepsilon}\int_0^T |(D^- u)_s|\, |s-\varepsilon-((s-\varepsilon)\vee 0)|\, ds \\
&\quad + \frac{1}{\varepsilon}\int_t^{(t+\varepsilon)\wedge T}\int_{r-\varepsilon}^{r} |D_r u_s| dsdr.
\end{aligned}$$

Consequently, (4.47) and the dominated convergence theorem imply

$$\begin{aligned}
&E\left(\sup_{t\in[0,T]}\left|\int_0^t\left(\frac{1}{\varepsilon}\int_s^{(s+\varepsilon)\wedge T} D_r u_s dr\right) ds - \int_0^t (D^- u)_s ds\right|\right) \\
&\leq \frac{2}{\varepsilon}\int_0^T\int_{(r-\varepsilon)\vee 0}^{r} E\left(|D_r u_s - (D^- u)_r|\right) dsdr \\
&\quad + \frac{1}{\varepsilon}\int_0^T \left(E|(D^- u)_s|\right) |s-\varepsilon-((s-\varepsilon)\vee 0)|\, ds + E\left(\sup_{t\in[0,T]}\int_t^{(t+\varepsilon)\wedge T} |(D^- u)_r| dr\right) \\
&\leq 2\int_0^T \sup_{(r-\varepsilon)\vee 0\leq s\leq r} E\left(|D_r u_s - (D^- u)_r|\right) dr \\
&\quad + \frac{1}{\varepsilon}\int_0^T \left(E|(D^- u)_s|\right) |s-\varepsilon-((s-\varepsilon)\vee 0)|\, ds \\
&\quad + E\left(\sup_{t\in[0,T]}\int_t^{(t+\varepsilon)\wedge T} |(D^- u)_r| dr\right) \to 0, \quad \text{as } \varepsilon\to 0.
\end{aligned}$$

Step 4: Finally, we give the missing details of the proof.

Let $q \in (2,4)$ such that $p = \frac{2q}{4-q}$. Then, Lemma 63 and Theorems 52 imply that there is a constant $C > 0$ such that

$$C^{-1}E\left(\sup_{t\in[0,T]}\left|\int_0^t\left(\frac{1}{\varepsilon}\int_{(r-\varepsilon)\vee 0}^r u_s ds - u_r\right)\delta_2 B_r\right|^q\right)$$
$$\leq E\left[\int_0^T\left|\frac{1}{\varepsilon}\int_{(r-\varepsilon)\vee 0}^r u_s ds - u_r\right|^p dr\right] + \left\|\left(\frac{1}{\varepsilon}\int_{(\cdot-\varepsilon)\vee 0}^{\cdot} u_s ds - u_{\cdot}\right)\right\|^2_{\mathbb{L}^F} \to 0, \quad \text{as } \varepsilon \to 0,$$

where the convergence to zero follows by proceeding as in the proofs of (5.24) and Lemma 63. Thus, the result is now a consequence of Steps 1-3 and (5.26). □

As a consequence of Proposition 101, we have that the forward integral also satisfies Proposition 60 and Lemma 48.

Proposition 102 *Let $u \in \mathbb{L}^F_{1-} \cap L^p(\Omega \times [0,T])$, for some $p > 2$, $(\mathcal{F}_t)_{t\in[0,T]}$ the filtration generated by B and all the P-negligible sets, and τ an $\mathcal{F}_t$-stopping time. Then, $u1_{[0,\tau]}$ belongs to $Dom\delta^-_B([0,t])$ and*

$$\int_0^s u_r dB^-_r\big|_{s=t\wedge\tau} = \int_0^t u_r 1_{[0,\tau]}(r) dB^-_r, \quad t \in [0,T].$$

Remark 80 *By Proposition 100, we have that $\int_0^{\cdot} u_r dB^-_r$ has continuous paths. Therefore, $\int_0^s u_r dB^-_r\big|_{s=t\wedge\tau}$ is well-defined, for all $t \in [0,T]$. Moreover, It is enough to assume that τ is a non-negative random variable since, unlike the Itô's integral, we do not need to consider adapted processes to the underlying filtration (see the proof of this result).*

Proof. Let $\eta > 0$ and $t \in [0,T]$. Then, we can apply Proposition 101 to establish

$$P\left[\left|\int_0^s u_r dB^-_r\big|_{s=t\wedge\tau} - \frac{1}{\varepsilon}\int_0^t u_s 1_{[0,\tau]}(s)\left(B_{(s+\varepsilon)\wedge T} - B_s\right) ds\right| > \eta\right]$$
$$= P\left[\left|\int_0^s u_r dB^-_r\big|_{s=t\wedge\tau} - \frac{1}{\varepsilon}\int_0^{t\wedge\tau} u_s\left(B_{(s+\varepsilon)\wedge T} - B_s\right) ds\right| > \eta\right]$$
$$\leq P\left[\sup_{\theta\in[0,T]}\left|\int_0^\theta u_r dB^-_r - \frac{1}{\varepsilon}\int_0^\theta u_s\left(B_{(s+\varepsilon)\wedge T} - B_s\right) ds\right| > \eta\right] \to 0, \quad \text{as } \varepsilon \to 0.$$

In consequence, the result holds due to Definition 43. □

Other applications of Theorem 67 are the Itô's formula and Fubini's theorem for the forward integral, which are stated in the following two results.

Theorem 68 *Consider the $\mathbb{R}$-valued stochastic process*

$$X_t = X_0 + \int_0^t u_s dB^-_s + \int_0^t v_s ds, \quad t \in [0,T],$$

where $X_0 \in \mathbb{D}^{1,2}_{loc}$, $u \in (\mathbb{D}^{2,2}(L^2([0,T])) \cap \mathbb{D}^{1,4}(L^2([0,T])) \cap \mathbb{L}^{1,2,f}_{1-})_{loc}$ is such that $\int_0^{\cdot} u_{i,s}\delta_2 B_s$ has continuous paths and $v, D^-u \in (\mathbb{D}^{1,2}(L^2([0,T])))_{loc}$. Let $f : \mathbb{R} \to \mathbb{R}$ be a twice continuously differentiable function. Then,

$$f(X_t) = f(X_0) + \int_0^t f'(X_s)v_s ds + \int_0^t f'(X_s)u_s dB^-_s + \frac{1}{2}\int_0^t f''(X_s)u_s^2 ds, \quad t \in [0,T]. \tag{5.27}$$

Remark 81 *Note that (5.27) agrees with the usual Itô's formula given by the classical stochastic calculus (see (3.21)). Moreover, in Frandoli and Russo [51] and some references therein, the Itô's formula has been considered for stochastic processes x having finite quadratic variation. That is, the integral $\varepsilon^{-1}\int_0^{\cdot}\left(x_{(s+\varepsilon)\wedge T}-x_s\right)^2 ds$ converges uniformly on $[0,T]$ in probability. This limit is denoted by $[x]_\cdot$. In this case, for $f\in\mathcal{C}^2(\mathbb{R})$, the integral $\int_0^{\cdot} f'(x_s)dx_s^-$ is well-defined and*

$$f(x_t)=f(x_0)+\int_0^t f'(x_s)dx_s^- +\frac{1}{2}\int_0^t f''(x_s)d[x]_s,\quad t\in[0,T].$$

Proof of Theorem 68: By Remark 79 and Theorem 54, we have

$$\begin{aligned} f(X_t) &= f(X_0)+\int_0^t f'(X_s)u_s\delta_2 B_s+\int_0^t f'(X_s)v_s ds+\int_0^t f'(X_s)(D^-u)_s ds \\ &\quad +\frac{1}{2}\int_0^t f''(X_s)u_s^2 ds+\int_0^t f''(X_s)\left(D^-X\right)_s u_s ds,\quad t\in[0,T]. \end{aligned}$$

The localization argument utilized in the proof of Theorem 54 implies that we can assume that $f'(X)u\in\mathbb{D}^{1,2}(L^2([0,T]))$, f' and f'' are bounded functions on $\mathbb{R}$, and that $D^-u\in\mathbb{D}^{1,2}(L^2([0,T]))$. In consequence, (4.49) yields that the last equality becomes

$$\begin{aligned} f(X_t) &= f(X_0)+\int_0^t f'(X_s)u_s\delta_2 dB_s+\int_0^t f'(X_s)v_s ds \\ &\quad +\int_0^t \left(D^-\left(f'(X_\cdot)u_\cdot\right)\right)_s ds+\frac{1}{2}\int_0^t f''(X_s)u_s^2 ds,\quad t\in[0,T]. \end{aligned}$$

Hence, using Remark 79 again, we get that (5.27) holds and, therefore, the proof is finished. □

Now, we consider Fubini's theorem for the forward integral.

Theorem 69 *Let μ be a σ-finite measure on the measurable space $(X,\mathcal{C})$. Assume that $\varphi:\Omega\times X\to L^2([0,T])$ is an $\mathcal{F}\otimes\mathcal{C}$-measurable function satisfying the following conditions:*

i) φ belongs to $L^1\left(X;\mathbb{D}^{1,2}(L^2([0,T]))\right)$.

ii) $(\omega,x)\mapsto\delta_2\left(\varphi(x)\right)$ is $\mathcal{F}\otimes\mathcal{C}$-measurable.

iii) The Bochner integral $\int_X\delta_2\left(\varphi(x)\right)\mu(dx)$ is well-defined w.p.1 and belongs to $L^2(\Omega)$.

iv) For μ-almost all $x\in X$, we have that $\varphi(x)\in\mathbb{L}_{1-}^{1,2,f}$ and

$$\lim_{n\to\infty}\int_X\int_0^T \sup_{0\vee(s-(1/n))\le t\le s} E\left|(D_s\varphi(x)(t))-(D^-\varphi(x))_s\right| ds\mu(dx)=0.$$

Then, $\int_0^T\int_X\varphi(x)(s)\mu(dx)dB_s^-$ and $\int_X\int_0^T\varphi(x)(s)dB_s^-\mu(dx)$ are well-defined, and they are the same. That is,

$$\int_0^T\int_X\varphi(x)(s)\mu(dx)dB_s^- =\int_X\int_0^T\varphi(x)(s)dB_s^-\mu(dx),\quad w.p.1.$$

Proof. We can apply Hypotheses *i)* and *iv)*, together with Remark 79, to establish that, for μ-almost all $x\in X$, we are able to write

$$\int_0^T\varphi(x)(s)dB_s^- =\delta_2\left(\varphi(x)\right)+\int_0^T\left(D^-\varphi(x)\right)_s ds.$$

Therefore, Theorem 48, Assumptions *i*)-*iv*) and Fubini's theorem give

$$\begin{aligned}
&\int_X \int_0^T \varphi(x)(s) dB_s^- \mu(dx) \\
&= \delta_2\left(\int_X \varphi(x)\mu(dx)\right) + \int_X \int_0^T \left(D^-\varphi(x)\right)_s ds\mu(dx) \\
&= \delta_2\left(\int_X \varphi(x)\mu(dx)\right) + \int_0^T \int_X \left(D^-\varphi(x)\right)_s \mu(dx)ds, \quad \text{w.p.1.} \qquad (5.28)
\end{aligned}$$

On the other hand, by Hypothesis *i*), we obtain that $t \mapsto \int_X \varphi(x)(t)\mu(dx)$ belongs to $\mathbb{D}^{1,2}(L^2([0,T]))$. Moreover, using Fubini's theorem again, we have,

$$\begin{aligned}
&\int_0^T \sup_{0\vee(s-(1/n))\le t\le s} E\left|D_s\left(\int_X \varphi(x)(t)\mu(dx)\right) ds - \int_X \left(D^-\varphi(x)\right)_s \mu(dx)\right| ds \\
&\le \int_0^T \int_X \sup_{0\vee(s-(1/n))\le t\le s} E\left|D_s\varphi(x)(t) - \left(D^-\varphi(x)\right)_s\right| \mu(dx)ds \\
&= \int_X \int_0^T \sup_{0\vee(s-(1/n))\le t\le s} E\left|D_s\varphi(x)(t) - \left(D^-\varphi(x)\right)_s\right| ds\mu(dx) \to 0,
\end{aligned}$$

as $n \to \infty$, where last convergence follows from Assumption *iv*). In consequence, we have proven that $t \mapsto \int_X \varphi(x)(t)\mu(dx)$ is in $\mathbb{L}_{1-}^{1,2,f}$ with

$$\left(D^- \int_X \varphi(x)\mu(dx)\right)_s = \int_X \left(D^-\varphi(x)\right)_s \mu(dx), \quad s \in [0;T].$$

Finally, Theorems 48 and 67, and (5.28) imply that the result is satisfied. □

Our next aim is to state two substitution formulae for the forward integral. To do so, we now introduce some notation: by $\mathcal{R}$, we denote the family of stochastic processes $u = \{u_t(x) : t \in [0,T],\ x \in \mathbb{R}\}$ parameterized by $x \in \mathbb{R}$, which are $\mathcal{P}_{[0,T]} \otimes \mathcal{B}(\mathbb{R})$-measurable. Here, $\mathcal{P}_{[0,T]}$ is the predictable σ-algebra on $\Omega \times [0,T]$. Consider the following subspaces of $\mathcal{R}$:

$$\begin{aligned}
\mathcal{R}_2 = &\left\{u \in \mathcal{R} : \int_0^T u_t(0)^2 dt < \infty,\ u_t(\cdot) \text{ is differentiable and} \right. \\
&\left. \int_{-n}^n \int_0^T u_t'(x)^2 dtdx < \infty \text{ for al } n \ge 1\right\}.
\end{aligned}$$

The following result is a substitution formula for the forward integral. Its proof is due to Russo and Vallois [190].

Proposition 103 *Let* $u \in \mathcal{R}$, $\delta > 1$ *such that:*

i) $E\left(\int_0^T |u_s(0)|^2 ds\right) < \infty$,

ii) For each $N > 0$ *and* $|x|, |y| \le N$, *we have*

$$E\left(\int_0^T |u_s(x) - u_s(y)|^2 ds\right) \le C_N |x-y|^\delta.$$

Then, the Itô's integral $\int_0^T u_s(x)dB_s$ *has a continuous version in* x *and, for every random variable* L, *the composition* $u(L)$ *belongs to* $Dom\,\delta_B^-([0,T])$ *and*

$$\int_0^T u_s(L)dB_s^- = \left(\int_0^T u_s(x)dB_s\right)_{x=L} \qquad w.p.1. \tag{5.29}$$

Remark 82 *From the proof of this result, we can see that (5.29) is true when we change* 0 *and* T *by* a *and* b, *respectively, where* $0 \le a < b \le T$. *Moreover, we can use the ideas developed in the proof of this result to change* B *by a semimartingale (see Russo and Vallois [191]).*

Proof. Observe that, without loss of generality, we can assume that L is a bounded random variable since Proposition 97. So, in the remaining of this proof, we suppose that $|L| \le n$ for some $n \ge 1$. Set

$$U(x) = \int_0^T u_s(x)dB_s, \quad x \in [-n,n],$$

and, for $\varepsilon > 0$,

$$V(x,\varepsilon) = \varepsilon^{-1}\int_0^T u_s(x)(B_{(s+\varepsilon)\wedge T} - B_s)ds = \int_0^T \left(\varepsilon^{-1}\int_{0\vee(r-\varepsilon)}^r u_s(x)ds\right)dB_r.$$

Then, Hypothesis *ii)* implies that, for $x, y \in \mathbb{R}$, $|x|, |y| \le n$,

$$\begin{aligned} &E\left((U(x)-U(y))^2\right) \\ &= E\left(\int_0^T (u_s(x)-u_s(y))^2 ds\right) \le C_{n,T}|x-y|^\delta \end{aligned} \tag{5.30}$$

and

$$\begin{aligned} &E\left((V(x,\varepsilon)-V(y,\varepsilon))^2\right) \\ &= E\left(\int_0^T \left(\varepsilon^{-1}\int_{0\vee(r-\varepsilon)}^r (u_s(x)-u_s(y))\,ds\right)^2 dr\right) \\ &\le E\left(\int_0^T \varepsilon^{-1}\int_{0\vee(r-\varepsilon)}^r (u_s(x)-u_s(y))^2\,dsdr\right) \le C_{n,T}|x-y|^\delta. \end{aligned} \tag{5.31}$$

Therefore, by Theorem 1, we have that $U(\cdot)$ and $V(\cdot,\varepsilon)$ have continuous versions, which are also denoted by $U(\cdot)$ and $V(\cdot,\varepsilon)$, respectively. Note that these continuous versions also satisfy inequalities (5.30) and (5.31), respectively. Consequently, for $\kappa \in (0, \delta-1)$, the Garsia, Rodemich and Rumsey lemma [59] (see Corollary 32 in the appendix) leads us to find two constants C_1 and C_2, independent of ε, and random variables Γ and Γ_ε such that

$$|U(x)-U(y)|^2 \le C_1\Gamma|x-y|^\kappa \quad \text{and} \quad |V(x,\varepsilon)-V(y,\varepsilon)|^2 \le C_1\Gamma_\varepsilon|x-y|^\kappa, \; x,y \in [-n,n],$$

with

$$E(\Gamma), E(\Gamma_\varepsilon) \le C_2.$$

Hence, for a simple and measurable random variable L_m such that $\|L-L_m\|_{L^\infty(\Omega)} \le m^{-1}$,

$$\begin{aligned} E\left((V(L,\varepsilon)-U(L))^2\right) &\le 8E\left((V(L,\varepsilon)-(V(L_m,\varepsilon))^2\right) + 8E\left((V(L_m,\varepsilon)-U(L_m))^2\right) \\ &\quad +8E\left((U(L_m)-U(L))^2\right) \\ &\le 16C_1C_2m^{-\kappa} + 8E\left((V(L_m,\varepsilon)-U(L_m))^2\right). \end{aligned}$$

Thus, we obtain that (5.29) holds since m is arbitrary and we can proceed as in the proof of Proposition 98 to show that $V(L_m, \varepsilon)$ goes to $U(L_m)$ in $L^2(\Omega)$, as ε tends to 0, which follows from the fact that L_m is a bounded simple random variable. □

Proposition 103 requires the existence of moments of the process $u_t(x)$ and its proof is based on Garsia-Rodemich-Rumsey lemma [59]. The version of this result that we utilize in last proof is stated in Barlow and Yor [18], and its proof is provided in Stroock and Varadhan [201]. For the convenience of the reader, we include this lemma and its proof in the Appendix (see Section 8.4). Moreover, If we assume that the random field $u_t(x)$ belongs to the class $\mathcal{R}_2$, then the substitution formula for the forward integral is still true, although we cannot apply Garsia-Rodemich-Rumsey lemma in this case.

Proposition 104 *Let $u \in \mathcal{R}_2$. Then, for every random variable L, $u(L)$ belongs to $Dom\,\delta_B^-([0,T])$ and*

$$\int_0^T u_s(L)dB_s^- = \left(\int_0^T u_s(x)dB_s\right)_{x=L} \quad w.p.1. \tag{5.32}$$

Proof: From Proposition 97, we can assume that the random variable L satisfies $|L| \leq n$ for some $n \geq 1$. For each positive integer $m \geq 1$, let τ_m be the stopping time

$$\tau_m = \inf\left\{t \geq 0 : \int_{-n}^{n} \int_0^t u_s'(x)^2 ds dx \geq m \text{ or } \int_0^t u_s(0)^2 ds \geq m\right\} \wedge T.$$

Replacing the random field $u_t(x)$ by $u_t(x)1_{\{t \leq \tau_m\}}$, we can assume that $\int_0^T u_s(0)^2 ds \leq m$ and $\int_{-n}^{n} \int_0^T u_s'(x)^2 ds dx \leq m$.

On the other hand, we have

$$\begin{aligned}
&\varepsilon^{-1} \int_0^T u_s(L)\left(B_{(s+\varepsilon)\wedge T} - B_s\right) ds \\
&= \varepsilon^{-1} \int_0^T u_s(0)\left(B_{(s+\varepsilon)\wedge T} - B_s\right) ds \\
&\quad + \varepsilon^{-1} \int_0^T \left(\int_0^L u_s'(y)dy\right)\left(B_{(s+\varepsilon)\wedge T} - B_s\right) ds = I_1 + I_2.
\end{aligned}$$

I_1 converges in $L^2(\Omega)$, as ε tends to zero, to the stochastic Itô's integral $\int_0^T u_s(0)dB_s$ because of Proposition 98.

Now, using Fubini's theorem, I_2 has the form

$$I_2 = \varepsilon^{-1} \int_0^L \left(\int_0^T u_s'(y)\left(B_{(s+\varepsilon)\wedge T} - B_s\right) ds\right) dy.$$

We claim that I_2 converges to $I = \int_0^L \left(\int_0^T u_s'(y)dW_s\right) dy$ in $L^2(\Omega)$, as ε goes to zero. Indeed, assume that $u \equiv 0$ on $\Omega \times [0,T]^c \times \mathbb{R}$. Then, Fubini's theorem implies

$$\begin{aligned}
E|I_2 - I|^2 &\leq \frac{1}{\varepsilon^2} E\left|\int_0^L \int_0^T \left(\int_{t-\varepsilon}^t \left[u_s'(y) - u_t'(y)\right] ds\right) dB_t dy\right|^2 \\
&\leq \frac{n}{\varepsilon} \int_{-n}^{n} E \int_0^T \int_{t-\varepsilon}^t \left[u_s'(y) - u_t'(y)\right]^2 ds dt dy. \qquad (5.33)
\end{aligned}$$

The right-hand side of (5.33) converges to zero as ε tends to zero because $u \in \mathcal{R}_2$. Finally, using Fubini's theorem again, we obtain

$$\int_0^T u_s(0)dB_s + \int_0^L \left(\int_0^T u_s'(y)dB_s\right)dy = \left(\int_0^T u_s(x)dB_s\right)_{x=L}.$$

□

In order to finish this section, we establish a result associated with the enlargement of a filtration. Towards this end, we first consider some auxiliary tool.

Lemma 64 *Let $v \in [0,T]$, $u \in \mathbb{L}^{1,2,f}$ and ϕ a $\mathcal{F} \otimes \mathcal{B}([0,T])$-measurable function such that $(1_{[v,T]}(\cdot)\phi(\cdot)) \in Dom\,\delta_2$. Then*

$$E\left(u_t \int_v^T \phi(s)\delta_2 B_s\right) = E\left(\int_v^T (D_s u_t)\phi(s)ds\right), \quad \textit{for a.a. } t \in [0,v].$$

Proof. From the definition of the space $\mathbb{L}^{1,2,f}$, there exists a sequence $\{u^{(n)} : n \in \mathbb{N}\}$ of $L^2([0,T])$-valued smooth random variables such that

$$E\left(\left|u_t^{(n)} - u_t\right|^2 + \int_t^T \left|D_s(u_t^{(n)} - u_t)\right|^2 ds\right) \to 0 \quad \text{as } n \to \infty, \text{ for a.a. } t \in [0,T].$$

Choose $t \in [0,T]$ satisfying this convergence. Hence, the duality relation (4.2) yields, for $n \in \mathbb{N}$,

$$E\left(u_t^{(n)} \int_v^T \phi(s)\delta_2 B_s\right) = E\left(\int_v^T (D_s u_t^{(n)})\phi(s)ds\right).$$

Therefore, we only need to take the limit as n goes to infinity to finish the proof. □

Now, we consider a random variable L in $\mathbb{D}^{1,2}$ such that, for a.a. $s \in [0,T]$, the process

$$I_\cdot(s,L) := 1_{[s,T]}(\cdot)1_{[\int_s^T (D_u L)^2 du > 0]}\left(\int_s^T (D_u L)^2 du\right)^{-1}(D_s L)(D_\cdot L)$$

belongs to Dom δ_2 and we assume that there exists a $\mathcal{P} \otimes \mathcal{B}(\mathbb{R})$-measurable random field $h = \{h_t(y) : t \in [0,T] \text{ and } y \in \mathbb{R}\}$ such that

$$E\left[\int_0^T I_u(s,L)\delta_2 B_u | \mathcal{F}_s \vee L\right] = h_s(L). \tag{5.34}$$

Here, the filtration $(\mathcal{F}_t)_{t\in[0,T]}$ is generated by B and the P-null sets.

In the following result, we use the convention $\sigma_s^\pi(y) = \sigma_s \pi_s(y)$, where σ is a process and π is a random field in $\mathcal{R}$.

Proposition 105 *Let σ be a bounded predictable process, $\pi \in \mathcal{R}_2$ and $L \in \mathbb{D}^{1,2}$ such that, for each $t \in [0,T]$, $\pi_t'(\cdot)$ is continuous, $1_{[0,t]}(\cdot)\sigma^\pi(L)$ is in $\mathbb{L}^F$, L is as in (5.34) with $h(L) \in L^2(\Omega \times [0,T])$ and*

$$E\left(\int_0^T (\sigma_s^\pi)'(L)^2 (D_s L)^2 ds + \left(\int_0^T (\sigma_s^\pi)'(L)^2 ds\right)\left(\int_0^T (D_s L)^2 ds\right)\right) < \infty.$$

Then,

$$E\left(\int_0^t \sigma_s^\pi(L)dB_s^-\right) = E\left(\int_0^t \sigma_s^\pi h_s(L)ds\right), \quad \textit{for } t \in [0,T].$$

Proof. Fix $t \in [0,T]$. By Theorem 67, Lemma 64, the definition of the process $I_.(s,L)$ and (5.34), we can write

$$\begin{aligned} E\left(\int_0^t \sigma_s^\pi(L)dB_s^-\right) &= E\left(\delta_2(1_{[0,t]}(\cdot)\sigma_\cdot^\pi(L)) + \int_0^t (\sigma_s^\pi)'(L)D_sLds\right) \\ &= E\left(\int_0^t (\sigma_s^\pi)'(L)D_sLds\right) \\ &= E\left(\int_0^t \int_s^T I_\theta(s,L)D_\theta(\sigma_s^\pi(L))d\theta ds\right) \\ &= E\left(\int_0^t \left(\int_s^T I_\theta(s,L)\delta_2 B_\theta\right)\sigma_s^\pi(L)ds\right) \\ &= E\left(\int_0^t \sigma_s^\pi(L)E\left[\int_s^T I_\theta(s,L)\delta_2 B_\theta|\mathcal{F}_s \vee L\right]ds\right) \\ &= E\left(\int_0^t \sigma_s^\pi(L)h_s(L)ds\right). \end{aligned}$$

Thus, the proof is complete. □

We are ready to state the result related to the enlargement of a filtration.

Theorem 70 *Under the assumptions of Proposition 105, we have that the process* $\{B_t - \int_0^t h_s(L)ds : t \in [0,T]\}$ *is a martingale with respect to the filtration* $(\sigma(\mathcal{F}_t \vee L))_{t\in[0,T]}$, *where* $\mathcal{F}_t$ *is the* σ*-algebra generated by* $\{B_s : s \in [0,t]\}$, *augmented with the sets of probability zero.*

Proof. Fix $a, b \in [0,T]$ with $a < b$. Let $f \in C^2(\mathbb{R})$ be a smooth function with compact support and F a smooth $\mathcal{F}_a$-measurable random variable. Set

$$\sigma = F1_{[a,b]} \quad \text{and} \quad \pi(x) = f(x)1_{[a,b]}.$$

Then, Proposition 105 yields

$$E\left(Ff(L)\int_a^b h_s(L)ds\right) = E\left(\int_a^b Ff(L)dB_s^-\right) = E\left(Ff(L)\left(B_b - B_a\right)\right),$$

which establishes

$$E\left(Ff(L)\left[B_b - \int_0^b h_s(L)ds\right]\right) = E\left(Ff(L)\left[B_a - \int_0^a h_s(L)ds\right]\right).$$

Therefore, the proof is complete. □

As an exercise, suppose that $L = \lambda B_T + (1-\lambda)\xi$, where $\lambda \in [0,1]$ and ξ is a standard normal random variable independent of the Brownian motion B. Show that L satisfies (5.34) with

$$h_s(y) = \frac{\lambda(y - \lambda B_s)}{\lambda^2(T-s) + (1-\lambda)^2}.$$

The reader interested in results associated with Theorem 70 can consult Itô [88], León et al. [116], Mansuy and Yor [141], and Protter [180] (and references therein).

5.3 Fractional Brownian motion case

We first point out that, in this section, we assume $B = \{B_t : t \in [0,T]\}$ is a fractional Brownian motion with Hurst parameter $H > 1/2$ because the forward integral $\int_a^b u_s dB_s^-$ is not well-defined in general when $H < 1/2$. Indeed, a simple argument is to take the expectation on the approximation of the forward integral and suppose that $u = B$. That is, Fubini's theorem and (1.44) imply that, for $0 \le a < T$,

$$\begin{aligned}
&E\left(\frac{1}{\varepsilon}\int_a^T B_s\left(B_{(s+\varepsilon)\wedge T} - B_s\right) ds\right)\\
&= \frac{1}{2\varepsilon}\int_a^T \left([(s+\varepsilon)\wedge T]^{2H} - s^{2H} - [(s+\varepsilon)\wedge T - s]^{2H}\right) ds\\
&= \frac{1}{2\varepsilon}\left(\int_a^T \left([(s+\varepsilon)\wedge T]^{2H} - s^{2H}\right) ds - \int_a^{T-\varepsilon} \varepsilon^{2H} ds - \int_{T-\varepsilon}^T (T-s)^{2H} ds\right) \to -\infty,
\end{aligned}$$

as $\varepsilon \downarrow 0$. However, the Stratonovich integral, which will be studied in Section 6, does not have this problem since the expectation of its approximation converges to a constant. It means, for ε small enough, we can use the dominated convergence theorem in order to get

$$\begin{aligned}
&E\left(\frac{1}{2\varepsilon}\int_a^T B_s\left(B_{(s+\varepsilon)\wedge T} - B_{(s-\varepsilon)\vee 0}\right) ds\right)\\
&= \frac{1}{2\varepsilon}\int_a^T \left([(s+\varepsilon)\wedge T]^{2H} - [(s-\varepsilon)\vee 0]^{2H}\right) ds\\
&\quad -\frac{1}{2\varepsilon}\int_a^T \left([(s+\varepsilon)\wedge T - s]^{2H} - [s - ((s-\varepsilon)\vee 0)]^{2H}\right) ds \qquad (5.35)\\
&= \frac{1}{2\varepsilon}\int_a^T \left([(s+\varepsilon)\wedge T]^{2H} - [(s-\varepsilon)\vee 0]^{2H}\right) ds\\
&\quad -\frac{1}{2\varepsilon}\int_{T-\varepsilon}^T (T-s)^{2H} ds + \frac{1}{2\varepsilon}\int_a^{a+\varepsilon} [s - ((s-\varepsilon)\vee 0)]^{2H} ds \to \frac{1}{2}\left(T^{2H} - a^{2H}\right),
\end{aligned}$$

as $\varepsilon \downarrow 0$. Note that in this analysis we have, for ε small enough,

$$\begin{aligned}
&\frac{1}{2\varepsilon}\int_{a+\varepsilon}^T [s - ((s-\varepsilon)\vee 0)]^{2H} ds - \frac{1}{2\varepsilon}\int_a^{T-\varepsilon} [(s+\varepsilon)\wedge T - s]^{2H} ds\\
&\quad = \frac{1}{2\varepsilon}\int_{a+\varepsilon}^T \varepsilon^{2H} ds - \frac{1}{2\varepsilon}\int_a^{T-\varepsilon} \varepsilon^{2H} ds = 0.
\end{aligned}$$

In other words, the two integrals in the first line of these equalities cancel each other (see (5.35)). In this way, we obtain that $E\left(\frac{1}{2\varepsilon}\int_a^T B_s\left(B_{(s+\varepsilon)\wedge T} - B_{(s-\varepsilon)\vee 0}\right) ds\right)$ converges to $\frac{1}{2}\left(T^{2H} - a^{2H}\right)$, as ε goes to zero. We use this fact to deal with the Stratonovich integral with respect to fractional Brownian motion in Section 6.

Now, we analyze Theorem 67 in the case that B is fractional Brownian motion with $H > 1/2$. Towards this end, we need to state the following auxiliary result.

Lemma 65 *Let* $u \in \mathbb{D}^{1,2}(|\mathcal{H}|)$, $\varepsilon > 0$ *and* $u_{\varepsilon,t} = \frac{1}{\varepsilon}\int_{(t-\varepsilon)\vee 0}^t u_s ds$, $t \in [0,T]$. *Then, there exists a constant* $C_H > 0$ *such that*

$$\|u_{\varepsilon,\cdot}\|^2_{\mathbb{D}^{1,2}(|\mathcal{H}|)} \le C_H \|u\|^2_{\mathbb{D}^{1,2}(|\mathcal{H}|)}.$$

Proof. Without loss of generality, we assume that $u \equiv 0$ on $\Omega \times [0,T]^c$ in order to simplify the notation. Thus, applying the change of variables $r = s - t$, we obtain

$$\int_{(t-\varepsilon)\vee 0}^{t} u_s ds = \int_{t-\varepsilon}^{t} u_s ds = \int_{-\varepsilon}^{0} u_{r+t} dr.$$

Hence, we can write

$$\begin{aligned} \|u_{\varepsilon,\cdot}\|_{|\mathcal{H}|}^2 &= H(2H-1)\int_0^T\int_0^T |u_{\varepsilon,r}|\,|u_{\varepsilon,s}|\,|r-s|^{2H-2}dsdr \\ &\leq \frac{C_H}{\varepsilon^2}\int_0^T\int_0^T\int_{-\varepsilon}^0\int_{-\varepsilon}^0 |u_{r+y}|\,|u_{s+\theta}|\,|r-s|^{2H-2}dyd\theta dsdr. \end{aligned}$$

Now, we can use the change of variables formula again, together with Fubini's theorem and our convention on u, to get

$$\begin{aligned} \|u_{\varepsilon,\cdot}\|_{|\mathcal{H}|}^2 &\leq \frac{C_H}{\varepsilon^2}\int_0^T\int_{-\varepsilon}^0\int_{-\varepsilon}^0 |u_{s+\theta}|\int_y^{T+y} |u_r|\,|r-s-y|^{2H-2}drdyd\theta ds \\ &\leq \frac{C_H}{\varepsilon^2}\int_0^T\int_0^T\int_{-\varepsilon}^0\int_{-\varepsilon}^0 |u_{s+\theta}|\,|u_r|\,|r-s-y|^{2H-2}dyd\theta drds \\ &\leq \frac{C_H}{\varepsilon^2}\int_0^T\int_0^T\int_{-\varepsilon}^0\int_{-\varepsilon}^0 |u_s|\,|u_r|\,|r-s-y+\theta|^{2H-2}dyd\theta drds \\ &\leq \frac{C_H}{\varepsilon^2}\int_0^T\int_0^T\int_{-\varepsilon}^0\int_{-2\varepsilon}^0 |u_s|\,|u_r|\,|r-s+y|^{2H-2}dyd\theta drds \\ &\leq \frac{C_H}{\varepsilon}\int_0^T\int_0^T\int_{-2\varepsilon}^{2\varepsilon} |u_s|\,|u_r|\,|r-s+y|^{2H-2}dyd\theta drds. \end{aligned} \tag{5.36}$$

Note that if $r - s > 4\varepsilon$, then

$$\frac{1}{\varepsilon}\int_{-2\varepsilon}^{2\varepsilon}(r-s+y)^{2H-2}dy \leq 4(r-s-2\varepsilon)^{2H-2} \leq 2^{4-2H}(r-s)^{2H-2},$$

where last inequality holds if and only if $(r-s)^{2-2H} < 2^{2-2H}(r-s-2\varepsilon)^{2-2H}$ if and only if $(r-s) > 4\varepsilon$. Similarly, for $s - r > 4\varepsilon$, we also have

$$\frac{1}{\varepsilon}\int_{-2\varepsilon}^{2\varepsilon}(s-r-y)^{2H-2}dy \leq 2^{4-2H}(s-r)^{2H-2}.$$

Now, we assume that $|r-s| \leq 4\varepsilon$. In this case, we are able to establish

$$\begin{aligned} \frac{1}{\varepsilon}\int_{-2\varepsilon}^{2\varepsilon}|r-s+y|^{2H-2}dy &= \frac{1}{\varepsilon}\int_{-2\varepsilon+r-s}^{2\varepsilon+r-s}|y|^{2H-2}dy \leq \frac{1}{\varepsilon}\int_{-6\varepsilon}^{6\varepsilon}|y|^{2H-2}dy \\ &= \frac{2}{\varepsilon}\int_0^{6\varepsilon} y^{2H-2}dy = \frac{2}{\varepsilon(2H-1)}(6\varepsilon)^{2H-1} \\ &\leq \frac{6^{2H-1}4^{3-2H}}{2H-1}(4\varepsilon)^{2H-2} \leq \frac{6^{2H-1}4^{3-2H}}{2H-1}|r-s|^{2H-2}. \end{aligned}$$

In consequence, we have proven that (5.36) allows us to conclude that there exists a constant $C_H > 0$ such that

$$\|u_{\varepsilon,\cdot}\|_{|\mathcal{H}|}^2 \leq C_H\,\|u\|_{|\mathcal{H}|}^2, \quad \text{with probability 1.}$$

Finally, proceeding similarly, it follows

$$\begin{aligned}
\|Du_{\varepsilon,\cdot}\|^2_{|\mathcal{H}|\times|\mathcal{H}|} &= [H(2H-1)]^2 \int_{[0,T]^2} |\theta-\eta|^{2H-2} \\
&\quad \times \left(\int_{[0,T]^2} |D_\eta u_{\varepsilon,r}|\,|D_\theta u_{\varepsilon,s}|\,|r-s|^{2H-2}\,drds \right) d\theta d\eta \\
&\leq C_H \int_{[0,T]^2} |\theta-\eta|^{2H-2} \\
&\quad \times \left(\int_{[0,T]^2} |D_\eta u_r|\,|D_\theta u_s|\,|r-s|^{2H-2}\,drds \right) d\theta d\eta \\
&= C_H \|Du\|^2_{|\mathcal{H}|\times|\mathcal{H}|}, \quad \text{with probability 1.}
\end{aligned}$$

So, it is now easy to finish the proof by taking expectations in the last two inequalities. □

Some parts of the proof of the next result are quite similar to those of the proof of the Theorem 67, so we only sketch them.

Theorem 71 *Let $u = \{u_t : t \in [0,T]\}$ be a stochastic process in $\mathbb{D}^{1,2}(|\mathcal{H}|)$ such that*

$$\int_0^T \int_0^T |D_s u_t|\,|t-s|^{2H-2}\,dsdt < \infty, \quad \textit{with probability 1.} \tag{5.37}$$

Then, for $0 \leq a < b \leq T$, u belongs to $Dom\delta_B^-([a,b])$ and

$$\int_a^b u_s dB_s^- = \delta_2\left(1_{[a,b]}(\cdot)u_\cdot\right) + H(2H-1)\int_a^b \int_0^T D_s u_t\,|t-s|^{2H-2}\,dsdt. \tag{5.38}$$

Remark 83 *Note that* (5.37) *holds if, for example, Du belongs to $L^q([0,T]^2)$ with probability 1, for some $q > 1/(2H-1)$.*

Proof. In order to simplify the notation, we assume, without loss of generality, that $a = 0$ and $b = T$ since we can change u by $u1_{[a,b]}$. In particular, we have that $u \equiv 0$ on $\Omega \times [0,T]^c$.

We consider $\varepsilon > 0$ and a sequence $\{u^{(n)} \in \mathcal{S}(L^2([0,T])) : n \in \mathbb{N}\} \subset \mathcal{E}(|\mathcal{H}|)$ of the form

$$u^{(n)} = \sum_{j=0}^{N-1} F_j 1_{(t_j,t_{j+1}]}, \quad \text{with } 0 = t_0 < t_1 < \ldots < t_N = T \text{ and } F_j \in \mathcal{S},$$

such that $u^{(n)}$ converges to u in $\mathbb{D}^{1,2}(|\mathcal{H}|)$, as $n \to \infty$. Then, Lemma 36 and Theorem 48 imply that, for $n \in \mathbb{N}$,

$$\begin{aligned}
&\frac{1}{\varepsilon}\int_0^T u_t^{(n)} \left(B_{(t+\varepsilon)\wedge T} - B_t\right) dt \\
&= \int_0^T \left(\frac{1}{\varepsilon}\int_{(r-\varepsilon)\vee 0}^r u_t^{(n)} dt \right) \delta_2 B_r + \frac{1}{\varepsilon}\int_0^T \left\langle Du_t^{(n)}, 1_{]t,(t+\varepsilon)\wedge T]} \right\rangle_{\mathcal{H}} dt.
\end{aligned} \tag{5.39}$$

Note that (1.122) and the change of variables formula $\theta = r_2 - t$ yield

$$\begin{aligned}
&\left| \int_0^T \langle Du_t, 1_{]t,(t+\varepsilon)\wedge T]} \rangle_{\mathcal{H}} dt \right| \\
&\leq C_H \int_0^T \int_0^T |D_r u_t| \int_0^{(t+\varepsilon)\wedge T - t} |\theta + t - r|^{2H-2} d\theta dr dt = C_H(I_1 + I_2)
\end{aligned}$$

with

$$\begin{aligned} I_1 &= \int_0^T \int_0^t |D_r u_t| \int_0^{(t+\varepsilon)\wedge T-t} |\theta+t-r|^{2H-2} d\theta dr dt \\ &\leq \int_0^T \int_0^t |D_r u_t| \int_0^{(t+\varepsilon)\wedge T-t} \theta^{2H-2} d\theta dr dt \leq C_{H,T} \int_0^T \int_0^T |D_r u_t| dr dt. \end{aligned}$$

Proceeding similarly with I_2, we can write

$$\begin{aligned} &\left(\int_0^T \left\langle D(u_t-u_t^{(n)}), 1_{]t,(t+\varepsilon)\wedge T]}\right\rangle_{\mathcal{H}} dt\right)^2 \\ &\leq C_{H,T}\left(\int_0^T \int_0^T |D_r(u_t-u_t^{(n)})| dr dt\right)^2 \\ &= C_{H,T}\int_{[0,T]^4} |D_{r_1}(u_{t_1}-u_{t_1}^{(n)})||D_{r_2}(u_{t_2}-u_{t_2}^{(n)})| dr_1 dt_1 dr_2 dt_2 \\ &\leq C_{H,T}\|u-u^{(n)}\|^2_{|\mathcal{H}|\otimes|\mathcal{H}|} \to 0 \quad \text{in } L^1(\Omega), \quad \text{as } n\to\infty. \end{aligned} \tag{5.40}$$

Also, as in the proof of Theorem 67, we have

$$\begin{aligned} &\left(\int_0^T \left|u_s^{(n)}-u_s\right| \left|B_{(s+\varepsilon)\wedge T}-B_s\right| ds\right)^2 \\ &\leq \left[\sup_{r\in[0,T]} |B_r|^2\right]\left(\int_0^T \left|u_s^{(n)}-u_s\right| ds\right)^2 \\ &\leq T^{2-2H}\left[\sup_{r\in[0,T]} |B_r|^2\right]\int_0^T\int_0^T \left|u_t^{(n)}-u_t\right|\left|u_s^{(n)}-u_s\right| |t-s|^{2H-2} ds dt \\ &= [H(2H-1)]^{-1} T^{2-2H}\left[\sup_{r\in[0,T]} |B_r|^2\right]\left\|u^{(n)}-u\right\|^2_{|\mathcal{H}|}. \end{aligned} \tag{5.41}$$

and, Proposition 62 and Lemma 65 allow us to see that there is a constant $C_H > 0$ such that

$$E\left(\left\{\int_0^T \left(\frac{1}{\varepsilon}\int_{(r-\varepsilon)\vee 0}^r \left[u_t^{(n)}-u_t\right] dt\right)\delta_2 B_r\right\}^2\right) \leq C_H \left\|u^{(n)}-u\right\|^2_{\mathbb{D}^{1,2}(|\mathcal{H}|)}.$$

In consequence, (5.40) and (5.41) give that (5.39) becomes

$$\begin{aligned} &\frac{1}{\varepsilon}\int_0^T u_t\left(B_{(t+\varepsilon)\wedge T}-B_t\right) dt \\ &= \int_0^T \left(\frac{1}{\varepsilon}\int_{(r-\varepsilon)\vee 0}^r u_t dt\right)\delta_2 B_r + \frac{1}{\varepsilon}\int_0^T \left\langle Du_t, 1_{]t,(t+\varepsilon)\wedge T]}\right\rangle_{\mathcal{H}} dt. \end{aligned}$$

We can prove that Proposition 62, Lemma 65 and the proof of Theorem 67 yield

$$\int_0^T \left(\frac{1}{\varepsilon}\int_{(r-\varepsilon)\vee 0}^r u_t dt\right)\delta_2 B_r \to \delta_2(u) \quad \text{in } L^2(\Omega), \ \text{as } \varepsilon\to 0,$$

Note that

$$\begin{aligned}
&\frac{1}{\varepsilon}\int_0^T \left\langle Du_t, 1_{]t,(t+\varepsilon)\wedge T]}\right\rangle_{\mathcal{H}} dt \\
&= \frac{H(2H-1)}{\varepsilon}\int_0^T\int_0^T D_s u_t \int_0^{(t+\varepsilon)\wedge T-t} |t-s+x|^{2H-2}dxdsdt \\
&\to H(2H-1)\int_0^T\int_0^T D_s u_t |s-t|^{2H-2}dsdt \quad \text{w.p.1}, \quad \text{as } \varepsilon\to 0,
\end{aligned}$$

if we can apply the dominated convergence theorem in the last claim. And this is true since $\varepsilon^{-1}\int_0^\varepsilon |t-s+x|^{2H-2}dx \le C_H|t-s|^{2H-2}$. So, if this inequality holds, we have verified that $\varepsilon^{-1}\int_0^T u_t\left(B_{(t+\varepsilon)\wedge T}-B_t\right)dt$ converges in probability, as $\varepsilon\to 0$, and that (5.38) is satisfied.

Finally, suppose that $0<\varepsilon<r-t$. Then,

$$\frac{1}{\varepsilon}\int_0^\varepsilon |t-s+x|^{2H-2}dx = \frac{1}{(2H-1)\varepsilon}\left[(r-t)^{2H-1}-(r-t-\varepsilon)^{2H-1}\right] = I(\varepsilon).$$

For $(r-t)/2<\varepsilon<(r-t)$, we have

$$I(\varepsilon) \le \frac{1}{(2H-1)\varepsilon}(r-t)^{2H-1} \le \frac{2}{(2H-1)}(r-t)^{2H-2}$$

and, for $0<\varepsilon<(r-t)/2$, the mean value theorem implies that $I(\varepsilon)\le (r-t-\varepsilon)^{2H-2}\le ((r-t)/2)^{2H-2}$. Missing cases are easier to prove. □

Now, in order to consider an extension of Theorem 71, we take advantage of Theorem 66 in the case that y is fractional Brownian motion B with Hurst parameter $H>1/2$. We emphasize that B satisfies Hypotheses **A1** and **A2** introduced in Section 5.1. Let f satisfy Hypothesis **B** and $G(x)=\int_0^x f(s)ds$, $x\in\mathbb{R}$. Then, G is the difference of two convex functions because f is the difference of two bounded and increasing functions, which implies that G has bounded right and left derivatives. Hence, we can use Coutin et al. [32] (Proposition 7) to establish

$$\begin{aligned}
\int_0^{B_t} f(s)ds = G(B_t) &= \int_0^t f(B_s-)\delta_2 B_s + \frac{1}{2}\int_{\mathbb{R}} L_t^a G''(da) \\
&= \int_0^t f(B_s)\delta_2 B_s + \frac{1}{2}\int_{\mathbb{R}} L_t^a G''(da),
\end{aligned}$$

where the last equality follows from the fact that the bounded variation functions have a countable number of discontinuities, L_t^a is the local time of B at a and G'' has distribution function $f(\cdot -)$. Consequently, now we can associate the forward integral with the divergence operator by means of Theorem 66. Namely, we have

$$\int_0^t f(B_s)dB_s^- = \int_0^t f(B_s)\delta_2 B_s + \frac{1}{2}\int_{\mathbb{R}} L_t^a G''(da).$$

In particular, Corollary 28 and [32] lead us to write the following version of Tanaka's formula:

$$B_t^+ = \int_0^t 1_{\{B_s>0\}}dB_s^- = \int_0^t 1_{\{B_s>0\}}\delta_2 B_s + \frac{1}{2}L_t^0. \tag{5.42}$$

On the other hand, from Theorem 71, we get that, under suitable conditions on u,

$$\int_0^t u_s dB_s^- = \int_0^t u_s\delta_2 B_s + Tr(Du),$$

where the term $Tr(Du)$ is given by

$$Tr(u) = P - \lim_{\varepsilon \downarrow 0} \varepsilon^{-1} \int_0^T < Du_s, 1_{[s,(s+\varepsilon)\wedge T]} >_{\mathcal{H}} ds.$$

The proof of this fact is based on Lemma 36. Therefore, roughly speaking, (5.42) suggests that $L_t^0/2$ could be interpreted as a trace term. One way to do this is by utilizing the approximation of the forward integral $F_\varepsilon(B_t^{(\varepsilon)}) - F_\varepsilon(B_0^{(\varepsilon)})$ with $f(x) = 1_{\{x>0\}}$ (see (5.12) and Lemma 94). Indeed, Lemma 36, together with the fact that $s \mapsto \varepsilon^{-1}\langle D.B_s^{(\varepsilon)}, 1_{]s,(s+\varepsilon)\wedge T]}(\cdot)\rangle_{\mathcal{H}}$ is bounded (see Lemma 95, (5.7) and (5.8)), implies

$$\begin{aligned} F_\varepsilon(B_t^{(\varepsilon)}) - F_\varepsilon(B_0^{(\varepsilon)}) &= \varepsilon^{-1} \int_0^t F_\varepsilon'(B_s^{(\varepsilon)})(B_{(s+\varepsilon)\wedge T} - B_s) ds \\ &= \varepsilon^{-1} \int_0^t \delta_2(F_\varepsilon'(B_s^{(\varepsilon)}) 1_{]s,(s+\varepsilon)\wedge T]}(\cdot)) ds \\ &\quad + \varepsilon^{-1} \int_0^t P_\varepsilon(B_s^{(\varepsilon)}) \langle D.B_s^{(\varepsilon)}, 1_{]s,(s+\varepsilon)\wedge T]}(\cdot)\rangle_{\mathcal{H}} ds. \end{aligned}$$

So, in this case, the last integral could be the approximation of a trace term and it turns out that this integral, for some $p > 1$, converges to $\frac{1}{2}L_t^0$ in $L^p(\Omega)$, which is stated by Lemmas 96 and 99. Moreover, also for some $p > 1$, $F_\varepsilon(B_t^{(\varepsilon)}) - F_\varepsilon(B_0^{(\varepsilon)})$ goes, in $L^p(\Omega)$, to either $\int_0^{B_t} f(x)\mathrm{d}x$, or $\int_0^t f(B_s) dB_s^-$ due to (5.12) and (5.13). Consequently, (5.42) implies that $\varepsilon^{-1} \int_0^t \delta_2(F_\varepsilon'(B_s^{(\varepsilon)}) 1_{]s,(s+\varepsilon)\wedge T]}(\cdot)) ds$ converges to $\int_0^t 1_{\{B_s>0\}} \delta_2 B_s$ in $L^p(\Omega)$, as $\varepsilon \to 0$, for some $p > 1$. In other words, $\frac{1}{2}L_t^0$ could be, indeed, interpreted as a trace term in (5.42).

On the other hand, by Theorem 56, Remark 67 and Theorem 71, we also have a maximal inequality for the forward integral with respect to fractional Brownian motion.

Theorem 72 *Let B be fBm with $H > 1/2$, $p > 2/H$ and $u = \{u_t : t \in [0,T]\}$ a process in $\mathbb{L}_{H-\varepsilon}^{1,p}$, for some $\varepsilon < H - (1/p)$. Moreover, assume that, for some $q > 1/(2H-1)$,*

$$E\left\{\int_0^T |(u_r)|^p dr + \int_0^T \left(\int_0^T |D_t u_r|^{\frac{1}{H}} dt\right)^{pH} dr + \left(\int_0^T \int_0^T |D_s u_t|^q ds dt\right)^{p/q}\right\} < \infty.$$

Then,

$$\begin{aligned} E\left(\sup_{t\in[0,T]} \left|\int_0^t u_s dB_s^-\right|^p\right) &\leq C_{H,p,q,\varepsilon,T} E\left[\left(\int_0^T |u_s|^{\frac{1}{H-\varepsilon}} ds\right)^{p(H-\varepsilon)}\right. \\ &\quad + \left(\int_0^T \left(\int_0^T |D_s u_r|^{\frac{1}{H}} dr\right)^{\frac{H}{H-\varepsilon}} ds\right)^{p(H-\varepsilon)} \\ &\quad \left. + \left(\int_0^T \int_0^T |D_s u_t|^q ds dt\right)^{p/q}\right]. \end{aligned}$$

Remark 84 *From (5.38) and Theorem 56, we have that $t \mapsto \int_0^t u_s dB_s^-$ is a continuous stochastic process. Thus, the expectation on the left-hand side of the last inequality is well-defined.*

Proof. The result is an immediate consequence of Theorems 56 and 71, and Remark 83. □

We can proceed as in the proof of Proposition 101 to show the following result. If the reader has problems to prove it, she/he is able to consult León and Tudor [124] to see details.

Proposition 106 *Assume that* $u \in \mathbb{L}^{1,2}_{H-\rho}$, *for some* $0 < \rho < H - \frac{1}{2}$, *and that the trace condition*

$$\int_0^T \int_0^T |D_s u_t| \, |t-s|^{2H-2} \, ds dt < \infty \quad \textit{with probability 1,}$$

holds. Then, the stochastic process $t \mapsto \varepsilon^{-1} \int_0^t u_s \left(B_{(s+\varepsilon)\wedge T} - B_s\right) ds$ *converges uniformly on* $[0,T]$ *in probability, as* $\varepsilon \to 0$, *and*

$$\int_0^t u_s dB_s^- = \int_0^t u_s \delta_2 B_s + H(2H-1) \int_0^t \int_0^T D_s u_r \, |r-s|^{2H-2} \, ds dr, \quad t \in [0,T].$$

As a consequence of this proposition, we state the following result (compare it with Proposition 102).

Proposition 107 *Let* τ *be a non-negative random variable. Then, under the assumptions of Proposition 106,* $u 1_{[0,\tau]}$ *belongs to* $Dom\delta_B^-([0,t])$ *and*

$$\int_0^s u_r dB_r^- \Big|_{s=t\wedge\tau} = \int_0^t u_r 1_{[0,\tau]}(r) dB_r^-, \quad \textit{for } t \in [0,T].$$

Remark 85 *Note that* $\int_0^s u_r dB_r^- \Big|_{s=t\wedge\tau}$ *is well-defined for every* $t \in [0,T]$ *since Proposition 106 implies that the forward integral* $\int_0^{\cdot} u_s dB_s^-$ *has continuous paths, with probability 1.*

Proof. The proof of this result follows that of Proposition 102, line by line. □

Now, we deal with an Itô's type formula for the forward integral with respect to fractional Brownian motion with a parameter bigger than 1/2.

Proposition 108 *Let* u *be a stochastic process with paths in* $\mathcal{C}_1^\lambda([0,T])$, *where* $\lambda + H > 1$, $F \in \mathcal{C}_b^2(\mathbb{R})$ *and* $x(t) = x(0) + \int_0^t u_s dB_s^-$, $t \in [0,T]$. *Then,*

$$F(x(t)) = F(x(0)) + \int_0^t F'(x(s)) \, u_s dB_s^-, \quad t \in [0,T].$$

Remark 86 *In particular, for* $t \in [0,T]$, *we have that* $F(B_t) = F(0) + \int_0^t F'(B_s) dB_s^-$. *This case has been considered by Russo and Vallois [191] for processes with finite quadratic variation. Namely, if* X *is a stochastic process with finite quadratic variation* $[X]$, *then*

$$f(X_t) = F(X_0) + \int_0^t F'(X_s) dX_s^- + \frac{1}{2} \int_0^t F''(X_s) d[X]_s.$$

Note that, for $H > 1/2$, $[B] \equiv 0$. *Here,* $[X]_t = P - \lim_{\varepsilon\downarrow 0} \int_0^t \frac{\left(X_{(s+\varepsilon)\wedge T} - X_s\right)^2}{\varepsilon} ds$.

In general, we can use the Itô's formula for the divergence operator with respect to B *(i.e. Theorem 61), together with Theorem 71, to analyze Itô's type formulae for the forward integral with respect to fractional Brownian motion* B *(with* $H > 1/2$*), as it is done for the Stratonovich integral in Theorem 77 and Proposition 118 below. Actually, Proposition 118 provides us an Itô's formula not only for the Stratonovich integral, but also for the forward one.*

Proof. The result follows from Propositions 49 and 96. □

Now, we consider the Fubini's theorem for the forward integral with respect to fractional Brownian motion similar to Theorem 69.

Theorem 73 *Let μ be a finite measure of the measurable space $(X, \mathcal{C})$. Assume that $\varphi : \Omega \times X \to \mathcal{H}$ is an $\mathcal{F} \otimes \mathcal{C}$-measurable function satisfying the following conditions:*

i) φ belongs to $L^2\left(X; \mathbb{D}^{1,2}(|\mathcal{H}|)\right)$.

ii) $(\omega, x) \mapsto \delta_2\left(\varphi(x)\right)$ is $\mathcal{F} \otimes \mathcal{C}$-measurable.

iii) The integral $\int_X \delta_2\left(\varphi(x)\right)\mu(dx)$ is well-defined w.p.1 and belongs to $L^2(\Omega)$.

iv) For μ-almost all $x \in X$, we have that $\varphi(x) \in \mathbb{D}^{1,2}(|\mathcal{H}|)$ and

$$\int_0^T \int_0^T |t-s|^{2H-2} \int_X |D_s\varphi(x)(t)|\mu(dx)dsdt < \infty$$

Then, $\int_0^T \int_X \varphi(x)(s)\mu(dx)dB_s^-$ and $\int_X \int_0^T \varphi(x)(s)dB_s^-\mu(dx)$ are well-defined, and they are the same. That is,

$$\int_0^T \int_X \varphi(x)(s)\mu(dx)dB_s^- = \int_X \int_0^T \varphi(x)(s)dB_s^-\mu(dx), \quad w.p.1.$$

Remark 87 *Following Royden [188] (Section 7.3), we can show that $\mathbb{D}^{1,2}(|\mathcal{H}|)$ is a Banach space. So the space $L^2\left(X; \mathbb{D}^{1,2}(|\mathcal{H}|)\right)$ is well-defined due to Section 8.3.*

Proof. Let $\{\varphi^{(n)} : n \in \mathbb{N}\}$ be a sequence of simple processes from X into $\mathbb{D}^{1,2}(|\mathcal{H}|)$ such that

$$\int_X \|\varphi(x) - \varphi^{(n)}(x)\|^2_{\mathbb{D}^{1,2}(|\mathcal{H}|)}\mu(dx) \to 0, \quad \text{as } n \to \infty. \tag{5.43}$$

The fact that μ is a finite measure implies that there is a constant $C > 0$ so that, for $n, m \in \mathbb{N}$,

$$\begin{aligned} E\left(\left\|\int_X \left[\varphi^{(m)}(x) - \varphi^{(n)}(x)\right]\mu(dx)\right\|^2_{\mathcal{H}}\right) &\leq C\int_X E\left(\left\|\varphi^{(m)}(x) - \varphi^{(n)}(x)\right\|^2_{\mathcal{H}}\right)\mu(dx) \\ &\leq C\int_X \left\|\varphi^{(m)}(x) - \varphi^{(n)}(x)\right\|^2_{\mathbb{D}^{1,2}(|\mathcal{H}|)}\mu(dx), \end{aligned}$$

which, together with (5.43), gives that the $L^2(\Omega; \mathcal{H})$-valued Bochner integral $\int_X \varphi(x)\mu(dx)$ exists and agrees with the $\mathbb{D}^{1,2}(|\mathcal{H}|)$-valued Bochner integral of φ with respect to μ since

$$\begin{aligned} E\left(\left\|\int_X \left[\varphi(x) - \varphi^{(n)}(x)\right]\mu(dx)\right\|^2_{\mathcal{H}}\right) &\leq \left\|\int_X \left[\varphi(x) - \varphi^{(n)}(x)\right]\mu(dx)\right\|^2_{\mathbb{D}^{1,2}(|\mathcal{H}|)} \\ &\leq C\int_X \left\|\varphi(x) - \varphi^{(n)}(x)\right\|^2_{\mathbb{D}^{1,2}(|\mathcal{H}|)}\mu(dx). \end{aligned}$$

Similarly, we can show that $D\left(\int_X \varphi(x)\mu(dx)\right) = \int_X D\left(\varphi(x)\right)\mu(dx)$, where the integral on the left-hand side is the $\mathbb{D}^{1,2}(|\mathcal{H}|)$-valued Bochner integral of φ and the one on the right-hand side is the $\mathcal{H} \otimes \mathcal{H}$-valued Bochner integral of $D(\varphi)$ both with respect to μ.

On the other hand, Hypotheses i) and iv), and Theorem 71 lead us to establish

$$\begin{aligned}\int_0^T \varphi(x)(s)dB_s^- &= \delta_2\left(\varphi(x)(\cdot)\right) \\ &\quad +H(2H-1)\int_0^T\int_0^T D_s\varphi(x)(t)\,|t-s|^{2H-2}\,dsdt, \quad \mu-\text{a.s.}\end{aligned}$$

Therefore, Theorem 48 and Assumptions i)-iv) imply that $\int_0^T \int_X \varphi(x)(s)\mu(dx)\delta_2 B_s$ is well-defined. Hence, Fubini's theorem and Theorem 48 give

$$\begin{aligned}\int_X\int_0^T \varphi(x)(s)dB_s^-\mu(dx) &= \delta_2\left(\int_X \varphi(x)(\cdot)\mu(dx)\right) + H(2H-1) \\ &\quad \times \int_0^T\int_0^T D_s\left(\int_X \varphi(x)(t)\mu(dx)\right)|t-s|^{2H-2}\,dsdt \\ &= \int_0^T\int_X \varphi(x)(s)\mu(dx)dB_s^-,\end{aligned}$$

where, in last equality, we use Theorem 71 again. The proof is now complete. □

Chapter 6

Stratonovich Integration

In order to motivate the use of Stratonovich integration, suppose that we have an absolutely continuous function $x : [a, b] \to \mathbb{R}$ of the form

$$x(t) = x_a + \int_a^t g(s)ds, \quad t \in [a, b].$$

Here. $x_a \in \mathbb{R}$ and $g \in L^1([a, b])$. It is well-known that a function is absolutely continuous if and only if it is the integral of its derivative plus a constant (see, for example, Royden [187]). Now, let f be a smooth function in $C^1(\mathbb{R})$. Then, the fundamental theorem of calculus implies

$$f(x(t)) = f(x_a) + \int_a^t f'(x(s))\, g(s)ds, \quad t \in [a, b]. \tag{6.1}$$

Unfortunately, the stochastic integral of Itô does not satisfy this important property because an extra term appears. Namely, consider the Itô process

$$X_t = x_a + \int_a^t g_s dW_s, \quad t \in [a, b],$$

where $W = \{W_t : t \in [a, b]\}$ is a $\mathcal{F}_t$-Brownian motion and g is now a $\mathcal{F}_t$-adapted and measurable process with square-integrable paths. Then, for $f \in C^2(\mathbb{R})$, the Itô's formula (i.e., Theorem 40) yields

$$f(X_t) = f(x_a) + \int_a^t f'(X_s)\, g(s)dW_s + \frac{1}{2}\int_a^t f''(X_s)\, g_s^2 ds, \quad t \in [a, b].$$

Therefore, if we define other stochastic integral of $f'(X.)g(\cdot)$ with respect to W as

$$\int_a^t f'(X_s)\, g(s) \circ dW_s = \int_a^t f'(X_s)\, g(s)dW_s + \frac{1}{2}\int_a^t f''(X_s)\, g_s^2 ds, \quad t \in [a, b]. \tag{6.2}$$

Then, this integral satisfies

$$\int_a^t f'(X_s)\, g(s) \circ dW_s = f(X_t) - f(X_a), \quad t \in [a, b]. \tag{6.3}$$

In other words, this stochastic integral meets the rule of the ordinary calculus (6.1). But $E\left(\int_a^t f'(X_s)\, g(s) \circ dW_s\right) \neq 0$ and it is not a martingale. We observe that Protter [180] (Theorems 29 and 34) shows that, in this case, we can write the last stochastic integral as

$$\begin{aligned}\int_a^t f'(X_s)\, g(s) \circ dW_s &= \int_0^t f'(X_s)\, g(s)dW_s \\ &\quad + \frac{1}{2}\left[f'(X.)1_{[a,b]}(\cdot), \int_0^{\cdot} 1_{[a,b]}(s)g(s)dW_s\right]_t, \quad t \in [a, b],\end{aligned}$$

DOI: 10.1201/9781003484912-6

where, on the right-hand side, the stochastic integrals are in the Itô's sense, and $\left[f'(X_\cdot)1_{[a,b]}(\cdot), \int_0^\cdot 1_{[a,b]}(s)g(s)dW_s\right]$ denotes the quadratic variation of the processes $f'(X_\cdot)1_{[a,b]}(\cdot)$ and $\int_0^\cdot 1_{[a,b]}(s)g(s)dW_s$ (see (3.44) for its definition).

In general, for two continuous semimartingales X and Y, the Fisk-Stratonovich integral of X with respect to Y, denoted by $\int_0^t X_s \circ dY_s$, is defined as

$$\int_0^t X_s \circ dY_s = \int_0^t X_s dY_s + \frac{1}{2}\left[X,Y\right]_t.$$

Here, the stochastic integral on the right-hand side is in the Itô's sense and to simplify the notation, we assume that $a=0$. The interested reader on stochastic integration with respect to semimartingales can consult the book by Protter [180]. In particular, in [180] (Theorem 22, Chapter V), we can find the following result.

Theorem 74 *Let $X = (X^{(1)},\ldots,X^{(n)})$ be a continuous n-dimensional semimartingale (i.e., $X^{(i)}$ is a real-valued continuous semimartingale, $i\in\{1,\ldots,n\}$) and $f:\mathbb{R}^n\to\mathbb{R}$ a function in $C^2(\mathbb{R}^n)$. Then,*

$$f(X_t) = f(X_0) + \sum_{i=1}^n \int_0^t \partial_{x_i} f(X_s)\circ dX_s^{(i)}, \quad t\in[0,b].$$

In Protter [180], it is pointed out that this result is an immediate consequence of the Itô's formula and the fact that

$$\sum_{i=1}^n \left[\partial_{x_i} f(X), X^{(i)}\right]_t = \frac{1}{2}\sum_{1\le i,j\le n}\int_0^t \partial^2_{x_i x_j} d\left[X^{(i)},X^{(j)}\right]_s, \quad t\in[0,b].$$

The reader interested in the case that X is a càdlàg semimartingale can consult [180] (Theorem 22, Chapter V).

In what follows, we consider some extensions of this integral, which will be called Stratonovich integral.

As we have already observed, the Stratonovich integral follows the rule of the ordinary calculus, which has allowed to establish links between stochastic and ordinary differential equations [42] even in the case that the initial condition is a random variable that may depend on the whole paths of the Brownian motion W and the coefficients are random (see, for example, Kohatsu-Higa and León [101]). Since the integrands are not adapted to the underlying filtration in this case, the main tool to establish this result is the Malliavin calculus based on the Itô-Ventzell formula [164] and the definition of the Stratonovich integral $\int_0^t u_s\circ dW_s$ as the limit in probability of the sequence

$$\sum_{i=0}^{n-1}\frac{1}{t_{i+1}-t_i}\left(\int_{t_i}^{t_{i+1}} u_s ds\right)\left(W_{t_{i+1}}-W_{t_i}\right), \quad \text{as } |\pi|\to 0 \tag{6.4}$$

whenever this limit exists. Here, $\pi=\{0=t_0<t_1<\ldots<t_n=t\}$ is a partition of $[0,t]$. The Stochastic calculus based on the Malliavin calculus for the Skorohod and Stratonovich integrals has been begun by Nualart and Pardoux [159]. In order to relate this integral to the divergence operator, we introduce the following definition. For u in the space $\mathbb{D}^{1,2}(L^2([0,T]))$, D^+X is the process in $L^q(\Omega\times[0,T])$, with $q\ge 1$, such that

$$\lim_{n\to\infty}\int_0^T \sup_{s\le t\le T\wedge(s+(1/n))} E\left(\left|D_s u_t - (D^+u)_s\right|^q\right) ds = 0. \tag{6.5}$$

Nualart [158] has shown that this definition of Stratonovich integral (i.e., the limit of (6.4)) is equal to the divergence operator plus an extra term. Namely,

$$\int_0^t u_s \circ dW_s = \delta_2(u) + \frac{1}{2}\int_0^t \left\{ \left(D^- u\right)_s + \left(D^+ u\right)_s \right\} ds, \tag{6.6}$$

where $D^- u$ is given by definition (4.47) and $D^+ u$ is defined in (6.5), both with $q = 1$. In particular, when u has the form

$$u_t = x + \int_0^t f_s ds + \int_0^t g_s dW_s, \quad t \in [0, T],$$

with f and g two $\mathcal{F}_t$-adapted and measurable processes such that $f \in L^1([0,T])$ and $g \in L^2([0,T])$ with probability 1, we have

$$\int_0^t u_s \circ dW_s = \int_0^t u_s dW_s + \frac{1}{2}\left[u, W\right]_t, \quad t \in [0, T],$$

due to Remark 55 and Proposition 70.

The Stratonovich integral with respect to semimartingales has also been taken into account to work with not necessarily adapted integrands to the underlying filtration. To do so, introduce the Stratonovich integral as the limit in probability of the sequence introduced in (6.4) when we change W by a continuous martingale. Under suitable conditions, Kohatsu-Higa et al. [102] have stated an existence and uniqueness result for semilinear Stratonovich SDEs driven by semimartingales. Here, the initial condition is a random variable, the coefficients are random and depend on time, and the diffusion coefficient is linear. The techniques applied in this paper are only based on the definition of the Stratonovich integral without using the Malliavin calculus (see Section 7.3.2).

As the forward integral (see Propositions 103 and 104), the Stratonovich integral also satisfies the substitution formula as we will see in Section 6.2 below. So, in the case that a SDE of Stratonovich type has coefficients independent of $(\omega, t) \in \Omega \times [0, T]$, the process $X_t = \varphi_t(X_0)$ is a solution of this stochastic equation, where the random variable X_0 is the initial condition and $\varphi_t(x)$ is the flow associated with this coefficients (see Millet et al. [148], or Theorem 6.1.1 in Nualart [156]).

The Stratonovich integral, with respect to a continuous process Y, of a process u with integrable paths has been also defined as the limit in probability of the sequence

$$\frac{1}{2}\sum_{i=0}^{n-1}\left(u_{t_{i+1}} + u_{t_i}\right)\left(Y_{t_{i+1}} - Y_{t_i}\right), \quad \text{as } |\pi| \to 0$$

if this limit exists. Here, $\pi = \{0 = t_0 < t_1 < \ldots < t_n = t\}$ is a partition of $[0, t]$. In order to explain the idea of this definition, suppose that Y is a continuous $\mathcal{F}_t$-semimartingale, and that u is a right-continuous with left-limits $\mathcal{F}_t$-adapted and measurable process such that $[u, Y]$ exists. Then,

$$\begin{aligned}
&\sum_{i=0}^{n-1}\frac{\left(u_{t_{i+1}} + u_{t_i}\right)}{2}\left(Y_{t_{i+1}} - Y_{t_i}\right) \\
&= \sum_{i=0}^{n-1} u_{t_i}\left(Y_{t_{i+1}} - Y_{t_i}\right) + \frac{1}{2}\sum_{i=0}^{n-1}\left(u_{t_{i+1}} - u_{t_i}\right)\left(Y_{t_{i+1}} - Y_{t_i}\right) \\
&\to \int_0^t u_{s-} dY_s + \frac{1}{2}\left[u, Y\right]_t, \quad \text{in probability, as } |\pi| \to 0.
\end{aligned} \tag{6.7}$$

Here, the stochastic integral of $u_{\cdot-}$ with respect to Y is in the Itô's sense again.

In the case that Y is the Brownian motion W, an integral of a process u with paths in $L^2([0,T])$ can be defined as

$$\int_0^T u_s \star dW_s = \sum_{n=1}^{\infty} \langle u, e_n\rangle_{L^2([0,T])} W(e_n)$$

if the series converges in probability and the limit is independent of the orthonormal complete system $\{e_i \in L^2([0,T]) : n \in \mathbb{N}\}$. Remember that $W(e_n)$ stands for the stochastic integral $\int_0^T e_n(s)dW_s$. Note that the idea of this integral takes into account property (5.2) (see also Proposition 109.i)). That is, it is well-know that the paths of u have the representation

$$u_t = \sum_{n=1}^{\infty} \langle u, e_n\rangle_{L^2([0,T])} e_n(t).$$

Hence, roughly speaking, we have

$$\begin{aligned} \int_0^T u_s \star dW_s &= \int_0^T \sum_{n=1}^{\infty} \langle u, e_n\rangle_{L^2([0,T])} e_n(s) \star dW_s = \sum_{n=1}^{\infty} \int_0^T \langle u, e_n\rangle_{L^2([0,T])} e_n(s) \star dW_s \\ &= \sum_{n=1}^{\infty} \langle u, e_n\rangle_{L^2([0,T])} W(e_n). \end{aligned}$$

This integral has been studied by several authors. Among them, we can mention Ogawa [166, 168, 167] (see also Nualart [158] and references therein). As Nualart [158] has pointed out, if u belongs to $\mathbb{D}^{1,2}(L^2([0,T]))$, then property (4.3) of the divergence operator implies

$$\langle u, e_n\rangle_{L^2([0,T])}\delta_2(e_n) = \delta_2\left(\langle u, e_n\rangle_{L^2([0,T])} e_n\right) + \int_0^T \int_0^T (D_s u_t) e_n(t) e_n(s) ds dt.$$

This equality allows us to analyze conditions that guarantee the existence of the integral $\int_0^T u_s \star dW_s$ and to study its relation with the divergence operator, as it is done by Nualart and Zakai [162] (Proposition 6.1). Moreover, Nualart and Zakai have shown that the existence of the integral $\int_0^T u_s \star dW_s$ gives that the Stratonovich integral of u with respect to W is well-defined and both integrals are the same.

The definition considered in the remainder of this book for the Stratonovich integral is the one introduced by Russo and Vallois [190]. Namely:

Definition 44 *Let X be a continuous process and u a measurable process with integrable paths on the interval $[0,T]$. The Stratonovich integral of u with respect to X, denoted by $\int_0^T u_s \circ dX_s$, is defined as the limit in probability*

$$\int_0^T u_s \circ dX_s = P - \lim_{\varepsilon \downarrow 0} \int_0^T u_s \frac{X_{(s+\varepsilon)\wedge T} - X_{(s-\varepsilon)\vee 0}}{2\varepsilon} ds,$$

provided this limit exists.

In order to explain the idea of the definition of this integral, we deal with an $\mathcal{F}_t$-Brownian motion W again. Then, if u is an $\mathcal{F}_t$-adapted and measurable process, Proposition 98,

Föllmer et al. [52], and Pardoux and Protter [173] imply

$$
\begin{aligned}
&P-\lim_{\varepsilon\downarrow 0}\int_0^T u_s\frac{W_{(s+\varepsilon)\wedge T}-W_{(s-\varepsilon)\vee 0}}{2\varepsilon}ds\\
&= P-\lim_{\varepsilon\downarrow 0}\int_0^T u_s\frac{W_{(s+\varepsilon)\wedge T}-W_s}{2\varepsilon}ds+P-\lim_{\varepsilon\downarrow 0}\int_0^T u_s\frac{W_s-W_{(s-\varepsilon)\vee 0}}{2\varepsilon}ds\\
&= \frac{1}{2}\int_0^T u_s dW_s+\frac{1}{2}\int_0^T u_s d^*W_s,
\end{aligned}
$$

where the first integral (resp. the second integral) on the right-hand side is the Itô's (resp. backward) integral of u with respect to W, which agrees with (6.7). Indeed, from Föllmer et al. [52], we have

$$
\begin{aligned}
&\frac{1}{2}\sum_{i=0}^{n-1}\left(u_{t_{i+1}}+u_{t_i}\right)\left(W_{t_{i+1}}-W_{t_i}\right)\\
&= \frac{1}{2}\sum_{i=0}^{n-1}u_{t_i}\left(W_{t_{i+1}}-W_{t_i}\right)+\frac{1}{2}\sum_{i=0}^{n-1}u_{t_{i+1}}\left(W_{t_{i+1}}-W_{t_i}\right)\\
&\to \frac{1}{2}\int_0^T u_s dW_s+\frac{1}{2}\int_0^T u_s d^*W_s \quad \text{in probability, as } |\pi|\to 0.
\end{aligned}
$$

6.1 Some properties of Stratonovich integral

Remember that in this chapter, the Stratonovich integral is given by Definition 44. Now, we refine this definition.

Definition 45 *Let $X=\{X_t : t\in[0,T]\}$ be a continuous process and $u=\{u_t : t\in[0,T]\}$ a measurable process with integrable paths on the interval $[0,T]$. We say that u is Stratonovich integrable with respect to X on the interval $[a,b]$, $0\le a<b\le T$, if the integral*

$$
\frac{1}{2\varepsilon}\int_a^b u_s\left(X_{(s+\varepsilon)\wedge T}-X_{(s-\varepsilon)\vee 0}\right)ds \tag{6.8}
$$

converges in probability, as $\varepsilon\downarrow 0$. The limit is denoted by $\int_a^b u_s\circ dX_s$ and it is called the Stratonovich integral of u with respect to X. Its domain is represented by $Dom\,\delta_X^S([a,b])$.

Note that the integral in (6.8) is well-defined since the process u has paths in $L^1([0,T])$ and X is pathwise bounded because it is continuous. As in Section 5.1, this definition leads us to state the following result.

Proposition 109 *Let X and u be as in Definition 45. Then we have:*

i) Suppose that F is a random variable and $u\in Dom\,\delta_X^S([a,b])$. Then, $Fu\in Dom\,\delta_X^S([a,b])$ and

$$
F\int_a^b u_s\circ dX_s=\int_a^b (Fu_s)\circ dX_s.
$$

ii) $u \in Dom\,\delta_X^S([a,b])$ *if and only if* $u1_{[a,b]} \in Dom\,\delta_X^S([0,T])$. *In this case,*

$$\int_a^b u_s \circ dX_s = \int_0^T u_s 1_{[a,b]}(s) \circ dX_s.$$

As in Proposition 96, the Stratonovich integral agrees with the Young integral if the integrand and the integrator are Hölder continuous such that the sum of their exponents is bigger than 1. Namely:

Proposition 110 *Let* $\kappa+\nu > 1$, *and* u *and* X *two processes with paths in* $C_1^\kappa([0,T];\mathbb{R})$ *and* $C_1^\nu([0,T];\mathbb{R})$, *respectively. Then, for* $a,b \in [0,T]$, $a < b$, *we have that* $u \in Dom\,\delta_X^S([a,b]) \cap Dom(\int_a^b \cdot d^Y X)$ *and*

$$\int_a^b u_s \circ dX_s = \int_a^b u_s d^Y X_s, \quad w.p.1,$$

where the right-hand side is the Young integral of u *with respect to* X.

We recall that the last Young integral is well-defined due to Theorem 33. The proof of this result is similar to that of Proposition 96. The details are left to the reader as an exercise.

Similar to the forward integral, the Stratonovich one has the local property. That is, for two measurable processes u and v belonging to Dom $\delta_X^S([a,b])$, we have

$$1_A \int_0^t u_s \circ dX_s = 1_A \int_0^t v_s \circ dX_s, \quad \text{w.p.1, for each } t \in [a,b],$$

whenever $u = v$ on $A \times [a,b]$. The proof of this fact follows the same lines of that of Lemma 60. Therefore, we can extend the domain of the Stratonovich integral apparently, as usual. However, we are able to apply the proof of Proposition 97 to establish the following result.

Proposition 111 *Assume that we have two stochastic processes* u *and* X *with integrable and continuous paths, respectively, such that there is a sequence* $\{(\Omega_n, u^{(n)}) : n \geq 1\} \subset \mathcal{F} \times Dom\,\delta_X^S([a,b])$ *satisfying*

i) $\Omega_n \nearrow \Omega$ *w.p.1.*

ii) $u = u^{(n)}$ *almost surely on* $\Omega_n \times [a,b]$.

Then, u *belongs to* $Dom\,\delta_X^S([a,b])$ *and*

$$\int_a^b u_s \circ dX_s = \int_a^b u_s^{(n)} \circ dX_s \quad \text{on } \Omega_n \text{ for all } n \in \mathbb{N} \text{ w.p.1.}$$

In order to finish this section, we suppose that $y = \{y_t : t \in [0,T]\}$ is a process satisfying Hypotheses **A1** and **A2**, and $f : \mathbb{R} \to \mathbb{R}$ is a function fulfilling Condition **B**. Remember that these assumptions are established in Section 5.

For $\varepsilon > 0$, instead of the process $y^{(\varepsilon)}$ introduced in (5.8), we work with the process

$$\tilde{y}_t^{(\varepsilon)} = \frac{1}{2\varepsilon} \int_{t-\varepsilon}^{t+\varepsilon} y_{(s\wedge T)\vee 0} ds, \quad t \in [0,T],$$

because we are dealing with the Stratonovich integral now. This process also satisfies (see (8.25))

$$|\tilde{y}_t^{(\varepsilon)} - y_t| \leq G_\rho \varepsilon^{H-\rho}, \quad t \in [0,T] \text{ and } \delta < H, \tag{6.9}$$

which is easy to prove utilizing Hypothesis **A1**. Also, as in (5.10), the fundamental theorem of calculus allows us to obtain

$$\begin{aligned}\int_{\tilde{y}_0^{(\varepsilon)}}^{\tilde{y}_t^{(\varepsilon)}} F_\varepsilon'(x)dx &= F_\varepsilon(\tilde{y}_t^{(\varepsilon)}) - F_\varepsilon(\tilde{y}_0^{(\varepsilon)}) \\ &= \varepsilon^{-1}\int_0^t F_\varepsilon'(\tilde{y}_s^{(\varepsilon)})(y_{(s+\varepsilon)\wedge T} - y_{(s-\varepsilon)\vee 0})ds \qquad (6.10)\end{aligned}$$

Now, we can state the following:

Theorem 75 *Under the assumptions of Theorem 66, we have*

$$\int_0^t f(y_s)\circ dy_s = \int_{y_0}^{y_t} f(x)dx, \quad t\in[0,T].$$

Remark 88 *In particular, by Theorem 66 and Remark 78, we have*

$$\int_0^t f(y_s)\circ dy_s = \int_0^t f(y_s)dy_s^- = \int_0^t f(y_s)dy_s, \quad t\in[0,T].$$

Here, the last integral on the right-hand side is the generalized Stieltjes integral given by Zähle [218], which is introduced in Section 2.3.

Proof. We observe that, proceeding as in the first part of the proof of Theorem 66 and using (6.9), we obtain

$$F_\varepsilon(\tilde{y}_t^{(\varepsilon)}) - F_\varepsilon(\tilde{y}_0^{(\varepsilon)}) \to \int_{y_0}^{y_t} f(x)dx \quad \text{in } L^2(\Omega), \text{ as } \varepsilon\downarrow 0.$$

Hence, from (6.10), we only need to show that the integral

$$\varepsilon^{-1}\int_0^t [f(y_s) - F_\varepsilon'(\tilde{y}_s^{(\varepsilon)})](y_{(s+\varepsilon)\wedge T} - y_{(s-\varepsilon)\vee})ds$$

tends to zero in probability, as $\varepsilon\downarrow 0$, in order to finish the proof. Towards this end, we observe that we can proceed as in the proof of Lemma 93 to get that there exists $C>0$ such that

$$\begin{aligned}\varepsilon^{-3/2+H-\rho}\left|\int_{\tilde{y}_s^{(\varepsilon)}-r}^{y_s-r} e^{-x^2/2\varepsilon}dx\right| &\le CG_\rho^{2n+2\rho}\varepsilon^{(-1+H-2\rho)+[2(H-\rho)-1]n+2\rho(H-\rho)} \\ &\quad +(n+1)(1+G_\rho^2)^{n-1}G_\rho\varepsilon^{-3/2+2H-2\rho}e^{-(y_s-r)^2/2\varepsilon} \\ &\quad +CG_\rho\varepsilon^{-2+2H-3\rho}\int_{s-\varepsilon}^{s+\varepsilon}\frac{1}{|y_{(u\wedge T)\vee 0}-r|^{1-2\rho}}du,\end{aligned}$$

for all $n\in\mathbb{N}$, $\varepsilon\in(0,1)$, $s\in[0,T]$ and $r\in\mathbb{R}$.

Finally, with this inequality in hand, we only have to follow the proof of Lemma 94 in order to see that the result that we are analyzing is satisfied. □

6.2 Brownian motion

In this section, we consider an $\mathcal{F}_t$-Brownian motion $B=\{B_t : t\in[0,T]\}$.

Our first goal here is to establish identity (6.6). To do so, we state the following auxiliary result, whose proof is similar to that of Lemma 63. The details are left to the reader as an exercise.

Lemma 66 *Let u be a process in $u \in \mathbb{D}^{1,2}(L^2([0,T]))$ and $\varepsilon > 0$. Then, the process $s \mapsto \frac{1}{2\varepsilon}\int_{(s-\varepsilon)\vee 0}^{(s+\varepsilon)\wedge T} u_r dr$ is also in $\mathbb{D}^{1,2}(L^2([0,T]))$ and*

$$\left\| \frac{1}{2\varepsilon}\int_{(\cdot-\varepsilon)\vee 0}^{(\cdot+\varepsilon)\wedge T} u_s ds \right\|_{1,2,L^2([0,T])} \leq \|u\|_{1,2,L^2([0,T])}.$$

In order to give meaning to the second sum on the right-hand side of (6.6), we introduce the following spaces.

Definition 46 *Let $u \in \mathbb{D}^{1,2}(L^2([0,T]))$ and $q \geq 1$. We say that u belongs to the space $\mathbb{D}^{1,2}_{q,+}(L^2([0,T]))$ (resp. $\mathbb{D}^{1,2}_{q,-}(L^2([0,T]))$ if there exists the process $D^+u = \{(D^+u)_t : t \in [0,T]\}$ (resp. $D^-u = \{(D^-u)_t : t \in [0,T]\}$) satisfying condition (6.5) (resp. (4.47)). Moreover, we set $\mathbb{D}^{1,2}_{q,c}(L^2([0,T])) = \mathbb{D}^{1,2}_{q,+}(L^2([0,T])) \cap \mathbb{D}^{1,2}_{q,-}(L^2([0,T]))$.*

Now, we are ready to see that our claim holds.

Theorem 76 *Let $u \in \left(\mathbb{D}^{1,2}_{1,c}(L^2([0,T]))\right)_{loc}$. Then, for $0 \leq a < b \leq T$, $u1_{[a,b]} \in (Dom\,\delta_2)_{loc} \cap Dom\,\delta^S_B([a,b])$ and*

$$\int_a^b u_s \circ dB_s = \delta_2(u1_{[a,b]}) + \frac{1}{2}\int_a^b \left((D^-u)_s + (D^+u)_s\right) ds. \tag{6.11}$$

Remark 89 *$\delta_2(u)$ is given by Definition 38, and the processes D^-u and D^+u are well-defined due to Definition 23 (see also (4.48)).*

Proof. By Propositions 22, 66 and 111, we can assume that u belongs to the space $\mathbb{D}^{1,2}_{1,c}(L^2([0,T]))$, $a = 0$ and $b = T$ without loss of generality.

Now, choose a sequence $\{u^{(n)} \in \mathcal{S}(L^2([0,T])) : n \in \mathbb{N}\}$ such that $\|u - u^{(n)}\|_{\mathbb{D}^{1,2}(L^2([0,T]))} \to 0$, as $n \to \infty$. With the help of this sequence and proceeding as in the proof of Theorem 67, we have that, for $\varepsilon > 0$,

$$\begin{aligned}
&\frac{1}{2\varepsilon}\int_0^T u_s\left(B_{(s+\varepsilon)\wedge T} - B_{(s-\varepsilon)\vee 0}\right) ds \\
&= \int_0^T \left(\frac{1}{2\varepsilon}\int_{(r-\varepsilon)\vee 0}^{(r+\varepsilon)\wedge T} u_s ds\right)\delta_2 B_r + \int_0^T \left(\frac{1}{2\varepsilon}\int_{(r-\varepsilon)\vee 0}^{(r+\varepsilon)\wedge T} D_r u_s ds\right) dr \\
&= \int_0^T \left(\frac{1}{2\varepsilon}\int_{(r-\varepsilon)\vee 0}^{(r+\varepsilon)\wedge T} u_s ds\right)\delta_2 B_r + \int_0^T \left(\frac{1}{2\varepsilon}\int_{(r-\varepsilon)\vee 0}^{r} D_r u_s ds\right) dr \\
&\quad + \int_0^T \left(\frac{1}{2\varepsilon}\int_r^{(r+\varepsilon)\wedge T} D_r u_s ds\right) dr.
\end{aligned}$$

Hence, we can show that the result holds by following the end of the proof of Theorem 67. To do so, we need to use Lemmas 36 and 66 instead of Lemmas 62 and 63, respectively. □

Now, the reader has to be able to show that if u is as in Theorem 76, the random variable given by (6.4) converges to the right-hand side of (6.6), as $|\pi| \to 0$. Moreover, in this case, for $\beta \in [0,1]$, the reader could prove

$$\sum_{i=0}^{n-1} \left(\beta u_{t_{i+1}} + (1-\beta) u_{t_i}\right) \left(B_{t_{i+1}} - B_{t_i}\right)$$
$$\to \delta_2(u) + \beta \int_0^t \left(D^+ u\right)_s ds + (1-\beta) \int_0^t \left(D^- u\right)_s ds, \quad \text{as } |\pi| \to 0. \tag{6.12}$$

Thus, we obtain the forward integral of u with respect to B if $\beta = 0$, and the Stratonovich one if $\beta = 1/2$ due to Theorems 67 and 76, respectively.

Consider a partition $\{0 = t_0 < t_1 < \ldots < t_n = T\}$ of $[0,T]$, $\beta \in [0,1]$ and $\alpha_i = (1-\beta)t_i + \beta t_{i+1}$, $i = 0, \ldots, n-1$. Calculate

$$L^2(\Omega) - \lim_{|\pi| \to 0} \sum_{i=0}^{n-1} B_{\alpha_i} \left(B_{t_{i+1}} - B_{t_i}\right)$$

as an exercise. Concerning this example, it is stated the following result in Protter [180] (Theorem V.30), where he used the concept of convergence uniformly on compacts in probability. That is, a sequence of processes $\{y^{(n)} : n \in \mathbb{N}\}$ converge to the process y uniformly on compacts in probability if, for each $t \in [0,T]$, the random variable $\sup_{0 \le s \le t} \left| y_s^{(n)} - y_s \right|$ converges to zero in probability.

Proposition 112 *Let X be an $\mathcal{F}_t$-semimartingale and Y a continuous $\mathcal{F}_t$-semimartingale such that $[X,Y] = \int_0^\cdot J_s ds$. Also, let $f \in C^1(\mathbb{R})$. Then,*

$$\lim_{|\pi| \to 0} \sum_{i=0}^{n-1} f(Y_{\alpha_i}) \left(B_{t_{i+1}} - B_{t_i}\right) = \int_{]0,T]} f(Y_s) dX_s + \beta \int_{]0,T]} f'(Y_s) d\,[X,Y]_s,$$

where the convergence is uniformly on compacts in probability.

Note that the last right-hand side is the Fisk-Stratonovich integral if $\beta = 1/2$.

By Proposition 112, (6.4) and (6.12), we get that the Stratonovich integral is the limit in probability of Riemann sums when we take, in each interval of the partition, either the value of the integrand at the middle point, or an average of the integrand.

As in the forward integral, we can take advantage of Theorem 76 to obtain estimates for the moments of the Stratonovich integral, which is convenient since limits in probability are difficult to control. For instance, an equivalent estimate to (5.25) is

$$E\left(\left(\int_a^b u_s \circ dB_s\right)^2\right)$$
$$\le 2c_2 \left\| u.1_{[a,b]}(\cdot) \right\|_{1,2,L^2([0,T])}^2 + 2(b-a) E\left(\int_a^b \left[(D^- u)_s + (D^+ u)_s\right]^2 ds\right),$$

for any $u \in \mathbb{D}_{2,c}^{1,2}(L^2([0,T]))$ and $0 \le a < b \le T$. Here, we apply Proposition 62 to calculate the second moment of the involved divergence.

We can also compare the forward and Stratonovich integrals with respect to Brownian motion.

Corollary 30 *Let* $u \in \left(\mathbb{D}^{1,2}_{1,c}(L^2([0,T]))\right)_{loc}$. *Then, for* $0 \le a < b \le T$, *u is in* $Dom\,\delta^-_B([a,b]) \cap Dom\,\delta^S_B([a,b])$ *and*

$$\int_a^b u_s \circ dB_s = \int_a^b u_s dB_s^- + \frac{1}{2}\int_a^b \left((D^+u)_s - (D^-u)_s\right) ds.$$

Proof. The result is an immediate consequence of (5.21), and Remark 79. □

Moreover, two immediate consequences of Theorem 76, together with Propositions 99 and 100, are the following two results.

Proposition 113 *Let u be a process in* $\mathbb{D}^{2,2}_{1.c}(L^2([0,T]))$, $p \in (2,4)$ *and* $q = 2p/(4-p)$ *such that* $u \in L^q(\Omega \times [0,T];\mathbb{R})$ *and* $[D^+u - D^-u] \in L^p(\Omega \times [0,T];\mathbb{R})$. *Then, there exists a constant C depending only on p and T such that*

$$\begin{aligned} &E\left(\sup_{t\in[0,T]}\left|\int_0^t u_s \circ dB_s\right|^p\right) \\ &\le\; CE\left[\int_0^T |u_s|^q ds + \int_{\Delta_1^T}\{(Du_t)_s\}^2 dsdt\right. \\ &\quad\left. + \int_{\Delta_2^T}\left\{\left(D^2u_t\right)_{r,s}\right\}^2 drdsdt + \int_0^T \left|\left(D^+u\right)_s - \left(D^-u\right)_s\right|^p ds\right], \end{aligned}$$

where Δ_1^T *and* Δ_2^T *are defined in* (4.31).

Proposition 114 *Let* $p > 2$ *and* $u \in \left(\mathbb{D}^{2,2}_{1,c}(L^2([0,T])) \cap L^p(\Omega \times [0,T])\right)_{loc}$. *Then, the process* $\{\int_0^t u_s \circ dB_s : t \in [0,T]\}$ *has continuous paths.*

Remark 90 *In Nualart [158], using Meyer's inequality, it is considered this result under different hypotheses. Namely, the result is true if u belongs to* $\mathbb{D}^{1,2}_{1,c}(L^2([0,T]))$ *and* $E\int_0^T\left(\int_0^T (D_s u_t)^2 ds\right)^{p/2} dt < \infty$, *for some* $p > 2$.

As in the forward integral, sometimes it is convenient to define the Stratonovich integral as the uniform limit on $[0,T]$ in probability of the integral $\frac{1}{2\varepsilon}\int_0^T u_s\left(B_{(s+\varepsilon)\wedge T} - B_{(s-\varepsilon)\vee 0}\right) ds$, as $\varepsilon \downarrow 0$. Since the proof of following result is quite similar to that of Proposition 101, we only sketch it in some parts.

Proposition 115 *Let u be a process in* $\mathbb{D}^{2,2}_{1,c}(L^2([0,T])) \cap L^p(\Omega \times [0,T])$, *for some* $p > 2$. *Then,* $(2\varepsilon)^{-1}\int_0^{\cdot} u_s\left(B_{(s+\varepsilon)\wedge T} - B_{(s-\varepsilon)\vee 0}\right) ds$ *converges to* $\int_0^{\cdot} u_s \circ dB_s$ *uniformly on* $[0,T]$ *in probability, as* $\varepsilon \to 0$.

Remark 91 *As in Proposition 101, we can use Theorem 76 to change the interval* $[0,T]$ *by* $[a,b]$, *for* $0 \le a < b \le T$.

Proof. As in (5.26), we can establish the equality

$$\begin{aligned}
&\frac{1}{2\varepsilon}\int_0^t u_s\left(B_{(s+\varepsilon)\wedge T}-B_{(s-\varepsilon)\vee 0}\right)ds\\
&= \int_0^{(t-\varepsilon)\vee 0}\left(\frac{1}{2\varepsilon}\int_{(r-\varepsilon)\vee 0}^{r+\varepsilon} u_s ds\right)\delta_2 B_r + \int_0^t\left(\frac{1}{2\varepsilon}\int_{(s-\varepsilon)\vee 0}^{(s+\varepsilon)\wedge T} D_r u_s dr\right)ds\\
&+\frac{1}{2\varepsilon}\int_{(t-2\varepsilon)\vee 0}^t u_s\left(B_{(s+\varepsilon)\wedge T}-B_{(t-\varepsilon)\vee 0}\right)ds\\
&-\frac{1}{2\varepsilon}\int_{(t-2\varepsilon)\vee 0}^t\int_{(t-\varepsilon)\vee 0}^{(s+\varepsilon)\wedge T} D_r u_s dr ds. \qquad (6.13)
\end{aligned}$$

The last two summands go to zero uniformly on $[0,T]$ in probability, as $\varepsilon\to 0$. In fact, we only need to proceed as in Steps 1 and 2 of the proof of Proposition 101.

For the second integral on the right-hand side of (6.13), we have

$$\begin{aligned}
&\left|\int_0^t\left(\frac{1}{2\varepsilon}\int_{(s-\varepsilon)\vee 0}^{(s+\varepsilon)\wedge T} D_r u_s dr\right)ds-\frac{1}{2}\int_0^t\left((D^-u)_s+(D^+u)_s\right)ds\right|\\
&\le \left|\int_0^t\left(\frac{1}{2\varepsilon}\int_s^{(s+\varepsilon)\wedge T} D_r u_s dr\right)ds-\frac{1}{2}\int_0^t (D^-u)_s ds\right|\\
&+\left|\int_0^t\left(\frac{1}{2\varepsilon}\int_{(s-\varepsilon)\vee 0}^{s} D_r u_s dr\right)ds-\frac{1}{2}\int_0^t (D^+u)_s ds\right|.
\end{aligned}$$

In consequence, we just have to continue as in Step 3 of the proof of Proposition 101.

Finally, observe that $t\mapsto\int_0^t u_s\delta_2 B_s$ has continuous paths due to Proposition 80 and $\sup_{t\in[0,T]}\left|\int_{(t-\varepsilon)\vee 0}^t u_s\delta_2 B_s\right| \le 2\sup_{t\in[0,T]}\left|\int_0^t u_s\delta_2 B_s\right| \in L^1(\Omega)$ because of Theorem 52. Therefore, the dominated convergence theorem leads us to conclude

$$E\left(\sup_{t\in[0,T]}\left|\int_{(t-\varepsilon)\vee 0}^t u_s\delta_2 B_s\right|\right)\to 0 \quad \text{as } \varepsilon\downarrow 0.$$

Thus, we can finish the proof by using Step 4 in the proof of Proposition 101. □

A consequence of Proposition 115 is the following result (compare it with Proposition 102).

Proposition 116 *Let u be as in Proposition 115 and $\tau:\Omega\to\mathbb{R}_+$ a random variable. Then, $u1_{[0,\tau]}$ belongs to Dom$\,\delta_B^S([0,T])$ and*

$$\int_0^s u_r\circ dB_r\Big|_{s=t\wedge\tau}=\int_0^t u_r 1_{[0,\tau]}(r)\circ dB_r, \quad t\in[0,T].$$

As in Proposition 102, this result makes sense because the process $\int_0^\cdot u_s\circ dB_s$ has continuous paths, which is an immediate consequence of Proposition 115.

Now, we see that the Stratonovich integral satisfies the "fundamental theorem of calculus" (6.3).

Theorem 77 *Let* $X_0 \in \mathbb{D}^{1,2}_{loc}$ *and* $u \in \left(\mathbb{D}^{2,2}_{1,c}(L^2([0,T])) \cap \mathbb{D}^{1,4}(L^2([0,T]))\right)_{loc}$ *such that* $D^-u, D^+u \in \left(\mathbb{D}^{1,2}(L^2([0,T]))\right)_{loc}$ *and the process* $\int_0^{\cdot} u_s \circ dB_s$ *is continuous. Moreover, consider a process* v *in* $\left(\mathbb{D}^{1,2}(L^2([0,T]))\right)_{loc}$ *and set*

$$X_t = X_0 + \int_0^t v_s ds + \int_0^t u_s \circ dB_s, \quad t \in [0,T].$$

Then, for $f \in C^2(\mathbb{R})$, *we have*

$$f(X_t) = f(X_0) + \int_0^t f'(X_s) v_s ds + \int_0^t f'(X_s) u_s \circ dB_s, \quad t \in [0,T].$$

Proof. In the proof of Theorem 68, it is pointed out that we can assume that $f \in C^2_b(\mathbb{R})$ and that $f'(X)u \in \mathbb{D}^{1,2}_{1,c}(L^2([0,T]))$ due to the localization procedure. Then, utilizing Theorem 68 again, we can write

$$\begin{aligned} f(X_t) &= f(X_0) + \int_0^t f'(X_s) u_s \delta_2 dB_s + \int_0^t f'(X_s) v_s ds + \frac{1}{2}\int_0^t f''(X_s) u_s^2 ds \\ &\quad + \frac{1}{2}\int_0^t f'(X_s)\left((D^-u)_s + (D^+u)_s\right) ds + \int_0^t f''(X_s)(D^-X)_s u_s ds, \quad t \in [0,T]. \end{aligned}$$

Additionally, proceeding as in (4.49), we can show that the equality

$$\left(D^+X\right)_t = u_t + \left(D^-X\right)_t, \quad t \in [0,T],$$

holds. Thus, Theorem 76 implies

$$\begin{aligned} f(X_t) &= f(X_0) + \int_0^t f'(X_s) u_s \delta_2 dB_s + \int_0^t f'(X_s) v_s ds \\ &\quad + \frac{1}{2}\int_0^t f'(X_s)\left((D^-u)_s + (D^+u)_s\right) ds \\ &\quad + \frac{1}{2}\int_0^t \left((D^- f'(X))_s + (D^+ f'(X))_s\right) u_s ds \\ &= f(X_0) + \int_0^t f'(X_s) u_s \delta_2 dB_s + \int_0^t f'(X_s) v_s ds \\ &\quad + \frac{1}{2}\int_0^t \left((D^- f'(X)u)_s + (D^+ f'(X)u)_s\right) ds \\ &= f(X_0) + \int_0^t f'(X_s) u_s \circ dB_s + \int_0^t f'(X_s) v_s ds, \quad t \in [0,T]. \end{aligned}$$

Therefore, the proof is finished. □

As in Theorem 69, we can state a Fubini's type theorem for the Stratonovich integral as follows.

Theorem 78 *Let* μ *be a* σ*-finite measure on the measurable space* $(X, \mathcal{C})$. *Assume that* $\varphi : \Omega \times X \to L^2([0,T])$ *is an* $\mathcal{F} \otimes \mathcal{C}$*-measurable function satisfying the following conditions:*

i) φ *belongs to* $L^1\left(X; \mathbb{D}^{1,2}(L^2([0,T]))\right)$.

ii) $(\omega, x) \mapsto \delta_2\left(\varphi(x)\right)$ *is* $\mathcal{F} \otimes \mathcal{C}$*-measurable.*

iii) The Bochner integral $\int_X \delta_2\left(\varphi(x)\right) \mu(dx)$ *is well-defined w.p.1 and belongs to* $L^2(\Omega)$.

iv) For μ-almost all $x \in X$, we have that $\varphi(x) \in \mathbb{D}^{1,2}_{1,c}(L^2([0,T]))$, and

$$\lim_{n\to\infty} \int_X \int_0^T \sup_{0\vee(s-(1/n))\leq t\leq s} E\left|(D_s\varphi(x)(t)) - (D^-\varphi(x))_s\right| ds\mu(dx) = 0$$

and

$$\lim_{n\to\infty} \int_X \int_0^T \sup_{s\leq t\leq T\wedge(s+(1/n))} E\left|(D_s\varphi(x)(t)) - (D^+\varphi(x))_s\right| ds\mu(dx) = 0.$$

Then, $\int_0^T \left(\int_X \varphi(x)(s)\mu(dx)\right) \circ dB_s$ and $\int_X \left(\int_0^T \varphi(x)(s) \circ dB_s\right) \mu(dx)$ are well-defined and

$$\int_0^T \left(\int_X \varphi(x)(s)\mu(dx)\right) \circ dB_s = \int_X \left(\int_0^T \varphi(x)(s) \circ dB_s\right) \mu(dx), \quad w.p.1.$$

Proof. We only need to proceed as in the proof of Theorem 69 in order to show that this Fubini's theorem holds. In this case, we must use Theorem 76 instead of Remark 79 and Theorem 67. □

Note that Propositions 113-115, and Theorems 77 and 78 are examples of the fact that the Stratonovich integral inherits some properties of the divergence operator through relation (6.11).

On the other hand, we now deal with the substitution formula for the Stratonovich integral with respect to Brownian motion (compare it with Nualart [158], Theorem 3.2.10). In the following result, we suppose that B is an $\mathcal{F}_t$-Brownian motion as we are assuming in this section.

Proposition 117 *Let $\{u_t(x) : t \in [0,T]$ and $x \in \mathbb{R}\}$ be a measurable and $\mathcal{F}_t$-adapted random field such that:*

i) $\int_0^T u_t(x) \circ dB_t$ is well-defined for all $x \in \mathbb{R}$.

ii) $\int_0^T E\left([u_t(0)]^2\right) dt < \infty$ and $t \mapsto u_t(x)$ is continuous in $L^2(\Omega)$, for every $x \in \mathbb{R}$.

iii) There exist $p \geq 2$ and $\alpha > 1$ such that, for $K > 0$, there is c_k satisfying

$$E\left(|u_t(x) - u_t(y)|^p\right) \leq c_k|x-y|^\alpha \tag{6.14}$$

and $E\left(|u_t(x) - u_t(y) - u_s(x) + u_s(y)|^p\right) \leq c_k|x-y|^\alpha|t-s|^{p/2}$, for all $|x|, |y| \leq K$ and $s,t \in [0,T]$.

Then, for any random variable F, we have that $u(F)$ is Stratonovich integrable and

$$\int_0^T u_t(F) \circ dB_s = \int_0^T u_t(x) \circ dB_s\Big|_{x=F}.$$

Proof. Let $\varepsilon > 0$. Therefore,

$$\begin{aligned} I_\varepsilon(x) &= \frac{1}{2\varepsilon}\int_0^T u_s(x)\left(B_{(s+\varepsilon)\wedge T} - B_{(s-\varepsilon)\vee 0}\right) ds \\ &= \frac{1}{2\varepsilon}\int_0^T u_{(s-\varepsilon)\vee 0}(x)\left(B_{(s+\varepsilon)\wedge T} - B_{(s-\varepsilon)\vee 0}\right) ds \\ &\quad +\frac{1}{2\varepsilon}\int_0^T \left(u_s(x) - u_{(s-\varepsilon)\vee 0}(x)\right)\left(B_{(s+\varepsilon)\wedge T} - B_{(s-\varepsilon)\vee 0}\right) ds \\ &:= I_{1,\varepsilon}(x) + I_{2,\varepsilon}(x), \quad x \in \mathbb{R}. \end{aligned} \tag{6.15}$$

Now, we divide the proof into several steps.

Step 1: Here, we see that, for each $x \in \mathbb{R}$, $I_{1,\varepsilon}(x)$ converges in probability to the Itô's integral $\int_0^T u_t(x)dB_t$, as $\varepsilon \downarrow 0$.

Theorem 48, Hypothesis *ii)* and the change of variables $v = s - \varepsilon$ imply that, for ε small enough,

$$\begin{aligned}
&I_{1,\varepsilon}(x)\\
&= \frac{1}{2\varepsilon}\int_0^T \left(\int_{(s-\varepsilon)\vee 0}^{(s+\varepsilon)\wedge T} u_{(s-\varepsilon)\vee 0}(x)dB_r\right)ds = \int_0^T \left(\frac{1}{2\varepsilon}\int_{(r-\varepsilon)\vee 0}^{(r+\varepsilon)\wedge T} u_{(s-\varepsilon)\vee 0}(x)ds\right)\delta_2 B_r\\
&= \int_0^\varepsilon \left(\frac{1}{2\varepsilon}\int_0^{r+\varepsilon} u_{(s-\varepsilon)\vee 0}(x)ds\right)\delta_2 B_r + \int_{T-\varepsilon}^T \left(\frac{1}{2\varepsilon}\int_{r-\varepsilon}^T u_{(s-\varepsilon)\vee 0}(x)ds\right)\delta_2 B_r\\
&\quad + \int_\varepsilon^{T-\varepsilon} \left(\frac{1}{2\varepsilon}\int_{r-\varepsilon}^{r+\varepsilon} u_{(s-\varepsilon)\vee 0}(x)ds\right)\delta_2 B_r = \int_0^\varepsilon \left(\frac{u_0(x)}{2} + \frac{1}{2\varepsilon}\int_0^r u_s(x)ds\right)\delta_2 B_r\\
&\quad + \int_{T-\varepsilon}^T \left(\frac{1}{2\varepsilon}\int_{r-2\varepsilon}^{T-\varepsilon} u_s(x)ds\right)\delta_2 B_r + \int_\varepsilon^{T-\varepsilon} \left(\frac{1}{2\varepsilon}\int_{r-2\varepsilon}^{r} u_{s\vee 0}(x)ds\right)\delta_2 B_r\\
&= I_{1,1,\varepsilon}(x) + I_{1,2,\varepsilon}(x) + I_{1,3,\varepsilon}(x). \qquad (6.16)
\end{aligned}$$

Note that $I_{1,1,\varepsilon}(x)$, $I_{1,2,\varepsilon}(x)$ and $I_{1,3,\varepsilon}(x)$ are Itô's integrals. Thus,

$$\begin{aligned}
E\left((I_{1,1,\varepsilon}(x))^2\right) &= E\int_0^\varepsilon \left(\frac{u_0(x)}{2} + \frac{1}{2\varepsilon}\int_0^r u_s(x)ds\right)^2 dr\\
&\le \varepsilon E\left(u_0(x)^2\right) + \frac{\sup_{0\le s\le T} E\left(u_s(x)^2\right)}{\varepsilon}\int_0^\varepsilon\int_0^r dsdr \to 0, \text{ as } \varepsilon \downarrow 0. \quad (6.17)
\end{aligned}$$

Similarly, we have

$$\begin{aligned}
E\left((I_{1,2,\varepsilon}(x))^2\right) &= E\int_{T-\varepsilon}^T \left(\frac{1}{2\varepsilon}\int_{r-2\varepsilon}^{T-\varepsilon} u_s(x)ds\right)^2 dr\\
&\le \frac{\sup_{0\le s\le T} E\left(u_s(x)^2\right)}{2\varepsilon}2\varepsilon^2 \to 0, \quad \text{as } \varepsilon \downarrow 0. \qquad (6.18)
\end{aligned}$$

Finally, in this step, we consider

$$\begin{aligned}
&E\left(\left(I_{1,3,\varepsilon}(x) - \int_0^T u_s(x)dB_s\right)^2\right)\\
&= E\Bigg\{\Bigg(\int_\varepsilon^{2\varepsilon}\left(\frac{1}{2\varepsilon}\int_{r-2\varepsilon}^r u_{s\vee 0}(x)ds\right)dB_r - \int_0^{2\varepsilon} u_s(x)dB_s - \int_{T-\varepsilon}^T u_s(x)dB_s\\
&\qquad + \int_{2\varepsilon}^{T-\varepsilon}\left(\frac{1}{2\varepsilon}\int_{r-2\varepsilon}^r (u_s(x) - u_r(x))ds\right)dB_r\Bigg)^2\Bigg\}\\
&\le CE\Bigg\{\int_\varepsilon^{2\varepsilon}\left(\frac{1}{2\varepsilon}\int_{r-2\varepsilon}^r u_{s\vee 0}(x)^2ds\right)dr + \int_0^{2\varepsilon} u_s(x)^2ds - \int_{T-\varepsilon}^T u_s(x)^2ds\\
&\qquad + \int_{2\varepsilon}^{T-\varepsilon}\left(\frac{1}{2\varepsilon}\int_{r-2\varepsilon}^r (u_s(x) - u_r(x))^2ds\right)dr\Bigg\} \to 0, \quad \text{as } \varepsilon \downarrow 0,
\end{aligned}$$

due to Hypothesis *ii*). Hence, the claim of this part of the proof holds because of (6.16)-(6.18).

Step 2: Now, we show that $I_{1,\varepsilon}(F)$ goes to $\int_0^T u_s(x)dB_s|_{x=F}$ in probability, as $\varepsilon \downarrow 0$.

In order to prove that this step is true, we apply Lemma 86 in the appendix. So, choose $x, y \in \mathbb{R}$ such that $|x|, |y| \leq M$. Then, (6.16), Burkholder-Davis-Gundy inequality and Hypothesis *iii*) lead us to write

$$\begin{aligned} E\left((I_{1,\varepsilon}(x) - I_{1,\varepsilon}(y)|^p\right) &\leq E\left\{\left|\int_0^T \left(\frac{1}{2\varepsilon}\int_{(r-2\varepsilon)\vee 0}^r (u_{s\vee 0}(x) - u_{s\vee 0}(y))ds\right) dB_r\right|^p\right\} \\ &\leq C_p E\left\{\left[\int_0^T \left(\frac{1}{2\varepsilon}\int_{(r-2\varepsilon)\vee 0}^r (u_{s\vee 0}(x) - u_{s\vee 0}(y))ds\right)^2 dr\right]^{p/2}\right\} \\ &\leq C_{p,T}\frac{1}{2\varepsilon}\int_0^T\int_{(r-2\varepsilon)\vee 0}^r E\left(|u_{s\vee 0}(x) - u_{s\vee 0}(y)|^p\right) dsdr \\ &\leq C_{p,T,M}|x-y|^\alpha. \end{aligned}$$

Consequently, the claim of this step is satisfied.

Step 3: The following aim is to prove that $I_{2,\varepsilon}$ introduced in (6.15) satisfies inequality (6.14), for some $p > 0$ and $\alpha > 1$.

Fix $\eta \in (0,1)$ such that $\eta\alpha > 1$, where α is given in (6.14). Then, Hölder inequality implies

$$\begin{aligned} \|I_{2,\varepsilon}(x) - I_{2,\varepsilon}(y)\|_{L^{\eta p}(\Omega)} &\leq \frac{1}{2\varepsilon}\int_0^T \big\|\left(u_s(x) - u_{(s-\varepsilon)\vee 0}(x) - u_s(y) + u_{(s-\varepsilon)\vee 0}(y)\right) \\ &\qquad \times \left(B_{(s+\varepsilon)\wedge T} - B_{(s-\varepsilon)\vee 0}\right)\big\|_{L^{\eta p}(\Omega)} ds \\ &\leq \frac{1}{2\varepsilon}\int_0^T \left\|u_s(x) - u_{(s-\varepsilon)\vee 0}(x) - u_s(y) + u_{(s-\varepsilon)\vee 0}(y)\right\|_{L^p(\Omega)} \\ &\qquad \times \left\|B_{(s+\varepsilon)\wedge T} - B_{(s-\varepsilon)\vee 0}\right\|_{L^{p\eta/(1-\eta)}(\Omega)} ds. \end{aligned}$$

Hence, (1.10) and Hypothesis *iii*) yield that, for $|x|, |y| \leq M$,

$$\begin{aligned} \|I_{2,\varepsilon}(x) - I_{2,\varepsilon}(y)\|_{L^{\eta p}(\Omega)} &\leq \frac{1}{\sqrt{2\varepsilon}}\int_0^T \left\|u_s(x) - u_{(s-\varepsilon)\vee 0}(x) - u_s(y) + u_{(s-\varepsilon)\vee 0}(y)\right\|_{L^p(\Omega)} ds \\ &\leq C_{M,T}|x-y|^{\alpha/p}. \end{aligned}$$

Now, it is easy to verify that the claim of this step is true.

Step 4: Finally, we finish the proof.

By (6.15), Hypothesis *i*) and Step 1, we have that, for $x \in \mathbb{R}$, $I_{2,\varepsilon}(x)$ converges in probability to $\int_0^T u_s(x) \circ dB_s - \int_0^T u_s(x)dB_s$, as $\varepsilon \downarrow 0$. In consequence, Steps 2 and 3, together with Lemma 86, give that

$$\begin{aligned} I_\varepsilon(F) = I_{1,\varepsilon}(F) + I_{2,\varepsilon}(F) &\to \int_0^T u_s(x)dB_s|_{x=F} \\ + \left(\int_0^T u_s(x) \circ dB_s|_{x=F} - \int_0^T u_s(x)dB_s|_{x=F}\right) &= \int_0^T u_s(x) \circ dB_s|_{x=F}. \end{aligned}$$

In this way, the proof is complete. □

6.3 Fractional Brownian motion

In the remainder of this chapter, we consider the case that $B = \{B_t : t \in [0,T]\}$ is fractional Brownian motion with Hurst parameter $H \in (0,1)$.

6.3.1 Case $H > 1/2$

Proceeding as in Theorem 71, Alòs and Nualart [13] have established that if the process $u = \{u_t : t \in [0,T]\}$ belongs to $\mathbb{D}^{1,2}(|\mathcal{H}|)$ and

$$\int_0^T \int_0^T |D_s u_t| \, |t-s|^{2H-2} dsdt < \infty, \quad \text{with probability 1,} \tag{6.19}$$

then, for $0 \le a < b \le T$, u is in $\text{Dom}\, \delta_B^{\text{S}}([a,b])$ and

$$\int_a^b u_s \circ dB_s = \delta_2\left(1_{[a,b]}(\cdot)u_\cdot\right) + H(2H-1)\int_a^b \int_0^T D_s u_t \, |t-s|^{2H-2} dsdt. \tag{6.20}$$

But, this is exactly what Theorem 71 states for the forward integral. In other words, we have the following result.

Theorem 79 *Let u be a process in $\mathbb{D}^{1,2}(|\mathcal{H}|)$ such that* (6.19) *holds. Then, for any $0 \le a < b \le T$, $u \in Dom\delta_B^-([a,b]) \cap Dom\,\delta_B^S([a,b])$ and* (6.20) *is satisfied. Moreover, in this case, the forward and Stratonovich integrals of u with respect to B are the same.*

An important consequence of Theorem 79 is that we can see easily that some results for the forward (resp. Stratonovich) integral are also satisfied for the Stratonovich (resp. forward) integral such as Theorem 72 and Proposition 106. An example showing that the forward integral inherits a property from the Stratonovich one is the following result. Towards this end, we consider property (1.142). That is, remember that a functions $f : [0,T] \times \mathbb{R} \to \mathbb{R}$ satisfies (1.142) if there exists a constant $C > 0$ such that

$$|f(s,x)| \le C \exp(\sigma x^2), \quad (s,x) \in [0,T] \times \mathbb{R}, \tag{6.21}$$

with $0 < \sigma < \left(\sup_{0 \le t \le T} R_B(t,t)\right)^{-1}/4$.

Proposition 118 *Let B be a fractional Brownian motion with $H > 1/2$ and $f : [0,T] \times \mathbb{R} \to \mathbb{R}$ a function in $\mathcal{C}^{1,2}(\mathbb{R})$ such that $f, \partial_x f, \partial_x^2 f$ and $\partial_t f$ satisfy inequality* (6.21). *Then, $\partial_x f(\cdot, B_\cdot)$ belongs to $Dom\,\delta_B^S([0,t])$ and*

$$f(t,B_t) = f(0,0) + \int_0^t \partial_t f(s,B_s)ds + \int_0^t \partial_x f(s,B_s) \circ dB_s, \quad t \in [0,T].$$

Remark 92 *We see in the proof of this proposition that $\partial_x f(\cdot, B_\cdot)$ is in the space $\mathbb{D}^{1,2}(|\mathcal{H}|)$. Thus, this result is also true for the forward integral, as Theorem 79 is also valid for this integral.*

Proof. Note that inequality (6.21) implies that $\partial_x f(\cdot, B_\cdot)$ belongs to $\mathbb{L}_H^{1,2}$ and, therefore, $\partial_x f(\cdot, B_\cdot)$ also belongs to $\mathbb{D}^{1,2}(|\mathcal{H}|)$ due to (1.129). Moreover, Theorem 59 allows us to write

$$\begin{aligned} f(t,B_t) &= f(0,0) + \int_0^t \partial_t f(s,B_s)ds + \int_0^t \partial_x f(s,B_s)\delta_2 B_s \\ &\quad + H\int_0^t \partial_x^2 f(s,B_s)s^{2H-1}ds, \quad t \in [0,T]. \end{aligned} \tag{6.22}$$

Finally, we have

$$\begin{aligned}\int_0^t \int_0^T D_s\partial_x f(r,B_r)|r-s|^{2H-2}dsdr &= \int_0^t \int_0^r \partial_x^2 f(r,B_r)|r-s|^{2H-2}dsdr \\ &= (2H-1)^{-1}\int_0^t \partial_x^2 f(r,B_r)r^{2H-1}dr,\end{aligned}$$

which, together with (6.20) and (6.22), allows us to deduce that the proof is complete. □

6.3.2 Case $H \in (1/4, 1/2)$

The purpose of this section is to use the techniques of the Malliavin calculus to analyze the Stratonovich integral with respect to some Gaussian-Volterra processes, which include the fractional Brownian motion with Hurst parameter H in $(1/4, 1/2)$. The Gaussian Volterra process is introduced in Definition 16. Namely, a stochastic process $B = \{B_t : t \in [0,T]\}$ is a Volterra process if it has the form

$$B_t = \int_0^t K(t,r)dW_r, \quad t \in [0,T].$$

Here, $W = \{W_t : t \in [0,T]\}$ is a Brownian motion and the kernel $K = \{K(t,s) : 0 < s < t \le T\}$ is square-integrable. That is, $K(t,\cdot) \in L^2([0,t])$, for every $t \in [0,T]$. As in Section 1.3, we suppose that (1.34) holds. It means,

$$\sup_{t\in[0,T]} \int_0^t K(t,s)^2 ds = \sup_{t\in[0,T]} R_B(t,t) < \infty.$$

Additionally, we assume that the kernel K satisfies the following two inequalities, with $\alpha \in (0,1/2)$ and $c > 0$ a constant:

i) $|K(t,s)| \le c\left((t-s)^{-\alpha} + s^{-\alpha}\right)$.

ii) $\left|\frac{\partial K}{\partial t}(t,s)\right| \le c(t-s)^{-1-\alpha}$.

It is worth mentioning that these two conditions are inspired by the kernel of fractional Brownian motion with $\alpha = \frac{1}{2} - H$ and $H \in (0,1/2)$, which is introduced in Section 1.4. Indeed, Theorem 9 and Lemma 54, together with Remark 69, imply the kernel in (4.76) satisfies Condition *i)*. Concerning Condition *ii)*, we apply Lemma 54 and (1.143) to show that inequality *ii)* also holds for fractional Brownian motion.

Now, we consider the seminorm on the set $\mathcal{E}([0,T])$ of step functions on $[0,T]$ (see Definition 13):

$$\|\varphi\|_K^2 = \int_0^T \varphi^2(s)K(T,s)^2 ds + \int_0^T \left(\int_s^T |\varphi(t)-\varphi(s)|\,(t-s)^{-1-\alpha}dt\right)^2 ds. \tag{6.23}$$

We denote by $\mathcal{H}_K$ the completion of $\mathcal{E}([0,T])$ with respect to this seminorm $\|\cdot\|_K^2$. Note that (1.37) yields that, for some constant $C > 0$,

$$\int_0^T \left((K^*\varphi)(s)\right)^2 ds \le C\,\|\varphi\|_K^2, \quad \text{for any step function } \varphi \in \mathcal{E}([0,T]).$$

Therefore, the space $\mathcal{H}_K$ is continuously contained in the reproducing kernel Hilbert space $\mathcal{H}$ of the Gaussian process B. Thus, we have the inclusion $\mathbb{D}^{1,2}(\mathcal{H}_K) \subset \mathbb{D}^{1,2}(\mathcal{H})$. Remember that a process $u = \{u_t : t \in [0,T]\}$ belongs to $\mathbb{D}^{1,2}(\mathcal{H}_K)$ if there exists a sequence $\{\varphi_n\}$ of bounded simple $\mathcal{H}_K$-valued processes of the form

$$\varphi_n = \sum_{j=0}^{n-1} F_j 1_{(t_j,t_{j+1}]}, \tag{6.24}$$

where F_j is a smooth random variable of the form $F_j = f_j(B_{s_1^j}, ..., B_{s^j_{m(j)}})$, with $f_j \in \mathcal{C}_b^\infty(\mathbb{R})$ (i.e., it is a bounded function with bounded partial derivatives), and $\{0 = t_0 < t_1 < ... < t_n = T\}$ is a partition of $[0,T]$ such that

$$E\left(\|u - \varphi_n\|_K^2\right) + E\left(\int_0^T \|D_r u - D_r \varphi_n\|_K^2 \, dr\right) \to 0, \quad \text{as } n \to \infty. \tag{6.25}$$

In particular, we have that if $u \in \mathbb{D}^{1,2}(\mathcal{H}_K)$, then u is also in $\operatorname{Dom}\delta_2^B$ and the transfer principle implies that $K^* u \in \mathbb{D}_W^{1,2}(L^2([0,T])$ and (1.39) holds.

In order to study the relation between the Stratonovich integrals δ_B^S and the divergence operator δ_2^B, we give the notion of trace. It means, we say that a process $u \in \mathbb{D}^{1,2}(\mathcal{H}_K)$ belongs to the space $\mathbb{D}_C^{1,2}(\mathcal{H}_K)$ if the limit in probability

$$TrDu := \lim_{\varepsilon \to 0} \frac{1}{2\varepsilon} \int_0^T \left\langle D^B u_s, \mathbf{1}_{[(s-\varepsilon)\vee 0,(s+\varepsilon)\wedge T]} \right\rangle_{\mathcal{H}} ds \tag{6.26}$$

exists. The random variable $TrDu$ is called the trace of Du.

Now, we state the relation between the operators δ_B^S and δ_2^B.

Theorem 80 *Assume that the kernel K of B satisfies Conditions i) and ii), and let u be a process in $\mathbb{D}_C^{1,2}(\mathcal{H}_K)$ such that*

$$E\int_0^T u_s^2 \left(s^{-2\alpha} + (T-s)^{-2\alpha}\right) ds < \infty \tag{6.27}$$

and

$$E\int_0^T \int_0^T (D_r u_s)^2 \left(s^{-2\alpha} + (T-s)^{-2\alpha}\right) dsdr < \infty. \tag{6.28}$$

Then $u \in Dom\,\delta_B^S$ and

$$\delta_B^S(u) = \delta_2^B(u) + TrDu.$$

Remark 93 *Remember that the fact that $\mathbb{D}^{1,2}(\mathcal{H}_K) \subset \mathbb{D}^{1,2}(\mathcal{H})$ implies that u also belongs to $Dom\,\delta_2^B$.*

In order to prove that last theorem is true, we need the following technical result, where we use the conventions $u_s \equiv 0$ for $s \notin [0,T]$ and

$$u_t^\varepsilon := \frac{1}{2\varepsilon} \int_{t-\varepsilon}^{t+\varepsilon} u_s ds, \quad \text{for } \varepsilon > 0,$$

Lemma 67 *Let u be a simple process of the form (6.24). Then u^ε converges to u in $\mathbb{D}^{1,2}(\mathcal{H}_K)$, as $\varepsilon \downarrow 0$.*

Proof. Assume that the process u is given by the right-hand side of (6.24). Then u is a bounded process. Hence, Condition i) on the kernel K and the dominated convergence theorem lead us to get

$$E\left(\int_0^T (u_s - u_s^\varepsilon)^2 K(T,s)^2 ds\right) \to 0 \quad \text{as } \varepsilon \downarrow 0. \tag{6.29}$$

Now, fix an index $i \in \{0, 1, ..., n-1\}$. Thus, (6.24) implies that $u_t - u_s = 0$ for $s, t \in [t_i, t_{i+1}]$. In consequence,

$$\begin{aligned}
&\int_{t_i}^{t_{i+1}} \left(\int_s^T |u_t^\varepsilon - u_s^\varepsilon - (u_t - u_s)| (t-s)^{-1-\alpha} dt\right)^2 ds \\
&\leq 2\int_{t_i}^{t_{i+1}} \left(\int_s^{t_{i+1}} |u_t^\varepsilon - u_s^\varepsilon| (t-s)^{-1-\alpha} dt\right)^2 ds \\
&\quad +2\int_{t_i}^{t_{i+1}} \left(\int_{t_{i+1}}^T |u_t^\varepsilon - u_s^\varepsilon - (u_t - u_s)| (t-s)^{-1-\alpha} dt\right)^2 ds \\
&= 2A_1(i,\varepsilon) + 2A_2(i,\varepsilon).
\end{aligned} \tag{6.30}$$

The convergence of the term $A_2(i, \varepsilon)$ to 0, as $\varepsilon \downarrow 0$, is a consequence of the dominated convergence theorem, and the facts that u is a bounded process and

$$|u_t^\varepsilon - u_s^\varepsilon - (u_t - u_s)| (t-s)^{-1-\alpha} \longrightarrow 0, \quad \text{for a.a. } 0 \leq s < t \leq T, \text{ as } \varepsilon \downarrow 0.$$

Concerning the term $A_1(i, \varepsilon)$, let $\varepsilon < \frac{1}{4} \min_{0 \leq i \leq n-1} |t_{i+1} - t_i|$. Then $u_t^\varepsilon - u_s^\varepsilon = 0$ if s and t belong to $[t_i + 2\varepsilon, t_{i+1} - 2\varepsilon]$. Therefore,

$$\begin{aligned}
E(A_1(i,\varepsilon)) \leq\ & 8\int_{t_i}^{t_i+2\varepsilon} \left(\int_s^{t_i+2\varepsilon} |u_t^\varepsilon - u_s^\varepsilon| (t-s)^{-1-\alpha} dt\right)^2 ds \\
&+8\int_{t_{i+1}-2\varepsilon}^{t_{i+1}} \left(\int_s^{t_{i+1}} |u_t^\varepsilon - u_s^\varepsilon| (t-s)^{-1-\alpha} dt\right)^2 ds \\
&+8\int_{t_i}^{t_i+2\varepsilon} \left(\int_{t_i+2\varepsilon}^{t_{i+1}} |u_t^\varepsilon - u_s^\varepsilon| (t-s)^{-1-\alpha} dt\right)^2 ds \\
&+8\int_{t_i+2\varepsilon}^{t_{i+1}-2\varepsilon} \left(\int_{t_{i+1}-2\varepsilon}^{t_{i+1}} |u_t^\varepsilon - u_s^\varepsilon| (t-s)^{-1-\alpha} dt\right)^2 ds.
\end{aligned}$$

The first two integrals go to zero because $|u_t^\varepsilon - u_s^\varepsilon| \leq \frac{c}{\varepsilon}|t-s|$. The convergence to zero, as $\varepsilon \downarrow 0$, of the last two addends follows from the fact that u_t^ε is bounded. In consequence, we have shown

$$E\|u - u^\varepsilon\|_K^2 \to 0, \quad \text{as } \varepsilon \to 0.$$

Finally, proceeding similarly, we are able to see the convergence

$$E\int_0^T \|D_r u - D_r u^\varepsilon\|_K^2 dr \to 0, \quad \text{as } \varepsilon \to 0.$$

Thus, the proof is complete. □

Now, we are ready to show that Theorem 80 is satisfied.

Proof of Theorem 80. By property (4.3) and Fubini's theorem for the divergence operator (i.e., Theorem 48), we have

$$
\begin{aligned}
&(2\varepsilon)^{-1}\int_0^T u_s\left(B_{(s+\varepsilon)\wedge T}-B_{(s-\varepsilon)\vee 0}\right)ds\\
&= (2\varepsilon)^{-1}\int_0^T \delta_2^B\left(u_s 1_{[(s-\varepsilon)\vee 0,(s+\varepsilon)\wedge T]}(\cdot)\right)ds\\
&\quad +(2\varepsilon)^{-1}\int_0^T \left\langle D_\cdot^B u_s, 1_{[(s-\varepsilon)\vee 0,(s+\varepsilon)\wedge T]}(\cdot)\right\rangle_{\mathcal{H}} ds\\
&= (2\varepsilon)^{-1}\int_0^T\left(\int_{(r-\varepsilon)\vee 0}^{(r+\varepsilon)\wedge T} u_s ds\right)\delta_2 B_r+(2\varepsilon)^{-1}\int_0^T \left\langle D_\cdot^B u_s, 1_{[(s-\varepsilon)\vee 0,(s+\varepsilon)\wedge T]}(\cdot)\right\rangle_{\mathcal{H}} ds\\
&= \int_0^T u_r^\varepsilon \delta_2 B_r+(2\varepsilon)^{-1}\int_0^T \left\langle D_\cdot^B u_s, 1_{[(s-\varepsilon)\vee 0,(s+\varepsilon)\wedge T]}(\cdot)\right\rangle_{\mathcal{H}} ds.
\end{aligned}
$$

By hypothesis, $u\in\mathbb{D}_C^{1,2}(\mathcal{H}_K)$. Therefore, the definition of the random variable $TrDu$ implies

$$
(2\varepsilon)^{-1}\int_0^T \left\langle D_\cdot^B u_s, 1_{[(s-\varepsilon)\vee 0,(s+\varepsilon)\wedge T]}(\cdot)\right\rangle_{\mathcal{H}} ds\to TrDu \quad \text{in probability, as } \varepsilon\downarrow 0.
$$

Therefore, we only need to show that $\int_0^T u_r^\varepsilon\delta_2 B_r$ converges to $\delta_2^B(u)$ in $L^2(\Omega)$, as $\varepsilon\downarrow 0$, in order to finish the proof. To do so, it is enough to see that u^ε goes to u in the space $\mathbb{D}^{1,2}(\mathcal{H}_K)$. Towards this end, fix $\delta>0$ and choose a bounded simple $\mathcal{H}_K$-valued simple processes φ as in (6.24) such that

$$
E\left(\|u-\varphi\|_K^2+\int_0^T \|D_r u-D_r\varphi\|_K^2\,dr\right)\le\delta \tag{6.31}
$$

(see (6.25)). Thus, Lemma 67 yields that, for ε small enough,

$$
\begin{aligned}
&E\left(\|u-u^\varepsilon\|_K^2+\int_0^T \|D_r(u-u^\varepsilon)\|_K^2\,dr\right)\\
&\le cE\left(\|u-\varphi\|_K^2+\int_0^T \|D_r(u-\varphi)\|_K^2\,dr\right)\\
&\quad +cE\left(\|\varphi-\varphi^\varepsilon\|_K^2+\int_0^T \|D_r(\varphi-\varphi^\varepsilon)\|_K^2\,dr\right)\\
&\quad +cE\left(\|\varphi^\varepsilon-u^\varepsilon\|_K^2+\int_0^T \|D_r(\varphi^\varepsilon-u^\varepsilon)\|_K^2\,dr\right)\\
&\le 2c\delta+cE\left(\|\varphi^\varepsilon-u^\varepsilon\|_K^2+\int_0^T \|D_r(\varphi^\varepsilon-u^\varepsilon)\|_K^2\,dr\right).
\end{aligned} \tag{6.32}
$$

Note that Fubini's theorem and Hölder inequality give

$$
\begin{aligned}
\int_0^T E\left((\varphi_s^\varepsilon-u_s^\varepsilon)^2\ K(T,s)^2\right)ds &= \int_0^T E\left(\left(\frac{1}{2\varepsilon}\int_{s-\varepsilon}^{s+\varepsilon}(\varphi_r-u_r)\,dr\right)^2\right)K(T,s)^2 ds\\
&\le \int_0^T E\left((\varphi_r-u_r)^2\right)\left(\frac{1}{2\varepsilon}\int_{(r-\varepsilon)\vee 0}^{(r+\varepsilon)\wedge T} K(T,s)^2 ds\right)dr.
\end{aligned}
$$

In order to continue with the proof, we observe that Property i) of the Kernel K allows us to establish

$$(2\varepsilon)^{-1}\int_{(r-\varepsilon)\vee 0}^{(r+\varepsilon)\wedge T} K(T,t)^2 dt \leq c\left[(T-r)^{-2\alpha}+r^{-2\alpha}\right].$$

For instance,

$$\int_0^{r+\varepsilon} t^{-2\alpha}dt = \frac{(r+\varepsilon)(r+\varepsilon)^{-2\alpha}}{1-2\alpha} \leq \frac{2\varepsilon}{1-2\alpha}r^{-2\alpha}, \quad r\in(0,\varepsilon).$$

If $r\in(\varepsilon,2\varepsilon)$, the fact that $0<1-2\alpha<1$ leads us to

$$\int_{r-\varepsilon}^{r+\varepsilon} t^{-2\alpha}dt = \frac{(r+\varepsilon)^{1-2\alpha}-(r-\varepsilon)^{1-2\alpha}}{1-2\alpha} \leq \frac{(2\varepsilon)^{1-2\alpha}}{1-2\alpha} \leq \frac{2\varepsilon}{1-2\alpha}r^{-2\alpha}.$$

For the case that $r\geq 2\varepsilon$, apply the mean value theorem. Hence, from (6.31), we obtain

$$\limsup_{\varepsilon\downarrow 0}\int_0^T E\left(\varphi_s^\varepsilon - u_s^\varepsilon\right)^2\ K(T,s)^2 ds \leq \int_0^T E\left(\varphi_s-u_s\right)^2\ K(T,s)^2 ds\ \leq\delta. \tag{6.33}$$

Acting similarly,

$$\begin{aligned}
&E\left(\int_0^T\left(\int_s^T\left|\varphi_t^\varepsilon-u_t^\varepsilon-\varphi_s^\varepsilon+u_s^\varepsilon\right|(t-s)^{-1-\alpha}dt\right)^2 ds\right)\\
&= \frac{1}{4\varepsilon^2}E\left(\int_0^T\left(\int_{-\varepsilon}^{\varepsilon}\int_s^T\left|(\varphi-u)_{t-\theta}-(\varphi-u)_{s-\theta}\right|(t-s)^{-1-\alpha}dtd\theta\right)^2 ds\right)\\
&= \frac{1}{4\varepsilon^2}E\left(\int_0^T\left(\int_{s-\varepsilon}^{s+\varepsilon}\int_r^{T+r-s}\left|(\varphi-u)_t-(\varphi-u)_r\right|(t-r)^{-1-\alpha}dtdr\right)^2 ds\right)\\
&\leq \frac{1}{2\varepsilon}E\left(\int_0^T\int_{s-\varepsilon}^{s+\varepsilon}\left(\int_r^{T+\varepsilon}\left|(\varphi-u)_t-(\varphi-u)_r\right|(t-r)^{-1-\alpha}dt\right)^2 drds\right)\\
&= \frac{1}{2\varepsilon}E\left(\int_{-\varepsilon}^{T+\varepsilon}\int_{(r-\varepsilon)\vee 0}^{(r+\varepsilon)\wedge T}\left(\int_r^{T+\varepsilon}\left|\varphi_t-u_t-\varphi_r+u_r\right|(t-r)^{-1-\alpha}dt\right)^2 dsdr\right)\\
&\leq E\left(\int_{-\varepsilon}^{T+\varepsilon}\left(\int_r^{T+\varepsilon}\left|\varphi_t-u_t-\varphi_r+u_r\right|(t-r)^{-1-\alpha}dt\right)^2 dr\right).
\end{aligned} \tag{6.34}$$

Consequently (6.32)-(6.34), together with a similar argument, give

$$\limsup_{\varepsilon\downarrow 0}E\left(\|\varphi^\varepsilon-u^\varepsilon\|_K^2\right)+\limsup_{\varepsilon\downarrow 0}E\left(\int_0^T\|D_r\left(\varphi^\varepsilon-u^\varepsilon\right)\|_K^2\,dr\right)\leq 4\delta.$$

Since δ is arbitrary, u^ε converges to u in the norm of the space $\mathbb{D}^{1,2}\left(\mathcal{H}_K\right)$, as $\varepsilon\downarrow 0$, which implies that $\int_0^T u_r^\varepsilon\delta_2 B_r$ tends in $L^2(\Omega)$ to $\delta_2^B(u)$. In this way, the proof is complete. □

Remark 94 *Theorem 80 is true for fractional Brownian motion and Riemann-Liouville fractional Brownian motion, which is introduced in Section 4.5, both with parameter* $H\in(0,1/2)$ *because their kernels satisfy Conditions i) and ii) of this section.*

On the other hand, our next aim is to provide the existence of the Stratonovich integrals of $f(B)$ with respect to the Gaussian process B by means of Theorem 80 in the case that B is either fractional Brownian motion, or Riemann-Liouville fractional process with $H \in (1/4, 1/2)$. Towards this end, we first study the following result.

Lemma 68 *Let $H \in (1/4, 1/2)$. Assume that the Kernel $K = \{K(t,s) : 0 \le s < t \le T\}$ satisfies Conditions i) and ii) of this section with $\alpha = \frac{1}{2} - H$. Then, there exists a constant $c_H > 0$ such that*

$$\|B_t - B_s\|_{L^2(\Omega)} \le c_H |t-s|^H, \quad s,t \in [0,T].$$

Proof. Let $s,t \in [0,T]$, $s < t$. Then,

$$\begin{aligned} E\left(|B_t - B_s|^2\right) &= \int_s^t K(t,r)^2 dr + 2\int_0^s K(t,r)\left[K(t,r) - K(s,r)\right] dr \\ &\quad + \int_0^s \left[K(s,r)^2 - K(t,r)^2\right] dr \\ &= \int_s^t K(t,r)^2 dr + 2\int_0^s K(t,r) \int_s^t \frac{\partial K}{\partial u}(u,r) du dr \\ &\quad - \int_0^s \left[K(s,r) + K(t,r)\right] \int_s^t \frac{\partial K}{\partial u}(u,r) du dr = I_1 + 2I_2 - I_3. \end{aligned} \quad (6.35)$$

Hence, Hypothesis $i)$ allows us to write

$$I_1 \le 2c^2 \int_s^t \left[(t-r)^{-2\alpha} + r^{-2\alpha}\right] dr = c_H\left[(t-s)^{2H} + t^{2H} - s^{2H}\right] \le c_H (t-s)^{2H}. \quad (6.36)$$

Here, we recall you that the constant c_H may change from line to line. Now, we use Hypotheses $i)$ and $ii)$ to obtain

$$\begin{aligned} |I_2| &\le c^2 \int_0^s \left[(t-r)^{-\alpha} + r^{-\alpha}\right] \int_s^t (u-r)^{-1-\alpha} du dr \\ &= c^2 \int_0^s (t-r)^{-\alpha} \int_s^t (u-r)^{-1-\alpha} du dr + c^2 \int_0^s r^{-\alpha} \int_s^t (u-r)^{-1-\alpha} du dr \\ &= I_{21} + I_{22}. \end{aligned} \quad (6.37)$$

For I_{21}, we have

$$\begin{aligned} |I_{21}| &\le c_H \int_0^s (t-r)^{-\alpha}\left[(s-r)^{-\alpha} - (t-r)^{-\alpha}\right] dr \\ &\le c_H \int_0^s \left[(s-r)^{-2\alpha} - (t-r)^{-2\alpha}\right] dr = c_H\left[s^{2H} - t^{2H} + (t-s)^{2H}\right] \\ &\le c_H (t-s)^{2H}. \end{aligned} \quad (6.38)$$

and, for I_{22}, we apply Lemma 87 and the change of variables $u = t - r$ to get

$$\begin{aligned} |I_{22}| &= c_H \int_0^s r^{-\alpha}\left[(s-r)^{-\alpha} - (t-r)^{-\alpha}\right] dr \\ &= c_H \left\{\int_0^s r^{-\alpha}(s-r)^{-\alpha} dr - \int_0^t r^{-\alpha}(t-r)^{-\alpha} dr + \int_s^t r^{-\alpha}(t-r)^{-\alpha} dr\right\} \\ &= c_H \left\{s^{2H} - t^{2H} + \int_0^{t-s} u^{-\alpha}(t-u)^{-\alpha}\right\} \\ &\le c_H \left\{s^{2H} - t^{2H} + \int_0^{t-s} u^{-\alpha}(t-s-u)^{-\alpha}\right\} \le (t-s)^{2H}. \end{aligned} \quad (6.39)$$

From (6.35), (6.37), (6.38) and (6.39), we consider the following calculations to estimate I_3:

$$\begin{aligned}
&\left|\int_0^s K(s,r)\int_s^t \frac{\partial K}{\partial u}(u,r)dudr\right| \\
&\leq c_H \int_0^s \left[(s-r)^{-\alpha}+r^{-\alpha}\right]\left[(s-r)^{-\alpha}-(t-r)^{-\alpha}\right]dr \\
&\leq c_H \left\{\int_0^s (s-r)^{-\alpha}\left[(s-r)^{-\alpha}-(t-r)^{-\alpha}\right]dr + I_{22}\right\}. \qquad (6.40)
\end{aligned}$$

Finally,

$$\begin{aligned}
&\int_0^s (s-r)^{-\alpha}\left[(s-r)^{-\alpha}-(t-r)^{-\alpha}\right]dr \\
&\leq \int_0^s \left[(s-r)^{-2\alpha}-(t-r)^{-2\alpha}\right]dr \leq c_H\left\{s^{2H}-t^{2H}+(t-s)^{2H}\right\} \\
&\leq c_H(t-s)^{2H}.
\end{aligned}$$

Consequently, the result follows from (6.35)-(6.40). □

By Proposition 29 and the proof of Proposition 30, we have that, under Assumptions *i)* and *ii)* in this section with $\alpha = \frac{1}{2} - H$, the process $f(B)$ belongs to $L^2(\Omega; \mathcal{H}_K)$ in the case that $H \in (1/4, 1/2)$ and $f \in C^1(\mathbb{R})$ satisfies condition (1.135). In order to apply Theorem 80 to study the existence of some Stratonovich integrals, our next purpose is to state the conditions that guarantee the process $f(B)$ is also in $\mathbb{D}^{1,2}(\mathcal{H}_K)$. To do so, we first establish the following auxiliary result, where we use the sequence $\{\varphi^{(n)} : n \in \mathbb{N}\}$ introduced in Proposition 29. That is,

$$\varphi^{(n)} = \sum_{i=0}^{2^n-1}\left[B_{t_i^n} 1_{(t_i^n, \bar{t}_i^n]} + B_{t_{i+1}^n} 1_{(\bar{t}_i^n, t_{i+1}^n]}\right],$$

with

$$t_i^n = \frac{Ti}{2^n} \quad \text{and} \quad \bar{t}_i^n = t_i^n + \frac{T}{2^{n+1}}, \qquad n \in \mathbb{N}.$$

The reason that we consider this sequence is that, in the proof of Proposition 29, we have already proven that $f(\varphi^{(n)})$ converges to $f(B)$ in the space $L^2(\Omega; \mathcal{H}_K)$, as $n \to \infty$.

Lemma 69 *Let $f \in C^1(\mathbb{R})$ be a function satisfying Condition* (1.135). *Then, for $s, r, t \in [0,T]$, $s < t$,*

$$\begin{aligned}
&D_r\left[f(B_t) - f(\varphi^{(n)}(t)) - \left(f(B_s) - f(\varphi^{(n)}(s))\right)\right] \\
&= \left[f'(B_t) - f'(\varphi^{(n)}(t)) - \left(f'(B_s) - f'(\varphi^{(n)}(s))\right)\right] 1_{[0,s]}(r) + J_2^n,
\end{aligned}$$

with

$$\begin{aligned}
J_2^n &= f'(B_t)1_{(s,t]}(r) - \sum_{i=0}^{2^n-1} f'(B_{t_i^n})1_{[t_i^n,s]}(r)1_{(t_i^n,\bar{t}_i^n]}(s)1_{(\bar{t}_i^n,T]}(t) \\
&\quad + \sum_{i=0}^{2^n-1} f'(B_{t_{i+1}^n})1_{[s,t_{i+1}^n]}(r)1_{(\bar{t}_i^n,t_{i+1}^n]}(s)1_{(t_{i+1}^n,T]}(t) \\
&\quad - \sum_{i=0}^{2^n-1} f'(B_{t_i^n})1_{[s,t_i^n]}(r)1_{[t_i^n,\bar{t}_i^n]}(t)1_{[0,t_i^n)}(s) \\
&\quad - \sum_{i=0}^{2^n-1} f'(B_{t_{i+1}^n})1_{[s,t_{i+1}^n]}(r)1_{(\bar{t}_i^n,t_{i+1}^n]}(t)1_{[0,\bar{t}_i^n)}(s).
\end{aligned}$$

Proof. Fix $s, r, t \in [0, T]$, with $s < t$. Therefore,

$$
\begin{aligned}
& D_r\left[f(B_t) - f(\varphi^{(n)}(t)) - \left(f(B_s) - f(\varphi^{(n)}(s))\right)\right] \\
&= \left[f'(B_t) - f'(\varphi^{(n)}(t)) - \left(f'(B_s) - f'(\varphi^{(n)}(s))\right)\right] 1_{[0,s]}(r) + f'(B_t)1_{(s,t]}(r) \\
&\quad - \sum_{i=0}^{2^n-1}\left[f'(B_{t_i^n})1_{[t_i^n,s]}(r)1_{(t_i^n,\bar{t}_i^n]}(s) - f'(B_{t_{i+1}^n})1_{[s,t_{i+1}^n]}(r)1_{(\bar{t}_i^n,t_{i+1}^n]}(s)\right] \\
&\quad - \sum_{i=0}^{2^n-1}\left[f'(B_{t_i^n})\left(1_{[0,t_i^n]}(r) - 1_{[0,s]}(r)\right)1_{(t_i^n,\bar{t}_i^n]}(t) + f'(B_{t_{i+1}^n})1_{[s,t_{i+1}^n]}(r)1_{(\bar{t}_i^n,t_{i+1}^n]}(t)\right] \\
&= \left[f'(B_t) - f'(\varphi^{(n)}(t)) - \left(f'(B_s) - f'(\varphi^{(n)}(s))\right)\right] 1_{[0,s]}(r) + J_2^n. \qquad (6.41)
\end{aligned}
$$

The term J_2^n can be decomposed as follows:

$$
\begin{aligned}
J_2^n &= f'(B_t)1_{(s,t]}(r) - \sum_{i=0}^{2^n-1} f'(B_{t_i^n})1_{[t_i^n,s]}(r)1_{(t_i^n,\bar{t}_i^n]^2}(s,t) \\
&\quad - \sum_{i=0}^{2^n-1} f'(B_{t_i^n})1_{[t_i^n,s]}(r)1_{(t_i^n,\bar{t}_i^n]}(s)1_{(\bar{t}_i^n,T]}(t) \\
&\quad + \sum_{i=0}^{2^n-1} f'(B_{t_{i+1}^n})1_{[s,t_{i+1}^n]}(r)1_{(\bar{t}_i^n,t_{i+1}^n]}(s)\left(1_{(\bar{t}_i^n,t_{i+1}^n]}(t) + 1_{(t_{i+1}^n,T]}(t)\right) \\
&\quad + \sum_{i=0}^{2^n-1} f'(B_{t_i^n})1_{[t_i^n,s]}(r)1_{[t_i^n,\bar{t}_i^n]^2}(s,t) - \sum_{i=0}^{2^n-1} f'(B_{t_i^n})1_{[s,t_i^n]}(r)1_{[t_i^n,\bar{t}_i^n]}(t)1_{[0,t_i^n)}(s) \\
&\quad - \sum_{i=0}^{2^n-1} f'(B_{t_{i+1}^n})1_{[s,t_{i+1}^n]}(r)1_{(\bar{t}_i^n,t_{i+1}^n]}(t)\left(1_{(\bar{t}_i^n,t_{i+1}^n]}(s) + 1_{[0,\bar{t}_i^n)}(s)\right).
\end{aligned}
$$

Hence, canceling terms, we have that the result is satisfied. □

Now, we are ready to see that $f(B)$ belongs to $\mathbb{D}^{1,2}(\mathcal{H}_K)$.

Proposition 119 *Assume that $H \in (1/4, 1/2)$. Let Hypotheses i) and ii) be satisfied with $\alpha = \frac{1}{2} - H$, and $f \in C^2(\mathbb{R})$ a function such that f, f' and f'' fulfill (1.135). Then, the process $f(B)$ is in the space $\mathbb{D}^{1,2}(\mathcal{H}_K)$.*

Remark 95 *By Propositions 30 and 31, we have that $f(B)$ belongs to $\mathbb{D}^{1,2}(\mathcal{H}_K)$ if B is either fractional Brownian motion, or Riemann-Liouville fractional Brownian motion (see Definition 25) with $H > 1/4$.*

Proof. Let $\{\varphi^{(n)} : n \in \mathbb{N}\}$ be the sequence utilized in Lemma 69. As we have already pointed out in the proof of Proposition 29, we show that $f(\varphi^{(n)})$ converges to $f(B)$ in the space $L^2(\Omega; \mathcal{H}_K)$, as $n \to \infty$. So, in order to finish the proof, it is only missing to see that

$$
E\int_0^T \left\|D_r f(B) - D_r f(\varphi^{(n)})\right\|_K^2 dr \to 0, \quad \text{as } n \to \infty,
$$

where $\|\cdot\|_K$ is introduced in (6.23). Towards this end, we first analyze the convergence of the double integral $\int_0^T \int_0^T \left(D_r f(B_s) - D_r f(\varphi^{(n)}(s))\right)^2 K(T,s)^2 ds dr$ using the definition of the sequence $\{f(\varphi^{(n)}) : n \in \mathbb{N}\}$. That is,

$$\begin{aligned}
&\int_0^T \int_0^T \left(D_r f(B_s) - D_r f(\varphi^{(n)}(s))\right)^2 K(T,s)^2 ds dr \\
&\leq 2\int_0^T \int_r^T \left(f'(B_s) - f'(\varphi^{(n)}(s))\right)^2 K(T,s)^2 ds dr \\
&\quad +2\int_0^T \int_0^T \Bigg[\sum_{i=0}^{2^n-1} f'(B_{t_i^n})\left(1_{[0,s]}(r) - 1_{[0,t_i^n]}(r)\right) 1_{(t_i^n,\bar{t}_i^n]}(s) \\
&\quad + \sum_{i=0}^{2^n-1} f'(B_{t_{i+1}^n})\left(1_{[0,s]}(r) - 1_{[0,t_{i+1}^n]}(r)\right) 1_{(\bar{t}_i^n,t_{i+1}^n]}(s)\Bigg]^2 K(T,s)^2 ds dr \\
&\leq 2\int_0^T \int_0^T \left(f'(B_s) - f'(\varphi^{(n)}(s))\right)^2 K(T,s)^2 ds dr \\
&\quad +2\int_0^T \int_0^T \Bigg[\sum_{i=0}^{2^n-1} f'(B_{t_i^n}) 1_{[t_i^n,s]}(r) 1_{(t_i^n,\bar{t}_i^n]}(s) \\
&\quad - \sum_{i=0}^{2^n-1} f'(B_{t_{i+1}^n}) 1_{[s,t_{i+1}^n]}(r) 1_{(\bar{t}_i^n,t_{i+1}^n]}(s)\Bigg]^2 K(T,s)^2 ds dr = I_1^n + I_2^n.
\end{aligned}$$

The convergence of $E(I_1^n)$ to zero follows as that in (1.138). The fact that $E(I_2^n) \to 0$ is a consequence of the dominated convergence theorem since the integrand of I_2^n goes to zero almost surely and it is bounded by $\left(\sup_{r\in[0,T]} |f'(B_r)|^2\right) K(T,s)^2$. Remember that $\sup_{r\in[0,T]} |f'(B_r)|$ belongs to $L^2(\Omega)$ due to Lemma 23. In consequence, we only need to establish

$$E\left\{\int_0^T \int_0^T \left(\int_s^T D_r\left[f(B_t) - f(\varphi^{(n)}(t)) - \left(f(B_s) - f(\varphi^{(n)}(s))\right)\right] \times (t-s)^{-1-\alpha} dt\right)^2 ds dr\right\} \to 0, \quad \text{as } n \to \infty, \tag{6.42}$$

to complete the proof.

Concerning the first term in (6.41), we have

$$\begin{aligned}
&E\left\{\int_0^T \int_0^T \left(\int_s^T \left[f'(B_t) - f'(\varphi^{(n)}(t)) - \left(f'(B_s) - f'(\varphi^{(n)}(s))\right)\right] \times 1_{[0,s]}(r)(t-s)^{-1-\alpha} dt\right)^2 ds dr\right\} \\
&\leq E\left\{\int_0^T \int_0^T \left(\int_s^T \left[f'(B_t) - f'(\varphi^{(n)}(t)) - \left(f'(B_s) - f'(\varphi^{(n)}(s))\right)\right] \times (t-s)^{-1-\alpha} dt\right)^2 ds dr\right\} \to 0, \quad \text{as } n \to \infty.
\end{aligned}$$

Note that the convergence to zero is an immediate consequence of the proof of Proposition 29 and Proposition 30, by writing f' instead of f.

For the second summand of J_2^2 in Lemma 69, we can write

$$\begin{aligned}
&\int_0^T\int_0^T\left(\int_s^T\sum_{i=0}^{2^n-1} f'(B_{t_i^n})1_{[t_i^n,s]}(r)1_{(t_i^n,\bar{t}_i^n]}(s)1_{(\bar{t}_i^n,T]}(t)(t-s)^{-1-\alpha}dt\right)^2 dsdr\\
&= \sum_{i=0}^{2^n-1} f'(B_{t_i^n})^2\int_{t_i^n}^{\bar{t}_i^n}\int_{t_i^n}^{s}\left(\int_{\bar{t}_i^n}^T(t-s)^{-1-\alpha}dt\right)^2 dsdr\\
&\le \sum_{i=0}^{2^n-1}\frac{f'(B_{t_i^n})^2}{\alpha^2}\int_{t_i^n}^{\bar{t}_i^n}\int_{t_i^n}^{s}(\bar{t}_i^n-s)^{-2\alpha}drds\\
&\le \sum_{i=0}^{2^n-1}\frac{f'(B_{t_i^n})^2T}{\alpha^2 2^{n+1}}\int_{t_i^n}^{\bar{t}_i^n}(\bar{t}_i^n-s)^{-2\alpha}ds\le c_H\left(\sup_{r\in[0,T]}|f'(B_r)|\right)(T2^{-(n+1)})^{2H}\\
&\to 0,\quad \text{in } L^2(\Omega),\quad \text{as } n\to\infty.
\end{aligned}$$

Proceeding similarly, we can show that last integral also goes to zero in $L^2(\Omega)$ when we change the second summand in J_2^n by the thirst one.

Now, the last term of J_2^n can be written as

$$-\sum_{i=0}^{2^n-1} f'(B_{t_{i+1}^n})1_{(\bar{t}_i^n,t_{i+1}^n]}(t)1_{[0,\bar{t}_i^n)}(s)\left(1_{(s,T]}(r)+1_{(t,t_{i+1}^n]}(r)\right)=J_{51}+J_{52}^n.$$

Note that

$$\begin{aligned}
&\int_0^T\int_0^T\left(\int_s^T J_{52}^n(t-s)^{-1-\alpha}dt\right)^2 dsdr\\
&\le \int_0^T\int_0^T\left(\int_s^T\sum_{i=0}^{2^n-1}|f'(B_{t_{i+1}^n})|1_{(\bar{t}_i^n,t_{i+1}^n]}(t)1_{[0,\bar{t}_i^n)}(s)1_{(\bar{t}_i^n,t_{i+1}^n]}(r)(t-s)^{-1-\alpha}dt\right)^2 dsdr\\
&= \sum_{i=0}^{2^n-1} f'(B_{t_{i+1}^n})^2\int_{\bar{t}_i^n}^{t_{i+1}^n}\int_0^{\bar{t}_i^n}\left(\int_{\bar{t}_i^n}^{t_{i+1}^n}(t-s)^{-1-\alpha}dt\right)^2 dsdr\\
&= \sum_{i=0}^{2^n-1}\frac{f'(B_{t_{i+1}^n})^2T}{\alpha^2 2^{n+1}}\int_0^{\bar{t}_i^n}\left((\bar{t}_i^n-s)^{-\alpha}-(t_{i+1}^n-s)^{-\alpha}\right)^2 ds\\
&= \sum_{i=0}^{2^n-1}\frac{f'(B_{t_{i+1}^n})^2T}{\alpha^2 2^{n+1}}\int_0^{\bar{t}_i^n}\left((\bar{t}_i^n-s)^{-2\alpha}-(t_{i+1}^n-s)^{-2\alpha}\right) ds\\
&\le c_H\left(\sup_{r\in[0,T]}|f'(B_r)|\right)(T2^{-(n+1)})^{2H}\to 0,\quad \text{in } L^2(\Omega),\ \text{as } n\to\infty.
\end{aligned}$$

Finally, we observe that J_{51} plus the fourth summand of J_2^n converges to $-f'(B_t)1_{(s,t]}(r)$ and it is bounded by $\left(\sup_{r\in[0,T]}|f'(B_r)|\right)1_{(s,t]}(r)$. Thus, the dominated convergence theorem allows us to finish the proof. □

Now, we are ready to analyze the Stratonovich integral of $f(B)$ via Theorem 80.

Example 3 *Here, we suppose that the Gaussian process $B=\{B_t : t\in[0,T]\}$ is fractional Brownian motion with Hurst parameter $H\in(1/4,1/2)$.*

For $f \in C^2(\mathbb{R})$ such that f, f' and f'' satisfy (1.135), we have that the process $f(B)$ is in the space $\mathbb{D}^{1,2}(\mathcal{H}_K)$ due to Remark 95. We claim that this process also belongs to $\mathbb{D}^{1,2}_C(\mathcal{H}_K)$. Indeed, we only need to see that the trace $TrDf(B)$ exists. To do so, we consider the following calculations:

$$\begin{aligned}
&(2\varepsilon)^{-1}\int_0^T f'(B_t)\left\langle 1_{[0,t]}, 1_{[(t-\varepsilon)\vee 0,(t+\varepsilon)\wedge T]}\right\rangle_{\mathcal{H}} dt \\
&= (2\varepsilon)^{-1}\int_0^T f'(B_t)\left(R(t,(t+\varepsilon)\wedge T) - R(t,(t-\varepsilon)\vee 0)\ \right) dt \\
&= (4\varepsilon)^{-1}\int_0^T f'(B_t)\left[((t+\varepsilon)\wedge T)^{2H} - ((t-\varepsilon)\vee 0)^{2H}\right. \\
&\qquad \left. -((t+\varepsilon)\wedge T - t)^{2H} + (t-((t-\varepsilon)\vee 0))^{2H}\right] dt
\end{aligned}$$

Since, for ε small enough,

$$\begin{aligned}
&\left|\int_0^T \left[((t+\varepsilon)\wedge T - t)^{2H} - (t-((t-\varepsilon)\vee 0))^{2H}\right] dt\right| \\
&= \left|\int_0^\varepsilon \left[\varepsilon^{2H} - t^{2H}\right] dt + \int_{T-\varepsilon}^T \left[(T-t)^{2H} - \varepsilon^{2H}\right] dt\right| \le \int_0^\varepsilon \varepsilon^{2H} dt + 2\int_{T-\varepsilon}^T \varepsilon^{2H} dt,
\end{aligned}$$

the dominated convergence theorem and (6.26) imply that $f(B_t)$ belongs to the space $\mathbb{D}^{1,2}_C(\mathcal{H}_K)$ with $TrDf(B) = H\int_0^T f'(B_s)s^{2H-1}ds$. Thus, as a consequence of Theorem 80 and the fact that $\sup_{r\in[0,T]}|f(B_r)| \in L^2(\Omega)$, we have that $f(B)$ belongs to $Dom\,\delta^S_B$ and

$$\int_0^T f(B_t)\circ dB_t = \delta^B_2\left(f(B)\right) + H\int_0^T f'(B_t)t^{2H-1}dt. \tag{6.43}$$

The following example shows that $f(B)$ belongs to the domain of the Stratonovich integral when B is the Riemann-Liouville Brownian process, which is introduced in Definition 25.

Example 4 *In this example, the process $B = \{B_t : t \in [0,T]\}$ is Riemann-Liouville fractional Brownian process with $\alpha < 1/4$ (i.e., $H > 1/4$).*

Let $f \in C^2(\mathbb{R})$ be such that f, f' and f'' satisfy (1.135), then the process $f(B)$ is also in the space $\mathbb{D}^{1,2}(\mathcal{H}_K)$ because of Remark 95. Thus, by Theorem 80, we only need to check that $trDf(B)$ exists in order to have that $f(B)$ belongs to the domain of the Stratonovich integral.

Remember that the kernel $\{K(t,s) : 0 \le s < t \le T\}$ is given by $K(t,s) = (t-s)^{-\alpha}$ in this case. So, the covariance function of the process B is given by

$$\begin{aligned}
R(t,s) &= \int_0^s (t-r)^{-\alpha}(s-r)^{-\alpha} dr = \int_0^s (t-s+r)^{-\alpha} r^{-\alpha} dr \\
&= s^{-2\alpha}\int_0^s \left(\frac{t-s+r}{s}\right)^{-\alpha}\left(\frac{r}{s}\right)^{-\alpha} dr = s^{1-2\alpha}G\left(\frac{t-s}{s}\right), \quad s < t,
\end{aligned}$$

with $G(t) = \int_0^1 (t+r)^{-\alpha} r^{-\alpha} dr$.

Now, we consider the study of the existence of $TrDf(B)$

$$\begin{aligned}
&(2\varepsilon)^{-1}\int_0^T f'(B_t)\left\langle 1_{[0,t]}, 1_{[(t-\varepsilon)\vee 0,(t+\varepsilon)\wedge T]}\right\rangle_{\mathcal{H}} dt \\
&= (2\varepsilon)^{-1}\int_0^T f'(B_t)\left(R\left((t+\varepsilon)\wedge T, t\right) - R\left(t,(t-\varepsilon)\vee 0\right)\right) dt.
\end{aligned}$$

Hence, previous equalities yield

$$\begin{aligned}
&(2\varepsilon)^{-1}\int_0^T f'(B_t)\left\langle 1_{[0,t]},1_{[(t-\varepsilon)\vee 0,(t+\varepsilon)\wedge T]}\right\rangle_{\mathcal{H}}dt\\
&= (2\varepsilon)^{-1}\int_\varepsilon^{T-\varepsilon} f'(B_t)\left(t^{1-2\alpha}-(t-\varepsilon)^{1-2\alpha}\right)G\left(\frac{\varepsilon}{t}\right)dt\\
&\quad +(2\varepsilon)^{-1}\int_\varepsilon^{T-\varepsilon} f'(B_t)(t-\varepsilon)^{1-2\alpha}\left(G\left(\frac{\varepsilon}{t}\right)-G\left(\frac{\varepsilon}{t-\varepsilon}\right)\right)dt\\
&\quad +(2\varepsilon)^{-1}\left(\int_0^\varepsilon f'(B_t)R(t+\varepsilon,t)\,dt\right.\\
&\quad \left.+\int_{T-\varepsilon}^T f'(B_t)\left(R(T,t)-R(t,t-\varepsilon)\right)dt\right) = I_{1,\varepsilon}+I_{2,\varepsilon}+I_{3,\varepsilon}. \qquad (6.44)
\end{aligned}$$

Observe that he term $I_{3,\varepsilon}$ converges to zero in $L^1(\Omega)$, as $\varepsilon\downarrow 0$. Also, by the dominated convergence theorem, the term $I_{1,\varepsilon}$ tends to $(\frac{1}{2}-\alpha)G(0)\int_0^T f'(B_t)t^{-2\alpha}\,dt$.

To study the convergence of $I_{2,\varepsilon}$, we choose $s,r>0$ and $\delta>0$ such that $2\alpha+\delta<1$, then we have

$$\left|\frac{d}{dr}\left(s^{-\alpha}(s+r)^{-\alpha}\right)\right| = \alpha s^{-\alpha}(s+r)^{-1-\alpha}\le \alpha s^{-1+\delta}r^{-2\alpha-\delta}.$$

Therefore $|G'(r)|\le \alpha r^{-2\alpha-\delta}\int_0^1 s^{-1+\delta}ds = c_\delta r^{-2\alpha-\delta}$. Hence, for $t\in[\varepsilon,T-\varepsilon]$, there is $\theta_{t,\varepsilon}\in\left(\frac{\varepsilon}{t},\frac{\varepsilon}{t-\varepsilon}\right)$ such that

$$\begin{aligned}
(2\varepsilon)^{-1}(t-\varepsilon)^{1-2\alpha}\left|G\left(\frac{\varepsilon}{t}\right)-G\left(\frac{\varepsilon}{t-\varepsilon}\right)\right| &\le c_\delta\varepsilon t^{-1}(t-\varepsilon)^{-2\alpha}(\theta_{t,\varepsilon})^{-2\alpha-\delta}\\
&\le c_\delta(t-\varepsilon)^{-2\alpha}\left(\frac{\varepsilon}{t}\right)^{1-2\alpha-\delta} \qquad (6.45)\\
&\le c_\delta(t-\varepsilon)^{-2\alpha}. \qquad (6.46)
\end{aligned}$$

Note that (6.45) allows us to get

$$(2\varepsilon)^{-1}(t-\varepsilon)^{1-2\alpha}\left|G\left(\frac{\varepsilon}{t}\right)-G\left(\frac{\varepsilon}{t-\varepsilon}\right)\right|1_{[\varepsilon,T-\varepsilon]}(t)\to 0,\quad \text{as } \varepsilon\downarrow 0,$$

which, together with (6.46) and the dominated convergence theorem, gives $I_{2,\varepsilon}\to 0$, in $L^1(\Omega)$, as $\varepsilon\downarrow 0$. Resuming, from Theorem 80, we have

$$\int_0^T f(B_s)\circ dB_s = \delta_2^B(f(B)) + (\frac{1}{2}-\alpha)G(0)\int_0^T f'(B_t)t^{-2\alpha}\,dt.$$

Finally, we show an example of a process that does not belong to the space $\mathbb{D}_C^{1,2}(\mathcal{H}_K)$.

Example 5 *Let f and B be as in Example 4, and $W=\{W_t : t\in[0,T]\}$ the Wiener process appearing in the integral representation of B as a Volterra Gaussian process. Thus, the goal of this example is to show that the process $u=W$ does not belong to the space $\mathbb{D}_C^{1,2}(\mathcal{H}_K)$. In other words, we prove that $trDf(W)$ does not exist. Towards this end, we consider the following calculations: by the transfer principle (4.93), we have*

$$\begin{aligned}
&(2\varepsilon)^{-1}\int_0^T \langle DW_t, 1_{[(t-\varepsilon)\vee 0,(t+\varepsilon)\wedge T]}\rangle_{\mathcal{H}}\, dt \\
&= (2\varepsilon)^{-1}\int_0^T \left\langle K^*(DW_t), K^*\left(1_{[(t-\varepsilon)\vee 0,(t+\varepsilon)\wedge T]}\right)\right\rangle_{L^2([0,T])} dt \\
&= \frac{1}{2\varepsilon}\left(\int_0^T\int_0^t ((t+\varepsilon)\wedge T-s)^{-\alpha}\, dsdt - \int_0^T\int_0^{(t-\varepsilon)\vee 0} ((t-\varepsilon)\vee 0-s)^{-\alpha}\, dsdt\right) \\
&= \frac{1}{2\varepsilon}(1-\alpha)^{-1}(2-\alpha)^{-1}\left(T^{2-\alpha}-2\varepsilon^{2-\alpha}-(T-\varepsilon)^{2-\alpha}\right) \\
&\quad -\frac{1}{2\varepsilon}(1-\alpha)^{-1}\varepsilon^{1-\alpha}(T-\varepsilon)+\frac{1}{2}(1-\alpha)^{-1}T^{1-\alpha},
\end{aligned}$$

which does not converge as $\varepsilon \downarrow 0$. This proves our claim.

Now, we analyze a version of Theorem 80 for RLfp.

Theorem 81 *Let B be Riemann-Liouville fractional process with parameter $\alpha < 1/4$ and $u \in \mathbb{D}^{1,2}(\mathcal{H}_K)$ such that, for $0 \le t_2 < t_1 \le T$,*

$$\left|D_s^W u_{t_1} - D_s^W u_{t_2}\right| 1_{[0,t_2]}(s) \le \bar{G}(t_1-t_2)^{\beta} \tag{6.47}$$

and

$$\left|D_s^W u_{t_1}\right| \le \bar{G}|t_1-s|^{\beta}, \quad s\in[0,T], \tag{6.48}$$

for some random variable $\bar{G}$ and $\beta \in (\alpha,1)$. Then, the process u also belongs to the space $\mathbb{D}_C^{1,2}(\mathcal{H}_K)$ and

$$\int_0^T u_s \circ dB_s = \delta_2^B(u) - \alpha\int_0^T\int_0^t \left(D_s^W u_t\right)(t-s)^{-1-\alpha} dsdt.$$

Remark 96 *By (4.92) and Theorem 63, we have that u belongs to the space $\mathbb{D}_W^{1,2}(L^2([0,T]))$.*

Proof. From the proof of Theorem 80, we only need to establish that the random variable $trDu$ exists and that it is equal to $-\alpha\int_0^T\int_0^t \left(D_s^W u_t\right)(t-s)^{-1-\alpha}dsdt$.

Let $\varepsilon > 0$. Then, (4.93) implies

$$\begin{aligned}
&\frac{1}{2\varepsilon}\int_0^T \left\langle D^B u_t, \mathbf{1}_{[(t-\varepsilon)\vee 0,(t+\varepsilon)\wedge T]}\right\rangle_{\mathcal{H}} dt \\
&= \frac{1}{2\varepsilon}\int_0^T\int_0^T \left(D_s^W u_t\right)\left(K(t+\varepsilon)\wedge T, s) - K((t-\varepsilon)\vee 0, s)\right) dsdt \\
&= \frac{1}{2\varepsilon}\int_0^T\int_0^T \left(D_s^W u_t\right)\left[((t+\varepsilon)\wedge T-s)^{-\alpha}1_{[0,(t+\varepsilon)\wedge T]}(s)\right. \\
&\qquad \left. -((t-\varepsilon)\vee 0-s)^{-\alpha}1_{[0,(t-\varepsilon)\vee 0]}(s)\right] dsdt.
\end{aligned} \tag{6.49}$$

Now, we will apply the dominated convergence theorem to study the convergence in probability of the last integral. Towards this end, observe that, for $\omega \in \Omega$ and $s, t \in (0,T)$, $s \neq t$, we have

$$\begin{aligned}
I_1 &= \frac{1}{2\varepsilon}\left(D_s^W u_t\right)\left[((t+\varepsilon)\wedge T-s)^{-\alpha}1_{[0,(t+\varepsilon)\wedge T]}(s) - ((t-\varepsilon)\vee 0-s)^{-\alpha}1_{[0,(t-\varepsilon)\vee 0]}(s)\right] \\
&\to -\alpha\left(D_s^W u_t\right)(t-s)^{-1-\alpha}1_{[0,t)}(s).
\end{aligned} \tag{6.50}$$

Moreover, (6.47) and (6.48) yield

$$\begin{aligned}
|I_1| &\leq \frac{1}{2\varepsilon}\left|D_s^W u_t\right|\left\{\left[((t-\varepsilon)\vee 0-s)^{-\alpha}-((t+\varepsilon)\wedge T-s)^{-\alpha}\right]1_{[0,(t-\varepsilon)\vee 0]}(s)\right.\\
&\quad \left.+((t+\varepsilon)\wedge T-s)^{-\alpha}1_{((t-\varepsilon)\vee 0,(t+\varepsilon)\wedge T]}(s)\right\}\\
&\leq \frac{1}{2\varepsilon}\left|D_s^W u_{(t-\varepsilon)\vee 0}\right|\left[((t-\varepsilon)\vee 0-s)^{-\alpha}-((t+\varepsilon)\wedge T-s)^{-\alpha}\right]1_{[0,(t-\varepsilon)\vee 0]}(s)\\
&\quad +\frac{\bar{G}\varepsilon^{\beta}}{2\varepsilon}\left[((t-\varepsilon)\vee 0-s)^{-\alpha}-((t+\varepsilon)\wedge T-s)^{-\alpha}\right]1_{[0,(t-\varepsilon)\vee 0]}(s)\\
&\quad +\frac{\bar{G}}{2\varepsilon}|t-s|^{\beta}((t+\varepsilon)\wedge T-s)^{-\alpha}1_{((t-\varepsilon)\vee 0,(t+\varepsilon)\wedge T]}(s)\\
&\leq \alpha\bar{G}((t-\varepsilon)\vee 0-s)^{\beta-\alpha-1}1_{[0,(t-\varepsilon)\vee 0]}(s)\\
&\quad +\frac{\bar{G}\varepsilon^{\beta}}{2\varepsilon}\left[((t-\varepsilon)\vee 0-s)^{-\alpha}-((t+\varepsilon)\wedge T-s)^{-\alpha}\right]1_{[0,(t-\varepsilon)\vee 0]}(s)\\
&\quad +\frac{\bar{G}\varepsilon^{\beta}}{2\varepsilon}((t+\varepsilon)\wedge T-s)^{-\alpha}1_{((t-\varepsilon)\vee 0,(t+\varepsilon)\wedge T]}(s)=J_1+J_2+J_3. \qquad (6.51)
\end{aligned}$$

Note that J_3 converges to zero, as $\varepsilon\to 0$, for all $\omega\in\Omega$ and $s\neq t$, and

$$\begin{aligned}
\int_0^T\int_0^T J_3 dsdt &= \frac{\bar{G}}{2\varepsilon^{1-\beta}}\int_0^T\int_{(t-\varepsilon)\vee 0}^{(t+\varepsilon)\wedge T}((t+\varepsilon)\wedge T-s)^{-\alpha}dsdt\\
&= \frac{\bar{G}}{2(1-\alpha)\varepsilon^{1-\beta}}\int_0^T((t+\varepsilon)\wedge T-(t-\varepsilon)\vee 0)^{1-\alpha}dt\to 0, \quad \text{as } \varepsilon\downarrow 0.
\end{aligned}$$

Similarly, J_2 converges to zero, as $\varepsilon\to 0$, for all $\omega\in\Omega$ and $s\neq t$, and

$$\begin{aligned}
&\int_0^T\int_0^T J_2 dsdt\\
&= \frac{\bar{G}}{2\varepsilon^{1-\beta}}\int_0^T\int_0^{(t-\varepsilon)\vee 0}\left[((t-\varepsilon)\vee 0-s)^{-\alpha}-((t+\varepsilon)\wedge T-s)^{-\alpha}\right]dsdt\\
&= \frac{\bar{G}}{2(1-\alpha)\varepsilon^{1-\beta}}\int_0^T\left[((t-\varepsilon)\vee 0)^{1-\alpha}-((t+\varepsilon)\wedge T)^{1-\alpha}\right.\\
&\qquad \left.+((t+\varepsilon)\wedge T-(t-\varepsilon)\vee 0)^{1-\alpha}\right]dt\to 0, \quad \text{as } \varepsilon\downarrow 0.
\end{aligned}$$

Also, it is easy to see that $J_1\to\alpha\bar{G}(t-s)^{\beta-\alpha-1}1_{[0,t)}(s)$, as $\varepsilon\to 0$, for all $\omega\in\Omega$ and $s\neq t$, and

$$\lim_{\varepsilon\downarrow 0}\int_0^T\int_0^T J_1 dsdt=\alpha\bar{G}\int_0^T\int_0^t(t-s)^{\beta-\alpha-1}dsdt.$$

Finally, the last six convergences (i.e., of J_1, J_2 and J_3, and their integrals), together with (6.49)-(6.50) and the dominated convergence theorem, imply that $trDu$ exists and it is equal to

$$-\alpha\int_0^T\int_0^t\left(D_s^W u_t\right)(t-s)^{-1-\alpha}dsdt.$$

Therefore, the proof is complete. □

Two processes that satisfy conditions (6.47) and (6.48) are

$$t\mapsto\int_0^t(t-r)^{\beta}dW_r \quad\text{and}\quad t\mapsto\int_0^t Y_r dr, \quad t\in[0,T].$$

Here, $\beta > \alpha$ and Y is an adapted process in $\mathbb{D}^{1,2}(\mathcal{H}_K)$ (see Corollary 9) such that $\sup_{s\in[0,T]}\int_s^T |D_s^W Y_r|^{1/(1-\beta)}dr$ is a finite random variable with probability 1. As an exercise to the reader, it is left to verify that both processes belongs to the space $\mathbb{D}^{1,2}(\mathcal{H}_K)$.

On the other hand, our next purpose is to analyze an Itô's type formula for the Stratonovich integral.

Remember that the process B satisfies Conditions *i)* and *ii)* at the beginning of this section. Moreover, we consider a process u satisfying:

iii) Let u be a process in the space $\mathbb{D}_W^{2,4}(\mathcal{H}_K)$ such that u and $D_r^W u$ are λ–Hölder continuous in the norm of the space $\mathbb{D}_W^{1,4}$, for some $\lambda > \alpha$ and each $r \in [0,T]$, and the function

$$\gamma_r = \sup_{0\le s\le T}\left\|D_r^W u_s\right\|_{1,4} + \sup_{0\le s<t\le T}\frac{\left\|D_r^W u_t - D_r^W u_s\right\|_{1,4}}{|t-s|^{\lambda}}$$

satisfies $\int_0^T \gamma_r^p dr < \infty$ for some $p > \frac{2}{1-4\alpha}$.

Under this condition, we establish the following integration by parts formula for the Stratonovich integral.

Theorem 82 *Suppose $\alpha < \frac{1}{4}$ and let u be an adapted process in $\mathbb{D}_W^{2,4}(\mathcal{H}_K)$ satisfying (6.27) and (6.28). Moreover, assume that Hypotheses i)-iii) of this section hold and that there is a process $(\nabla u) \in \mathbb{D}_W^{1,2}(L^2([0,T]))$ such that the limit in probability*

$$\int_0^T \left|(\nabla u)_s - \frac{1}{2\varepsilon}\left\langle D^B u_s, \mathbf{1}_{[(s-\varepsilon)\vee 0,(s+\varepsilon)\wedge T]}\right\rangle_{\mathcal{H}}\right| ds \to 0 \tag{6.52}$$

is satisfied. Let $X_t = \int_0^t u_s \circ dB_s$ and $F \in C_b^2(\mathbb{R})$, then the process $s \mapsto F'(X_s)u_s$ is Stratonovich integrable with respect to B and

$$F(X_t) = F(0) + \int_0^t F'(X_s)u_s \circ dB_s.$$

Proof. From Theorem 80, we have

$$X_t = \int_0^t u_s\delta_2 B_s + \int_0^t (\nabla u)_s\, ds, \quad t\in[0,T].$$

Then, by a straightforward extension of Theorem 60, we can establish that $F'(X_s)u_s$ is Skorohod integrable with respect to B and, for $t\in[0,T]$,

$$\begin{aligned} F(X_t) &= F(0) + \int_0^t F'(X_s)u_s\delta_2 B_s \\ &+ \int_0^t F''(X_s)\,u_s\left(\int_0^s \frac{\partial K}{\partial s}(s,r)\left(\int_0^s D_r^W (K_s^* u)_\theta\,\delta_2 W_\theta\right)dr\right)ds \\ &+ \frac{1}{2}\int_0^t F''(X_s)\frac{\partial}{\partial s}\left(\int_0^s (K_s^* u)_r^2\, dr\right)ds + \int_0^t F'(X_s)\,(\nabla u)_s\, ds \\ &+ \int_0^t F''(X_s)u_s\int_0^s\left(\int_r^s D_r^W (\nabla u)_\theta\, d\theta\right)\frac{\partial K}{\partial s}(s,r)\, drds. \end{aligned}$$

Therefore, in order to finish the proof, we only need to check that the limit in probability

$$\lim_{\varepsilon\downarrow 0}\frac{1}{2\varepsilon}\int_0^t \left\langle D^B\left(F'(X_s)u_s\right), \mathbf{1}_{[(s-\varepsilon)\vee 0,(s+\varepsilon)\wedge T]}\right\rangle_{\mathcal{H}} ds$$

exists and that it is equal to

$$\int_0^t F''(X_s)\,u_s\left(\int_0^s \frac{\partial K}{\partial s}(s,r)\left(\int_0^s D_r^W (K_s^* u)_\theta\,\delta_2 W_\theta\right)dr\right)ds$$
$$+\frac{1}{2}\int_0^t F''(X_s)\frac{\partial}{\partial s}\left(\int_0^s (K_s^* u)_r^2\,dr\right)ds+\int_0^t F'(X_s)\,(\nabla u)_s\,ds$$
$$+\int_0^t F''(X_s)u_s\int_0^s\left(\int_r^s D_r^W (\nabla u)_\theta\,d\theta\right)\frac{\partial K}{\partial s}(s,r)\,drds,\quad t\in[0,T].$$

Towards this end, we consider

$$\begin{aligned}
&\frac{1}{2\varepsilon}\int_0^t \left\langle D^B\left(F'(X_s)u_s\right),\mathbf{1}_{[(s-\varepsilon)\vee 0,(s+\varepsilon)\wedge T]}\right\rangle_{\mathcal{H}}ds\\
=&\ \frac{1}{2\varepsilon}\int_0^t F'(X_s)\left\langle D^B u_s,\mathbf{1}_{[(s-\varepsilon)\vee 0,(s+\varepsilon)\wedge T]}\right\rangle_{\mathcal{H}}ds\\
&+\frac{1}{2\varepsilon}\int_0^t F''(X_s)u_s\left\langle D^B X_s,\mathbf{1}_{[(s-\varepsilon)\vee 0,(s+\varepsilon)\wedge T]}\right\rangle_{\mathcal{H}}ds\\
=&\ \frac{1}{2\varepsilon}\int_0^t F'(X_s)\left\langle D^B u_s,\mathbf{1}_{[(s-\varepsilon)\vee 0,(s+\varepsilon)\wedge T]}\right\rangle_{\mathcal{H}}ds\\
&+\frac{1}{2\varepsilon}\int_0^t F''(X_s)u_s\left\langle D^B\left(\int_0^s u_r\delta_2 B_r\right),\mathbf{1}_{[(s-\varepsilon)\vee 0,(s+\varepsilon)\wedge T]}\right\rangle_{\mathcal{H}}ds\\
&+\frac{1}{2\varepsilon}\int_0^t F''(X_s)u_s\left\langle D^B\left(\int_0^s (\nabla u)_r\,dr\right),\mathbf{1}_{[(s-\varepsilon)\vee 0,(s+\varepsilon)\wedge T]}\right\rangle_{\mathcal{H}}ds\\
=&\ T_1^\varepsilon+T_2^\varepsilon+T_3^\varepsilon.
\end{aligned}$$

We have that the first term converges to $\int_0^t F'(X_s)\,(\nabla u)_s\,ds$ in probability due to (6.52), for every $t\in[0,T]$.

For T_2^ε, we apply the transfer principle:

$$\begin{aligned}
T_2^\varepsilon =&\ \frac{1}{2\varepsilon}\int_0^t F''(X_s)u_s\left[\int_0^{s+\varepsilon} D_\theta^W\left(\int_0^s (K_s^*u)_r\,\delta_2 W_r\right)K(s+\varepsilon,\theta)\,d\theta\right.\\
&\left.-\int_0^{s-\varepsilon} D_\theta^W\left(\int_0^s (K_s^*u)_r\,\delta_2 W_r\right)K(s-\varepsilon,\theta)\,d\theta\right]ds\\
=&\ \frac{1}{2\varepsilon}\int_0^t F''(X_s)u_s\left[\int_0^s (K_s^*u)_\theta\,K(s+\varepsilon,\theta)\,d\theta\right.\\
&\left.-\int_0^{s-\varepsilon}(K_s^*u)_\theta\,K(s-\varepsilon,\theta)\,d\theta\right]ds\\
&+\frac{1}{2\varepsilon}\int_0^t F''(X_s)u_s\left[\int_0^{s+\varepsilon}\left(\int_0^s D_\theta^W (K_s^*u)_r\,\delta_2 W_r\right)K(s+\varepsilon,\theta)\,d\theta\right.\\
&\left.-\int_0^{s-\varepsilon}\left(\int_0^s D_\theta^W (K_s^*u)_r\,\delta_2 W_r\right)K(s-\varepsilon,\theta)\,d\theta\right]ds\\
=&\ T_{2,1}^\varepsilon+T_{2,2}^\varepsilon.
\end{aligned}$$

In order to analyze $T_{2,1}^\varepsilon$, we consider the following calculations. Using the definition of K_s^*u, which is given in (4.73), we can write

$$\begin{aligned}
&\frac{1}{2\varepsilon}\left[\int_0^s (K_s^* u)_\theta\, K(s+\varepsilon,\theta)\,d\theta - \int_0^{s-\varepsilon} (K_s^* u)_\theta\, K(s-\varepsilon,\theta)\,d\theta\right] \\
&= \frac{1}{2\varepsilon}\left[\int_0^s K(s,\theta)u_\theta K(s+\varepsilon,\theta)\,d\theta - \int_0^{s-\varepsilon} K(s,\theta)u_\theta K(s-\varepsilon,\theta)\,d\theta\right] \\
&\quad + \frac{1}{2\varepsilon}\left[\int_0^s \left(\int_\theta^s \frac{\partial K}{\partial r}(r,\theta)(u_r - u_\theta)dr\right) K(s+\varepsilon,\theta)\,d\theta\right. \\
&\qquad \left. - \int_0^{s-\varepsilon}\left(\int_\theta^s \frac{\partial K}{\partial r}(r,\theta)(u_r - u_\theta)dr\right) K(s-\varepsilon,\theta)\,d\theta\right].
\end{aligned}$$

We add and subtract u_s in the first two integrals on the right-hand side of this equality to see that last expression is equal to

$$\begin{aligned}
&\frac{u_s}{2\varepsilon}\left[R(s,s+\varepsilon) - R(s,s-\varepsilon)\right] \\
&\quad + \frac{1}{2\varepsilon}\left[\int_0^T K(s,\theta)\,(u_\theta - u_s)\left[K(s+\varepsilon,\theta) - K(s-\varepsilon,\theta)\right]d\theta\right] \\
&\quad + \frac{1}{2\varepsilon}\int_0^s \left(\int_\theta^s \frac{\partial K}{\partial r}(r,\theta)(u_r - u_\theta)dr\right)\left[K(s+\varepsilon,\theta) - K(s-\varepsilon,\theta)\right]d\theta.
\end{aligned}$$

Hence, now it is easy to see that $T_{2,1}^{\varepsilon}$ converges in $L^1(\Omega)$ to

$$\begin{aligned}
&H\int_0^t F''(X_s)u_s^2\, s^{2H-1}ds + \frac{1}{2}\int_0^t F''(X_s)u_s\left(\int_0^s (u_\theta - u_s)\frac{\partial K^2}{\partial s}(s,\theta)\,d\theta\right)ds \\
&\quad + \int_0^t F''(X_s)u_s \int_0^s\left(\int_\theta^s \frac{\partial K}{\partial r}(r,\theta)(u_r - u_\theta)dr\right)\frac{\partial K}{\partial s}(s,\theta)\,d\theta ds \\
&= \frac{1}{2}\int_0^t F''(X_s)\frac{\partial}{\partial s}\int_0^s (K_s^* u)_\theta^2\,d\theta.
\end{aligned}$$

The term $T_{2,2}^{\varepsilon}$ goes in $L^1(\Omega)$ to

$$\int_0^t F''(X_s)u_s\left(\int_0^s \frac{\partial K}{\partial s}(s,\theta)\left(\int_0^s D_\theta^W (K_s^* u)_r\, \delta_2 W_r\right)d\theta\right)ds.$$

Finally, we deal with the term T_3^{ε}. Using the transfer principle again, we can state

$$\begin{aligned}
T_3^{\varepsilon} &= \frac{1}{2\varepsilon}\int_0^t F''(X_s)u_s\left[\int_0^T\left(\int_0^s D_\theta^W (\nabla u)_r\,dr\right)K(s+\varepsilon,\theta)\,d\theta\right. \\
&\qquad \left. - \int_0^T\left(\int_0^s D_\theta^W (\nabla u)_r\,dr\right)K(s-\varepsilon,\theta)\,d\theta\right]ds \\
&\to \int_0^t F''(X_s)u_s\int_0^s\left(\int_0^s D_r^W (\nabla u)_\theta\,d\theta\right)\frac{\partial K}{\partial s}(s,r)\,drds,
\end{aligned}$$

in $L^1(\Omega)$, as $\varepsilon \downarrow 0$. Thus, the proof is complete. □

On the other hand, concerning Definition 29, Stratonovich integral with respect to fractional Brownian motion B with Hurst parameter $H \in (1/3,1)$ has been related to a rough integral. Namely, let $\mathcal{D}_B(\mathbb{R})$ be the set of all controlled processes (Y, Y') by B. That is, for $s, t \in [0, T]$, $s < t$, we have

$$Y_t - Y_s = Y_s^{'}(B_t - B_s) + R_{st}^Y, \tag{6.53}$$

with

$$P - \lim_{\varepsilon \downarrow 0} \frac{1}{\varepsilon} \int_0^t R_{s((s+\varepsilon)\wedge T)}^Y \left(B_{(s+\varepsilon)\wedge T} - B_s\right) ds = 0. \tag{6.54}$$

Also consider the two-parameter process $\mathbb{B}_{st} = \frac{1}{2}(B_t - B_s)^2$, which satisfies the Chen's relation $(\delta\mathbb{B})_{sut} = (\delta B)_{su}(\delta B)_{ut}$. Thus, for the fBm B, the rough integral of $(Y, Y') \in \mathcal{D}_B(\mathbb{R})$ with respect to $\mathbf{B} = (B, \mathbb{B})$ is defined by

$$\int_0^T Y_s d\mathbf{B}_s = P - \lim_{\varepsilon \downarrow 0} \frac{1}{\varepsilon} \int_0^T \left(Y_s \left[B_{(s+\varepsilon)\wedge T} - B_s\right] + Y_s^{'} \mathbb{B}_{s((s+\varepsilon)\wedge T)}\right) ds, \tag{6.55}$$

if this limit in probability exists.

Under suitable assumptions on the derivative DY' in the Malliavin calculus sense (see Theorem 83 below), Ohashi and Russo [169] have shown that $(Y, Y') \in \mathcal{D}_B(\mathbb{R})$ is rough integrable with respect to $\mathbf{B}$ (see (6.55)) if and only if Y is Stratonovich integrable with respect to B (according to Definition 44). In this case, both integrals are the same. It means,

$$\int_0^T Y_s d\mathbf{B}_s = \int_0^T Y_s \circ dB_s. \tag{6.56}$$

The main ingredients to prove that this equality holds are the following:

i) For $s \in \mathbb{R}$ and $\tilde{s} = s \wedge T$,

$$\partial_s R(s, T) = \begin{cases} H\left[\tilde{s}^{2H-1} + (T - \tilde{s})^{2H-1}\right], & \text{if } s < T, \\ 0, & \text{if } s > T. \end{cases}$$

Therefore, for every $s \in [0, T]$, $R(dx, s) = \partial_x R(x, s)dx$ is a finite non-negative measure with support on $[0, T]$.

ii) The Schwartz distribution $\partial_{s_1,s_2}^2 R(s_1, s_2)(s_1 - s_2)$ is the Radon measure

$$\bar{\mu}(ds_1, ds_2) = H(2H-1)|s_1 - s_2|^{2H-1}\text{sgn}(s_1 - s_2)1_{[0,T]^2\setminus D}(s_1, s_2)ds_1 ds_2,$$

where $D = \{(s, t) \in \mathbb{R}_+^2 : s = t\}$.

Equality (6.56) is a particular case of the problem studied in [169], which considers d-dimensional singular covariance Gaussian processes X instead of B. In what follows, we explain under what conditions (6.56) is still valid when we change B to X.

In the remaining of this section, we consider a d-dimensional singular covariance Gaussian process $X = \{X_t : t \in [0, T]\}$ such that its covariance function R satisfies:

(A_R) For every $s \in [0, T]$, the Schwartz distribution (see Rudin [189]) $\partial_x R(x, s)$ is such that $R(dx, s) = \partial_x R(x, s)dx$ is a finite non-negative measure with compact support on $[0, T]$.

(B_R) The product of the Schwartz distribution $\partial_{s_1,s_2}^2 R(s_1, s_2)$ with the function $(s_1, s_2) \mapsto (s_1 - s_2)$ is a real Radon measure, which is represented by $\bar{\mu}$.

(C_R) $i)$ The Schwartz distribution $\partial^2_{s_1,s_2}R(s_1,s_2)$ is a sigma-finite non-positive measure μ, which is absolutely continuous with respect to the Lebesgue measure on $\mathbb{R}^2_+\setminus D$ such that its Radon-Nikodym derivative satisfies

$$\left|\partial^2_{s_1,s_2}R(s_1,s_2)\right| \leq C|s_1-s_2|^\alpha + \phi(s_1,s_2), \quad (s_1,s_2)\in[0,T]^2\setminus D,$$

where $C>0$, $\alpha\in(-3/2,-1)$ and $\phi:[0,T]^2\setminus D\to\mathbb{R}_+$ is a symmetric p-integrable function, for some $p>1$.

$ii)$ $E\left(|X_t-X_s|^2_{\mathbb{R}^d}\right)\leq C|t-s|^{\alpha+2}$, for $s,t\in[0,T]$.

$iii)$ For $(s_1,s_2)\in[0,T]^2\setminus D$, we have

$$|R(s_1,T)-R(s_2,T)|\leq C|s_1-s_2|^{\alpha+2}.$$

$iv)$ There is a non-increasing integrable function $\psi:[0,T]\to\mathbb{R}_+$ such that, for $a,b,c,d\in[0,T]$,

1. $\int_a^b|\phi(r,c)|dr\leq C|b-a|^{(\alpha+2)/2}\psi(c)$,
2. $\int_c^d\psi(r)dr\leq C|d-c|^{(\alpha+2)/2}$,
3. $s\mapsto s^{(\alpha+2)/2}\psi(s)$ belongs to $L^1([0,T])$.

By convention, any function f defined on $[0,T]$ is extended to $\mathbb{R}$ as $s\mapsto f((s\vee 0)\wedge T)$. We also use the notation

$$I(f)=\int_0^\infty f(s)dX_s := -\int_0^\infty \langle X_s, df(s)\rangle_{\mathbb{R}^d},$$

for every function $f\in C_0^1(\mathbb{R}_+;\mathbb{R}^d)$.

Under Hypotheses (A_R)-(C_R), Kruk and Russo [104] have established

$$E\left(I(f)^2\right)=\|f\|^2_{\tilde{L}_R(\mathbb{R}^d)}, \quad f\in C_0^1(\mathbb{R}_+;\mathbb{R}^d),$$

where $\tilde{L}_R(\mathbb{R}^d)$ is the linear space of all Borel-measurable functions $f:\mathbb{R}_+\to\mathbb{R}^d$ such that

(i) $\int_0^\infty |f(s)|^2_{\mathbb{R}^d}|R|(ds,\infty)<\infty$,

(ii) $\int_{\mathbb{R}^2_+\setminus D}|f(s_1)-f(s_2)|^2_{\mathbb{R}^d}|\mu|(ds_1,ds_2)<\infty$.

and

$$\|f\|^2_{\tilde{L}_R(\mathbb{R}^d)}=\int_0^\infty|f(s)|^2_{\mathbb{R}^d}R(ds,\infty)-\frac{1}{2}\int_{\mathbb{R}^2_+\setminus D}|f(s_1)-f(s_2)|^2_{\mathbb{R}^d}\mu(ds_1,ds_2).$$

In particular, $I: C_0^1(\mathbb{R}_+;\mathbb{R}^d)\to L^2(\Omega)$ can be uniquely extended to a linear isometry $I:\tilde{L}_R(\mathbb{R}^d)\to L^2(\Omega)$, which satisfies

$$\int_0^\infty 1_{[0,t]}dX = X_t \quad\text{and}\quad R(s,t)=\left\langle 1_{[0,t]},1_{[0,s]}\right\rangle_{\tilde{L}_R(\mathbb{R}^d)}.$$

Thus, we can introduce a Malliavin calculus based on $\tilde{L}_R(\mathbb{R}^d)$ by considering the space $\mathcal{S}$ as the family of all the random variables F of the form

$$F=f(I(\varphi_1),\ldots,I(\varphi_n))$$

with $f\in C_b^\infty(\mathbb{R}^n)$ and $\varphi_1,\ldots,\varphi_n\in\tilde{L}_R(\mathbb{R}^d)$.

The next step to deal with rough integration with respect to X is to take into account the two-parameter process

$$\mathbb{X}^{i,j}_{st} = \begin{cases} \int_s^t \left(X^i_r - X^i_s\right) \circ dX^j_r, & \text{if } i \neq j, \\ \frac{1}{2}\left(X^i_t - X^i_s\right)^2, & \text{otherwise.} \end{cases}$$

Using that Stratonovich integral is a linear operator, we have that $\mathbb{X}$ satisfies Chen's relation $\left(\delta\mathbb{X}^{i,j}\right)_{sut} = \left(\delta X^i\right)_{su}\left(\delta X^j\right)_{ut}$. Moreover, by proceeding as in the proof of Theorem 67, the process $\mathbb{X}$ is well-defined since, for $i \neq j$, $D^j(X^i_u - X^i_s) = 0$, which follows from the fact that X^i and X^j are independent. Actually, we get $\mathbb{X}^{i,j}_{st} = \delta_2^{X^j}\left((X^i_\cdot - X^i_s)1_{[s,t]}\right)$, $i \neq j$. The reader can see Ohashi and Russo [169] for details.

Now, $(Y,Y') \in \mathcal{D}_X(\mathbb{R}^d)$ if $(Y,Y') \in \mathcal{L}(\mathbb{R}^d,\mathbb{R}) \times \mathcal{L}(\mathbb{R}^d,\mathcal{L}(\mathbb{R}^d,\mathbb{R}))$, and (6.53) and (6.54) hold when we change B by X. Therefore, we can use (6.55) in order to define the rough integral of $(Y,Y') \in \mathcal{D}_X(\mathbb{R}^d)$ with respect to $\mathbf{X} = (X,\mathbb{X})$ by writing $\mathbf{X}$, X and $\mathbb{X}$ instead of $\mathbf{B}$, B and $\mathbb{B}$, respectively.

The main result of the paper [169] is the following theorem. The reader can consult this paper for the proof of this result.

Theorem 83 *Assume that (A_R)-(C_R) are satisfied with $-4/3 < \alpha < -1$. Let $(Y,Y') \in \mathcal{D}_X(\mathbb{R}^d)$, where Y' satisfies:*

1. *$s \mapsto D_v Y'_s$ is continuous w.p.1 on $(0,T) \setminus \{v\}$ for almost all v.*
2. *There is $p > 2$ such that $s \mapsto Y'_s$ is a $\mathbb{D}^{1,p}(\mathbb{R}^{d\times d})$-valued continuous function and*

$$\sup_{0\le s\le T} E\left(|Y'_s|^p_{\mathbb{R}^{d\times d}}\right) + \sup_{0\le s,t\le T} E\left(|D_t Y'_s|^p_{\mathbb{R}^{d^3}}\right) < \infty.$$

3. *There exists $q > 2$ such that*

$$\int_0^T \int_{v_2}^T \sup_{s\ge v_1 \text{ or } s<v_2} \|D_{v_1} Y'_s - D_{v_2} Y'_s\|^q_{L^q(\Omega;\mathbb{R}^{d^3})} |\partial^2 R(v_1,v_2)|^{q/2} dv_1 dv_2 < \infty.$$

Then, $(Y,Y') \in \mathcal{D}_X(\mathbb{R}^d)$ is rough integrable with respect to $\mathbf{X}$ if and only if Y is Stratonovich integrable with respect to X. In this case, we have

$$\int_0^T Y_s d\mathbf{X}_s = \int_0^T Y_s \circ dX_s.$$

6.3.3 Case $H < 1/2$

Throughout this section, the Gaussian process $B = \{B_t : t \in [0,T]\}$ is fractional Brownian motion with Hurst parameter $H \in (0,1/2)$, unlike previous section where $H \in (1/4,1/2)$. The purpose of this section is to define a stochastic integral of the Stratonovich type and to state a relation between this integral and the extended divergence, which is introduced in Definition 40. In order to avoid confusion with the classical divergence operator with respect to B, we denote the extended divergence by $\delta_{\mathcal{T}}$ in what follows.

In this section, we work with $\mathcal{T} : \mathcal{H} \subset L^2([0,T]) \to L^2([0,T])$ and the set $\mathcal{S}_{\mathcal{T}}$, which are defined in Sections 4.3.1 and 1.5.5 , respectively. Namely, the space $\mathcal{S}_{\mathcal{T}}$ is introduced in (1.110), and the operator $\mathcal{T}$ is given by (1.119) and (1.120). This operator $\mathcal{T}$ satisfies Hypotheses (H1), (H2) and (H3) considered in Section 4.2, which are needed to introduce the extended divergence operator $\delta_{\mathcal{T}}$.

We are ready to define our extended Stratonovich integral with respect to fractional Brownian motion. Remember that we are dealing with the case that $H \in (0,1/2)$. The following definition is inspired by that of Russo and Vallois [190], and (4.61).

Definition 47 *Let $u = \{u_t : t \in [0,T]\}$ be a measurable process with integrable paths such that $E\left(\left(\int_0^T |u_t| dt\right)^p\right) < \infty$ for some $p > 2$. We say that u belongs to $\mathcal{D}(\delta_S^{B^H})$ if there exists a square-integrable random variable $\int_0^T u_t \circ dB_t^H$ such that*

$$\left\langle F, \int_0^T u_t \circ dB_t^H \right\rangle_{L^2(\Omega)} = \lim_{\varepsilon \downarrow 0} \left\langle F, \frac{1}{2\varepsilon} \int_0^T u_s \left(B_{(s+\varepsilon)\wedge T} - B_{(s-\varepsilon)\vee 0}\right) ds \right\rangle_{L^2(\Omega)}, \tag{6.57}$$

provided this limit exists for every $F \in \mathcal{S}_T$. In this case, $\int_0^T u_t \circ dB_t^H$ is called the weak Stratonovich integral of u with respect to the fBm B.

Remarks 14 *(i) Note that $\int_0^T u_s(B_{(s+\varepsilon)\wedge T} - B_{(s-\varepsilon)\vee 0})ds$ is a square-integrable random variable since $\sup_{0 \le s \le T} |B_s|$ is in $L^p(\Omega)$, for any $p \ge 1$.*

(ii) Hypothesis (H2) in Section 4.2 implies that there is at most one square-integrable random variable $\int_0^T u_t \circ dB_t^H$ such that (6.57) holds for every $F \in \mathcal{S}_T$.

(iii) Let $G \in \mathcal{S}_T$ be a bounded random variable and $u \in \mathcal{D}(\delta_S^{B^H})$. Then, Gu also belongs to $\mathcal{D}(\delta_S^{B^H})$ and $\int_0^T Gu_t \circ dB_t^H = G\int_0^T u_t \circ dB_t^H$.

(iv) Consider a process $u = \{u_t : t \in [0,T]\}$ with β-Hölder continuous paths such that $\beta + H > 1$ and $||u||_{\infty,[0,T]} + ||u||_{\beta,[0,T]}$ is in $L^p(\Omega)$, for some $p > 2$. Then, by Proposition 110, we have that $u \in \mathcal{D}(\delta_S^{B^H})$ and $\int_0^T u_t \circ dB_t^H$ agrees with the integral given by Young [216] of u with respect to B.

Remember that Russo and Vallois [190] (see Definition 45) have defined the symmetric integral of u with respect to B as

$$\lim_{\varepsilon \downarrow 0} \left(\frac{1}{2\varepsilon} \int_0^T u_s \left(B_{(s+\varepsilon)\wedge T} - B_{(s-\varepsilon)\vee 0}\right) ds \right), \tag{6.58}$$

where the limit is in probability. In Cheridito and Nualart [27], it has been pointed out that B is in the domain of this integral, but B^2 is not in this domain whenever $H \in (0, 1/6]$. Moreover, from Remark 95, we have that B belongs to $\mathbb{D}^{1,2}(\mathcal{H}_K)$ for $H \in (1/4, 1/2)$ (compare it with Proposition 85), and (6.43) establishes a relation between the symmetric integral in [190], and the classical divergence operator with respect to B. In particular, (6.43) establishes that the relation

$$\int_0^T B_s \circ dB_s = \delta_2(B) + \frac{1}{2}T^{2H} \tag{6.59}$$

is satisfied for $H \in (1/4, 1/2)$. Here, δ_2 is the classical divergence operator with respect to B.

In our case (i.e., for the weak Stratonovich integral), using the ideas developed in the proof of Theorem 80, we have the following result.

Proposition 120 *Let $H < \frac{1}{2}$. Then, B belongs to $\mathcal{D}(\delta_S^{B^H})$ and*

$$\int_0^T B_s \circ dB_s^H = \delta_T(B) + \frac{1}{2}T^{2H}.$$

Remark 97 *Remember that δ_T is defined in (4.61) and that its domain is denoted by Dom δ^*. By Proposition 85, we have that $B \in (Dom\,\delta^* \setminus Dom\,\delta_2)$, for $H < 1/4$, since the definition of δ_2 requires that $B \in \mathcal{H}$ (see (4.2)). However, (6.59) holds even for $H < 1/4$ if we write δ_T instead of δ_2. It means, we now utilize that $B \in Dom\,\delta^*$. Note that this proposition, together with Theorem 55, Section 1.5.6.1 and Proposition 84, implies*

$$\int_0^t B_s \circ dB_s^H = \frac{1}{2}\left(I_2(1 \otimes 1) + T^{2H}\right) = \frac{1}{2}(B_T)^2$$

is true even for $H < 1/4$. Furthermore, in Theorem 85 below, in particular, we see that $p(B) \in \mathcal{D}(\delta_S^{B^H})$, for any real polynomial function p. The proof of this fact does not require that the integrand is in Dom δ^*. We also observe that the proof of Proposition 120, together with the one of Theorem 85, explains how we can handle the existence of a Stratonovich integral introduced in Definition 47.

Proof of Proposition 120. Let $\varepsilon > 0$. Then, Lemma 36 leads us to write

$$\begin{aligned}
&\int_0^T B_s\left(B_{(s+\varepsilon)\wedge T} - B_{(s-\varepsilon)\vee 0}\right) ds \\
&\quad= \int_0^T \delta_2\left(B_s 1_{[(s-\varepsilon)\vee 0,(s+\varepsilon)\wedge T]}(\cdot)\right) ds + \int_0^T \left\langle 1_{[0,s]}, 1_{[(s-\varepsilon)\vee 0,(s+\varepsilon)\wedge T]}\right\rangle_{\mathcal{H}} ds \\
&\quad= \int_0^T \delta_2\left(B_s 1_{[(s-\varepsilon)\vee 0,(s+\varepsilon)\wedge T]}(\cdot)\right) ds \\
&\qquad+ \int_0^T \left(R(s,(s+\varepsilon)\wedge T) - R(s,(s-\varepsilon)\vee 0)\right) ds. \qquad (6.60)
\end{aligned}$$

Hence, (1.44), Remarks 8.*iv*) and Fubini's theorem (i.e., Theorem 48) give that, for any $F \in \mathcal{S}_T$,

$$\begin{aligned}
&\frac{1}{2\varepsilon}E\left(F\int_0^T B_s\left(B_{(s+\varepsilon)\wedge T} - B_{(s-\varepsilon)\vee 0}\right) ds\right) \\
&\quad= \frac{1}{2\varepsilon}E\left(\int_0^T \left\langle \mathcal{D}_T F(\cdot), B_s 1_{[(s-\varepsilon)\vee 0,(s+\varepsilon)\wedge T]}(\cdot)\right\rangle_{L^2([0,T])} ds\right) \\
&\qquad+\frac{1}{4\varepsilon}E\left(F\int_0^T \Big[((s+\varepsilon)\wedge T)^{2H} - ((s-\varepsilon)\vee 0)^{2H}\right. \\
&\qquad\qquad\left. - ((s+\varepsilon)\wedge T - s)^{2H} + (s-(s-\varepsilon)\vee 0)^{2H}\Big] ds\right) \\
&\quad= E\left(\int_0^T D_T F(r)\left(\frac{1}{2\varepsilon}\int_{(r-\varepsilon)\vee 0}^{(r+\varepsilon)\wedge T} B_s ds\right) dr\right) \\
&\qquad+\frac{1}{4\varepsilon}E\left(F\int_0^T \Big[((s+\varepsilon)\wedge T)^{2H} - ((s-\varepsilon)\vee 0)^{2H}\right. \\
&\qquad\qquad\left. - ((s+\varepsilon)\wedge T - s)^{2H} + (s-(s-\varepsilon)\vee 0)^{2H}\Big] ds\right),
\end{aligned}$$

which goes to $E\left(\int_0^T D_T F(s) B_s ds + HF\int_0^T s^{2H-1}ds\right)$, as $\varepsilon \downarrow 0$. In consequence, Proposition 84 and (4.61) yield that B belongs to $\mathcal{D}(\delta_S^{B^H})$ and

$$\int_0^T B_s \circ dB_s^H = \delta_T(B) + \frac{1}{2}T^{2H}.$$

Therefore, the proof is finished. □

Remark 98 *Remember that, at the beginning of Section 5.3, we have already explained why the forward integral $\int_0^T B_s dB_s^-$ does not exist, while the Stratonovich integral $\int_0^T B_s \circ dB_s$ is well-defined in the case that $H < 1/2$. This shows the importance of the study of Stratonovich type integrals with respect to fractional Brownian motion with Hurst parameter less than $1/2$.*

Following the proof of Proposition 120, we obtain a version of Theorem 80 for $H < 1/2$. Namely:

Theorem 84 *Let $p > 2$ and $u \in \left[\mathbb{D}^{1,2}(L^2([0,T])) \cap L^p(\Omega \times [0,T]) \cap (Dom\,\delta^*)\right]$ a process such that*

i) For each $\varepsilon > 0$ small enough, we have

$$\left(u_s(B_{(s+\varepsilon)\wedge T} - B_{(s-\varepsilon)\vee 0}) - \langle Du_s, 1_{[(s-\varepsilon)\vee 0,(s+\varepsilon)\wedge T]}\rangle_{\mathcal{H}}\right) \in L^2(\Omega),$$

for almost all $s \in [0,T]$.

ii) There exists a square-integrable random variable $TrDu$ such that

$$E\left(FTrDu\right) = \lim_{\varepsilon\downarrow 0}\frac{1}{2\varepsilon}E\left(F\int_0^T \left\langle Du_s, 1_{[(s-\varepsilon)\vee 0,(s+\varepsilon)\wedge T]}\right\rangle_{\mathcal{H}}\right),$$

for every $F \in \mathcal{S}_T$.

Then, u belongs to $\mathcal{D}(\delta_S^{B^H})$ and

$$\int_0^T u_s \circ dB_s^H = \delta_T(u) + TrDu.$$

Proof. We only need to change B_s and $1_{[0,s]}$ by u_s and Du_s, respectively, in the proof of Proposition 120 to show that the result is true. Indeed, remember that $DB_s = 1_{[0,s]}$. □

Before establishing an Itô's type formula for the Stratonovich stochastic integral given in Definition 47, we introduce some notations.

The set $C_e^{1,2}([0,T]\times\mathbb{R})$ stands for all functions f in $C^{1,2}([0,T]\times\mathbb{R})$ such that

$$\max\left\{|f(t,x)|, |\partial_t f(t,x)|, |\partial_x f(t,x)|, |\partial_x^2 f(t,x)|\right\} \le c\exp\left(C|x|\right),$$

for $(t,x) \in [0,T]\times\mathbb{R}$ and some positive constants c and C. Also, we use the notations

$$B_t^{H,\varepsilon} := \frac{1}{2\varepsilon}\int_0^t \left(B_{(s+\varepsilon)\wedge T} - B_{(s-\varepsilon)\vee 0}\right) ds, \quad \text{for } t\in[0,T] \text{ and } \varepsilon > 0, \tag{6.61}$$

and

$$\int_0^t u_s \circ dB_s^H := \int_0^T (u_s 1_{[0,t]}(s)) \circ dB_s^H, \quad t \in [0,T].$$

Now, we are ready to establish an Itô's type formula for the weak Stratonovich integral. In order to appreciate the main ideas of Definition 47, we provide some details of the proof of this Itô's formula in Section 6.3.3.1 as auxiliary results. Moreover, in this way, we avoid a long and tedious proof.

Theorem 85 *Let $f \in C_e^{1,2}([0,T]\times\mathbb{R})$. Then, $\partial_x f(\cdot, B_\cdot)1_{[0,t]}(\cdot) \in \mathcal{D}(\delta_S^{B^H})$ and*

$$f(t,B_t) = f(0,0) + \int_0^t \partial_t f(s,B_s)ds + \int_0^t \partial_x f(s,B_s)\circ dB_s^H, \quad \textit{for } t\in[0,T]. \tag{6.62}$$

Remark 99 *The symmetric integral in [190] of B^2 with respect to B (see (6.58)) does not exist for $H \le 1/6$ according to Cheridito and Nualart [27]. But, as a consequence of Theorem 85, we have that the integral $\int_0^T p(B_s)\circ dB_s^H$ is well-defined, for any polynomial function $p:\mathbb{R}\to\mathbb{R}$.*

Proof. Let $t\in[0,T]$ and $\varepsilon>0$. Then, the fundamental theorem of calculus leads to state

$$\begin{aligned}
&f(t,B_t^{H,\varepsilon})\\
&= f(0,0) + \int_0^t \partial_t f(s,B_s^{H,\varepsilon})ds + \frac{1}{2\varepsilon}\int_0^t \partial_x f(s,B_s^{H,\varepsilon})\left(B_{(s+\varepsilon)\wedge T} - B_{(s-\varepsilon)\vee 0}\right)ds\\
&= f(0,0) + \int_0^t \partial_t f(s,B_s^{H,\varepsilon})ds + \frac{1}{2\varepsilon}\int_0^t \partial_x f(s,B_s)\left(B_{(s+\varepsilon)\wedge T} - B_{(s-\varepsilon)\vee 0}\right)ds\\
&\quad +\frac{1}{2\varepsilon}\int_0^t \left(\partial_x f(s,B_s^{H,\varepsilon}) - \partial_x f(s,B_s)\right)\left(B_{(s+\varepsilon)\wedge T} - B_{(s-\varepsilon)\vee 0}\right)ds.
\end{aligned} \tag{6.63}$$

Therefore, by applying Lemmas 36 and 71, Propositions 21 and 23, [155] (property (4.13)) and (6.66)-(6.68), we get

$$\begin{aligned}
I_t^\varepsilon &:= \frac{1}{2\varepsilon}\int_0^t \left(\partial_x f(s,B_s^{H,\varepsilon}) - \partial_x f(s,B_s)\right)\left(B_{(s+\varepsilon)\wedge T} - B_{(s-\varepsilon)\vee 0}\right)ds\\
&= \frac{1}{2\varepsilon}\int_0^t \delta_2\left(\left(\partial_x f(s,B_s^{H,\varepsilon}) - \partial_x f(s,B_s)\right)1_{[(s-\varepsilon)\vee 0,(s+\varepsilon)\wedge T]}\right)ds\\
&\quad +\frac{1}{2\varepsilon}\int_0^t \left\langle \partial_x^2 f(s,B_s^{H,\varepsilon})DB_s^{H,\varepsilon} - \partial_x^2 f(s,B_s)1_{[0,s]}, 1_{[(s-\varepsilon)\vee 0,(s+\varepsilon)\wedge T]}\right\rangle_{\mathcal{H}} ds.
\end{aligned} \tag{6.64}$$

Now, we consider the first term on the right-hand side of the last equality. From Remark 8.*iv*), we can show that, for $F\in\mathcal{S}_T$,

$$\begin{aligned}
&E\left(\frac{F}{2\varepsilon}\int_0^t \delta_2\left(\left(\partial_x f(s,B_s^{H,\varepsilon}) - \partial_x f(s,B_s)\right)1_{[(s-\varepsilon)\vee 0,(s+\varepsilon)\wedge T]}\right)ds\right)\\
&= E\left(\frac{1}{2\varepsilon}\int_0^t \left\langle D_T F, \left(\partial_x f(s,B_s^{H,\varepsilon}) - \partial_x f(s,B_s)\right)1_{[(s-\varepsilon)\vee 0,(s+\varepsilon)\wedge T]}\right\rangle_{L^2([0,T])}ds\right).
\end{aligned} \tag{6.65}$$

Thus, using Hölder inequality, we have

$$\begin{aligned}
&E\left(\frac{F}{2\varepsilon}\int_0^t \delta_2\left(\left(\partial_x f(s,B_s^{H,\varepsilon}) - \partial_x f(s,B_s)\right)1_{[(s-\varepsilon)\vee 0,(s+\varepsilon)\wedge T]}\right)ds\right)\\
&\le \left(E\int_0^t \left(\partial_x f(s,B_s^{H,\varepsilon}) - \partial_x f(s,B_s)\right)^2 ds\right)^{1/2}\\
&\quad\times\left(E\int_0^t\left(\frac{1}{2\varepsilon}\int_{(s-\varepsilon)\vee 0}^{(s+\varepsilon)\wedge T} D_T F(r)dr\right)^2 ds\right)^{1/2}\\
&\le \left(E\int_0^t \left(\partial_x f(s,B_s^{H,\varepsilon}) - \partial_x f(s,B_s)\right)^2 ds\right)^{1/2}\\
&\quad\times\left(E\int_0^t \frac{1}{2\varepsilon}\int_{(s-\varepsilon)\vee 0}^{(s+\varepsilon)\wedge T} \left(D_T F(r)\right)^2 drds\right)^{1/2}.
\end{aligned}$$

Hence, Fubini's theorem implies

$$E\left(\frac{F}{2\varepsilon}\int_0^t \delta_2\left(\left(\partial_x f(s,B_s^{H,\varepsilon})-\partial_x f(s,B_s)\right)1_{[(s-\varepsilon)\vee 0,(s+\varepsilon)\wedge T]}\right)ds\right)$$
$$\leq\left(E\int_0^t\left(\partial_x f(s,B_s^{H,\varepsilon})-\partial_x f(s,B_s)\right)^2ds\right)^{1/2}\left(E\int_0^{(t+\varepsilon)\wedge T}\left(D_{\mathcal{T}}F(r)\right)^2dr\right)^{1/2}.$$

In consequence, using that f is a function in $C_e^{1,2}([0,T]\times\mathbb{R})$, [155] (property (4.13)), (6.68), (6.64), and Lemmas 72 and 73 below, we obtain that $\lim_{\varepsilon\downarrow 0}E\left(FI_t^\varepsilon\right))=0$, for every $F\in\mathcal{S}_T$. Therefore, due to (6.63) and Lemma 70, $s\mapsto\partial_x f(s,B_s^H)1_{[0,t]}(s)$ belongs to $\mathcal{D}(\delta_S^{B^H})$ and (6.62) is true. In other words, the proof is complete. □

Note that we could use the duality relation (4.2) between the derivative and the classical divergence operators on the left-hand side of (6.65) to see that I_t^ε goes to zero, as $\varepsilon\to 0$. But, in this way, this left-hand side would have the form

$$E\left(\frac{1}{2\varepsilon}\int_0^t\langle DF,\left(\partial_x f(s,B_s^{H,\varepsilon})-\partial_x f(s,B_s)\right)1_{[(s-\varepsilon)\vee 0,(s+\varepsilon)\wedge T]}\rangle_{\mathcal{H}}\,ds\right).$$

In general, the analysis of the convergence of this quantity is not easy to handle because of the inner product in the Hilbert space $\mathcal{H}$ as Lemmas 72 and 73 below show. Moreover, we know that the first summand on the right-hand side of (6.60) times $(2\varepsilon)^{-1}$ does not converge if we use the classical duality relation (4.2) since B does not belong to $\mathcal{H}$ for $H<1/4$, as Proposition 85 establishes. The advantage of applying the duality relation (4.61) for the extended divergence operator is that the inner product in $\mathcal{H}$ becomes the one of $L^2([0,T])$, which is well-known and has tools to work with convergences there, as the study of the right-hand side of (6.65) exhibits.

6.3.3.1 Auxiliary results

The purpose of this section is to provide the auxiliary tool to give the missing details in the proof of Theorem 85.

Lemma 70 *Let $f\in C_e^{1,2}([0,T]\times\mathbb{R})$ and $t\in[0,T]$. Then,*

$$f(t,B_t)=\lim_{\varepsilon\downarrow 0}f(t,B_t^{H,\varepsilon})\quad \text{and}\quad \int_0^t\partial_t f(s,B_s)ds=\lim_{\varepsilon\downarrow 0}\int_0^t\partial_t f(s,B_s^{H,\varepsilon})ds,$$

in $L^2(\Omega)$.

Proof. Let $t\in(0,T]$ and $0<\varepsilon<t$. Then, (6.61) and the change of variables formula imply

$$B_t^{H,\varepsilon}:=\frac{1}{2\varepsilon}\left(\int_{t-\varepsilon}^{t+\varepsilon}B_{s\wedge T}ds-\int_0^\varepsilon B_sds\right).\tag{6.66}$$

Therefore, the continuity of B yields that $B_t^{H,\varepsilon}$ goes to B_t, as $\varepsilon\downarrow 0$, w.p.1. Moreover, for $0\leq t\leq\varepsilon$, we have

$$B_t^{H,\varepsilon}=\frac{1}{2\varepsilon}\int_\varepsilon^{t+\varepsilon}B_{s\wedge T}ds,\tag{6.67}$$

which, together with (6.66), gives

$$\sup_{s\in[0,T]}|B_s^{H,\varepsilon}|\leq 2\sup_{s\in[0,T]}|B_s|.\tag{6.68}$$

Thus, the result is a consequence of the facts that, for any $c>0$, $\exp(c\sup_{s\in[0,T]}|B_s|)\in L^2(\Omega)$ (see Theorem 4.2 in [155]) and $f\in C_e^{1,2}([0,T]\times\mathbb{R})$, and the dominated convergence theorem. □

Lemma 71 *Let $a, b \in [0, T]$ and $\varepsilon \geq 0$ be such that $a \leq b + \varepsilon$. Then, $\int_a^{b+\varepsilon} B^H_{s\wedge T} ds$ is in $\mathbb{D}^{1,2}$ and*

$$\left\langle D \int_a^{b+\varepsilon} B_{s\wedge T} ds, \phi \right\rangle_{\mathcal{H}} = \int_a^{b+\varepsilon} \langle 1_{[0,s\wedge T]}, \phi \rangle_{\mathcal{H}} ds, \quad \text{for } \phi \in \mathcal{H}.$$

Proof. The continuity of B and $(\sup_{s\in[0,T]} |B_s|) \in L^2(\Omega)$ imply

$$\int_a^{b+\varepsilon} B_{s\wedge T} ds = \lim_{|\pi|\to 0} \sum_{i=0}^{n-1} B_{t_i \wedge T}(t_{i+1} - t_i),$$

where the limit is in $L^2(\Omega)$ and $\pi = \{a = t_0 < t_1 < \ldots < t_n = b + \varepsilon\}$ is a partition of $[a, b+\varepsilon]$. Consequently, $\int_a^{b+\varepsilon} B_{s\wedge T} ds$ is a square-integrable random variable in the chaos of order 1 and, therefore, it is in $\mathbb{D}^{1,2}$.

Finally, Fubini's theorem, together with the duality relation (4.2) between the derivative and the divergence operators yields that, for $\phi \in \mathcal{H}$,

$$\begin{aligned}\left\langle D \int_a^{b+\varepsilon} B_{s\wedge T} ds, \phi \right\rangle_{\mathcal{H}} &= E\left(\left\langle D \int_a^{b+\varepsilon} B_{s\wedge T} ds, \phi \right\rangle_{\mathcal{H}}\right) = E\left(\delta(\phi) \int_a^{b+\varepsilon} B_{s\wedge T} ds\right) \\ &= \int_a^{b+\varepsilon} E\left(\delta_2(\phi) B_{s\wedge T}\right) ds = \int_a^{b+\varepsilon} \langle 1_{[0,s\wedge T]}, \phi \rangle_{\mathcal{H}} ds.\end{aligned}$$

The proof is now complete. □

Lemma 72 *Let $f \in C^{1,2}_e([0, T] \times \mathbb{R})$. Then,*

$$\lim_{\varepsilon\downarrow 0} E\left(\left(\frac{1}{2\varepsilon}\int_0^\varepsilon |\langle \partial_x^2 f(s, B^{H,\varepsilon}_s) DB^{H,\varepsilon}_s - \partial_x^2 f(s, B_s) 1_{[0,s]}, 1_{[(s-\varepsilon)\vee 0,(s+\varepsilon)\wedge T]} \rangle_{\mathcal{H}}| \, ds\right)^2\right) = 0.$$

Proof. Let $\varepsilon < T/2$. Then, (6.67) leads to write

$$B^{H,\varepsilon}_t = \frac{1}{2\varepsilon} \int_\varepsilon^{t+\varepsilon} B_s ds, \quad t \in [0, \varepsilon].$$

Hence, the fact that $f \in C^{1,2}_e([0, T] \times \mathbb{R})$, (1.44), (6.68) and Lemma 71 imply

$$\begin{aligned}&\frac{1}{2\varepsilon}\int_0^\varepsilon |\langle \partial_x^2 f(s, B^{H,\varepsilon}_s) DB^{H,\varepsilon}_s - \partial_x^2 f(s, B_s) 1_{[0,s]}, 1_{[(s-\varepsilon)\vee 0,(s+\varepsilon)\wedge T]} \rangle_{\mathcal{H}}| \, ds \\ &= \frac{1}{2\varepsilon}\int_0^\varepsilon \left| \partial_x^2 f(s, B^{H,\varepsilon}_s) \frac{1}{2\varepsilon} \int_\varepsilon^{s+\varepsilon} \langle 1_{[0,u]}, 1_{[0,s+\varepsilon]} \rangle_{\mathcal{H}} du - \partial_x^2 f(s, B_s) \langle 1_{[0,s]}, 1_{[0,s+\varepsilon]} \rangle_{\mathcal{H}} \right| ds \\ &= \frac{1}{2\varepsilon}\int_0^\varepsilon \left| \partial_x^2 f(s, B^{H,\varepsilon}_s) \frac{1}{2\varepsilon} \int_\varepsilon^{s+\varepsilon} R(u, s+\varepsilon) du - \partial_x^2 f(s, B_s) R(s, s+\varepsilon) \right| ds \\ &\leq \frac{c}{\varepsilon} \exp(C \sup_{s\in[0,T]} |B_s|) \int_0^\varepsilon \left(\frac{1}{2\varepsilon} \int_\varepsilon^{s+\varepsilon} |R(u, s+\varepsilon)| du + |R(s, s+\varepsilon)| \right) ds \\ &\leq c \exp(C \sup_{s\in[0,T]} |B_s|) \varepsilon^{2H},\end{aligned}$$

which tends to 0 in $L^2(\Omega)$, as $\varepsilon \downarrow 0$, since $\exp(C \sup_{s\in[0,T]} |B^H_s|) \in L^2(\Omega)$ due to [155] (Theorem 4.2). In consequence, the result is satisfied. □

Lemma 73 *Let $f \in C_e^{1,2}([0,T]\times\mathbb{R})$ and $t \in [0,T]$. Then,*

$$\lim_{\varepsilon\downarrow 0} E\left(\left(\frac{1}{2\varepsilon}\int_{t\wedge\varepsilon}^{t} |\langle \partial_x^2 f(s,B_s^{H,\varepsilon})DB_s^{H,\varepsilon} - \partial_x^2 f(s,B_s)1_{[0,s]}, 1_{[(s-\varepsilon)\vee 0,(s+\varepsilon)\wedge T]}\rangle_{\mathcal{H}}|\, ds\right)^2\right) = 0.$$

Proof. Let $0<\varepsilon<t/2\le T/2$. Then, from (6.66), we get

$$\begin{aligned}
&\frac{1}{2\varepsilon}\int_{\varepsilon}^{t} |\langle \partial_x^2 f(s,B_s^{H,\varepsilon})DB_s^{H,\varepsilon} - \partial_x^2 f(s,B_s)1_{[0,s]}, 1_{[s-\varepsilon,(s+\varepsilon)\wedge T]}\rangle_{\mathcal{H}}|\, ds\\
&\le \frac{1}{2\varepsilon}\int_{\varepsilon}^{t} \left|\left\langle \partial_x^2 f(s,B_s^{H,\varepsilon})D\left\{\frac{1}{2\varepsilon}\int_{s-\varepsilon}^{s+\varepsilon} B_{u\wedge T}du\right\} - \partial_x^2 f(s,B_s)1_{[0,s]}, 1_{[s-\varepsilon,(s+\varepsilon)\wedge T]}\right\rangle_{\mathcal{H}}\right| ds\\
&\quad + \frac{1}{4\varepsilon^2}\int_{\varepsilon}^{t} \left|\left\langle \partial_x^2 f(s,B_s^{H,\varepsilon})D\int_0^{\varepsilon} B_u du, 1_{[s-\varepsilon,(s+\varepsilon)\wedge T]}\right\rangle_{\mathcal{H}}\right| ds = I_1^\varepsilon + I_2^\varepsilon. \qquad (6.69)
\end{aligned}$$

Now, we divide the proof into three parts.

Step 1: Here, we consider the convergence to zero of I_2^ε in $L^2(\Omega)$, as $\varepsilon\downarrow 0$.

From Lemma 71 and (6.68), we have that, for some constant $C>0$,

$$\begin{aligned}
I_2^\varepsilon &= \frac{1}{4\varepsilon^2}\int_{\varepsilon}^{t}\left|\partial_x^2 f(s,B_s^{H,\varepsilon})\int_0^{\varepsilon}(R(u,(s+\varepsilon)\wedge T) - R(u,s-\varepsilon))\,du\right| ds\\
&\le \frac{C\exp(C\sup_{s\in[0,T]}|B_s|)}{\varepsilon^2}\left(\int_{\varepsilon}^{2\varepsilon}\int_0^{\varepsilon} |R(u,(s+\varepsilon)\wedge T) - R(u,s-\varepsilon)|\,duds\right.\\
&\quad \left.+\int_{2\varepsilon}^{t}\int_0^{\varepsilon} |R(u,(s+\varepsilon)\wedge T) - R(u,s-\varepsilon)|\,duds\right)\\
&= \frac{C\exp(C\sup_{s\in[0,T]}|B_s|)}{\varepsilon^2}\left(I_{2,1}^\varepsilon + I_{2,2}^\varepsilon\right). \qquad (6.70)
\end{aligned}$$

Note that the covariance of fractional Brownian motion (1.44) gives

$$I_{2,1}^\varepsilon \le C\varepsilon^{2H}\int_{\varepsilon}^{2\varepsilon}\int_0^{\varepsilon} duds = C\varepsilon^{2+2H} \qquad (6.71)$$

and, for $u\in(0,\varepsilon)$ and $v\in(\varepsilon,T]$, $|\partial_v R(v,u)| = H\left((v-u)^{2H-1} - v^{2H-1}\right)$ which, together with the mean value theorem, leads us to

$$\begin{aligned}
I_{2,2}^\varepsilon &\le C\varepsilon\int_{2\varepsilon}^{t}\int_0^{\varepsilon}\left\{(s-2\varepsilon)^{2H-1} - ((s+\varepsilon)\wedge T)^{2H-1}\right\} duds\\
&= C\varepsilon^2\int_{2\varepsilon}^{t}\left\{(s-2\varepsilon)^{2H-1} - ((s+\varepsilon)\wedge T)^{2H-1}\right\} ds. \qquad (6.72)
\end{aligned}$$

Thus, for $0<t<T$ and ε small enough, we establish

$$\begin{aligned}
I_{2,2}^\varepsilon &\le C\varepsilon^2\int_{2\varepsilon}^{t}\left\{(s-2\varepsilon)^{2H-1} - (s+\varepsilon)^{2H-1}\right\} ds\\
&= C\varepsilon^2\left[(t-2\varepsilon)^{2H} - (t+\varepsilon)^{2H} + (3\varepsilon)^{2H}\right] \le C\varepsilon^{2+2H}. \qquad (6.73)
\end{aligned}$$

Also, we have to consider the case that $t=T$. But, (6.72) yields that, for ε small enough,

$$\begin{aligned}
I_{2,2}^\varepsilon &\le C\varepsilon^2\int_{2\varepsilon}^{T-\varepsilon}\left\{(s-2\varepsilon)^{2H-1} - (s+\varepsilon)^{2H-1}\right\} ds\\
&\quad + C\varepsilon^2\int_{T-\varepsilon}^{T}\left\{(s-2\varepsilon)^{2H-1} - (T)^{2H-1}\right\} ds\\
&= C\varepsilon^2\left|\frac{1}{2H}\left((T-2\varepsilon)^{2H} - (T)^{2H} + (3\varepsilon)^{2H}\right) - T^{2H-1}\varepsilon\right| \le C\varepsilon^{2+2H}. \qquad (6.74)
\end{aligned}$$

Therefore, we have proved that I_2^ε converges to zero in $L^2(\Omega)$, as $\varepsilon \to 0$, because of inequalities (6.70)-(6.74) and Nourdin [155] (Theorem 4.2).

Step 2: Now, we show that I_1^ε tends to zero in $L^2(\Omega)$, as $\varepsilon \to 0$.

By (6.69),

$$\begin{aligned} I_1^\varepsilon &\leq \frac{1}{2\varepsilon}\int_\varepsilon^t \left|\left\langle \left(\partial_x^2 f(s,B_s^{H,\varepsilon}) - \partial_x^2 f(s,B_s)\right) 1_{[0,s]}, 1_{[s-\varepsilon,(s+\varepsilon)\wedge T]}\right\rangle_{\mathcal{H}}\right| ds \\ &+\frac{1}{2\varepsilon}\int_\varepsilon^t \left|\left\langle \partial_x^2 f(s,B_s^{H,\varepsilon})\left(D\left\{\frac{1}{2\varepsilon}\int_{s-\varepsilon}^{s+\varepsilon} B_{u\wedge T}du\right\} - 1_{[0,s]}\right), 1_{[s-\varepsilon,(s+\varepsilon)\wedge T]}\right\rangle_{\mathcal{H}}\right| ds \\ &= I_{1,1}^\varepsilon + I_{1,2}^\varepsilon. \end{aligned} \tag{6.75}$$

We first deal with $I_{1,1}^\varepsilon$ when $t < T$. Then, for ε small enough, the mean value theorem and (1.44) yield

$$\begin{aligned} I_{1,1}^\varepsilon &= \frac{1}{2\varepsilon}\int_\varepsilon^t \left|\left(\partial_x^2 f(s,B_s^{H,\varepsilon}) - \partial_x^2 f(s,B_s)\right)\left(R(s,s+\varepsilon) - R(s,s-\varepsilon)\right)\right| ds \\ &= \frac{1}{4\varepsilon}\int_\varepsilon^t \left|\partial_x^2 f(s,B_s^{H,\varepsilon}) - \partial_x^2 f(s,B_s)\right| \left((s+\varepsilon)^{2H} - (s-\varepsilon)^{2H}\right) ds \\ &\leq H\int_\varepsilon^t \left|\partial_x^2 f(s,B_s^{H,\varepsilon}) - \partial_x^2 f(s,B_s)\right| (s-\varepsilon)^{2H-1} ds. \end{aligned} \tag{6.76}$$

Similarly, for $t = T$ and ε small enough, the mean value theorem implies

$$\begin{aligned} I_{1,1}^\varepsilon &= \frac{1}{4\varepsilon}\int_\varepsilon^{T-\varepsilon} \left|\partial_x^2 f(s,B_s^{H,\varepsilon}) - \partial_x^2 f(s,B_s)\right| \left((s+\varepsilon)^{2H} - (s-\varepsilon)^{2H}\right) ds \\ &+\frac{1}{4\varepsilon}\int_{T-\varepsilon}^T \left|\partial_x^2 f(s,B_s^{H,\varepsilon}) - \partial_x^2 f(s,B_s)\right| \left(T^{2H} - (s-\varepsilon)^{2H} + \varepsilon^{2H} - (T-s)^{2H}\right) ds \\ &\leq H\int_\varepsilon^T \left|\partial_x^2 f(s,B_s^{H,\varepsilon}) - \partial_x^2 f(s,B_s)\right| (s-\varepsilon)^{2H-1} ds \\ &+H\int_{T-\varepsilon}^T \left|\partial_x^2 f(s,B_s^{H,\varepsilon}) - \partial_x^2 f(s,B_s)\right| \left((T-s)^{2H-1} + (s-\varepsilon)^{2H-1}\right) ds, \end{aligned}$$

which, together with [155] (Theorem 4.2), $f \in C_e^{1,2}([0,T]\times\mathbb{R})$ and (6.76), allows us to conclude that $I_{1,1}^\varepsilon$ converges to zero in $L^2(\Omega)$, as $\varepsilon \to 0$. Hence, to see that Step 2 is also satisfied, we must show that $I_{1,2}^\varepsilon$ also goes to zero in $L^2(\Omega)$, as $\varepsilon \downarrow 0$, since (6.75). To do so, we first suppose that $t < T$. In this case, we apply Lemma 71 and the convention $G = C\exp(C\sup_{s\in[0,T]}|B_s^H|)$ to prove that, for ε small enough,

$$\begin{aligned} I_{1,2}^\varepsilon &\leq \frac{G}{4\varepsilon^2}\int_\varepsilon^t \left|\int_{s-\varepsilon}^{s+\varepsilon} [R(u,s+\varepsilon) - R(u,s-\varepsilon) - R(s,s+\varepsilon) + R(s,s-\varepsilon)]\,du\right| ds \\ &\leq C\frac{G}{\varepsilon^2}\int_\varepsilon^t \left|\int_{s-\varepsilon}^{s+\varepsilon} \left[(s+\varepsilon-u)^{2H} - (u-(s-\varepsilon))^{2H}\right] du\right| ds = 0. \end{aligned} \tag{6.77}$$

Furthermore, for $t = T$ and ε small enough, we have

$$\begin{aligned} I_{1,2}^\varepsilon \leq C\frac{G}{\varepsilon^2}\int_{T-\varepsilon}^T \Bigg|\int_{s-\varepsilon}^{s+\varepsilon} &[R(u\wedge T,(s+\varepsilon)\wedge T) \\ &-R(u\wedge T,s-\varepsilon) - R(s,(s+\varepsilon)\wedge T) + R(s,s-\varepsilon)]\,du\Bigg| ds. \end{aligned}$$

Therefore, (1.44) allows us to establish

$$\begin{aligned} I_{1,2}^{\varepsilon} &\leq C\frac{G}{\varepsilon^2}\int_{T-\varepsilon}^{T}\left|\int_{s-\varepsilon}^{s+\varepsilon}\left[(T-u\wedge T)^{2H}\right.\right. \\ &\qquad \left.\left. -(u\wedge T-(s-\varepsilon))^{2H}+\varepsilon^{2H}-(T-s)^{2H}\right]du\right|ds \\ &\leq C\frac{G}{\varepsilon^2}\int_{T-\varepsilon}^{T}\int_{s-\varepsilon}^{s+\varepsilon}\varepsilon^{2H}duds = CG\varepsilon^{2H}. \end{aligned}$$

Thus, (6.77) gives that $I_{1,2}^{\varepsilon}\to 0$ in $L^2(\Omega)$, as ε goes to zero.
Step 3: Finally, (6.69), and Steps 1 and 2 yield that the proof is complete. □

Chapter 7

Stochastic Differential Equations

Currently, there exists a great interest in the study of the theory of probability due in large part to the applications of stochastic differential equations in various areas of scientific knowledge such as physics, biology, financial mathematics, among others. Roughly speaking, in general, a differential equation of the form

$$dX_t = b(t, X_t)dt, \qquad t \in (a, b],$$

models certain phenomenon, which is of our interest. Therefore, the solutions to this equation tell us how the phenomenon in question evolves in time. However, the analysis of these type of equations requires that all the parameters involved are completely known (for instance, the initial condition, coefficients, external perturbations of the system, etc.). But, sometimes this is impossible because of the nature of the problem that is being studied, the scant information available, the cost involved, or simply because they are obtained experimentally. In consequence, in general, it is necessary to do a "random correction" to the equation in question. That is, we need to introduce a parameter in a set where a probability measure is defined and whose elements represent the possible factors that influence the system so the solution X gives all the possible realizations of the phenomenon in question. The way to incorporate these elements is allowing the coefficients to be random and to disturb the equation by means of a "random noise". Hence, we can take the central limit theorem into account to consider stochastic differential equations of the form

$$dX_t = b(t, X_t)dt + \sigma(t, X_t)dB_t, \qquad t \in (a, b], \tag{7.1}$$

where $0 \leq a < b \leq T$ and $B = \{B_t : t \in [0, T]\}$ is a suitable Gaussian process defined on a complete probability space since the central limit theorem states that "the superposition of a large number of small independent disturbances produce a Gaussian noise".

Note that equation (7.1) is a symbolic expression and it may be interpreted in different ways. Namely, one way to see that X is a solution of equation (7.1) is that X is a stochastic process satisfying

$$X_t = \xi + \int_0^t b(s, X_s)ds + \int_0^t \sigma(s, X_s)dB_s, \quad \text{w.p.1, for each } t \in [0, T].$$

Here ξ is a given random variable. Even in this case, the meaning of the solution to this equation depends on the interpretation of the stochastic integral that we may employ.

The main aim of this section is to analyze equation (7.1) using the different definitions of stochastic integral considered in this book. In some cases, we only explain the ideas of the construction of the unique solution to (7.1) and indicate where the reader can consult the details in order to simplify the exposition and not to give elaborated and boring features, which may complicate the understanding of the main theory in the study of stochastic differential equations.

DOI: 10.1201/9781003484912-7

7.1 Stochastic differential equations of Itô type

In the literature, a process used to stand for the disturbance in equation (7.1) is the Brownian motion or Wiener process W since it is a Gaussian process with independent and stationary increments whose generalized derivative is white noise. Gaussian white noise is a stationary generalized process and a good model of "noise". It means white noise represents stationary and rapidly fluctuating phenomena (see, for example, Arnold [16] or Hida [75]). However, white noise does not exist in the traditional sense. For all the above, it is important to consider stochastic differential equations (SDEs) driven by Brownian motion W, which are written as

$$\begin{aligned} dX_t &= b(t, X_t)dt + \sigma(t, X_t)dW_t, \qquad t \in [a, T], \\ X_a &= \xi. \end{aligned} \tag{7.2}$$

Here the initial condition ξ is a random variable and $a \in [0, T)$.

In this section, $(\mathcal{F}_t)_{t\in[0,T]}$ stands for a filtration of complete sub-σ-algebras on a complete probability space $(\Omega, \mathcal{F}, P)$ (see Definition 1) and $W = \{W_t : t \in [0, T]\}$ is a m-dimensional $\mathcal{F}_t$-Brownian motion as it is introduced in Definition 36 (see also Definition 11). All the processes contemplated in this section are $\mathcal{F}_t$-adapted and measurable as functions on $\Omega \times [a, T]$. This hypothesis is a natural one because the σ-algebra $\mathcal{F}_t$ represents the information we have up to time t of a phenomenon or problem that we are analyzing in general. So, in equation (7.2), the initial condition ξ is a $\mathcal{F}_a$-measurable d-dimensional random vector, and the coefficients

$$b : \Omega \times [a, T] \times \mathbb{R}^d \to \mathbb{R}^d \quad \text{and} \quad \sigma : \Omega \times [a, T] \times \mathbb{R}^d \to \mathcal{L}(\mathbb{R}^m, \mathbb{R}^d)$$

are $\mathcal{F} \otimes \mathcal{B}([a, T]) \otimes \mathcal{B}(\mathbb{R}^d)$-measurable random fields such that, for $x \in \mathbb{R}^d$, $b(\cdot, x)$ and $\sigma(\cdot, x)$ are $\mathcal{F}_t$-adapted, measurable, and $\mathbb{R}^d$ and $\mathcal{L}(\mathbb{R}^m, \mathbb{R}^d)$-valued random processes, respectively. Here, $\mathcal{L}(\mathbb{R}^m, \mathbb{R}^d)$ denotes the family of all linear operators from $\mathbb{R}^m$ to $\mathbb{R}^d$.

As in Theorem 41, we have that equation (7.2) means that, for $i \in \{1, \ldots, d\}$,

$$X_t^{(i)} = \xi^{(i)} + \int_a^t b^{(i)}(s, X_s)ds + \sum_{j=1}^m \int_a^t \sigma^{i,j}(s, X_s)dW^{(j)}, \quad t \in [a, T], \tag{7.3}$$

where the matrix $\left(\sigma^{i,j}(s, X_s)\right)_{\substack{1\le i\le d\\ 1\le j\le m}}$ denotes the linear operator $\sigma(s, X_s)$. In this way, now we can introduce what a solution of equation (7.3) means:

Definition 48 *A (strong) solution X to equation* (7.2) *is a continuous, $\mathbb{R}^d$-valued, $\mathcal{F}_t$-adapted and measurable process such that:*

i) $P[X_a = \xi] = 1$.

ii) $b(\cdot, X_\cdot)$ and $\sigma(\cdot, X_\cdot)$ are $\mathcal{F}_t$-adapted, and vectors-valued stochastic processes.

iii) $P\left[\int_a^t \{|b(s, X_s)| + |\sigma(s, X_s)|^2\}ds < \infty\right] = 1$, for $t \in [a, T]$.

iv) $X_t = \xi + \int_a^t b(s, X_s)ds + \int_a^t \sigma(s, X_s)dW_s$ w.p.1, for each $t \in [a, T]$.

Remark 100 *The set of probability 1 in Statement iv) depends on $t \in [0, T]$. However, from Proposition 1, we have that X and the right-hand side of Statement iv) are indistinguishable* *since they are continuous $\mathbb{R}^d$-valued processes that are a modification of each other. So, the set of probability 1 is independent of the point t. Furthermore, Statement ii) yields that the integral $\int_a^t b(s, X_s)ds$ is introduced pathwise. That is, $b(\cdot, X_\cdot)$ belongs to $L^1([a,T])$ with probability 1 and the integral is defined ω by ω. Concerning $\int_a^t \sigma(s, X_s)dW_s$, it is the stochastic integral of $1_{[a,t]}(\cdot)\sigma(\cdot, X_\cdot)$ with respect to the Brownian motion W in the Itô sense, which is well-defined because $\sigma(\cdot, X_\cdot)$ is an $\mathcal{F}_t$-adapted and $\mathbb{R}^d$-valued process with square-integrable paths (see equality* (3.12) *and Proposition 56).*

As an example, consider the linear SDE of Itô type

$$\begin{aligned} dX_t &= b_t X_t dt + \sigma_t X_t dB_t, \qquad t \in [0,T], \\ X_0 &= x. \end{aligned} \tag{7.4}$$

Here $x \in \mathbb{R}$, B is a one-dimensional Brownian motion and $b, \sigma : \Omega \times [0,T] \to \mathbb{R}$ are two measurable and $\mathcal{F}_t$-adapted processes such that $b \in L^1([0,T])$ and $\sigma \in L^2([0,T])$ with probability 1. Proceeding as in Examples 2, we can apply Itô's formula to show that the unique solution to equation (7.4) is the process $t \mapsto x \exp\left(\int_0^t b_s ds + \int_0^t \sigma_s dB_s - \frac{1}{2}\int_0^t \sigma_s^2 ds\right)$.

It is worth mentioning that Itô's formula has also allowed SDEs to be related to ordinary differential equations [42] (see also Section 7.3), and differential equations of parabolic and elliptic type [96].

A useful tool to deal with the uniqueness for the solution of equation (7.2) is the so-called Gronwall's lemma:

Proposition 121 (Gronwall's lemma) *Let $C > 0$ be a constant, and f and g two functions in $L^1([a,T])$ such that*

$$f(t) \le g(t) + C\int_a^t f(s)ds, \quad t \in [a,T]. \tag{7.5}$$

Then,

$$f(t) \le g(t) + C\int_a^t \exp\{C(t-s)\}\, g(s)ds, \quad t \in [a,T].$$

In particular, if $g \equiv M$ with M a constant, we have

$$f(t) \le M\exp\{C(t-a)\}, \quad t \in [a,T].$$

Remark 101 *Note that $f \equiv 0$ whenever $f \ge 0$ and $g \equiv 0$.*

Proof. Set $\alpha(t) = \int_a^t f(s)ds$, $t \in [a,T]$. Then, the function $\beta(t) = \alpha(t)e^{-Ct}$ is absolutely continuous on the interval $[a,T]$. Thus, the fundamental theorem of calculus yields

$$\beta(t) - \beta(a) = \int_a^t \beta'(s)ds = \int_a^t \left(f(s) - C\alpha(s)\right)e^{-Cs}ds, \quad t \in [a,T].$$

Hence (7.5) and the fact that $\alpha(a) = 0$ imply

$$\alpha(t) \le e^{Ct}\int_a^t g(s)e^{-Cs}ds, \quad t \in [a,T],$$

and, therefore, using inequality (7.5) again, we obtain

$$f(t) \le g(t) + C\alpha(t) \le g(t) + C\int_a^t g(s)e^{C(t-s)}ds, \quad t \in [a,T].$$

Finally, assume that $g \equiv M$. Then, last inequality allows us to establish

$$f(t) \le M + CM \int_a^t e^{C(t-s)} ds = M + M\left(e^{C(t-a)} - 1\right) = Me^{C(t-a)}, \quad t \in [a,T].$$

Thus, the proof is complete. □

We have already mentioned that Gronwall's lemma (i.e., Proposition 121) is a useful tool to deal with the uniqueness of the solutions of stochastic differential equations. For instance, assume that $\sigma \equiv 0$ in equation (7.2) and that $b : [a,T] \times \mathbb{R} \to \mathbb{R}$ is a Lipschitz function in x, uniformly on $t \in [a,T]$. That is, there is a constant $C > 0$ such that

$$|b(t,x) - b(t,y)| \le C|x-y|, \quad \text{for } t \in [a,T] \text{ and } x, y \in \mathbb{R}.$$

In this case, if we have that $f, g : [a,T] \times \to \mathbb{R}$ are two solutions, then we have

$$|f(t) - g(t)| \le \int_0^t |b(s,f(s)) - b(s,g(s))|\, ds \le C \int_0^t |f(s) - g(s)|\, ds, \quad t \in [a,T].$$

Therefore, Remark 101 leads us to conclude that $f - g \equiv 0$, as we wanted to prove. In other words, we have stated the uniqueness of equation (7.2) in the case that $\sigma \equiv 0$.

Now, we establish the classical result for the existence of a unique solution to equation (7.2). More general results can be found in [96, 180].

In the following proof, as in Proposition 54.*ii*), we use

$$E\left(\left|\int_a^t \sigma(s,X_s)dW_s\right|^2\right) = E\left(\int_a^t |\sigma(s,X_s)|^2 ds\right), \quad t \in [a,T],$$

whenever $\int_a^t E\left(|\sigma(s,X_s)|^2\right) ds < \infty$. The proof of this fact is left to the reader as an exercise.

Theorem 86 *Let $b : \Omega \times [a,T] \times \mathbb{R}^d \to \mathbb{R}^d$ and $\sigma : \Omega \times [a,T] \times \mathbb{R}^d \to \mathcal{L}(\mathbb{R}^m, \mathbb{R}^d)$ be two measurable random fields satisfying:*

i) (Local Lipschitz condition) For each $N > 0$, there exists a constant $K_N > 0$ such that, for $|x|$, $|y| \le N$, we have

$$|b(t,x) - b(t,y)| + |\sigma(t,x) - \sigma(t,y)| \le K_N|x-y|, \quad t \in [a,T]. \tag{7.6}$$

ii) (Linear growth) There is a constant $K > 0$ such that

$$|b(t,x)|^2 + |\sigma(t,x)|^2 \le K\left(1 + |x|^2\right), \quad x \in \mathbb{R} \text{ and } t \in [a,T].$$

iii) For $x \in \mathbb{R}$, $s \mapsto b(s,x)$ and $s \mapsto \sigma(s,x)$ are two vector-valued and $\mathcal{F}_t$-adapted processes.

Then, the stochastic differential equation (7.2) has a unique solution.

Remarks 15 *i) The uniqueness of the solution to the SDE (7.2) means that if we have two solutions according to Definition 48, then they are indistinguishable.*

ii) The unique solution to (7.2) belongs to $L^2(\Omega \times [a,T])$ if $\xi \in L^2(\Omega; \mathbb{R}^d)$ and K_N in (7.6) is independent of N.

iii) We will see in the proof of this result that both Conditions i) and ii) are used to construct the solution to equation (7.2) and that we only need Condition i) to establish the uniqueness of the solution of (7.2). Later on, we give an example where the solution blows up according to an $\mathcal{F}_t$-stopping time if Hypothesis ii) is omitted. Moreover, observe that the coefficients of SDE (7.4) do not satisfy Assumptions i) and ii) in general. However, the one-dimensional linear stochastic differential equation (7.4) has a unique solution due to Itô's formula, as we have already shown in Examples 2.

Proof. Note that if X is a continuous, $\mathbb{R}^d$-valued and $\mathcal{F}_t$-adapted process, then for $\pi = \{a = t_0 < t_1 < \ldots < t_n = T\}$, we get that $X_t = \lim_{|\pi|\to 0} \sum_{i=0}^{n-1} X_{t_i} 1_{]t_i,t_{i+1}]}(t)$ for all $t \in [a,T]$, w.p.1. In consequence, now it is easy to prove that $b(\cdot, X.)$ and $\sigma(\cdot, X.)$ are also vector-valued and $\mathcal{F}_t$-adapted processes.

Now, the proof is divided into several steps.

Step 1. Here we assume that the initial condition ξ belongs to $L^2(\Omega)$, and that b and σ are Lipschitz functions on $\mathbb{R}^d$, uniformly in $t \in [a,T]$. It means, for all $x, y \in \mathbb{R}^d$ and $\omega \in \Omega$, inequality (7.6) holds when we change K_N by a constant $K > 0$ independent of N.

We first assume that X and Y are two solutions in $L^2(\Omega \times [0,T])$ to the stochastic differential equation (7.2). Then, Hölder inequality and Proposition 54 imply

$$E\left(|X_t - Y_t|^2\right) \leq 4K^2\left[(T-a)+1\right]\int_a^t E\left(|X_s - Y_s|^2\right) ds, \quad t \in [a,T].$$

Thus, Gronwall's lemma (i.e., Proposition 121) gives that $X = Y$ in the space $L^2(\Omega \times [a,T])$. In this way, we have obtained that the SDE (7.2) has at most one solution.

On the other hand, to show the existence of a solution to (7.2), we only need to proceed as in the deterministic case (i.e., $\sigma \equiv 0$ and ξ is independent of $\omega \in \Omega$). That is, we define inductively the processes

$$X^{(0)} \equiv \xi$$

and, for $n \geq 1$,

$$X_t^{(n)} = \xi + \int_a^t b(s, X_s^{(n-1)})ds + \int_a^t \sigma(s, X_s^{(n-1)})dW_s, \quad t \in [a,T].$$

Note that these processes are square-integrable due to the linear growth of the coefficients. Moreover, through induction on n, the fact that $\xi \in L^2(\Omega)$, Theorem 100 and the Lipschitz condition on the coefficients, we can show that there exists a constant $C > 0$ independent of n such that the inequality

$$E\left(\sup_{r\in[a,t]}\left(X_r^{(n)} - X_r^{(n-1)}\right)^2\right) \leq C\int_a^t E\left(\left|X_s^{(n-1)} - X_s^{(n-2)}\right|^2\right) ds, \quad t \in [a,T],$$

holds for any $n \in \mathbb{N}$. Consequently, we have, for $t \in [a,T]$,

$$\begin{aligned} &E\left(\sup_{r\in[a,t]}\left|X_r^{(n)} - X_r^{(n-1)}\right|^2\right) \\ &\quad \leq C^{n-1}\int_a^t \frac{(t-s)^{n-2}}{(n-2)!} E\left|X_s^{(1)} - X_s^{(0)}\right|^2 ds \leq LC^{n-1}\int_a^t \frac{(t-s)^{n-2}}{(n-2)!} ds, \end{aligned}$$

with

$$L = E\left(\sup_{r\in[a,t]}\left|X_r^{(1)} - X_r^{(0)}\right|^2\right) \leq 4K^2\left[(T-a)+1\right]\int_a^T (1 + E\left(|\xi|^2\right) ds.$$

Therefore, we are able to establish that, for $n < m$,

$$\begin{aligned}
&\left\| \sup_{t\in[a,T]} \left(X_t^{(m)} - X_t^{(n)}\right) \right\|_{L^2(\Omega)} \\
&= \left\| \sup_{t\in[a,T]} \left[\sum_{k=n}^{m-1} \left(X_t^{(k+1)} - X_t^{(k)}\right) \right] \right\|_{L^2(\Omega)} \leq \sum_{k=n}^{m-1} \left\| \sup_{t\in[a,T]} \left(X_t^{(k+1)} - X_t^{(k)}\right) \right\|_{L^2(\Omega)} \\
&\leq L \sum_{k=n}^{m-1} \frac{C^{k/2}(T-a)^{k/2}}{\sqrt{k!}} \to 0, \quad \text{as } n, m \to \infty.
\end{aligned}$$

In other words, we have shown that $\{X^{(n)} : n \in \mathbb{N}\}$ is a Cauchy sequence in $L^2(\Omega \times [a,T])$. In order to see that the limit X of this sequence is a continuous, $\mathbb{R}^d$-valued and $\mathcal{F}_t$-adapted process, we utilize the inequality

$$\sum_{n=1}^{\infty} P\left(\left[\sup_{t\in[a,T]} \left| X_t^{(n+1)} - X_t^{(n)} \right| > n^{-2} \right] \right) \leq L \sum_{n=1}^{\infty} n^4 \frac{C(T-a)^n}{n!} < \infty,$$

together with the Borel-Cantelly lemma, to have that

$$X_t^{(0)} + \sum_{k=0}^{n} \left(X_t^{(k)} - X_t^{(k-1)}\right) = X_t^{(n)}$$

converges uniformly on $t \in [a,T]$ to X. That is, we have also proven that the process X is a continuous, $\mathcal{F}_t$-adapted and $\mathbb{R}^d$-valued process. Moreover, the Lipschitz condition of the coefficients implies that this $\mathbb{R}^d$-valued process is a solution to the stochastic differential equation (7.2).

Step 2. Now, the initial condition ξ is a $\mathcal{F}_a$-measurable random variable with values in $\mathbb{R}^d$, and the coefficients b and σ are still Lipschitz functions on $\mathbb{R}^d$, uniformly in $t \in [a,T]$.

We first deal with the uniqueness for the solution to (7.2) again. Towards this end, consider two solutions X and Y of the SDE (7.2) and, for $n \in \mathbb{N}$, introduce the function $\varphi_n : [a,T] \to [0,1]$ defined by

$$\varphi_n(t) = \begin{cases} 1, & \text{if } \left(\sup_{a\leq s\leq t} |X_s|\right) \vee \left(\sup_{a\leq s\leq t} |Y_s|\right) \leq n, \\ 0, & \text{otherwise.} \end{cases}$$

It is not difficult to see that φ_n is an F_t-adapted process since X and Y are continuous, F_t-adapted and $\mathbb{R}^d$-valued stochastic processes. Also, we have $\varphi_n(t) = \varphi_n(t)\varphi_n(s)$, for $a \leq s \leq t \leq T$. Then, from the local property of the Itô's integral (i.e., Proposition 55) and proceeding as in the uniqueness established in Step 1, we have that there is a constant $C > 0$ satisfying

$$\begin{aligned}
E\left(\varphi_n(t)\,|X_t - Y_t|^2\right) &\leq 2E\left(\varphi_n(t) \left| \int_a^t \varphi_n(s)\,(b(s,X_s) - b(s,Y_s))\,ds \right|^2 \right. \\
&\qquad \left. +2\varphi_n(t) \left| \int_a^t \varphi_n(s)\,(\sigma(s,X_s) - \sigma(s,Y_s))\,dW_s \right|^2 \right) \\
&\leq C \int_a^t E\left(\varphi_n(s)\,|X_s - Y_s|^2\right) ds, \quad t \in [a,T].
\end{aligned}$$

Note that $t \mapsto \varphi_n(t)\,|X_t - Y_t|^2$ belongs to $L^1(\Omega \times [a,T])$ because of the definition of the function φ_n. Thus, the Gronwall's lemma allows us to show

$$E\left(\varphi_n(t)\,|X_t - Y_t|^2\right) = 0, \quad t \in [a,T].$$

Hence, we get that $\varphi_n(t)(X_t - Y_t) = 0$ with probability 1 for each $n \in \mathbb{N}$ and $t \in [a,T]$. Now, we use the continuity of the paths of X and Y to conclude that X is a modification of Y. In this way, the continuity of X and Y, together with Proposition 1, yields that these two processes are indistinguishable. In consequence, we obtain the uniqueness of the solution for the SDE (7.2).

Now, we consider the existence of the solution. For $n \in \mathbb{N}$, we use the convention $\xi^{(n)} = \xi 1_{[|\xi|\le n]}$. It is clear that $\xi^{(n)} \in L^2(\Omega)$ and we denote by $Y^{(n)} \in L^2(\Omega \times [a,T])$ the continuous and $\mathcal{F}_t$-adapted solution of the SDE

$$Y_t^{(n)} = \xi^{(n)} + \int_a^t b(s, Y_s^{(n)})ds + \int_a^t \sigma(s, Y_s^{(n)})dW_s, \quad t \in [a,T].$$

Remember that $Y^{(n)}$ is well-defined because of Step 1. We proceed as in Step 1, and utilize that $\left(\xi^{(n)} - \xi^{(\tilde{n})}\right) 1_{[|\xi|\le n]} = 0$ for $\tilde{n} > n$, Proposition 58 and Theorem 100 to obtain that, for $t \in [a,T]$,

$$\begin{aligned} E\left(\sup_{r\in[a,t]}\left|Y_r^{(n)} - Y_r^{(\tilde{n})}\right|^2 1_{[|\xi|\le n]}\right) &\le C\int_a^t E\left(\left|Y_s^{(n)} - Y_s^{(\tilde{n})}\right|^2 1_{[|\xi|\le n]}\right) ds \\ &\le C\int_a^t E\left(\sup_{r\in[a,s]}\left|Y_r^{(n)} - Y_r^{(\tilde{n})}\right|^2 1_{[|\xi|\le n]}\right) ds. \end{aligned}$$

Iterating this inequality, we also get

$$\begin{aligned} &\int_a^t E\left(\sup_{r\in[a,s]}\left|Y_r^{(n)} - Y_r^{(\tilde{n})}\right|^2 1_{[|\xi|\le n]}\right) ds \\ &\quad\le C\int_a^t \left[\int_0^s E\left(\sup_{r\in[a,u]}\left|Y_r^{(n)} - Y_r^{(\tilde{n})}\right|^2 1_{[|\xi|\le n]}\right) du\right] ds. \end{aligned}$$

Thus, these two inequalities, together with Gronwall's lemma, yield

$$E\left(\sup_{r\in[a,t]}\left|Y_r^{(n)} - Y_r^{(\tilde{n})}\right|^2 1_{[|\xi|\le n]}\right) = 0, \quad t \in [a,T],$$

and

$$\int_a^T E\left(\sup_{r\in[a,s]}\left|Y_r^{(n)} - Y_r^{(\tilde{n})}\right|^2 1_{[|\xi|\le n]}\right) ds = 0,$$

which imply

$$\begin{aligned} &P\left[\sup_{r\in[a,t]}\left|Y_r^{(n)} - Y_r^{(\tilde{n})}\right|^2 > 0\right] \\ &\quad + P\left[\int_0^T \left(\sup_{r\in[a,s]}\left|Y_r^{(n)} - Y_r^{(\tilde{n})}\right|^2\right) ds > 0\right] \le 2P\left[|\xi| > n\right], \quad t \in [a,T]. \end{aligned}$$

In this way, we have proven that the sequence $\{Y^{(n)} : n \in \mathbb{N}\}$ converges uniformly in probability to an $\mathbb{R}^d$-valued, continuous and $\mathcal{F}_t$-adapted process Y such that the sequence

$$\int_a^T \left|Y_s^{(n)} - Y_s\right|^2 ds$$

also converges in probability to zero, as $n \to \infty$. In order to finish this step, we notice that Y is a solution of (7.2) due to inequality (3.17).

Step 3. Finally, we prove the general case.

We observe that the proof of the uniqueness of the solution to the SDE (7.2) in Step 2 also gives the one for the solution of (7.2) under the assumptions of this step.

Now, we figure out a solution of the SDE in consideration. To do so, we use Spivak [200] (Exercise 2-26) to choose a function $\varphi : \mathbb{R}^d \to [0,1]$ in $C_b^\infty(\mathbb{R}^d, \mathbb{R})$ such that

$$\varphi(x) = \begin{cases} 1, & \text{if } |x| \le 1, \\ 0, & \text{if } |x| \ge 2. \end{cases}$$

Furthermore, for $n \in \mathbb{N}$, set $\varphi_n = \varphi(x/n)x$ and

$$X_t^{(n)} = \xi + \int_a^t b(s, \varphi_n(X_s^{(n)}))ds + \int_a^t \sigma(s, \varphi_n(X_s^{(n)}))dW_s, \quad t \in [a,T].$$

Note that this equation has a unique continuous and $\mathcal{F}_t$-adapted solution since Step 2. Introduce

$$\lambda_n = \inf\{t \ge a : \left|X_t^{(n)}\right| \ge n\} \quad \text{and} \quad \tilde{\lambda}_n = \inf\{t \ge a : \left|X_t^{(n+1)}\right| \ge n\},$$

which are $\mathcal{F}_t$-stopping times (see Lemma 34). Then, we are able to apply Proposition 60 to establish

$$\begin{aligned} X_{t\wedge\lambda_n\wedge\tilde{\lambda}_n}^{(n)} &= \xi + \int_a^{t\wedge\lambda_n\wedge\tilde{\lambda}_n} b(s, \varphi_n(X_s^{(n)}))ds + \int_a^{t\wedge\lambda_n\wedge\tilde{\lambda}_n} \sigma(\varphi_n(X_s^{(n)}))dW_s \\ &= \xi + \int_a^t b(s, \varphi_n(X_{s\wedge\lambda_n\wedge\tilde{\lambda}_n}^{(n)}))1_{[0,\lambda_n\wedge\tilde{\lambda}_n]}(s)ds \\ &\quad + \int_a^t \sigma(s, \varphi_n(X_{s\wedge\lambda_n\wedge\tilde{\lambda}_n}^{(n)}))1_{[0,\lambda_n\wedge\tilde{\lambda}_n]}(s)dW_s, \quad t \in [a,T], \end{aligned}$$

and

$$\begin{aligned} X_{t\wedge\lambda_n\wedge\tilde{\lambda}_n}^{(n+1)} &= \xi + \int_a^{t\wedge\lambda_n\wedge\tilde{\lambda}_n} b(s, \varphi_{n+1}(X_s^{(n+1)}))ds + \int_a^{t\wedge\wedge\lambda_n\tilde{\lambda}_n} \sigma(\varphi_{n+1}(X_s^{(n+1)}))dW_s \\ &= \xi + \int_a^t b(s, \varphi_{n+1}(X_s^{(n+1)}))1_{[0,\lambda_n\wedge\tilde{\lambda}_n]}(s)ds \\ &\quad + \int_a^t \sigma(s, \varphi_{n+1}(X_s^{(n+1)}))1_{[0,\lambda_n\wedge\tilde{\lambda}_n]}(s)dW_s \\ &= \xi + \int_a^t b(s, \varphi_n(X_{s\wedge\lambda_n\wedge\tilde{\lambda}_n}^{(n+1)}))1_{[0,\lambda_n\wedge\tilde{\lambda}_n]}(s)ds \\ &\quad + \int_a^t \sigma(s, \varphi_n(X_{s\wedge\lambda_n\wedge\tilde{\lambda}_n}^{(n+1)}))1_{[0,\lambda_n\wedge\tilde{\lambda}_n]}(s)dW_s, \quad t \in [a,T]. \end{aligned}$$

Consequently, the last two stochastic differential equations and Step 2 allow us to conclude

$$X_t^{(n)} = X_t^{(n+1)}, \quad \text{for all } t \in [a, \lambda_n \wedge \tilde{\lambda}_n], \text{ w.p.1.}$$

In particular, we also get $\lambda_n = \tilde{\lambda}_n$ and $\lambda_n \leq \lambda_{n+1}$. Therefore, by Proposition 55, the process $X_t = X_t^{(n)}$, for $0 \leq t \leq \lambda_n$, satisfies the SDE

$$X_t = \xi + \int_a^t b(s, X_s)ds + \int_a^t \sigma(s, X_s)dW_s.$$

Hence, we only need to show that $\lim_{n\to\infty} \lambda_n > T$ with probability 1. Towards this end, we use that ξ is $\mathcal{F}_a$-measurable to get that $Y_t^{(n)} := X_t^{(n)} e^{-|\xi|}$ is such that

$$\begin{aligned} Y_t^{(n)} &= \xi e^{-|\xi|} + \int_a^t b\left(s, \varphi_n(X_s^{(n)})\right) e^{-|\xi|}ds + \int_a^t \sigma\left(s, \varphi_n(X_s^{(n)})\right) e^{-|\xi|}dW_s \\ &= \xi e^{-|\xi|} + \int_a^t b\left(s, \varphi_n(Y_s^{(n)}e^{|\xi|})\right) e^{-|\xi|}ds + \int_a^t \sigma\left(s, \varphi_n(Y_s^{(n)}e^{|\xi|})\right) e^{-|\xi|}dW_s, \end{aligned}$$

for $t \in [a, T]$. Note that the coefficients of this equation are Lipschitz with linear growth. Moreover, the constant of the linear growth is K, which is the one in the Hypothesis of this result and therefore is independent of n. Indeed, for all $x \in \mathbb{R}^d$ and $n \in \mathbb{N}$,

$$|b(s, \varphi_n(xe^{|\xi|})e^{-|\xi|}|^2 + |\sigma(s, xe^{|\xi|})e^{-|\xi|}|^2 \leq Ke^{-2|\xi|}\left(1 + |\varphi_n(xe^{|\xi|})|^2\right) \leq K\left(1 + |x|^2\right).$$

In consequence, proceeding as in Step 1 (i.e., using the Gronwall's lemma and the linear growth), and applying that $\exp(-|\varphi|)|\varphi| \leq 1$, we can state that there is a constant $C_{a,T}$ independent of n such that $E\left(\sup_{t\in[a,T]} \left|Y_t^{(n)}\right|^2\right) \leq C_{a,T} < \infty$.

Finally, Chebyshev's inequality yields that, for $\delta > 0$,

$$\begin{aligned} P\left([\lambda_n \leq T]\right) &= P\left(\left[\sup_{t\in[a,T]} \left|X_t^{(n)}\right| \geq n\right]\right) = P\left(\left[\sup_{t\in[a,T]} \left|Y_t^{(n)}\right| \geq ne^{-|\xi|}\right]\right) \\ &\leq P\left(\left[\sup_{t\in[a,T]} \left|Y_t^{(n)}\right| \geq n\delta\right]\right) + P\left(\left[e^{-|\xi|} \leq \delta\right]\right) \\ &\leq \frac{C_{a,T}}{n^2\delta^2} + P\left([|\xi| \geq -\log(\delta)]\right). \end{aligned}$$

Thus, the result is true due to $\lim_{\delta\to 0}\lim_{n\to\infty} P([\lambda_n \leq T]) = 0$, $\{\lambda_n : n \in \mathbb{N}\}$ being a non-decreasing sequence of stopping times and the Fatou's lemma. □

In the last proof, we use the linear growth of the coefficients to show that the SDE (7.2) has a unique solution X such that $sup_{t\in[a,T]}|X_t| < \infty$ with probability 1. A natural question is the following: What happens if the coefficients do not satisfy linear growth? To answer this question, we use the Osgood criterion [171] for ordinary differential equations. Namely, consider the equation

$$v(t) = c + \int_a^t b(v(s))ds, \quad t \geq a. \tag{7.7}$$

Here, $c \in \mathbb{R}$ and $b : \mathbb{R} \to (0, \infty)$ is a continuous function. Note that if a solution v exists on the interval $[a, a + \varepsilon]$, for some $\varepsilon > 0$, then v has a positive derivative. Therefore, the existence of a unique local solution to equation (7.7) that could exploit in finite time follows from the Osgood criterion [171]. Namely, (7.7) implies that for $t > a$,

$$t - a = \int_a^t \frac{v'(s)}{b(v(s))}ds = \int_c^{v(t)} \frac{ds}{b(s)},$$

where last equality is a consequence of the change of variables formula. In consequence, since b is positive on $\mathbb{R}$, the function $F(x) = \int_c^x \frac{ds}{b(s)}$, $x \in \mathbb{R}$, is increasing. Moreover, $v(t) = F^{-1}(t-a)$. Hence, the solution v of equation (7.7) exploits in finite time if and only if $F(\infty) < \infty$. In this case, the time of the explosion is $a + \int_c^\infty \frac{ds}{b(s)}$. Indeed, suppose that t_0 is the first time that $v(t_0) = \infty$. Then, the fact that v is increasing gives

$$\int_c^\infty \frac{ds}{b(s)} = \lim_{t \uparrow t_0} \int_c^{v(t)} \frac{ds}{b(s)} = \lim_{t \uparrow t_0} (t-a) = t_0 - a.$$

Reciprocally, for $t_0 = a + \int_c^\infty \frac{ds}{b(s)}$, we have that $v(t_0) = \infty$, otherwise

$$t_0 = a + \int_c^{v(t_0)} \frac{ds}{b(s)} < a + \int_c^\infty \frac{ds}{b(s)} = t_0,$$

which is a contradiction.

Observe that if b satisfies the linear growth, then there exists a constant $C > 0$ such that

$$\infty = \int_c^\infty \frac{ds}{C(1+|s|)} \leq \int_0^\infty \frac{ds}{b(s)}.$$

Resuming, equation (7.7) has a unique continuous solution that could exploit in finite time.

On the other hand, now suppose that $b(c) = 0$, $b(t) > 0$ and $\int_c^t \frac{ds}{b(s)} < \infty$ for any $t > c$. Remember that we are assuming that b is a continuous function. Then, in this case, we still have that $F^{-1}(t-a)$ is a solution of (7.7) such that $v(t) > c$ on the interval $(a, a+\varepsilon)$ for some $\varepsilon > 0$. But, for $\eta > 0$, the function

$$v_\eta(s) = \begin{cases} c, & \text{if } a \leq s \leq a+\eta, \\ F^{-1}(t-(a+\eta)), & \text{if } t \geq a+\eta, \end{cases} \tag{7.8}$$

and $v \equiv c$ are also solutions of equation (7.7).

In the book by Agarwal and Lakshmikantham [3], the reader can find results on the uniqueness and nonuniqueness of ordinary differential equation. For instance, the following result is stated in this book.

Proposition 122 *Let $b : [0, \infty) \to \mathbb{R}$ be a continuous and non-decreasing function such that $b(0) = 0$, $b(t) > 0$, for $t > 0$, and $\int_0^\cdot \frac{ds}{b(s)} = \infty$. Also, let $\varphi : [0, T] \to \mathbb{R}_+ \cup \{0\}$ be a continuous function such that $\varphi(t) \leq \int_0^t b(\varphi(s))ds$, $t \in [0, T]$. Then, $\varphi = 0$ on the interval $[0, T]$.*

Unlike (7.8) (i.e., the case $\int_c^\cdot \frac{ds}{b(s)} < \infty$), equation (7.7) has a unique solution if $a = c = 0$ and b is as in the last proposition.

As example, for $c > 0$, we first consider the equation

$$\varphi(t) = c + \int_0^t \varphi(s)^2 ds, \quad t \geq 0.$$

In this case, $F(x) = \int_c^x \frac{ds}{s^2} = c^{-1} - x^{-1}$, $x > c$. Hence, the unique solution of the last equation is $\varphi(t) = 1/(c^{-1} - t)$, which blows up at $t = 1/c$.

The second equation is

$$\varphi(t) = \int_0^t \varphi(s)^{1/2} ds, \quad t \geq 0.$$

This equation has many solutions according to (7.8). These solutions do not explode in finite time because $s \mapsto s^{1/2}$ has linear growth on $[0, \infty)$.

We now return to the stochastic case. That is, we give a stochastic differential equation of the form (7.2) that has a unique solution that explodes in finite time. Towards this end, consider the process

$$Y_t = \left(a^{-1} - W_t\right)^{-1}, \quad 0 \leq t \leq \lambda_a.$$

Here, $W = \{W_t : t \geq 0\}$ is a one-dimensional Brownian motion, $a > 0$ and λ_a is the stopping time $\inf\{t > 0 : W_t > a^{-1}\}$. Then, Itô's formula (3.21) with $f(x) = x^{-1}$ implies

$$Y_t = a + \int_0^t Y_s^2 dW_s + \int_0^t Y_s^3 ds, \quad 0 \leq t < \lambda_a. \tag{7.9}$$

In fact, as we have done in several occasions, to prove that this equality is satisfied, we choose a function $\varphi : \mathbb{R} \to [0,1]$ in $C_b^\infty(\mathbb{R})$ such that

$$\varphi(x) = \begin{cases} 1, & \text{if } |x| \leq 1, \\ 0, & \text{if } |x| \geq 2, \end{cases}$$

and, for $n \in \mathbb{N}$, we use the notation $\varphi_n(x) = \varphi(x/n)x$ and introduce the stopping time $\lambda_{a,n} = \inf\{t > 0 : (a^{-1} - W_t)^{-1} > n\}$. Then, on the stochastic interval $[0, \lambda_{a,n}]$, the process $\varphi_n(Y)$ is equal to Y and satisfies the stochastic differential equation (7.9) due to Proposition 55 (i.e., the local property of the Itô's integral). In this way, we have also proven that the process Y is the unique solution to the SDE (7.9) that blow up in finite time with explosion time λ_a. This fact is explained by the following result, whose proof can be found in the book by Mckean [145] (see also Protter [180]).

Proposition 123 *Let b and σ be two function in $C^1(\mathbb{R})$ and $c \in \mathbb{R}$. Then, there exists a stopping time $0 \leq \lambda \leq \infty$ such that the SDE*

$$X_t = c + \int_0^t b(X_s)ds + \int_0^t \sigma(X_s)dW_s, \quad t < \lambda,$$

has a unique solution such that X_λ is equal to either ∞, or $-\infty$ if $\lambda < \infty$.

Note that if b, or σ does not have a bounded derivative, then we do not have that both coefficients satisfy the linear growth condition.

An important tool of stochastic calculus to know if the last stopping time λ is finite with positive probability is the so-called Feller test (see, for example, Karatzas and Shreve [96]). The reader can consult de Pablo et al. [36] for applications of SDEs with blow up. In the case that b is non-decreasing and positive, and $\sigma \equiv 1$, the Feller test is equivalent to Osgood criterion [171] as it is established in León and Villa [125]. That is, in this case, the solution explodes in finite time if and only if $\int_c^\infty \frac{ds}{b(s)} < \infty$.

Unfortunately, the distribution of the explosion time of a certain SDE is not easy to calculate. One way to do it is by using linear second-order ordinary differential equations. Indeed, Feller [49] has pointed out the Laplace transformation of this distribution is a bounded solution to some related ordinary differential equations. A generalization of this result is given in León et al. [121] (see also León et al. [120]). Moreover, Dávila et al. [35] have analyzed some numerical schemes to approximate the time of explosion of SDEs. Extension of the Feller test and Osgood criterion for SDEs have been studied by León et al. [121, 125, 119] (see also Foondun and Nualart [53]).

On the other hand, in Theorem 86, we have found conditions that guarantee the equation (7.2) has a unique solution X. But, in practice, it is impossible to know X explicitly in any situation. Thus, the analysis of equation (7.2) also requires to apply techniques to approximate the solution X, which are called numerical methods. That is, it is necessary to study and use discrete-time schemes to estimate X. The achievement of a numerical method depends on the computer's implementation of it and the rate of increase of the error $|X_t - x(t,h)|$, where $x(t,h)$ represents the approximation of X with the step side of h. For details, the reader can consult the books [68, 76, 99, 100]. Reference [76] considers Itô versus Stratonovich calculus.

7.2 Anticipating stochastic differential equations

Throughout this section, $(\Omega, \mathcal{F}, P)$ and $W = \{W_t : t \in [0,T]\}$ represent the canonical Wiener space and the canonical Wiener process, which are introduced in Section 1.2.1. Also, we deal with the space $\mathbb{D}^{1,\infty}$ defined as the restriction of $\mathbb{D}^{1,2}$ to those random variables F for which the norm

$$\|F\|_{1,\infty} = \|F\|_{L^\infty(\Omega)} + \||DF|_{L^2([0,T])}\|_{L^\infty(\Omega)} < \infty.$$

Buckdahn [24, 25] has shown that any elements in $\mathbb{D}^{1,\infty}$ can be approximated in $\mathbb{D}^{1,2}$ by sequence of smooth functionals that are uniformly bounded in $\mathbb{D}^{1,\infty}$.

Now, we consider the stochastic differential equation

$$X_t = \eta + \int_0^t b_s X_s ds + \int_0^t \sigma_s X_s \delta_1 W_s, \quad t \in [0,T]. \tag{7.10}$$

Here, $\sigma \in \mathbb{D}^{1,\infty}(L^2([0,T]))$ (see (4.1)), $b \in L^2([0,T]; L^\infty(\Omega))$ and $\eta \in L^\infty(\Omega)$. Note that we are not assuming that neither η is $\mathcal{F}_0$-measurable, nor b and σ are $\mathcal{F}_t$-adapted processes. In consequence, we are not dealing with an Itô's type equation because the initial condition may depend on all the information that generates the process W as well as σ and b. In this case, the stochastic integral is the divergence operator with respect to the Canonical Wiener process W. So, we have the following:

Definition 49 *A measurable process $X = \{X_t : t \in [0,T]\}$ is a solution to equation* (7.10) *if*

(i) $b.X.$ belongs to $L^1([0,T])$ with probability 1.

(ii) The process $\sigma.X.1_{[0,t]}$ is in $Dom\,\delta_1$, for all $t \in [0,T]$.

(iii) Equality (7.10) *holds with probability 1 for each $t \in [0,T]$.*

Note that the set of probability 1 in Statement *iii)* depends on $t \in [0,T]$ and remember that δ_1 is defined in (4.2).

In order to give the close expression for the unique solution to equation (7.10), we need to consider two families of transformation on Ω. These transformations are solutions of the equations

$$T_t\omega = \omega + \int_0^{t\wedge\cdot} \sigma_s(T_s\omega) ds, \quad \omega \in \Omega \text{ and } t \in [0,T], \tag{7.11}$$

and

$$A_{s,t}\omega = \omega - \int_{s\wedge\cdot}^{t\wedge\cdot} \sigma_r(A_{r,t}\omega) dr, \quad \omega \in \Omega \text{ and } 0 \le s \le t \le T. \tag{7.12}$$

Note that $T_t\omega$ and $A_{s,t}\omega$ belong to Ω, for $\omega \in \Omega$, since the elements of this space are continuous functions in $C_0([0,T])$. (see Section 1.2.1).

In Buckdahn [24, 25], it is stated the following two results.

Proposition 124 *Assume that σ belongs to the space $L^2([0,T];\mathbb{D}^{1,\infty})$. Then, there exists a unique family $\{T_t : t \in [0,T]\}$ of absolutely continuous transformations $T_t : \Omega \to \Omega$ that is the solution to equation (7.11) such that the process $\sigma(T) = \{\sigma_t(T_t), t \in [0,T]\}$ is in $L^2([0,T];\mathbb{D}^{1,\infty})$, and, for each $t \in [0,T]$, the transformation T_t is invertible whose inverse A_t has the density*

$$
\begin{aligned}
\mathcal{L}_t &= \frac{dP \circ [A_t]^{-1}}{dP} = \exp\left\{ -\int_0^t \sigma_s(T_s)\delta_2 W_s - \frac{1}{2}\int_0^t \sigma_s(T_s)^2 ds \right. \\
&\quad \left. - \int_0^t \int_0^s (D_r\sigma_s)(T_s) D_s[\sigma_r(T_r)] dr\, ds \right\}.
\end{aligned}
$$

The second result in [24, 25] is the following:

Proposition 125 *Suppose that $\sigma \in L^2([0,T];\mathbb{D}^{1,\infty})$. Then, equation (7.12) has a unique solution $\{A_{s,t} = T_s \circ A_{0.t} : 0 \le s \le t \le T\}$ in the space of absolutely continuous transformations. Moreover, the inverse transformation $A_{s,t}^{-1} = T_t \circ A_{0,s}$ has the density*

$$
\begin{aligned}
L_{s,t} &= \exp\left\{ \int_s^t \sigma_r(A_{r,t})\delta_2 W_r - \frac{1}{2}\int_s^t \sigma_r(A_{r,t})^2 dr \right. \\
&\quad \left. - \int_s^t \int_r^t (D_u\sigma_r)(A_{r,t}) D_r[\sigma_u(A_{u,t})] du dr \right\}, \quad 0 \le s \le t \le T.
\end{aligned}
$$

As consequence of Propositions 124 and 125, we get that

$$
T_t(A_{0,u}\omega) = A_{t,u}\omega \quad \text{and} \quad A_{s,t}(T_t\omega) = T_s\omega,
$$

for all $\omega \in \Omega$ and $0 \le s \le t \le u \le T$. In particular, if $u = t$ and $s = 0$, we obtain

$$
T_t(A_{0,t}\omega) = A_{0,t}(T_t\omega) = \omega, \quad \omega \in \Omega.
$$

That is, $A_t = A_{0,t}$ is the inverse of T_t on Ω. Moreover, we also have Propositions 124 and 125 imply

$$
E\left(F(A_t)\right) = E\left(F\mathcal{L}_t\right) \quad \text{and} \quad E\left(F\right) = E\left(F(A_{s,t})L_{s,t}\right), \tag{7.13}
$$

where $t \in [0,T]$ and $F \in L^\infty(\Omega)$.

Now, we use the convention $L_t = L_{0,t}$, $t \in [0,T]$. The following result related to the existence and uniqueness of the solution of equation (7.10) is stated in Buckdahn [24, 25]. So, we only sketch its proof.

Theorem 87 *Let $\sigma \in L^2([0,T];\mathbb{D}^{1,\infty})$, $b \in L^1([0,T];L^\infty(\Omega))$ and $\eta \in L^\infty(\Omega)$. Then, the process X defined by*

$$
X_t = \eta(A_t)\exp\left\{ \int_0^t b_s(T_sA_t) ds \right\} L_t, \quad t \in [0,T], \tag{7.14}
$$

belongs to $L^1(\Omega \times [0,T])$ and is a solution to equation (7.10). Conversely, assume that σ, $b \in L^\infty(\Omega \times [0,T])$ and $D\sigma \in L^\infty(\Omega \times [0,T]^2)$. Also, assume that $Y \in L^1(\Omega \times [0,T])$ is a process such that $\sigma Y 1_{[0,t]} \in Dom\,\delta_1$, for $t \in [0,T]$, and it satisfies equation (7.10). Then, Y is the process given by (7.14).

Remark 102 *If the process σ is also adapted to the filtration generated by $W = \{W_t : t \in [0,T]\}$, it is easy to see that the transformation A_t takes the form $A_t\omega = \omega - \int_0^{t\wedge\cdot} \sigma_s(\omega)ds$. Therefore,*

$$L_t = \exp\left\{\int_0^t \sigma_s dW_s - \frac{1}{2}\int_0^1 \sigma_s^2 ds\right\}.$$

Furthermore, if the process b is also adapted, the process

$$X_t = \eta(A_t)\exp\left\{\int_0^t \sigma_s dW_s - \frac{1}{2}\int_0^t \sigma_s^2 ds + \int_0^t b_s ds\right\}, \quad t \in [0,T],$$

is the solution of equation (7.10) in this case. In consequence, for a deterministic η, the solution coincides with the classical result for linear stochastic differential equations of Itô's type. This last fact can be also proven via the Itô's formula (see Example 2.ii)). Note that, in the general case, we cannot use Itô's formula to show that Theorem 87 holds.

Proof. Fix $t \in [0,T]$.

We first consider the existence of a solution to equation (7.10). To do so, let X be the process given by (7.14) and G in $\mathcal{S}$. Then, it is not difficult to show that $X \in L^1(\Omega\times[0,T])$, and $X_t \in L^1(\Omega)$. Therefore, (7.14) and (7.13) imply

$$\begin{aligned} E\left(\int_0^t \sigma_s X_s D_s G ds\right) &= E\left[\int_0^t \sigma_s\eta(A_s)L_s\exp\left\{\int_0^s b_r(T_rA_s)dr\right\}D_sGds\right] \\ &= E\left[\int_0^t \sigma_s(T_s)\eta\exp\left\{\int_0^s b_r(T_r)dr\right\}(D_sG)(T_s)ds\right]. \end{aligned}$$

Hence, using the fact that $\frac{d}{ds}G(T_s) = \sigma_s(T_s)(D_sG)(T_s)$ and the integration by parts formula, we are able to write

$$\begin{aligned} E\left(\int_0^t \sigma_s X_s D_s G ds\right) &= E\left(\int_0^t \eta\exp\left\{\int_0^s b_r(T_r)dr\right\}\frac{d}{ds}G(T_s)ds\right) \\ &= E\left(\eta\exp\left\{\int_0^t b_s(T_s)ds\right\}G(T_t) - \eta G\right) \\ &\quad -E\left(\int_0^t \eta b_s(T_s)\exp\left\{\int_0^s b_r(T_r)dr\right\}G(T_s)ds\right) \\ &= E\left(\eta(A_t)\exp\left\{\int_0^t b_s(T_sA_t)ds\right\}L_tG\right) - E(\eta G) \\ &\quad -E\left(\int_0^t \eta(A_s)b_s\exp\left\{\int_0^s b_r(T_rA_s)dr\right\}L_sGds\right) \\ &= E(X_tG) - E(\eta G) - E\left(\int_0^t b_sX_sdsG\right). \end{aligned}$$

Using that the random variable $X_t - \eta - \int_0^t b_sX_sds$ belongs to $L^1(\Omega)$, we obtain that $\sigma X 1_{[0,t]}$ is a process in $\operatorname{Dom}\delta_1$ and $\int_0^t \sigma_sX_s\delta_1W_s = X_t - \eta - \int_0^t b_sX_sds$. Thus, process X given by (7.14) is a solution to equation (7.10).

Finally, we deal with the uniqueness of the solution for equation (7.10). So, consider a process $Y \in L^1(\Omega\times[0,T])$ that is a solution of equation (7.10). Since $D\sigma \in L^\infty(\Omega\times[0,T]^2)$, Buckdahn [24, 25] has shown that there exists a sequence $\{\sigma^n : n \in \mathbb{N}\}$ of measurable processes such that $\sigma_t^n \in \mathcal{S}$, $\|\sigma - \sigma^n\|_{\mathbb{D}^{1,2}(L^2([0,T]))} \to 0$, as $n \to \infty$, and

$$\frac{dG(A_t^n)}{dt} = -\sigma_t^n D_tG(A_t^n), \quad \text{for } G \in \mathcal{S},$$

where $\{T^n_s : s \in [0,T]\}$ and $\{A^n_s : s \in [0,T]\}$ are the transformations associated with σ^n through Propositions 124 and 125, respectively. In consequence, for $G \in \mathcal{S}$, (7.10) yields

$$E\left(Y_t G(A^n_t)\right) = E\left(\eta G(A^n_t)\right) + E\left(\int_0^t \sigma_s Y_s D_s[G(A^n_t)]ds\right) + E\left(\int_0^t b_s Y_s G(A_t)ds\right). \quad (7.15)$$

Thus, we can apply the fact that $G(A^n_t) = G + \int_0^t \frac{d}{ds}G(A^n_s)ds$ and Fubini's theorem to establish

$$\begin{aligned} E\left(Y_t G(A^n_t)\right) &= E(\eta G) - E\left(\int_0^t \eta\sigma^n_s D_s[G(A^n_s)]ds\right) + E\left(\int_0^t \sigma_s Y_s D_s[G(A^n_s)]ds\right) \\ &\quad - E\left(\int_0^t \int_0^r \sigma_s Y_s D_s(\sigma^n_r D_r[G(A^n_r)])dsdr\right) + E\left(\int_0^t b_s Y_s G(A^n_s)ds\right) \\ &\quad - E\left(\int_0^t \int_0^r b_s Y_s \sigma^n_r D_r[G(A^n_r)]dsdr\right). \end{aligned}$$

Hence, (7.15), with $\sigma^n_r D_r[G(A^n_r)]$ instead of $G(A^n_t)$, leads us to conclude

$$E\left(Y_t G(A^n_t)\right) = E(\eta G) + E\left(\int_0^t \left[\sigma_s - \sigma^n_s\right] Y_s D_s\left[G(A^n_s)\right]ds\right) + E\left(\int_0^t b_s Y_s G(A^n_s)ds\right).$$

Consequently, taking the limit as $n \to \infty$, we can write

$$E\left(Y_t G(A_t)\right) = E(\eta G) + E\left(\int_0^t b_s Y_s G(A_s)ds\right).$$

Therefore, (7.13) allows us to get

$$E[Y_t(T_t)\mathcal{L}_t G] = E[\eta G] + E\left[G\int_0^t b_s(T_s)Y_s(T_s)\mathcal{L}_s ds\right].$$

In this way, we have proven that

$$Y_t(T_t)\mathcal{L}_t = \eta + \int_0^t b_s(T_s)Y_s(T_s)\mathcal{L}_s ds \quad \text{for each } t \in [0,T], \text{ w.p.1},$$

due to $\mathcal{S}$ being dense in $L^\infty(\Omega)$. In order to complete the proof, define

$$\tilde{Y}_t = \eta + \int_0^t b_s(T_s)Y_s(T_s)\mathcal{L}_s ds \quad \text{for } t \in [0,T] \text{ and } \omega \in \Omega,$$

which is a continuous modification of $Y_\cdot(T_\cdot)\mathcal{L}_\cdot$. Therefore, Proposition 1 and Fubini's theorem imply

$$\tilde{Y}_t = \eta + \int_0^t b_s(T_s)\tilde{Y}_s ds, \quad \text{with probability 1, for all } t \in [0,T].$$

That is, $Y_t(T_t)\mathcal{L}_t = \tilde{Y}_t = \eta\exp\left(\int_0^t b_s(T_s)ds\right)$, for each $t \in [0,T]$, with probability 1. The proof is finished now because of (7.13). □

Remarks 16 *The anticipating linear stochastic differential equation* (7.10) *has been also considered in another situation. Namely:*

i) *León et al. [115] have used the techniques shown in the proof of Theorem 87 to analyze the semilinear stochastic differential equation*

$$X_t = \eta + \int_0^t b(s, X_s)ds + \int_0^t \sigma_s X_s \delta_1 W_s, \quad t \in [0, T].$$

Here, b is a measurable and $\mathcal{F}_t$-adapted random field, and σ is a measurable and $\mathcal{F}_t$-adapted stochastic process. In this case, the solution X has the form

$$X_t = L_{0,t} Z_t(A_t, \eta(A_t)), \quad t \in [0, T], \tag{7.16}$$

where the random field Z is the unique solution to

$$Z_t(\omega, x) = x + \int_0^t L_{0,s}^{-1}(T_t\omega) b(s, L_{0,s}(T_t\omega) Z_s(\omega, x), T_s\omega)ds, \quad t \in [0, T].$$

ii) *The case that σ is a deterministic function in $L^2([0,T])$ and b is a measurable random field, the stochastic differential equation in the previous statement is studied by Nualart [158].*

iii) *Let $B = \{B_t : t \in [0,T]\}$ be fractional Brownian motion with Hurst parameter $H \in (0,1)$ and consider the stochastic differential equation*

$$X_t = \eta + \int_0^t b(s, X_s)ds + \int_0^t \sigma_s X_s \delta B_s, \quad t \in [0, T]. \tag{7.17}$$

Here, the stochastic integral is the divergence operator given by (4.2) if $H > 1/2$, and it is the extension of the divergence operator introduced in Definition 40 if $H < 1/2$ (see Section 4.3.1 for details).

In this case, Jien and Ma [90] analyze equation (7.17) on the canonical fractional Wiener space $(\Omega, \mathcal{F}, P_H)$, where $\Omega = C_0([0,T])$, $\mathcal{F}$ is the topological σ-algebra on Ω and P_H is the unique probability measure such that $B_t(\omega) = \omega(t)$ is fractional Brownian motion. In this space, these authors define the transformations $T_t : \Omega \to \Omega$ and $A_{s,t} : \Omega \to \Omega$ as the solutions to the equations

$$(T_t\omega)(s) = \omega_s + \int_0^{t\wedge s} K(s,r)\sigma_r(T_r\omega)dr, \quad t, s \in [0, T],$$

and

$$(A_{s,t}\omega)(v) = \omega_s - \int_{s\wedge v}^{t\wedge v} K(v,r)\sigma_r(A_{r,t}\omega)dr, \quad t, s, v \in [0, T].$$

In these equations, the kernel $K = \{K(t,s) : 0 \le s < t \le T\}$ is given by (1.66), or (1.67), according to the value of the Hurst parameter H. Proceeding as in the proof of Theorem 87, Jien and Ma [90] have established that the process X given by (7.16) is still the solution to equation (7.17). In other words, Jien and Ma [90] have proven that the Buckdahn [24, 25] method is still useful to analyze anticipating fractional stochastic differential equations.

iv) *Concerning anticipating stochastic differential equations driven by a Lévy process, León et al. [114] have dealt with the linear stochastic differential equation*

$$\begin{aligned} X_t &= X_0 + \int_0^t b_s X_s ds + \int_0^t a_s X_s \delta_1 W_s \\ &\quad + \int_0^t \int_{\mathbb{R}\setminus\{0\}} v_s(y) X_{s-}\, d\tilde{N}(s,y), \quad 0 \le t \le T, \end{aligned} \tag{7.18}$$

on the Canonical space $\Omega_W \otimes \Omega_N$*. Here,* X_0 *is a random variable, and* a*,* b *and* $v(y)$*, for any* $y \in (\mathbb{R} \setminus \{0\})$*, are measurable stochastic processes. In this statement, all the processes are defined on the canonical Lévy space*

$$(\Omega, \mathcal{F}, P) = (\Omega_W \otimes \Omega_N, \mathcal{F}_W \otimes \mathcal{F}_N, P_W \otimes P_N),$$

on $[0, T]$*, where* $(\Omega_W, \mathcal{F}_W, P_W)$ *is the canonical Wiener space defined in Section 1.2.1 and* $(\Omega_N, \mathcal{F}_N, P_N)$ *is the canonical Lévy space for a pure jump Lévy process* N *with Lévy measure* ν *as it is introduced by Solé et al. [199]. The stochastic integral with respect to the canonical Wiener process* W *(resp. the canonical compensated Poisson random measure* $d\tilde{N}(t,y) := dN(t,y) - dt\,\nu(dy)$*) is given by the duality relation* (4.2) *(resp. is defined pathwise).*

In order to apply the techniques developed by Buckdahn [24, 25], León et al. [114] study the Girsanov type transformations $\{T_t : \Omega \to \Omega_W : 0 \le t \le T\}$ *and* $\{A_{s,t} : \Omega \to \Omega_W : 0 \le s \le t \le T\}$ *given as the solutions of the equations*

$$(T_t\,\omega)_\cdot = \omega'_\cdot + \int_0^{t\wedge\cdot} a_s(T_s\omega, \omega'')\, ds$$

and

$$(A_{s,t}\,\omega)_\cdot = \omega'_\cdot - \int_{s\wedge\cdot}^{t\wedge\cdot} a_r(A_{r,t}\omega, \omega'')\, dr,$$

respectively, with $\omega = (\omega', \omega'') \in \Omega$*. Under suitable condition, we can use the results of Buckdahn [24, 25] to show*

$$E\left[F(A_{s,t}\omega, \omega'')L_{s,t}(\omega)\right] = E\left[F\right] \quad \textit{and} \quad E\left[F(A_{s,t}\omega, \omega'')\right] = E\left[F\mathcal{L}_{s,t}\right],$$

where $F \in L^\infty(\Omega)$*,*

$$\begin{aligned} L_{s,t}(\omega) &= \exp\left\{\int_s^t a_r(A_{r,t}\omega, \omega'')\delta_2 W_r - \frac{1}{2}\int_s^t a_r^2(A_{r,t}\omega, \omega'')dr \right.\\ &\left. - \int_s^t \int_r^t (D_u^W a_r)(A_{r,t}\omega, \omega'') D_r^W\left[a_u(A_{u,t}\omega, \omega'')\right] du dr\right\} \end{aligned}$$

is the density of $A_{s,t}^{-1}$ *and*

$$\begin{aligned} \mathcal{L}_{s,t}(\omega) &= \exp\left\{-\int_s^t a_r(T_t A_r\omega, \omega'')\delta_2 W_r - \frac{1}{2}\int_s^t a_r^2(T_t A_r\omega, \omega'')dr \right.\\ &\left. - \int_s^t \int_s^r (D_u^W a_r)(T_t A_r\omega, \omega'') D_r^W\left[a_u(T_t A_u\omega, \omega'')\right] du dr\right\}. \end{aligned}$$

In consequence, the ideas in the proof of Theorem 87 allow us to see that the process

$$\begin{aligned} X_t &= X_0(A_{0,t}) \exp\left\{\int_0^t b_s(A_{s,t})\, ds\right\} L_{0,t} \exp\left\{-\int_0^t \int_{\mathbb{R}\setminus\{0\}} v_s(y, A_{s,t})\, \nu(dy)\, ds\right\} \\ &\times \prod_{s\le t, y\in(\mathbb{R}\setminus\{0\})} \Big[1 + v_s(y, A_{s,t})\Delta N(s,y)\Big], \quad t \in [0, T], \end{aligned}$$

is the unique solution in $L^1(\Omega\times[0,T])$ *of equation* (7.18) *such that the pathwise integral* $\int_0^T \int_{\mathbb{R}\setminus\{0\}} |v_s(y) X_{s-}| dN(s,y)$ *belongs to* $L^1(\Omega)$*.*

7.3 Stochastic differential equation of Stratonovich type

In this section, we study the existence of a unique solution to autonomous stochastic differential equations of Stratonovich type, which are driven by either Brownian motion, a semimartingale, or fractional Brownian motion. To do so, as a main tool, we use the Doss-Sussmann transformation [42, 204, 203]. This transformation is a link between ordinary and stochastic differential equations.

7.3.1 Brownian motion

Let $W = \{W_t : t \in [0,T]\}$ be Brownian motion. Then, we can use Doss-Sussmann transformation [42, 204, 203], together with the Itô's formula for the Stratonovich integral (i.e., Theorem 74), to study the existence of a unique solution to the stochastic differential equation

$$X_t = x_0 + \int_0^t a(X_s) \circ dW_s + \int_0^t b(X_s)\, ds, \quad t \in [0,T], \tag{7.19}$$

with $x_0 \in \mathbb{R}$ and $a, b : \mathbb{R} \to \mathbb{R}$ two functions such that $a \in C_b^2(\mathbb{R})$ (i.e., it is a bounded function with two bounded continuous derivatives) and b is Lipschitz on $\mathbb{R}$. To do so, we first analyze the solution α to the ordinary differential equation

$$\begin{aligned} \frac{\partial \alpha}{\partial x}(y,x) &= a(\alpha(y,x)), \\ \alpha(y,0) &= y. \end{aligned} \tag{7.20}$$

This equation has a unique solution since a is Lipschitz on $\mathbb{R}$ by hypothesis. Moreover, Hartman [73] implies that α is in $C^2(\mathbb{R}^2)$.

We will need the following properties of the solution α of the equation (7.20), which are established in Doss [42].

Lemma 74 *Let $a \in C_b^2(\mathbb{R}^2)$. Then, for $x, x_1, y \in \mathbb{R}$., we have:*

i) $\alpha(y,x) = \alpha(\alpha(y,x_1), x - x_1)$.

ii) $y = \alpha(\alpha(y,x), -x)$.

iii) $\partial_y \alpha(y,x) = \exp\left(\int_0^x a'(\alpha(y,s))ds\right)$.

iv) $(\partial_x \alpha)(y,-x) = a(y)(\partial_y \alpha)(y,-x)$.

Proof. The right and left-hand sides of *i)* are two solutions of equation (7.20) such that both are equal to $\alpha(y,x_1)$ at $x = x_1$. Hence, the uniqueness of the solution of (7.20) yields that both solutions are the same.

Changing x and x_1 by 0 and x, respectively, Statement *ii)* follows from *i)*.

From (7.20) and Hartman [73], we have

$$\partial_y \alpha(y,x) = 1 + \int_0^x a'(\alpha(y,s))\partial_y \alpha(y,s)ds, \quad (y,x) \in \mathbb{R}^2.$$

Thus, Statement *iii)* also holds.

Finally, Assertion *i)* gives that

$$\begin{aligned} 0 &= \partial_{x_1} \alpha(y,-x) = \partial_{x_1}\left[\alpha(\alpha(y,x_1), -x - x_1)\right] \\ &= (\partial_y \alpha)(\alpha(y,x_1), -x - x_1)a(\alpha(y,x_1)) - (\partial_x \alpha)(\alpha(y,x_1), -x - x_1)). \end{aligned}$$

In consequence, with $x_1 = 0$, we have that Statement *iv*) is true and, therefore, the proof is complete. □

In order to continue with the study of (7.19), Doss [42] also considers the equation

$$\begin{aligned} Y_t &= x_0 + \int_0^t \left(\frac{\partial \alpha}{\partial y} (Y_u, W_u) \right)^{-1} b(\alpha(Y_u, W_u))\, du \\ &= x_0 + \int_0^t \exp\left(- \int_0^{W_u} a'(\alpha(Y_u, z))\, dz \right) b(\alpha(Y_u, W_u))\, du, \quad t \in [0,T]. \quad (7.21) \end{aligned}$$

Here, this equation is defined pathwise (i.e., ω by ω) and the second equality follows from Lemma 74. Moreover, we assume that b belongs to $C_b^1(\mathbb{R})$. Note that this equation has a unique solution due to the function

$$y \mapsto \exp\left(- \int_0^{W_u} a'(\alpha(y,z)dz \right) b(\alpha(y, W_u))$$

being Lypschitz for almost all $\omega \in \Omega$ since a' and b are Lypschitz and $\sup_{t\in[0,T]} |W_t| < \infty$, with probability 1.

Now, we are ready to study the existence and uniqueness of the solution to equation (7.19).

Theorem 88 *Let $a \in C_b^2(\mathbb{R})$ and $b \in C_b^1(\mathbb{R})$. Then, equation (7.19) has a unique solution X, which is an $\mathcal{F}_t^W$-semimartingale. Moreover, this solution has the form*

$$X_t = \alpha(Y_t, W_t), \quad t \in [0,T]. \tag{7.22}$$

Proof. Let X be the stochastic process introduced in (7.22) and fix $t \in [0,T]$. Note that this process is an $\mathcal{F}_t^W$-semimartingale since Itô's formula (see Theorem 40). Then, the Itô's formula for the Stratonovich integral (i.e., Theorem 74) leads us to write

$$X_t = \alpha(x_0, 0) + \int_0^t \partial_y \alpha(Y_s, W_s) dY_s + \int_0^t \partial_x \alpha(Y_s, W_s) \circ dW_s.$$

In consequence (7.20) and Lemma 74.*iii*) imply that X is a solution to the stochastic differential equation (7.19).

Reciprocally, assume that X is an $\mathcal{F}_t^W$-semimartingale such that it is also a solution of equation (7.19). Then, applying Theorem 74 again to the process $D = \alpha(X, -W)$, we have

$$D_t = x_0 + \int_0^t \partial_y \alpha(X_s, -W_s) \circ dX_s - \int_0^t \partial_x \alpha(X_s, -W_s) \circ dW_s.$$

Therefore, using Lemma 74 again, we are able to establish

$$\begin{aligned} D_t &= x_0 + \int_0^t [\partial_y \alpha(X_s, -W_s) a(X_s) - \partial_x \alpha(X_s, -W_s)] \circ dW_s \\ &\quad + \int_0^t \partial_y \alpha(X_s, -W_s) b(X_s) ds = \int_0^t \partial_y \alpha(X_s, -W_s) b(X_s) ds. \quad (7.23) \end{aligned}$$

Observe that Lemma 74.*ii*) gives that $\alpha(D_s, W_s) = \alpha(\alpha(X_s, -W_s), W_s) = X_s$ and

$$\partial_y \alpha(X_s, -W_s) = (\partial_y \alpha(\alpha(X_s, -W_s), W_s))^{-1} = (\partial_y \alpha(D_s, W_s))^{-1}.$$

Hence, Lemma 74.*iii*) and (7.23) yield

$$D_t = x_0 + \int_0^t \exp\left(-\int_0^{W_s} a'\left(\alpha\left(D_s, z\right)\right) dz\right) b\left(\alpha\left(D_s, W_s\right)\right) ds.$$

In this way, we have proven that $\alpha(X, -W) = Y$, where Y is the solution to equation (7.21). Finally, by using Lemma 74.*ii*) again, we get

$$\alpha(Y, W) = \alpha(\alpha(X, -W), W) = X,$$

which implies the uniqueness for the solution to equation (7.19). Thus, the proof is now complete. □

Doss [42] also considers the Stratonovich stochastic differential equation on $\mathbb{R}^d$

$$X_t = x_0 + \int_0^t b(X_s)\, ds + \sum_{j=1}^{k} \int_0^t a^j(X_s) \circ dW_s^j, \quad t \in [0, T]. \tag{7.24}$$

Here, $W = \{W_t : t \in [0, T]\}$ is a k-dimensional Brownian motion, $x_0 \in \mathbb{R}^d$ and $b, a^j : \mathbb{R}^d \to \mathbb{R}^d$, with $j \in \{1, \ldots, k\}$. Under this assumptions, the pathwise equation (7.21) has the form

$$Y_t = x_0 + \sum_{i=1}^{d} \int_0^t \frac{\partial \alpha}{\partial y^i}(\alpha(Y_s, W_s), -W_s) b^i(\alpha(Y_s, W_s)) ds, \quad t \in [0, T].$$

Doss [42] has shown that the process X introduced in (7.22) is still a solution to (7.24) if a satisfies the Frobenius condition. That is, for all $m, j \in \{1, \ldots, k\}$ and $i \in \{1, \ldots, d\}$,

$$\sum_{l=1}^{d} a_{lm} \frac{\partial a_{ij}}{\partial x^l} = \sum_{l=1}^{d} a_{lj} \frac{\partial a_{im}}{\partial x^l}. \tag{7.25}$$

Unfortunately, the Frobenius's condition is rarely satisfied in the case that $k > 1$. The reason this condition is required is that the equation (7.20) has a unique solution if and only if the Frobenius condition is satisfied (see Dieudonné [40]).

We have analyzed equation (7.24) with x_0, a^j and b deterministic so far. However, we can handle the random case as it is done in Kohatsu-Higa and León [101]. Namely, now, x_0 is an $\mathbb{R}^d$-valued random variable and $b, a^j : \Omega \times \mathbb{R}^d \to \mathbb{R}^d$ are random fields such that for some $p \geq 4$, and for all $i \in \{1, \ldots, d\}$ and $j \in \{1, \ldots, k\}$, we have

(*i*) $b^i(x), a_{ij}(x) \in \mathbb{D}^{1,p}$, for every $x \in \mathbb{R}^d$.

(*ii*) $(\omega, x) \mapsto b^i(x)$, $(\omega, x) \mapsto a_{ij}(x)$, $(\omega, t, x) \mapsto D_t b^i(x)$ and $(\omega, t, x) \mapsto D_t a_{ij}(x)$ are measurable maps.

(*iii*) $b^i, a_{ij} \in C^1(\mathbb{R}^d)$ and $D_t b^i, D_t a_{ij} \in C^0(\mathbb{R}^d, \mathbb{R}^k)$, for almost all $(\omega, t) \in \Omega \times [0, T]$.

Under this assumptions, Kohatsu-Higa and León [101] have stated the following result.

Theorem 89 *For $i \in \{1, \ldots, d\}$ and $j \in \{1, \ldots, k\}$, assume:*

1. *For $(\omega, t) \in \Omega \times [0, T]$, $a_{ij} \in C^2(\mathbb{R}^d)$ and $D_t a_{ij} \in C^1(\mathbb{R}^d, \mathbb{R}^k)$.*

2. *There exists $M \in \cap_{p \geq 2} \mathbb{D}^{1,p}$ such that:*

   (*i*) $\operatorname*{ess\,sup}_{r \in [0,T]} E|D_r M|^{16} < \infty$ *and* $DM \in \cap_{p \geq 2} L^p(\Omega \times [0, T]; \mathbb{R}^k)$.

(ii) $|a_{ij}(0)|,\ |b^i(0)| \leq M$.

(iii) b^i *and its partial derivatives are random Lipschitz with Lipschitz random variable* M.

(iv) a_{ij} *and its partial derivatives up to order 2 are random Lipschitz with Lipschitz random variable* M.

3. *There exists* $\xi \in \cap_{p\geq 2} L^p(\Omega \times [0,T])$ *such that, for all* $r \in [0,T]$,

   (i) $\operatorname{esssup}_{t\in[0,T]} E|\xi_t|^{16} < \infty$.

   (ii) $|D_r a_{ij}(0)|,\ |D_r b^i(0)| \leq \xi_r$.

   (iii) $D_r a_{ij}$*, its partial derivatives and* $D_r b^i$ *are random Lipschitz with Lipschitz random constant* ξ_r.

4. $x_0^i \in \cap_{p\geq 2}\mathbb{D}^{1,p}$ *and* $\operatorname{esssup}_{r\in[0,T]} E|D_r x_0^i|^{16} < \infty$.

5. *a satisfies the Frobenius condition* (7.25).

Then, equation (7.24) *has a unique solution given by the process* X *introduced in* (7.22).

Remarks 17 *i) Here, a measurable process* u *with integrable paths is said to be Stratonovich integrable with respect* W^j *if the sequence* (6.4) *converges in probability when we write* W^j *instead of* W.

ii) The process X *is a local solution of equation* (7.24). *That is, the* $\mathbb{R}^d$*-valued stochastic process* X *is continuous and there exists a sequence* $\{\beta_n : n \in \mathbb{N}\}$ *of random variables such that:*

a) $[\beta_n = 1] \uparrow \Omega$ *w.p.1.*

b) $\beta_n X, \beta_n a^j(X) \in (\mathbb{D}_C^{1,8}(L^2([0,T])))^d$*, for* $j \in \{1, ..., k\}$.

c) $\beta_n X_t = \beta_n x_0 + \int_0^t \beta_n b(X_s)ds + \sum_{j=1}^{k}\int_0^t \beta_n a^j(X_s) \circ dW_s^j$*, with probability 1.*

iii) In this case, we cannot apply Itô's formula for the Stratonovich integral (i.e., Theorem 74) because a and b are random fields, and x_0 *is a random variable that is not necessarily* $\mathcal{F}_0^W$*-measurable. So, in order to see that Theorem 89 holds, we have to use the techniques of Malliavin calculus based on a chain rule for random fields and the Itô-Ventzell formula, which is established by Ocone and Pardoux [164]. The Itô-Ventzell formula is a useful Itô's type one to deal with equations with stochastic fields as coefficients and random initial conditions.*

7.3.2 Semimartingales

The purpose of this section is to explain how Kohatsu-Higa et al. [102] have extended Theorem 89 to the case that Brownian motion W is changed by a continuous semimartingale. Towards this end, they use the classical stochastic calculus based on a generalized Stratonovich integral instead of the techniques of the Malliavin calculus as it is done in the analysis of Theorem 89. In this way, Kohatsu-Higa et al. [102] have obtained simple proofs of the results.

Let $Z = \{Z_t : t \in [0,T]\}$ be a continuous semimartingale defined on a filtered probability space $(\Omega, \mathcal{F}, (\mathcal{F}_t)_{t\in[0,T]}, P)$ satisfying the usual conditions. Suppose that this semimartingale (see [180]) has the canonical decomposition of the form

$$Z_t = M_t + A_t, \quad t \in [0,T],$$

where $M = \{M_t : t \in [0,T]\}$ is a continuous local martingale such that $M_0 = 0$ and there is an $\mathcal{F}_t$-adapted process $m \in L^1(\Omega \times [0,T])$ satisfying $[M]_t = \int_0^t m_s ds$, and $A = \{A_t : t \in [0,T]\}$ is a process with continuous and bounded variation paths. See (3.44) for the definition of the quadratic variation $[M]$ of M.

Given a partition $\pi = \{0 = t_0 < t_1 < \cdots < t_n = T\}$, we introduce the polygonal approximation of the martingale M associated to π by

$$M_t^\pi = \int_0^t \sum_{i=0}^{n-1} \frac{M_{t_{i+1}} - M_{t_i}}{t_{i+1} - t_i} \mathbf{1}_{(t_i,t_{i+1}]}(s)ds, \quad t \in [0,T].$$

Now, we give the notion of the Stratonovich integral with respect to M that is used in Kohatsu-Higa et al. [102].

Definition 50 *Let $u = \{u_t : t \in [0,T]\}$ be a measurable stochastic process with integrable paths. The Stratonovich integral of u with respect to M, denoted by $\int_0^t u_s \circ dM_s$, is the continuous stochastic process defined by*

$$\lim_{|\pi|\downarrow 0} P\left(\sup_{0\le t\le T} \left| \int_0^t u_s \circ dM_s - \int_0^t u_s \dot{M}_s^\pi ds \right| > \epsilon \right) = 0,$$

for all $\epsilon > 0$, if this limit exists. Here, $\dot{M}_s^\pi$ stands for $\frac{dM_s^\pi}{ds}$.

Note that this definition is nothing else than the one introduced in (6.4) when we change the convergence in probability by the uniform convergence in probability on intervals.

Concerning the semimartingale Z, we consider the stochastic differential equation

$$X_t = X_0 + \int_0^t b(X_s)dA_s + \int_0^t a(X_s) \circ dM_s, \quad t \in [0,T], \tag{7.26}$$

where X_0 is a random variable. Observe that we do not assume it to be $\mathcal{F}_0$-measurable. A continuous stochastic process $X = \{X_t : t \in [0,T]\}$ is a solution to the stochastic differential equation (7.26) if $a(X)$ is Stratonovich integrable with respect to M in the sense of Definition 50 and (7.26) holds for each $t \in [0,T]$ with probability 1. In fact, the set of probability 1 where (7.26) is satisfied is independent of t since Proposition 1.

In order to state the existence result for the solution of the equation (7.26), we need to study the following auxiliary result.

Lemma 75 *Let $Y = \{Y_t : t \in [0,T]\}$ be a continuous process and $\pi = \{0 = t_0 < t_1 < \cdots < t_n = T\}$ a partition of $[0,T]$. Then,*

$$\lim_{|\pi|\downarrow 0} P\left(\sup_{0\le t\le T} \left| \int_0^t Y_s (M_s^\pi - M_s) \dot{M}_s^\pi ds \right| > \epsilon \right) = 0,$$

for all $\epsilon > 0$.

Proof. We choose the partition $\{0 = s_0 < s_1 < \ldots < s_m = T\}$ with $s_j = \frac{jT}{m}$. So, we have

$$\begin{aligned}
&\left|\int_0^t Y_s(M_s^\pi - M_s)\dot{M}_s^\pi ds\right| \\
&\leq \sum_{j=0}^{m-1}\left|Y_{s_j}\int_{s_j\wedge t}^{s_{j+1}\wedge t}(M_s^\pi - M_s)\dot{M}_s^\pi ds\right| \\
&\quad + \left|\sum_{j=0}^{m-1}\int_{s_j\wedge t}^{s_{j+1}\wedge t}(Y_s - Y_{s_j})(M_s^\pi - M_s)\dot{M}_s^\pi ds\right| := I_t^1 + I_t^2.
\end{aligned}$$

We claim that $\lim_{|\pi|\downarrow 0} P(\sup_{0\leq t\leq T}|I_t^1| > \epsilon) = 0$ for all $\epsilon > 0$. Indeed,

$$\int_{s_j\wedge t}^{s_{j+1}\wedge t}(M_s^\pi - M_s)\dot{M}_s^\pi ds = \frac{1}{2}\left((M_{s_{j+1}\wedge t}^\pi)^2 - (M_{s_j\wedge t}^\pi)^2\right) - \int_{s_j\wedge t}^{s_{j+1}\wedge t} M_s\dot{M}_s^\pi ds.$$

Then, applying Yor [215] and using the absolute continuity of the quadratic variation $[M]$, we conclude that $\int_{s_j\wedge t}^{s_{j+1}\wedge t} M_s\dot{M}_s^\pi ds$ converges uniformly in probability on $[0,T]$ to

$$\frac{1}{2}\left((M_{s_{j+1}\wedge t})^2 - (M_{s_j\wedge t})^2\right), \quad \text{as } |\pi|\downarrow 0.$$

For the term I^2, we have that the inequality

$$|I_t^2| \leq \left(\sup_{|s-t|\leq\frac{T}{m}}|Y_s - Y_t|\right)\int_0^T |M_s^\pi - M_s|\,|\dot{M}_s^\pi|ds$$

holds. Thus,

$$\lim_{m\to\infty}\lim_{|\pi|\downarrow 0} P(\sup_{0\leq t\leq T}|I_t^2| > \epsilon) = 0, \quad \text{for all } \epsilon > 0. \tag{7.27}$$

Indeed,

$$\begin{aligned}
\int_0^T |\dot{M}_s^\pi||M_s^\pi - M_s|ds &\leq \sum_{i=0}^{n-1}\frac{|\Delta_i M|}{\Delta_i t}\int_{t_i}^{t_{i+1}}\left(\frac{|\Delta_i M|}{\Delta_i t}(s - t_i) + |M_s - M_{t_i}|\right)ds \\
&\leq \frac{1}{2}\sum_{i=0}^{n-1}|\Delta_i M|^2 + \sum_{i=0}^{n-1}\sup_{s\in[t_i,t_{i+1}]}(M_s - M_{t_i})^2,
\end{aligned}$$

with $\Delta_i t = t_{i+1} - t_i$ and $\Delta_i M = M_{t_{i+1}} - M_{t_i}$. Therefore, Burkholder-Davis-Gundy and Chevyshev inequalities give

$$\lim_{K\uparrow\infty}\sup_\pi P\left\{\int_0^T |\dot{M}_s^\pi||M_s^\pi - M_s|ds > K\right\} = 0, \tag{7.28}$$

which allows us to conclude that (7.27) is true. The proof is now complete. □

We are ready to state the existence result for the solution of equation (7.26), which has been established by Kohatsu-Higa et al. [102].

Theorem 90 *Let X_0 be a random variable and $b, a : \Omega\times\mathbb{R}\to\mathbb{R}$ measurable random fields such that $b\in C_b^1(\mathbb{R})$ and $a\in C_b^2(\mathbb{R})$ for each $\omega\in\Omega$. Then, equation (7.26) has a solution X, which is given by*

$$X_t = \alpha(Y_t, W_t), \quad t\in[0,T],$$

where α is the pathwise solution of equation (7.20) *and the process $Y = \{Y_t : t \in [0,T]\}$ is the unique solution of the pathwise equation*

$$Y_t = X_0 + \int_0^t f(Y_s, M_s)dA_s, \quad t \in [0,T],$$

with

$$f(y,x) = b(\alpha(y,x)) \exp(-\int_0^x a'(\alpha(y,z))dz), \quad x, y \in \mathbb{R}.$$

Remark 103 *Unlike Theorem 89, this result requires fewer assumptions because we do not use the techniques of Malliavin calculus and the Itô-Ventzell formula in its proof, but instead, we now give a proof based on the classical stochastic calculus and Definition 50 of the Stratonovich integral.*

Proof. In order to simplify the notation, we omit $\omega \in \Omega$ in the coefficients as usual.

Set $X_t = \alpha(Y_t, M_t)$ and $X_t^\pi = \alpha(Y_t, M_t^\pi)$ for the partition $\pi = \{0 = t_0 < t_1 < \cdots < t_n = T\}$ of $[0,T]$. Then, (7.20) implies

$$\lim_{|\pi|\downarrow 0} \sup_{0 \le t \le T} |X_t - X_t^\pi| \le \|a\|_\infty \lim_{|\pi|\downarrow 0} \sup_{0 \le t \le T} |M_t - M_t^\pi| = 0.$$

for all $\omega \in \Omega$. Using Lemma 74.*iii*) and (7.20), the process X^π can be decomposed as

$$\begin{aligned} X_t^\pi &= X_0 + \int_0^t a(X_s^\pi)\dot{M}_s^\pi ds \\ &+ \int_0^t b(\alpha(Y_s, M_s)) \exp\left(\int_{M_s}^{M_s^\pi} a'(\alpha(Y_s, r))dr\right) dA_s := A_t^1 + A_t^2, \end{aligned}$$

with

$$\begin{aligned} A_t^1 &= \int_0^t a(X_s^\pi)\dot{M}_s^\pi ds = \int_0^t (a(X_s^\pi) - a(X_s))\dot{M}_s^\pi ds \\ &+ \int_0^t a(X_s)\dot{M}_s^\pi ds := B_t^1 + B_t^2, \quad t \in [0,T]. \end{aligned}$$

Note that the dominated convergence theorem gives

$$\lim_{|\pi|\downarrow 0} \sup_{0 \le t \le T} |A_t^2 - (X_0 + \int_0^t b(X_s)dA_s)| = 0,$$

for all $\omega \in \Omega$. Consequently, we have

$$\lim_{|\pi|\downarrow 0} \sup_{0 \le t \le T} |A_t^1 - (X_t - X_0 - \int_0^t b(X_s)dA_s)| = 0,$$

for all $\omega \in \Omega$. Hence, from Definition 50, we only need to show that B^1 goes to zero, uniformly in probability on [0,T], as $|\pi|$ tends to zero, in order to prove that the process X is a solution of equation(7.26). To do so, we write

$$\begin{aligned} a(X_s^\pi) - a(X_s) &= a(\alpha(Y_s, M_s^\pi)) - a(\alpha(Y_s, M_s)) = a'a(X_s)(M_s^\pi - M_s) \\ &+ \frac{1}{2}(a(a')^2 + a^2a'')(\alpha(Y_s, \xi_s))(M_s^\pi - M_s)^2, \end{aligned}$$

where ξ_s is between M_s^π and M_s. Therefore, applying that a is in $C_b^2(\mathbb{R})$, we obtain

$$\begin{aligned}
&\left| \int_0^t (a(a')^2 + a^2a'')(\alpha(Y_s,\xi_s))(M_s^\pi - M_s)^2 \dot{M}_s^\pi ds \right| \\
&\leq C(\omega) \int_0^t |\dot{M}_s^\pi|(M_s^\pi - M_s)^2 ds \\
&\leq C(\omega) \left(\sup_{|u-v|\leq|\pi|} |M_u - M_v| \right) \int_0^t |\dot{M}_s^\pi||M_s^\pi - M_s| ds,
\end{aligned}$$

which converges to zero, uniformly in probability on $[0,T]$, as $|\pi| \downarrow 0$, because of (7.28).

Finally, the convergence to zero of the term $\int_0^t a'a(X_s)(M_s^\pi - M_s)\dot{M}_s^\pi ds$ follows from Lemma 75. In this way, the proof is finished. □

Now, we state the conditions that guarantee the uniqueness of the solution to equation (7.26).

Let $\mathcal{A}$ represent the set of continuous processes X such that there exists a sequence $\{M^n : n \in \mathbb{N}\}$ of processes with absolutely continuous trajectories satisfying:

(a) $\|M^n - M\|_{\infty,[0,T]} \to 0$, as $n \to \infty$, with probability 1.

(b) The unique solution X^n of the equation

$$X_t^n = X_0 + \int_0^t b(X_s^n)dA_s + \int_0^t \sigma(X_s^n)\dot{M}_s^n ds, \quad t \in [0,T],$$

converges to X with probability 1, for each $t \in [0,T]$.

Theorem 91 *Equation* (7.26) *has a unique solution in the class* $\mathcal{A}$.

Remark 104 *Unlike Theorem 89, the uniqueness of the solutions is proven in the family* $\mathcal{A}$ *because we do not use the Itô-Ventzell formula in the proof of this result.*

Proof. We first assume that the process $X \in \mathcal{A}$ is a solution of equation (7.26).

Let f and Y be the function and the process defined in Theorem 90, respectively. Also, let $|A|$ be the total variation of the process A. Then, (7.20), Lemma 74 and Condition (b) lead us to prove that, for $n \in \mathbb{N}$, $X^n = \alpha(Y^n, M^n)$ where Y^n is the unique solution to the pathwise equation

$$Y_t^n = X_0 + \int_0^t f(Y_s^n, M_s^n)dA_s, \quad t \in [0,T].$$

Hence, using the particular expression of the function f and the mean value theorem, together with (7.20), Lemma 74 and Hypothesis (a), we have that there exists a random variable K such that

$$\begin{aligned}
|Y_t - Y_t^n| &\leq \int_0^t |f(Y_s, M_s) - f(Y_s^n, M_s^n)| \, d|A|_s \\
&\leq K \int_0^t |Y_s - Y_s^n| d|A|_s + K \int_0^t |M_s - M_s^n| d|A|_s \\
&\leq K \int_0^t |Y_s - Y_s^n| d|A|_s + KT|A|_T \|M^n - M\|_{\infty,[0,T]}.
\end{aligned}$$

In consequence, Gronwall's lemma (i.e., Proposition 121) yields that $\|Y^n - Y\|_{\infty,[0,T]} \to 0$, with probability 1. Therefore, by definition of $\mathcal{A}$, one has that $X = \alpha(Y, M)$, Thus, the uniqueness for the solution of equation (7.26) follows.

Finally, proceeding as before, the solution X to (7.26) defined in Theorem 90 belongs to the family $\mathcal{A}$ since

$$\|M - M^\pi\|_{\infty,[0,T]} \leq 2 \sup_{|u-v|\leq|\pi|} |M_u - M_v| .$$

□

In the multidimensional case, we can obtain similar results to those of Theorems 90 and 91 under the Frobenius condition (7.25). The details are left to the reader as an exercise.

The techniques applied to study Theorems 90 and 91 are also used by Kohatsu-Higa et al. [102] to consider the semilinear stochastic differential equation of Stratonovich type

$$X_t = X_0 + \int_0^t b(s, X_s)dA_s + \int_0^t a_s X_s \circ dM_s, \quad t \in [0, T]. \tag{7.29}$$

Here, the coefficients $b : \Omega \times [0, T] \times \mathbb{R} \to \mathbb{R}$ and $a : \Omega \times [0, T] \to \mathbb{R}$ are measurable functions such that a is Stratonovich integrable with respect to the martingale M and b is Lypschitz, uniformly in time, for each $\omega \in \Omega$. That is, there exists a random variable K such that $|b(\omega, s, x) - b(\omega, s, y)| \leq K(\omega)|x - y|$, for all $\omega \in \Omega$, $s \in [0, T]$ and $x, y \in \mathbb{R}$. Moreover, this random variable is such that $|b(\omega, s, 0)| \leq K(\omega)$, $(\omega, s) \in \Omega \times [0, T]$ (i.e., b also has linear growth).

The existence and uniqueness result for the solution to equation (7.29) is carried out in the class $\mathcal{A}$ of all the continuous process X such that aX is Stratonovich integrable with respect to M and, for every $t \in [0, T]$ and $\varepsilon > 0$,

$$\lim_{|\pi|\downarrow 0} \lim_{|\pi'|\downarrow 0} P\left(\left| \int_0^t a_s X_s \exp\left(-\int_0^s a_r \dot{M}_r^\pi dr \right) \left[\dot{M}_s^{\pi'} - \dot{M}_s^\pi \right] ds \right| > \varepsilon \right) = 0.$$

Under suitable conditions on the coefficient a, Kohatsu-Higa et al. [102] have established that equation (7.29) has a unique solution X belonging to the family $\mathcal{A}$, which is given by the unique solution of the pathwise equation

$$X_t = \exp(\int_0^t a_s \circ dM_s) X_0 + \int_0^t \exp(\int_u^t a_s \circ dM_s) b(u, X_u) dA_u, \quad t \in [0, T].$$

7.3.3 Fractional Brownian motion

In this section, $B = \{B_t : t \in [0, T]\}$ is fractional Brownian motion with Hurst parameter $H \in (0, 1)$.

As we have already pointed out in Section 1.4, fractional Brownian motion B is not a semimartingale when $H \neq 1/2$. Therefore, we cannot use the classical Itô's calculus to deal with stochastic differential equations of the form

$$X_t = x + \int_0^t b(s, X_s)\, ds + \int_0^t \sigma(s, X_s)\, dB_s, \quad t \in [0, T], \tag{7.30}$$

where $\sigma, b : \Omega \times [0, T] \times \mathbb{R}$ are suitable measurable random fields, since the above stochastic integral does not exist in the classical Itô sense. Thus, this expression is symbolic and it could be interpreted in different ways according to the definition that we use of stochastic integral, as the integrals that we have studied in this book.

In the case $H \in (1/2, 1)$, it is reasonable to interpret (7.30) as a pathwise ordinary differential equation (i.e., path–by–path) because, in this case, B has both zero quadratic variation (see Lin [127]) and β-Hölder continuous paths, for every $\beta < H$. That is, it is natural to interpret the stochastic integral in (7.30) as a pathwise Riemann-Stieltjes one, which is introduced in Section 2. Remember that $\int_0^T X_s dB_s$ exists as a pathwise Riemann–Stieltjes integral for any λ-Hölder continuous stochastic process X with $\lambda > 1 - H$. This pathwise equation has been considered by different researchers under special assumptions on the diffusion coefficient σ (see, for instance, Kleptsyna et al. [98], Lin [127] and Ruzmaikina [192]). In order to improve this trajectory by trajectory approach, Nualart and Răşcanu [160] (see Section 7.4 below), and Zähle [219] have considered equation (7.30) through the extension of the Riemann-Stieltjes integral given by Zähle [218] (see Section 2.3 for its definition). Here, σ satisfies some suitable Hölder conditions, or it is smooth enough. Their proofs are based on contraction principles for the Zähle's generalized integral with integrands in some Besov spaces. A pathwise ordinary differential equation with σ discontinuous has been analyzed by Garzón et al. [60, 61] (see Section 7.3.3.1 and 7.3.3.3 below), and Torres and Viitasaari [207] . In Section 7.4, we deal with fractional stochastic differential equations and equations with power coefficients, both driven by B.

Also, for $H > 1/2$, the stochastic differential equation (7.30) has been studied when the stochastic integral is either the forward one (see Section 7.3.3.1), or the divergence operator (see Section 7.2).

For $H < 1/2$, the existence of a unique solution to fractional stochastic differential equations of the form (7.30) has been contemplated, for example, in the following situations:

(*i*) $H > 1/4$ and the stochastic integral is the symmetric one given by Russo and Vallois [190] (see Definition 44), which is a limit in probability. This is done by Alòs et al. [7] using the Malliavin calculus.

(*ii*) Coutin and Qian [33] have considered the case $H > 1/4$ and the integral is defined by means of the rough paths theory (see Section 2.2.3).

(*iii*) $H > 1/6$ and the symmetric integral defined by Russo and Vallois [190] is given as a uniform limit in probability. The results can be found in Gradinaru et al. [67] and Nourdin [154].

(*iv*) $H \leq 1/6$ and the integral is a renormalized Stratonovich integral. This is also stated in [67] and [154].

v) $H \in (0, 1)$ and the stochastic integral is the divergence operator, it is studied by Jien and Ma [90].

vi) León [110] has given some examples for $H < 1/2$ and the integral is the extension of the Stratonovich integral introduced in Section 6.3.3.

On the other hand, in order to continue with the analysis of applications of the Doss-Sussmann transformation, the main purpose of this section is to study the existence and uniqueness for the solution of the stochastic differential equation

$$X_t = x + \int_0^t b(X_s)\,ds + \int_0^t a(X_s) \circ dB_s, \quad t \in [0, T]. \tag{7.31}$$

Here, $a, b : \mathbb{R} \to \mathbb{R}$ are bounded and measurable functions. The meaning of the solution to the stochastic differential equation (7.31) is the following:

Definition 51 *We say that a process $X = \{X_t : t \in [0, T]\}$ is a solution to* (7.31) *if the integrals of the right-hand side of this equation are well-defined and the equality in* (7.31) *holds.*

In this definition, "the stochastic integral exists" means that the involved integrand belongs to the domain of the definition of stochastic integral that we are considering.

7.3.3.1 Case $H > 1/2$

Here, we suppose that $H > 1/2$ and the stochastic integral with respect to B is the Riemann-Stieltjes integral, or forward integral. Remember that Stratonovich integral agrees with Riemann-Stieltjes and forward integrals under suitable conditions (see, for instance, Propositions 96 and 110, (5.38) and (6.20))

Now, assume that the involved stochastic integral is in the Riemann-Stieltjes sense, and that $a \in C_b^2(\mathbb{R})$ and $b \in C_b^1(\mathbb{R})$, with probability 1.

An extension of Theorem 35 to $f \in C_b^2(\mathbb{R}^2)$, which is left to the reader as an exercise, implies that the unique solution to equation (7.31) is the process

$$X_t = \alpha(Y_t, B_t), \quad t \in [0,T],$$

where, $\alpha : \Omega \times \mathbb{R}^2 \to \mathbb{R}$ and the process Y are given by (7.20) and (7.21), respectively, ω by ω. Here, we write B instead of W in equation (7.21). Indeed, to show this, we only need to proceed as in the proof of Theorem 88. Remember that we can study equation (7.31) in the multidimensional case under the Frobenius condition (7.25). That is, we have to change the k-dimensional Brownian motion W by a k-dimensional fractional Brownian motion B in (7.24).

On the other hand, it is also possible to deal with a discontinuous coefficient $a : \mathbb{R} \to \mathbb{R}$ in equation (7.31). In fact, Garzón et al. [61] consider the case $b \equiv 0$ and

$$a(x) = a_+ 1_{\{x \geq 0\}} + a_- 1_{\{x<0\}}, \quad \text{with } a+, a_- > 0,\ a_+ \neq a_-$$

and the stochastic integral is the extension of the Riemann-Stieltjes integral introduced in Section 2.3. In this case, we have the following result.

Proposition 126 *Let $\beta \in (1-H, H)$, $\beta \vee \frac{1}{2} < \gamma < H$ and $\varepsilon > 0$ small enough. Then, equation (7.31) has a unique solution $X \in C_1^\gamma([0,T])$ such that:*

i) $|X|^{-1} \in L^{\beta+\varepsilon/\gamma}([0,T])$, with probability 1.

ii) For $t \in [0,T]$, X_t has a density p_t such that $\sup_{y\in\mathbb{R}} p_\cdot(y)$ belongs to $L^1([0,T])$.

Moreover, the solution X is given by $X_t = \Lambda^{-1}(\Lambda(x) + B_t)$, $t \in [0,T]$, where $\Lambda : \mathbb{R} \to \mathbb{R}$ is the Lamperti transformation

$$\Lambda(x) = \int_0^x \frac{ds}{a(s)}, \quad x \in \mathbb{R}.$$

Remarks 18 *i) Since Λ is a strictly increasing function on $\mathbb{R}$, we have that Λ^{-1} is well-defined.*

ii) The main tool to prove the existence of the solution is Proposition 52 and, for the uniqueness of the solution, the function a is approximated by suitable continuous functions.

On the other hand, we observe that we can use the ideas developed by Kohatsu-Higa et al. [102] in the study of equation (7.29) to see that, under suitable conditions on the coefficients a and b, the semilinear stochastic differential equation of forward type

$$X_t = X_0 + \int_0^t b(s, X_s)ds + \int_0^t a_s X_s dB_s^-, \quad t \in [0,T], \tag{7.32}$$

and the pathwise equation

$$X_t = \exp(\int_0^t a_s dB_s^-)X_0 + \int_0^t \exp(\int_u^t a_s dB_s^-)b(u, X_u)du, \quad t \in [0,T],$$

have the same solutions. Hence, equation (7.32) has a unique solution, which is given by the solution of the last equation. The reader can see León and Tudor [124] for details.

In order to finish this section, we note that numerical methods for stochastic differential equations driven by fBm have also been analyzed by several authors using different approaches. For example, the reader can see the papers [78, 80, 81, 149]. Moreover, for SDEs with singular drift, we refer to the papers [39, 77, 220, 221].

7.3.3.2 Case $1/4 < H < 1/2$

In this section, we first use the techniques of the Malliavin calculus to study the existence of a solution to equation (7.31) when $H \in (1/4, 1/2)$. The Stratonovich integral in the following result is introduced in Definition 44.

Proposition 127 *Let $H \in (1/4, 1/2)$, $a \in \mathcal{C}_b^2(\mathbb{R})$ and $b \in \mathcal{C}_b^1(\mathbb{R})$. Then, the process*

$$X_t = \alpha\left(Y_t, B_t\right), \quad t \in [0,T],$$

is a solution to equation (7.31). Here, α is the solution to equation (7.20) and Y satisfies the pathwise integral equation

$$Y_t = x + \int_0^t \left(\partial_y \alpha\left(Y_s, B_s\right)\right)^{-1} b\left(\alpha\left(Y_s, B_s\right)\right) ds, \quad t \in [0,T].$$

Proof. Let $\varepsilon > 0$ and fix $t \in [0,T]$. We use the notations

$$B_t^{\varepsilon} = \frac{1}{2\varepsilon}\int_0^t \left(B_{(s+\varepsilon)\wedge T} - B_{(s-\varepsilon)\vee 0}\right) ds \quad \text{and} \quad X_t^{\varepsilon} = \alpha\left(Y_t, B_t^{\varepsilon}\right).$$

From the fundamental theorem of calculus, it follows

$$\begin{aligned} X_t^{\varepsilon} &= x + \frac{1}{2\varepsilon}\int_0^t a\left(\alpha\left(Y_s, B_s^{\varepsilon}\right)\right)\left(B_{(s+\varepsilon)\wedge T} - B_{(s-\varepsilon)\vee 0}\right) ds \\ &\quad + \int_0^t \left(\partial_y \alpha\left(Y_s, B_s^{\varepsilon}\right)\right)\left(\partial_y \alpha\left(Y_s, B_s\right)\right)^{-1} b\left(\alpha\left(Y_s, B_s\right)\right) ds \\ &= x + \frac{1}{2\varepsilon}\int_0^t a\left(\alpha\left(Y_s, B_s\right)\right)\left(B_{(s+\varepsilon)\wedge T} - B_{(s-\varepsilon)\vee 0}\right) ds \\ &\quad + \frac{1}{2\varepsilon}\int_0^t \left[a\left(\alpha\left(Y_s, B_s^{\varepsilon}\right)\right) - a\left(\alpha\left(Y_s, B_s\right)\right)\right]\left(B_{(s+\varepsilon)\wedge T} - B_{(s-\varepsilon)\vee 0}\right) ds \\ &\quad + \int_0^t \left(\partial_y \alpha\left(Y_s, B_s^{\varepsilon}\right)\right)\left(\partial_y \alpha\left(Y_s, B_s\right)\right)^{-1} b\left(\alpha\left(Y_s, B_s\right)\right) ds. \end{aligned} \tag{7.33}$$

Now, the proof is divided into three steps.
Step 1. The fact that a and b are bounded, together with Lemma 74.*iii*), leads us to show that the last integral in (7.33) converges to $\int_0^t b\left(\alpha\left(Y_s, B_s\right)\right) ds$ with probability 1, as $\varepsilon \downarrow 0$. Moreover, (7.20) allows us to see that X^{ε} goes to X_t, as $\varepsilon \downarrow 0$, also with probability 1.
Step 2. Now, we prove that the process $a\left(\alpha\left(Y_\cdot, B_\cdot\right)\right)$ belongs to $\operatorname{Dom}\delta_B^{\mathrm{S}}([0,t])$ (i.e., it is Stratonovich integrable).

Observe that Lemma 74 implies

$$Y_s = x + \int_0^s f(Y_u, B_u) du, \quad s \in [0,t], \tag{7.34}$$

where $f(y,x) = \exp\left(-\int_0^x a'(\alpha(y,s))\,ds\right) b(\alpha(y,x))$. Fix $N \in \mathbb{N}$ and let φ_N be an infinitely differentiable function with compact support such that $\varphi_N(x) = x$ for $|x| \leq N$. Consider the function $f_N(y,x) := \varphi_N(x) f(y,x)$ and the solution Y^N to (7.34) when we replace f by f_N. Note that the processes Y and Y^N coincide on the set

$$\Omega_N = \left\{\omega \in \Omega : \sup_{s \leq T} |B_s| < N\right\}.$$

Taking into account that $\Omega = \cup_{N=1}^{\infty} \Omega_N$ because of the continuity of B, we only need to show that $a\left(\alpha\left(Y_\cdot^N, B_\cdot\right)\right)$ belongs to $\operatorname{Dom} \delta_B^{\mathrm{S}}([0,t])$, for each N, due to Proposition 111. It is clear that Y^N is in the set $\mathbb{D}_W^{1,2}(\mathcal{H})$ and Proposition 24 yields that, for $0 \leq r < s \leq T$,

$$D_r^W Y_s^N = \int_r^s \frac{\partial f_N}{\partial x}(Y_u^N, B_u) K(u,r)\,du + \int_r^s \frac{\partial f_N}{\partial y}(Y_u^N, B_u)\left(D_r^W Y_u^N\right) du.$$

Here, W is the Brownian motion introduced in Theorem 9. Consequently, Gronwall's inequality allows us to prove that there is a constant $C_N > 0$ such that

$$\left|D_r^W Y_s^N\right| < C_N \int_r^s K(u,r)\,du < C_N (s-r)^{1-\tilde{\alpha}} r^{-\tilde{\alpha}}, \quad 0 \leq r < s \leq T,$$

where last inequality follows from the fact that the kernel of fractional Brownian motion satisfies Condition $i)$ in Section 6.3.2 with $\tilde{\alpha} = \frac{1}{2} - H$. Hence, by Proposition 21, we obtain that $a\left(\alpha\left(B, Y^N\right)\right) \in \mathbb{D}_W^{1,2}(\mathcal{H})$ and

$$\begin{aligned} D_r^W\left[a\left(\alpha\left(Y_s^N, B_s\right)\right)\right] &= \mathbf{1}_{[0,s]}(r)\left(a'\left(\alpha\left(Y_s^N, B_s\right)\right) a\left(\alpha\left(Y_s^N, B_s\right)\right) K(s,r)\right. \\ &\quad \left. + a'\left(\alpha\left(Y_s^N, B_s\right)\right) \frac{\partial \alpha}{\partial y}\left(Y_s^N, B_s\right) D_r^W Y_s^N\right). \end{aligned}$$

Moreover, we have that, proceeding as in the proof of Proposition 119, the process $a\left(\alpha\left(B, Y^N\right)\right)$ is also in the space $\mathbb{D}^{1,2}(\mathcal{H}_{\mathcal{K}})$. Indeed, this follows from the fact that (7.34) and

$$D_r Y_s^N = \int_r^s \frac{\partial f_N}{\partial x}(Y_u^N, B_u) \mathbf{1}_{[0,u]}(r)\,du + \int_r^s \frac{\partial f_N}{\partial y}(Y_u^N, B_u)\left(D_r Y_u^N\right) du, \quad r < s,$$

give that $|Y_{s_1}^N - Y_{s_2}^N| \leq C_N |s_2 - s_1|$ and $|D_r\left(Y_{s_1}^N - Y_{s_2}^N\right)| \leq C_N |s_2 - s_1|$, for $r, s_1, s_2 \in [0,T]$. Therefore, in order to conclude the proof of this step, we only need to see the existence of the trace of $D\left(a\left(\alpha\left(B, Y^N\right)\right)\right)$ because of Theorem 80.

In order to simplify the notation, we set $A(y,x) = a(\alpha(y,x))$. Thus, we can write

$$\begin{aligned} &\frac{1}{2\varepsilon}\int_0^t \left\langle Da\left(\alpha\left(Y_s^N, B_s\right)\right), \mathbf{1}_{[(s-\varepsilon)\vee 0,(s+\varepsilon)\wedge T]}\right\rangle_{\mathcal{H}} ds \\ &= \frac{1}{2\varepsilon}\int_0^t \left\langle \frac{\partial A}{\partial x}\left(Y_s^N, B_s\right) \mathbf{1}_{[0,s]}, \mathbf{1}_{[(s-\varepsilon)\vee 0,(s+\varepsilon)\wedge T]}\right\rangle_{\mathcal{H}} ds \\ &\quad + \frac{1}{2\varepsilon}\int_0^t \left\langle \frac{\partial A}{\partial y}\left(Y_s^N, B_s\right) DY_s^N, \mathbf{1}_{[(s-\varepsilon)\vee 0,(s+\varepsilon)\wedge T]}\right\rangle_{\mathcal{H}} ds \\ &= \frac{1}{2\varepsilon}\int_0^t \frac{\partial A}{\partial x}\left(Y_s^N, B_s\right)\left[R((s+\varepsilon)\wedge T, s) - R((s-\varepsilon)\vee 0, s)\right] ds \\ &\quad + \frac{1}{2\varepsilon}\int_0^t \frac{\partial A}{\partial y}\left(Y_s^N, B_s\right)\left\langle DY_s^N, \mathbf{1}_{[(s-\varepsilon)\vee 0,(s+\varepsilon)\wedge T]}\right\rangle_{\mathcal{H}} ds. \end{aligned} \tag{7.35}$$

As in (6.43), the first integral on the right-hand side of this equality converges to

$$H \int_0^t \frac{\partial A}{\partial x} \left(Y_s^N, B_s\right) s^{2H-1} ds, \quad \text{w.p.1.}$$

Concerning the convergence of the last summand in (7.35), we apply Proposition 24 again to get

$$\begin{aligned}
&\frac{1}{2\varepsilon} \int_0^t \frac{\partial A}{\partial y} \left(Y_s^N, B_s\right) \left\langle DY_s^N, \mathbf{1}_{[(s-\varepsilon)\vee 0,(s+\varepsilon)\wedge T]} \right\rangle_{\mathcal{H}} ds \\
&= \frac{1}{2\varepsilon} \int_0^t \frac{\partial A}{\partial y} \left(Y_s^N, B_s\right) \left[\int_0^{(s+\varepsilon)\wedge T} D_\theta^W Y_s^N K\left((s+\varepsilon)\wedge T, \theta\right) d\theta \right. \\
&\qquad \left. - \int_0^{(s-\varepsilon)\vee 0} D_\theta^W Y_s^N K\left((s-\varepsilon)\vee 0, \theta\right) d\theta \right] ds.
\end{aligned}$$

Using the estimate $\left|D_\theta^W Y_s^N\right| < C_N (s-\theta)^{1-\tilde{\alpha}} \theta^{-\tilde{\alpha}}$, this term tends to

$$\int_0^t \frac{\partial A}{\partial y} \left(Y_s^N, B_s\right) \left(\int_0^s D_\theta Y_s^N \frac{\partial K}{\partial s} (s,\theta)\, d\theta \right) ds, \quad \text{with probability 1.}$$

Step 3. By (7.33), and Steps 1 and 2, It remains to check that

$$\frac{1}{2\varepsilon} \int_0^t \left[a\left(\alpha\left(Y_s^N, B_s^\varepsilon\right)\right) - a\left(\alpha\left(Y_s^N, B_s\right)\right) \right] \left(B_{(s+\varepsilon)\wedge T} - B_{(s-\varepsilon)\vee 0}\right) ds$$

converges in probability to zero, as $\varepsilon \downarrow 0$. In fact, from previous steps, we have already known that it converges in probability. So, it is sufficient to see that the limit is zero. Towards this end, let G be a smooth functional. Then, the duality relation (4.2) implies

$$\begin{aligned}
&\frac{1}{2\varepsilon} E\left[G \int_0^t \left[A\left(Y_s^N, B_s^\varepsilon\right) - A\left(Y_s^N, B_s\right) \right] \left(B_{(s+\varepsilon)\wedge T} - B_{(s-\varepsilon)\vee 0}\right) ds \right] \\
&= \frac{1}{2\varepsilon} E\left[G \int_0^t \left[A\left(Y_s^N, B_s^\varepsilon\right) - A\left(Y_s^N, B_s\right) \right] \delta_2^B \left(1_{[(s-\varepsilon)\vee 0,(s+\varepsilon)\wedge T]}\right) ds \right] \\
&= \frac{1}{2\varepsilon} E\left[\int_0^t \left\langle D\left[G\left(A\left(Y_s^N, B_s^\varepsilon\right) - A\left(Y_s^N, B_s\right)\right) \right], 1_{[(s-\varepsilon)\vee 0,(s+\varepsilon)\wedge T]} \right\rangle_{\mathcal{H}} ds \right] \\
&= \frac{1}{2\varepsilon} E\left[\int_0^t \left[A\left(Y_s^N, B_s^\varepsilon\right) - A\left(Y_s^N, B_s\right) \right] \left\langle DG, 1_{[(s-\varepsilon)\vee 0,(s+\varepsilon)\wedge T]} \right\rangle_{\mathcal{H}} ds \right] \\
&+ \frac{1}{2\varepsilon} E\left[G \int_0^t \frac{\partial A}{\partial x} \left(Y_s^N, B_s^\varepsilon\right) \left\langle DB_s^\varepsilon - \mathbf{1}_{[0,s]}, 1_{[(s-\varepsilon)\vee 0,(s+\varepsilon)\wedge T]} \right\rangle_{\mathcal{H}} ds \right] \\
&+ \frac{1}{2\varepsilon} E\left[G \int_0^t \left(\frac{\partial A}{\partial x} \left(Y_s^N, B_s^\varepsilon\right) - \frac{\partial A}{\partial x} \left(Y_s^N, B_s\right) \right) \left\langle 1_{[0,s]}, 1_{[(s-\varepsilon)\vee 0,(s+\varepsilon)\wedge T]} \right\rangle_{\mathcal{H}} ds \right] \\
&+ \frac{1}{2\varepsilon} E\left[G \int_0^t \left(\frac{\partial A}{\partial y} \left(Y_s^N, B_s^\varepsilon\right) - \frac{\partial A}{\partial y} \left(Y_s^N, B_s\right) \right) \left\langle DY_s^N, 1_{[(s-\varepsilon)\vee 0,(s+\varepsilon)\wedge T]} \right\rangle_{\mathcal{H}} ds \right].
\end{aligned}$$

Finally, Lemmas 72 and 73 (with $f(x) = x^2$), Proposition 24 and the dominated convergence theorem lead us to establish that the last expression goes to zero, as $\varepsilon \downarrow 0$. The proof is now complete. □

Regarding numerical methods for fractional SDE, Araya et al. [15] have applied Doss-Sussmann representation of the solution and an approximation of this solution via a first-order Taylor expansion. Also, several authors have used rough paths theory to study the rate of increase of the error. For example, we refer to [83, 111, 130] and references therein.

7.3.3.3 Case $H < 1/2$

Now, it is the turn of the case that the Hurst parameter H is less than one half. Thus, the Stratonovich integral that we use in this section is given by Definition 47.

As an example, we deal with the existence of a unique solution to the fractional linear stochastic differential equation

$$X_t = x_0 + \int_0^t b(s)X_s ds + \int_0^t \sigma X_s \circ dB_s^H, \quad t \in [0,T], \tag{7.36}$$

with $x_0, \sigma \in \mathbb{R}$ and $b \in L^1([0,T])$.

We apply Theorem 85 to state the following result.

Proposition 128 *Let $b : [0,T] \to \mathbb{R}$ be a continuous function. Then, the process*

$$X_t = x_0 \exp\left(\int_0^t b(s)ds + \sigma B_t\right), \quad t \in [0,T], \tag{7.37}$$

is a solution to equation (7.36).

Remark 105 *Note that X is a continuous process that belongs to $L^p(\Omega \times [0,T])$, for any $p \geq 2$.*

Proof. The result is an immediate consequence of Theorem 85. Indeed, we only need to observe that the function

$$f(t,x) = x_0 \exp\left(\int_0^t b(s)ds + \sigma x\right), \quad (t,x) \in [0,T] \times \mathbb{R},$$

belongs to $C_e^{1,2}([0,T] \times \mathbb{R})$. □

In order to show the uniqueness for the solution to equation (7.36), we now study some properties of the process X introduced in (7.37). Towards this end, we proceed as in Section 7.3.2.

We use the notation

$$Y_t^\varepsilon = x_0 + \int_0^t b(s)Y_s ds + \frac{1}{2\varepsilon}\int_0^t \sigma Y_s (B_{(s+\varepsilon)\wedge T} - B_{(s-\varepsilon)\vee 0})ds, \quad t \in [0,T],$$

where $\varepsilon > 0$ and Y is a process with integrable paths. Moreover, we choose a function $\psi : \mathbb{R} \to [0,1]$ in $C_b^\infty(\mathbb{R})$ such that

$$\psi(x) = \begin{cases} 1, & \text{if } |x| \leq 1, \\ 0, & \text{if } |x| \geq 2. \end{cases}$$

For $m \in \mathbb{N}$, $\psi_m : \mathbb{R} \to \mathbb{R}$ denotes the function $\psi_m(x) = \psi(x/m)x$, and $\{F_n : n \in \mathbb{N}\} \subset \mathcal{S}(L^2([0,T]))$ is a sequence that converges to B in $L^2(\Omega \times [0,T])$ and almost surely, where

$$F_n = \sum_{i=1}^{N_n} f_{i,n}(B(\phi_{1,n}), \ldots, B(\phi_{i_n,n}))g_{i,n}, \tag{7.38}$$

with $g_{i,n} \in C^1([0,T])$ and $f_{i,n}(B(\phi_{1,n}), \ldots, B(\phi_{i_n,n})) \in \mathcal{S}_T$. Note that there exists such a sequence due to $B \in L^2(\Omega; L^2([0,T]))$ and Hypothesis (H2) in Section 4.2. Without loss of generality, we can assume that $F_n(0) = 0$. In fact, we can write $F_n\tilde{\psi}_n$ instead of F_n, where $\tilde{\psi}_n : \mathbb{R}_+ \to [0,1]$ is a function in $C_b^\infty(\mathbb{R}_+)$ such that

$$\tilde{\psi}_n(t) = \begin{cases} 1, & \text{if } t \geq \frac{2}{n}, \\ 0, & \text{if } 0 \leq t \leq \frac{1}{n}. \end{cases}$$

Now. we study some auxiliary tool.

Lemma 76 *Let X be the process introduced in (7.37). Then,*

$$\lim_{m\to\infty}\lim_{n\to\infty}\lim_{\varepsilon\downarrow 0} E\left(FX_t^\varepsilon \exp\left(-\sigma\psi_m(F_n(t))\right)\right) = E\left(FX_t \exp(-\sigma B_t)\right), \tag{7.39}$$

for almost all $t\in[0,T]$, for all $F\in\mathcal{S}_T$.

Remark 106 *The set $\{t\in[0,T] : (7.39) \text{ holds}\}$ is independent of the random variable F.*

Proof. Since $F\exp\left(-\sigma\psi_m(F_n(t))\right)$ belongs to $\mathcal{S}_T$, for $t\in[0,T]$ and $n,m\in\mathbb{N}$, we have

$$\lim_{\varepsilon\downarrow 0} E\left(FX_t^\varepsilon \exp\left(-\sigma\psi_m(F_n(t))\right)\right) = E\left(FX_t \exp\left(-\sigma\psi_m(F_n(t))\right)\right).$$

Therefore, now it is not difficult to finish the proof using the definitions of ψ_m and F_n. □

Lemma 77 *Let X be the process given by (7.37). Then,*

$$\lim_{m\to\infty}\lim_{n\to\infty}\lim_{\varepsilon\downarrow 0} E\left[\frac{F}{2\varepsilon}\int_0^t X_s \exp\left(-\sigma\psi_m(F_n(s))\right)\left(B_{(s+\varepsilon)\wedge T} - B_{(s-\varepsilon)\vee 0}\right) ds \right.$$
$$\left. -F\int_0^t X_s^\varepsilon \exp\left(-\sigma\psi_m(F_n(s))\right)\psi_m'(F_n(s))F_n'(s)ds\right] = 0,$$

for almost all $t\in[0,T]$, for all $F\in\mathcal{S}_T$.

Proof. Let $t\in[0,T]$, $\varepsilon>0$ and $n,m\in\mathbb{N}$. Thus, the fundamental theorem of calculus implies

$$\begin{aligned} X_t^\varepsilon \exp\left(-\sigma\psi_m(F_n(t))\right) &= x_0 + \int_0^t b(s)X_s \exp\left(-\sigma\psi_m(F_n(s))\right) ds \\ &\quad + \frac{\sigma}{2\varepsilon}\int_0^t X_s \exp\left(-\sigma\psi_m(F_n(s))\right)\left(B_{(s+\varepsilon)\wedge T} - B_{(s-\varepsilon)\vee 0}\right) ds \\ &\quad -\sigma\int_0^t X_s^\varepsilon \exp\left(-\sigma\psi_m(F_n(s))\right)\psi_m'(F_n(s))F_n'(s)ds. \end{aligned} \tag{7.40}$$

Hence, from Lemma 76, we only need to see

$$\lim_{m\to\infty}\lim_{n\to\infty} E\left[F\int_0^t b(s)X_s \exp\left(-\sigma\psi_m(F_n(s))\right) ds\right] = E\left[F\int_0^t b(s)X_s \exp\left(-\sigma B_s\right) ds\right]. \tag{7.41}$$

To do so, we observe that Hölder inequality implies

$$\begin{aligned} & E\left[|F|\int_0^t |b(s)X_s|\left|\exp\left(-\sigma\psi_m(F_n(s))\right) - \exp\left(-\sigma B_s\right)\right| ds\right] \\ \leq\ & C\left(E\int_0^T |FX_s|^2 ds\right)^{1/2}\left(E\int_0^T \left(\exp\left(-\sigma\psi_m(F_n(s))\right) - \exp\left(-\sigma B_s\right)\right)^2 ds\right)^{1/2} \\ \leq\ & C\left(E\int_0^T \left(\exp\left(-\sigma\psi_m(F_n(s))\right) - \exp\left(-\sigma\psi_m(B_s)\right)\right)^2 ds\right)^{1/2} \\ & +C\left(E\int_0^T \left(\exp\left(-\sigma\psi_m(B_s)\right) - \exp\left(-\sigma B_s\right)\right)^2 ds\right)^{1/2}. \end{aligned}$$

Therefore, (7.41) is true. In this way, we have that the proof is complete. □

Now, we imitate the ideas developed in Section 7.3.2 to show that (7.36) has at most one solution. Towards this end, we let $\mathcal{A}$ represent the family of all the processes Y such that (*i*) Y is a continuous process in $L^p(\Omega \times [0,T])$, for some $p > 2$.

(*ii*) There is a sequence $\{F_n : n \in \mathbb{N}\} \subset \mathcal{S}(L^2([0,T]))$ that goes to B in $L^2(\Omega \times [0,T])$ and almost surely. Moreover, we assume that F_n is as in (7.38), with $g_{i,n} \in C^1([0,T])$, $g_{i,n}(0) = 0$ and $f_{i,n}(B(\phi_{1,n}), \ldots, B(\phi_{i_n,n})) \in \mathcal{S}_T$.

(*iii*) For almost all $t \in [0,T]$,

$$\lim_{m\to\infty} \lim_{n\to\infty} \lim_{\varepsilon\downarrow 0} E\left(FY_t^\varepsilon \exp\left(-\sigma\psi_m(F_n(t))\right)\right) = E\left(FY_t \exp(-\sigma B_t)\right),$$

for all $F \in \mathcal{S}_T$.

(*iv*) For almost all $t \in [0,T]$,

$$\lim_{m\to\infty} \lim_{n\to\infty} \lim_{\varepsilon\downarrow 0} E\left[\frac{F}{2\varepsilon}\int_0^t Y_s \exp\left(-\sigma\psi_m(F_n(s))\right)\left(B_{(s+\varepsilon)\wedge T} - B_{(s-\varepsilon)\vee 0}\right) ds\right.$$
$$\left. -F\int_0^t Y_s^\varepsilon \exp\left(-\sigma\psi_m(F_n(s))\right)\psi_m^{'}(F_n(s))F_n^{'}(s)ds\right] = 0,$$

for all $F \in \mathcal{S}_T$.

We are ready to establish the uniqueness of the solution to equation (7.36).

Proposition 129 *Let Y be a solution of equation (7.36) in $\mathcal{A}$. Then, $Y = X$ in $L^p(\Omega \times [0,T])$, where X is introduced in* (7.36).

Proof. Note that (7.40) is also satisfied if we write Y and Y^ε instead of X and X^ε, respectively. In consequence, applying the definition of family $\mathcal{A}$ and proceeding as in the proof of Lemma 77, we are able to show

$$E\left(FY_t \exp(-\sigma B_t)\right) = E\left[F\left(x_0 + \int_0^t b(s)Y_s \exp\left(-\sigma B_s\right) ds\right)\right],$$

for almost all $t \in [0,T]$, for all $F \in \mathcal{S}_T$. Finally, due to $\mathcal{S}_T$ being a dense set of $L^2(\Omega)$, we can write

$$Y_t \exp(-\sigma B_t) = x_0 + \int_0^t b(s)Y_s \exp\left(-\sigma B_s\right) ds, \quad \text{w.p.1, for almost all } t \in [0,T].$$

Hence, the continuity of the process Y implies that $Y_t \exp(-\sigma B_t) = x_0 \exp(\int_0^t b(s)ds)$. Therefore, the proof is complete. □

Proceeding as in this section, León [110] has also studied the existence of a unique solution for the reduced stochastic differential equation

$$X_t = x + \int_0^t \sigma(X_s) \circ dB_s^H, \quad t \in [0,T], \tag{7.42}$$

with $\sigma \in C_b^2(\mathbb{R})$. Now, the main tool is the solution α of the ordinary differential equation (7.20). That is, we use the Doss-Sussmann transformation [42, 204, 203]. In this case, the unique solution of equation (7.42) is the continuous process $X_t = \alpha(x, B_t)$, $t \in [0,T]$.

On the other hand, as in the case $H > 1/2$, we can consider equation (7.42) with a discontinuous $\sigma : \mathbb{R} \to \mathbb{R}$. For instance, we can choose the function

$$\sigma(x) = \sigma_+ 1_{\{x \geq 0\}} + \sigma_- 1_{\{x<0\}}, \quad \text{with } \sigma_+, \sigma_- > 0 \text{ and } \sigma_- \neq \sigma_+.$$

For this case, the solution is the process $X_t = \Lambda^{-1}(\Lambda(x) + B_t)$, $t \in [0,T]$, where $\Lambda : \mathbb{R} \to \mathbb{R}$ is the Lamperti transformation again. That is, $\Lambda(x) = \int_0^x \frac{ds}{\sigma(s)}$, $x \in \mathbb{R}$. The reader interested in this result can consult Garzón et al. [60]. The difficulty in proving this result is that, in this case, the Stratonovich integral introduced in Definition 47 is not easy to handle since it is given as a weak limit.

7.4 Young stochastic differential equations

In this section, the involved driving noise $x = \{x_t : t \in [0,T]\}$ is a Hölder continuous function with exponent bigger than one half and the stochastic integral is either Young integral, or the extended Young integral, which are introduced in Section 2. Note that the paths of fractional Brownian motion, or Riemann-Liouville fractional process with Hurst parameter $H > 1/2$ can be used as driving noise.

We first consider the Young delay equation of the form

$$\begin{aligned} y_t &= \xi_0 + \int_0^t f(\mathcal{Z}_u^y)\, dx_u, \quad t \in [0,T], \\ \mathcal{Z}_0^y &= \xi. \end{aligned} \tag{7.43}$$

Here, the integral is the Young one studied in Section 2.2 (see (2.47)), $\gamma \in (1/2, 1)$, x is a function in $\mathcal{C}_1^\gamma([0,T]; \mathbb{R}^d)$, the initial condition ξ belongs to the space $\mathcal{C}_1^\gamma([-h,0]; \mathbb{R}^n)$, with $h > 0$, and the coefficient f is a function $f : \mathcal{C}_1^\lambda([-h,0]; \mathbb{R}^n) \to \mathbb{R}^{n\times d}$, where $1/2 < \lambda < \gamma$. Moreover, for a solution y to equation (7.43), we mean a function in the space

$$\mathcal{C}_{\xi,0,T}^\lambda(\mathbb{R}^n) := \{\zeta \in \mathcal{C}_1^\lambda([-h,T]; \mathbb{R}^n) : \zeta = \xi \text{ on } [-h,0]\}$$

and $\mathcal{Z}_t^y : [-h,0] \to \mathbb{R}^n$ is the function $y_{t+\cdot}$. The meaning of (7.43) in terms of the components of involved functions is

$$y_t^i = \xi_0^i + \sum_{j=1}^d \int_0^t f(\mathcal{Z}_u^y)^{i,j}\, dx_u^j, \quad i \in \{1, \dots, n\},$$

where $f(\mathcal{Z}_u^y)$ is the $n \times d$-matrix $\left(f(\mathcal{Z}_u^y)^{i,j}\right)_{\substack{1\le i\le n \\ 1\le j\le d}}$.

Applying a fixed point argument, León and Tindel [123] have stated the following existence and uniqueness result for equation (7.43). Here, the main tool to analyze the contraction argument is the inequality in Theorem 33.

Theorem 92 *Assume that there exist a positive constant M and $\lambda \in (1/2, \gamma)$ such that*

$$|f(\zeta_1)| \le M \left(1 + \sup_{\theta\in[-h,0]} |\zeta_1(\theta)|\right) \tag{7.44}$$

and

$$|f(\zeta_2) - f(\zeta_1)| \le M \left(\sup_{\theta\in[-h,0]} |\zeta_2(\theta) - \zeta_1(\theta)| \right),$$

for $\zeta_1, \zeta_2 \in \mathcal{C}_1^\lambda([-h,0]; \mathbb{R}^n)$. Then, the delay equation (7.43) has a unique solution in $\mathcal{C}_{\xi,0,T}^\lambda(\mathbb{R}^n)$.

Remark 107 *León and Tindel [123] have assumed that $|f(\zeta)| \le M$, for all $\zeta \in \mathcal{C}_1^\lambda([-h,0]; \mathbb{R}^n)$, instead of the linear growth* (7.44)*. But, it is easy to show that this theorem is still true under Condition* (7.44).

Note that the Hypotheses in the theorem include the following two examples:

i) $f(\mathcal{Z}_t^y) = \sigma(\mathcal{Z}_t^y(h_1), \dots, \mathcal{Z}_t^y(h_k))$, where $k \in \mathbb{N}$, $h_1, \dots, h_k \in [-h,0]$ and $\sigma : \mathbb{R}^{n\times k} \to \mathbb{R}^{n\times d}$ is a Lipschitz function with linear growth.

ii) $f(\mathcal{Z}_t^y) = \sigma\left(\int_{-h}^0 y_{t-\theta}\mu(d\theta)\right)$, where μ is a finite signed measure on $[-h,0]$ and $\sigma : \mathbb{R}^n \to \mathbb{R}^{n\times d}$ is a Lipschitz function with linear growth.

On the other hand, another application of the fixed point theorem to study differential equation of Young type is the paper by Nualart and Răşcanu [160]. The equation that they consider has the form

$$y_t = y_0 + \int_0^t b(s,y_s)ds + \int_0^t \sigma(s,y_s)dx_s, \quad t\in[0,T]. \tag{7.45}$$

Now, the integral with respect the noise $x : [0,T] \to \mathbb{R}^d$ is the extension of Young integral defined by the fractional calculus, which is studied in Section 2.3, $y_0 \in \mathbb{R}^n$, $b : [0,T]\times\mathbb{R}^n \to \mathbb{R}^n$ and $\sigma : [0,T]\times\mathbb{R}^n \to \mathbb{R}^{n\times d}$.

Nualart and Răşcanu [160] have worked under the following assumptions on the coefficients.

(1) $(t,x) \mapsto \sigma(t,x)$ is continuously differentiable in x, there exist $\beta,\delta\in(0,1]$ and, for each $N\geq 0$, there is a constant $M_N > 0$ such that:

   i) $|\sigma(t,x)-\sigma(t,y)| \leq M_0|x-y|$, for $x,y\in\mathbb{R}^n$ and $t\in[0,T]$.

   ii) $|\partial_{x_i}\sigma(t,x)-\partial_{y_i}\sigma(t,y)| \leq M_N|x-y|^\delta$, for $|x|,|y|\leq N$, $t\in[0,T]$ and $i=1,\ldots,n$.

   iii) $|\sigma(t,x)-\sigma(s,x)| + |\partial_{x_i}\sigma(t,x)-\partial_{x_i}\sigma(s,x)| \leq M_0|t-s|^\beta$, for $x\in\mathbb{R}^n$, $s,t\in[0,T]$ and $i=1,\ldots,n$.

(2) There exists $b_0\in L^p([0,T];\mathbb{R}^n)$, with $p\geq 2$, and, for $N\geq 0$, there is a constant $L_N>0$ satisfying:

   i) $|b(t,x)-b(t,y)| \leq L_N|x-y|$, for $|x|,|y|\leq N$ and $t\in[0,T]$.

   ii) $|b(t,x)| \leq L_0|x| + b_0(t)$, for $x\in\mathbb{R}^n$ and $t\in[0,T]$.

(3) There are $\gamma\in[0,1]$ and $K_0>0$ such that $|\sigma(t,x)|\leq K_0(1+|x|^\gamma)$, for $x\in\mathbb{R}^n$ and $t\in[0,T]$.

As we have already pointed out, Nualart and Răşcanu [160] have established the following result.

Theorem 93 *Let* $0<\alpha<\min\left\{\frac{1}{2},\beta,\frac{\delta}{1+\delta}\right\}$. *Suppose that Assumptions (1)-(3) above hold and that* x *is an* $\mathbb{R}^d$*-valued Hölder continuous function with exponent* $1-\alpha+\varepsilon$, *for some* $\varepsilon>0$. *Then, equation* (7.45) *has a unique* $(1-\alpha)$*-Hölder continuous solution* y.

Remark 108 *As an immediate consequence, equation* (7.45) *has a unique solution if* x *is a path of either fractional Brownian motion, or the Riemann-Liouville fractional process with Hurst parameter* $H>1/2$ *satisfying* $1-H<\alpha<\min\left\{\frac{1}{2},\beta,\frac{\delta}{1+\delta}\right\}$.

Actually, in [160], the noise x belongs to the space $W_T^{1-\alpha,\infty}(0,T;\mathbb{R}^d)$ and the fixed point argument is carried out on the space $W_0^{\alpha,\infty}(0,T;\mathbb{R}^n)$. In order to explain this and to simplify the notation, suppose that $d=n=1$.

$W_T^{1-\alpha,\infty}(0,T)$ stands for the space of all the measurable functions $x:[0,T]\to\mathbb{R}$ such that

$$\|x\|_{1-\alpha,\infty,T} := \sup_{0<s<t<T}\left(\frac{|x_t-x_s|}{(t-s)^{1-\alpha}} + \int_s^t \frac{|x_r-x_s|}{(r-s)^{2-\alpha}}dr\right) < \infty,$$

and $W_0^{\alpha,\infty}(0,T)$ represents the space of measurable functions $f:[0,T]\to\mathbb{R}$ such that

$$\|f\|_{\alpha,\infty} := \sup_{t\in[0,T]}\left(|f(t)| + \int_0^t \frac{|f(t)-f(s)|}{(t-s)^{\alpha+1}}ds\right).$$

Therefore, Definition 30 of the Young integral $\int_0^t f_s dx_s$ implies

$$\left|\int_0^t f_s dx_s\right| \le \left(\sup_{0<s<t}\left|\left(D_{t-}^{1-\alpha}x_{t-}\right)(s)\right|\right)\int_0^t \left|\left(D_{0+}^{\alpha}f\right)(s)\right|ds, \quad t\in[0,T].$$

Using this inequality and the notation $G_t(f)=\int_0^t f_s dx_s$, Nualart and Răşcanu [160] have obtained the estimation

$$\begin{aligned} &|G_t(f)| + \int_0^t \frac{|G_t(f)-G_s(f)|}{(t-s)^{\alpha+1}}ds \\ &\le\; C_{\alpha,T}\|x\|_{1-\alpha,\infty,T}\int_0^t \left[(t-r)^{-2\alpha}+r^{-\alpha}\right]\left(|f(r)| + \int_0^r \frac{|f(r)-f(s)|}{(r-s)^{\alpha+1}}ds\right)dr. \end{aligned}$$

Hence, it is easy to understand how the fixed point theorem is applied in the study of equation (7.45) .

Now, we consider the semilinear fractional differential equation of Young type

$$x_t = \xi_t + \int_0^t (t-s)^{\beta-1}\, g(s,x_s)ds + \int_0^t (t-s)^{\alpha-1}x_s f(s)d\theta_s, \quad t\in[0,T]. \tag{7.46}$$

Here, θ is a γ-Hölder continuous function with $\gamma>2/3$, $g:[0,T]\times\mathbb{R}\to\mathbb{R}$ is a Lipschitz continuous function with linear growth, uniformly in $[0,T]$, $\alpha,\beta\in(2/3,1)$ and f is a Hölder continuous function on $[0,T]$ with a suitable exponent. The last integral in (7.46) is given by Definition 30 (i.e., it is the extension of the Young integral introduced via the fractional calculus).

In León and Márquez-Carreras [112], equation (7.46) is analyzed for two types of initial conditions. Namely:

(ξ1) $\xi:[0,T]\to\mathbb{R}$ is a measurable function such that $t\mapsto t^{1-\alpha}\xi_t$ belongs to $\mathcal{C}_1^{2\alpha-1+\varepsilon}([0,T])$, for some ε small enough .

(ξ2) The function ξ belongs to $\mathcal{C}_1^{1-\alpha}([0,T])$.

Notice that the function $\xi_t=t^{\alpha-1}\tilde{\xi}_t$ satisfies Condition (ξ1) if $\tilde{\xi}\in\mathcal{C}_1^{2\alpha-1+\varepsilon}([0,T])$, for some $\varepsilon>0$. But, ξ is a discontinuous function at 0 when $\tilde{\xi}_0\neq 0$.

For $0<\varepsilon<\alpha$ and $1<\alpha+\beta$, the main tool to deal with equation (7.46) is the fixed point theorem again, together with the techniques of the fractional calculus and the identities

$$\begin{aligned} (t-r)^{\alpha-1}1_{[0,t]}(r) &= \frac{\Gamma(\alpha)}{\Gamma(\alpha-\varepsilon)\Gamma(\varepsilon)}\int_r^T (t-s)^{\varepsilon-1}(s-r)^{\alpha-\varepsilon-1}1_{[0,t]}(s)ds \\ &= \frac{\Gamma(\alpha)}{\Gamma(\varepsilon)}I_{T-}^{\alpha-\varepsilon}\left((t-\cdot)^{\varepsilon-1}1_{[0,t]}(\cdot)\right)(r), \quad r\in[0,T], \\ (t-r)^{\alpha-1} &= \frac{\Gamma(\alpha)}{\Gamma(\varepsilon)}I_{t-}^{\alpha-\varepsilon}\left((t-\cdot)^{\varepsilon-1}\right)(r), \quad r\in[0,T], \end{aligned}$$

and, for $x\in(a,b)$, $0\le a<b\le T$,

$$I_{a+}^{\alpha+\beta-1}\left((b-\cdot)^{-\beta}(\cdot-a)^{-\alpha}\right)(x) = (b-a)^{1-\alpha-\beta}\frac{\Gamma(1-\alpha)}{\Gamma(\beta)}(x-a)^{\beta-1}(b-x)^{\alpha-1}.$$

Note that the first three equalities are derived from (1.56) and Lemma 87. The fourth one is an immediate consequence of Samko et al. [193] (equalities (1.74), (1.75) and (2.46)).

Under Condition ($\xi 1$), the fixed point argument is applied to the operator

$$\mathcal{M}(\rho)_t = \xi_t + \int_0^t (t-s)^{\beta-1}\, g(s,\rho_s)ds + \int_0^t (t-s)^{\alpha-1}\rho_s d\theta_s, \quad t \in [0,T], \tag{7.47}$$

defined on the space

$$\mathcal{L}_T = \left\{\rho : [0,T] \to \mathbb{R} : (t \mapsto t^{1-\alpha}\rho_t) \in \mathcal{C}^{2\alpha-1+\tilde{\varepsilon}}([0,T]), \text{ for } \tilde{\varepsilon} < \varepsilon\right\}.$$

This family is a normed space, whose norm is

$$|||\rho|||_{2\alpha-1+\tilde{\varepsilon},[0,T]} = \| \cdot^{1-\alpha} \rho_\cdot \|_{2\alpha-1+\tilde{\varepsilon},[0,T]} + \| \cdot^{1-\alpha} \rho_\cdot \|_{\infty,[0,T]}.$$

We are ready to state the first existence and uniqueness for the solution of (7.46).

Proposition 130 *Suppose that $\gamma > \beta > \alpha$, $\alpha \in (1-\frac{\gamma}{2}, 2/3)$, $f \in \mathcal{C}_1^{2\alpha-1+\varepsilon}([0,T])$ and that the initial condition is as in Condition ($\xi 1$). Then, equation* (7.46) *has a unique solution in the space $\mathcal{L}_T$.*

On the other hand, for $\beta > 1-\alpha$, we have that $\mathcal{M} : \mathcal{C}_1^{1-\alpha}([0,T]) \to \mathcal{C}_1^{1-\alpha}([0,T])$, as it is proven in [112] (Lemma 3.3). Remember that the operator $\mathcal{M}$ is introduced in (7.47). In consequence, in the case that the initial condition satisfies Condition ($\xi 2$), we can also utilize the fixed point theorem to state the following result.

Proposition 131 *Assume that $1 > \beta \geq 1-\alpha$, $\alpha \in \left(1-\frac{\gamma}{2}, \gamma\right)$, $f \in \mathcal{C}_1^{1-\alpha}([0,T])$ and that Condition ($\xi 2$) holds. Then, equation*(7.46) *has a unique solution in the space $\mathcal{C}_1^{1-\alpha}([0,T])$.*

We observe that it is natural to have a function as an initial condition in equation (7.46). Indeed, suppose that we are studying the Volterra-type equation

$$y_t = a + \int_0^t (t-s)^{-\alpha}\Psi(y_s)d\theta_s, \quad t \in [0,T],$$

where the initial condition is a constant $a \in \mathbb{R}$, and that we know the solution of this equation up to a time $T_0 < T$. Therefore, now we need to analyze the equation

$$\begin{aligned} y_t &= a + \int_0^{T_0} (t-s)^{-\alpha}\Psi(y_s)d\theta_s + \int_{T_0}^t (t-s)^{-\alpha}\Psi(y_s)d\theta_s \\ &= \xi_t + \int_{T_0}^t (t-s)^{-\alpha}\Psi(y_s)d\theta_s, \quad t \in [T_0,T], \end{aligned}$$

to figure out the solution y on the whole interval $[0,T]$. But, in this equation, $\xi_\cdot = a + \int_0^{T_0}(\cdot - s)^{-\alpha}\Psi(y_s)d\theta_s$ is a function.

We also point out that considering Condition ($\xi 1$) (resp. ($\xi 2$)), we deal with an equation in the sense of Riemann-Liouville (resp. Caputo). For more information about this subject when $\theta_s = s$, we refer to Podlubny [179].

Remark 109 *Let $B = \{B_t : t \in [0,T]\}$ be a d-dimensional fractional Brownian motion with Hurst parameter $H \in (1/2, 1)$ defined on a complete probability space $(\Omega, \mathcal{F}, P)$. Consider the m-dimensional stochastic differential equation*

$$x_t = x_0 + \sum_{j=1}^d \int_0^t \sigma^j(x_s)dB_s^j, \quad t \in [0,T], \tag{7.48}$$

where $x_0 \in \mathbb{R}^m$. If σ is Hölder continuous with exponent $\kappa > \frac{1}{H} - 1$, then there exists a solution x, which has γ-Hölder continuous trajectories, for any $\gamma < H$. This has been proved by Lyons [131] using the Young's integral and p-variation estimates (see Section 2.1). An extension of this result where there is a measurable drift with linear growth is given by Duncan and Nualart [45]. In both cases, the solutions are figure out via a Picard iteration procedure.

For the case that $\kappa \leq \frac{1}{H} - 1$, León et al. [118] use the extension of the Young integral given by the fractional calculus, which is introduced in Section 2.3, to build a solution to equation (7.48). *For instance, suppose that $m = d = 1$, $x_0 = 0$ and $\sigma(\xi) = C|\xi|^\kappa$. Then, the process $y_t = \phi^{-1}(B_t)$, $t \in [0,T]$, is a solution different than zero to equation* (7.48), *where $\phi(\xi) = \int_0^\xi \frac{dx}{\sigma(x)}$. Also note that $y \equiv 0$ is another solution to* (7.48). *In particular, this shows that, under the weak assumptions on σ in the last three cases, we cannot expect the uniqueness for the solution, which requires σ to be differentiable with Hölder continuous partial derivatives of orders bigger than $\frac{1}{H} - 1$ (see [131, 160]).*

7.5 Linear fractional differential equations of Skorohod type

In this section, $B = \{B_t : t \in [0,T]\}$ is fractional Brownian motion with Hurst parameter $H \in (0, 1/2)$ and we use the convention

$$\alpha = \frac{1}{2} - H.$$

Here, we study the chaos decomposition of the solution to the linear stochastic differential equation of the form

$$X_t = \eta + \int_0^t a(s)X_s ds + \int_0^t b(s)X_s \delta_2 B_s, \quad t \in [0,T]. \tag{7.49}$$

In this equation, η is a square-integrable random variable having the chaos decomposition

$$\eta = \sum_{n=0}^{\infty} I_n(\eta_n), \tag{7.50}$$

$a \in L^2([0,T])$ and $b \in \mathcal{H}$, which is defined in (1.83). That is,

$$\mathcal{H} = \left\{ f \in L^2([0,T]) : f(s) = s^\alpha I_{T-}^\alpha \left(u^{-\alpha}\phi_f(u)\right)(s) \text{ for some } \phi_f \in L^2([0,T]) \right\}.$$

The stochastic integral is the extension of the divergence operator, which is analyzed in Sections 4.2 and 4.3.1

The main purpose of this section is to apply Theorems 12 and 55 to establish the existence of a unique continuous solution to the linear fractional differential equation (7.49) through the chaos decomposition approach. This chaotic expansion procedure is initiated in Shiota [196] for $H = 1/2$ (i.e., B is Brownian motion).

First assume that equation (7.49) has a solution X in $L^2(\Omega \times [0,T])$ having the chaos decomposition

$$X_t = \sum_{n=0}^{\infty} I_n(f_n^t), \quad f_n \in \mathcal{H}^{\odot n} \otimes L^2([0,T]). \tag{7.51}$$

Then the uniqueness of the chaotic representation (7.51) established in Theorem 12, together with the characterization of the extension of the divergence operator via the chaos representation (i.e., Theorem 55), implies

$$f_0^t = \eta_0 \exp(\int_0^t a(s)ds), \quad t \in [0,T],$$

and, for $t, t_1, \ldots, t_n \in [0,T]$ and $n \geq 1$,

$$\begin{aligned} f_n^t(t_1,\ldots,t_n) &= \eta_n(t_1,\ldots,t_n) + \int_0^t a(s) f_n^s(t_1,\ldots,t_n)ds \\ &+ \frac{1}{n}\sum_{j=1}^n b(t_j) f_{n-1}^{t_j}(\hat{t}_j) 1_{[0,t]}(t_j). \end{aligned} \tag{7.52}$$

Here $f_{n-m}^t(\hat{t}_{i_1},\ldots,\hat{t}_{i_m})$ means that the function f_{n-m}^t is evaluated in t's other than $t_{i_1},\ldots,t_{i_m}$. Therefore, by induction on $n \in \mathbb{N}$, we have

$$\begin{aligned} & f_n^t(t_1,\ldots,t_n) \\ &= \exp\left(\int_0^t a(s)ds\right)\Bigg[\eta_n(t_1,\ldots,t_n) + \sum_{j=1}^n \sum_{\Delta_{j,n}} \frac{(n-j)!}{j!n!} b^{\otimes j}(t_{i_1},\ldots,t_{i_j}) \\ &\quad \times \eta_{n-j}(\hat{t}_{i_1},\ldots,\hat{t}_{i_j}) 1_{[0,t]^j}(t_{i_1},\ldots,t_{i_j})\Bigg]. \end{aligned} \tag{7.53}$$

where

$$\Delta_{j,n} = \{\{i_1,\ldots,i_j\} \subset \{1,\ldots,n\} : i_k \neq i_\ell \text{ if } k \neq \ell\}.$$

Consequently, equation (7.49) has at most one solution in $L^2(\Omega \times [0,T])$.

Conversely, suppose that the functions f_n given by (7.53) satisfy the following conditions:

1. For every $n \geq 0$, $f_n \in \mathcal{H}^{\odot n} \otimes L^2([0,T])$.
2. The process $Y_t = \sum_{n=0}^{\infty} I_n(f_n^t)$ is in $L^2(\Omega \times [0,T])$. That is,

$$\sum_{n=0}^{\infty} n! \int_0^T ||f_n^t||^2_{\mathcal{H}^{\otimes n}} dt < \infty.$$

3. For almost all $t \in [0,T]$, $bY1_{[0,t]}$ belongs to Dom δ^* (see Definition 40 and Theorem 55).

Then process Y is a solution in $L^2(\Omega \times [0,T])$ of equation (7.49).

This approach is the main tool to establish the following existence and uniqueness result for the solution to (7.49), where we use the following notation: ϕ_b is introduced in (1.83), $C(p,q)$ denotes the norm of the fractional integral I_{T-}^{α} as a bounded linear operator from $L^p([0,T])$ into $L^q([0,T])$. Remember that Theorem 7 explains us how big q can be. Also, we utilize the convention

$$B_{H,p,t} = 1 + C_H\left(B_{T,p}||\phi_b 1_{[0,t]}||_{L^p([0,T])} + ||\phi_b 1_{[0,t]}||_{L^2([0,T])}\right)^2$$

with

$$B_{T,p} = \frac{T^{(p-2)/2p}}{\Gamma(1-\alpha)} C(p, p/(1-\alpha p)) \left(\frac{p-2(1-\alpha p)}{p-2}\right)^{(p-2(1-p\alpha))/2p}.$$

Theorem 94 *Let η be as in (7.50), $p \in (2, 1/\alpha)$, $a \in L^2([0,T])$, $b \in \mathcal{H}$, $\phi_b \in L^p([0,T])$ and*

$$\sum_{k=0}^{\infty}(k+1)!||\eta_k||^2_{\mathcal{H}^{\otimes k}}(\sup_{t\in[0,T)} B_{H,\tilde{p},t})^k < \infty \tag{7.54}$$

for some $\tilde{p} \in (2,p)$. Then, the process $X_t = \sum_{n=0}^{\infty} I_n(f_n^t)$, where $t \in [0,T]$ and f_n is given by (7.53), is the unique solution to equation (7.49) in $L^2(\Omega \times [0,T])$. Additionally, suppose

$$\sum_{k=1}^{\infty} e^{k\theta} k^{k/2}||\eta_k||_{\mathcal{H}^{\otimes k}} < \infty, \tag{7.55}$$

for some θ satisfying $(1+e^{2\theta})(\frac{p-2}{2p} \wedge \alpha \wedge \frac{(1-\alpha p)}{p}) > 1$. Then, the process X has a continuous version in the space $L^2(\Omega \times [0,T])$.

Remarks 19 *i) In Lemma 78 below, we show that $b1_{[0,t]} \in \mathcal{H}$ and estimate $||\phi_{b1_{[0,t]}}||_{L^{\tilde{p}}([0,T])}$, for $t \in [0,T]$.*

ii) Examples of random variables η that satisfy Hypothesis (7.54) are the following:

a) η has finite chaos decomposition. That is, there is $m \in \mathbb{N}$ such that $\eta = \sum_{n=0}^{m} I_n(\eta_n)$.

b) The chaos decomposition of η has exponential growth. It means, there is a positive constant C such that, for all $n \in \mathbb{N}$,

$$||\eta_n||_{\mathcal{H}^{\otimes n}} \leq \frac{C^n}{n!}.$$

c) There exists $\varepsilon > 0$ such that $\sum_{k=0}^{\infty}(k!)^{1+\varepsilon}||\eta_k||^2_{\mathcal{H}^{\otimes k}} < \infty$.

iii) From (7.53), the solution X of equation (7.49) has the form

$$X_t = \exp(\int_0^t a(s)ds)Y_t, \quad t \in [0,T],$$

where Y is the solution in $L^2(\Omega \times [0,T])$ to the equation

$$Y_t = \eta + \int_0^t b(s)Y_s\delta_2 B_s, \quad t \in [0,T].$$

iv) We have that (7.54) is satisfied (resp. (7.55) is false) if

$$||\eta_k||_{\mathcal{H}^{\otimes k}} = \left[k^2(k+1)!(\sup_{t\in[0,T)} B_{H,\tilde{p},t})^k\right]^{-1/2}$$

(resp. and $e^\theta \geq \sup_{t\in[0,T)} B_{H,\tilde{p},t}$).

v) (7.55) does not necessarily give that (7.54) is satisfied if θ is such that $e^{\theta+\varepsilon} \leq (\sup_{t\in[0,T)} B_{H,\tilde{p},t})^{1/2}$ for some $\varepsilon > 1/2$. In fact, in the case that

$$||\eta_k||_{\mathcal{H}^{\otimes k}} = (\sup_{t\in[0,T)} B_{H,\tilde{p},t})^{-k/2}((k+1)!)^{-1/2},$$

we have that (7.55) holds, but (7.54) is not true.

vi) Initial conditions as those in Statement ii) meet both Conditions (7.54) *and* (7.55).

The proof of Theorem 94 is mainly based on Theorem 7 and the following two auxiliary results.

Lemma 78 *Let $f \in \mathcal{H}$ be such that $\phi_f \in L^p([0,T])$, for some $p \in (2, 1/\alpha)$, and $t \in [0,T]$. Then, $f1_{[0,t]}$ also belongs to $\mathcal{H}$ and*

$$||\phi_{f1_{[0,t]}}||_{L^{p'}([0,T])} \leq C_{\alpha,p',p,t}||\phi_f||_{L^p([0,T])},$$

where $p' \in [2,p)$ and $C_{\alpha,p',p,t} = t^{(p-p')/p'p} + \frac{C(p,p(1-\alpha p)^{-1})}{\Gamma(1-\alpha)}(\int_0^t (t-s)^{-p'\alpha q} ds)^{1/p'q}$ with $q = \frac{p}{p-p'(1-\alpha p)}$.

Remarks 20 *i) Note that $1 - p'\alpha q > 0$.*

ii) Remember that $C(p, p(1-\alpha p)^{-1})$ is the norm of the linear operator $I^\alpha_{T-} : L^p([0,T]) \to L^{p(1-\alpha p)^{-1}}([0,T])$ (according to Theorem 7).

Proof. Fix $t \in [0,T]$. Then, Proposition 17 yields

$$\begin{aligned}(D^\alpha_{T-}(u^{-\alpha} f(u) 1_{[0,t]}(u)))(s) &= 1_{[0,t]}(s)\Big[s^{-\alpha}\phi_f(s) \\ &\quad + \frac{\alpha}{\Gamma(1-\alpha)}\int_t^T \frac{u^{-\alpha} f(u)}{(u-s)^{1+\alpha}} du\Big], \quad s \in [0,T].\end{aligned}$$

Then, (1.83) gives

$$\phi_{f1_{[0,t]}}(s) = 1_{[0,t]}(s)\left[\phi_f(s) + \frac{\alpha s^\alpha}{\Gamma(1-\alpha)}\int_t^T \frac{u^{-\alpha} f(u)}{(u-s)^{1+\alpha}} du\right], \quad s \in [0,T]. \tag{7.56}$$

Now, we use the inequality

$$\alpha \int_t^r (r-u)^{\alpha-1} u^{-\alpha} (u-s)^{-\alpha-1} du \leq t^{-\alpha}(t-s)^{-\alpha}(r-s)^{-1}(r-t)^\alpha, \quad 0 < s < t < r,$$

which follows from the changes of variables $z = (r-u)(u-s)^{-1}$ and $y = (t-s)z$, and the fact that $y \mapsto \frac{(t-s)r+ys}{y+t-s}$ is a decreasing function on $[0, r-t]$, to establish

$$\begin{aligned}&||(\cdot)^\alpha 1_{[0,t]}(\cdot)\int_t^T \frac{u^{-\alpha}|f(u)|}{(u-\cdot)^{1+\alpha}} du||_{L^{p'}([0,T])} \\ &\leq \frac{1}{\alpha}||1_{[0,t]}(\cdot)(t-\cdot)^{-\alpha}\frac{1}{\Gamma(\alpha)}\int_t^T \frac{|\phi_f(r)|}{(r-\cdot)^{1-\alpha}} dr||_{L^{p'}([0,T])}.\end{aligned}$$

Thus, Theorem 7 allows us to conclude that, for $q = \frac{p}{p-p'(1-\alpha p)}$,

$$\begin{aligned}&\left\|(\cdot)^\alpha 1_{[0,t]}(\cdot)\int_t^T \frac{u^{-\alpha}|f(u)|}{(u-\cdot)^{1+\alpha}} du\right\|_{L^{p'}([0,T])} \\ &\leq \frac{C(p,p(1-\alpha p)^{-1})}{\alpha}\left(\int_0^t (t-s)^{-p'\alpha q} ds\right)^{1/p'q} ||\phi_f||_{L^p([0,T])}.\end{aligned}$$

Therefore, from (7.56), we have that the proof is complete. □

Lemma 79 *Let f be a function in the space $\mathcal{H}$ and $g : [0,T] \to \mathbb{R}$ a β-Hölder continuous function with $\beta > \alpha$. Then, the function gf is also in $\mathcal{H}$.*

Proof: Applying equality (1.61) again, we get that, for $s \in [0,T]$,

$$(D^\alpha_{T-}(u^{-\alpha}g(u)f(u)))(s) = g(s)(D^\alpha_{T-}(u^{-\alpha}f(u)))(s) + \frac{\alpha}{\Gamma(1-\alpha)}\int_s^T u^{-\alpha}f(u)\frac{g(s)-g(u)}{(u-s)^{\alpha+1}}du.$$

Finally, the Hölder continuity of g, Fubini's theorem and Hölder inequality imply that $s \mapsto s^\alpha \int_s^T u^{-\alpha}|f(u)|\frac{|g(s)-g(u)|}{(u-s)^{\alpha+1}}du$ is a square-integrable function. Indeed,

$$\begin{aligned}
&\int_0^T \left(s^\alpha \int_s^T u^{-\alpha}|f(u)|\frac{|g(s)-g(u)|}{(u-s)^{\alpha+1}}du\right)^2 ds \\
&\leq \int_0^T \left(\int_s^T |f(u)|\frac{|g(s)-g(u)|}{(u-s)^{\alpha+1}}du\right)^2 ds \leq C\int_0^T \left(\int_s^T |f(u)|(u-s)^{\beta-\alpha-1}du\right)^2 ds \\
&\leq C\int_0^T\int_s^T |f(u)|^2(u-s)^{\beta-\alpha-1}duds = C\int_0^T |f(u)|^2\int_0^u (u-s)^{\beta-\alpha-1}dsdu < \infty.
\end{aligned}$$

Hence, from (1.83), we have that the result is satisfied. □

Remark 110 *The approach explained in this section is also used by León and Pérez-Abreu [122] to consider equation (7.49) in the case that $H = 1/2$ (i.e., the process B is Brownian motion), $b \in L^2([0,T])$ and a is a process in the chaos of order 1. In this case the relation (7.52) becomes much more complicated and quite difficult to solve due to the multiplication formula in Lemma 12. In fact, the kernel f_n now depends on the kernels f_{n-1} and f_{n+1}. As an application, León and Márquez-Carreras [113] study some types of stability for the solution to the SDE (7.49).*

7.6 Stochastic heat equation

In this section, we study the existence and uniqueness of the solution to the stochastic partial differential equation

$$\frac{\partial u}{\partial t} = A_t u + f(t,x,u(t,x)) + g(t,x,u(t,x))\frac{\partial^2 W}{\partial t\partial x}, \tag{7.57}$$

with Dirichlet boundary conditions $u(t,0) = u(t,1) = 0$. Here, the initial condition u_0 is a continuous function on $[0,1]$ such that $u_0(0) = u_0(1) = 0$, W is an isonormal Gaussian process defined on the Hilbert space $L^2([0,T]\times[0,1])$ (see Definition 19) and A is a random second order differential operator of the form

$$A_t = a(t,x)\frac{\partial^2}{\partial x^2} + b(t,x)\frac{\partial}{\partial x}, \quad (t,x) \in [0,T]\times[0,1]. \tag{7.58}$$

Here, the coefficients $f, g : \Omega \times [0,T]\times[0,1]\times\mathbb{R} \to \mathbb{R}$ are measurable functions satisfying Lipschitz and linear growth type conditions, and a and b are measurable random fields. Also, we assume that these four functions are adapted to the filtration generated by W. To do so, we first introduce the framework and the notation that we use in this section.

For $t \in [0,T]$, we denote the set $[0,t] \times [0,1]$ and the Borel subsets of $[0,T] \times [0,1]$ by I_t and $\mathcal{B}(I_T)$, respectively. Consider a zero-mean Gaussian family of random variables $W = \{W(A) : A \in \mathcal{B}(I_T)\}$ defined on a complete probability space $(\Omega, \mathcal{F}, P)$ such that

$$E(W(A)W(B)) = \mu(A \cap B),$$

where μ is the Lebesgue measure on I_T. Throughout this section, $\mathcal{K}$ stands for the Hilbert space $L^2(I_T)$ and, for $h \in \mathcal{K}$, $W(h) = \int_{I_T} h(s,x)dW(s,x)$. Note that, as in the definition of the Wiener integral given in Section 1.2.2, we can define last stochastic integral as

$$\begin{aligned} \int_{I_T} h(s,x)dW(s,x) &= L^2(\Omega) - \lim_{n\to 0} \int_{I_T} h_n(s,x)dW(s,x) \\ &:= L^2(\Omega) - \lim_{n\to 0} \sum_{j=1}^{N_n} a_{j,n} W(A_{j,n}), \end{aligned}$$

where $h_n = \sum_{j=1}^{N_n} a_{j,n} 1_{A_{j,n}}$ is a simple function that converges to h in $\mathcal{K}$, as $n \to \infty$. Indeed, we only need to observe that

$$E\left(\left(\int_{I_T} h_n(s,x)dW(s,x)\right)^2\right) = \int_{I_T} h_n^2(s,x)d\mu(s,x).$$

In conclusion, $W = \{W(h) : h \in \mathcal{K}\}$ is an isonormal Gaussian process. Therefore, now, the smooth and cylindrical random variables F in the space $\mathcal{S}$ have the form

$$F = f(W(h_1), \ldots, W(h_n)),$$

with $f \in C_b^\infty(\mathbb{R}^n)$ and $h_1, \ldots, h_n \in \mathcal{K}$. In consequence, for this random variable,

$$D_{t,x}F = \sum_{i=1}^{n} \frac{\partial f}{\partial x_i}(W(h_1), \ldots, W(h_n))h_i(t,x), \quad (t,x) \in I_T.$$

Hence, the domain of the divergence operator δ_2 (denoted by Dom δ_2) is the space of random fields u in $L^2(\Omega \times I_T)$ such that there exists a random variable $\delta_2(u) \in L^2(\Omega)$ satisfying the duality relation

$$E(\delta_2(u)F) = E\left(\int_{I_T} (D_{t,x}F)\, u(t,x)d\mu(t,x)\right), \quad \text{for every } F \in \mathcal{S}.$$

We assume that the σ-algebra $\mathcal{F}$ is generated by W and the P-null sets. $W(t,x) := W([0,t] \times [0,x])$ defines a two-parameter Wiener process. Thus, for each $t \in [0,T]$, $\mathcal{F}_t$ denotes the σ-algebra generated by $\{W(s,x) : (s,x) \in I_t\}$ and the P-null sets. We say that a random field $u = \{u(t,x) : (t,x) \in I_T\}$ is $\mathcal{F}_t$-adapted if $u(t,x)$ is $\mathcal{F}_t$-measurable for all $(t,x) \in I_T$.

Now, the forward integral with respect to the two-parameter Wiener process W is introduced in the following way. Let $u = \{u(t,x) : (t,x) \in I_T\}$ be a measurable random field with integrable trajectories. That is, $\int_{I_T} |u(t,x)|d\mu(t,x) < \infty$ with probability 1. For every $n > 1$, we introduce the approximation

$$I^n(u) = \frac{2^{2n}}{T} \int_{I_T} u(t,x)W(J^{n+}(t,x))d\mu(t,x),$$

where

$$J^{n+}(t,x) = [t, (t + 2^{-n}T) \wedge T] \times [x, (x + 2^{-n}) \wedge 1].$$

Definition 52 *Let u be a random field in $L^1(I_T)$ with probability 1. We define the forward integral of u with respect to W as the limit in probability*

$$\int_{I_T} u(t,x)dW^-_{t,x} := P - \lim_{n\to\infty} I^n(u),$$

provided the limit exists. In this case, we say that u belongs to the domain of the forward integral (denoted by $Dom\,\delta^-$, for short).

On the other hand, the stochastic partial differential equation (7.57) is a symbolic expression and it could be interpreted in different ways (see, for example, Friedman [56] and Walsh [210]). In order to contemplate one of them, we consider the fundamental solution $\Gamma = \{\Gamma_{t,s}(x,y) : 0 \le s < t \le T \text{ and } x, y \in [0,1]\}$ of the random second order differential operator A in (7.58), which satisfies

$$\begin{aligned} \frac{\partial \Gamma_{t,s}}{\partial t} &= A_t \Gamma_{t,s}, \quad 0 \le s < t \le T, \\ \Gamma_{t,s}(0,y) &= \Gamma_{t,s}(1,y) = 0, \quad 0 \le s < t \le T \text{ and } y \in [0,1], \\ \lim_{t\downarrow s} \int_0^1 \Gamma_{t,s}(x,y) f(y) dy &= f(x), \quad \text{for } f \in \mathcal{C}([0,1]). \end{aligned}$$

It is well-known that Γ can be constructed using the parametrix method studied in Friedman [56]. This method is applied in Alòs et al. [6] to analyze some properties of Γ like estimates for Γ, and its derivatives $D\Gamma$ and $D^2\Gamma$ in the Malliavin calculus sense. In this framework, the required conditions on the stochastic operator A to deal with the parametrix method in [6] are the following:

(A1) The coefficients $a, b : \Omega \times I_T \to \mathbb{R}$ are measurable and $\mathcal{F}_t$-adapted random fields (*adaptability*).

(A2) There exist constants $0 < c_1 < c_2$ such that (*uniform ellipticity*)

$$c_1 \le a(t,x) \le c_2, \quad s,t \in [0,T] \text{ and } x, y \in [0,1].$$

(A3) The random fields a and b are continuous and bounded in I_T, and satisfy the following Hölder type conditions:

$$|a(t,x) - a(s,y)| \le K(|t-s|^{\frac{\alpha}{2}} + |x-y|^{\alpha}) \quad \text{and} \quad |b(t,x) - b(t,y)| \le K|x-y|^{\alpha},$$

for some constants $K > 0$ and $\alpha \in (0,1]$, and for all $s,t \in [0,T]$ and $x,y \in [0,1]$ (*Hölder continuity*).

(A4) For each $(t,x) \in I_T$, the random variables $a(t,x)$ and $b(t,x)$ are in the space $\mathbb{D}^{2,2}$.

(A5) There exists a non-negative random field $\psi = \{\psi(t,x) : (t,x) \in I_T\}$ such that $\int_0^1 \psi^p(t,x)dx < C$, for some constant $C > 0$ and $p > 8$, and, for every (t,x), (s,y) and (r,v) in I_T,

$$|D_{r,v}a(t,x)| + |D_{r,v}b(t,x)| \le \psi(r,v)$$

and

$$|D_{r,v}(a(t,x) - a(s,y))| \le \psi(r,v)(|t-s|^{\frac{\alpha}{2}} + |x-y|^{\alpha}).$$

(A6) There is a non-negative measurable function $L = \{L(r,z,s,y) : (r,z),(s,y) \in I_T\}$ such that $\int_0^1\int_0^1 L^p(r,z,s,y)drdy < C$, for some constant $C > 0$ and $p > 0$, and

$$|D_{r,z}D_{\tilde{r},\tilde{z}}a(t,x)| + |D_{r,z}D_{\tilde{r},\tilde{z}}b(t,x)| \leq L(r,z,\tilde{r},\tilde{z})$$

and

$$|D_{r,z}D_{\tilde{r},\tilde{z}}(a(t,x) - a(s,y))| \leq L(r,z,\tilde{r},\tilde{z})(|t-s|^{\frac{\alpha}{2}} + |x-y|^{\alpha}),$$

for every (t,x), (s,y), (r,z), and $(\tilde{r},\tilde{z})$ in I_T.

Under Assumptions (A1)-(A6), Alòs et al. [6] have established that the fundamental solution $\Gamma = \{\Gamma_{s,t}(x,y) : 0 \leq s < t \leq T \text{ and } x,y \in [0,1]\}$ enjoys the following properties:

(P1) For each $0 \leq s < t \leq T$, $x,y \in [0,1]$, $\Gamma_{t,s}(x,y)$ is an $\mathcal{F}_t$-measurable random variable in $\mathbb{D}^{2,2}$ such that

$$|\Gamma_{t,s}(x,y)| \leq \tilde{a}(t-s)^{1/2}\exp\left(-(x-y)^2/[\tilde{b}(t-s)]\right), \quad \text{for some constants } \tilde{a},\tilde{b} > 0,$$

and

$$\|D\Gamma_{t,\cdot}(x,\cdot)\|_{L^2(I_t^2\times\Omega)} + \|D^2\Gamma_{t,\cdot}(x,\cdot)\|_{L^2(I_t^3\times\Omega)} < \infty.$$

(P2) $\Gamma_{t,s}(x,y)$ is pathwise continuous on the set $\{(s,y,t,x) : 0 \leq s < t \leq T \text{ and } x,y \in [0,1]\}$.

(P3) For all $0 \leq s < r < t \leq T$ and $x,y \in [0,1]$,

$$\Gamma_{t,s}(x,y) = \int_0^1 \Gamma_{t,r}(x,z)\Gamma_{r,s}(z,y)dz.$$

(P4) There exists a version of the derivative $D_{r,z}\Gamma_{t,s}(x,y)$ such that for all $\omega \in \Omega$, $x,z \in [0,1]$ and $0 < s < t \leq T$, the limit

$$D^-_{s,z}\Gamma_{t,s}(x,\cdot) = L^2([0,1]) - \lim_{\epsilon\downarrow 0} D_{s,z}\Gamma_{t,s-\epsilon}(x,\cdot)$$

exists. Moreover, $D^-_{s,z}\Gamma_{t,s}(x,y)$ is in $\mathbb{D}^{1,2}$ for each $0 < s < t$ and $x,y,z \in [0,1]$.

(P5) For all $x \in [0,1]$, $0 \leq s < t \leq T$ and p as in (A5), we have

$$E\left[\int_0^1 \left(\int_0^1 |D^-_{s,y}\Gamma_{t,s}(x,u)|^{\frac{p}{p-1}}du\right)^{p-1} dy\,\middle|\,\mathcal{F}_s\right] \leq C(t-s)^{-\delta},$$

for some constants $C > 0$ and $\delta \in [0,1)$.

(P6) For all $x,z \in [0,1]$, $0 \leq r < s < t \leq T$ and p as in (A5),

$$E\left[\int_0^1 \left(\int_0^1\int_0^1 |D_{r,z}D^-_{s,y}\Gamma_{t,s}(x,u)|^{\frac{p}{p-1}}dudz\right)^{p-1} dy\,\middle|\,\mathcal{F}_s\right] \leq C(t-s)^{-\delta},$$

for some constants $C > 0$ and $\delta \in [0,1)$.

(P7) For all $0 \leq r < s < t \leq T$ and $x,y,z \in [0,1]$, the equality

$$D_{s,y}\Gamma_{t,r}(x,z) = \int_0^1 \left[D^-_{s,y}\Gamma_{t,s}(x,u)\right]\Gamma_{s,r}(u,z)du.$$

holds.

(P8) Let p be as in (A5). Then, there exist a version of the derivative $D_{r,z}\Gamma_{t,s}(x,y)$ and an element $D^*\Gamma = \{D^*_{r,z}\Gamma_{t,r}(x,z) : (r,z) \in I_t\}$ in $L^{\frac{p}{p-1}}(\Omega \times I_t)$, for each $(t,x) \in I_T$, such that

$$\int_{I_t} \left(\sup_{(s,y)\in J^{m-}(r,z)} E|D_{r,z}\Gamma_{t,s}(x,y) - D^*_{r,z}\Gamma_{t,r}(x,z)|^{\frac{p}{p-1}} \right) d\mu(r,z) \to 0,$$

as $m \to \infty$. Here, $J^{m-}(t,x) = [(t - 2^{-m}T) \vee 0, t] \times [(x - 2^{-m}) \vee 0, x]$.

With the last properties on Γ, Alòs et al. [6] have obtained the following relation between the forward integral and the divergence operator with respect to W, which is one of the main tools to study the stochastic partial differential equation (7.57).

Proposition 132 *Assume that Hypotheses (A1)-(A6) hold. Let $p > 8$ be as in (A5) and $\phi = \{\phi(s,y); (s,y) \in I_T\}$ an $\mathcal{F}_t$-adapted random field in $L^p(\Omega \times I_T)$. Then, for almost all $(t,x) \in I_T$,*

$$\{\Gamma_{t,s}(x,y)\phi(s,y)1_{[0,t]}(s), (s,y) \in I_T\}$$

belongs to $Dom\,\delta \cap Dom\,\delta^-$ and

$$\int_{I_t} \Gamma_{t,s}(x,y)\phi(s,y)dW^-_{s,y} = \int_{I_t} \Gamma_{t,s}(x,y)\phi(s,y)\delta_2 W_{s,y} + \int_{I_t} \left[D^*_{s,y}\Gamma_{t,s}(x,y)\right] \phi(s,y)d\mu(s,y).$$

This proposition allows us to establish the second main tool of Alòs et al. [6] to deal with the stochastic partial differential equation (7.57).

Proposition 133 *Under the assumptions of Proposition 132, we have that there is a constant $C_{p,T,\Gamma} > 0$ that only depends on T, p and Γ such that*

$$E\left(\sup_{(t,x)\in I_T} |\int_{I_t} \Gamma_{t,s}(x,y)\phi(s,y)dW^-_{s,y}|^p \right) \le C_{p,T,\Gamma} \int_{I_T} E|\phi(s,y)|^p d\mu(s,y).$$

Continuing with the analysis of equation (7.57), we now give the assumptions on the coefficients $f, g : \Omega \times [0,T] \times [0,1] \times \mathbb{R} \to \mathbb{R}$ of this equation.

(C1) The random fields f and g are measurable and $\mathcal{F}_t$-adapted such that

$$|f(t,x,y) - f(t,x,z)| \le L|y-z|, \qquad |f(t,x,y)| \le K(1+|y|),$$

$$|g(t,x,y) - g(t,x,z)| \le L|y-z| \quad \text{and} \quad |g(t,x,y)| \le K(1+|y|),$$

for all $t \in [0,T]$, $x \in [0,1]$ and $y,z \in \mathbb{R}$, and for some constants $K, L > 0$.

We are ready to introduce the notion of two solutions of the stochastic partial differential equation (7.57). Towards, this end, we consider the adjoint of the operator A_t in (7.58), which is formally written as

$$A_t^*\varphi = \frac{\partial^2}{\partial x^2}\left(a(t,\cdot)\varphi\right) - \frac{\partial}{\partial x}\left(b(t,\cdot)\varphi\right).$$

In order to define the meaning of weak solution for equation (7.57), we also assume that the operator A in (7.58) satisfies:

(A7) There exists a dense subset $\mathcal{H}_0$ in $L^2([0,1])$ such that $\mathcal{H}_0$ is included in the domain of $A_t^*(\omega)$, for all $(\omega,t) \in \Omega \times [0,T]$.

We are ready to introduce two interpretations of solution to equation (7.57).

Definition 53 *Let* $u = \{u(t,x) : (t,x) \in I_T\}$ *be an* $\mathcal{F}_t$*-adapted and continuous random field. We say that* u *is a weak solution to equation* (7.57) *if and only if, for every* $\varphi \in \mathcal{H}_0$ *and* $t \in [0,T]$*, we have*

$$\begin{aligned}\int_0^1 u(t,x)\varphi(x)dx &= \int_0^1 u_0(x)\varphi(x)dx + \int_{I_t} (A_s^*\varphi)(x)u(s,x)d\mu(s,x) \\ &+ \int_{I_t} f(s,x,u(s,x))\varphi(x)d\mu(s,x) + \int_{I_t} g(s,x,u(s,x))\varphi(x)dW_{s,x} \ \ w.p.1.\end{aligned}$$

The second interpretation of solution to the stochastic partial differential equation (7.57) is the following integral equation:

Definition 54 *Let* $u = \{u(t,x) : (t,x) \in I_T\}$ *be an* $\mathcal{F}_t$*-adapted and continuous random field. We say that* u *is a mild solution to equation* (7.57) *if and only if, for all* $t \in [0,T]$ *and* $x \in [0,1]$*, we have*

$$\begin{aligned}u(t,x) &= \int_0^1 \Gamma_{t,0}(x,y)u_0(y)dy + \int_{I_t} \Gamma_{t,s}(x,y)f(s,y,u(s,y))d\mu(s,y) \\ &+ \int_{I_t} \Gamma_{t,s}(x,y)g(s,y,u(s,y))dW_{s,y}^- \ \ w.p1.\end{aligned}$$

The following result is the relation between the mild and weak solutions of equation (7.57).

Theorem 95 (Alòs et al. [6]) *Suppose that Assumptions (A1)-(A7) and (C1) are satisfied. Then, equation* (7.57) *has a unique mild solution* $u = \{u(t,x) : (t,x) \in I_T\}$ *in* $L^2(\Omega \times I_T)$*. Moreover, this random field* u *is a weak solution of equation* (7.57).

Observe that all elements involved in the stochastic partial differential equation (7.57) are adapted to the filtration generated by W. Therefore, the analysis of this equation must be carried out through classical stochastic calculus of Itô. However, the stochastic integral

$$\int_{I_t} \Gamma_{t,s}(x,y)g(s,y,u(s,y))dW_{s,y}$$

has no meaning in the Itô's calculus sense since $\Gamma_{t,s}(x,y)$ is $\mathcal{F}_t$-measurable. That is, in this case, the mild solution has no meaning in the classical Itô's calculus. In consequence, this is a problem where all involved elements are adapted to the underlying filtration that is solved using the techniques of the Malliavin calculus because Itô's calculus is not useful for this problem.

Using the framework of semigroup theory, Theorem 95 has been considered by León and Nualart [117]. Moreover, the study of the equivalence of solutions to evolution equations with random generators via the Malliavin calculus was first considered in León [108].

Also, using the semigroup techniques and Itô's calculus, Chojnowska-Michalik [28] has studied the equivalence of several solutions in the case that $A_t \equiv A$ is a deterministic infinitesimal generator of a C_0-semigroup and the driven noise is a particular semimartingale. This analysis has been extended by León et al. [66, 107] to the case that $\{A_t : t \in [0,T]\}$ is a deterministic family of infinitesimal generators of C_0-semigroups on a real-separable Hilbert space (see Pazy [175]), and the driven noise is a general semimartingale that may depend on the solution.

Chapter 8

Appendix

Here, we analyze several auxiliary tools, which will help the reader to understand some proofs and results in this book.

8.1 Monotone class lemma

Although we are assuming that the reader is familiar with measure theory, in this chapter we study two forms of the monotone class lemma to keep this book self-contained.

In this section, B denotes a non-empty set.

Definition 55 *i) A family $\mathcal{A} \neq \emptyset$ of subsets of B is a π-system if $A_1, A_2 \in \mathcal{A}$ implies that $A_1 \cap A_2$ also belongs to $\mathcal{A}$.*

ii) A class $\mathcal{L}$ of subsets of B is a λ-system if:

a) $B \in \mathcal{L}$.

b) If $A_1, A_2 \in \mathcal{L}$, $A_1 \subset A_2$, then $A_2 \setminus A_1$.

c) Let $\{A_n \in \mathcal{L}, n \in \mathbb{N}\}$ be such that $A_n \subset A_{n+1}$. Then, $\cup_{n\in\mathbb{N}} A_n \in \mathcal{L}$.

In order to state the main results of this section, we need to establish the following tool:

Lemma 80 *Let $\mathcal{L}$ be a λ-system of B and $A_1, A_2 \in \mathcal{L}$ such that $A_1 \cap A_2 = \emptyset$. Then, $A_1 \cup A_2 \in \mathcal{L}$.*

Proof. Let $A \in \mathcal{L}$. Then, Definitions 55.*ii*.a) and 55.*ii*.b) yield $A^c \in \mathcal{L}$. Thus, using Definition 55.*ii*.b) again, we get that $A_2^c \setminus A_1$ also belongs to $\mathcal{L}$ due to $A_1 \subset A_2^c$. Finally, the fact that $A_1 \cup A_2 = (A_2^c \setminus A_1)^c$ implies the result. □

Lemma 81 *Let $\mathcal{L}$ be a family that is both a π-system and a λ-system. Then, $\mathcal{L}$ is also a σ-algebra.*

Proof. From Definition 55.*ii*.c), we only need to prove that $A_1 \cup A_2 \in \mathcal{L}$ if $A_1, A_2 \in \mathcal{L}$. But, $A_1 \cup A_2 = A_1 \cup (A_2 \setminus A_1 \cap A_2)$. Hence, the fact that $\mathcal{L}$ is a π-system and Lemma 80 imply that the result is true. □

Now, we are ready to establish the first main result of this section.

Theorem 96 (π and λ-systems) *Let $\mathcal{A}$ be a π-system and $\mathcal{L}$ a λ-system such that $\mathcal{A} \subset \mathcal{L}$. Then, $\sigma(\mathcal{A}) \subset \mathcal{L}$.*

Proof. Let $\mathcal{L}(\mathcal{A})$ be the λ-system generated by $\mathcal{A}$ (i.e., the intersection of all the λ-systems that contain $\mathcal{A}$). Note that $\mathcal{L}(\mathcal{A})$ is well-defined because the intersection of λ-systems is also a λ-system and 2^B is a λ-system. Consequently, $\mathcal{A} \subset \mathcal{L}(\mathcal{A}) \subset \mathcal{L}$. Thus, we only need

DOI: 10.1201/9781003484912-8

to see that $\mathcal{L}(\mathcal{A})$ is a π-system because, in this case, Lemma 81 implies that it is also a σ-algebra. Towards this end, choose $A \in \mathcal{A}$ and set $\mathcal{A}_1 = \{C \subset B : A \cap C \in \mathcal{L}(\mathcal{A})\}$. Since $\mathcal{A}$ is a π-system, it is easy to see that $\mathcal{A}_1$ is a λ-system that contains $\mathcal{A}$. Hence, we have that $\mathcal{L}(\mathcal{A}) \subset \mathcal{A}_1$, which yields that $A \cap C \in \mathcal{L}(\mathcal{A})$, for every $A \in \mathcal{A}$ and $C \in \mathcal{L}(\mathcal{A})$.

Finally, let $A \in \mathcal{L}(\mathcal{A})$. Then, proceeding similarly, we can show that

$$\mathcal{A}_2 = \{C \subset B : A \cap C \in \mathcal{L}(\mathcal{A})\}$$

is a λ-system such that $\mathcal{L}(\mathcal{A}) \subset \mathcal{A}_2$, which establishes that $\mathcal{L}(\mathcal{A})$ is a π-system. Therefore, the proof is complete. □

As an example, we consider the following result, which is used in the proof of the Fubini's theorem for Itô's integral with respect to Brownian motion.

Lemma 82 *Let $f : \Omega \times [0,T] \times \mathbb{R} \to \mathbb{R}$ be an $\mathcal{F} \otimes \mathcal{B}([0,T]) \otimes \mathcal{B}(\mathbb{R})$-measurable random field and $X : \Omega \times [0,T] \to \mathbb{R}$ a measurable stochastic process. Then, $t \mapsto f(t, X_t)$ is also a measurable process.*

Proof. Observe that it is enough to prove that the process $t \mapsto 1_A(t, X_t)$ is measurable, for every $A \in \mathcal{F} \otimes \mathcal{B}([0,T]) \otimes \mathcal{B}(\mathbb{R})$ due to any measurable random field being the limit of linear combinations of processes of this form. So, we introduce the λ-system

$$\mathcal{L} = \{A \in \mathcal{F} \otimes \mathcal{B}([0,T]) \otimes \mathcal{B}(\mathbb{R}) : 1_A(\cdot, X_\cdot) \text{ is a measurable process}\}$$

and the π-system

$$\mathcal{A} = \{A \in \mathcal{F} \otimes \mathcal{B}([0,T]) \otimes \mathcal{B}(\mathbb{R}) : A = F \times D \times C\}.$$

Thus, the result is a consequence of Theorem 96, and the facts that $\sigma(\mathcal{A}) = \mathcal{F} \otimes \mathcal{B}([0,T]) \otimes \mathcal{B}(\mathbb{R})$ and that, for $A = F \times D \times C$ in $\mathcal{A}$, the process $1_A(t, X_t) = 1_{F \times D}(t) 1_C(X_t)$ is measurable since it is the product of two measurable processes. □

Now, we state the second main result of this section.

Theorem 97 (monotone class lemma) *Let $B \neq \emptyset$, $\mathcal{A}$ a π-system of subsets of B and $\mathcal{G}$ a linear family of functions from B into $\mathbb{R}$ such that:*

i) $1 \in \mathcal{G}$.

ii) $1_A \in \mathcal{G}$, *for every* $A \in \mathcal{A}$.

iii) If, for $n \in \mathbb{N}$, $\xi_n \in \mathcal{G}$, $\xi_n \geq 0$, $\xi_n \uparrow \xi$ and ξ is a bounded function, then $\xi \in \mathcal{G}$.

Then, $\mathcal{G}$ contains all the bounded and $\sigma\{\mathcal{A}\}$-measurable functions.

Proof. By hypothesis, the family $\mathcal{L} = \{A \subset B : 1_A \in \mathcal{G}\}$ is a λ-system that contains $\mathcal{A}$. Thus, Theorem 96 implies that $\sigma\{\mathcal{A}\} \subset \mathcal{L}$. Therefore, the linearity of $\mathcal{G}$ gives that all the simple $\sigma\{\mathcal{A}\}$-measurable functions also belong to $\mathcal{G}$. Consequently, $\mathcal{G}$ contains all the non-negative bounded and $\sigma\{\mathcal{A}\}$-measurable functions η because, in this case, there exists a sequence $\{\eta_n : n \in \mathbb{N}\}$ of simple $\sigma\{\mathcal{A}\}$-measurable functions such that $\eta_n \uparrow \eta$, as $n \to \infty$. Finally, for a $\sigma\{\mathcal{A}\}$-measurable bounded function η, the result follows from the decomposition $\eta = \eta^+ - \eta^-$. □

8.2 Progressively measurable modifications

Now, we give the auxiliary tool needed to prove that Theorem 2 holds.

Lemma 83 *Let Y be a random variable in $L^1(\Omega)$ and $(\mathcal{F}_t)_{t\in[0,T]}$ a filtration on $(\Omega, \mathcal{F}, P)$. Then, there exists a progressively measurable modification of the stochastic process $t \mapsto E(Y|\mathcal{F}_t)$.*

Proof. The martingale convergence theorem (see Dellacherie and Meyer [38]) implies

$$\lim_{s\downarrow t} E(Y|\mathcal{F}_s) = E(Y|\mathcal{F}_{t+}) \text{ and } \lim_{s\uparrow t} E(Y|\mathcal{F}_s) = E(Y|\mathcal{F}_{t-}) \tag{8.1}$$

in $L^1(\Omega)$ and w.p.1. Here, $\mathcal{F}_{t-} = \sigma\left(\cup_{s<t}\mathcal{F}_s\right)$. Consequently, the function $t \mapsto E(Y|\mathcal{F}_t)$ is regulated from the interval $[0,T]$ into $L^1(\Omega)$. Thus, Dieudonné [40] (Result (7.6.1)) gives that there exists a countable subset $\mathcal{C}$ of $[0,T]$ such that the last function is continuous on $[0,T]\setminus\mathcal{C}$. In consequence, from (8.1), the progressively measurable process

$$Y_t^{(n)} = 1_{\{0\}}(t)E(Y|\mathcal{F}_0)+\sum_{i=0}^{n-1} E(Y|\mathcal{F}_{\frac{iT}{n}})1_{(\frac{iT}{n},\frac{(i+1)T}{n}]\setminus\mathcal{C}}(t)+\sum_{s\in(\mathcal{C}\setminus\{0\})} 1_{\{s\}}(t)E(Y|\mathcal{F}_s),\ t\in[0,T],$$

converges to $E(Y|\mathcal{F}_t)$ w.p.1, for each $t\in[0,T]$.

Finally, the fact that

$$\begin{aligned}\Gamma &:= \left\{(\omega,t)\in\Omega\times[0,T] : \lim_{n\to\infty} Y_t^{(n)}(\omega) \text{ exists}\right\}\\ &= \bigcap_{n=1}^{\infty}\bigcup_{m=1}^{\infty}\bigcap_{p,q=m}^{\infty}\left\{(\omega,t)\in\Omega\times[0,T] : |Y_t^{(p)}(\omega)-Y_t^{(q)}(\omega)|<\frac{1}{n}\right\}\end{aligned}$$

yields that 1_Γ is a progressively measurable process and, therefore, the process $\tilde{Y}_t(\omega) = \lim_{n\to\infty} Y_t^{(n)}(\omega)1_\Gamma(\omega,t)$ is the modification that we are looking for. That is, the proof is complete. □

As a consequence of the last lemma, we have the following:

Corollary 31 *Let $A\in\mathcal{F}\otimes\mathcal{B}([0,T])$ and $(\mathcal{F}_t)_{t\in[0,T]}$ a filtration on $(\Omega,\mathcal{F},P)$. Then, the process $t\mapsto E(1_A(\cdot,t)|\mathcal{F}_t)$ has a progressively measurable modification.*

Proof. We first assume that A is a measurable rectangle of the for $F\times I$, with $I\in\mathcal{B}([0,T])$ and $F\in\mathcal{F}$. Since, for $t\in[0,T]$, $E(1_A(\cdot,t)|\mathcal{F}_t) = 1_I(t)E(1_F|\mathcal{F}_t)$, then, Lemma 83 implies that the result is true in this case.

Finally, the result is also satisfied for any $A\in\mathcal{F}\otimes\mathcal{B}([0,T])$ due to a π and λ-systems argument (see Theorem 96) because the set of all the measurable rectangles is a π-system and the family

$$\Lambda = \{B\in\mathcal{F}\otimes\mathcal{B}([0,T]) : E(1_B(\cdot,t)|\mathcal{F}_t) \text{ has a progressively measurable modification}\}$$

is a λ system. Indeed, note that, in order to see that this last claim holds, we only need to show that if $A_n\uparrow A$, with $A_n\in\Lambda$, then A is also a set in Λ. But, for $n\in\mathbb{N}$, the definition of Λ allows us to find a progressively measurable modification $Y^{(n)}$ of $E(1_{A_n}|\mathcal{F}_t)$ such that $Y_t^{(n)}$ is nondecreasing for each t fixed. Consequently, the process Y defined as $\lim_{n\to\infty} Y^{(n)}$ on the set where this limit exists, and equal to zero otherwise, is a progressively measurable version of $E(1_A|\mathcal{F}_t)$ (for details see the proof of Lemma 83). Hence, the result is true. □

Lemma 84 *Let X be a measurable process and $\varepsilon > 0$. Then, there exists a measurable process Y with countable range such that*

$$\sup_{(\omega,t)\in\Omega\times[0,T]} |X_t(\omega) - Y_t(\omega)| < \varepsilon. \tag{8.2}$$

Proof. Let $\{y_n \in \mathbb{R} : n \in \mathbb{N}\}$ be a dense set of $\mathbb{R}$ and $B_\varepsilon(y_n)$ the open ball centered at y_n and radius ε. Then, the process $Y = \sum_{n=1}^{\infty} y_n 1_{\Gamma_n}$, with $\Gamma_n = \{(\omega, t) \in \Omega \times [0,T] : X_t(\omega) \in \left(B_\varepsilon(y_n) \setminus \cup_{i=1}^{n-1} B_\varepsilon(y_i)\right)\}$, satisfies inequality (8.2). Thus, the proof is complete. □

8.3 Bochner integral

Let $(X, \mathcal{C})$ be a measurable space, Y a Banach space and $f : X \to Y$. In this case, we can deal with different definitions of measurability for f. Among them, we can consider the following:

i) The function f is called measurable if $F^{-1}(C) \in \mathcal{B}(Y)$, for every $C \in \mathcal{C}$.

ii) The function f is said to be $\mathcal{C}$-strongly measurable if there exists a sequence $\{\varphi_n : X \to Y : n \in \mathbb{N}\}$ of simple functions of the form

$$\varphi_n = \sum_{i=1}^{N} y_{i,n} 1_{C_{i,n}} \quad \text{with } y_{i,n} \in Y \text{ and } C_{i,n} \in \mathcal{C}, \tag{8.3}$$

such that φ_n converges to f pointwise, as $n \to \infty$.

By Hytönen et al. [85] (Corollary 1.1.10), we have that a strongly measurable function is also a measurable one. Furthermore, in the case that Y is a separable Banach space, a function is strongly measurable if and only if it is measurable. This last fact is known as Pettis's measurability theorem.

Now, we assume that μ is a measure on $(X, \mathcal{C})$. Thus, we can consider other interpretation of measurability for f. The so-called strongly μ-measurability. It means, f is strongly μ-measurable if there is a sequence $\{\varphi_n : X \to Y : n \in \mathbb{N}\}$ of simple functions that converges to f μ-almost surely.

The Bochner integral of the simple function φ_n defined in (8.3) with respect to μ is given as

$$\int_X \varphi_n(x)\mu(dx) := \sum_{i=1}^{N} y_{i,n}\mu(C_{i,n}).$$

It is easy to see that this definition is independent of the representation of φ_n as a simple function. This fact is left as an exercise to the reader. The Bochner integral is extended to strongly μ-measurable functions:

Definition 56 *A strongly μ-measurable function $f : X \to Y$ is Bochner integrable with respect to the measure μ if there exists a sequence $\{\varphi_n : X \to Y : n \in \mathbb{N}\}$ of simple functions such that*

$$\int_X |\varphi_n(x) - f(x)|_Y \, \mu(dx) \to 0, \quad \text{as } n \to \infty. \tag{8.4}$$

In this case, for $C \in \mathcal{C}$, we defined the Bochner integral of f on C with respect to μ as

$$\int_C f(x)\mu(dx) = \lim_{n\to\infty} \int_C \varphi_n(x)\mu(dx).$$

It is also easy to show that the Bochner integral $\int_C f(x)\mu(dx)$ is well-defined and independent of the sequence $\{\varphi_n : n \in \mathbb{N}\}$ since $\left\{\int_C \varphi_n(x)\mu(dx) : n \in \mathbb{N}\right\}$ is a Cauchy sequence in Y and $\left|\int_C f(x)\mu(dx)\right|_Y \leq \int_C |f(x)|_Y \mu(dx)$, which follow from the definition of the Bochner integral for simple functions. Indeed, for $n, m \in \mathbb{N}$,

$$\left|\int_C (\varphi_n(x) - \varphi_m(x))\,\mu(dx)\right|_Y \leq \int_X |\varphi_n(x) - \varphi_m(x)|_Y\,\mu(dx)$$

and

$$\begin{aligned}\left|\int_C f(x)\mu(dx)\right|_Y &= \lim_{n\to\infty}\left|\int_C \varphi_n(x)\mu(dx)\right|_Y \leq \lim_{n\to\infty}\int_C |\varphi_n(x)|_Y\,\mu(dx) \\ &= \int_C |f(x)|_Y\,\mu(dx), \end{aligned} \tag{8.5}$$

where last equality is a consequence of the inequality

$$\left|\int_C |f(x)|_Y\,\mu(dx) - \int_C |\varphi_n(x)|_Y\,\mu(dx)\right| \leq \int_C |f(x) - \varphi_n(x)|_Y\,\mu(dx).$$

The following result is a consequence of Definition 56.

Lemma 85 *Let $\tilde{Y}$ be another Banach space, $C \in \mathcal{C}$, $f : X \to Y$ a Bochner integrable function on C with respect to μ and $T : Y \to \tilde{Y}$ a bounded linear operator. Then, $T(f)$ is also Bochner integrable on C with respect to μ and*

$$\int_C T(f(x))\mu(dx) = T\left(\int_C f(x)\mu(dx)\right).$$

Proof. Let $\{\varphi_n : X \to Y : n \in \mathbb{N}\}$ be a sequence of simple functions satisfying (8.4). Then, $\{T(\varphi_n) : n \in \mathbb{N}\}$ is a sequence of $\tilde{Y}$-valued simple functions on X such that

$$\int_C |T(\varphi_n(x)) - T(f(x))|_{\tilde{Y}}\,\mu(dx) \leq C\int_X |\varphi_n(x) - f(x)|_Y\,\mu(dx).$$

Thus, $T(f)$ is a Bochner integrable function on C with respect to μ .

Finally, the definition of the Bochner integral for simple function and (8.5) yield

$$\begin{aligned}\int_C T(f(x))\,\mu(dx) &= \lim_{n\to\infty}\int_C T(\varphi_n(x))\,\mu(dx) = \lim_{n\to\infty} T\left(\int_C \varphi_n(x)\mu(dx)\right) \\ &= T\left(\int_C f(x)\mu(dx)\right).\end{aligned}$$

Consequently, the proof is complete. □

8.4 Garsia-Rodemich-Rumsey lemma

The purpose of this section is to establish the Garsia-Rodemich-Rumsey lemma [59], which is used to get the substitution property for the forward and Stratonovich integrals analyzed in Sections 5 and 6. Actually, we are interested in a consequence of this lemma that is related to Kolmogorov's continuity criterion, which is analyzed in Theorem 1. Namely, we can study the modulus of continuity of stochastic processes by means of the moments of their increments.

Now, we study the version of Garsia-Rodemich-Rumsey lemma that we need in this book.

Proposition 134 *Let $\Psi, p : [0,\infty) \to [0,\infty)$ be two continuous and strictly increasing functions such that $\Psi(0) = p(0) = 0$ and $\lim_{x\to\infty}\Psi(x) = \infty$. Also, let $T > 0$ and $\varphi \in C([0,T];\mathbb{R}^d)$ so that*

$$\int_0^T\int_0^T \Psi\left(\frac{|\varphi(t)-\varphi(s)|}{p(|t-s|)}\right) dsdt \le C,$$

for some constant $C > 0$. Then,

$$|\varphi(t)-\varphi(s)| \le 8\int_0^{t-s} \Psi^{-1}\left(\frac{4C}{r^2}\right) p(dr), \quad \textit{for } 0 \le s < t \le T.$$

Proof. Here, we use the convention

$$I(t) = \int_0^T \Psi\left(\frac{|\varphi(t)-\varphi(s)|}{p(|t-s|)}\right) ds, \quad t \in [0,T].$$

Our next goal is to use mathematical induction on n to build two non-increasing sequences $\{t_n : n \in (\mathbb{N}\cup\{0\})\}$ and $\{d_n : n \in (\mathbb{N}\cup\{-1,0\})\}$ as follows. Set $d_{-1} = T$ and $t_0 \in (0, d_{-1})$ such that $I(t_0) \le C/T$. Note that such a t_0 exists due to $\int_0^T I(t)dt \le C$ by hypothesis. So, given t_{n-1}, we use that the function p is a continuous non-decreasing function such that $p(0) = 0$ to define the point d_{n-1} as the one that satisfies $p(d_{n-1}) = p(t_{n-1})/2$. The point t_n is chosen as a point that satisfies $t_n \in (0, d_{n-1})$,

$$I(t_n) \le \frac{2C}{d_{n-1}} \quad \text{and} \quad \Psi\left(\frac{|\varphi(t_n)-\varphi(t_{n-1})|}{p(|t_n-t_{n-1}|)}\right) \le \frac{2I(t_{n-1})}{d_{n-1}}. \tag{8.6}$$

Note that we can choose such a t_n since the set of points in $(0, d_{n-1})$ for which the first (resp. second) inequality fails has Lebesgue measure less than $d_{n-1}/2$. Indeed, let $A = \{t \in (0, d_{n-1}) : I(t) > 2C/d_{n-1}\}$. Then,

$$C \ge \int_0^T I(t)dt \ge \int_A I(t)dt > \lambda(A)2C/d_{n-1},$$

which implies our claim for the firs inequality in (8.6). For the second inequality, we proceed as follows. If $I(t_{n-1}) = 0$, then $\Psi\left(\frac{|\varphi(\cdot)-\varphi(t_{n-1})|}{p(|\cdot-t_{n-1}|)}\right) = 0$, almost surely. In this case, our claim is true because $\lambda(A) < d_{n-1}/2$. Moreover, if $I(t_{n-1}) \ne 0$, then, the set

$$B = \left\{t \in (0, d_{n-1}) : \Psi\left(\frac{|\varphi(t)-\varphi(t_{n-1})|}{p(|t-t_{n-1}|)}\right) > \frac{2I(t_{n-1})}{d_{n-1}}\right\}$$

is such that

$$I(t_{n-1}) > \int_B 2I(t_{n-1})/d_{n-1}ds = \lambda(B)2I(t_{n-1})/d_{n-1}.$$

Therefore our claim is satisfied. That is, we can choose $t_n \in (0, d_{n-1})$ such that inequalities in (8.6) hold. Note that, by construction, together with the fact that p is strictly increasing, we also have

$$2p(d_{n+1}) = p(t_{n+1}) \le p(d_n) = p(t_n)/2,$$

which yields $d_{n+1} < t_{n+1} \le d_n < t_n$, $t_n \le p(t_0)/2^n \le p(T)/2^n$ and

$$p(t_n - t_{n+1}) \le p(t_n) = 2p(d_n) = 4\left(p(d_n) - \frac{1}{2}p(d_n)\right) \le 4\left(p(d_n) - p(d_{n+1})\right).$$

Hence, $t_n \downarrow 0$, as $n \to \infty$, and (8.6) gives

$$\begin{aligned} |\varphi(t_0) - \varphi(0)| &\le \sum_{n=0}^{\infty} |\varphi(t_n) - \varphi(t_{n+1})| \le \sum_{n=0}^{\infty} \Psi^{-1}\left(2I(t_n)/d_n\right) p(t_n - t_{n+1}) \\ &\le 4\sum_{n=0}^{\infty} \Psi^{-1}\left(4C/(d_{n-1}d_n)\right)\left(p(d_n) - p(d_{n+1})\right) \\ &\le 4\sum_{n=0}^{\infty} \Psi^{-1}\left(4C/d_n^2\right)\left(p(d_n) - p(d_{n+1})\right) \\ &\le 4\sum_{n=0}^{\infty} \int_{d_{n+1}}^{d_n} \Psi^{-1}\left(4C/u^2\right) p(du) \le 4\int_0^T \Psi^{-1}\left(4C/u^2\right) p(du). \quad (8.7) \end{aligned}$$

Now we repeat the above procedure with the function $\varphi(T-\cdot)$ instead of φ. The corresponding sequence $\{(\tilde{t_n}, \tilde{d}_{n-1}) : n \in \mathbb{N} \cap \{0\}\}$ is built as above. But, with $(\tilde{t_0}, \tilde{d}_{-1}) = (T - t_0, T)$. Thus, proceeding as before, we also get the inequality

$$|\varphi(T) - \varphi(t_0)| \le 4\int_0^T \Psi^{-1}\left(4C/u^2\right) p(du),$$

which, together with (8.7), leads us to establish

$$|\varphi(T) - \varphi(0)| \le 8\int_0^T \Psi^{-1}\left(4C/u^2\right) p(du). \tag{8.8}$$

Now, for $0 \le s \le t \le T$, we apply this inequality to the functions

$$\bar{\varphi}(\cdot) = \varphi\left(s + \frac{t-s}{T}\cdot\right) \quad \text{and} \quad \bar{p}(\cdot) = p\left(\frac{t-s}{T}\cdot\right).$$

in the following way. For these functions, the change of variables formula $\tilde{u} = s + \frac{t-s}{T}u$ yields

$$\begin{aligned} &\int_0^T \int_0^T \Psi\left(\frac{|\bar{\varphi}(u) - \bar{\varphi}(v)|}{\bar{p}(|u-v|)}\right) dudv \\ &= \left(\frac{T}{t-s}\right)^2 \int_s^t \int_s^t \Psi\left(\frac{|\varphi(u) - \varphi(v)|}{p(|u-v|)}\right) dudv \le C\left(\frac{T}{t-s}\right)^2 := \bar{C}. \end{aligned}$$

In consequence, (8.8) allows us to conclude

$$|\varphi(t) - \varphi(s)| = |\bar{\varphi}(T) - \bar{\varphi}(0)| \le 8\int_0^T \Psi^{-1}\left(4\bar{C}/u^2\right) \bar{p}(du).$$

Thus, the proof is finished by noting that Proposition 32 gives, for any partition $\pi = \{0 = t_0 < \ldots < t_n = T\}$ of the interval $[0, T]$,

$$\begin{aligned} \int_0^T \Psi^{-1}\left(4\bar{C}/u^2\right) \bar{p}(du) &= \lim_{|\pi| \to 0} \sum_{i=0}^{n-1} \Psi^{-1}\left(4\bar{C}/t_{i+1}^2\right) p\left(\left(\frac{t-s}{T}t_{i+1}\right) - \left(\frac{t-s}{T}t_i\right)\right) \\ &= \int_0^{t-s} \Psi^{-1}\left(4C/u^2\right) p(du) \end{aligned}$$

due to $\{0 = \frac{t-s}{T}t_0 < \frac{t-s}{T}t_1 < \ldots < \frac{t-s}{T}t_n = t-s\}$ being a partition of the interval $[0, t-s]$. $\square$

An immediate consequence of Proposition 134 is the following result.

Corollary 32 *Let $X = \{X_t : t \in [0,T]\}$ be a continuous process such that, for some $\gamma, C > 0$ and $\alpha > 1$,*

$$E\left(|X_t - X_s|^\gamma\right) \le C|t-s|^\alpha, \quad s,t \in [0,T].$$

Then, for $\kappa \in (0, \alpha - 1)$, we have that there exists a constant $C = C_{\gamma,\kappa}$ such that

$$|X_t - X_s|^\gamma \le C|t-s|^\kappa \Gamma \quad \text{with} \quad \Gamma = \int_0^T \int_0^T \frac{|X_t - X_s|^\gamma}{|t-s|^{\kappa+2}} dsdt.$$

Moreover, if $E\left(|X_{t_0}|^\gamma\right) < \infty$, we have that $E\left(\sup_{t\in[0,T]} |X_t|^\gamma\right) < \infty$.

Proof. we only need to apply Proposition 134 with $\Psi(x) = x^\gamma$ and $p(x) = x^{(\kappa+2)/\gamma}$, $x \in [0, \infty)$. $\square$

We need the following consequence of Corollary 32 to study the substitution formula for the Stratonovich integral, as it is done in Nualart [158].

Lemma 86 *Let $u_n = \{u_n(x) : x \in \mathbb{R}\}$ and $u = \{u(x) : x \in \mathbb{R}\}$ be random fields such that $u_n(x)$ converges in probability to $u(x)$, as $n \to \infty$, for each $x \in \mathbb{R}$. Moreover, assume that, for $M > 0$, there exist $C_M > 0$, $p > 0$ and $\alpha > 1$ satisfying*

$$E\left(|u_n(x) - u_n(y)|^p\right) \le C_M |x-y|^\alpha, \quad \text{for all } |x|, |y| \le M \text{ and } n \in \mathbb{N}. \tag{8.9}$$

Then, for every random variable F, we obtain that $P - \lim_{n\to\infty} u_n(F) = u(F)$.

Proof. Note that inequality (8.9) is true when we write u instead of u_n by Fatou's lemma.

We first assume that $|F| \le M$, for some $M > 0$. Fix an arbitrary $\varepsilon > 0$. Hence, we can find a simple random variable F_ε of the form $F_\varepsilon = \sum_{i=0}^n a_i 1_{\Omega_i}$ such that the family $\{\Omega_1, \ldots, \Omega_n\}$ is a measurable partition of Ω, $a_i \in \mathbb{R}$ and $\|F - F_\varepsilon\|_{L^\infty(\Omega)} < \varepsilon$. Then, from Corollary 32, we have that, for $m \in (0, \alpha - 1)$, there are a constant $c > 0$, and random variables Γ_n and Γ such that $E(\Gamma_n), E(\Gamma) < c$,

$$|u_n(x) - u_n(y)|^p \le |x-y|^m \Gamma_n \quad \text{and} \quad |u(x) - u(y)|^p \le |x-y|^m \Gamma.$$

In consequence, for $\eta > 0$,

$$\begin{aligned}
&P\left[|u_n(F) - u(F)| > \eta\right] \\
&\quad \le \; P\left[|u_n(F) - u_n(F_\varepsilon)| > \eta/4\right] + P\left[|u_n(F_\varepsilon) - u(F_\varepsilon)| > \eta/4\right] + P\left[|u(F_\varepsilon) - u(F)| > \eta/4\right] \\
&\quad \le \; \frac{2(4^p)c\varepsilon^m}{\eta^p} + P\left[|u_n(F_\varepsilon) - u(F_\varepsilon)| > \eta/4\right].
\end{aligned}$$

Since $u_n(F_\varepsilon) - u(F_\varepsilon) = \sum_{i=1}^n \left(u_n(a_i) - u(a_i)\right) 1_{\Omega_i}$, the result holds for any bounded random variable F.

Now, consider and arbitrary random variable F. Thus,

$$\begin{aligned}
&P\left[|u_n(F) - u(F)| > \eta\right] \\
&\quad \le \; P\left[1_{[|F|\le M]} |u_n(F) - u(F)| > \eta/2\right] + P\left[1_{[|F|>M]} |u_n(F) - u(F)| > \eta/2\right] \\
&\quad \le \; P\left[1_{[|F|\le M]} \left|u_n(F 1_{[|F|\le M]}) - u(F 1_{[|F|\le M]})\right| > \eta/2\right] + P\left[|F| > M\right] \\
&\quad \le \; P\left[\left|u_n(F 1_{[|F|\le M]}) - u(F 1_{[|F|\le M]})\right| > \eta/2\right] + P\left[|F| > M\right].
\end{aligned}$$

Therefore, the result is satisfied because it is true for bounded random variables, as we have already established. $\square$

8.5 Fractional calculus

In this section, we analyze some facts on fractional integrals and derivatives that we need to understand the extension of the Young's integral introduced by Zähle [218] (see Section 2.3).

In Section 1.4.1, it is considered the gamma function defined only on $\mathbb{R}_+$. This function is given by the integral

$$\Gamma(\alpha) = \int_0^\infty x^{\alpha-1}e^{-x}dx, \quad \alpha > 0. \tag{8.10}$$

Hence, the integration by parts formula yields

$$\Gamma(\alpha+1) = -x^\alpha e^{-x}\Big|_{x=0}^{x=\infty} + \alpha \int_0^\infty x^{\alpha-1}e^{-x}dx = \alpha\Gamma(\alpha). \tag{8.11}$$

In particular, $\Gamma(n) = (n-1)!$, for $n \in \mathbb{N}$, since (8.10) implies that $\Gamma(1) = 1$. The gamma function is relate to the beta function B, which is defined by the integral

$$B(\alpha,\mu) = \int_0^1 x^{\alpha-1}(1-x)^{\mu-1}dx, \quad \alpha,\mu > 0.$$

Lemma 87 *Let $a, b \in \mathbb{R}$, $a < b$, and $\alpha, \mu > 0$. Then,*

$$\int_a^b (b-s)^{\mu-1}(s-a)^{\alpha-1}ds = B(\alpha,\mu)(b-a)^{\alpha+\mu-1} = \frac{\Gamma(\alpha)\Gamma(\mu)}{\Gamma(\alpha+\mu)}(b-a)^{\alpha+\mu-1}.$$

Proof. We first consider the change of variables $y = (s-a)/(b-a)$ to obtain

$$\int_a^b (b-s)^{\mu-1}(s-a)^{\alpha-1}ds = (b-a)^{\alpha+\mu-1}\int_0^1 s^{\alpha-1}(1-s)^{\mu-1}ds.$$

Consequently, we only need to show that $B(\alpha,\mu) = \frac{\Gamma(\alpha)\Gamma(\mu)}{\Gamma(\alpha+\mu)}$ in order to finish the proof. Towards this end, we observe that the definition of Γ leads us to establish

$$\Gamma(\alpha)\Gamma(\mu) = \int_0^\infty \int_0^\infty e^{-(x+y)}x^{\alpha-1}y^{\mu-1}dxdy.$$

So, we can now apply the change of variables formula $x = st$ and $y = t(1-s)$ to this double integral. To do so, observe that $x + y = t$ and that the Jacobian of this transformation is $-t$. Thus,

$$\begin{aligned}\Gamma(\alpha)\Gamma(\mu) &= \int_0^1 \int_0^\infty e^{-t}(st)^{\alpha-1}t^{\mu-1}(1-s)^{\mu-1}tdtds \\ &= \int_0^\infty e^{-t}t^{\alpha+\mu-1}dt\left(\int_0^1 s^{\alpha-1}(1-s)^{\mu-1}ds\right) = \Gamma(\alpha+\mu)B(\alpha,\mu).\end{aligned}$$

Therefore, the proof is complete. □

We will also need the following equality.

Lemma 88 *Let $c < a < b$ and $\mu, \nu > 0$. Then,*

$$\int_a^b \frac{(y-a)^{\mu-1}(b-y)^{\nu-1}}{(y-c)^{\mu+\nu}}dy = B(\mu,\nu)\frac{(b-a)^{\mu+\nu-1}}{(b-c)^\mu(a-c)^\nu}.$$

Proof. We only need to apply the change of variables formula

$$y - c = \frac{(b-c)(a-c)}{(b-a)t + a - c}$$

in order to prove that the result is true. In fact, note that we have

$$y - a = \frac{(a-c)(b-a)(1-t)}{(b-a)t + a - c} \quad \text{and} \quad b - y = \frac{(b-c)(b-a)}{(b-a)t + a - c}.$$

□

Now, we deal with the well-posedness of Abel equation

Proposition 135 *Let f be a function in $L^1([a,b])$ and $\alpha \in (0,1)$. Then,*

i) The Abel equation

$$\frac{1}{\Gamma(\alpha)} \int_a^x \varphi(y)(x-y)^{\alpha-1} dy = f(x), \quad x \in (a,b), \tag{8.12}$$

has a solution $\varphi \in L^1([a,b])$ if and only if the left-sided fractional integral $I_{a+}^{1-\alpha} f$ is absolutely continuous on $[a,b]$ and $(I_{a+}^{1-\alpha} f)(a) = 0$.

ii) The equation

$$\frac{1}{\Gamma(\alpha)} \int_x^b \varphi(y)(y-x)^{\alpha-1} dy = f(x), \quad x \in (a,b),$$

has a solution φ in $L^1([a,b])$ if and only if the right-sided fractional integral $I_{b-}^{1-\alpha} f$ is absolutely continuous on $[a,b]$ and $(I_{b-}^{1-\alpha} f)(b) = 0$.

Moreover, in both cases, the solutions are unique.

Proof. The proof of Statement *ii)* is quite similar to that of Assertion *i)*. So, we only show that Statement *i)* holds.

We have already known that $I_{a+}^{1-\alpha} f$ is absolutely continuous on $[a,b]$ and $(I_{a+}^{1-\alpha} f)(a) = 0$ if the Abel equation (8.12) has a solution because this was established for equation (1.58) in Section 1.4.1.

Sufficiency of i): By hypothesis, the function

$$\varphi(x) = \frac{1}{\Gamma(1-\alpha)} \frac{d}{dx} \int_a^x \frac{f(y)dy}{(x-y)^\alpha}, \quad \text{for almost all } x \in [a,b],$$

belongs to $L^1([a,b])$. We now show that this function is a solution of (8.12). To do so, set

$$g(x) := \frac{1}{\Gamma(\alpha)} \int_a^x \frac{\varphi(y)}{(x-y)^{1-\alpha}} \quad \text{for almost all } x \in [a,b].$$

But, the analysis done for equation (1.58) implies that $x \mapsto \left(I_{a+}^{1-\alpha} g\right)(x)$ is an absolutely continuous function and

$$\varphi(x) = \frac{1}{\Gamma(1-\alpha)} \frac{d}{dx} \int_a^x \frac{g(y)dy}{(x-y)^\alpha}, \quad \text{for almost all } x \in [a,b].$$

Therefore, utilizing that $x \mapsto \left(I_{a+}^{1-\alpha} f\right)(x)$ is also an absolutely continuous function, we get

$$\left(I_{a+}^{1-\alpha} g\right)(x) - \left(I_{a+}^{1-\alpha} f\right)(x) = c, \quad \text{for } x \in [a,b].$$

Note that $c = 0$ since $\left(I_{a+}^{1-\alpha} f\right)(a) = 0$ by hypothesis and $\left(I_{a+}^{1-\alpha} g\right)(a) = 0$ because of the study carried out to equation (1.58). Thus, the uniqueness for the solution of equation (1.58) yields that $f = g$ and that the Abel equation (8.12) has a unique solution in $L^1([a,b])$. The proof is now complete. □

Now, based on Marchaud's type fractional derivatives (see the right-hand sides of (1.63) and (1.64)), we study necessary and sufficient conditions for a function to belong to either $I_{a+}^{\alpha}(L^p)$, or $I_{b-}^{\alpha}(L^p)$. These conditions are related with the existence of the limits in (1.64), or in (1.63), respectively. Towards this end, we establish, for $\varepsilon > 0$, the notations

$$\left(D_{a+}^{\alpha,\varepsilon} f\right)(t) = \frac{1}{\Gamma(1-\alpha)}\left(\frac{f(t)}{(t-a)^{\alpha}} + \alpha\int_a^{t-\varepsilon}\frac{f(t)-f(r)}{(t-r)^{1+\alpha}}\,dr\right), \quad t \in [a,b],$$

and

$$\left(D_{b-}^{\alpha,\varepsilon} f\right)(t) = \frac{1}{\Gamma(1-\alpha)}\left(\frac{f(t)}{(b-t)^{\alpha}} + \alpha\int_{t+\varepsilon}^{b}\frac{f(t)-f(r)}{(r-t)^{1+\alpha}}\,dr\right), \quad t \in [a,b]. \tag{8.13}$$

Here, we use the convention $f \equiv 0$ on $[a,b]^c$. Remember that we are assuming that $a, b \in \mathbb{R}$.

Theorem 98 *Let p be in $[1,\infty)$ and $\alpha \in (0,1)$. Then, f belongs to $I_{a+}^{\alpha}(L^p)$ if and only if $f \in L^p([a,b])$ and $D_{a+}^{\alpha,\varepsilon} f$ converges in $L^p([a,b])$, as $\varepsilon \to 0$. Moreover, in this case, we have that $f = I_{a+}^{\alpha}\varphi$, where φ is the limit of $D_{a+}^{\alpha,\varepsilon} f$ in $L^p([a,b])$, as $\varepsilon \to 0$.*

Remark 111 *Note that if the function $t \mapsto \frac{f(t)}{(t-a)^{\alpha}}$ belongs to $L^p([a,b])$ and the integral $t \mapsto \int_a^{t-\varepsilon}\frac{f(t)-f(r)}{(t-r)^{1+\alpha}}\,dr$ converges in $L^p([a,b])$, as $\varepsilon \to 0$. Then,*

$$\varphi(t) = \frac{1}{\Gamma(1-\alpha)}\left(\frac{f(t)}{(t-a)^{\alpha}} + \alpha L^p([a,b]) - \lim_{\varepsilon\to 0}\int_a^{t-\varepsilon}\frac{f(t)-f(r)}{(t-r)^{1+\alpha}}\,dr\right).$$

Furthermore, for $p = 1$, φ has been contemplated in Corollaries 7 and 8, and Proposition 17.

Proof. Here, we make use of the functions

$$k(t) = \frac{1}{\Gamma(\alpha)}\left(t^{\alpha-1} - (t-1)^{\alpha-1}1_{\{t>1\}}\right) \quad \text{and} \quad \mathcal{K}(t) = \frac{\alpha}{t\Gamma(1-\alpha)}\int_0^t k(u)du,$$

for $t > 0$. Note that the function

$$\mathcal{K}(t) = \frac{t^{\alpha} - [(t-1)\vee 0]^{\alpha}}{t\Gamma(1-\alpha)\Gamma(\alpha)} > 0, \quad \text{for } t > 0,$$

belongs to $L^1(\mathbb{R}_+)$ because the mean value theorem gives

$$0 < \mathcal{K}(t) \leq C_{\alpha}(t-1)^{\alpha-2}, \quad \text{for } t > 1.$$

We recall that, in this proof, any function $g : [a,b] \to \mathbb{R}$ is extended to $\mathbb{R}$ by defining $g(t) = 0$ for all $t \in [a,b]^c$.

Necessity: In this part of the proof, we assume that $f = I_{a+}^{p}\varphi$ with $\varphi \in L^p([a,b])$.

Let $x \in [a+\varepsilon, b]$ and $t \in [\varepsilon, x-a]$. Then, the change of variables $u = x - y$ allows us to get

$$\begin{aligned} f(x) - f(x-t) &= \frac{1}{\Gamma(\alpha)} \left(\int_a^x \frac{\varphi(y)}{(x-y)^{1-\alpha}} dy - \int_a^{x-t} \frac{\varphi(y)}{(x-t-y)^{1-\alpha}} dy \right) \\ &= \frac{1}{\Gamma(\alpha)} \left(\int_0^{x-a} \frac{\varphi(x-u)}{u^{1-\alpha}} du - \int_t^{x-a} \frac{\varphi(x-u)}{(u-t)^{1-\alpha}} du \right) \\ &= t^{\alpha-1} \int_0^{x-a} \varphi(x-u) k(u/t) du. \end{aligned}$$

Consequently, for $x \in [a+\varepsilon, b]$, from Fubini's theorem and the change of variables $v = \frac{u}{t}$, we obtain

$$\begin{aligned} &\int_a^{x-\varepsilon} \frac{f(x) - f(t)}{(x-t)^{1+\alpha}} dt \\ &= \int_\varepsilon^{x-a} \frac{f(x) - f(x-t)}{t^{1+\alpha}} dt = \int_0^{x-a} \varphi(x-u) \int_\varepsilon^{x-a} t^{-2} k(u/t) dt du \\ &= \int_0^{x-a} \frac{\varphi(x-u)}{u} \int_{u/(x-a)}^{u/\varepsilon} k(v) dv du \\ &= \int_0^{x-a} \frac{\varphi(x-u)}{u} \left(\int_0^{u/\varepsilon} k(v) dv - \frac{1}{\Gamma(\alpha)} \int_0^{u/(x-a)} v^{\alpha-1} dv \right) du \\ &= \int_0^{x-a} \frac{\varphi(x-u)}{u} \int_0^{u/\varepsilon} k(v) dv du - \frac{1}{\alpha\Gamma(\alpha)(x-a)^\alpha} \int_0^{x-a} \frac{\varphi(x-u)}{u^{1-\alpha}} du. \end{aligned}$$

Thus, applying first the change of variables formula $y = x - u$ and then the one $y = u/\varepsilon$, we can write

$$\begin{aligned} \left(D_{a+}^{\alpha,\varepsilon} f\right)(x) &= \frac{\alpha}{\Gamma(1-\alpha)} \int_0^{x-a} \frac{\varphi(x-u)}{u} \int_0^{u/\varepsilon} k(v) dv du \\ &= \frac{\alpha}{\Gamma(1-\alpha)} \int_0^{(x-a)/\varepsilon} \frac{\varphi(x-\varepsilon u)}{u} \int_0^u k(v) dv du \\ &= \frac{\alpha}{\Gamma(1-\alpha)} \int_0^\infty \frac{\varphi(x-\varepsilon u)}{u} \int_0^u k(v) dv du \\ &= \int_0^\infty \varphi(x-\varepsilon u) \mathcal{K}(u) du, \quad x \in [a+\varepsilon, b], \end{aligned} \tag{8.14}$$

where, in the penultimate equality, we make use of the convention $\varphi = 0$ on $[a,b]^c$.

For $a \le x < a + \varepsilon$, we can establish

$$\begin{aligned} \left(D_{a+}^{\alpha,\varepsilon} f\right)(x) &= \frac{\alpha f(x)}{\Gamma(1-\alpha)} \int_a^{x-\varepsilon} \frac{1}{(x-t)^{1+\alpha}} dt + \frac{f(x)}{\Gamma(1-\alpha)(x-a)^\alpha} \\ &= \frac{f(x)}{\varepsilon^\alpha \Gamma(1-\alpha)} = \frac{1}{\varepsilon^\alpha \Gamma(1-\alpha)\Gamma(\alpha)} \int_a^x \frac{\varphi(u)}{(x-u)^{1-\alpha}} du. \end{aligned} \tag{8.15}$$

Now, we deal with the convergence of $D^{\alpha,\varepsilon}_{a+}f$ in $L^p([a,b])$. Note that the change of variables formula $u=x+a-y$ and (8.5) imply

$$
\begin{aligned}
&\frac{1}{\varepsilon^\alpha}\left(\int_a^{a+\varepsilon}\left|\int_a^x \frac{\varphi(y)}{(x-y)^{1-\alpha}}dy\right|^p dx\right)^{1/p}\\
&= \frac{1}{\varepsilon^\alpha}\left(\int_a^{a+\varepsilon}\left|\int_a^x \frac{\varphi(x+a-y)}{(y-a)^{1-\alpha}}dy\right|^p dx\right)^{1/p}\\
&\leq \frac{1}{\varepsilon^\alpha}\left(\int_a^{a+\varepsilon}\left[\int_a^{a+\varepsilon} \frac{|\varphi(x+a-y)|}{(y-a)^{1-\alpha}}dy\right]^p dx\right)^{1/p}\\
&\leq \frac{1}{\varepsilon^\alpha}\int_a^{a+\varepsilon}\frac{\|\varphi(\cdot+a-y)\|_{L^p([a,a+\varepsilon])}}{(y-a)^{1-\alpha}}dy \leq \frac{\|\varphi\|_{L^p([a,a+\varepsilon])}}{\varepsilon^\alpha}\int_a^{a+\varepsilon}\frac{dy}{(y-a)^{1-\alpha}}\\
&= \frac{1}{\alpha}\|\varphi\|_{L^p([a,a+\varepsilon])}\to 0, \quad \text{as } \varepsilon\to 0. \qquad (8.16)
\end{aligned}
$$

In order to continuous with our analysis, remember that we saw $\mathcal{K}\in L^1(\mathbb{R}_+)$ at the beginning of this proof. So, (8.5) and (8.14) lead us to write

$$
\begin{aligned}
&\left(\int_{a+\varepsilon}^b\left|\left(D^{\alpha,\varepsilon}_{a+}f\right)(x)-\varphi(x)\int_0^\infty \mathcal{K}(u)du\right|^p dx\right)^{1/p}\\
&= \left(\int_{a+\varepsilon}^b\left|\int_0^\infty \mathcal{K}(u)\left(\varphi(x-\varepsilon u)-\varphi(x)\right)du\right|^p dx\right)^{1/p}\\
&\leq \int_0^\infty \mathcal{K}(u)\left\|\varphi(\cdot-\varepsilon u)-\varphi(\cdot)\right\|_{L^p([a,b])}du\to 0, \quad \text{as } \varepsilon\to 0.
\end{aligned}
$$

Hence, (8.15) and (8.16) allow us to conclude that $\left(D^{\alpha,\varepsilon}_{a+}f\right)(\cdot)\to\varphi(\cdot)\int_0^\infty\mathcal{K}(u)du$ in $L^p([a,b])$, as $\varepsilon\to 0$.

Finally, we have $\int_0^\infty \mathcal{K}(u)du=1$ because $D^{\alpha,\varepsilon}_{a+}f\to\varphi$ in $L^q([a,b])$, as $\varepsilon\to 0$, for any $q\in(1,1/\alpha)$, if $f\in C^1([a,b])$ (see Proposition 17 and (1.64)).

Sufficiency: Now, we assume that $D^{\alpha,\varepsilon}_{a+}f$ goes to φ in $L^p([a,b])$.

Observe that, in order to prove that the result is satisfied, we only need to show that $f=L^p([a,b])-\lim_{\varepsilon\to 0}I^\alpha_{a+}(D^{\alpha,\varepsilon}_{a+}f)$ due to (1.55) implying

$$
\left\|I^\alpha_{a+}\left(D^{\alpha,\varepsilon}_{a+}f-\varphi\right)\right\|_{L^p([a,b])}\leq C\left\|D^{\alpha,\varepsilon}_{a+}f-\varphi\right\|_{L^p([a,b])}.
$$

Fubini's theorem yields, for $x\in(a+\varepsilon,b)$,

$$
\begin{aligned}
&\alpha\int_{a+\varepsilon}^x\frac{1}{(x-y)^{1-\alpha}}\int_a^{y-\varepsilon}\frac{f(y)-f(t)}{(y-t)^{1+\alpha}}dtdy\\
&= \alpha\int_{a+\varepsilon}^x\frac{1}{(x-y)^{1-\alpha}}\left(\int_a^{y-\varepsilon}\frac{f(y)}{(y-t)^{1+\alpha}}dt-\int_a^{y-\varepsilon}\frac{f(t)}{(y-t)^{1+\alpha}}dt\right)dy\\
&= \int_{a+\varepsilon}^x\frac{f(y)}{(x-y)^{1-\alpha}}\left[\varepsilon^{-\alpha}-(y-a)^{-\alpha}\right]dy-\alpha\int_a^{x-\varepsilon}f(t)\int_{t+\varepsilon}^x(x-y)^{\alpha-1}(y-t)^{-\alpha-1}dydt.
\end{aligned}
$$

Therefore, using the definitions of $D^{\alpha,\varepsilon}_{a+}f$ and I^α_{a+}, we obtain that, for $x\in(a+\varepsilon,b)$,

$$
\begin{aligned}
I^\alpha_{a+}\left(D^{\alpha,\varepsilon}_{a+}f\right)(x) &= \frac{1}{\Gamma(1-\alpha)\Gamma(\alpha)}\int_a^x\frac{f(y)}{(x-y)^{1-\alpha}\varepsilon^\alpha}dy\\
&\quad -\frac{\alpha}{\Gamma(1-\alpha)\Gamma(\alpha)}\int_a^{x-\varepsilon}f(t)\int_{t+\varepsilon}^x\frac{(x-y)^{\alpha-1}}{(y-t)^{\alpha+1}}dydt.
\end{aligned}
$$

Consequently, from Lemmas 87 and 88, and (8.11), we get

$$
\begin{aligned}
& I^{\alpha}_{a+}\left(D^{\alpha,\varepsilon}_{a+}f\right)(x) \\
&= \frac{1}{\Gamma(1-\alpha)\Gamma(\alpha)\varepsilon^{\alpha}}\left(\int_a^x \frac{f(y)}{(x-y)^{1-\alpha}}dy - \alpha B(1,\alpha)\int_a^{x-\varepsilon} f(t)\frac{(x-t-\varepsilon)^{\alpha}}{x-t}dt\right) \\
&= \frac{1}{\Gamma(1-\alpha)\Gamma(\alpha)\varepsilon^{\alpha}}\left(\int_a^x \frac{f(y)}{(x-y)^{1-\alpha}}dy - \int_a^{x-\varepsilon} f(y)\frac{(x-y-\varepsilon)^{\alpha}}{x-y}dy\right), \quad x \in [a+\varepsilon, b].
\end{aligned}
$$

Thus, the change of variables formula $y = x - \varepsilon t$ yields

$$
\begin{aligned}
& I^{\alpha}_{a+}\left(D^{\alpha,\varepsilon}_{a+}f\right)(x) \\
&= \frac{1}{\Gamma(1-\alpha)\Gamma(\alpha)}\left(\int_0^{(x-a)/\varepsilon} \frac{f(x-\varepsilon t)t^{\alpha}}{t}dt - \int_1^{(x-a)/\varepsilon} \frac{f(x-\varepsilon t)(t-1)^{\alpha}}{t}dt\right) \\
&= \int_0^{(x-a)/\varepsilon} f(x-\varepsilon t)\mathcal{K}(t)dt = \int_0^{\infty} f(x-\varepsilon t)\mathcal{K}(t)dt, \quad x \in [a+\varepsilon, b],
\end{aligned}
$$

which, together with the equality

$$
I^{\alpha}_{a+}\left(D^{\alpha,\varepsilon}_{a+}f\right)(x) = \frac{1}{\Gamma(1-\alpha)\varepsilon^{\alpha}} I^{\alpha}_{a+}(f)(x), \quad x \in (a, a+\varepsilon),
$$

and (8.16), allows us to conclude that $f = L^p([a,b]) - \lim_{\varepsilon\to 0} I^{\alpha}_{a+}(D^{\alpha,\varepsilon}_{a+}f)$. In consequence, the proof is complete. □

Proceeding as in the proof of Theorem 98, we can also show that the following result holds, where we utilize the notation (8.13).

Theorem 99 *Let $p \in [1,\infty)$ and $\alpha \in (0,1)$. Then, f belongs to $I^{\alpha}_{b-}(L^p)$ if and only if $f \in L^p([a,b])$ and $D^{\alpha,\varepsilon}_{b-}f$ converges in $L^p([a,b])$, as $\varepsilon \to 0$. In this case, we have that $f = I^{\alpha}_{b-}\psi$, where $\psi = L^p([a,b]) - \lim_{\varepsilon\to 0} D^{\alpha,\varepsilon}_{b-}f$.*

8.6 Fractional integral of some discontinuous integrands with respect to Hölder continuous processes

The aim of this section is to provide the auxiliary tool needed to understand the proof of Proposition 52.

We begin with the existence of the fractional integral in Proposition 52.

Proposition 136 *Let g and f be two processes with paths in $C^{\alpha}_1([a,b])$ and $C^{\beta}_1([a,b])$, respectively, where $\alpha + \beta > 1$ and g satisfies Assumption (2.72). Also let $G : \mathbb{R} \to \mathbb{R}$ be a function with locally finite variation. Then, the pathwise fractional integral $\int_a^b G(g(s))df(s)$ exists.*

Remark 112 *Remember that the equality*

$$
\left(\int_a^b G(g(s))df(s)\right)(\omega) = \int_a^b G(g(\omega,s))df(\omega,s), \quad \omega \in \Omega,
$$

holds.

Proof. Let $\gamma \in (1-\beta, \alpha)$. Then, by (1.65), we obtain that f^{b-} belongs to $I_{b-}^{1-\gamma}(L^q)$ with probability 1, for any $q > 1$, since f has β-Hölder continuous paths. Thus, in order to finish the proof, we only need to show that $G(g) \in I_{a+}^{\gamma}(L^p)$ with probability 1, for some $p > 1$ (see Definition 30). Towards this end, we will show that, for $p > 1$ close enough to 1,

$$\left(t \mapsto \frac{|G(g(t)+)|}{(t-a)^{\gamma}} + \gamma \int_a^t \frac{|G(g(t)+) - G(g(r)+)|}{(t-r)^{1+\gamma}} dr \right) \in L^p([a,b]),$$

which implies that $G(g) \in I_{a+}^{\gamma}(L^p)$ due to Condition (2.72). Indeed, the set J_G of discontinuities of G is at most countable because G is a locally bounded variation function on $\mathbb{R}$ (see Lemma 108). Therefore, Fubini's theorem and (2.72) yield

$$E\left(\int_a^b 1_{J_G}(g(s)) ds \right) = \int_a^b E\left[1_{J_G}(g(s))\right] ds = 0.$$

Consequently, we can assume that G is right-continuous without loss of generality.

Now, we divide the proof into several steps.

Step 1: Here we suppose that G is a non-decreasing function on $\mathbb{R}$ and show that, for $p > 1$ and the closure I_g of the image of g, we have

$$\begin{aligned} &\int_a^b \left(\int_a^t \frac{|G(g(t)) - G(g(r))|}{(t-r)^{1+\gamma}} dr \right)^p dt \\ &\leq 2(b-a)^{p-1} \left(p(1+\gamma) - 1\right)^{-1} \mu_G(I_g)^{p-1} \|g\|_{\alpha,[a,b]}^{(p(1+\gamma)-1)/\alpha} \\ &\quad \times \int_a^b \int_{I_g} |g(t) - y|^{(1-p(1+\gamma))/\alpha} \mu_G(dy) dt, \end{aligned} \tag{8.17}$$

where μ_G is the associated measure of the locally bounded variation function G (see Remark 30). Observe that the definitions of μ_G and I_g, together with Condition (2.72), imply that, for $t, r \in [a,b]$,

$$\begin{aligned} |G(g(t)) - G(g(r))|^p &= |G(g(t)) - G(g(r))|^{p-1} |G(g(t)) - G(g(r))| \\ &\leq \mu_G(I_g)^{p-1} |G(g(t)) - G(g(r))| \\ &= \mu_G(I_g)^{p-1} \left(\int_{I_g} 1_{(g(r),g(t)]}(y) \mu_G(dy) + \int_{I_g} 1_{(g(t),g(r)]}(y) \mu_G(dy) \right) \\ &= \mu_G(I_g)^{p-1} \left(\int_{I_g} 1_{(g(r),g(t))}(y) \mu_G(dy) + \int_{I_g} 1_{(g(t),g(r))}(y) \mu_G(dy) \right), \end{aligned}$$

for almost all $(\omega, t, r) \in \Omega \times [a,b]^2$. Hence, Hölder inequality gives

$$\begin{aligned} &\int_a^b \left(\int_a^t \frac{|G(g(t)) - G(g(r))|}{(t-r)^{1+\gamma}} dr \right)^p dt \\ &\leq (b-a)^{p-1} \int_a^b \int_a^t \frac{|G(g(t)) - G(g(r))|^p}{(t-r)^{p(1+\gamma)}} dr dt \\ &\leq \mu_G(I_g)^{p-1} (b-a)^{p-1} \left(\int_a^b \int_a^t \int_{I_g} \frac{1_{(g(r),g(t))}(y)}{(t-r)^{p(1+\gamma)}} \mu_G(dy) dr dt \right. \\ &\quad \left. + \int_a^b \int_a^t \int_{I_g} \frac{1_{(g(t),g(r))}(y)}{(t-r)^{p(1+\gamma)}} \mu_G(dy) dr dt \right) \\ &= \mu_G(I_g)^{p-1} (b-a)^{p-1} \left(\int_{I_g} J_1(y) \mu_G(dy) + \int_{I_g} J_2(y) \mu_G(dy) \right). \end{aligned} \tag{8.18}$$

Note that, for $r < t$ and $y \in (g(r), g(t))$, there exists $s_{r,t} \in (r,t)$ such that $g(s_{r,t}) = y$ and $g(u) > y$, for $u \in (s_{r,t}, t)$. In consequence,

$$|g(t) - y| = |g(t) - g(s_{r,t})| \leq \|g\|_{\alpha,[a,b]} |t - s_{r,t}|^{\alpha}, \tag{8.19}$$

which leads us to obtain

$$\begin{aligned} 1_{\{g(t)>y\}} \int_a^t \frac{1_{\{g(r)<y\}}}{(t-r)^{p(1+\gamma)}} dr &\leq \frac{(t - s_{r,t})^{1-p(1+\gamma)}}{p(1+\gamma) - 1} \\ &\leq \frac{\|g\|_{\alpha,[a,b]}^{(p(1+\gamma)-1)/\alpha} |g(t) - y|^{(1-p(1+\gamma))/\alpha}}{p(1+\gamma) - 1}. \end{aligned} \tag{8.20}$$

Therefore, from (8.18), we get

$$\int_{I_g} J_1(y)\mu_G(dy) \leq \frac{\|g\|_{\alpha,[a,b]}^{(p(1+\gamma)-1)/\alpha}}{p(1+\gamma) - 1} \int_a^b \int_{I_g} |g(t) - y|^{(1-p(1+\gamma))/\alpha} \mu_G(dy)dt.$$

Likewise, this inequality is also true if we write J_2 instead of J_1 since $1_{(g(t),g(r))}(y) = 1_{(-g(r),-g(t))}(-y)$. That is, for J_2, we only need to change g and y by $-g$ and $-y$, respectively. In this way, we have proven that (8.17) is true if G is a non-decreasing function.
Step 2: It is well-known that $G = G_1 - G_2$, where G_1 and G_2 are two non-decreasing functions on $\mathbb{R}$. Thus, (8.17) implies that

$$\begin{aligned} &\int_a^b \left(\int_a^t \frac{|G(g(t)) - G(g(r))|}{(t-r)^{1+\gamma}} dr \right)^p dt \\ &\leq 2^{p+2}(b-a)^{p-1} \left(p(1+\gamma) - 1\right)^{-1} |\mu_G|(I_g)^{p-1} \|g\|_{\alpha,[a,b]}^{(p(1+\gamma)-1)/\alpha} \\ &\quad \times \int_a^b \int_{I_g} |g(t) - y|^{(1-p(1+\gamma))/\alpha} |\mu_G|(dy)dt, \end{aligned}$$

with $|\mu_G| = \mu_{G_1} + \mu_{G_2}$.
Step 3: Here, we show that the left-hand side of (8.17) is finite for $p \in (1, \frac{1+\alpha}{1+\gamma})$, w.p.1.

Fix $p \in (1, \frac{1+\alpha}{1+\gamma})$ and let $N \in \mathbb{N}$. Then, Condition (2.72) and Fubini's theorem yield

$$\begin{aligned} &E\left(\int_a^b \int_{[N,N]} |g(t) - y|^{(1-p(1+\gamma))/\alpha} |\mu_G|(dy)dt \right) \\ &= \int_{[N,N]} \int_a^b E\left(1_{\{|g(t)-y|\leq 1\}} |g(t) - y|^{(1-p(1+\gamma))/\alpha}\right) dt|\mu_G|(dy) \\ &\quad + \int_{[N,N]} \int_a^b E\left(1_{\{|g(t)-y|>1\}} |g(t) - y|^{(1-p(1+\gamma))/\alpha}\right) dt|\mu_G|(dy) \\ &\leq \int_{[N,N]} \int_a^b \int_{-1}^1 |x|^{(1-p(1+\gamma))/\alpha} p_t(x+y)dxdt|\mu_G|(dy) + (b-a)|\mu_G|([N,N]) \\ &\leq \int_{[N,N]} \int_a^b \hat{p}_t \int_{-1}^1 |x|^{(1-p(1+\gamma))/\alpha} dxdt|\mu_G|(dy) + (b-a)|\mu_G|([N,N]) < \infty. \end{aligned} \tag{8.21}$$

Hence, Step 2 allows us to conclude

$$1_{\{[\|g\|_{\infty,[a,b]} \leq N]\}} \int_a^b \left(\int_a^t \frac{|G(g(t)) - G(g(r))|}{(t-r)^{1+\gamma}} dr \right)^p dt < \infty, \quad \text{w.p.1},$$

which, together with the fact that g has α-Hölder-continuous paths, gives that the claim of this step is satisfied.

Step 4: Finally, we observe that $t \mapsto \frac{|G(g(t))|}{(t-a)^\gamma}$ belongs to $L^p([a,b])$, for $p \leq 1/\gamma$, with probability 1 since

$$\frac{|G(g(t))|}{(t-a)^\gamma} \leq \frac{\|G(g)\|_{\infty,[a,b]}}{(t-a)^\gamma}$$

and $\|G(g)\|_{\infty,[a,b]} < \infty$ with probability 1, which follows from the facts that g has α-Hölder-continuous paths and G is a locally bounded variation function. Thus, by Step 3 and proceeding as in (1.65), we have that $G(g) \in I^\gamma_{a+}(L^p)$, for $1 < p < \frac{1+\alpha}{1+\gamma} \wedge \frac{1}{\gamma}$. In order words, the proof is complete. □

On the other hand, in order to show that Proposition 52 hold, we will need the following three auxiliary results due to Chen et al. [26].

Lemma 89 *Let $G : \mathbb{R} \to \mathbb{R}$ be a bounded, non-decreasing and right-continuous function, $\varphi \in C^\infty(\mathbb{R})$ a nonnegative function with support $[-1,0]$ and $\int_{\mathbb{R}} \varphi(s)ds = 1$, and $G_n = G*\varphi_n$, where $n \in \mathbb{N}$ and $\varphi_n(x) = n\varphi(nx)$. Then, $G_n \to G$ pointwise and, for any bounded and continuous function h on $\mathbb{R}$, $\int_{\mathbb{R}} h(y)\mu_{G_n}(dy) \to \int_{\mathbb{R}} h(y)\mu_G(dy)$, as $n \to \infty$.*

Proof. We have that, for $n \in \mathbb{N}$ and $x \in \mathbb{R}$, $G_n(x) = \int_{\mathbb{R}} G(x-z)\varphi_n(z)dz$. Thus, $G_n(x) \to G(x)$, as $n \to \infty$, because G is right-continuous and $\varphi_n(z) = 0$ for $z \in (-1/n, 0)^c$. Moreover, G_n is a non-decreasing and bounded function due to G being so. Therefore, $G_n(-\infty+)$ and $G_n(\infty-)$ exist and they agree with $G(-\infty+)$ and $G(\infty-)$, respectively, because of the dominated convergence theorem and the definition of G_n. In this way, we can define $G_n(-\infty) = G_n(-\infty+) = G(-\infty+)$ and $G_n(\infty) = G_n(\infty-) = G(\infty-)$. The measure μ_{G_n} is defined as $\mu_{G_n}((a,b]) = G_n(b) - G_n(a)$, for $-\infty \leq a < b \leq \infty$. In particular, we get $\mu_{G_n}(\mathbb{R}) = \mu_G(\mathbb{R})$ and

$$\lim_{n\to\infty} \frac{\mu_{G_n}((a,b])}{\mu_{G_n}(\mathbb{R})} = \frac{\mu_G((a,b])}{\mu_G(\mathbb{R})}.$$

Consequently, the result is now a consequence of Portmanteau theorem (see, for instance, Theorem 2.1 and Section 2 in Billingsley [21]). □

Now, we state the second auxiliary result needed to show Proposition 52.

Lemma 90 *Let $\alpha \in (0,1)$ and $g : \Omega \times [a,b] \to \mathbb{R}$ a process with α-Hölder continuous paths such that, for almost $t \in [a,b]$, the random variable $g(t)$ is absolutely continuous with density satisfying Assumption (2.72). Then, for $\gamma \in (0,\alpha)$ and $q \geq \gamma/\alpha$, the functions*

$$\Phi(y) = E\left(\left[1 + \|g\|_{\alpha,[a,b]}\right]^{-q} \int_a^b \int_a^t \frac{1_{\{g(s)<y<g(t)\}}}{(t-s)^{1+\gamma}} dsdt\right)$$

and

$$\Psi(y) = E\left(\left[1 + \|g\|_{\alpha,[a,b]}\right]^{-q} \int_a^b \int_a^t \frac{1_{\{g(t)<y<g(s)\}}}{(t-s)^{1+\gamma}} dsdt\right), \quad y \in \mathbb{R},$$

are bounded and continuous.

Proof. We use the notation $\Phi(y) = E(\phi(y))$. Likewise (8.19) and (8.20), we obtain

$$\phi(y) \leq \frac{\|g\|^{\gamma/\alpha}_{\alpha,[a,b]}}{\gamma\left(1 + \|g\|_{\alpha,[a,b]}\right)^q} \int_a^b |g(t) - y|^{-\gamma/\alpha} dt.$$

Therefore, as in (8.21), we can utilize Condition (2.72) to establish

$$\Phi(y) \leq \frac{1}{\gamma} \int_a^b \left(1 + \frac{2}{1-\gamma\alpha^{-1}} \hat{p}_t\right) dt, \quad y \in \mathbb{R}. \tag{8.22}$$

Hence, using (2.72) again, the function Φ is bounded on $\mathbb{R}$.

Now, we see that Φ is right-continuous. Towards this end, we choose $y \in \mathbb{R}$ and $\varepsilon > 0$, and we observe that the equality

$$1_{\{g(s)<y+\varepsilon<g(t)\}} - 1_{\{g(s)<y<g(t)\}} = 1_{\{y\le g(s)<y+\varepsilon<g(t)\}} - 1_{\{g(s)<y<g(t)\le y+\varepsilon\}}$$

implies

$$\Phi(y+\varepsilon) - \Phi(y) = E\left(\int_a^b \int_a^t \Phi_1(s,t,\varepsilon)dsdt\right) - E\left(\int_a^b \int_a^t \Phi_2(s,t,\varepsilon)dsdt\right), \tag{8.23}$$

where

$$\Phi_1(s,t,\varepsilon) = \left[1 + \|g\|_{\alpha,[a,b]}\right]^{-q} \frac{1_{\{y\le g(s)<y+\varepsilon<g(t)\}}}{(t-s)^{1+\gamma}}$$

and

$$\Phi_2(s,t,\varepsilon) = \left[1 + \|g\|_{\alpha,[a,b]}\right]^{-q} \frac{1_{\{g(s)<y<g(t)\le y+\varepsilon\}}}{(t-s)^{1+\gamma}}.$$

Note that the fact that $g(t)$ has a density for almost all $t \in [a,b]$ leads to conclude that, for almost all $(\omega, s, t) \in \Omega \times \{(s,t) \in [a,b]^2 : s \le t\}$,

$$\Phi_1(s,t,\varepsilon) \to 0 \quad \text{and} \quad \Phi_2(s,t,\varepsilon) \to 0, \quad \text{as } \varepsilon \downarrow 0. \tag{8.24}$$

The following step is to verify that $E\left(\int_a^b \int_a^t \Phi_1(s,t,\varepsilon)dsdt\right)$ goes to zero as $\varepsilon \downarrow 0$. To do so, we fix $p \in (1, \frac{1+\alpha}{1+\gamma})$ so that $1/p > 1+\gamma-\alpha q$. This p satisfies that $\tilde{q} \ge \tilde{\gamma}/\alpha$ an $0 < \tilde{\gamma} < \alpha$, with $\tilde{q} = pq$ and $\tilde{\gamma} = (1+\gamma)p - 1$. Consequently, (8.22) allows us to get

$$\begin{aligned} E\left(\int_a^b \int_a^t \Phi_1(s,t,\varepsilon)^p dsdt\right) &\le E\left(\left[1 + \|g\|_{\alpha,[a,b]}\right]^{-\tilde{q}} \int_a^b \int_a^t \frac{1_{\{g(s)<y+\varepsilon<g(t)\}}}{(t-s)^{1+\tilde{\gamma}}} dsdt\right) \\ &\le \frac{1}{\tilde{\gamma}} \int_a^b \left(1 + \frac{2}{1-\tilde{\gamma}\alpha^{-1}} \hat{p}_t\right) dt. \end{aligned}$$

Since the last quantity is independent of ε, (8.24) and Kallenberg [94] (Proposition 3.12 and Exercise 3.6) yield that $\{\Phi_1(\cdot,\varepsilon) : \varepsilon > 0\}$ is uniformly integrable and, therefore,

$$E\left(\int_a^b \int_a^t \Phi_1(s,t,\varepsilon)dsdt\right) \to 0, \quad \text{as } \varepsilon \to 0.$$

We also have that $E\left(\int_a^b \int_a^t \Phi_2(s,t,\varepsilon)dsdt\right) \to 0$, as $\varepsilon \to 0$, due to the inequality

$$\Phi_2(s,t,\varepsilon) \le \left[1 + \|g\|_{\alpha,[a,b]}\right]^{-q} \frac{1_{\{g(s)<y<g(t)\}}}{(t-s)^{1+\gamma}},$$

(8.22), (8.24) and the dominated convergence theorem. Hence, (8.23) allows us to conclude that Φ is right-continuous on $\mathbb{R}$.

Next step is to demonstrate that Φ is also left-continuous. So, we fix $y \in \mathbb{R}$ again and observe that the equality

$$1_{\{g(s)<y-\varepsilon<g(t)\}} - 1_{\{g(s)<y<g(t)\}} = 1_{\{g(s)<y-\varepsilon<g(t)\le y\}} - 1_{\{y-\varepsilon\le g(s)<y<g(t)\}}$$

holds. This equality allows us to prove that Φ is left-continuous by following the aforementioned approach in the proof of the fact that Φ is right-continuous.

Finally, in order to verify that Ψ is also a bounded and continuous function on $\mathbb{R}$, we only need to change the process g by $\tilde{g} = -g$. The proof is now complete. □

Now, we state the third auxiliary result needed to deal with Proposition 52.

Lemma 91 *Let $G : \mathbb{R} \to \mathbb{R}$ be a right-continuous function of locally bounded variation, φ_n as in Lemma 89 and g a process as in Lemma 90. Then, for $\gamma \in (0, \alpha)$, we have*

$$P - \lim_{n\to\infty} \int_a^b \int_a^t \frac{|G_n(g(t)) - G_n(g(s)) - (G(g(t)) - G(g(s)))|}{(t-s)^{1+\gamma}} ds = 0.$$

Remark 113 *In the proof of Proposition 136, we have established that*

$$t \mapsto \int_a^t \frac{|G(g(t)) - G(g(s))|}{(t-s)^{1+\gamma}} ds$$

belongs to $L^p([a,b])$, with probability 1, for some $p > 1$. Moreover, observe that $G_n(g)$ has α-Hölder continuous paths since φ_n has a bounded derivative.

Proof. Remember that $G_n = G * \varphi_n$ and $\gamma \in (0, \alpha)$. For $y \in \mathbb{R}$, we set

$$h(y) = E\left(\left[1 + \|g\|_{\alpha,[a,b]}\right]^{-q} \int_a^b \int_a^t \frac{1_{\{g(s)<y<g(t)\}} + 1_{\{g(t)<y<g(s)\}}}{(t-s)^{1+\gamma}} dsdt \right).$$

Note that h is a continuous function by Lemma 90. Then,

$$E\left(\left[1 + \|g\|_{\alpha,[a,b]}\right]^{-q} \int_a^b \int_a^t \frac{|G_n(g(t)) - G_n(g(s))|}{(t-s)^{1+\gamma}} dsdt \right) = \int_{\mathbb{R}} h(y) \mu_{G_n}(dy)$$

and

$$E\left(\left[1 + \|g\|_{\alpha,[a,b]}\right]^{-q} \int_a^b \int_a^t \frac{|G(g(t)) - G(g(s))|}{(t-s)^{1+\gamma}} dsdt \right) = \int_{\mathbb{R}} h(y) \mu_G(dy).$$

With these two last equalities in mind, we now divide the proof into three parts.
Step 1: Here, we suppose that G is a right-continuous, non-decreasing and bounded function on $\mathbb{R}$. In this case, the definition of G_n and the right continuity of G (see Lemma 89) allows us to get

$$\frac{G_n(g(t)) - G_n(g(s))}{(t-s)^{1+\gamma}} \to \frac{G(g(t)) - G(g(s))}{(t-s)^{1+\gamma}},$$

for almost all $(\omega, t, s) \in \Omega \times \{(t,s) \in [a,b]^2 : s < t\}$, and Lemmas 89 and 90 imply

$$E\left(\left[1 + \|g\|_{\alpha,[a,b]}\right]^{-q} \int_a^b \int_a^t \frac{|G_n(g(t)) - G_n(g(s))|}{(t-s)^{1+\gamma}} ds \right)$$
$$\to E\left(\left[1 + \|g\|_{\alpha,[a,b]}\right]^{-q} \int_a^b \int_a^t \frac{|G(g(t)) - G(g(s))|}{(t-s)^{1+\gamma}} ds \right),$$

as $n \to \infty$. Hence, Kallenberg [94] (Lemma 1.32) yields

$$\lim_{n\to\infty} \left[1 + \|g\|_{\alpha,[a,b]}\right]^{-q} \frac{G_n(g(t)) - G_n(g(s))}{(t-s)^{1+\gamma}} = \left[1 + \|g\|_{\alpha,[a,b]}\right]^{-q} \frac{G(g(t)) - G(g(s))}{(t-s)^{1+\gamma}}$$

in $L^1(\Omega \times \{(t,s) \in [a,b]^2 : s \leq t\})$ So, the result holds under the assumptions of this step.
Step 2: Suppose that G is a right-continuous function with bounded variation on $\mathbb{R}$. In consequence, there exist $G^{(1)}$ and $G^{(2)}$ two non-decreasing and bounded function on $\mathbb{R}$ such

that $G = G^{(1)} - G^{(2)}$ (see Lemma 106). Moreover, by Condition (2.72), we can assume that they are right-continuous without loss of generality. Thus,

$$\begin{aligned}
&\int_a^b \int_a^t \frac{|G_n(g(t)) - G_n(g(s)) - (G(g(t)) - G(g(s)))|}{(t-s)^{1+\gamma}} ds \\
\leq & \int_a^b \int_a^t \frac{\left|G_n^{(1)}(g(t)) - G_n^{(1)}(g(s)) - \left(G^{(1)}(g(t)) - G^{(1)}(g(s))\right)\right|}{(t-s)^{1+\gamma}} ds \\
+ & \int_a^b \int_a^t \frac{\left|G_n^{(2)}(g(t)) - G_n^{(2)}(g(s)) - \left(G^{(2)}(g(t)) - G^{(2)}(g(s))\right)\right|}{(t-s)^{1+\gamma}} ds.
\end{aligned}$$

Now, it is easy to see that the result is also true in this case because of Step 1.
Step 3: Finally, we assume that G is a right-continuous function with locally bounded variation. In this case, we consider, for $m \in \mathbb{N}$, the event $A_m = \left[\|g\|_{\infty,[a,b]} \leq m\right]$ and the function $\tilde{G}^{(m)} = G\left((m \wedge \cdot) \vee (-m)\right)$. Therefore, Step 2 implies

$$\begin{aligned}
&P\left(\left[\int_a^b \int_a^t \frac{|G_n(g(t)) - G_n(g(s)) - (G(g(t)) - G(g(s)))|}{(t-s)^{1+\gamma}} ds > \varepsilon, A_m\right]\right) \\
= \ & P\left(\left[\int_a^b \int_a^t \frac{\left|\tilde{G}_n^{(m)}(g(t)) - \tilde{G}_n^{(m)}(g(s)) - \left(\tilde{G}^{(m)}(g(t)) - \tilde{G}^{(m)}(g(s))\right)\right|}{(t-s)^{1+\gamma}} ds > \varepsilon, A_m\right]\right) \\
\leq \ & P\left(\left[\int_a^b \int_a^t \frac{\left|\tilde{G}_n^{(m)}(g(t)) - \tilde{G}_n^{(m)}(g(s)) - \left(\tilde{G}^{(m)}(g(t)) - \tilde{G}^{(m)}(g(s))\right)\right|}{(t-s)^{1+\gamma}} ds > \varepsilon\right]\right) \\
& \to 0, \quad \text{as } \varepsilon \to 0.
\end{aligned}$$

In consequence, we conclude that the result is satisfied because $\Omega = \cup_{m\in\mathbb{N}} A_m$, with probability 1 (see the proof of Proposition 97). □

8.7 Forward integral of some discontinuous integrands with respect to Hölder continuous processes

Here, we state some auxiliary tools that we need for the analysis of Theorem 66 in order to improve the presentation of Section 5 and to avoid a long and tedious proof of this theorem.

The first auxiliary result is to figure out the distance between y and $y^{(\varepsilon)}$, which is introduced in (5.8).

Lemma 92 *Let $y = \{y_t : t \in [0,T]\}$ be a stochastic process satisfying Hypothesis* **A1** *and $y^{(\varepsilon)}$ given in* (5.8). *Then,*

$$|y_s^{(\varepsilon)} - y_s| \leq G_\delta \varepsilon^{H-\delta}, \quad \textit{for all } \varepsilon > 0, \ s \in [0,T] \textit{ and } \delta < H, \tag{8.25}$$

where G_δ is introduced in (5.5).

Proof. From (5.5) and (5.8), we can write

$$\begin{aligned} |y_s^{(\varepsilon)} - y_s| &= |\frac{1}{\varepsilon}\int_s^{s+\varepsilon} y_{u\wedge T}du - \frac{1}{\varepsilon}\int_s^{s+\varepsilon} y_s du| \leq \frac{1}{\varepsilon}G_\delta \int_s^{s+\varepsilon} |u\wedge T - s|^{H-\delta}du \\ &\leq G_\delta|(s+\varepsilon)\wedge T - s|^{H-\delta} \leq G_\delta \varepsilon^{H-\delta}. \end{aligned}$$

Thus, (8.25) is satisfied. □

As a consequence of Lemma 92, we also have that, for $r, u \in [0,T]$ and $\delta < H$,

$$|y_r^{(\varepsilon)} - y_u^{(\varepsilon)}| \leq |y_r^{(\varepsilon)} - y_r| + |y_u^{(\varepsilon)} - y_u| + |y_r - y_u| \leq 2G_\delta \varepsilon^{H-\delta} + G_\delta |r-u|^{H-\delta}. \quad (8.26)$$

The purpose of the following two results is to confirm that the convergence established in (5.13) is valid.

Lemma 93 *Let $\rho > 0$ be such that $2(H-\rho) > 1$ and assume that Hypothesis* **A1** *holds. Then, there exists a constant $C > 0$ satisfying*

$$\begin{aligned} \varepsilon^{-3/2+H-\rho}\left|\int_{y_s^{(\varepsilon)}-r}^{y_s-r} e^{-x^2/2\varepsilon}dx\right| &\leq CG_\rho^{2n+2\rho}\varepsilon^{(-1+H-2\rho)+[2(H-\rho)-1]n+2\rho(H-\rho)} \\ &\quad +(n+1)(1+G_\rho^2)^{n-1}G_\rho\varepsilon^{-3/2+2H-2\rho}e^{-(y_s-r)^2/2\varepsilon} \\ &\quad +CG_\rho\varepsilon^{-2+2H-3\rho}\int_s^{s+\varepsilon}\frac{1}{|y_{u\wedge T}-r|^{1-2\rho}}du, \end{aligned}$$

for all $n \in \mathbb{N}$, $\varepsilon \in (0,1)$, $s \in [0,T]$ and $r \in \mathbb{R}$.

Proof. Fix $s \in [0,T]$ and $r \in \mathbb{R}$. We observe that, in order to prove this result, we have to consider only two cases. Namely, $(y_s - r)(y_s^{(\varepsilon)} - r) < 0$ and $(y_s - r)(y_s^{(\varepsilon)} - r) > 0$ because we can proceed as in the case that both terms have the same sign if the product is zero. So, now we divide the proof into two parts.

Step1: Here, we suppose that $(y_s - r)(y_s^{(\varepsilon)} - r) < 0$.

In order to fix ideas, we assume that $y_s - r > 0$ and $y_s^{(\varepsilon)} - r < 0$. After the analysis of this case, the reader will be easily able to see that the result is also true if $y_s - r < 0$ and $y_s^{(\varepsilon)} - r > 0$.

Choose $\rho < H$ and $\varepsilon > 0$. Then, (8.25) leads us to establish

$$|(y_s - r) - (y_s^{(\varepsilon)} - r)| = y_s - y_s^{(\varepsilon)} \leq G_\rho \varepsilon^{H-\rho},$$

which implies

$$y_s - r \leq G_\rho\varepsilon^{H-\rho} \quad \text{and} \quad -(y_s^{(\varepsilon)} - r) \leq G_\rho \varepsilon^{H-\rho}. \quad (8.27)$$

Notice that we have

$$\begin{aligned} &\varepsilon^{-3/2+H-\rho}\int_{y_s^{(\varepsilon)}-r}^{y_s-r} e^{-x^2/2\varepsilon}dx \\ &= \varepsilon^{-3/2+H-\rho}\left[\int_{y_s^{(\varepsilon)}-r}^{0} e^{-x^2/2\varepsilon}dx + \int_0^{y_s-r} e^{-x^2/2\varepsilon}dx\right] \\ &= \varepsilon^{-3/2+H-\rho}\left[\int_0^{-(y_s^{(\varepsilon)}-r)} e^{-x^2/2\varepsilon}dx + \int_0^{y_s-r} e^{-x^2/2\varepsilon}dx\right]. \quad (8.28) \end{aligned}$$

We first analyze the second integral in (8.28). The triangle inequality, (8.27) and the mean valued theorem yield

$$\begin{aligned}
&\varepsilon^{-3/2+H-\rho}\left|\int_0^{y_s-r} e^{-x^2/2\varepsilon}dx\right| \\
&\leq \varepsilon^{-3/2+H-\rho}\left(\int_0^{y_s-r}\left|e^{-x^2/2\varepsilon}-e^{-(y_s-r)^2/2\varepsilon}\right|dx+\int_0^{y_s-r} e^{-(y_s-r)^2/2\varepsilon}dx\right) \\
&\leq \varepsilon^{-3/2+H-\rho}\left(\int_0^{y_s-r}\left|\frac{x^2}{2\varepsilon}-\frac{(y_s-r)^2}{2\varepsilon}\right|e^{-x^2/2\varepsilon}dx+\int_0^{y_s-r} e^{-(y_s-r)^2/2\varepsilon}dx\right) \\
&\leq \varepsilon^{-3/2+H-\rho}\left(G_\rho^2\varepsilon^{2(H-\rho)-1}\int_0^{y_s-r} e^{-x^2/2\varepsilon}dx+\int_0^{y_s-r} e^{-(y_s-r)^2/2\varepsilon}dx\right). \qquad (8.29)
\end{aligned}$$

Proceeding similarly (i.e., repeating last argument), for $\rho > 0$ such that $2(H-\rho) > 1$ and $\varepsilon < 1$, we can write

$$\begin{aligned}
&\varepsilon^{-3/2+H-\rho}\left|\int_0^{y_s-r} e^{-x^2/2\varepsilon}dx\right| \\
&\leq \varepsilon^{-3/2+H-\rho}\left(G_\rho^4\varepsilon^{[2(H-\rho)-1]2}\int_0^{y_s-r} e^{-x^2/2\varepsilon}dx\right. \\
&\quad \left.+G_\rho^2\varepsilon^{2(H-\rho)-1}\int_0^{y_s-r} e^{-(y_s-r)^2/2\varepsilon}dx+\int_0^{y_s-r} e^{-(y_s-r)^2/2\varepsilon}dx\right) \\
&\leq \varepsilon^{-3/2+H-\rho}\left(G_\rho^4\varepsilon^{[2(H-\rho)-1]2}\int_0^{y_s-r} e^{-x^2/2\varepsilon}dx+2(G_\rho^2+1)\int_0^{y_s-r} e^{-(y_s-r)^2/2\varepsilon}dx\right).
\end{aligned}$$

Thus, induction on n, (8.27) and (8.29) allow us to show

$$\begin{aligned}
&\varepsilon^{-3/2+H-\rho}\int_0^{y_s-r} e^{-x^2/2\varepsilon}dx \\
&\leq \varepsilon^{-3/2+H-\rho}\left(G_\rho^{2n}\varepsilon^{[2(H-\rho)-1]n}\int_0^{y_s-r} e^{-x^2/2\varepsilon}dx+n(G_\rho^2+1)^{n-1}\int_0^{y_s-r} e^{-(y_s-r)^2/2\varepsilon}dx\right) \\
&\leq \varepsilon^{-3/2+H-\rho}\left(G_\rho^{2n}\varepsilon^{[2(H-\rho)-1]n}\int_0^{y_s-r} e^{-x^2/2\varepsilon}dx+n(G_\rho^2+1)^{n-1}G_\rho\varepsilon^{H-\rho}e^{-(y_s-r)^2/2\varepsilon}\right) \\
&\leq CG_\rho^{2n+2\rho}\varepsilon^{-1+H-2\rho+[2(H-\rho)-1]n+2\rho(H-\rho)} \\
&\quad +n(G_\rho^2+1)^{n-1}G_\rho\varepsilon^{-3/2+2H-2\rho}e^{-(y_s-r)^2/2\varepsilon}, \qquad (8.30)
\end{aligned}$$

where last inequality follows from the fact that (8.37) below and (8.27) give that

$$\frac{1}{\varepsilon^{1/2-\rho}}\int_0^{y_s-r} e^{-x^2/2\varepsilon}dx \leq C\int_0^{y_s-r}\frac{dx}{x^{1-2\rho}} = C\frac{(y_s-r)^{2\rho}}{2\rho} \leq CG_\rho^{2\rho}(\varepsilon^{H-\rho})^{2\rho}$$

is satisfied.

Now, we deal with the first integral in (8.28).

If $-(y_s^{(\varepsilon)}-r) \leq y_s - r$, then

$$\varepsilon^{-3/2+H-\rho}\int_0^{-(y_s^{(\varepsilon)}-r)} e^{-x^2/2\varepsilon}dx \leq \varepsilon^{-3/2+H-\rho}\int_0^{y_s-r} e^{-x^2/2\varepsilon}dx. \qquad (8.31)$$

In the case that $y_s - r \le -(y_s^{(\varepsilon)} - r)$, we apply (8.27) again to establish

$$\begin{aligned}
&\varepsilon^{-3/2+H-\rho}\int_0^{-(y_s^{(\varepsilon)}-r)} e^{-x^2/2\varepsilon}dx \\
\le\ & \varepsilon^{-3/2+H-\rho}\left[\int_0^{y_s-r} e^{-x^2/2\varepsilon}dx + \int_{y_s-r}^{-(y_s^{(\varepsilon)}-r)} e^{-x^2/2\varepsilon}dx\right] \\
\le\ & \varepsilon^{-3/2+H-\rho}\left[\int_0^{y_s-r} e^{-x^2/2\varepsilon}dx + e^{-(y_s-r)^2/2\varepsilon}\left|-(y_s^{(\varepsilon)}-r)-(y_s-r)\right|\right] \\
\le\ & \varepsilon^{-3/2+H-\rho}\int_0^{y_s-r} e^{-x^2/2\varepsilon}dx + G_\rho\varepsilon^{-3/2+2H-2\rho}e^{-(y_s-r)^2/2\varepsilon}.
\end{aligned}$$

In consequence, (8.28), (8.30) and (8.31) give

$$\begin{aligned}
\varepsilon^{-3/2+H-\rho}\int_{y_s^{(\varepsilon)}-r}^{y_s-r} e^{-x^2/2\varepsilon}dx \ \le\ & CG_\rho^{2n+2\rho}\varepsilon^{(-1+H-2\rho)+[2(H-\rho)-1]n+2\rho(H-\rho)} \\
& +(n+1)(1+G_\rho^2)^{n-1}G_\rho\varepsilon^{-3/2+2H-2\rho}e^{-(y_s-r)^2/2\varepsilon}.
\end{aligned}$$

Step 2: In this part of the proof, we suppose that $(y_s - r)(y_s^{(\varepsilon)} - r) > 0$.

As in Step 1, without loss of generality, we can assume that $y_s - r > 0$ and $y_s^{(\varepsilon)} - r > 0$ since it is able to show the missing situation analogously.

For $y_s - r < y_s^{(\varepsilon)} - r$, we make use of Lemma 92 to get

$$\begin{aligned}
\varepsilon^{-3/2+H-\rho}\int_{y_s-r}^{y_s^{(\varepsilon)}-r} e^{-x^2/2\varepsilon}dx \ \le\ & \varepsilon^{-3/2+H-\rho}e^{-(y_s-r)^2/2\varepsilon}\left|y_s - y_s^{(\varepsilon)}\right| \\
\le\ & G_\rho\varepsilon^{-3/2+2H-2\rho}e^{-(y_s-r)^2/2\varepsilon}. \qquad (8.32)
\end{aligned}$$

But, for $y_s^{(\varepsilon)} - r < y_s - r$, (5.8) implies

$$y_s^{(\varepsilon)} - r = \int_s^{s+\varepsilon}(y_{u\wedge T} - r)\frac{du}{\varepsilon}. \qquad (8.33)$$

Note that if there is some $\hat{u} \in (s, s+\varepsilon]$ such that $y_{\hat{u}\wedge T} - r = 0$, we have

$$y_s - r = y_s - r - (y_{\hat{u}\wedge T} - r) = y_{s\wedge T} - y_{\hat{u}\wedge T} \le G_\rho\varepsilon^{H-\rho}.$$

Hence, (8.30) proves

$$\begin{aligned}
\varepsilon^{-3/2+H-\rho}\left|\int_{y_s^{(\varepsilon)}-r}^{y_s-r} e^{-x^2/2\varepsilon}dx\right| &\le 2\varepsilon^{-3/2+H-\rho}\int_0^{y_s-r} e^{-x^2/2\varepsilon}dx \\
&\le CG_\rho^{2n+2\rho}\varepsilon^{-1+H-2\rho+[2(H-\rho)-1]n+2\rho(H-\rho)} \\
&\quad + n(G_\rho^2+1)^{n-1}G_\rho\varepsilon^{-3/2+2H-2\rho}e^{-(y_s-r)^2/2\varepsilon}. \qquad (8.34)
\end{aligned}$$

Therefore, in the remaining of proof, we can assume that $y_{u\wedge T} - r > 0$, for all $u \in [s, s+\varepsilon]$.

Observe that Hölder inequality yields

$$|y_s^{(\varepsilon)} - r|^{1-2\rho} = (\int_s^{s+\varepsilon}|y_{u\wedge T} - r|\frac{du}{\varepsilon})^{1-2\rho} \ge \int_s^{s+\varepsilon}|y_{u\wedge T} - r|^{1-2\rho}\frac{du}{\varepsilon},$$

which, together with Jensen inequality, allows us to obtain

$$\frac{1}{|y_s^{(\varepsilon)} - r|^{1-2\rho}} \le \left(\int_s^{s+\varepsilon} |y_{u\wedge T} - r|^{1-2\rho}\frac{du}{\varepsilon}\right)^{-1} \le \frac{1}{\varepsilon}\int_s^{s+\varepsilon} \frac{du}{|y_{u\wedge T} - r|^{1-2\rho}}. \tag{8.35}$$

Thus, by (8.25) and (8.37), we have

$$\begin{aligned}
&\varepsilon^{-3/2+H-\rho}\int_{y_s^{(\varepsilon)}-r}^{y_s-r} e^{-x^2/2\varepsilon}dx \\
\le\ & \varepsilon^{-3/2+H-\rho}e^{-(y_s^{(\varepsilon)}-r)^2/2\varepsilon}|y_s - y_s^{(\varepsilon)}| \le G_\rho\varepsilon^{-3/2+2H-2\rho}e^{-(y_s^{(\varepsilon)}-r)^2/2\varepsilon} \\
\le\ & CG_\rho\varepsilon^{-1+2H-3\rho}\frac{1}{|y_s^{(\varepsilon)} - r|^{1-2\rho}} \le CG_\rho\varepsilon^{-2+2H-3\rho}\int_s^{s+\varepsilon}\frac{1}{|y_{u\wedge T} - r|^{1-2\rho}}du.
\end{aligned}$$

Finally, utilizing this inequality, (8.32), (8.34) and Step 1, the proof is complete. □

Lemma 94 *The assumptions of Theorem 66 imply that, for some $p > 1$ and all $t \in [0, T]$,*

$$\varepsilon^{-1}\int_0^t F'_\varepsilon(y_s^{(\varepsilon)})(y_{(s+\varepsilon)\wedge T} - y_s)ds \to \int_0^t f(y_s)dy_s^-, \quad \textit{in } L^p(\Omega), \textit{ as } \varepsilon \downarrow 0.$$

Proof. Fix $t \in [0, T]$. Note that, by (5.12), we only need to see that the integral $\varepsilon^{-1}\int_0^t [f(y_s) - F'_\varepsilon(y_s^{(\varepsilon)})](y_{(s+\varepsilon)\wedge T} - y_s)ds$ goes to zero in probability as $\varepsilon \downarrow 0$. So, we observe

$$\begin{aligned}
&\varepsilon^{-1}\int_0^t [f(y_s) - F'_\varepsilon(y_s^{(\varepsilon)})](y_{(s+\varepsilon)\wedge T} - y_s)ds \\
=\ & \varepsilon^{-1}\int_0^t [f(y_s) - F'_\varepsilon(y_s)](y_{(s+\varepsilon)\wedge T} - y_s)ds \\
&+\varepsilon^{-1}\int_0^t [F'_\varepsilon(y_s) - F'_\varepsilon(y_s^{(\varepsilon)})](y_{(s+\varepsilon)\wedge T} - y_s)ds = I_1 + I_2.
\end{aligned} \tag{8.36}$$

Now, we divide the proof into two parts in order to study the convergence of I_1 and I_2 in $L^p(\Omega)$.
Step 1: Here, we show that, for some $p > 1$, I_1 in (8.36) tends to zero in $L^p(\Omega)$, as $\varepsilon \downarrow 0$.

Note that the function $g(x) = x^\alpha e^{-ax}$, $x > 0$, has a maximum at $x = \frac{\alpha}{a}$, which leads us to state

$$\left(\frac{1}{\varepsilon}\right)^\alpha e^{-x^2/4\varepsilon} \le \left(\frac{4\alpha}{x^2}\right)^\alpha e^{-\alpha}. \tag{8.37}$$

Hence, Hölder inequality, Fubini's theorem, Hypotheses **A1, A2** and **B**, Remark 77, Lemma

61 and inequality (5.5) give

$$\begin{aligned}
E(|I_1|^p) &\le \varepsilon^{(-1+H-\rho)p} E\left[G_\rho^p \left(\int_0^t |f(y_s+) - F_\varepsilon'(y_s)|ds\right)^p\right] \\
&\le C\varepsilon^{(-1+H-\rho)p} E\left[G_\rho^p \left(\int_0^t \int_{\mathbb{R}} e^{-(y_s-x)^2/4\varepsilon} |\mu_f|(dx)ds\right)^p\right] \\
&= C\varepsilon^{\rho p} E\left[G_\rho^p \left(\int_0^t \int_{\mathbb{R}} \left(\frac{1}{\varepsilon}\right)^{1+2\rho-H} e^{-(y_s-x)^2/4\varepsilon} |\mu_f|(dx)ds\right)^p\right] \\
&\le C\varepsilon^{\rho p} E\left[G_\rho^p \left(\int_0^t \int_{\mathbb{R}} \frac{|\mu_f|(dx)}{(y_s-x)^{2(1+2\rho-H)}} ds\right)^p\right] \\
&\le C\varepsilon^{\rho p} (E(G_\rho^{p(1+\eta)/\eta}))^{\eta/(1+\eta)} \\
&\quad \times \left(\int_{\mathbb{R}} E\int_0^t \frac{ds}{|y_s-x|^{2(1+\eta)(1+2\rho-H)p}} |\mu_f|(dx)\right)^{1/(1+\eta)} \le C\varepsilon^{\rho p},
\end{aligned}$$

where the last inequality is true if ρ and η are small enough, and p is close enough to 1 such that $2(1+\eta)(1+2\rho-H)p < 1$. Therefore, the claim of this step holds.

Step 2: Now, we show that I_2 in (8.36) converges to zero in $L^p(\Omega)$, as $\varepsilon \downarrow 0$, for some $p > 1$.

Let $\rho > 0$ such that $2(H-\rho) > 1$. Then, (5.5), (5.7) and Hypothesis **B** yield

$$\begin{aligned}
|I_2| &\le \varepsilon^{-1} \int_0^t |F_\varepsilon'(y_s) - F_\varepsilon'(y_s^{(\varepsilon)})||y_{(s+\varepsilon)\wedge T} - y_s|ds \\
&\le G_\rho \varepsilon^{-1+H-\rho} \int_0^t \left| \int_{\mathbb{R}} f(y_s - x)P_\varepsilon(x)dx - \int_{\mathbb{R}} f(y_s^{(\varepsilon)} - x)P_\varepsilon(x)dx \right| ds \\
&\le G_\rho \varepsilon^{-1+H-\rho} \int_0^t \int_{\mathbb{R}} \left| f((y_s-x)+) - f((y_s^{(\varepsilon)} - x)+) \right| P_\varepsilon(x)dxds \\
&= G_\rho \varepsilon^{-1+H-\rho} \int_0^t \int_{\mathbb{R}} \left| \int_{\mathbb{R}} 1_{((y_s-x)\wedge(y_s^{(\varepsilon)}-x),(y_s-x)\vee(y_s^{(\varepsilon)}-x)]}(r)d\mu_f(r) \right| P_\varepsilon(x)dxds \\
&\le G_\rho \varepsilon^{-1+H-\rho} \int_0^t \int_{\mathbb{R}} \int_{\mathbb{R}} 1_{((y_s-r)\wedge(y_s^{(\varepsilon)}-r),(y_s-r)\vee(y_s^{(\varepsilon)}-r)]}(x)P_\varepsilon(x)dx|\mu_f|(dr)ds \\
&= G_\rho \varepsilon^{-1+H-\rho} \int_0^t \int_{\mathbb{R}} \int_{(y_s-r)\wedge(y_s^{(\varepsilon)}-r)}^{(y_s-r)\vee(y_s^{(\varepsilon)}-r)} P_\varepsilon(x)dx|\mu_f|(dr)ds \\
&= G_\rho \varepsilon^{-1+H-\rho} \int_0^t \int_{\mathbb{R}} \left| \int_{y_s^{(\varepsilon)}-r}^{y_s-r} P_\varepsilon(x)dx \right| |\mu_f|(dr)ds \\
&\le CG_\rho \varepsilon^{-3/2+H-\rho} \int_{\mathbb{R}} \int_0^t \left| \int_{y_s^{(\varepsilon)}-r}^{y_s-r} e^{-x^2/2\varepsilon}dx \right| ds|\mu_f|(dr).
\end{aligned}$$

Thus, Lemma 93 implies that there is $C > 0$ such that, for all $n \in \mathbb{N}$ and $\varepsilon \in (0,1)$,

$$\begin{aligned}
|I_2| &\le C\int_{\mathbb{R}} \int_0^t G_\rho^{2n+2\rho+1} \varepsilon^{(-1+H-2\rho)+[2(H-\rho)-1]n+2\rho(H-\rho)} ds|\mu_f|(dr) \\
&\quad + C\int_{\mathbb{R}} \int_0^t (n+1)(1+G_\rho^2)^{n-1} G_\rho^2 \varepsilon^{-3/2+2H-2\rho} e^{-(y_s-r)^2/2\varepsilon} ds|\mu_f|(dr) \\
&\quad + C\int_{\mathbb{R}} \int_0^t G_\rho^2 \varepsilon^{-2+2H-3\rho} \int_s^{s+\varepsilon} \frac{1}{|y_{u\wedge T} - r|^{1-2\rho}} duds|\mu_f|(dr) = I_{2,1} + I_{2,2} + I_{2,3}.
\end{aligned}$$

For n large enough, we have

$$I_{2,1} \leq CG_\rho^{2n+2\rho+1}\varepsilon^{(-1+H-2\rho)+[2(H-\rho)-1]n+2\rho(H-\rho)},$$

whose right-hand side goes to zero in $L^p(\Omega)$ as $\varepsilon \downarrow 0$, for any $p > 1$ and $n \in \mathbb{N}$ such that $H - 2\rho + [2(H-\rho)-1]n + 2\rho(H-\rho) > 1$.

From (8.37), we get

$$I_{2,2} \leq C(n+1)(1+G_\rho^2)^{n-1}G_\rho^2\varepsilon^{-1+2H-3\rho}\int_{\mathbb{R}}\int_0^t \frac{ds}{|y_s - r|^{1-2\rho}}|\mu_f|(dx),$$

which, together with Remark 77, yields that, for ρ small enough, $I_{2,2}$ converges to zero in $L^p(\Omega)$, as $\varepsilon \downarrow 0$, for some $p > 1$, by proceeding as in the analysis of $I_{2,3}$ below.

Finally, Fubini's theorem and Hölder inequality give

$$\begin{aligned}
E\left(|I_{2,3}|^p\right) &\leq C\varepsilon^{(-1+2H-3\rho)p}E\left(G_\rho^{2p}\left(\int_{\mathbb{R}}\int_0^{t+\varepsilon}\frac{1}{|y_{u\wedge T}-r|^{1-2\rho}}du|\mu_f|(dr)\right)^p\right)\\
&\leq C\varepsilon^{(-1+2H-3\rho)p}E\left(G_\rho^{2p}\int_{\mathbb{R}}\int_0^{t+\varepsilon}\frac{1}{|y_{u\wedge T}-r|^{(1-2\rho)p}}du|\mu_f|(dr)\right)\\
&\leq C\varepsilon^{(-1+2H-3\rho)p}\left(E\left(G_\rho^{2p(1+\eta)/\eta}\right)\right)^{\eta/(1+\eta)}\\
&\quad\times\left(E\left(\int_{\mathbb{R}}\int_0^{t+\varepsilon}\frac{1}{|y_{u\wedge T}-r|^{(1-2\rho)(1+\eta)p}}du|\mu_f|(dr)\right)\right)^{1/(1+\eta)}\\
&\to 0, \quad \text{as } \varepsilon \downarrow 0,
\end{aligned}$$

for $1+\eta$ and p close enough to 1. In fact, for $t < T$ and $\varepsilon < T - t$, Remark 2.v) allows us to obtain

$$\begin{aligned}
&E\left(\int_{\mathbb{R}}\int_0^{t+\varepsilon}\frac{1}{|y_{u\wedge T}-r|^{(1-2\rho)(1+\eta)p}}du|\mu_f|(dr)\right)\\
&= E\left(\int_{\mathbb{R}}\int_0^{t+\varepsilon}\frac{1}{|y_u - r|^{(1-2\rho)(1+\eta)p}}du|\mu_f|(dr)\right) < \infty
\end{aligned}$$

and, for $t = T$, Hypothesis **A2** and Remark 77 imply

$$\begin{aligned}
&E\left(\int_{\mathbb{R}}\int_0^{T+\varepsilon}\frac{1}{|y_{u\wedge T}-r|^{(1-2\rho)(1+\eta)p}}du|\mu_f|(dr)\right)\\
&= E\left(\int_{\mathbb{R}}\int_0^{T}\frac{1}{|y_{u\wedge T}-r|^{(1-2\rho)(1+\eta)p}}du|\mu_f|(dr)\right)\\
&\quad+\varepsilon E\left(\int_{\mathbb{R}}\frac{1}{|y_T-r|^{(1-2\rho)(1+\eta)p}}|\mu_f|(dr)\right)\\
&= E\left(\int_{\mathbb{R}}\int_0^{T}\frac{1}{|y_{u\wedge T}-r|^{(1-2\rho)(1+\eta)p}}du|\mu_f|(dr)\right)\\
&\quad+\varepsilon\left(\int_{\mathbb{R}}\int_{\mathbb{R}}\frac{p_T(z)}{|z-r|^{(1-2\rho)(1+\eta)p}}dz|\mu_f|(dr)\right) < \infty.
\end{aligned}$$

Therefore, the proof is finished. □

Now, we state the auxiliary tool that we need to understand some results in Section 5.3, where the process y is a fractional Brownian motion with Hurst parameter $H > 1/2$. So, we consider the space $\mathcal{H}$ related to fractional Brownian motion, which is introduced in Examples 1.iii).

Lemma 95 *Let y be fractional Brownian motion, $\varepsilon > 0$ and $s \in [0,T]$. Then,*

$$\begin{aligned}\left\langle D.y_s^{(\varepsilon)}, 1_{]s,(s+\varepsilon)\wedge T]}(\cdot)\right\rangle_{\mathcal{H}} &= \frac{((s+\varepsilon)\wedge T)^{2H} - s^{2H}}{2} \\ &\quad - \int_s^{s+\varepsilon} \frac{((s+\varepsilon)\wedge T - u\wedge T)^{2H} - (u\wedge T - s)^{2H}}{2\varepsilon} du.\end{aligned}$$

In particular, if $s+\varepsilon \le T$, we have that $\left\langle D.y_s^{(\varepsilon)}, 1_{]s,(s+\varepsilon)\wedge T]}(\cdot)\right\rangle_{\mathcal{H}} = \frac{(s+\varepsilon)^{2H}-s^{2H}}{2}$.

Proof. (1.44) and (5.8) imply

$$\begin{aligned}&\left\langle D.y_s^{(\varepsilon)}, 1_{]s,(s+\varepsilon)\wedge T]}(\cdot)\right\rangle_{\mathcal{H}} \\ &= \left\langle D.\left(\frac{1}{\varepsilon}\int_s^{s+\varepsilon} y_{u\wedge T} du\right), 1_{]s,(s+\varepsilon)\wedge T]}(\cdot)\right\rangle_{\mathcal{H}} = \frac{1}{\varepsilon}\left\langle \int_s^{s+\varepsilon} 1_{[0,u\wedge T]}(\cdot) du, 1_{]s,(s+\varepsilon)\wedge T]}(\cdot)\right\rangle_{\mathcal{H}} \\ &= \int_s^{s+\varepsilon} \frac{((s+\varepsilon)\wedge T)^{2H} - s^{2H} - ((s+\varepsilon)\wedge T - u\wedge T)^{2H} + (u\wedge T - s)^{2H}}{2\varepsilon} du.\end{aligned}$$

Hence, the proof is complete. □

Lemma 96 *Let y be a fractional Brownian motion with Hurst parameter $H > 1/2$ and $t \in [0,T]$. Then,*

$$\frac{1}{\varepsilon}\int_0^t P_\varepsilon(y_s)\left\langle D.y_s^{(\varepsilon)}, 1_{]s,(s+\varepsilon)\wedge T]}(\cdot)\right\rangle_{\mathcal{H}} ds \to \frac{1}{2}L_t^0 \quad \text{in } L^2(\Omega), \text{ as } \varepsilon \downarrow 0, \tag{8.38}$$

where L_t^0 is the local time of y at 0.

Proof. Fix $t \in [0,T]$. Note that, in order to prove that the result holds, we only need to see that

$$\int_0^t P_\varepsilon(y_s)\left(\varepsilon^{-1}\left\langle D.y_s^{(\varepsilon)}, 1_{]s,(s+\varepsilon)\wedge T]}(\cdot)\right\rangle_{\mathcal{H}} - Hs^{2H-1}\right) ds \to 0, \quad \text{in } L^2(\Omega),$$

as ε goes to zero, due to Coutin et al. [32] (Proposition 2).

For $\varepsilon > 0$ small enough, we have that Lemma 95 leads us to show

$$\begin{aligned}&\int_0^t P_\varepsilon(y_s)\left(\varepsilon^{-1}\left\langle D.y_s^{(\varepsilon)}, 1_{]s,(s+\varepsilon)\wedge T]}(\cdot)\right\rangle_{\mathcal{H}} - Hs^{2H-1}\right) ds \\ &\le \int_0^t P_\varepsilon(y_s)\left|\frac{(s+\varepsilon)^{2H} - s^{2H}}{2\varepsilon} - Hs^{2H-1}\right| ds \\ &\quad + \int_{t-\varepsilon}^t P_\varepsilon(y_s)\left|\int_s^{s+\varepsilon} \frac{((s+\varepsilon)\wedge T - u\wedge T)^{2H} - (u\wedge T - s)^{2H}}{2\varepsilon^2} du\right| ds \\ &= I_1 + I_2.\end{aligned} \tag{8.39}$$

The mean value theorem, and the facts that there is a constant $C > 0$ such that, for $x \neq 0$, $P_\varepsilon(x) \le (\sqrt{2\pi})^{-1}\frac{1}{\sqrt{\varepsilon}} \wedge \frac{C}{|x|}$ and $H > 1/2$ give that, for $\delta > 0$ small enough,

$$\begin{aligned}|I_2| &\le C\int_{t-\varepsilon}^t P_\varepsilon(y_s)\int_s^{s+\varepsilon}\frac{(s+\varepsilon)\wedge T - s}{2\varepsilon^2} duds \le C\int_{t-\varepsilon}^t P_\varepsilon(y_s) ds \\ &\le C(\sqrt{\varepsilon})^{-\delta-1/2}\int_{t-\varepsilon}^t |y_s|^{\delta-1/2} ds \le C(\varepsilon)^{-\delta/2}\varepsilon^{1/4}\left(\int_{t-\varepsilon}^t |y_s|^{2\delta-1} ds\right)^{1/2}.\end{aligned}$$

Hence, Lemma 1 yields

$$\begin{aligned} E\left(|I_2|^2\right) &\leq C\varepsilon^{-\delta}\varepsilon^{1/2}\int_{t-\varepsilon}^{t}\frac{ds}{s^{H(1-2\delta)}} = C\varepsilon^{1/2-\delta}\left(t^{1-H+2H\delta}-(t-\varepsilon)^{1-H+2H\delta}\right) \\ &\leq C\varepsilon^{1/2-\delta}\varepsilon^{1-H+2H\delta}\to 0, \quad \text{in } L^2(\Omega), \quad \text{as } \varepsilon\downarrow 0. \end{aligned}$$

Finally, for $s\in[0,t]$, Taylor's theorem yields that $(s+\varepsilon)^{2H}\leq s^{2H}+2Hs^{2H-1}\varepsilon+H(2H-1)s^{2H-2}\varepsilon^2$, which, together with (8.39), establishes

$$|I_1| \leq C\varepsilon\int_0^t P_\varepsilon(y_s)s^{2H-2}ds \leq C\sqrt{\varepsilon}t^{2H-1}\to 0, \quad \text{in } L^2(\Omega) \text{ as } \varepsilon\downarrow 0.$$

Therefore, the result is true. □

Lemma 97 *Let y be fractional Brownian motion with Hurst parameter $H>1/2$ and $s\in[0,T]$ such that $0<y_s\leq G_\rho\varepsilon^{H-\rho}$, for $0<\rho<(2H-1)/2$. Then,*

$$\int_0^{y_s} e^{-u^2/2\varepsilon}udu \leq G_\rho^{2n}\varepsilon^{n(2H-2\rho-1)}\int_0^{y_s} e^{-u^2/2\varepsilon}udu + nG_\rho^2\left(G_\rho^2+1\right)^{n-1}e^{-y_s^2/2\varepsilon}\varepsilon^{2H-2\rho},$$

for all $n\in\mathbb{N}$ and $\varepsilon\in(0,1)$.

Proof. By the mean value theorem, we have

$$\begin{aligned} \int_0^{y_s} e^{-u^2/2\varepsilon}udu &\leq \int_0^{y_s}\left|e^{-u^2/2\varepsilon}-e^{-y_s^2/2\varepsilon}\right|udu + \int_0^{y_s} e^{-y_s^2/2\varepsilon}udu \\ &\leq \frac{1}{2\varepsilon}\int_0^{y_s}\left|u^2-y_s^2\right|e^{-u^2/2\varepsilon}udu + \int_0^{y_s} e^{-y_s^2/2\varepsilon}udu \\ &\leq G_\rho^2\varepsilon^{2(H-\rho)-1}\int_0^{y_s} e^{-u^2/2\varepsilon}udu + G_\rho^2\varepsilon^{2H-2\rho}e^{-y_s^2/2\varepsilon}. \end{aligned}$$

Now, it is easy to finish the proof using induction on $n\in\mathbb{N}$. □

Lemma 98 *Let y be fractional Brownian motion with Hurst parameter $H>1/2$, $\varepsilon\in(0,1)$, $\rho\in(0,H-1/2)$ and $s\in[0,T]$. Then,*

$$\begin{aligned} |P_\varepsilon(y_s)-P_\varepsilon(y_s^{(\varepsilon)})| &\leq CG_\rho^{2(n+1)}\varepsilon^{n(2H-2\rho-1)-3/2+2H-2\rho} \\ &\quad +(n+1)G_\rho^2\left(G_\rho^2+1\right)^{n-1}e^{-y_s^2/2\varepsilon}\varepsilon^{-3/2+2H-2\rho} \\ &\quad +CG_\rho\varepsilon^{H-1/2-\rho}\left(\frac{1}{y_s^{1-\rho}}+\varepsilon^{-1}\int_s^{s+\varepsilon}\frac{1}{|y_{u\wedge T}|^{1-2\rho}}du\right). \end{aligned}$$

Proof. Fix $s\in[0,t]$, $0<\rho<H-1/2$ and $\varepsilon\in(0,1)$. Therefore,

$$|P_\varepsilon(y_s)-P_\varepsilon(y_s^{(\varepsilon)})| = \varepsilon^{-3/2}\left|\int_{y_s\wedge y_s^{(\varepsilon)}}^{y_s\vee y_s^{(\varepsilon)}}\frac{e^{-u^2/2\varepsilon}}{\sqrt{2\pi}}udu\right|. \tag{8.40}$$

Now, we divide the proof into two cases.
Step 1: Here, we assume that $y_sy_s^{(\varepsilon)}<0$. We only analyze the case that $y_s^{(\varepsilon)}<0<y_s$ because the other one follows similarly.

We have that (8.25) gives that $y_s, |y_s^{(\varepsilon)}| \leq G_\rho \varepsilon^{H-\rho}$. Consequently,

$$\begin{aligned}
&|P_\varepsilon(y_s) - P_\varepsilon((y_s^{(\varepsilon)})| \\
&\leq \varepsilon^{-3/2} \left| \int_{y_s^{(\varepsilon)}}^{y_s} e^{-x^2/2\varepsilon} x dx \right| \leq \varepsilon^{-3/2} \left[\left| \int_{y_s^{(\varepsilon)}}^{0} e^{-x^2/2\varepsilon} x dx \right| + \int_0^{y_s} e^{-x^2/2\varepsilon} x dx \right] \\
&= \varepsilon^{-3/2} \left[\int_0^{-y_s^{(\varepsilon)}} e^{-x^2/2\varepsilon} x dx + \int_0^{y_s} e^{-x^2/2\varepsilon} x dx \right]. \qquad (8.41)
\end{aligned}$$

We first study the first integral in (8.41). For $-y_s^{(\varepsilon)} \leq y_s$, we get

$$\varepsilon^{-3/2} \int_0^{-y_s^{(\varepsilon)}} e^{-x^2/2\varepsilon} x dx \leq \varepsilon^{-3/2} \int_0^{y_s} e^{-x^2/2\varepsilon} x dx.$$

Also, for $y_s \leq -y_s^{(\varepsilon)}$, we obtain

$$\begin{aligned}
\varepsilon^{-3/2} \int_0^{-y_s^{(\varepsilon)}} e^{-x^2/2\varepsilon} x dx &\leq \varepsilon^{-3/2} \left[\int_0^{y_s} e^{-x^2/2\varepsilon} x dx + \int_{y_s}^{-y_s^{(\varepsilon)}} e^{-x^2/2\varepsilon} x dx \right] \\
&\leq \varepsilon^{-3/2} \left[\int_0^{y_s} e^{-x^2/2\varepsilon} x dx + e^{-y_s^2/2\varepsilon} G_\rho |-y_s^{(\varepsilon)} - y_s| \varepsilon^{H-\rho} \right] \\
&\leq \varepsilon^{-3/2} \int_0^{y_s} e^{-x^2/2\varepsilon} dx + 2G_\rho^2 \varepsilon^{-3/2+2H-2\rho} e^{-y_s^2/2\varepsilon}.
\end{aligned}$$

Thus, Lemma 97 and (8.41) allow us to deduce

$$\begin{aligned}
\left| P_\varepsilon(y_s) - P_\varepsilon(y_s^{(\varepsilon)}) \right| &\leq CG_\rho^{2n} \varepsilon^{n(2H-2\rho-1)-3/2} \int_0^{y_s} e^{-u^2/2\varepsilon} u du \\
&\quad + nG_\rho^2 \left(G_\rho^2 + 1 \right)^{n-1} e^{-y_s^2/2\varepsilon} \varepsilon^{-3/2+2H-2\rho} + 2G_\rho^2 \varepsilon^{-3/2+2H-2\rho} e^{-y_s^2/2\varepsilon} \\
&\leq CG_\rho^{2(n+1)} \varepsilon^{n(2H-2\rho-1)-3/2+2H-2\rho} \\
&\quad + nG_\rho^2 \left(G_\rho^2 + 1 \right)^{n-1} e^{-y_s^2/2\varepsilon} \varepsilon^{-3/2+2H-2\rho} + 2G_\rho^2 \varepsilon^{-3/2+2H-2\rho} e^{-y_s^2/2\varepsilon}.
\end{aligned}$$

Step 2: Now, we deal with the case that $y_s y_s^{(\varepsilon)} > 0$.

We only need to consider the case that $y_s > 0$ and $y_s^{(\varepsilon)} > 0$. Otherwise, we use the change of variable $v = -u$ and proceed similarly. Also, without loss of generality, we can suppose that $y > 0$ on $[s, s+\varepsilon]$ since (8.25) gives that we are in the situation of Step 1 if there is $u \in [s, s+\varepsilon]$ such that $y_u = 0$. Hence, by (8.25), (8.37), (8.35) and (8.40), we are able to establish

$$\begin{aligned}
\left| P_\varepsilon(y_s) - P_\varepsilon(y_s^{(\varepsilon)}) \right| &= C\varepsilon^{-3/2} \int_{y_s \wedge y_s^{(\varepsilon)}}^{y_s \vee y_s^{(\varepsilon)}} e^{-u^2/2\varepsilon} \frac{u^{2-\rho}}{u^{1-\rho}} du \\
&\leq C\varepsilon^{-1/2-\rho/2} \int_{y_s \wedge y_s^{(\varepsilon)}}^{y_s \vee y_s^{(\varepsilon)}} \frac{du}{u^{1-\rho}} \\
&\leq C\varepsilon^{-1/2-\rho/2} \left(\frac{1}{y_s^{1-\rho}} + \frac{1}{(y_s^{(\varepsilon)})^{1-\rho}} \right) |y_s - y_s^{(\varepsilon)}| \\
&\leq CG_\rho \varepsilon^{H-1/2-\rho} \left(\frac{1}{y_s^{1-\rho}} + \frac{1}{(y_s^{(\varepsilon)})^{1-\rho}} \right) \\
&\leq CG_\rho \varepsilon^{H-1/2-\rho} \left(\frac{1}{y_s^{1-\rho}} + \varepsilon^{-1} \int_s^{s+\varepsilon} \frac{1}{|y_{u\wedge T}|^{1-2\rho}} du \right).
\end{aligned}$$

In consequence, the result is satisfied. □

Lemma 99 *Let y be fractional Brownian motion with Hurst parameter $H > 1/2$ and $t \in [0,T]$. Then, for some $p > 1$,*

$$\int_0^t \left(P_\varepsilon(y_s) - P_\varepsilon(y_s^{(\varepsilon)})\right) \left\langle D.y_s^{(\varepsilon)}, 1_{]s,(s+\varepsilon)\wedge T]}(\cdot)\right\rangle_{\mathcal{H}} ds \to 0 \quad \textit{in } L^p(\Omega), \textit{ as } \varepsilon \downarrow 0.$$

Proof. Fix $t \in [0,T]$ and choose $\varepsilon \in (0,1)$. Consequently, Lemma 95 establishes that there is $C > 0$ (independent of s and ε) such that $\left|\langle D.y_s^{(\varepsilon)}, 1_{]s,(s+\varepsilon)\wedge T]}(\cdot)\rangle_{\mathcal{H}}\right| \leq C$, which implies

$$\left|\int_0^t \left(P_\varepsilon(y_s) - P_\varepsilon(y_s^{(\varepsilon)})\right) \left\langle D.y_s^{(\varepsilon)}, 1_{]s,(s+\varepsilon)\wedge T]}(\cdot)\right\rangle_{\mathcal{H}} ds\right| \leq C \int_0^t \left|P_\varepsilon(y_s) - P_\varepsilon(y_s^{(\varepsilon)})\right| ds.$$

Finally, from Lemma 98 and proceeding as in Step 2 of Lemma 94, we see that the result holds. □

8.8 Maximal inequalities

Here, we deal with Doob's maximal inequalities in order to see that the Itô's integral with respect to Brownian motion has continuous paths. These inequalities also play an important role in the study of convergences of either, martingales, submartingales, or supermartingales (see, for instance, [22, 38, 180]).

Definition 57 *Let $\mathbb{I}$ be a subset of $\mathbb{R}_+$ and $(\mathcal{F}_t)_{t\in\mathbb{I}}$ a filtration of the underlying probability space. We say that an $\mathcal{F}_t$-adapted process $X : \Omega \times \mathbb{I} \to \mathbb{R}$ is an $\mathcal{F}_t$-submartingale if and only if*

i) X_t belongs to $L^1(\Omega)$, for all $t \in \mathbb{I}$.

ii) $X_s \leq E[X_t|\mathcal{F}_s]$, for every pair $s,t \in \mathbb{I}$ such that $s \leq t$.

Also, a process Y is called an $\mathcal{F}_t$-supermartingale if $-Y$ is an $\mathcal{F}_t$-submartingale, and a process that is both an $\mathcal{F}_t$-submartingale and an $\mathcal{F}_t$-supermartingale is said to be an $\mathcal{F}_t$-martingale.

Proposition 137 *Let $X : \Omega \times \mathbb{N} \to \mathbb{R}$ be a nonnegative $\mathcal{F}_n$-submartingale and $m \in \mathbb{N}$. Then,*

i) For $\varepsilon > 0$, we have

$$P\left(\left[\max_{0\leq k\leq m} X_k \geq \varepsilon\right]\right) \leq \frac{1}{\varepsilon}\int_{\{\max_{0\leq k\leq m} X_k \geq \varepsilon\}} X_m dP \leq \frac{E(X_m)}{\varepsilon}.$$

ii) If $X_k \in L^p(\Omega)$, for some $p > 1$ and all $k = 1,\ldots,m$, we get

$$\left\|\max_{0\leq k\leq m} X_k\right\|_{L^p(\Omega)} \leq \frac{p}{p-1}\|X_m\|_{L^p(\Omega)}.$$

Remark 114 *Note that if $(X_k)_{k=1}^m$ is an $\mathcal{F}_k$-martingale, then $(|X_k|)_{k=1}^m$ is a nonnegative $\mathcal{F}_k$-submartingale by Jensen inequality for conditional expectation. So, in this case, Statements i) and ii) hold when we write $|X.|$ instead of $X.$.*

Proof. We use the notation $X_n^* = \max_{0\le k\le n} X_k$.

We first deal with Assertion *i*). Let $\tau = \inf\{n \in \mathbb{N} : X_n \ge \varepsilon\} \wedge m$. Then, it is easy to see that τ is an $\mathcal{F}_n$-stopping time and $[X_m^* \ge \varepsilon] = [X_\tau \ge \varepsilon]$ (see Definition 34). Hence, the fact that X is an $\mathcal{F}_n$-submartingale implies

$$\begin{aligned}
\varepsilon P([X_m^* \ge \varepsilon]) &= \varepsilon P([X_\tau \ge \varepsilon]) \le \int_{[X_\tau\ge\varepsilon]} X_\tau dP \\
&= \sum_{n=1}^{m} \int_{[X_\tau\ge\varepsilon]\cap[\tau=n]} X_\tau dP = \sum_{n=1}^{m} \int_{[X_n\ge\varepsilon]\cap[\tau=n]} X_n dP \\
&\le \sum_{n=1}^{m} \int_{[X_n\ge\varepsilon]\cap[\tau=n]} X_m dP = \int_{[X_\tau\ge\varepsilon]} X_m dP = \int_{[X_m^*\ge\varepsilon]} X_m dP \le E(X_m),
\end{aligned}$$

where the last inequality is true since X_m is a nonnegative random variable. Thus, Statement *i*) is satisfied.

Now, we consider Declaration *ii*). Let $F_{(X_m^*)^p}$ be the distribution function of the random variable $(X_m^*)^p$. Then, from Fubini's theorem and the change of variables formula applied twice, we can deduce

$$\begin{aligned}
E((X_m^*)^p) &= \int_{[0,\infty)} x F_{(X_m^*)^p}(dx) = \int_{[0,\infty)} \int_0^x dy F_{(X_m^*)^p}(dx) \\
&= \int_0^\infty \int_{\{y<x<\infty\}} F_{(X_m^*)^p}(dx) dy = \int_0^\infty P([(X_m^*)^p > y]) dy \\
&= \int_0^\infty P\left(\left[X_m^* > y^{1/p}\right]\right) dy = p\int_0^\infty y^{p-1} P([X_m^* > y]) dy.
\end{aligned}$$

Therefore, Statement *i*), Fubini's theorem and Hölder inequality yield

$$\begin{aligned}
E((X_m^*)^p) &\le p\int_0^\infty y^{p-2} \int_{[X_m^*>y]} X_m dP dy = p\int_\Omega X_m \int_0^{X_m^*} y^{p-2} dy dP \\
&= \frac{p}{p-1}\int_\Omega X_m (X_m^*)^{p-1} dP \le \frac{p}{p-1} \|X_m\|_{L^p(\Omega)} \|X_m^*\|_{L^p(\Omega)}^{p-1}.
\end{aligned}$$

Note that if $E((X_m^*)^p) = 0$, then there is nothing to show. So, we can assume that $E((X_m^*)^p) > 0$. Thus, the last inequality gives that the proof is complete. □

Observe that if $X : \Omega\times\mathbb{I} \to \mathbb{R}$ is a right-continuous process, where $\mathbb{I} = [a, b]$ is an interval of $\mathbb{R}$, then $\sup_{t\in\mathbb{I}} X_t = \sup_{t\in D} X_t$, for every dense set D of $\mathbb{I}$, which contains the point b. This fact allows us to state the following extension of Proposition 137.

Theorem 100 (Doob's $L^p(\Omega)$-maximal inequality) *Let $\mathbb{I} \subset \mathbb{R}$ be an interval, $X : \Omega \times \mathbb{I} \to \mathbb{R}$ a right-continuous $\mathcal{F}_t$-martingale, or a right-continuous nonnegative $\mathcal{F}_t$-submartingale and $X^* = \sup_{t\in\mathbb{I}} |X_t|$. Then,*

i) for $p \ge 1$ and $\varepsilon > 0$, we have that $\varepsilon^p P([X^* > \varepsilon]) \le \sup_{t\in\mathbb{I}} E(|X_t|^p)$.

ii) for $p > 1$, we obtain that $\|X^*\|_{L^p(\Omega)} \le \frac{p}{p-1} \sup_{t\in\mathbb{I}} \|X_t\|_{L^p(\Omega)}$.

Remark 115 *We have*

$$\sup_{t\in\mathbb{I}} E(|X_t|^p) = \begin{cases} E(|X_b|^p), & \text{if } \mathbb{I} = [a, b], \\ \lim_{t\uparrow b} E(|X_t|^p), & \text{if } \mathbb{I} = [a, b). \end{cases}$$

Proof. Notice that the Jensen inequality implies that $|X|^p$ is a nonnegative $\mathcal{F}_t$-submartingale. Thus we can apply Proposition 137 to this process. So, consider a dense set D of $\mathbb{I}$ and a sequence $\{D_n \subset D : n \in \mathbb{N}\}$ of finite and increasing sets such that $D = \cup_{n=1}^{\infty} D_n$. D and D_n have the point b if $\mathbb{I} = [a, b]$. Hence, the facts that $t \mapsto E(|X_t|^p)$ is a nondecreasing function,

$$P([X^* > \varepsilon]) = \lim_{n\to\infty} P\left(\left[\sup_{t\in D_n} |X_t| > \varepsilon\right]\right) \quad \text{and} \quad E((X^*)^p) = \lim_{n\to\infty} E\left(\sup_{t\in D_n} |X_t|^p\right),$$

together with Proposition 137, give that the result is true. □

8.9 Outer measures

In this section, we study some facts that we need to understand the construction of the canonical Wiener space introduced in Section 1.2.1.

In order to fix the notation, we give the following:

Definition 58 *An outer measure on a nonempty set X is a set function $\mu^* : \mathcal{P}(X) \to [0, \infty]$ such that:*

a) $\mu^*(\emptyset) = 0$.

b) $\mu^*(A) \leq \mu^*(B)$ *if* $A \subset B \subset X$.

c) Let $\{A_i \subset X : i \in \mathbb{N}\}$ be a sequence in $\mathcal{P}(X)$. Then, $\mu^*(\cup_{i\in\mathbb{N}} A_i) \leq \sum_{i=1}^{\infty} \mu^*(A_i)$.

Properties *b)* and *c)* of μ^* are called monotonicity and countable subadditivity, respectively.

Now, we give an example of an outer measure, which is induced by a set function ρ.

Proposition 138 *Let $\mathcal{E} \subset \mathcal{P}(X)$ be a family of subset of X such that*

a) $\emptyset \in \mathcal{E}$.

b) There is a sequence $\{E_i \in \mathcal{E} : i \in \mathbb{N}\}$ such that $X = \cup_{i\in\mathbb{N}} E_i$.

Also let $\rho : \mathcal{E} \to [0, \infty]$ be such that $\rho(\emptyset) = 0$. Set, for $A \in \mathcal{P}(X)$,

$$\rho^*(A) := \inf\left\{\sum_{i=1}^{\infty} \rho(E_i) : A \subset \cup_{i\in\mathbb{N}} E_i, \text{ with } E_i \in \mathcal{E} \text{ and } i \in \mathbb{N}\right\}. \tag{8.42}$$

Then ρ^* is an outer measure on X.

Remark 116 *The Hypotheses ensure that the infimum in the definition of ρ^* is not taken over the empty set. Moreover, ρ^* is called the outer measure induced by ρ.*

Proof. It is easy to see that Properties *a)* and *b)* of Definition 58 hold. Indeed, $\{\emptyset\}$ is a covering of $\emptyset$ and any covering of B is also a covering of the set A. In order to see Property *c)* is satisfied, let $\varepsilon > 0$, then, for every $n \in \mathbb{N}$, there is a sequence $\{E_{i,n} \in \mathcal{E} : i \in \mathbb{N}\}$ such that $A_n \subset \cup_{i\in\mathbb{N}} E_{i,n}$ and $\sum_{i=1}^{\infty} \rho(E_{i,n}) \leq \rho^*(A_n) + 2^{-n}\varepsilon$. Hence,

$$\rho^*(\cup_{n\in\mathbb{N}} A_n) \leq \sum_{n=1}^{\infty}\sum_{i=1}^{\infty} \rho(E_{i,n}) \leq \sum_{n=1}^{\infty} \left(\rho^*(A_n) + 2^{-n}\varepsilon\right) \leq \varepsilon + \sum_{n=1}^{\infty} \rho^*(A_n),$$

which implies that the proof is complete. □

Given an outer measure μ^*, $\mathcal{F}^*$ represents all the Carathéodory measurable sets (or μ^*-measurable sets). It means, $E \in \mathcal{P}(X)$ belongs to $\mathcal{F}^*$ if and only if we have

$$\mu^*(A) = \mu^*(E \cap A) + \mu^*(E^c \cap A), \quad \text{for any } A \in \mathcal{P}(X).$$

For example, $\mathcal{F}^*$ contains all the set E such that $\mu^*(E) = 0$ since the definition of outer measure yields that, for $A \in \mathcal{P}(X)$,

$$\mu^*(A) \leq \mu^*(E \cap A) + \mu^*(E^c \cap A) \leq \mu^*(E) + \mu^*(E^c \cap A) = \mu^*(E^c \cap A) \leq \mu^*(A).$$

The following result is well-known in measure theory (see, for instance, Royden [187]) and it is used to extend a measure defined on an algebra $\mathcal{A}$ to a measure on the σ-algebra $\sigma(\mathcal{A})$.

Proposition 139 (Carathéodory extension theorem) *i) Let μ^* be an outer measure on X. Then $(X, \mathcal{F}^*, \mu^*)$ is a complete measure space.*

ii) Let ρ^* be given by (8.42), where $\mathcal{E}$ is an algebra and ρ is a measure of it, then $\mathcal{E} \subset \mathcal{F}^*$ and $\rho^*(E) = \rho(E)$, for $E \in \mathcal{E}$. Moreover, If ρ is finite (or σ-finite), then $\rho^*|_{\sigma(\mathcal{E})}$ is the only extension of ρ to a measure on $\sigma(\mathcal{E})$.

The following auxiliary result will be utilized to introduce the Wiener measure.

Lemma 100 *Let $(X, \mathcal{F}, \mu)$ be a finite measure space, μ^* the outer measure induced by μ and $Y \subset X$. Then, the following statements are equivalents:*

i) Let $E_1, E_2 \in \mathcal{F}^*$. Then, $E_1 \cap Y = E_2 \cap Y$ implies that $\mu^*(E_1) = \mu^*(E_2)$.

ii) $\mu^*(Y) = \mu(X)$.

Proof. Assume that Statement *i)* is satisfied. Let $\mathcal{O}$ be a countable union of elements of $\mathcal{F}$ covering Y, $E_1 = \mathcal{O}$ and $E_2 = X$. Hence, $E_1 \cap Y = E_2 \cap Y$ and, by Proposition 139, $E_1, E_2 \in \mathcal{F}^*$. Consequently, our hypothesis gives that $\mu^*(\mathcal{O}) = \mu(X)$ and, therefore, the definition of μ^* implies that Statement *ii)* is also true.

Conversely, suppose that Statement *ii)* holds. Let $E_1, E_2 \in \mathcal{F}^*$ be such that $E_1 \cap Y = E_2 \cap Y$ and $E = E_1 \setminus (E_1 \cap E_2)$. Then, using Proposition 139 again, we obtain $E \in \mathcal{F}^*$ and $E \cap Y = \emptyset$, which implies

$$\mu^*(X) \geq \mu^*(E \cup Y) = \mu^*((E \cup Y) \cap E) + \mu^*((E \cup Y) \cap E^c) = \mu^*(E) + \mu^*(Y).$$

This, together with Statement *ii)* and Proposition 139, allows us to conclude

$$0 = \mu^*(E) = \mu^*(E_1) - \mu^*(E_1 \cap E_2).$$

That is, $\mu^*(E_1) = \mu^*(E_1 \cap E_2)$. Finally, we can also show that $\mu^*(E_2) = \mu^*(E_1 \cap E_2)$ by changing E by $E_2 \setminus (E_1 \cap E_2)$. □

8.9.1 Outer measure on a space of functions

Now, for $0 < t_1 < \ldots < t_n \leq T$, we introduce the probability measures on $\mathcal{B}(\mathbb{R}^n)$ given by

$$\mu_{t_1,\ldots,t_n}(B) := \int_B \prod_{i=1}^{n} \left(\frac{1}{\sqrt{2\pi(t_i - t_{i-1})}} \exp\left(-\frac{(u_i - u_{i-1})^2}{2(t_i - t_{i-1})} \right) \right) du_1 \cdots du_n,$$

with $B \in \mathcal{B}(\mathbb{R}^n)$ and $t_0, u_0 = 0$.

In order to see that the set of all probability measures of the last form is consistent, we give the following semigroup property.

Lemma 101 *Let $0 \leq s < r < t \leq T$ and $x, y \in \mathbb{R}$. Then,*

$$\frac{1}{\sqrt{2\pi(t-r)}\sqrt{2\pi(r-s)}} \int_{\mathbb{R}} \exp\left(-\frac{(x-u)^2}{2(t-r)}\right) \exp\left(-\frac{(u-y)^2}{2(r-s)}\right) du$$
$$= \frac{1}{\sqrt{2\pi(t-s)}} \exp\left(-\frac{(x-y)^2}{2(t-s)}\right).$$

Proof. We use the change of variables formula to prove the result holds. To do so, set

$$v = \sqrt{\frac{t-s}{t-r}} \frac{u-y}{\sqrt{2(r-s)}} + \sqrt{\frac{r-s}{t-r}} \frac{y-x}{\sqrt{2(t-s)}}. \tag{8.43}$$

Hence

$$v^2 = \frac{t-s}{t-r}\frac{(u-y)^2}{2(r-s)} + \frac{r-s}{t-r}\frac{(y-x)^2}{2(t-s)} + \frac{(u-y)(y-x)}{t-r}.$$

Consequently,

$$\begin{aligned} v^2 + \frac{(y-x)^2}{2(t-s)} &= \frac{t-s}{t-r}\frac{(u-y)^2}{2(r-s)} + \frac{t-s}{t-r}\frac{(y-x)^2}{2(t-s)} + \frac{(u-y)(y-x)}{t-r} \\ &= \frac{(u-y)^2}{2(r-s)} + \frac{1}{t-r}\left(\frac{(u-y)^2}{2} + \frac{(y-x)^2}{2} + (u-y)(y-x)\right) \\ &= \frac{(u-y)^2}{2(r-s)} + \frac{(u-x)^2}{2(t-r)}. \end{aligned}$$

Finally, we can apply the change of variables formula given by (8.43) in order to finish the proof. □

In particular, Lemma 101 implies that $\mu_{t_1,\ldots,t_n,s_1,\ldots,s_m}(B \times \mathbb{R}^m) = \mu_{t_1,\ldots,t_n}(B)$ and

$$\mu_{t_1,\ldots,t_n,k_1,\ldots,k_r,s_1,\ldots,s_m}(B \times \mathbb{R}^r \times A) = \mu_{t_1,\ldots,t_n,s_1,\ldots,s_m}(B \times A),$$

for every $B \in \mathcal{B}(\mathbb{R}^n)$, $A \in \mathcal{B}(\mathbb{R}^m)$ and $0 < t_1 < \ldots < t_n < k_1 < \ldots < k_r < s_1, \ldots, s_m \leq T$. Hence, if $\mathbb{R}^{(0,T]}$ represents the set of all the functions from $(0,T]$ into $\mathbb{R}$, the Kolmogorov extension theorem (see, for example, Karatzas and Shreve [96], Theorem 2.2.2) implies that there exists a probability measure μ_X defined on the measurable space $(\{0\} \times \mathbb{R}^{(0,T]}, \mathcal{B}(\{0\} \times \mathbb{R}^{(0,T]}))$ such that $\mu_X(C) = \mu_{t_1,\ldots,t_n}(B)$, where C is the cylinder

$$C = \{x \in \{0\} \times \mathbb{R}^{(0,T]} : (x(t_1), \ldots, x(t_n)) \in B\}, \tag{8.44}$$

$B \in \mathcal{B}(\mathbb{R}^n)$ and $0 < t_1 < \ldots < t_n \leq T$.

In what follows, we will use μ_X to define a measure P^W on the σ-algebra of Borel of the space $C_0([0,T])$ of all the real-valued continuous functions ω defined on $[0,T]$ such that $\omega(0) = 0$. Namely, P^W satisfies

$$P^W(\tilde{C}) = \mu_{t_1,\ldots,t_n}(B) = \mu_X(C), \tag{8.45}$$

where $\tilde{C}$ is given by the right-hand side of (8.44) when we write $C_0([0,T])$ instead of $\{0\} \times \mathbb{R}^{(0,T]}$. That is, $\tilde{C} = C \cap C_0([0,T])$. In other words, we need to consider sets of the form $C_Y = C \cap Y$, with C a cylinder and $Y \subset \{0\} \times \mathbb{R}^{(0,T]}$. Towards this end, we introduce the following: let $0 < t_1 < \ldots < t_n \leq T$ and $Y \subset \{0\} \times \mathbb{R}^{(0,T]}$, then $\mathcal{B}_{t_1,\ldots,t_n}$ (resp. $\mathcal{B}^Y_{t_1,\ldots,t_n}$) denotes the family of all the cylinders of the form (8.44) (resp. of the form $C \cap Y$, where C is as in (8.44)) and $q^Y_{t_1,\ldots,t_n} : Y \to \mathbb{R}^n$ is defined as $q^Y_{t_1,\ldots,t_n}(x) = (x(t_1), \ldots, x(t_n))$.

Lemma 102 *Let $0 < t_1 < \ldots < t_n \le T$ and $Y \subset \{0\} \times \mathbb{R}^{(0,T]}$ be such that the function $q^Y_{t_1,\ldots,t_n}$ is onto. Then, there exists a one-to-one correspondence between the elements of $\mathcal{B}(\mathbb{R}^n)$ and those of $\mathcal{B}^Y_{t_1,\ldots,t_n}$.*

Remark 117 *Note that if Y only contains the function identically zero. Then,*

$$\{x \in Y : (x(t_1), \ldots, x(t_n)) \in \mathbb{R}^n\} = \{x \in Y : (x(t_1), \ldots, x(t_n)) \in [0, \infty)^n\} = Y.$$

Proof. Since $q^Y_{t_1,\ldots,t_n} : Y \to \mathbb{R}^n$ is onto, then we have

$$q^Y_{t_1,\ldots,t_n}\left(\left(q^Y_{t_1,\ldots,t_n}\right)^{-1}(B)\right) = B, \quad \text{for } B \in \mathcal{B}(\mathbb{R}^n).$$

Thus, $\left(q^Y_{t_1,\ldots,t_n}\right)^{-1}(B_1) \neq \left(q^Y_{t_1,\ldots,t_n}\right)^{-1}(B_2)$ for $B_1, B_2 \in \mathcal{B}(\mathbb{R}^n)$ such that $B_1 \neq B_2$. The proof is now finished. □

Now, we are ready to give the first step to see that P^W introduced in equality (8.45) is well-defined.

Given $Y \subset \{0\} \times \mathbb{R}^{(0,T]}$ and motivated by (8.45), we could try to define a probability measure μ_Y on $\sigma\left(\cup_{n\in\mathbb{N}} \cup_{\{t_1,\ldots,t_n : 0<t_1<\ldots<t_n\le T\}} \mathcal{B}^Y_{t_1,\ldots,t_n}\right)$ such that $\mu_Y(C \cap Y) = \mu_X(C)$, for every cylinder C of the form (8.44). However, μ_Y may not be well-defined as Remark 117 shows.

Lemma 103 *Let $Y \subset \{0\} \times \mathbb{R}^{(0,T]}$ be such that, for every $0 < t_1 < \ldots < t_n \le T$ and $(y_1, \ldots, y_n) \in \mathbb{R}^n$, there exists $x \in Y$ so that $(x(t_1), \ldots, x(t_n)) = (y_1, \ldots, y_n)$. Then, μ_Y is a well-defined probability measure on the σ-algebra $\mathcal{B}^Y_{t_1,\ldots,t_n}$, for any $0 < t_1 < \ldots < t_n \le T$. Moreover, if $\mu^*_X(Y) = 1$, where μ^*_X is the outer measure induced by μ_X, we have that μ_Y is a well-defined probability measure of the algebra*

$$\mathcal{A}_Y := \cup_{n\in\mathbb{N}} \cup_{\{t_1,\ldots,t_n : 0<t_1<\ldots<t_n\le T\}} \mathcal{B}^Y_{t_1,\ldots,t_n}.$$

Remark 118 *Proceeding as in the proof of Lemma 2, we have that $\mathcal{A}_Y$ is an algebra on $\{0\} \times \mathbb{R}^{(0,T]}$.*

Proof. The first part of the result is an immediate consequence of Lemma 102. So, now, we assume that $\mu^*_X(Y) = 1$. We now see that μ_Y is well-defined on $\mathcal{A}_Y$. To do so, let $J \in \mathcal{A}_Y$ such that $J = I_1 \cap Y = I_2 \cap Y$, where I_1, I_2 are two cylinders of $\{0\} \times \mathbb{R}^{(0,T]}$. Then, the fact that $\mu^*_X(Y) = 1$, together with Lemma 100 and Proposition 139, implies that $\mu_X(I_1) = \mu_X(I_2)$. Thus, μ_Y is well-defined on $\mathcal{A}_Y$. Hence, in order to finish the proof, we only need to see that μ_Y is σ-additive on $\mathcal{A}_Y$.

Let $\{J_n : n \in \mathbb{N}\} \subset \mathcal{A}_Y$ such that $J_k \cap J_n = \emptyset$, for $k \neq n$, and $\cup_{n\in\mathbb{N}} J_n \in \mathcal{A}_Y$. Then, there exists a sequence $\{I_n : n \in \mathbb{N} \cup \{\infty\}\}$ of cylinders of $\{0\} \times \mathbb{R}^{(0,T]}$ such that

$$I_\infty \cap Y = \cup_{n\in\mathbb{N}} J_n = (\cup_{n\in\mathbb{N}} I_n) \cap Y. \tag{8.46}$$

We claim that $I_k \cap I_n = \emptyset$, $k \neq n$. Indeed, without loss of generality we can assume that

$$I_i = \{x \in \{0\} \times \mathbb{R}^{(0,T]} : (x(t_1), \ldots, x(t_m)) \in B_i\}, \quad i = k, n,$$

for some $0 < t_1 < \ldots < t_m \le T$ and $B_1, B_2 \in \mathcal{B}(\mathbb{R}^m)$. Note that if there is $(y_1, \ldots, y_m) \in B_k \cap B_n$, then, from our hypotheses, we can find $x \in Y$ such that $(x(t_1), \ldots, x(t_m)) = (y_1, \ldots, y_m)$. Consequently, $x \in J_k \cap J_n$, which is a contradiction. Therefore, our claim is true.

Finally, equality (8.46) and Lemma 100 give

$$\mu_Y(\cup_{n\in\mathbb{N}} J_n) = \mu_X(I_\infty) = \mu_X(\cup_{n\in\mathbb{N}} I_n) = \sum_{n=i}^{\infty} \mu_X(I_n) = \sum_{n=i}^{\infty} \mu_Y(J_n),$$

where we have used that μ_X is a measure on $(\{0\}\times\mathbb{R}^{(0,T]}, \mathcal{B}(\{0\}\times\mathbb{R}^{(0,T]}))$. Thus, the proof is complete □

The next goal is to show that $\mu_X^*(C_0([0,T])) = 1$. Towards this end, we continue analyzing some auxiliary results.

Lemma 104 *Let I be the cylinder $\{w \in \{0\}\times\mathbb{R}^{(0,T]} : |w(\frac{k}{2^n}T) - w(\frac{k-1}{2^n}T)| > a\frac{1}{2^{\alpha n}}\}$ with $a, \alpha > 0$, $n \in \mathbb{N}$ and $k \in \{1, \dots, 2^n\}$. Then,*

$$\mu_X(I) \leq \sqrt{\frac{2}{\pi}} \frac{2^{n(\alpha-1/2)}T^{1/2}}{a} \exp\left(-\frac{a^2}{2T}2^{n(1-2\alpha)}\right).$$

Proof. Proceeding as in (1.19), we have that $\mu_X(I) = \sqrt{\frac{2}{\pi T}}\sqrt{2^n}\int_{a\frac{1}{2^{\alpha n}}}^{\infty} \exp\left(-2^n\frac{x^2}{2T}\right)dx$. Thus, the change of variables formula $u = \sqrt{2^n/T}x$ implies

$$\begin{aligned}\mu_X(I) &= \sqrt{\frac{2}{\pi}}\int_{aT^{-1/2}(2^{-n})^{\alpha-\frac{1}{2}}}^{\infty} \exp\left(-\frac{u^2}{2}\right)du \\ &\leq \sqrt{\frac{2}{\pi}}\int_{aT^{-1/2}(2^{-n})^{\alpha-\frac{1}{2}}}^{\infty} \frac{uT^{1/2}}{a(2^{-n})^{\alpha-\frac{1}{2}}} \exp\left(-\frac{u^2}{2}\right)du.\end{aligned}$$

Now, it is easy to see that the result is satisfied. □

In the following result, we utilize the notation

$$H_\alpha(a) = \left\{w \in \{0\}\times\mathbb{R}^{(0,T]} : \text{there are } s_1, s_2 \in \Pi \text{ such that } |w(s_2) - w(s_1)| > a|s_2 - s_1|^\alpha\right\},$$

where Π is introduced in the proof of Theorem 1, and μ_X^* denotes the outer measure induced by $\mu_X : \mathcal{B}(\{0\}\times\mathbb{R}^{(0,T]}) \to [0,1]$, which is defined in Proposition 138.

Lemma 105 *Let $a > 0$ and $\alpha \in (0, \frac{1}{2})$ such that $2^{1-\delta} < \exp((2T)^{-1}a^2\delta)$, with $\delta = \frac{1}{2} - \alpha$. Then,*

$$\mu_X^*\left(H_\alpha\left(\frac{2a}{1-2^{-\alpha}}\right)\right) \leq \sqrt{\frac{2T}{\pi}} \frac{1}{a\left(1 - 2^{1-\delta}\exp(-\frac{1}{2T}a^2\delta)\right)}.$$

Proof. By Remark 8, we have that $H_\alpha\left(\frac{2a}{1-2^{-\alpha}}\right) \subset \cup_{n=1}^{\infty}\cup_{k=1}^{2^n} I_{k,n}$, with $I_{k,n} = \{w \in \{0\}\times\mathbb{R}^{(0,T]} : |w(\frac{k}{2^n}T) - w(\frac{k-1}{2^n}T)| > a\frac{1}{2^{\alpha n}}\}$. Thus, from Lemma 104, we get

$$\begin{aligned}\mu_X^*\left(H_\alpha\left(\frac{2a}{1-2^{-\alpha}}\right)\right) &\leq \sum_{n=1}^{\infty}\sum_{k=1}^{2^n}\mu_X(I_{k,n}) \\ &\leq \sum_{n=1}^{\infty}\sum_{k=1}^{2^n}\sqrt{\frac{2T}{\pi}}\frac{2^{n(\alpha-1/2)}}{a}\exp\left(-\frac{a^2}{2T}2^{n(1-2\alpha)}\right) \\ &= \sqrt{\frac{2T}{\pi}}\frac{1}{a}\sum_{n=1}^{\infty}2^n 2^{n(\alpha-1/2)}\exp\left(-\frac{a^2}{2T}2^{n(1-2\alpha)}\right) \\ &= \sqrt{\frac{2T}{\pi}}\frac{1}{a}\sum_{n=1}^{\infty}2^{n(1-\delta)}\exp\left(-\frac{a^2}{2T}2^{2n\delta}\right).\end{aligned}$$

Hence, the inequality $2^x \geq \frac{x}{2}$, for $x \geq 0$, implies

$$\mu_X^*\left(H_\alpha\left(\frac{2a}{1-2^{-\alpha}}\right)\right) \leq \sqrt{\frac{2T}{\pi}}\frac{1}{a}\sum_{n=0}^{\infty}\left[2^{1-\delta}\exp\left(-\frac{a^2}{2T}\delta\right)\right]^n = \sqrt{\frac{2T}{\pi}}\frac{1}{a}\frac{1}{1-2^{1-\delta}\exp\left(-\frac{a^2}{2T}\delta\right)}.$$

It means, the proof is complete. □

Now, we are ready to establish that $\mu_X^*(C_0([0,T])) = 1$.

Proposition 140 *Let $\alpha \in (0, \frac{1}{2})$ and C_α the family of real-valued Hölder continuous functions on $[0,T]$ with exponent α. Then, $\mu_X^*(C_\alpha) = 1$, and, consequently, $\mu_X^*(C_0([0,T])) = 1$.*

Proof. We note that it is enough to see that if there is $\varepsilon \in (0,1)$ such that $\mu_X^*(C_\alpha) = 1 - \varepsilon$, then we have that $\mu_X^*((\{0\} \times \mathbb{R}^{(0,T]}) \setminus H_\alpha(a)) < 1 - \varepsilon/8$, for any $a > 0$. Indeed, by Lemma 105, we have that $\mu_X^*(H_\alpha(a)) \leq \varepsilon/16$, for a large enough. Thus,

$$1 = \mu_X^*(\{0\} \times \mathbb{R}^{(0,T]}) \leq \mu_X^*(H_\alpha(a)) + \mu_X^*((\{0\} \times \mathbb{R}^{(0,T]}) \setminus (H_\alpha(a)) \leq \frac{\varepsilon}{16} + 1 - \frac{\varepsilon}{8} < 1,$$

which is a contradiction.

Now, assume that $\mu_X^*(C_\alpha) = 1 - \varepsilon$, for some $\varepsilon \in (0,1)$. Then, there exists a sequence $\{I_n : n \in \mathbb{N}\}$ of cylinders of $\{0\} \times \mathbb{R}^{(0,T]}$ such that

$$C_\alpha \subset \cup_{n=1}^{\infty} I_n \quad \text{and} \quad \sum_{n=1}^{\infty} \mu_X(I_n) < 1 - \frac{\varepsilon}{2}. \tag{8.47}$$

For $n \in \mathbb{N}$, we have that I_n has the form

$$I_n = \{x \in \{0\} \times \mathbb{R}^{(0,T]} : (x(t_{n,1}), \ldots, x(t_{n,m_n})) \in B_n\},$$

for some $B_n \in \mathcal{B}(\mathbb{R}^{m_n})$ and $0 < t_{n,1} < \ldots < t_{n,m_n} \leq T$. Hence $\mu_X(I_n) = \mu_{t_{n,1},\ldots,t_{n,m_n}}(B_n)$. Thus, the fact that $\mu_{t_{n,1},\ldots,t_{n,m_n}}$ is a probability measure on $\mathcal{B}(\mathbb{R}^{m_n})$ implies that there is a family $\{B_{n,k} \in \mathcal{B}(\mathbb{R}^{m_n}) : k \in \mathbb{N}\}$ such that $B_{n,k} \cap B_{n,j} = \emptyset$, $k \neq j$, $B_{n,k} = (b_{n,k,1}, c_{n,k,1}] \times \ldots \times (b_{n,k,m_n}, c_{n,k,m_n}]$, $I_n \subset \cup_{k \in \mathbb{N}} B_{n,k}$ and $\sum_{j=1}^{\infty} \mu_{t_{n,1},\ldots,t_{n,m_n}}(B_{n,j}) \leq \mu_X(I_n) + 2^{-n-2}\varepsilon$. Therefore, by (8.47), we can assume that

$$C_\alpha \subset \cup_{n=1}^{\infty} I_n \quad \text{and} \quad \sum_{n=1}^{\infty} \mu_X(I_n) < 1 - \frac{\varepsilon}{4},$$

with

$$I_n = \{x \in \{0\} \times \mathbb{R}^{(0,T]} : (x(t_{n,1}), \ldots, x(t_{n,m_n})) \in (b_{n,1}, c_{n,1}] \times \ldots \times (b_{n,m_n}, c_{n,m_n}]\}.$$

Consequently, for $a > 0$ fixed and $x \in ((\{0\} \times \mathbb{R}^{(0,T]}) \setminus H_\alpha(a))$, there exist $n \in \mathbb{N}$ and a unique $\tilde{x} \in C_\alpha$ such that $\tilde{x} \in I_n$ (i.e., $b_{n,j} < x(t_{n,j}) \leq c_{n,j}$), $\tilde{x}(s) = x(s)$, for every $s \in \Pi$, and $|\tilde{x}(t) - \tilde{x}(s)| \leq a|t-s|^\alpha$, $t, s \in [0,T]$. So, $x \in I_n^{(a)}$, where

$$\begin{aligned} I_n^{(a)} = \{&x \in \{0\} \times \mathbb{R}^{(0,T]} : \\ &x(s_{n,j}) \in [b_{n,j} - a|t_{n,j} - s_{n,j}|^\alpha, c_{n,j} + a|t_{n,j} - s_{n,j}|^\alpha],\ j = 1, \ldots, m_n\}. \end{aligned}$$

Here, $s_{n,j}$, $j = 1, \ldots, m_n$, is chosen such that $\mu_X(I_n^{(a)}) \leq \mu_X(I_n) + 2^{-3-n}\varepsilon$. Hence, we can conclude that $((\{0\} \times \mathbb{R}^{(0,T]}) \setminus H_\alpha(a)) \subset \cup_{n=1}^{\infty} I_n^{(a)}$ and

$$\mu_X^*((\{0\} \times \mathbb{R}^{(0,T]}) \setminus H_\alpha(a)) \leq \sum_{n=1}^{\infty} \mu_X(I_n^{(a)}) \leq 1 - \frac{\varepsilon}{4} + \frac{\varepsilon}{8} \sum_{n=1}^{\infty} 2^{-n} = 1 - \frac{\varepsilon}{8},$$

as we wanted to prove. □

8.10 Functions of bounded variation

Now we state two auxiliary results that we need in Section 2. Here, $g : [a, b] \to \mathbb{R}$ is a function of bounded variation on the interval $[a, b]$. That is, $\mathrm{var}_1(g; [a, b]) < \infty$ (see (2.4)).

Lemma 106 *A function $g : [a, b] \to \mathbb{R}$ has bounded variation if and only if there are two non-decreasing functions $g_1, g_2 : [a, b] \to \mathbb{R}$ such that $g = g_1 - g_2$.*

Remark 119 *Note that $g = (g_1 + c) - (g_2 + c)$, where $c \in \mathbb{R}$. Thus, the representation of g as a difference of two non-decreasing functions is not unique.*

Proof. Assume that g is a function of bounded variation on $[a, b]$. Therefore, by Lemma 24, we have that $x \mapsto \mathrm{var}_1(g, [a, x])$ is a non-decreasing function on $[a, b]$. Set

$$g_1(x) = \mathrm{var}_1(g, [a, x]) \quad \text{and} \quad g_2(x) = g_1(x) - g(x), \quad \text{for } x \in [a, b].$$

We claim that g_2 is a non-decreasing function on $[a, b]$, which implies that g is the difference of two non-decreasing functions. Indeed, let $x, y \in [a, b]$ be such that $x < y$. Then, using Lemma 24 again, we can write $g_2(y) = g_1(x) + \mathrm{var}_1(g, [x, y]) - g(y)$ and, consequently,

$$g_2(y) - g_2(x) = \mathrm{var}_1(g, [x, y]) - (g(y) - g(x)) \geq 0.$$

it means, the function g_2 is also non-decreasing.

Conversely, if $g_1, g_2 : [a, b] \to \mathbb{R}$ are two non-decreasing functions, then we are able to establish

$$\mathrm{var}_1(g_1 - g_2, [a, b]) \leq \mathrm{var}_1(g_1, [a, b]) + \mathrm{var}_1(g_2, [a, b]) = g_1(b) - g_1(a) + g_2(b) - g_2(a) < \infty.$$

That is, the proof is complete. □

Lemma 107 *Let $g : [a, b] \to \mathbb{R}$ be a function of bounded variation and $f(\cdot) = var_1(g, [a, \cdot])$. Then, for $x \in [a, b]$,*

$$f(x+) - f(x) = |g(x+) - g(x)| \quad \textit{and} \quad f(x) - f(x-) = |g(x) - g(x-)|.$$

Remark 120 *Remember that Lemma 25 gives that g is a regulated function. Also, by Lemma 24, we have that $x \mapsto var_1(g, [a, x])$ is a non-decreasing function on $[a, b]$, which yields that f is also a regulated function.*

Proof. Fix $x \in (a, b]$. From Lemma 24, we have that, for $y \in (a, x)$,

$$f(x) - f(y) = \mathrm{var}_1(g, [y, x]) \geq |g(x) - g(y)|.$$

Consequently, we can conclude, taking the limit as $y \uparrow x$,

$$f(x) - f(x-) \geq |g(x) - g(x-)|. \tag{8.48}$$

On the other hand, let $\varepsilon > 0$ be an arbitrary number and choose $\delta > 0$ such that

$$|g(x-) - g(y)| < \varepsilon/2, \quad \text{for } y \in (x - \delta, x).$$

Also choose a partition $\{a = x_0 < x_1 < \ldots < x_n = x\}$ of $[a, x]$ so that

$$x_{n-1} > x - \frac{\delta}{2} \quad \text{and} \quad f(x) - \sum_{i=1}^{n} |g(x_i) - g(x_{i-1})| < \varepsilon/2.$$

Therefore, from the last three inequalities, we get

$$\begin{aligned} f(x) - \sum_{i=1}^{n-1} |g(x_i) - g(x_{i-1})| &< \frac{\varepsilon}{2} + \sum_{i=1}^{n} |g(x_i) - g(x_{i-1})| - \sum_{i=1}^{n-1} |g(x_i) - g(x_{i-1})| \\ &= \frac{\varepsilon}{2} + |g(x) - g(x_{n-1})| \leq \frac{\varepsilon}{2} + |g(x) - g(x-)| \\ &\quad + |g(x-) - g(x_{n-1})| \leq |g(x) - g(x-)| + \varepsilon. \end{aligned}$$

Hence, it follows that, for $y \in (x_{n-1}, x)$,

$$f(x) - f(y) \leq f(x) - f(x_{n-1}) \leq f(x) - \sum_{i=1}^{n-1} |g(x_i) - g(x_{i-1})| \leq |g(x) - g(x-)| + \varepsilon.$$

In other words, letting $y \uparrow x$, we can conclude that $f(x) - f(x-) \leq |g(x) - g(x-)| + \varepsilon$, which, together with (8.48) and the fact that ε is arbitrary, implies that

$$f(x) - f(x-) = |g(x) - g(x-)| . \tag{8.49}$$

Now let x be in $[a, b)$. Note that, by last equality, we only need to show that $f(x+) - f(x) = |g(x+) - g(x)|$ to finish the proof. Towards this end, we introduce the function $\tilde{g} : [-b, -a] \to \mathbb{R}$ given by $\tilde{g}(x) = g(-x)$. It is easy to show that $\mathrm{var}_1(g, [c, d]) = \mathrm{var}_1(\tilde{g}, [-d, -c])$ if $[c, d] \subset [a, b]$, and $g(x+) - g(x) = -[\tilde{g}(-x) - \tilde{g}((-x)-)]$. So, Lemma 24 and equality (8.49) applied to the function $\tilde{g}$ allow to write

$$\begin{aligned} f(x+) - f(x) &= \lim_{\delta \downarrow 0} (\mathrm{var}_1(g, [a, x + \delta]) - \mathrm{var}_1(g, [a, x])) \\ &= \lim_{\delta \downarrow 0} \mathrm{var}_1(g, [x, x + \delta]) = \lim_{\delta \downarrow 0} \mathrm{var}_1(\tilde{g}, [-x - \delta, -x]) \\ &= \lim_{\delta \downarrow 0} (\mathrm{var}_1(\tilde{g}, [-b, -x]) - \mathrm{var}_1(\tilde{g}, [-b, -x - \delta])) \\ &= |\tilde{g}(-x) - \tilde{g}((-x)-)| = |g(x+) - g(x)| . \end{aligned}$$

Thus, the result is satisfied. □

The following result deals with the discontinuities of the bounded variation functions.

Lemma 108 *Let $g : [a, b] \to \mathbb{R}$ be a function of bounded variation. Then, g is continuous except at countably many points.*

Proof. By Lemma 25, we have already known that the function g is regulated. Also, from Lemma 106, g is the difference of two non-decreasing functions. Therefore, without loss of generality, we can assume that g is a non-decreasing function. So, it is easy to see that, for $x \in (a, b)$, we have that $g(x-) \leq g(x) \leq g(x+)$. In consequence g is discontinuous at x if and only if $g(x-) < g(x+)$. Now, choose $x, y \in (a, b)$, $x < y$, such that g is discontinuous at these points. Then, $(g(x-), g(x+)) \cap (g(y-), g(y+)) = \emptyset$. Hence, we can find two different rational numbers r_x and r_y such that $r_x \in (g(x-), g(x+))$ and $r_y \in (g(y-), g(y+))$. Thus, the proof is complete. □

8.11 Completion of a metric space

In this section, we see that a metric space (X, d) has a completion $(\tilde{X}, \tilde{d})$, which is a complete metric space. The fact that X can be considered as a dense set of $\tilde{X}$ gives that $\tilde{X}$ is a linear space if X is so.

Definition 59 *A completion of a metric space (X, d) is a pair consisting of a complete metric space $(\tilde{X}, \tilde{d})$ and an isometry $\varphi : X \to \tilde{X}$ such that $\varphi(X)$ is dense in $\tilde{X}$.*

In order to see that there is a completion for (X, d), a metric space $(\tilde{X}, \tilde{d})$ is built as follows. On the set of all the Cauchy sequences in X, it is considered the equivalence relation $\sim$ given by

$$\{x_n : n \in \mathbb{N}\} \sim \{y_n : n \in \mathbb{N}\} \quad \text{if and only if} \quad \lim_{n\to\infty} d(x_n, y_n) = 0.$$

We write $\{x_n\}$ instead of $\{x_n : n \in \mathbb{N}\}$ to simplify the notation. Given a Cauchy sequence $\{x_n\}$ of X, $[\{x_n\}]$ stands for the equivalence class that contains the sequence $\{x_n\}$ and $\tilde{X}$ is the set of all the equivalence classes:

$$\tilde{X} = \{[\{x_n\}] : \{x_n\} \text{ is a Cauchy sequence in } X\}. \tag{8.50}$$

We will identify an element $x \in X$ with $[x]$, where $[x]$ is the equivalence class of the constant sequence $\{x, x, x, \ldots\}$.

Now, we figure out a metric $\tilde{d}$ on $\tilde{X}$. Towards this end, note that if $\{x_n\}$ and $\{y_n\}$ are two Cauchy sequences in X, then the triangle inequality leads to

$$\begin{aligned} |d(x_n, y_n) - d(x_m, y_m)| &\leq |d(x_n, y_n) - d(x_n, y_m)| + |d(x_n, y_m) - d(x_m, y_m)| \\ &\leq d(y_n, y_m) + d(x_m, x_n), \end{aligned} \tag{8.51}$$

which implies that $\{d(x_n, y_n) : n \in \mathbb{N}\}$ is a Cauchy sequence of $\mathbb{R}$. Hence, we can define

$$\tilde{d}\left([\{x_n\}], [\{y_n\}]\right) := \lim_{n\to\infty} d(x_n, y_n). \tag{8.52}$$

We claim that $\tilde{d}$ is well-defined. Indeed, choose $\{x_n^{(1)}\} \in [\{x_n\}]$ and $\{y_n^{(1)}\} \in [\{y_n\}]$. So, proceeding as in (8.51), we obtain

$$\left| d(x_n, y_n) - d(x_n^{(1)}, y_n^{(1)}) \right| \leq d(x_n, x_n^{(1)}) + d(y_n, y_n^{(1)}) \to 0, \quad \text{as } n \to \infty,$$

which gives that $\lim_{n\to\infty} d(x_n, y_n) = \lim_{n\to\infty} d(x_n^{(1)}, y_n^{(1)})$ and therefore $\tilde{d}$ is well-defined. In order words, we have that $(\tilde{X}, \tilde{d})$ is a metric space. Moreover, if $(X, \|\cdot\|)$ (resp. $(X, \langle\cdot,\cdot\rangle_X)$) is a normed space (resp. a pre-Hilbert space), then $\tilde{X}$ is a linear space with the operations

$$\{x_n\} + \{y_n\} := \{\, x_n + y_n\} \quad \text{and} \quad k\{x_n\} := \{kx_n\},$$

where $\{x_n\}$ and $\{y_n\}$ are two Cauchy sequences of X, and k is a scalar in the underlying field related to X. Note that $x_n + y_n$ an kx_n are the addition an the multiplication by a scalar defined in X. Now, $d(x, y) = \|x - y\|$ and $(\tilde{X}, |\cdot|)$ (resp. $(\tilde{X}, \langle\cdot,\cdot\rangle)$) is also a normed space (resp. a Hilbert space) with $|[\{x_n\}]| = \lim_{n\to\infty} \|x_n\|$. Also note that $\{x_n\} \in [0]$ if $\|x_n\|$ goes to zero as $n \to \infty$.

Now, we are ready to state the main result of this section.

Theorem 101 *Let (X, d) be a metric space. Then, it has a completion $(\tilde{X}, \tilde{d})$. This completion is unique determined up to isometry. Moreover, if X is a linear normed space, then $\tilde{X}$ is a Banach space.*

Remark 121 *The uniqueness of the completion means that if we have two completions $(\varphi_1, (\tilde{X}_1, \tilde{d}_1))$ and $(\varphi_2, (\tilde{X}_2, \tilde{d}_2))$, then there exists a unique isometry $\psi : \tilde{X}_1 \to \tilde{X}_2$ such that $\psi\varphi_1 = \varphi_2$.*

Proof. Let us see that the space $(\tilde{X}, \tilde{d})$ introduced in (8.50) and (8.52) is complete. Thus, choose a Cauchy sequence $\{\tilde{x}_k\} = \left\{\left[\{x_n^{(k)} : n \in \mathbb{N}\}\right]\right\}$ of $\tilde{X}$ and $\varepsilon > 0$. So there is $N_\varepsilon > 0$ such that $\tilde{d}(\tilde{x}_k, \tilde{x}_m) < \varepsilon$, for $k, m > N_\varepsilon$. Similarly, using that $\{x_n^{(k)} : n \in \mathbb{N}\}$ is a Cauchy sequence of X, for every $k \in \mathbb{N}$, there exists $n_k > 0$ such that $d(x_m^{(k)}, x_{n_k}^{(k)}) < \frac{1}{k}$, for $m > n_k$. Hence, we can consider the sequence $\tilde{x} = \left\{x_{n_k}^{(k)} : k \in \mathbb{N}\right\}$. We claim that $\tilde{x}$ belongs to $\tilde{X}$ and that $\tilde{x}_k$ goes to $\tilde{x}$ in $(\tilde{X}, \tilde{d})$. Indeed,

$$
\begin{aligned}
d\left(x_{n_k}^{(k)}, x_{n_m}^{(m)}\right) &= \tilde{d}\left(\left[x_{n_k}^{(k)}\right], \left[x_{n_m}^{(m)}\right]\right) \le \tilde{d}\left(\left[x_{n_k}^{(k)}\right], \tilde{x}_k\right) + \tilde{d}\left(\tilde{x}_k, \tilde{x}_m\right) + \tilde{d}\left(\tilde{x}_m, \left[x_{n_m}^{(m)}\right]\right) \\
&\le \frac{1}{k} + \tilde{d}\left(\tilde{x}_k, \tilde{x}_m\right) + \frac{1}{m} \le \frac{1}{k} + \varepsilon + \frac{1}{m}, \quad \text{for } k, m > N_\varepsilon,
\end{aligned}
$$

which implies that $\tilde{x}$ belongs to $\tilde{X}$. Also notice that this inequality yields

$$
\begin{aligned}
\tilde{d}\left(\tilde{x}, \tilde{x}_k\right) &\le \tilde{d}\left(\tilde{x}, \left[x_{n_k}^{(k)}\right]\right) + \tilde{d}\left(\left[x_{n_k}^{(k)}\right], \tilde{x}_k\right) \le \tilde{d}\left(\tilde{x}, \left[x_{n_k}^{(k)}\right]\right) + \frac{1}{k} \\
&= \lim_{m \to \infty} d\left(x_{n_m}^{(m)}, x_{n_k}^{(k)}\right) + \frac{1}{k} \le \frac{2}{k} + \varepsilon, \quad \text{for } k > N_\varepsilon.
\end{aligned}
$$

Therefore our claim is satisfied and, consequently, $(\tilde{X}, \tilde{d})$ is a complete metric space.

Now introduce $\varphi : X \to \tilde{X}$ given by $\varphi(x) = [x]$. It is easy to see that X and $\varphi(X)$ are isometric spaces. Also, if $x \in \tilde{X}$, then the sequence $\{[x_{n_k}] : k \in \mathbb{N}\}$ goes to x, where n_k is such that $d\left(x_m, x_{n_k}\right) < 1/k$, for $m > n_k$. It means, $\varphi(X)$ is dense in $\tilde{X}$,

Finally, we deal with the uniqueness of the completion: suppose that $\left(\varphi_1, (\tilde{X}_1, \tilde{d}_1)\right)$ and $\left(\varphi_2, (\tilde{X}_2, \tilde{d}_2)\right)$ are two completions of (X, d). Since φ_1 is a $1-1$ function due to φ_1 being an isometry, we have that $\psi = \varphi_2 \circ \varphi_1^{-1} : \varphi_1(X) \to \varphi_2(X)$ is a surjective isometry. Thus it can be extended to an isometry $\tilde{\psi} : \tilde{X}_1 \to \tilde{X}_2$. In the same way, we can prove that there is also an isometry $\hat{\psi} : \tilde{X}_2 \to \tilde{X}_1$ such that $\hat{\psi} \circ \varphi_2 = \varphi_1$. These two isometries allow us to write $\hat{\psi} \circ \tilde{\psi} \circ \varphi_1 = \varphi_1$ and $\tilde{\psi} \circ \hat{\psi} \circ \varphi_2 = \varphi_2$, which give $\hat{\psi} \circ \tilde{\psi} = i_{\varphi_1(X)}$ and $\tilde{\psi} \circ \hat{\psi} = i_{\varphi_2(X)}$. Here, $i_{\varphi_1(X)}$ and $i_{\varphi_2(X)}$ are the identity operators on $\varphi_1(X)$ and $\varphi_2(X)$, respectively. But, the fact that $\varphi_1(X)$ and $\varphi_2(X)$ are dense in $\tilde{X}_1$ and $\tilde{X}_2$, respectively, implies $\hat{\psi} \circ \tilde{\psi} = i_{\tilde{X}_1}$ and $\tilde{\psi} \circ \hat{\psi} = i_{\tilde{X}_2}$. Now, it is not difficult to conclude that $\tilde{\psi}$ is the unique isometry from $\tilde{X}_1$ into $\tilde{X}_2$ such that $\tilde{\psi} \circ \varphi_1 = \varphi_2$. □

8.12 Tensor product of Hilbert spaces

The purpose of this section is to define the tensor product of two Hilbert spaces and study some properties of it, which are needed in this book. As the reader can see, this construction is easily extended to the tensor product of finitely many Hilbert spaces. There are different ways of introducing the tensor product of Hilbert spaces. Here we follow the approach given by Reed and Simon [182] because it is a quick way to understand some of the results established in this monograph.

Consider two real separable Hilbert spaces $\mathcal{H}$ and $\mathcal{K}$ with inner products $\langle \cdot, \cdot \rangle_{\mathcal{H}}$ and $\langle \cdot, \cdot \rangle_{\mathcal{K}}$, and norms $\| \cdot \|_{\mathcal{H}}$ and $\| \cdot \|_{\mathcal{K}}$, respectively. Let $h \in \mathcal{H}$ and $k \in \mathcal{K}$. Set the mapping $h \otimes k : \mathcal{H} \times \mathcal{K} \to \mathbb{R}$ given by

$$
h \otimes k(x, y) = \langle h, x \rangle_{\mathcal{H}} \langle k, y \rangle_{\mathcal{K}}, \quad (x, y) \in \mathcal{H} \times \mathcal{K}. \tag{8.53}
$$

It turns out that $h \otimes k$ is a bilinear mapping. That is, for $x_1, x_2 \in \mathcal{H}$, $y_1, y_2 \in \mathcal{K}$ and $r \in \mathbb{R}$, we have:

i) $h \otimes k(x_1 + x_2, y_1) = h \otimes k(x_1, y_1) + h \otimes k(x_2, y_1)$.

ii) $h \otimes k(x_1, y_1 + y_2) = h \otimes k(x_1, y_1) + h \otimes k(x_1, y_2)$.

iii) $h \otimes k(rx_1, y_1) = h \otimes k(x_1, ry_1) = r\,(h \otimes k(x_1, y_1))$.

Note that $h \otimes k$ can be also seen as a linear map $h \otimes k : \mathcal{H} \to \mathcal{K}$ defined by

$$h \otimes k(x) = \langle h, x\rangle_{\mathcal{H}} k, \quad x \in \mathcal{H},$$

which satisfies that $\sum_{i=1}^{\infty} \|h \otimes k(x_i)\|_{\mathcal{K}}^2 = \|h\|_{\mathcal{H}}^2 \|k\|_{\mathcal{K}}^2$, for all basis $\{x_i : i \in \mathbb{N}\}$ of the Hilbert space $\mathcal{H}$. It means, it is a Hilbert-Schmidt operator from $\mathcal{H}$ into $\mathcal{K}$.

Note that (8.53) implies that, for $h_1, \ldots, h_m \in \mathcal{H}$ and $k_1, \ldots, k_n \in \mathcal{K}$,

$$\left(\sum_{i=1}^{m} h_i\right) \otimes \left(\sum_{j=1}^{n} k_j\right) = \sum_{i=1}^{m} \sum_{j=1}^{n} h_i \otimes k_j \tag{8.54}$$

and, for $r \in \mathbb{R}$,

$$r\,(h_1 \otimes k_1) = (rh_1) \otimes k_1 = h_1 \otimes (rk_1) \tag{8.55}$$

hold.

The inner product of the tensor product of $\mathcal{H}$ and $\mathcal{K}$, denoted by $\langle \cdot, \cdot \rangle_{\mathcal{H}\otimes\mathcal{K}}$, is the one that satisfies

$$\langle h \otimes k, x \otimes y\rangle_{\mathcal{H}\otimes\mathcal{K}} = \langle h, x\rangle_{\mathcal{H}} \langle k, y\rangle_{\mathcal{K}}, \quad h, x \in \mathcal{H} \text{ and } k, y \in \mathcal{K}. \tag{8.56}$$

This is extended to $\mathrm{Span}(\{h \otimes k : h \in \mathcal{H},\, k \in \mathcal{K}\})$ by linearity. Note that an element in $\mathrm{Span}(\{h \otimes k : h \in \mathcal{H},\, k \in \mathcal{K}\})$ can have several representation due to (8.54) and (8.55). So it is necessary to see that $\langle \cdot, \cdot \rangle_{\mathcal{H}\otimes\mathcal{K}}$ is a well-defined and positive definite bilinear form. Before doing it, we give the following auxiliary result.

Lemma 109 *Let $h_1, \ldots, h_n \in \mathcal{H}$ and $k_1, \ldots, k_n \in \mathcal{K}$. Then, $\sum_{i=1}^{n} h_i \otimes k_i = 0$ if and only if there is a $n \times n$ matrix $C = (c_{i,\ell})$, with $c_{i,\ell} \in \mathbb{R}$, such that*

$$(h_1, \ldots, h_n)C = 0 \quad \textit{and} \quad C(k_1, \ldots, k_n)^T = (k_1, \ldots, k_n)^T,$$

where $(k_1, \ldots, k_n)^T$ stands for the transpose of the vector $(k_1, \ldots, k_n)$.

Proof. We first suppose that $\sum_{i=1}^{n} h_i \otimes k_i = 0$. Consider an orthonormal basis $\{y_1, \ldots, y_r\}$ of the linear subspace of $\mathcal{K}$ generated by the set $\{k_1, \ldots, k_n\}$. Consequently, there are two finite sequences $\{a_{i,j} \in \mathbb{R} : 0 \leq i \leq n \text{ and } 0 \leq j \leq r\}$ and $\{b_{j,\ell} \in \mathbb{R} : 0 \leq j \leq r \text{ and } 0 \leq \ell \leq n\}$ such that, for $i \in \{1, \ldots, n\}$ and $j \in \{1, \ldots, r\}$,

$$k_i = \sum_{j=1}^{r} a_{i,j} y_j \quad \text{and} \quad y_j = \sum_{\ell=1}^{n} b_{j,\ell} k_\ell.$$

Therefore, for the $n \times n$ matrix $(c_{i,\ell}) := (a_{i,j})(b_{j,\ell})$, we get

$$k_i = \sum_{j=1}^{r} a_{i,j} \left(\sum_{\ell=1}^{n} b_{j,\ell} k_\ell\right) = \sum_{\ell=1}^{n} c_{i,\ell} k_\ell$$

and

$$0 = \sum_{i=1}^{n} h_i \otimes k_i = \sum_{i=1}^{n} h_i \otimes \left(\sum_{j=1}^{r} a_{i,j} y_j \right) = \sum_{j=1}^{r} x_j \otimes y_j,$$

with $x_j = \sum_{i=1}^{n} a_{i,j} h_i$. Consequently, the fact that $\{y_1, \ldots, y_r\}$ is an orthonormal system in $\mathcal{K}$ allows us to deduce

$$0 = \left(\sum_{j=1}^{r} x_j \otimes y_j \right) (x_\alpha, y_\alpha) = \sum_{j=1}^{r} \langle x_j, x_\alpha \rangle_{\mathcal{H}} \langle y_j, y_\alpha \rangle_{\mathcal{K}} = \|x_\alpha\|_{\mathcal{H}}^2,$$

for $\alpha = 1, \ldots, r$. Thus,

$$\sum_{i=1}^{n} c_{i,\ell} h_i = \sum_{i=1}^{n} \left(\sum_{\alpha=1}^{r} a_{i,\alpha} b_{\alpha,\ell} \right) h_i = \sum_{\alpha=1}^{r} b_{\alpha,\ell} x_\alpha = 0, \quad \ell = 1, \ldots, n.$$

In other words, the necessity is true.

Finally we deal with the sufficient. So, using the definition of the matrix C and (8.54), we have

$$\begin{aligned} \sum_{i=1}^{n} h_i \otimes k_i &= \sum_{i=1}^{n} h_i \otimes \left(\sum_{\ell=1}^{n} c_{i,\ell} k_\ell \right) = \sum_{i=1}^{n} \sum_{\ell=1}^{n} c_{i,\ell} h_i \otimes k_\ell \\ &= \sum_{\ell=1}^{n} \left(\sum_{i=1}^{n} c_{i,\ell} h_i \right) \otimes k_\ell = \sum_{\ell=1}^{n} 0 \otimes k_\ell = 0. \end{aligned}$$

Now, the proof is complete. □

Proposition 141 $\langle \cdot, \cdot \rangle_{\mathcal{H} \otimes \mathcal{K}}$ *is a well-defined and a positive definite bilinear form on Span*$(\{h \otimes k : h \in \mathcal{H},\, k \in \mathcal{K}\})$.

Proof. We first show that $\langle \cdot, \cdot \rangle_{\mathcal{H} \otimes \mathcal{K}}$ is well-defined. So, choose $\mu, \eta \in \mathrm{Span}(\{h \otimes k : h \in \mathcal{H},\, k \in \mathcal{K}\})$ such that they have the representations

$$\mu = \sum_{i=1}^{n} h_i \otimes k_i = \sum_{j=1}^{m} \tilde{h}_j \otimes \tilde{k}_j \quad \text{and} \quad \eta = \sum_{i=1}^{p} x_i \otimes y_i = \sum_{j=1}^{q} \tilde{x}_j \otimes \tilde{y}_j$$

as elements of $\mathrm{Span}(\{h \otimes k : h \in \mathcal{H},\, k \in \mathcal{K}\})$. Hence, we must prove

$$\left\langle \sum_{i=1}^{n} h_i \otimes k_i, \sum_{i=1}^{p} x_i \otimes y_i \right\rangle_{\mathcal{H} \otimes \mathcal{K}} = \left\langle \sum_{j=1}^{m} \tilde{h}_j \otimes \tilde{k}_j, \sum_{j=1}^{q} \tilde{x}_j \otimes \tilde{y}_j \right\rangle_{\mathcal{H} \otimes \mathcal{K}},$$

which is true if the equalities

$$\begin{aligned} &\left\langle \sum_{i=1}^{n} h_i \otimes k_i - \sum_{j=1}^{m} \tilde{h}_j \otimes \tilde{k}_j, \sum_{i=1}^{p} x_i \otimes y_i \right\rangle_{\mathcal{H} \otimes \mathcal{K}} \\ = &\left\langle \sum_{j=1}^{m} \tilde{h}_j \otimes \tilde{k}_j, \sum_{j=1}^{q} \tilde{x}_j \otimes \tilde{y}_j - \sum_{i=1}^{p} x_i \otimes y_i \right\rangle_{\mathcal{H} \otimes \mathcal{K}} = 0 \end{aligned}$$

are satisfied. Thus, it is enough to prove that $\langle \sum_{i=1}^n h_i \otimes k_i, h \otimes k \rangle_{\mathcal{H}\otimes\mathcal{K}} = 0$, for $h \otimes k \in \mathcal{H}\otimes\mathcal{K}$ and $\sum_{i=1}^n h_i \otimes k_i = 0$. Therefore, assume that $\sum_{i=1}^n h_i \otimes k_i = 0$. Then, by Lemma 109, there exists an $n \times n$ matrix C such that

$$\begin{aligned}
\left\langle \sum_{i=1}^n h_i \otimes k_i, h \otimes k \right\rangle_{\mathcal{H}\otimes\mathcal{K}} &= \sum_{i=1}^n \langle h_i, h\rangle_{\mathcal{H}} \langle k_i, k\rangle_{\mathcal{K}} = \sum_{i=1}^n \langle h_i, h\rangle_{\mathcal{H}} \left\langle \sum_{j=1}^n c_{i,j} k_j, k \right\rangle_{\mathcal{K}} \\
&= \sum_{i=1}^n \sum_{j=1}^n c_{i,j} \langle h_i, h\rangle_{\mathcal{H}} \langle k_j, k\rangle_{\mathcal{K}} = \sum_{j=1}^n \left\langle \sum_{i=1}^n c_{i,j} h_i, h \right\rangle_{\mathcal{H}} \langle k_j, k\rangle_{\mathcal{K}} \\
&= \sum_{j=1}^n \langle 0, h\rangle_{\mathcal{H}} \langle k_j, k\rangle_{\mathcal{K}} = 0.
\end{aligned}$$

Consequently, $\langle \cdot, \cdot \rangle_{\mathcal{H}\otimes\mathcal{K}}$ is well-defined on $\mathrm{Span}(\{h \otimes k : h \in \mathcal{H},\ k \in \mathcal{K}\})$.

On the other hand, choose $\mu = \sum_{i=1}^p h_i \otimes k_i$. In this case, we obtain that μ also has the representation

$$\mu = \sum_{i=1}^n \sum_{j=1}^m d_{i,j} \tilde{h}_i \otimes \tilde{k}_j,$$

where $d_{i,j} \in \mathbb{R}$, and $\{\tilde{h}_1, \ldots, \tilde{h}_n\}$ and $\{\tilde{k}_1, \ldots, \tilde{k}_m\}$ are orthonormal bases of the spaces generated by the sets $\{h_1, \ldots, h_p\}$ and $\{k_1, \ldots, k_p\}$, respectively. Hence, the first part of the proof implies

$$\langle \mu, \mu \rangle_{\mathcal{H}\otimes\mathcal{K}} = \sum_{i=1}^n \sum_{j=1}^m \sum_{\ell=1}^n \sum_{\nu=1}^m d_{i,j} d_{\ell,\nu} \left\langle \tilde{h}_i, \tilde{h}_\ell \right\rangle_{\mathcal{H}} \left\langle \tilde{k}_j, \tilde{k}_\nu \right\rangle_{\mathcal{K}} = \sum_{i=1}^n \sum_{j=1}^m d_{i,j}^2 \geq 0.$$

Moreover, $\langle \mu, \mu \rangle_{\mathcal{H}\otimes\mathcal{K}} = 0$ if and only if $\mu = 0$. it means, $\langle \cdot, \cdot \rangle_{\mathcal{H}\otimes\mathcal{K}}$ is a positive definite on $\mathrm{Span}(\{h \otimes k : h \in \mathcal{H},\ k \in \mathcal{K}\})$. □

We introduce the tensor product of $\mathcal{H}$ and $\mathcal{K}$ using Theorem 101:

Definition 60 *The tensor product $\mathcal{H} \otimes \mathcal{K}$ of the real separable Hilbert spaces $\mathcal{H}$ and $\mathcal{K}$ is the completion of $(\mathrm{Span}(\{h \otimes k : h \in \mathcal{H},\ k \in \mathcal{K}\}), \langle \cdot, \cdot \rangle_{\mathcal{H}\otimes\mathcal{K}})$. The inner product in $\mathcal{H} \otimes \mathcal{K}$ is still denoted by $\langle \cdot, \cdot \rangle_{\mathcal{H}\otimes\mathcal{K}}$.*

As a consequence of the last definition, we can state the following result.

Proposition 142 *Let $\mathbb{I}, \mathbb{J} \subset \mathbb{N}$. Suppose that $\{h_i : i \in \mathbb{I}\}$ and $\{k_j : i \in \mathbb{J}\}$ are complete orthonormal systems of $\mathcal{H}$ and $\mathcal{K}$, respectively. Then $\{h_i \otimes k_j : (i,j) \in \mathbb{I} \times \mathbb{J}\}$ is an orthonormal basis of $\mathcal{H} \otimes \mathcal{K}$.*

Proof. It is clear that (8.56) implies $\{h_i \otimes k_j : (i,j) \in \mathbb{I} \times \mathbb{J}\}$ is an orthonormal system in $\mathcal{H} \otimes \mathcal{K}$. So, we only need to show that the closure of $\mathrm{Span}(\{h_i \otimes k_j : (i,j) \in \mathbb{I} \times \mathbb{J}\})$ contains $\mathrm{Span}(\{h \otimes k : h \in \mathcal{H},\ k \in \mathcal{K}\})$. Towards this end, let $h \in \mathcal{H}$ and $k \in \mathcal{K}$, then there exist $\{a_i \in \mathbb{R} : i \in \mathbb{I}\}$ and $\{b_j \in \mathbb{R} : j \in \mathbb{J}\}$ such that

$$h = \sum_{i \in \mathbb{I}} a_i h_i,\ \|h\|_{\mathcal{H}}^2 = \sum_{i \in \mathbb{I}} a_i^2 \quad \text{and} \quad k = \sum_{i \in \mathbb{J}} b_j k_i,\ \|k\|_{\mathcal{K}}^2 = \sum_{i \in \mathbb{J}} b_j^2.$$

Consequently,

$$\begin{aligned}
&\left\| h \otimes k - \sum_{\substack{i\in\mathbb{I}, j\in\mathbb{J}\\ i,j\le n}} a_i b_j h_i \otimes k_j \right\|_{\mathcal{H}\otimes\mathcal{K}} \\
\le & \left\| h \otimes \left(k - \sum_{\substack{j\in\mathbb{J}\\ j\le n}} b_j k_j \right) \right\|_{\mathcal{H}\otimes\mathcal{K}} + \left\| \left(h - \sum_{\substack{i\in\mathbb{I}\\ i\le n}} a_i h_i \right) \otimes \left(\sum_{\substack{j\in\mathbb{J}\\ j\le n}} b_j k_j \right) \right\|_{\mathcal{H}\otimes\mathcal{K}} \\
\le & \left\| k - \sum_{\substack{j\in\mathbb{J}\\ j\le n}} b_j k_j \right\|_{\mathcal{K}} \left(\sum_{i\in\mathbb{I}} a_i^2 \right)^{1/2} + \left\| h - \sum_{\substack{i\in\mathbb{I}\\ i\le n}} a_i h_i \right\|_{\mathcal{H}} \left(\sum_{j\in\mathbb{J}} b_j^2 \right)^{1/2} \to 0 \text{ as } n \to \infty.
\end{aligned}$$

Hence, it is easy to see that $\mathrm{Span}(\{h \otimes k : h \in \mathcal{H},\ k \in \mathcal{K}\})$ is contained in the closure of $\mathrm{Span}(\{h_i \otimes k_j : i, j \in \mathbb{N}\})$. $\square$

As an example, we assume that (X, μ) and (Y, ν) are either two complete, or two σ-finite measure spaces such that the Hilbert spaces $L^2(X, \mu)$ and $L^2(Y, \nu)$ are real separable. So we can choose bases $\{f_i : i \in \mathbb{I}\}$ and $\{g_j : j \in \mathbb{J}\}$ of $L^2(X, \mu)$ and $L^2(Y, \nu)$, respectively. Here, $\mathbb{I}$ and $\mathbb{J}$ are as in Proposition 142. Thus, from Proposition 142, $\{f_i \otimes g_j : (i, j) \in \mathbb{I} \times \mathbb{J}\}$ is a complete orthonormal system of $L^2(X, \mu) \otimes L^2(Y, \nu)$. Therefore, $\psi : L^2(X, \mu) \otimes L^2(Y, \nu) \to L^2(X \times Y, \mu \times \nu)$ given by $\psi\left(\sum_{i\in\mathbb{I}, j\in\mathbb{J}} c_{i,j} f_i \otimes g_j\right) = \sum_{i\in\mathbb{I}, j\in\mathbb{J}} c_{i,j} f_i g_j$ is an isometry because $\{f_i g_j : (i, j) \in \mathbb{I} \times \mathbb{J}\}$ is a complete orthonormal system of $L^2(X \times Y, \mu \otimes \nu)$ and

$$\left\| \sum_{i\in\mathbb{I}, j\in\mathbb{J}} c_{i,j} f_i \otimes g_j \right\|_{\mathcal{H}\otimes\mathcal{K}} = \sum_{i\in\mathbb{I}, j\in\mathbb{J}} c_{i,j}^2 = \left\| \sum_{i\in\mathbb{I}, j\in\mathbb{J}} c_{i,j} f_i g_j \right\|_{L^2(X\times Y, \mu\otimes\nu)}$$

holds. Thus, Definitions 59 and 60 yield

$$L^2(X, \mu) \otimes L^2(Y, \nu) = L^2(X \times Y, \mu \times \nu). \tag{8.57}$$

Similarly, for $n \ge 1$, we can show that

$$\begin{aligned}
& L^2(X, \mu; \mathbb{R}^n) \otimes L^2(Y, \nu; \mathbb{R}^n) \\
= \ & L^2\left((X \times \{1, \dots, n\}) \times (Y \times \{1, \dots, n\}), (\mu \times \delta) \times (\nu \times \delta)\right),
\end{aligned} \tag{8.58}$$

where δ is the measure on $(\{1, \dots, n\}, 2^{\{1,\dots,n\}})$ such that $\delta(\{\ell\}) = 1$ for all $\ell = 1, \dots, n$. In this case, for the bases $\{f_i : i \in \mathbb{I}\}$ and $\{g_j : j \in \mathbb{J}\}$ of $L^2(X, \mu; \mathbb{R}^n)$ and $L^2(Y, \nu; \mathbb{R}^n)$, respectively, the isometry ψ is now given by

$$\psi\left(\sum_{i\in\mathbb{I}, j\in\mathbb{J}} c_{i,j} f_i \otimes g_j \right) = \sum_{i\in\mathbb{I}, j\in\mathbb{J}} c_{i,j} \bar{f}_i \bar{g}_j. \tag{8.59}$$

Here $\bar{f}_i(x,\ell) = f_i^{(\ell)}(x)$ and $\bar{g}_j(y,\ell) = g_j^{(\ell)}(y)$, for $\ell = 1,\dots,n$, $x \in X$ and $y \in Y$. Indeed, we only need to observe

$$\begin{aligned}
\langle \bar{f}_i, \bar{f}_{\bar{i}}(x)\rangle_{L^2(X\times\{1,\dots,n\},\mu\times\delta)} &= \int_{X\times\{1,\dots,n\}} \bar{f}_i(x,\ell)\bar{f}_{\bar{i}}(x,\ell)\mu\times\delta(d(x,\ell)) \\
&= \sum_{\ell=1}^{n}\int_X \bar{f}_i(x,\ell)\bar{f}_{\bar{i}}(x,\ell)\mu(dx) = \sum_{\ell=1}^{n}\int_X f_i^{(\ell)}(x) f_{\bar{i}}^{(\ell)}(x)\mu(dx) \\
&= \int_X \langle f_i(x), f_{\bar{i}}(x)\rangle_{\mathbb{R}^n}\,\mu(dx) = \langle f_i, f_{\bar{i}}(x)\rangle_{L^2(X,\mu;\mathbb{R}^n)},
\end{aligned}$$

which, in particular, implies that the family $\{\bar{f}_i : i \in \mathbb{I}\}$ is a basis of $L^2\left((X\times\{1,\dots,n\},\mu\times\delta\right)$.

Now, consider another two real separable Hilbert spaces $\mathcal{H}_1$ and $\mathcal{K}_1$, and two bounded linear operators $A:\mathcal{H}\to\mathcal{H}_1$ and $B:\mathcal{K}\to\mathcal{K}_1$. Set

$$(A\otimes B)(h\otimes k) = A(h)\otimes B(k), \quad \text{for } h\in\mathcal{H} \text{ and } k\in\mathcal{K}, \tag{8.60}$$

and then extend it by linearity on Span($\{h\otimes k : h\in\mathcal{H},\, k\in\mathcal{K}\}$). Proceeding as in the proof of Proposition 141 (i.e., using Lemma 109), we have that this extension is well-defined. The proof of this fact is left to the reader as an exercise.

Proposition 143 *There is a unique bounded linear operator $A\otimes B : \mathcal{H}\otimes\mathcal{K}\to\mathcal{H}_1\otimes\mathcal{K}_1$ such that $\|A\otimes B\| = \|A\|\|B\|$ and (8.60) holds.*

Proof. We use the notations $I_{\mathcal{H}}$ and $I_{\mathcal{K}}$ for the identity operators on $\mathcal{H}$ and $\mathcal{K}$, respectively. Then, (8.60) gives

$$(A\otimes B)(h\otimes k) = (A\otimes I_{\mathcal{K}})\left((I_{\mathcal{H}}\otimes B)(h,k)\right), \quad \text{for } h\in\mathcal{H} \text{ and } k\in\mathcal{K}.$$

Choose an element $\mu = \sum_{i=1}^{p} h_i\otimes k_i$ of Span($\{h\otimes k : h\in\mathcal{H},\, k\in\mathcal{K}\}$). Note that μ also has the representation $\mu = \sum_{i=1}^{n}\sum_{j=1}^{m} d_{i,j}\tilde{h}_i\otimes\tilde{k}_j$, where $\{\tilde{h}_1,\dots,\tilde{h}_n\}$ and $\{\tilde{k}_1,\dots,\tilde{k}_m\}$ are orthonormal bases of the linear spaces generated by $\{h_1,\dots,h_p\}$ and $\{k_1,\dots,k_p\}$, respectively. Then,

$$\begin{aligned}
&\|(I_{\mathcal{H}}\otimes B)(\mu)\|^2_{\mathcal{H}\otimes\mathcal{K}_1} \\
&= \left\langle \sum_{i=1}^{n}\sum_{j=1}^{m} d_{i,j}\tilde{h}_i\otimes B\left(\tilde{k}_j\right), \sum_{\ell=1}^{n}\sum_{q=1}^{m} d_{\ell,q}\tilde{h}_\ell\otimes B\left(\tilde{k}_q\right)\right\rangle_{\mathcal{H}\otimes\mathcal{K}_1} \\
&= \sum_{i=1}^{n}\sum_{j,q=1}^{m} d_{i,j}d_{i,q}\left\langle B\left(\tilde{k}_j\right), B\left(\tilde{k}_q\right)\right\rangle_{\mathcal{K}_1} = \sum_{i=1}^{n}\left\| B\left(\sum_{j=1}^{m} d_{i,j}\tilde{k}_j\right)\right\|^2_{\mathcal{K}_1} \\
&\le \|B\|^2\sum_{i=1}^{n}\left\|\left(\sum_{j=1}^{m} d_{i,j}\tilde{k}_j\right)\right\|^2_{\mathcal{K}_1} = \|B\|^2\sum_{j=1}^{m}\sum_{j=1}^{m} d_{i,j}^2 = \|B\|^2\|\mu\|^2_{\mathcal{H}\otimes\mathcal{K}},
\end{aligned}$$

which gives $\|I_{\mathcal{H}}\otimes B\| \le \|B\|$. Proceeding similarly, we can establish that $\|A\otimes I_{\mathcal{K}}\| \le \|A\|$. Hence, on Span($\{h\otimes k : h\in\mathcal{H},\, k\in\mathcal{K}\}$),

$$\|A\otimes B\| = \|(A\otimes I_{\mathcal{K}})(I_{\mathcal{H}}\otimes B)\| \le \|A\|\|B\|. \tag{8.61}$$

Conversely, let $\varepsilon > 0$. Then, there are $h_\varepsilon\in\mathcal{H}$ and $k_\varepsilon\in\mathcal{K}$ such that

$$\|A(h_\varepsilon)\|_{\mathcal{H}_1} \ge (\|A\|-\varepsilon)\|h_\varepsilon\|_{\mathcal{H}} \quad\text{and}\quad \|B(k_\varepsilon)\|_{\mathcal{K}_1} \ge (\|B\|-\varepsilon)\|k_\varepsilon\|_{\mathcal{K}}.$$

Therefore,

$$\begin{aligned}\|(A\otimes B)(h_\varepsilon\otimes k_\varepsilon)\|_{\mathcal{H}_1\otimes\mathcal{K}_1} &= \|A(h_\varepsilon)\|_{\mathcal{H}_1}\|B(k_\varepsilon)\|_{\mathcal{K}_1}\geq(\|A\|-\varepsilon)(\|B\|-\varepsilon)\|h_\varepsilon\|_{\mathcal{H}}\|k_\varepsilon\|_{\mathcal{K}}\\ &= (\|A\|-\varepsilon)(\|B\|-\varepsilon)\|h_\varepsilon\otimes k_\varepsilon\|_{\mathcal{H}\otimes\mathcal{K}}.\end{aligned}$$

Since ε is arbitrary, (8.61) implies that $A\otimes B$ is a bounded linear operator on Span($\{h\otimes k: h\in\mathcal{H},\, k\in\mathcal{K}\}$) satisfying $\|A\otimes B\|=\|A\|\|B\|$.

Finally, the result follows from the fact that Span($\{h\otimes k: h\in\mathcal{H},\, k\in\mathcal{K}\}$) is a dense set of $\mathcal{H}\otimes\mathcal{K}$. □

Also, we have the following auxiliary result in the construction of Wiener-Itô multiple integrals.

Proposition 144 *Let ξ be an element of $\mathcal{H}\otimes\mathcal{K}$. Then, there is a unique continuous linear operator $\Xi:\mathcal{H}\to\mathcal{K}$ such that*

$$\langle\Xi(h),k\rangle_{\mathcal{K}}=\langle\xi,h\otimes k\rangle_{\mathcal{H}\otimes\mathcal{K}},\quad \text{for all } h\otimes k\in\mathcal{H}\otimes\mathcal{K}. \tag{8.62}$$

Proof. Let $h\in\mathcal{H}$. Then, the map $k\mapsto\langle\xi,h\otimes k\rangle_{\mathcal{H}\otimes\mathcal{K}}$ is a functional on $\mathcal{K}$ (i.e., it is a continuous bounded linear operator from $\mathcal{K}$ into $\mathbb{R}$), whose norm is bounded by $\|h\|_{\mathcal{H}}\|\xi\|_{\mathcal{H}\otimes\mathcal{K}}$. Thus, the Riesz representation theorem implies that there is a unique linear operator $\Xi:\mathcal{H}\to\mathcal{K}$ with norm bounded by $\|\xi\|_{\mathcal{H}\otimes\mathcal{K}}$ such that (8.62) holds. □

8.13 Hermite polynomials

In this section, we introduce the Hermite polynomials and some of their properties, which are used in this book. For $\rho>0$, the family of Hermite polynomials $\{H_n(\cdot,\rho):n\in\mathbb{N}\cup\{0\}\}$ is a complete orthonormal system of $L^2(\mathbb{R};\nu_\rho)$, with

$$\nu_\rho(dx)=(2\pi\rho)^{-1/2}\exp\left(-x^2/(2\rho)\right)dx.$$

The Hermite polynomials, named after the French mathematician Chales Hermite, arise as solutions of the simple harmonic oscillator of quantum mechanics (see, for instance, Faddeev and Yakubovskiĭ [48]).

Now, we proceed to introduce the family $\{H_n(\cdot,\rho):n\in\mathbb{N}\cup\{0\}\}$. Towards this end, for $t\geq 0,\rho>0$ and $x\in\mathbb{R}$, we use the power series of the exponential function to establish

$$\exp\left(tx-\frac{\rho t^2}{2}\right)=\sum_{n=0}^{\infty}\frac{t^n}{n!}\left(x-\frac{t\rho}{2}\right)^n=\sum_{n=0}^{\infty}\frac{t^n}{n!}\sum_{k=0}^{n}\binom{n}{k}2^{-k}(-\rho)^k x^{n-k}t^k.$$

Thus, the change $m=n+k$ leads us to write

$$\begin{aligned}\exp\left(tx-\frac{\rho t^2}{2}\right) &= \sum_{n=0}^{\infty}\sum_{m=n}^{2n}2^{n-m}(-\rho)^{m-n}t^m\frac{x^{2n-m}}{(2n-m)!(m-n)!}\\ &= \sum_{m=0}^{\infty}t^m\sum_{m/2\leq n\leq m}2^{n-m}(-\rho)^{m-n}\frac{x^{2n-m}}{(2n-m)!(m-n)!},\end{aligned}$$

where the interchange of the order of the double sum (i.e., last equality) follows from the fact that

$$\sum_{n=0}^{\infty}\sum_{m=n}^{2n}2^{n-m}\left|(-\rho)^{m-n}t^m\frac{x^{2n-m}}{(2n-m)!(m-n)!}\right|\leq\sum_{n=0}^{\infty}\frac{t^n}{n!}\left(|x|+\frac{t\rho}{2}\right)^n<\infty.$$

Therefore, making the change $\ell = m - n$, we get

$$\exp\left(tx - \frac{\rho t^2}{2}\right) = \sum_{m=0}^{\infty} \frac{t^m}{m!} \sum_{\ell=0}^{[m/2]} 2^{-\ell}(-\rho)^{\ell} \frac{m!}{(m-2\ell)!\ell!} x^{m-2\ell} := \sum_{m=0}^{\infty} \frac{t^m}{m!} H_m(x,\rho). \quad (8.63)$$

That is, the Hermite polynomials with parameter ρ are introduced as the coefficients of the power series of $t \mapsto \exp(tx - \frac{\rho t^2}{2})$, which is called the generating function of the Hermite polynomials. Consequently, the last equality yields that $H_n(\cdot,\rho)$ is a polynomial of order n given by

$$H_n(x,\rho) = \sum_{k=0}^{[n/2]} 2^{-k} \frac{n!}{k!(n-2k)!} (-\rho)^k x^{n-2k}, \quad x \in \mathbb{R}. \quad (8.64)$$

$H_n(\cdot,\rho)$ is called the Hermite polynomial of order n with parameter ρ. Hence, we easily obtain, for $x \in \mathbb{R}$,

$$H_{n+1}'(x,\rho) = (n+1)H_n(x,\rho) \quad (8.65)$$

and $\rho^{-n}H_n(\rho x, \rho^2) = H_n(x,1)$. Similarly, (8.63) implies

$$\begin{aligned}(x-\rho t)\exp\left(tx - \frac{\rho t^2}{2}\right) &= \frac{d\exp\left(tx - \frac{\rho t^2}{2}\right)}{dt} \\ &= \sum_{m=1}^{\infty} \frac{t^{m-1}}{(m-1)!} H_m(x,\rho) = \sum_{m=0}^{\infty} \frac{t^m}{m!} H_{m+1}(x,\rho).\end{aligned}$$

Hence, comparing the terms of $t^m/m!$, we obtain the recurrence relations $H_1(x,\rho) = xH_0(x,\rho)$ and

$$H_{m+1}(x,\rho) = xH_m(x,\rho) - \rho m H_{m-1}(x,\rho), \quad m \geq 1. \quad (8.66)$$

Also, from (8.63) and the Taylor series expansion for power series, we have

$$H_n(x,\rho) = \frac{d^n \exp\left(tx - \frac{\rho t^2}{2}\right)}{dt^n}\Big|_{t=0} = e^{x^2/2\rho} \frac{d^n \exp\left(-\frac{1}{2}\rho\left(t - \frac{x}{\rho}\right)^2\right)}{dt^n}\Big|_{t=0}.$$

Since

$$\begin{aligned}\frac{d\exp\left(-\frac{1}{2}\rho\left(t-\frac{x}{\rho}\right)^2\right)}{dt} &= -\rho(t - \frac{x}{\rho})\exp\left(-\frac{1}{2}\rho\left(t - \frac{x}{\rho}\right)^2\right) \\ &= -\rho\frac{d\exp\left(-\frac{1}{2}\rho\left(t-\frac{x}{\rho}\right)^2\right)}{dx},\end{aligned}$$

we can apply induction on n to show that the Rodrigues formula

$$H_n(x,\rho) = e^{x^2/(2\rho)}(-\rho)^n \frac{d^n e^{-x^2/(2\rho)}}{dx^n}, \quad x \in \mathbb{R} \text{ and } n \in \mathbb{N} \cup \{0\}$$

holds.

Now, we are ready to show that the Hermite polynomials is an orthogonal system.

Theorem 102 *Let $\rho > 0$. Then, $\{H_n(\cdot,\rho)/\sqrt{\rho^n n!} : n \in \mathbb{N} \cup \{0\}\}$ is an orthonormal system in $L^2(\mathbb{R};\nu_\rho)$, where $\nu_\rho(dx) = (2\pi\rho)^{-1/2}\exp\left(-x^2/(2\rho)\right)dx$.*

Proof. We first deal with the orthogonality of $\{H_n(\cdot,\rho) : n \in \mathbb{N} \cup \{0\}\}$. So, choose $n \geq 1$. Therefore, the Rodrigues formula yields

$$\int_{\mathbb{R}} H_n(x,\rho)e^{-x^2/2\rho}dx = (-\rho)^n \int_{\mathbb{R}} \frac{d^n e^{-x^2/2\rho}}{dx^n}dx = 0.$$

That is, $H_n(\cdot,\rho)$ is orthogonal to $H_0(\cdot,\rho) \equiv 1$ in $L^2(\mathbb{R};\nu_\rho)$, for every $n \in \mathbb{N}$.

Now, set $f_n(x) = \exp(-x^2/4\rho)H_n(x,\rho)$. Then, (8.65) gives

$$\begin{aligned} f_n'(x) &= \left(H_n'(x,\rho) - \frac{x}{2\rho}H_n(x,\rho)\right)\exp(-x^2/4\rho) \\ &= \left(nH_{n-1}(x,\rho) - \frac{x}{2\rho}H_n(x,\rho)\right)\exp(-x^2/4\rho) = nf_{n-1}(x) - \frac{x}{2\rho}f_n(x). \end{aligned}$$

Hence, we can use (8.66) to establish

$$\begin{aligned} f_n''(x) &= -\frac{1}{2\rho}f_n(x) - \frac{x}{2\rho}f_n'(x) + nf_{n-1}'(x) \\ &= -\frac{1}{2\rho}f_n(x) - \frac{x}{2\rho}\left(nf_{n-1}(x) - \frac{x}{2\rho}f_n(x)\right) + n\left((n-1)f_{n-2}(x) - \frac{x}{2\rho}f_{n-1}(x)\right) \\ &= \left(\frac{x^2}{4\rho^2} - \frac{1}{2\rho}\right)f_n(x) - \frac{n}{\rho}\left(xf_{n-1}(x) - (n-1)\rho f_{n-2}(x)\right) \\ &= \left(\frac{x^2}{4\rho^2} - \frac{1}{2\rho} - \frac{n}{\rho}\right)f_n(x), \quad x \in \mathbb{R}, \end{aligned}$$

which leads us to get that $\frac{d\left(f_n'(x)f_m(x) - f_m'(x)f_n(x)\right)}{dx} + \frac{m-n}{\rho}f_n(x)f_m(x) = 0$. Consequently, integrating this equality, we obtain for $n \neq m$,

$$\frac{n-m}{\rho}\int_{\mathbb{R}} H_n(x,\rho)H_m(x,\rho)e^{-x^2/2\rho}dx = \int_{\mathbb{R}} \frac{d\left(f_n'(x)f_m(x) - f_m'(x)f_n(x)\right)}{dx}dx = 0.$$

Thus, we have proven that $\{H_n(\cdot,\rho)/\sqrt{\rho^n n!} : n \in \mathbb{N} \cup \{0\}\}$ is an orthogonal system in $L^2(\mathbb{R};\nu_\rho)$. In order to show that it is also an orthonormal system, we apply equality (8.66) twice to see that the equality

$$\begin{aligned} & H_m(x,\rho)\left[H_m(x,\rho) - xH_{m-1}(x,\rho) + \rho(m-1)H_{m-2}(x,\rho)\right] \\ &= H_{m-1}(x,\rho)\left[H_{m+1}(x,\rho) - xH_m(x,\rho) + \rho m H_{m-1}(x,\rho)\right], \quad m \geq 2, \end{aligned}$$

is satisfied. Thus, the fact that $H_1(x,\rho) = x$ gives

$$\begin{aligned} \int_{\mathbb{R}} H_m^2(x,\rho)e^{-x^2/2\rho}dx &= \rho m \int_{\mathbb{R}} H_{m-1}^2(x,\rho)e^{-x^2/2\rho}dx \\ &= m!\rho^{m-1}\int_{\mathbb{R}} H_1^2(x,\rho)e^{-x^2/2\rho}dx = m!\rho^m\sqrt{2\pi\rho}. \end{aligned}$$

In consequence, the theorem is true. □

In order to state that $\{H_n(\cdot,\rho)/\sqrt{\rho^n n!} : n \in \mathbb{N} \cup \{0\}\}$ is a complete orthonormal system in $L^2(\mathbb{R};\nu_\rho)$, we use the following result concerning Fourier transform. For its proof, the reader can see the book by Andrews et al. [14] (Section 6.5).

Lemma 110 *Let $f \in L^2(\mathbb{R};\nu_\rho)$ be such that $\int_{\mathbb{R}} x^n f(x)\nu_\rho(dx) = 0$, for $n \in \mathbb{N} \cup \{0\}$. Then, $f = 0$ almost surely.*

Remark 122 *This result implies that the Fourier transform of $g(x) := f(x)\exp(-x^2/2)$ is zero. That is,*

$$\hat{g}(x) = \int_{\mathbb{R}} e^{ixy} f(y)\exp(-y^2/2)dy = 0, \quad \textit{for almost all } x \in \mathbb{R}.$$

Now, we can show that the family $\{H_n(\cdot,\rho)/\sqrt{\rho^n n!} : n \in \mathbb{N}\cup\{0\}\}$ is a basis of $L^2(\mathbb{R};\nu_\rho)$.

Theorem 103 *Let $\rho > 0$ and $f \in L^2(\mathbb{R};\nu_\rho)$. Then, the function f has the unique series expansion*

$$f(x) = \sum_{m=0}^{\infty} f_n \frac{H_n(x,\rho)}{\sqrt{\rho^n n!}} \ \textit{in } L^2(\mathbb{R};\nu_\rho), \quad \textit{with} \quad f_n = \frac{1}{\sqrt{\rho^n n!}}\int_{\mathbb{R}} f(x)H_n(x,\rho)d\nu_\rho(dx).$$

Moreover, $\int_{\mathbb{R}} f(x)^2 d\nu_\rho(dx) = \sum_{m=0}^{\infty} f_n^2$.

Proof. We observe that (8.64) implies that, for $x \in \mathbb{R}$, $H_0(x,\rho) \equiv 1$, $H_1(x,\rho) = x$ and $H_2(x,\rho) = x^2 - \rho$, which implies that $x^2 = H_2(x,\rho) + \rho H_0(x,\rho)$. Hence, by induction on n and (8.64), the monomial x^n belongs to Span($\{H_0(x,\rho),\ldots,H_n(x,\rho)\}$). Consequently,

$$\int_{\mathbb{R}} x^n f(x)d\nu_\rho(dx) = 0, \quad \text{for all } n \in \mathbb{N}\cup\{0\},$$

whenever

$$\int_{\mathbb{R}} H_n(x,\rho)f(x)d\nu_\rho(dx) = 0, \quad \text{for all } n \in \mathbb{N}\cup\{0\}.$$

Thus, from Lemma 110, we have that the orthogonal complement of $\{H_n(\cdot,\rho)/\sqrt{\rho^n n!} : n \in \mathbb{N}\cup\{0\}\}$ is the set $\{0\}$. Therefore, this family is a complete orthonormal system of $L^2(\mathbb{R};\nu_\rho)$ due to Theorem 102 and, in consequence, the proof is complete. □

8.14 The Ornstein-Uhlenbeck semigroup

In this section, we study some auxiliary results on the Ornstein-Uhlenbeck semigroup that we need to understand the proof of Theorem 52. So, here we suppose that $\mathcal{H} = L^2([0,T];\mathbb{R}^d)$. For a detailed exposition, the reader can consult Nualart [158].

Let $W = \{W(h) : h \in \mathcal{H}\}$ be the isonormal Gaussian process given by (4.20), which is related to the d-dimensional Brownian motion $B = \{B_t : t \in [0,T]\}$, $p \geq 1$ and $\mathcal{F}$ the σ-algebra generated by W (or by B). Then, we can apply the Doob-Dynkin lemma (see [181]) to prove that, for $F \in L^p(\Omega,\mathcal{F})$, there exists a measurable mapping $\Phi_F : \mathbb{R}^{\mathcal{H}} \to \mathbb{R}$ such that $F = \Phi_F(W)$, where $\mathbb{R}^{\mathcal{H}}$ is the space of all the functions from $\mathcal{H}$ into $\mathbb{R}$. The σ-algebra of $\mathbb{R}^{\mathcal{H}}$ is generated by the sets

$$\mathcal{A}_{h_1,\ldots,h_n}^{\mathcal{C}_1,\ldots,\mathcal{C}_n} := \left\{\Phi \in \mathbb{R}^{\mathcal{H}} : (\Phi(h_1),\ldots,\Phi(h_n)) \in \mathcal{C}\right\}, \tag{8.67}$$

where $h_1,\ldots,h_n \in \mathcal{H}$, $n \in \mathbb{N}$ and $\mathcal{C} = \mathcal{C}_1 \times \cdots \times \mathcal{C}_n \in \mathcal{B}(\mathbb{R}^n)$. Also, let $\tilde{W} = \{\tilde{W}(h) : h \in \mathcal{H}\}$ be the isonormal Gaussian process associated with an independent copy $\tilde{B} = \{\tilde{B}_t : t \in [0,T]\}$ of B. Thus, we can suppose that W and $\tilde{W}$ are defined on the probability space $(\Omega \times \tilde{\Omega}, \mathcal{F}\otimes\tilde{\mathcal{F}}, P \times \tilde{P})$. Therefore, for $t > 0$, we can consider, on the last probability space, the isonormal Gaussian process

$$G_t = \{G_t(h) := e^{-t}W(h) + \sqrt{1-e^{-2t}}\tilde{W}(h) : h \in \mathcal{H}\}$$

satisfying $E\left(G_t(h_1)G_t(h_2)\right) = \langle h_1, h_2\rangle_{\mathcal{H}}$, for $h_1.h_2 \in \mathcal{H}$.

Before introducing the Ornstein-Uhlenbeck semigroup, we state the following:

Lemma 111 *Let $\Phi : \mathbb{R}^{\mathcal{H}} \to \mathbb{R}$ be a bounded and measurable function and $t > 0$. Then,*

$$E(\Phi(W)) = E\tilde{E}(\Phi(G_t)). \tag{8.68}$$

Proof. We will use Theorem 97 (i.e., the monotone class lemma) to show that the result is true.

Set $\mathcal{G} = \{\Phi : \mathbb{R}^{\mathcal{H}} \to \mathbb{R} : (8.68) \text{ is satisfied}\}$. It is clear that this is a linear family of functions that contains $\Phi \equiv 1$. That is, Statement *i)* in Theorem 97 is satisfied in this case.

In order to consider Statement *ii)* of the class monotone lemma, let $h_1, \ldots, h_n \in \mathcal{H}$ and $\mathcal{C} = \mathcal{C}_1 \times \cdots \times \mathcal{C}_n \in \mathcal{B}(\mathbb{R}^n)$. Then, $(W(h_1), \ldots, W(h_n))$ and $(G_t(h_1), \ldots, G_t(h_n))$ are two Gaussian random vectors having zero mean vector and the same covariance matrix. Consequently, Section 1.1.1 implies that these random vectors have the same distribution function, which allows us to conclude

$$E\left(1_{\{(W(h_1),\ldots,W(h_n))\in\mathcal{C}\}}\right) = E\tilde{E}\left(1_{\{(G_t(h_1),\ldots,G_t(h_n))\in\mathcal{C}\}}\right).$$

It means, Statement *ii)* of Theorem 97 is valid with

$$\mathcal{A} = \left\{\mathcal{A}_{h_1,\ldots,h_n}^{\mathcal{C}_1,\ldots,\mathcal{C}_n} : h_i \in \mathcal{H},\ \mathcal{C}_i \in \mathcal{B}(\mathbb{R}), n \in \mathbb{N} \text{ and } 1 \le i \le n\right\},$$

where $\mathcal{A}_{h_1,\ldots,h_n}^{\mathcal{C}_1,\ldots,\mathcal{C}_n}$ is introduced in (8.67).

Finally, Statement *iii)* in Theorem 97 holds due to the monotone convergence theorem. □

Now, for $t \ge 0$ and $p \ge 1$, we introduce the Ornstein-Uhlenbeck operator $T_t : L^p(\Omega, \mathcal{F}) \to L^p(\Omega, \mathcal{F})$ given by

$$T_t(F) = \tilde{E}\left(\Phi_F\left(e^{-t}W + \sqrt{1-e^{-2t}}\tilde{W}\right)\right).$$

As an example, consider $F \in \mathcal{S}$ having the form $F = f(W(h_1), \ldots, Wh_n))$. Then,

$$T_t(F) = \tilde{E}\left[f\left(e^{-t}W(h_1) + \sqrt{1-e^{-2t}}\tilde{W}(h_1), \ldots, e^{-t}W(h_n) + \sqrt{1-e^{-2t}}\tilde{W}(h_n)\right)\right].$$

Note that T_t is a linear operator. Actually, T_t is a contraction on $L^p(\Omega, \mathcal{F})$. Indeed, Lemma 111 and Jensen's inequality lead us to obtain

$$\begin{aligned} E\left(|T_t(F)|^p\right) &= E\left(\left|\tilde{E}\left(\Phi_F\left(e^{-t}W + \sqrt{1-e^{-2t}}\tilde{W}\right)\right)\right|^p\right) \\ &\le E\tilde{E}\left(\left|\Phi_F\left(e^{-t}W + \sqrt{1-e^{-2t}}\tilde{W}\right)\right|^p\right) = E\left(|\Phi_F(W)|^p\right) = E\left(|F|^p\right), \end{aligned}$$

which gives that our claim is satisfied.

In order to see that $\{T_t : t \ge 0\}$ is a C_0-semigroup on $L^2(\Omega, \mathcal{F})$, we state the following result. Towards this end, we recall that the operator $D : \mathcal{S} \subset L^2(\Omega, \mathcal{F}) \to L^2(\Omega; \mathcal{H})$ is analyzed in Section 4.1 and that the Hermite polynomial $H_n(\cdot, 1)$ of order $n \in \mathbb{N}$ is defined in Section 8.13.

Proposition 145 *Let $h \in \mathcal{H}$ be such that $\|h\|_{\mathcal{H}} = 1$, $t > 0$ and $n \in \mathbb{N}$. Then, $T_t(H_n(W(h), 1))$ belongs to $\mathbb{D}^{1,2}$ and*

$$\begin{aligned} D\left[T_t(H_n(W(h), 1))\right] &= D\left[\tilde{E}\left(H_n\left(e^{-t}W(h) + \sqrt{1-e^{-2t}}\tilde{W}(h), 1\right)\right)\right] \\ &= ne^{-t}T_t(H_{n-1}(W(h), 1))h. \end{aligned}$$

Proof. Let Σ be the covariance matrix of the Gaussian vector

$$\left(\int_0^T h_s^{(1)} d\tilde{B}_s^{(1)}, \dots, \int_0^T h_s^{(d)} d\tilde{B}_s^{(d)}\right)$$

(see Subsection 1.1.1), $m > 0$ and $\pi_k = \{-m = t_0^k < t_1^k < \ldots < t_{i_k}^k = m\}$ a partition of the interval $[-m, m]$ such that $|\pi_k| < 1/k$. Set

$$\bar{\pi}_k = \left\{[s_{j_1}^1, s_{j_1+1}^1] \times \cdots \times [s_{j_d}^d, s_{j_d+1}^d] : j_\ell \in \{0, 1, \dots, i_k - 1\} \text{ and } s_{j_\ell}^i \in \pi_k\right\}.$$

Then, using that $H_n(\cdot, 1)$ is a polynomial on $\mathbb{R}$, we have

$$\begin{aligned}
&\int_{[-m,m]^d} H_n\left(W(h)e^{-t} + \sqrt{1-e^{-2t}}\sum_{i=1}^d x_i, 1\right) \exp\left(-\sum_{i=1}^d \Sigma_{ii}^{-1} x_i^2/2\right) dx \\
&= L^2(\Omega) - \lim_{k\to\infty} \sum_{j_1,\dots,j_d=0}^{i_k-1} H_n\left(W(h)e^{-t} + \sqrt{1-e^{-2t}}\sum_{i=1}^d s_{j_i}^i, 1\right) \\
&\qquad \times \exp\left(-\sum_{i=1}^d \Sigma_{ii}^{-1}(s_{j_i}^i)^2/2\right) \prod_{\ell=1}^d (s_{j_\ell+1}^\ell - s_{j_\ell}^\ell).
\end{aligned}$$

Hence, (8.65) and the fact that D is a closed operator on $L^2(\Omega)$ imply that the last integral belongs to $\mathbb{D}^{1,2}$ and

$$\begin{aligned}
&D\left(\int_{[-m,m]^d} H_n\left(W(h)e^{-t} + \sqrt{1-e^{-2t}}\sum_{i=1}^d x_i, 1\right) \exp\left(-\sum_{i=1}^d \Sigma_{ii}^{-1} x_i^2/2\right) dx\right) \\
&= ne^{-t}\left(\int_{[-m,m]^d} H_{n-1}\left(W(h)e^{-t} + \sqrt{1-e^{-2t}}\sum_{i=1}^d x_i, 1\right)\right. \\
&\qquad \left.\times \exp\left(-\sum_{i=1}^d \Sigma_{ii}^{-1} x_i^2/2\right) dx\right) h.
\end{aligned}$$

Therefore, using that the operator D is closed on $L^2(\Omega)$ again and Proposition 4, we get that the proof is complete. □

Now, we are ready to characterize the semigroup $\{T_t : t \geq 0\}$ through the chaos decomposition (see Theorem 12).

Proposition 146 *Let $F \in L^2(\Omega)$ have the chaos decomposition*

$$F = E(F) + \sum_{n=1}^\infty I_n(h_n), \quad \text{in } L^2(\Omega),$$

with $h_n \in \mathcal{H}^{\odot n}$, for all $n \geq 1$. Then, for $t \geq 0$, $T_t(F)$ has the chaos decomposition

$$T_t(F) = E(F) + \sum_{n=1}^\infty e^{-nt} I_n(h_n). \tag{8.69}$$

Proof. Let $e(\alpha)$ be given by (1.96) and $t \geq 0$. Then, (1.97) implies

$$\begin{aligned}
T_t\left(I_n(e(\alpha))\right) &= \tilde{E}\left(\prod_{i=1}^{\hat{\alpha}} H_{\alpha_i}\left(W(e_{j_i})e^{-t} + \sqrt{1-e^{-2t}}\tilde{W}(e_{j_i}), 1\right)\right) \\
&= \prod_{i=1}^{\hat{\alpha}} \tilde{E}\left(H_{\alpha_i}\left(W(e_{j_i})e^{-t} + \sqrt{1-e^{-2t}}\tilde{W}(e_{j_i}), 1\right)\right),
\end{aligned}$$

where last equality follows from the fact that $\{e_{j_i} : i = 1, \ldots, \hat{\alpha}\}$ is an orthonormal family in $\mathcal{H}$. Hence, Propositions 21 and 145 allow us to establish that $T_t\left(I_n(e(\alpha))\right)$ belongs to $\mathbb{D}^{n,2}$ and

$$E\left[D^k T_t\left(I_n(e(\alpha))\right)\right] = \begin{cases} 0, & \text{if } 1 \leq k < n, \text{ or } k > n, \\ n! e^{-nt} e(\alpha), & \text{if } k = n. \end{cases}$$

Therefore, Theorem 46, and Remarks 22.v) and 57 imply

$$T_t\left(I_n(e(\alpha))\right) = e^{-nt}(I_n(e(\alpha)).$$

Consequently, (1.97), Theorem 12 and the fact that T_t is a contraction on $L^2(\Omega, \mathcal{F})$ yield that the result is satisfied. □

Two immediate consequences of Theorem 46 and Proposition 146 are the following two results.

Corollary 33 *Let $F \in L^2(\Omega)$ and $t > 0$. Then, $T_t(F)$ belongs to $\mathbb{D}^{n,2}$, for every $n \in \mathbb{N}$. Moreover, $T_t : L^2(\Omega) \to L^2(\Omega)$ is a C_0-semigroup.*

Proof. Let $F \in L^2(\Omega)$ have the chaos decomposition

$$F = E(F) + \sum_{n=1}^{\infty} I_n(h_n).$$

Hence, (8.69) and Theorem 46, together with the fact that $x \mapsto p(x)e^{-xt}$ is a bounded function on $\mathbb{R}_+$ for any polynomial $p : \mathbb{R} \to \mathbb{R}$, yield that $T_t(F) \in \mathbb{D}^{n,2}$, for any $n \in \mathbb{N}$.

Finally, using (8.69) again, we have that $T_t : L^2(\Omega) \to L^2(\Omega)$ is a C_0-semigroup. Thus, the proof is complete. □

Corollary 34 *Let $F \in \mathbb{D}^{1,2}$ and $t > 0$. Then, $T_t(DF) = e^t D\left(T_t(F)\right)$ and there exists a constant C independent of t and F such that $E\left[|D(T_t(F))|^2_{\mathcal{H}}\right] \leq \frac{C}{2t} E\left(|F|^2\right)$.*

Proof. The equality is an immediate consequence of (8.69) and the inequality follows from (1.109) and Theorem 46. □

We also have the following proposition.

Proposition 147 *Let $p \geq 1$ and $F \in L^p(\Omega)$. Then, $\lim_{t \to 0} E\left(|T_t(F) - F|^p\right) = 0$.*

Proof. Let $\{F_n \in \mathcal{S} : n \in \mathbb{N}\}$ be a sequence that converges to F in $L^p(\Omega)$, as $n \to \infty$. Therefore, by making use of the fact that T_t is a contraction on $L^p(\Omega)$, we have

$$\begin{aligned} E\left(|T_t(F) - F|^p\right) &\leq 3^p E\left(|T_t(F) - T_t(F_n)|^p\right) + 3^p E\left(|T_t(F_n) - F_n|^p\right) + 3^p E\left(|F_n - F|^p\right) \\ &\leq 4^p E\left(|F_n - F|^p\right) + 3^p E\left(|T_t(F_n) - F_n|^p\right). \end{aligned}$$

For this reason, the dominated convergence theorem, Corollary 34 and the fact that F_n is a smooth random variable in $\mathcal{S}$ imply

$$\limsup_{t \to 0} E\left(|T_t(F) - F|^p\right) \leq 8E\left(|F_n - F|^p\right), \quad \text{for every } n \in \mathbb{N},$$

which gives that the result holds. □

Bibliography

[1] N.H. Abel. Auflösung einer mechanischen Aufgabe. *J. Reine Angew. Math.*, 1:153–157, 1826.

[2] N.H. Abel. Solutions de quelques problèmes à l'aide d'int'egrales défines. In *Oeuvres complètes*, èd. 1, pages 11–27. Grondahi & Sons, Christiania, 1881.

[3] R.P. Agarwal and V. Lakshmikantham. *Uniqueness and nonuniqueness criteria for ordinary differential equations*, volume 6 of *Series in Real Analysis.* World Scientific Publishing Co., Inc., River Edge, NJ, 1993.

[4] E. Alòs and D. García Lorite. *Malliavin Calculus in Finance: Theory and Practice.* Financial Mathematics Series. Chapman & Hall/CRC, 2021.

[5] E. Alòs and J.A. León. On the curvature on the smile in stochastic volatility models. *SIAM J. Financial Math.*, 8(1):373–399, 2017.

[6] E. Alòs, J.A. León, and D. Nualart. Stochastic heat equation with random coefficients. *Probab. Theory Relat. Fields*, 115:41–94, 1999.

[7] E. Alós, J.A. León, and D. Nualart. Stochastic Stratonovich calculus for fractional Brownian motion with Hurst parameter less than 1/2. *Taiwanese J. Math.*, 5(3):609–632, 2001.

[8] E. Alòs, J.A. León, and J. Vives. On the short-time behavior of the implied volatility for jump-diffusion models with stochastic volatility. *Finance Stoch.*, 11:571–589, 2007.

[9] E. Alòs, O. Mazet, and D. Nualart. Stochastic calculus with respect to fractional Brownian motion with Hurst parameter lesser than $\frac{1}{2}$. *Stochastic Process. Appl.*, 86(1):121–139, 2000.

[10] E. Alòs, O. Mazet, and D. Nualart. Stochastic calculus with respect to Gaussian processes. *Ann. Probab.*, 29(2):766–801, 2001.

[11] E. Alòs and D. Nualart. A maximal inequality for the Skorohod integral. In *Stochastic differential and difference equations (Győr, 1996)*, volume 23 of *Progr. Systems Control Theory*, pages 241–251. Birkhäuser Boston, Boston, MA, 1997.

[12] E. Alòs and D. Nualart. An extension of Itô's formula for anticipating processes. *J. Theoret. Probab.*, 11(2):493–514, 1998.

[13] E. Alòs and D. Nualart. Stochastic integration with respect to the fractional Brownian motion. *Stoch. Stoch. Rep.*, 75(3):129–152, 2003.

[14] G.E. Andrews, R. Askey, and R. Roy. *Special Functions.* Cambridge University Press, 1999.

[15] H. Araya, J.A. León, and S. Torres. Numerical scheme for stochastic differential equations driven by fractional Brownian motion with $1/4 < H < 1/2$. *J. Theoret. Probab.*, 33(3):1211–1237, 2020.

[16] L. Arnold. *Stochastic differential equations: theory and applications.* Wiley-Interscience [John Wiley & Sons], New York-London-Sydney, 1974. Translated from the German.

[17] L. Bachelier. Théorie de la spéculation. *Annales Scientifiques de l'É.N.S, 3e Série*, 17:21–86, 1900.

[18] M.T. Barlow and M. Yor. Semi-martingale inequalities via the Garsia-Rodemich-Rumsey lemma, and applications to local times. *J. Funct. Anal.*, 49(2):198–229, 1982.

[19] M.A. Berger and V.J. Mizel. An extension of the stochastic integral. *Ann. Probab.*, 10(2):435–450, 1982.

[20] F. Biagini, Y. Hu, B. Ø ksendal, and T. Zhang. *Stochastic Calculus for Fractional Brownian Motion and Applications.* Springer, 2008.

[21] P. Billingsley. *Convergence of Probability Measures.* John Wiley & Sons, Inc., New York-London-Sydney, 1968.

[22] T. Bojdecki. *Teoría General de Procesos Estocásticos e Integración Estocástica.* Aportaciones Matemáticas, Sociedad Matemática Mexicana, 1995.

[23] R. Brown, F.R.S. Hon, and M.R.S.E. & R.I. Acad. V.P.L.S. A brief account of microscopical observations made in the months of June, July and August 1827, on the particles contained in the pollen of plants; and on the general existence of active molecules in organic and inorganic bodies. *Philos. Mag. Series 2*, 4(21):161–173, 1828.

[24] R. Buckdahn. Linear Skorohod stochastic differential equations. *Probab. Theory Related Fields*, 90(2):223–240, 1991.

[25] R. Buckdahn. Anticipative Girsanov transformations and Skorohod stochastic differential equations. *Mem. Amer. Math. Soc.*, 111(533):viii+88, 1994.

[26] Z. Chen, L. Leskelä, and L. Viitasaari. Pathwise Stieltjes integrals of discontinuously evaluated stochastic processes. *Stoch. Process Their Appl.*, 129(8):2723–2757, 2019.

[27] P. Cheridito and D. Nualart. Stochastic integral of divergence type with respect to fractional Brownian motion with Hurst parameter $H \in (0, \frac{1}{2})$. *Ann. Inst. H. Poincaré Probab. Statist.*, 41(6):1049–1081, 2005.

[28] A. Chojnowska-Michalik. Stochastic differential equations in Hilbert spaces. In *Probability theory (Papers, VIIth Semester, Stefan Banach Internat. Math. Center, Warsaw, 1976)*, volume 5 of *Banach Center Publ.*, pages 53–74. PWN, Warsaw, 1979.

[29] K.L. Chung and J.L. Doob. Optionality and Measurability. *Am. J. Math.*, 87(2):397–424, 1965.

[30] J.M.C. Clark. The representation of functionals of Brownian motion by stochastic integrals. *Ann. Math. Statist.*, 41:1282–1295, 1970.

[31] F. Comte and E. Renault. Long memory in continuous-time stochastic volatility models. *Math. Finance*, 8(4):291–323, 1998.

[32] L. Coutin, D. Nualart, and C.A. Tudor. Tanaka formula for the fractional Brownian motion. *Stochastic Process. Appl.*, 94(2):301–315, 2001.

[33] L. Coutin and Z. Qian. Stochastic analysis, rough path analysis and fractional Brownian motions. *Probab. Theory Related Fields*, 122(1):108–140, 2002.

[34] H. Cramér. A contribution to the theory of stochastic processes. In *Proceedings of the Second Berkeley Symposium on Mathematical Statistics and Probability, 1950*, pages 329–339. University of California Press, Berkeley and Los Angeles, Calif., 1951.

[35] J. Dávila, J.F. Bonder, J.D. Rossi, P. Groisman, and M. Sued. Numerical analysis of stochastic differential equations with explosions. *Stoch. Anal. Appl.*, 23(4):809–825, 2005.

[36] A. De Pablo, R. Ferreira, F. Quirós, and J.L. Vázquez. Blow-up. el problema matemático de explosión para ecuaciones y sistemas de reacción-difusión. *Bol. Soc. Esp. Mat. Apl.*, 32:75–111, 2005.

[37] C. Dellacherie and P.A. Meyer. *Probabilities and Potential.* North-Holland, 1978.

[38] C. Dellacherie and P.A. Meyer. *Probabilities and Potential B. Theory of Martingales.* North-Holland, Mathematics studies 72, 1982.

[39] G. Di Nunno, Y. Mishura, and A. Yurchenko-Tytarenko. Drift-implicit Euler scheme for sandwiched processes driven by Hölder noises. *Numer. Algorithms*, 93(2):459–491, 2023.

[40] J. Dieudonné. *Foundations of Modern Analysis.* Academic Press, 1969.

[41] J.L. Doob. *Stochastic processes.* John Wiley & Sons, Inc., New York; Chapman & Hall, Limited, London, 1953.

[42] H. Doss. Liens entre équations différentielles stochastiques et ordinaires. *Ann. Inst. H. Poincaré Sect. B (N.S.)*, 13(2):99–125, 1977.

[43] P. Doukhan, G. Oppenheim, and M.S. Taqqu. *Theory and Applications of Long-Range Dependence.* Birkhäuser, 2003.

[44] R.M. Dudley. Wiener functionals as Itô integrals. *Ann. Probab.*, 5(1):140–141, 1977.

[45] T. Duncan and D. Nualart. Existence of strong solutions and uniqueness in law for stochastic differential equations driven by fractional Brownian motion. *Stoch. Dyn.*, 9(3):423–435, 2009.

[46] A. Dvoretzky, P. Erdös, and S. Kakutani. Nonincrease everywhere of the brownian motion process. In *Proceedings of the Fourth Berkeley Symposium on Mathematical Statistics and Probability. Vol. 2*, pages 103–116. University of California Press, 1961.

[47] A. Einstein. On the movement of small particles suspended in a stationary liquid demanded by the molecular-kinetic theory of heat. *Annalen der Physik*, 17:549–560, 1905.

[48] L.D. Faddeev and O.A. Yakubovskiĭ. *Lectures on Quantum Mechanics for Mathematics Students.* Student Mathematical Library 47, American Mathematical Society, 2009.

[49] W. Feller. Diffusion processes in one dimension. *Trans. Amer. Math. Soc.*, 77:1–31, 1954.

[50] F. Ferrari. Wayl and Marchaud derivatives: A forgotten history. *Mathematics*, 6(6):25 pp., 2018.

[51] F. Flandoli and F. Russo. Generalized integration and stochastic ODEs. *Ann. Probab.*, 30(1):270–292, 2002.

[52] H. Föllmer, P. Protter, and A.N. Shiryayev. Quadratic covariation and an extension of Itô's formula. *Bernoulli*, 1(1-2):149–169, 1995.

[53] M. Foondun and E. Nualart. The Osgood condition for stochastic partial differential equations. *Bernoulli*, 27(1):295–311, 2021.

[54] E. Fournié, J.-M. Lasry, J. Lebuchoux, and P.-L. Lions. Application of Malliavin calculus to Monte Carlo methods in finance II. *Finance Stoch.*, 5(2):201–236, 2001.

[55] E. Fournié, J.-M. Lasry, J. Lebuchoux, P.-L. Lions, and N. Touzi. Application of Malliavin calculus to Monte Carlo methods in finance. *Finance Stoch.*, 3(4):391–412, 1999.

[56] A. Friedman. *Partial differential equations of parabolic type.* Prentice-Hall, Inc., Englewood Cliffs, NJ, 1964.

[57] P.K. Friz and M. Hairer. *A course on rough paths.* Universitext. Springer, Cham, 2014. With an introduction to regularity structures.

[58] P.K. Friz and N.B. Victoir. *Multidimensional stochastic processes as rough paths*, volume 120 of *Cambridge Studies in Advanced Mathematics.* Cambridge University Press, Cambridge, 2010. Theory and applications.

[59] A. Garsia, E. Rodemich, and H. Rumsey. A real variable lemma and the continuity of paths of some Gaussian processes. *Indiana Univ. Math. J.*, 20(6):565–578, 1970/71.

[60] J. Garzón, J.A. León, J.A. Lozada, and S. Torres. A fractional stochastic differential equation with discontinuous diffusion driven by fbm with hurst parameter less than 1/2. *Preprint*, 2023.

[61] J. Garzón, J.A. León, and S. Torres. Fractional stochastic differential equation with discontinuous diffusion. *Stoch. Anal. Appl.*, 35(6):1113–1123, 2017.

[62] J. Gatheral, T. Jaisson, and M. Rosenbaum. Volatility is rough. *Quant. Finance*, 18(6):933–949, 2018.

[63] B. Gaveau and P. Trauber. L'intégrale stochastique comme opérateur de divergence dans l'espace fonctionnel. *J Funct. Anal.*, 46(2):230–238, 1982.

[64] I.I. Gikhman and A.V. Skorokhod. *Introduction to the theory of random processes.* Dover Publications, Inc., Mineola, NY, 1996. Translated from the 1965 Russian original, Reprint of the 1969 English translation, With a preface by Warren M. Hirsch.

[65] L.G. Gorostiza. Análisis de sistemas sometidos a perturbaciones estocásticas. *Ciencia*, 35:33–43, 1984.

[66] L.G. Gorostiza and J.A. León. A stochastic Fubini theorem and equivalence of extended solutions of stochastic evolution equations in Hilbert space. In *Random partial differential equations (Oberwolfach, 1989)*, volume 102 of *Internat. Ser. Numer. Math.*, pages 85–94. Birkhäuser, Basel, 1991.

[67] M. Gradinaru, I. Nourdin, F. Russo, and P. Vallois. m-order integrals and generalized Itô's formula: the case of a fractional Brownian motion with any Hurst index. *Ann. Inst. H. Poincaré Probab. Statist.*, 41(4):781–806, 2005.

[68] C. Graham and D. Talay. *Stochastic simulation and Monte Carlo methods*, volume 68 of *Stochastic Modelling and Applied Probability*. Springer, Heidelberg, 2013. Mathematical foundations of stochastic simulation.

[69] M.T. Greene and B.D. Fielitz. Long-term dependence in common stock returns. *Journal of Financial Economics*, 4(3):339–349, 1977.

[70] M. Gubinelli. Controlling rough paths. *J Funct. Anal.*, 216(1):86–140, 2004.

[71] M. Gubinelli and S. Tindel. Rough evolution equation. *Ann. Probab.*, 38(1):1–75, 2008.

[72] G.H. Hardy and J.E. Littlewood. Some properties of fractional integrals. I. *Math. Z.*, 27(1):565–606, 1928.

[73] P. Hartman. *Ordinary differential equations.* John Wiley & Sons, Inc., New York-London-Sydney, 1964.

[74] U.G. Haussmann. On the integral representation of functionals of Itô processes. *Stochastics*, 3(1):17–27, 1979.

[75] T. Hida. *Brownian Motion.* Stochastic Modelling and Applied Probability, Springer New York, 1980.

[76] D.J. Higham and P.E. Kloeden. *An introduction to the numerical simulation of stochastic differential equations.* Society for Industrial and Applied Mathematics (SIAM), Philadelphia, PA, [2021] ©2021.

[77] J. Hong, C. Huang, M. Kamrani, and X. Wang. Optimal strong convergence rate of a backward Euler type scheme for the Cox-Ingersoll-Ross model driven by fractional Brownian motion. *Stochastic Process. Appl.*, 130(5):2675–2692, 2020.

[78] J. Hong, C. Huang, and X. Wang. Optimal rate of convergence for two classes of schemes to stochastic differential equations driven by fractional Brownian motions. *IMA J. Numer. Anal.*, 41(2):1608–1638, 2021.

[79] C. Houdré and J. Villa. An example of infinite dimensional quasi-helix. In *Stochastic models (Mexico City, 2002)*, volume 336 of *Contemp. Math.*, pages 195–201. Amer. Math. Soc., Providence, RI, 2003.

[80] Y. Hu, Y. Liu, and D. Nualart. Rate of convergence and asymptotic error distribution of Euler approximation schemes for fractional diffusions. *Ann. Appl. Probab.*, 26(2):1147–1207, 2016.

[81] Y. Hu, Y. Liu, and D. Nualart. Crank-Nicolson scheme for stochastic differential equations driven by fractional Brownian motions. *Ann. Appl. Probab.*, 31(1):39–83, 2021.

[82] Y. Hu and D. Nualart. Rough path analysis via fractional calculus. *Trans. Amer. Math. Soc.*, 361(5):2689–2718, 2009.

[83] C. Huang. Optimal convergence rate of modified Milstein scheme for SDEs with rough fractional diffusions. *J. Differential Equations*, 344:325–351, 2023.

[84] H.E. Hurst. Long-term storage capacity of reservoirs. *ASCE Transactions*, 116(1):770–799, 1951.

[85] T. Hytönen, J. van Neerven, M. Veraar, and L. Weis. *Analysis in Banach space. Volume I: Martingales and Littlewood-Paley Theory.* Springer, 2016.

[86] N. Ikeda and S. Watanabe. *Stochastic Differential Equations and Diffusion Processes.* North-Holland Publishing Company, 1981.

[87] K. Itô. Multiple Wiener integrals. *J. Math. Soc. Japan*, 3(1):157–169, 1951.

[88] K. Itô. Extension of stochastic integrals. In *Proceedings of the International Symposium on Stochastic Differential Equations (Res. Inst. Math. Sci., Kyoto Univ., Kyoto, 1976)*, pages 95–109. Wiley, New York-Chichester-Brisbane, 1978.

[89] J. Jacod. *Calcul Stochastique et Problèmes de Martingales.* Lecture Notes in Mathematics 714 Springer, 1979.

[90] Y.-J. Jien and J. Ma. Stochastic differential equations driven by fractional Brownian motions. *Bernoulli*, 15(3):846–870, 2009.

[91] Y.M. Kabano. On extended stochastic integrals. *Theory Probab. Its Applications*, 20(4):710–722, 1976.

[92] S. Kaden and J. Potthoff. Progressive stochastic processes and applications to the Itô integral. *Stoch. Anal. Appl.*, 22(4):843–865, 2004.

[93] I. Kaj and M.S. Taqqu. Convergence to fractional Brownian motion and to the Telecom process: the integral representation approach. *Prog. Probab.*, 60:383–427, 2008.

[94] O. Kallenberg. *Foundations of Modern Probability.* Springer, 1997.

[95] I. Karatzas, D.L. Ocone, and J. Li. An extension of Clark's formula. *Stochastics Stochastics Rep.*, 37(3):127–131, 1991.

[96] I. Karatzas and S.E. Shreve. *Brownian Motion and Stochastic Calculus. Second Edition.* Springer-Verlag, 1998.

[97] T. Kawada and N. Kôno. A remark on nowhere differentiability of sample functions of Gaussian processes. *Proc. Japan Acad.*, 47(suppl, suppl. II):932–934, 1971.

[98] M.L. Kleptsyna, P.E. Kloeden, and V.V. An. Existence and uniqueness theorems for stochastic differential equations with fractal Brownian motion. *Problemy Peredachi Informatsii*, 34(4):51–61, 1998.

[99] P.E. Kloeden and E. Platen. *Numerical solution of stochastic differential equations*, volume 23 of *Applications of Mathematics (New York).* Springer-Verlag, Berlin, 1992.

[100] P.E. Kloeden, E. Platen, and H. Schurz. *Numerical solution of SDE through computer experiments.* Universitext. Springer-Verlag, Berlin, 1994. With 1 IBM-PC floppy disk (3.5 inch; HD).

[101] A. Kohatsu-Higa and J.A. León. Anticipating stochastic differential equations of Stratonovich type. *Appl. Math. Optim.*, 36(3):263–289, 1997.

[102] A. Kohatsu-Higa, J.A. León, and D. Nualart. Stochastic differential equations with random coefficients. *Bernoulli*, 3(2):233–245, 1997.

[103] A.N. Kolmogorov. Wienersche Spiralen und einige andere interessante Kurven im Hilbertschen Raum. *C. R. (Doklady)*, 26:115–118, 1940.

[104] I. Kruk and F. Russo. Malliavin-Skorohod calculus and Paley-Wiener integral for covariance singular processes. *inria-00540914*, pages 1–120, 2010.

[105] H. Kunita. Stochastic differential equations and stochastic flows of diffeomorphisms. In *École d'été de probabilités de Saint-Flour, XII—1982*, volume 1097 of *Lecture Notes in Mathematics*, pages 143–303. Springer, 1984.

[106] P. Lei and D. Nualart. Stochastic calculus for Gaussian processes and application to hitting times. *Commun. Stoch. Anal.*, 6(3):379–402, 2012.

[107] J.A. León. Stochastic evolution equations with respect to semimartingales in Hilbert space. *Stochastics Stochastics Rep.*, 27(1):1–21, 1989.

[108] J.A. León. On equivalence of solutions to stochastic differential equations with anticipating evolution systems. *Stochastic Anal. Appl.*, 8(3):363–387, 1990.

[109] J.A. León. Stochastic Fubini theorem for semimartingales in Hilbert space. *Canadian J. Math.*, XLII(5):890–901, 1990.

[110] J.A. León. Stratonovich type integration with respect to fractional Brownian motion with Hurst parameter less than 1/2. *Bernoulli*, 26(3):2436–2462, 2020.

[111] J.A. León, Y. Liu, and S. Tindel. Euler scheme for SDEs driven by fractional Brownian motions: Malliavin differentiability and uniform upper-bound estimates. *Stochastic Process. Appl.*, 175:Paper No. 104412, 20, 2024.

[112] J.A. León and D. Márquez-Carreras. Semilinear fractional stochastic differential equations driven by a γ-Hölder continuous signal with $\gamma > 2/3$. *Stoch. Dyn.*, 21(1):Paper No. 2050039, 29, 2021.

[113] J.A. León and D. Márquez-Carreras. Stability of stochastic bilinear equations with anticipating drift in the first Wiener chaos. *Preprint*, 2024.

[114] J.A. León, D. Márquez-Carreras, and J. Vives. Anticipating linear stochastic differential equations driven by a Lévy process. *Electron. J. Probab.*, 17:no. 89, 26, 2012.

[115] J.A. León, D. Márquez-Carreras, and J. Vives. Stability of some anticipating semilinear stochastic differential equations of skorohod type. *J. Dyn. Differ. Equ. Preprint*, 2023.

[116] J.A. León, R. Navarro, and D. Nualart. An anticipating calculus approach to the utility maximization of an insider. *Math. Financ.*, 13(1):171–185, 2003.

[117] J.A. León and D. Nualart. Stochastic evolution equations with random generators. *Ann. Probab.*, 26(1):149–186, 1998.

[118] J.A. León, D. Nualart, and S. Tindel. Young Differential equations with power type nonlinearities. *Stoch. Process. Their Appl.*, 127:3042–3067, 2017.

[119] J.A. León and L. Peralta. Some Feller and Osgood type criteria for semilinear stochastic differential equations. *Stoch. Dyn.*, 17(2):1750011, 19, 2017.

[120] J.A. León, L. Peralta, and J. Villa Morales. An example of explosion of a stochastic differential equation with random initial condition. In *Models in statistics and probability. III (Spanish)*, volume 47 of *Aportaciones Mat. Comun.*, pages 139–147. Inst. Mat. , UNAM, México, 2014.

[121] J.A. León, L. Peralta Hernández, and J. Villa-Morales. On the distribution of explosion time of stochastic differential equations. *Bol. Soc. Mat. Mexicana (3)*, 19(2):125–138, 2013.

[122] J.A. Leon and V. Perez-Abreu. Strong solutions of stochastic bilinear equations with anticipating drift in the first Wiener chaos. In *Stochastic processes*, pages 235–243. Springer, New York, 1993.

[123] J.A. León and S. Tindel. Malliavin calculus for fractional delay equations. *J. Theoret. Probab.*, 25(3):854–889, 2012.

[124] J.A. León and C. Tudor. Semilinear fractional stochastic differential equations. *Bol. Soc. Mat. Mexicana (3)*, 8(2):205–226, 2002.

[125] J.A. León and J. Villa. An Osgood criterion for integral equations with applications to stochastic differential equations with an additive noise. *Statist. Probab. Lett.*, 81(4):470–477, 2011.

[126] S.C. Lim and V.M. Sithi. Asymptotic properties of the fractional Brownian motion of Riemann-Liouville type. *Phys. Lett. A*, 206(5-6):311–317, 1995.

[127] S.J. Lin. Stochastic analysis of fractional Brownian motions. *Stochastics Stochastics Rep.*, 55(1-2):121–140, 1995.

[128] G. Lindgren. *Stationary stochastic processes. Theory and applications.* Text in Statistical Science. Chapman & Hall/CRC, 2013.

[129] J. Liouville. Memoire sur quelques questions de geometrie et de mecanique, et sur un noveau genre pour respondre ces questions. *Journal de l'École Polytechnique*, 13:1–69, 1832.

[130] Y. Liu and S. Tindel. First-order Euler scheme for SDEs driven by fractional Brownian motions: the rough case. *Ann. Appl. Probab.*, 29(2):758–826, 2019.

[131] T. Lyons. Differential equations driven by rough signals. I. An extension of an inequality of L. C. Young. *Math. Res. Lett.*, 1(4):451–464, 1994.

[132] T. Lyons and Z. Qian. *System control and rough paths.* Oxford Mathematical Monographs. Oxford University Press, Oxford, 2002. Oxford Science Publications.

[133] T.J. Lyons. Differential equations driven by rough signals. *Rev. Mat. Iberoamericana*, 14(2):215–310, 1998.

[134] T.J. Lyons. System control and rough paths. In *Numerical methods and stochastics (Toronto, ON, 1999)*, volume 34 of *Fields Inst. Commun.*, pages 91–99. Amer. Math. Soc., Providence, RI, 2002.

[135] J. Maas. Malliavin calculus and decoupling inequalities in Banach spaces. *J. Math. Anal. Appl.*, 363(2):383–398, 2010.

[136] P. Malliavin. Stochastic calculus of variations and hypoelliptic operators. In *Proceedings of the International Symposium on Stochastic Differential Equations, Kyoto 1976*, pages 195–263. Academic Press, 1978.

[137] B.B. Mandelbrot. *Fractals and scaling in finance.* Selected Works of Benoit B. Mandelbrot. Springer-Verlag, New York, 1997. Discontinuity, concentration, risk, Selecta Volume E, With a foreword by R. E. Gomory.

[138] B.B. Mandelbrot and J.W. Van Ness. Fractional Brownian motions, fractional noises and applications. *SIAM Review*, 10(4):422–437, 1968.

[139] B.B. Mandelbrot and J.R. Wallis. Computer experiments with fractional Gaussian noises. *Water Resour. Res.*, 5(1):228–241, 1969.

[140] B.B. Mandelbrot and J.R. Wallis. Some long-run properties of geophysical records. *Water Resour. Res.*, 5(2):321–340, 1969.

[141] R. Mansuy and M. Yor. *Random times and enlargements of filtrations in a Brownian setting*, volume 1873 of *Lecture Notes in Mathematics.* Springer-Verlag, Berlin, 2006.

[142] M.A. Marchaud. *Sur les Dérivées et Sur les Différences des Fonctions de Variables Réelles.* NUMDAM, [place of publication not identified], 1927.

[143] M.A. Marchaud. Sur les dérivées et sur les différences des fonctions de variables réelles. *Journal de Mathématiques Pures et Appliquées*, 9:337–425, 1927.

[144] T. Matsuda and N. Perkowski. An extension of the stochastic sewing lemma and applications to fractional stochastic calculus. *Forum Math. Sigma*, 12:Paper No. e52, 52, 2024.

[145] Jr. H.P. McKean. *Stochastic integrals.* Probability and Mathematical Statistics, No. 5. Academic Press, New York-London, 1969.

[146] E.J. McShane. Extension of range of functions. *Bull. New Ser. Am. Math. Soc.*, 40(12):837–842, 1934.

[147] P.A. Meyer. *Probability and potentials.* Blaisdell Publishing Co. Ginn and Co., Waltham, Mass.-Toronto, Ont.-London, 1966.

[148] A. Millet, D. Nualart, and M. Sanz. Large deviations for a class of anticipating stochastic differential equations. *Ann. Probab.*, 20(4):1902–1931, 1992.

[149] Y. Mishura and G. Shevchenko. The rate of convergence for Euler approximations of solutions of stochastic differential equations driven by fractional Brownian motion. *Stochastics*, 80(5):489–511, 2008.

[150] Y.S. Mishura. *Stochastic Calculus for Fractional Brownian Motion and Related Processes.* Springer-Verlag, 2008.

[151] O. Mocioalca and F. Viens. Skorohod integration and stochastic calculus beyond the fractional Brownian scale. *J. Funct. Anal.*, 222(2):385–434, 2005.

[152] G.M. Molchan and Ju.I. Golosov. Gaussian stationary processes with asymptotically a power spectrum. (russian). *Dokl. Akad. Nauk SSSR*, 184(3):546–549, 1969.

[153] J. Neveu. *Bases mathématiques du calcul des probabilités.* Masson et Cie, Éditeurs, Paris, 1970. Préface de R. Fortet, Deuxième édition, revue et corrigée.

[154] I. Nourdin. A simple theory for the study of SDEs driven by a fractional Brownian motion, in dimension one. In *Séminaire de probabilités XLI*, volume 1934 of *Lecture Notes in Math.*, pages 181–197. Springer, Berlin, 2008.

[155] I. Nourdin. *Selected aspects of fractional Brownian motion*, volume 4 of *Bocconi & Springer Series*. Springer, Milan; Bocconi University Press, Milan, 2012.

[156] D. Nualart. Analysis on Wiener space and anticipating stochastic calculus. In *Lectures on probability theory and statistics (Saint-Flour, 1995)*, volume 1690 of *Lecture Notes in Math.*, pages 123–227. Springer, Berlin, 1998.

[157] D. Nualart. Stochastic integration with respect to fractional brownian motion and applications. In *Stochastic Models*, pages 3–39. Conteporary Mathematics 39 Amer. Math. Soc., 2003.

[158] D. Nualart. *The Malliavin Calculus and Related Topics*. Springer, 2006.

[159] D. Nualart and E. Pardoux. Stochastic calculus with anticipating integrands. *Probability Theory and Related Fields*, 78:535–581, 1988.

[160] D. Nualart and A. Răşcanu. Differential equations driven by fractional Brownian motion. *Collect. Math.*, 53(1):55–81, 2002.

[161] D. Nualart and S. Tindel. A construction of the rough path above fractional Brownian motion using Volterra's representation. *Ann. Probab.*, 39(3):1061–1096, 2011.

[162] D. Nualart and M. Zakai. On the relation between the Stratonovich and Ogawa integrals. *Ann. Probab.*, 17(4):1536–1540, 1989.

[163] D. Ocone. Malliavin's calculus and stochastic integral representations of functionals of diffusion processes. *Stochastics*, 12(3-4):161–185, 1984.

[164] D. Ocone and E. Pardoux. A generalized Itô-Ventzell formula. Application to a class of anticipating stochastic differential equations. *Annales de l'Institut Henri Poincare (B)*, 25(1):39–71, 1989.

[165] D.L. Ocone and I. Karatzas. A generalized Clark representation formula, with applications to optimal portfolios. *Stochastics Stochastics Rep.*, 34(3-4):187–220, 1991.

[166] S. Ogawa. Quelques propriétés de l'intégrale stochastique du type noncausal. *Japan J. Appl. Math.*, 1(2):405–416, 1984.

[167] S. Ogawa. Une remarque sur l'approximation de l'intégrale stochastique du type noncausal par une suite des intégrales de Stieltjes. *Tohoku Math. J. (2)*, 36(1):41–48, 1984.

[168] S. Ogawa. The stochastic integral of noncausal type as an extension of the symmetric integrals. *Japan J. Appl. Math.*, 2(1):229–240, 1985.

[169] A. Ohashi and F. Russo. Rough paths and symmetric-Stratonovich integrals driven by singular covariance Gaussian processes. *Bernoulli*, 30(2):1197–1230, 2024.

[170] M. Ondreját and J. Seidler. On existence of progressively measurable modifications. *Electronic Communications in Probability*, 18(20):1–6, 2013.

[171] W.F. Osgood. Beweis der Existenz einer Lösung der Differentialgleichung $\frac{dy}{dx} = f(x, y)$ ohne Hinzunahme der Cauchy-Lipschitz'schen Bedingung. *Monatsh. Math. Phys.*, 9(1):331–345, 1898.

[172] R.E.A.C. Paley, N. Wiener, and A. Zygmund. Notes on random functions. *Mathematische Zeitschrift*, 37:647–668, 1933.

[173] É. Pardoux and P. Protter. A two-sided stochastic integral and its calculus. *Probab. Theory Related Fields*, 76(1):15–49, 1987.

[174] W.J. Park. On the equivalence of Gaussian processes with factorable covariance functions. *Proc. Amer. Math. Soc.*, 32:275–279, 1972.

[175] A. Pazy. *Semigroups of linear operators and applications to partial differential equations*, volume 44 of *Applied Mathematical Sciences*. Springer-Verlag, New York, 1983.

[176] V. Pipiras and M.S. Taqqu. Integration questions related to fractional Brownian motion. *Probab. Theory Related Fields*, 118(2):251–291, 2000.

[177] V. Pipiras and M.S. Taqqu. Are classes of deterministic integrands for fractional Brownian motion on an interval complete? *Bernoulli*, 7(6):873–897, 2001.

[178] G. Pisier. Riesz transforms: a simpler analytic proof of p. a. meyer's inequality. In *Séminaire de probabilités de Strasbourg, XXII-1988*, volume 1321 of *Lecture Notes in Mathematics*, pages 485–501. Springer, 1988.

[179] I. Podlubny. *Fractional Differential Equations*, volume 198 of *Mathematics in Science and Engineering*. Academic Press, Inc., San Diego, CA, 1999. An introduction to fractional derivatives, fractional differential equations, to methods of their solution and some of their applications.

[180] P.E. Protter. *Stochastic Integrationand Differential Equations. Second Edition.* Springer, 2004.

[181] M.M. Rao and R.J. Swift. *Probability Theory with Applications. Second Edition.* Springer, 2006.

[182] M. Reed and B. Simon. *Methods of Modern Mathematical Physics. I: Functional Analysis, Revised and Enlarged Edition.* Academic Press, INC., 1980.

[183] B. Riemann. Versuch einer allgemeinen auffassung der integration und differentiation. In *Gesammelte Mathematische Werke und Wissenchaftlicher*, Leipzig: Teubner, pages 33–344. 1876.

[184] L.C.G. Rogers. Arbitrage with fractional Brownian motion. *Math. Finance*, 7(1):95–105, 1997.

[185] B. Ross. A brief history and exposition of the fundamental theory of fractional calculus. In *Fractional Calculus and Its Applications*, volume 457 of *Lecture Notes in Mathematics*, pages 1–36. Springer, Berlin, Heidelberg, 1975.

[186] B. Ross. The development of fractional calculus 1695-1900. *Historia Mathematica*, 4(1):75–89, 1977.

[187] H.L. Royden. *Real Analysis. Third Edition.* Macmillan Publishing Company and Collier Macmillan Publising, 1988.

[188] H.L. Royden and P.M. Fitzpatrick. *Real analysis.* China Machine Press, fourth edition, 2010.

[189] W. Rudin. *Principles of Mathematical Analysis. Third Edition.* McGraw-Hill, Inc, 1976.

[190] F. Russo and P. Vallois. Forward, backward and symmetric stochastic integration. *Probab. Theory Relat. Fields*, 97(3):403–421, 1993.

[191] F. Russo and P. Vallois. Elements of stochastic calculus via regularization. In *Séminaire de Probabilités XL*, pages 147–185. Springer, 2007.

[192] A.A. Ruzmaikina. Stieltjes integrals of Hölder continuous functions with applications to fractional Brownian motion. *J. Statist. Phys.*, 100(5-6):1049–1069, 2000.

[193] S.G. Samko, A.A. Kilbas, and O.I. Marichev. *Fractional Integrals and Derivatives. Theory and Applications.* Gordon and Breach Science Publishers, 1993.

[194] G. Samorodnitsky. Long range dependence. *Foundations and Trends® in Stochastic Systems*, 1(3):163–257, 2007.

[195] G. Samorodnitsky and M.S. Taqqu. *Stable Non-Gaussian Random Processes.* Chapman & Hall, 1994.

[196] Y. Shiota. A linear stochastic integral equation containing the extended Itô integral. *Math. Rep. Toyama Univ.*, 9:43–65, 1986.

[197] A.N. Shiryayev. *Probability*, volume 95 of *Graduate Texts in Mathematics.* Springer-Verlag, New York, 1984. Translated from the Russian by R. P. Boas.

[198] A.V. Skorohod. On a generalization of a stochastic integral. *Theory Probab. its Appl.*, 20(2):219–233, 1975.

[199] J.Ll. Solé, F. Utzet, and J. Vives. Canonical Lévy process and Malliavin calculus. *Stochastic Process. Appl.*, 117(2):165–187, 2007.

[200] M. Spivak. *Calculus on Manifolds.* Addison-Wesley Publishing Company, 1965.

[201] D.W. Stroock and S.R. Srinivasa Varadhan. *Multidimensional diffusion processes.* Classics in Mathematics. Springer-Verlag, Berlin, 2006. Reprint of the 1997 edition.

[202] H. Sugita. On a characterization of the Sobolev spaces over an abstract Wiener space. *Kyoto J. Math.*, 25(4):717–725, 1985.

[203] H.J. Sussmann. An interpretation of stochastic differential equations as ordinary differential equations which depend on the sample point. *Bull. Amer. Math. Soc.*, 83(2):296–298, 1977.

[204] H.J. Sussmann. On the gap between deterministic and stochastic ordinary differential equations. *Ann. Probability*, 6(1):19–41, 1978.

[205] H.J. Ter Horst. Riemann-Stieltjes and Lebesgue-Stieltjes integrability. *Am. Math. Mon.*, 91(9):551–559, 1984.

[206] P. Todorovic. *An introduction to stochastic processes and their applications.* Springer Series in Statistics: Probability and its Applications. Springer-Verlag, New York, 1992.

[207] S. Torres and L. Viitasaari. Stochastic differential equations with discontinuous diffusion coefficients. *Theory Probab. Math. Statist.*, (109):159–175, 2023.

[208] N.M. Towghi. Multidimensional extension of L. C. Young's inequality. *JIPAM. J. Inequal. Pure Appl. Math.*, 3(2):Article 22, 13, 2002.

[209] H.G. Tucker. *A Graduate Course in Probability.* Academic Press, 1967.

[210] J.B. Walsh. An introduction to stochastic partial differential equations. In *École d'été de probabilités de Saint-Flour, XIV—1984*, volume 1180 of *Lecture Notes in Math.*, pages 265–439. Springer, Berlin, 1986.

[211] H. Weyl. Bemerkungen zum Begriff des Differentialquotienten gebrochener Ordnung. *Zürich. Naturf. Ges.*, 62:296–302, 1917.

[212] N. Wiener. Differential space. *J. Math. & Phys. Sci.*, 2:131–174, 1923.

[213] N. Wiener. The quadratic variation of a function and its Fourier coefficients. *J. Math. & Phys. Sci.*, 3(2):72–94, 1924.

[214] N. Wiener. The homogeneous chaos. *Am. J. Math.*, 60(4):897–936, 1938.

[215] M. Yor. Sur quelques approximations d'intégrales stochastiques. In *Séminaire de Probabilités, XI (Univ. Strasbourg, Strasbourg, 1975/1976)*, Lecture Notes in Math., Vol. 581, pages 518–528. Springer, Berlin-New York, 1977.

[216] L.C. Young. An inequality of the Hölder type, connected with Stieljes integration. *Acta Mathematica*, 67:251–282, 1936.

[217] L.C. Young. General inequalities for Stieltjes integrals and the convergence of Fourier series. *Math. Ann.*, 115(1):581–612, 1938.

[218] M. Zähle. Integration with respect to fractal functions and stochastic calculus I. *Probab. Theory Relat. Fields*, 111:333–374, 1998.

[219] M. Zähle. Integration with respect to fractal functions and stochastic calculus. II. *Math. Nachr.*, 225:145–183, 2001.

[220] S.-Q. Zhang and C. Yuan. Stochastic differential equations driven by fractional Brownian motion with locally Lipschitz drift and their implicit Euler approximation. *Proc. Roy. Soc. Edinburgh Sect. A*, 151(4):1278–1304, 2021.

[221] H. Zhou, Y. Hu, and J. Zhao. Numerical method for singular drift stochastic differential equation driven by fractional Brownian motion. *J. Comput. Appl. Math.*, 447:Paper No. 115902, 13, 2024.

Index

For Product Safety Concerns and Information please contact our EU representative GPSR@taylorandfrancis.com
Taylor & Francis Verlag GmbH, Kaufingerstraße 24, 80331 München, Germany

www.ingramcontent.com/pod-product-compliance
Lightning Source LLC
Chambersburg PA
CBHW080125220326
41598CB00032B/4963

* 9 7 8 1 0 3 2 7 7 8 1 2 9 *